EIGENVALUES OF INHOMOGENEOUS STRUCTURES

Unusual Closed-Form Solutions

EIGENVALUES OF INHOMOGENEOUS STRUCTURES

Unusual Closed-Form Solutions

Isaac Elishakoff

J. M. Rubin Foundation Distinguished Professor
Florida Atlantic University
Boca Raton, Florida

CRC Press
Taylor & Francis Group
Boca Raton London New York

CRC Press is an imprint of the
Taylor & Francis Group, an **informa** business

CRC Press
Taylor & Francis Group
6000 Broken Sound Parkway NW, Suite 300
Boca Raton, FL 33487-2742

© 2005 by Issac Elishakoff
CRC Press is an imprint of Taylor & Francis Group, an Informa business

First issued in paperback 2019

No claim to original U.S. Government works

ISBN-13: 978-0-367-45427-2 (pbk)
ISBN-13: 978-0-8493-2892-3 (hbk)

Visit the Taylor & Francis Web site at
http://www.taylorandfrancis.com

and the CRC Press Web site at
http://www.crcpress.com

Library of Congress Card Number 2004051927

Library of Congress Cataloging-in-Publication Data

Elishakoff, Isaac.
    Eigenvalues of inhomogenous structures : unusual closed-form solutions /
      Isaac Elishakoff.
    p. cm.
    Includes bibliographical references and index.
    ISBN 0-8493-2892-6 (alk. paper)
    1. Structural dynamics–Mathematical models. 2. Buckling
      (Mechanics)–Mathematical models. 3. Eigenvalues. I. Title.

TA654.E495 2004
624.1'76'015118–dc22

                2004051927

Dedicated to IFMA, Institute Français de Mécanique Avancée, France whose superbly educated engineers have been most instrumental in helping to bring this book to fruition; without their dedication it would take many more years to achieve this humble goal of communicating to the engineers, scientists and students the infinite number of closed-form solutions.

Isaac Elishakoff

Other Books from Professor Isaac Elishakoff

Textbook

I. Elishakoff, *Probabilistic Methods in the Theory of Structures*, Wiley—Interscience, New York, 1983 (second edition: Dover Publications, Mineola, New York, 1999).

Monographs

Y. Ben-Haim and I. Elishakoff, *Convex Models of Uncertainty in Applied Mechanics*, Elsevier Science Publishers, Amsterdam, 1990.

G. Cederbaum, I. Elishakoff, J. Aboudi and L. Librescu, *Random Vibrations and Reliability of Composite Structures*, Technomic Publishers, Lancaster, PA, 1992.

I. Elishakoff, Y. K. Lin and L. P. Zhu, *Probabilistic and Convex Modeling of Acoustically Excited Structures*, Elsevier Science Publishers, Amsterdam, 1994.

I. Elishakoff, *The Courage to Challenge*, 1st Books Library, Bloomington, IN, 2000.

I. Elishakoff, Y. W. Li, and J. H. Starnes, Jr., *Nonclassical Problems in the Theory of Elastic Stability*, Cambridge University Press, 2001.

I. Elishakoff and Y. J. Ren, *Large Variation Finite Element Method for Stochastic Problems*, Oxford University Press, 2003.

I. Elishakoff, *Safety Factors and Reliability: Friends or Foes?*, Kluwer Academic Publishers, Dordrecht, 2004.

Edited Volumes

I. Elishakoff and H. Lyon, (eds.), *Random Vibration - Status and Recent Developments*, Elsevier Science Publishers, Amsterdam, 1986.

I. Elishakoff and H. Irretier, (eds.), *Refined Dynamical Theories of Beams, Plates and Shells, and Their Applications*, Springer Verlag, Berlin, 1987.

I. Elishakoff, J. Arbocz, Ch. D. Babcock, Jr. and A. Libai, (eds.), *Buckling of Structures-Theory and Experiment*, Elsevier Science Publishers, Amsterdam, 1988.

S. T. Ariaratnam, G. Schuëller and I. Elishakoff, (eds.), *Stochastic Structural Dynamics-Progress in Theory and Applications*, Elsevier Applied Science Publishers, London, 1988.

C. Mei, H. F. Wolfe and I. Elishakoff, (eds.), *Vibration and Behavior of Composite Structures*, ASME Press, New York, 1989.

F. Casciati, I. Elishakoff and J. B. Roberts, (eds.), *Nonlinear Structural Systems under Random Conditions*, Elsevier Science Publishers, Amsterdam, 1990.

D. Hui and I. Elishakoff, (eds.), *Impact and Buckling of Structures*, ASME Press, New York, 1990.

A. K. Noor, I. Elishakoff and G. Hulbert, (eds.), *Symbolic Computations and Their Impact on Mechanics*, ASME Press, New York, 1990.

Y. K. Lin and I. Elishakoff, (eds.), *Stochastic Structural Dynamics-New Theoretical Developments*, Springer Verlag, Berlin, 1991.

I. Elishakoff and Y. K. Lin, (eds.), *Stochastic Structural Dynamics-New Applications*, Springer, Berlin, 1991.

I. Elishakoff (ed.), *Whys and Hows in Uncertainty Modeling*, Springer Verlag, Vienna, 2001.

A. P. Seyranian and I. Elishakoff (eds.), *Modern Problems of Structural Stability*, Springer Verlag, Vienna, 2002.

Contents

Foreword

This is a most remarkable and thorough review of the efforts that have been made to find closed-form solutions in the vibration and buckling of all manner of elastic rods, beams, columns and plates. The author is particularly, but not exclusively, concerned with variations in the stiffness of structural members. The resulting volume is the culmination of his studies over many years.

What more can be said about this monumental work, other than to express admiration? The author's solutions to particular problems will be very valuable for testing the validity and accuracy of various numerical techniques. Moreover, the study is of great academic interest, and is clearly a labor of love. The author is to be congratulated on this work, which is bound to be of considerable value to all interested in research in this area.

Dr. H.D. Conway
Professor Emeritus
Department of Theoretical and Applied Mechanics
Cornell University

It is generally believed that closed-form solutions exist for only a relatively few, very simple cases of bars, beams, columns, and plates. This monograph is living proof that there are, in fact, not just a few such solutions. Even in the current age of powerful numerical techniques and high-speed, large-capacity computers, there are a number of important uses for closed-form solutions:

- for preliminary design (often optimal)
- as bench-mark solutions for evaluating the accuracy of approximate and numerical solutions
- to gain more physical insight into the roles played by the various geometric and/or loading parameters

This book is fantastic. Professor Elishakoff is to be congratulated not only for pulling together a number of solutions from the international literature, but also for contributing a large number of solutions himself. Finally, he has explained in a very interesting fashion the history behind many of the solutions.

Dr. Charles W. Bert
Benjamin H. Perkinson Professor Emeritus
Aerospace and Mechanical Engineering
The University of Oklahoma

Prologue

The puzzled reader may wonder why such a seemingly self-promotional adjective like "unusual" was put in the title. Let us explain immediately that it was borrowed from a definitive article (Conway, 1981) by Dr. H.D. Conway, Professor Emeritus of Cornell University, in which he derived a closed-form solution for the bending of a circular plate on an elastic foundation, its thickness varying according to a specific law. His solution is in strong contrast to that for uniform circular plates, which involves transcendental Webber functions. The above adjective was fully justified and the author feels its use is equally called for in this monograph as well.

The present monograph summarizes the efforts and results in the search for closed-form solutions in vibration and buckling of inhomogeneous rods, beams, columns and plates, undertaken by the author, his colleagues and his associates over several years.

The exact solution for the buckling of columns with uniform cross-section was obtained by Leonhard Euler. In his famous paper of 1744, he derived the expression for the buckling load of a pinned column through nonlinear analysis of the elastica. Later, he took up the linear approach, and the celebrated formula for the uniform pinned column was re-derived, accordingly, in his paper of 1759 where he also posed the problem of buckling of inhomogeneous columns, with variable cross-sections. The stiffness $D(x)$ was taken to vary as

$$D(x) = D_0(a + bx/L)^m \tag{P.1}$$

where x is the coordinate along the column axis, L is the length of the beam, D_0 is the stiffness at the origin $x = 0$, a and b are real numbers, and m is a positive integer. Whereas the exact solution for the general positive integer m had to wait for over 150 years, to be solved in terms of Bessel functions by Dinnik (1912), Euler obtained solutions in terms of *elementary* (but not closed-form) functions for the particular cases $m = 2$ and $m = 4$. Namely, he derived the following buckling loads:

$$P_{cr} = D_0 \frac{b^2}{L^2} \left[\pi^2 \left(\ln \frac{a+b}{a} \right)^{-2} + \frac{1}{4} \right] \quad \text{for } m = 2$$

$$P_{cr} = \pi^2 D_0 a^2 (a+b)^2 / L^2 \quad \text{for } m = 4 \tag{P.2}$$

In the eighteenth century, Euler's (1759) solution was the only closed-form expression that was derived for buckling of an homogeneous structure.

Apparently, in the nineteenth century, the only closed-form solution for buckling was obtained by Engesser (1893). In the twentieth century three studies were reported, by Duncan (1937), Rzhanitsyn (1951) and by Elishakoff and Rollot (1999) all on buckling of columns. This monograph appears at the beginning of a new century and new millennium and presents an *infinte* number of closed-form solutions both in buckling and vibration. Engesser (1893) considered a column with the following moment of inertia:

$$D(x) = 4D_0 x(L - x)/L^2 \tag{P.3}$$

and postulated the mode shape

$$W(x) = 4x(L - x)A/L^2 \tag{P.4}$$

which, substituted into the governing differential equation of second order for a column, pinned at both ends,

$$D(x)\frac{d^2W}{dx^2} + PW = 0 \tag{P.5}$$

yields the following buckling load

$$P_{cr} = 8D_0/L^2 \tag{P.6}$$

An analogous expression for the buckling load was derived by Rzhanitsyn (1955) in a different problem.

Another solution of this kind was derived by Duncan (1937) in his interesting study on the Boobnov–Galerkin method as applied to vibration and buckling. Such a study on the approximate method would be an odd place to search for closed-form solutions, at the first glance. The variation of the flexural rigidity was taken as

$$D(x)\left[1 - \tfrac{7}{3}(x/L)^2\right]D_0 \tag{P.7}$$

The mode shape was taken as

$$W(x) = 7(x/L) - 10(x/L)^3 + 3(x/L)^5 \tag{P.8}$$

yielding, after substitution in Eq. (P.5), the buckling load

$$P_{cr} = \tfrac{60}{7}D_0/L^2 \tag{P.9}$$

Duncan (1937) apparently wanted to have a benchmark solution for inhomogeneous columns. Some interesting similarities can be noted between Engesser's and Duncan's examples. For *specific* columns with variable cross-section the mode shape is represented by *polynomial* functions, while for those with uniform cross-sections one needs to solve a *transcendental* equation.

Even in the latter with the exact solution available, the vibration frequency and the buckling load are given in terms of the number π and the accuracy of the result depends on that of the latter. Thus, for uniform structures in the easiest case the solution is obtained in terms of an *irrational* number, while for the variable cross-section case a *rational* expression is derived.

In the case of a non-homogeneous column Engesser (1893), Duncan (1937) and Rzhanitsyn (1955) provided the solution in terms of rational expressions for the buckling loads. This fact, although not of primary importance in the modern age of computers, is still remarkable. It can actually be viewed as a curiosity, and the relevant material can be included in courses of strength of materials and mechanics of solids. The lesson to be learned is, obviously, that one must be encouraged to seek simple solutions, and that such a pursuit may even be rewarded.

The idea that led to the present monograph is as follows. It occurred to the author, there might be *other* solutions of this kind. The following task was posed: find polynomial expressions for the stiffness of a column or a beam, such that the associated buckling modes of vibration mode shapes will also constitute polynomial expressions. Thus, our first study appeared (Elishakoff and Rollot, 1999), devoted to the buckling problems. This initial success stimulated scientific appetite and led to a more refined task: find a method that could bring out, in some systematic way, closed-form solutions for vibration frequencies and buckling loads.

The method for obtaining some *infinite set* of closed-form solutions is expounded in what follows. We appreciate the collaboration with Messrs. Olivier Rollot, Suleyman Candan, Zakoua Guédé, Roland Becquet, Denis Meyer of the French Institute of Advanced Mechanical Engineering, France, (whom the dedication refers to) Dr. Joseph Neuringer of Boca Raton, and Professor Menahem Baruch of the Technion–Israel Institute of Technology. Sincere gratitude is also expressed to Professor Maurice Lemaire of the IFMA, Aubiere, France. Discussions with Professor Yehuda Stavsky of the Technion–Israel Institute of Technology, on a possible application of the proposed method for rectangular plates is gratefully recorded, along with comments made by Professor Ivo Calió during the short course given at the Department of Structural Engineering, University of Catania, Italy, in June 2000. Table 2.1 was prepared by Mr. Pierfranceco Cacciola, Mr. Alessandro Palmeri of the University of Messina, and by Ms. Anna Sidoti of the University of Palermo. Tables 2.2–2.5 were prepared by Mr. Massimo Marletta and Mr. Franscesco Vincipova of the Department of Civil and Environmental Engineering of University of Catania, Italy. Table 4.5 was calculated by Suleyman Candan and Roland Becquet, IFMA, France. The author is extremely grateful to several colleagues for careful reading of the author's review paper (Elishakoff 2000h), which subsequently became the backbone of the first chapter of this monograph, and providing numerous constructive suggestions for its improvements. These colleagues are (in alphabetical order): Professor Menahem Baruch of the Technion–Israel Institute of Technology; Professor Charles W. Bert of the University of Oklahoma; Professor Rama B. Bhat of Concordia University, Canada; Professor Moshe Eisenberger of the Technion–Israel Institute of Technology; Professor Cornelius O. Horgan of the University of Virginia; Professor Graham M.L. Gladwell of the University of Waterloo, Canada; Professor Patricio A.A. Laura, Institute of Applied Mechanics, Bahia Blanca, Argentina; Professor Arthur W. Leissa of the Ohio State University; Professor Yitshak M. Ram of the Louisiana State University. Last but not least, grateful thanks are expressed to Professor Hans P.W. Gottlieb of Griffith University, Australia for providing some useful references that escaped my attention. The skillful typing of the manuscript by Messrs. Zakoua Guédé, Roland Becquet, Denis Meyer, Roberta Santoro, Cristina Gentilini, Rosemarie Ciucchi, Adawna Smith and Melissa Morris is gratefully appreciated. The efforts of the anonymous reviewers of several of our papers, published on this topic, are appreciated. All comments were very constructive, and the author is grateful for that. Two persons deserve infinite thanks: These are Dr. Joel A, Storch of the Aerospace Corporation, Los Angeles for painstakingly checking the numerical derivations, providing a list of typing and numerical errors, and possible simplifications of the analytic expressions, and Professor Charles W. Bert of the University of Oklahoma who went through the entire manuscript and magnanimously edited the text. I do not know how to thank them. I appreciate encouraging comments by Professor Emeritus Alexander R. Khvoles of Jerusalem, Israel. Each of them has been a godsend.

After completion of this book, in the process of its continuous improvement, I was fortunate also to cooperate with Dr. Ivo Calió of Catania, Italy; Professor Menahem Baruch of Haifa, Israel; Dr. Giulia Catellani of Modena, Italy; Professor Yitshak Ram of Baton Rouge, USA; Mr. James Endres of Deerfield Beach, USA; Professor Giuseppe Ruta of Rome, Italy; Ms. Roberta Santoro of Palermo, Italy; Ms. Cristina Gentilini of Bologna, Italy and Professors L. Wu and Q.-S. Wang of Anhui, People's Republic of China, on the topics touched upon in this book. It was decided not to include some results of these studies in this book for the author himself does not like overly thick books; in addition, some of these studies will soon be published in scientific journals. Many thanks to several anonymous reviewers for their encouragement and constructive comments all of which were implemented. The harmonious cooperation with Mrs. Cindi Carelli, Mrs. Jessica Vakili, Mrs. Andrea and Mrs. Veena of the CRC Press is gratefully recorded.

One more person deserves special thanks: Professor Werner Soedel, of Purdue University. When Mr. Rollot and the author were preparing the first article referred to above we intended to submit it for faster publication, to the *Journal of Sound and Vibration*, as a Letter to the Editor. (To digress, the *Letters to the Editor* section was initiated several decades ago by Professor Phil Doak — the Editor-in-Chief of this celebrated journal for several decades — as a means for rapid communication of ideas, which otherwise have to wait — sometimes for years — until, sometimes, the author's patience and/or the interest are perhaps exhausted.) Professor Soedel — Americas Editor — commented that the manuscript did not contain vibration data, and asked whether we cared to generalize it. Mr. Rollot quickly conjectured that a beam with constant density and constant modulus of elasticity but with variable cross-section would not allow polynomial mode shapes. Then, the idea came to this author that we ought to consider a more general case of variable stiffness and inertial coefficients, where there was no apparent reason for exclusion of a polynomial solution. Accordingly, the statement was added that a manuscript on the vibration aspect was under way, which was true — albeit "*in embryo.*" This led, as a chain reaction, to further investigations, of which this monograph is the final result. The conclusion, this time again, to regard a half-empty glass, as the one that is half-full, was drawn by us. The author is happy to arrive at this truism, which may seem to be drawn from the Ann Landers' encouraging advice columns. We hope that the search for the closed-form solutions will be given some additional boost by this monograph.

The significance of the closed-form solutions cannot be overestimated. Not only are they of extreme intrinsic interest, but, they also serve as benchmarks against which the accuracy of various approximate solutions (the Rayleigh–Ritz, Boobnov–Galerkin, finite differences, finite elements, differential quadrature and others) can be ascertained. As regards the intrinsic interest, it stems *inter alia* from the fact that inhomogeneity, desirable or undesirable is invariably present in a realistic structure. Of even greater importance is the fact that an inhomogeneous structure can be designed with prescribed specific properties — for example, a beam whose natural frequency is at the desired level, or is below the excitation level; or a column whose buckling load exceeds a prescribed level, or a structure whose fundamental frequency is outside the range where the excitation spectral density takes significant values. Since inhomogeneity provides, by its very nature, additional freedom in the choice of parameters, it is likely that many design problems will find simpler and possibly less time consuming solutions. Throughout this monograph we extensively utilized computerized symbolic algebra, user-friendly "computer assistants" MATHEMATICA® and MAPLE®, allowing for the blessed modern realization of a dream: analytical derivations on the computer. For the detailed description of the application of symbolic algebra to the buckling problems the interested readers can consult the recent monograph by Elishakoff, Li and Starnes (2001a). One more comment appears to be relevant: one computational analyst maintained, quite surprisingly, prominent during a conversation that analytical solutions, since they pertain to specialized cases have no future. Yet, as Niels Bohr remarked, "It is difficult to predict, especially the future." Analytical methods appear to this writer as a lamp which provides light for future numerical investigations.

The author will be most welcome to receive, through e-mail, elishako@fau.edu, through Fax 561-297-2825, or through regular mail, critiques, which are bound to be constructive, and will do his best to reply to them as feedback for the interested readers.

Isaac Elishakoff
Boca Raton
Eve of New Year 5765
September 2004

1

Introduction: Review of Direct, Semi-inverse and Inverse Eigenvalue Problems

This chapter presents a selective review (with emphasis on the word "selective") of the direct, semi-inverse and inverse eigenvalue problem for structures described by a differential equation with variable coefficients. It gives only a taste of the extensive research that has been conducted since 1759, when Leonhard Euler posed, apparently for the first time, a boundary value problem. Since then numerous studies have been conducted for rods, Bernoulli–Euler beams, Bresse–Timoshenko beams, Kirchhoff–Love and Mindlin–Reissner plates and shells, and structures analyzed via finer, high-order theories. This selective review classifies the solutions as belonging to one of three main classes: (1) direct problems, (2) semi-inverse problems, (3) inverse problems. In addition, some new closed-form solutions are reported that have been obtained via posing an inverse vibration problem. Due to the huge body of literature, the author limits himself to discussing classic theories of structures.

1.1 Introductory Remarks

The vibration and buckling eigenvalue problems in engineering may be roughly categorized as belonging to one of the following three classes: (1) direct or forward problems, (2) inverse or backward problems, (3) semi-inverse or semi-direct problems. Direct problems are associated with the determination of the vibration frequencies and/or the buckling loads of structures with specified configuration; namely, in the vibration context, direct problems call for evaluation of the natural frequencies of rods, beam, plates or shells of specified, uniform or non-uniform cross-section (in the case of rods and beams) or thickness (in the case of plates and shells), for both homogeneous structures, i.e., those with constant elastic properties and mass density, and inhomogeneous ones, in which the elastic properties and/or the material density vary with the coordinates. There are numerous methods of solving direct eigenvalue problems. These methods are conveniently subdivided into the exact and the approximate ones. Exact solutions may be available for both uniform and non-uniform homogeneous or inhomogeneous structures.

For completeness, we will recapitulate some simple examples reported in the literature.

1.2   Vibration of Uniform Homogeneous Beams

Consider first the vibration of the uniform homogeneous Bernoulli–Euler beams. The vibration of such beams is governed by the the following differential equation:

$$D\frac{\partial^4 w}{\partial x^4} + \rho A \frac{\partial^2 w}{\partial t^2} = 0 \tag{1.1}$$

where $D = EI$ is the flexural rigidity, E is the modulus of elasticity, I the moment of inertia, ρ the mass density, A the cross-sectional area, $w(x,t)$ the deflection (transverse displacement), x the axial coordinate and t the time. We are looking for the free harmonic vibrations with frequency ω, representing the displacement $w(x,t)$ as

$$w(x,t) = W(x)e^{i\alpha x} \tag{1.2}$$

where $W(x)$ is the mode shape. The differential equation (1.1) becomes

$$D\frac{d^4 W}{dx^4} - \rho A \omega^2 W = 0 \tag{1.3}$$

It is instructive to introduce the parameter

$$k^4 = \frac{\rho A \omega^2}{D} \tag{1.4}$$

so that Eq. (1.3) takes the form

$$\frac{d^4 W}{dx^4} - k^4 W = 0 \tag{1.5}$$

The solution $W(x)$ can be put in the form

$$W(x) - C_1 \sin(kx) + C_2 \cos(kx) + C_3 \sinh(kx) + C_4 \cosh(kx) \tag{1.6}$$

The mode shape $W(x)$ must satisfy the associated boundary conditions. For the end that is pinned (denoted by P),

$$W = 0 \qquad \frac{d^2 W}{dx^2} = 0 \tag{1.7}$$

For the end that is clamped (denoted by C),

$$W = 0 \qquad \frac{dW}{dx} = 0 \tag{1.8}$$

For the free end (denoted by F),

$$\frac{d^2W}{dx^2} = 0 \qquad \frac{d^3W}{dx^3} = 0 \tag{1.9}$$

whereas for the guided end (denoted by G),

$$\frac{dW}{dx} = 0 \qquad \frac{d^3W}{dx^3} = 0 \tag{1.10}$$

Note that the term "guided" was suggested to the author by Bert (2000b). Usually, textbooks do not discuss the guided end boundary condition. Some of the exceptions are the texts by Dimarogonas (1996) and Inman (1995), who refer to the end with boundary condition in Eq. (1.10) as the "guided" end (see also Natke, 1989; Korenev and Rabinovich, 1972). Satisfaction of the boundary conditions at both ends yields four homogeneous equations with four unknowns C_1, C_2, C_3 and C_4. Requiring that they be non-trivial, i.e.,

$$C_1^2 + C_2^2 + C_3^2 + C_4^2 \neq 0 \tag{1.11}$$

yields characteristic equations that give, for beams under different boundary conditions, associated non-trivial fundamental natural frequencies,

$$
\begin{aligned}
P\text{–}P: &\quad \sin(kL) = 0 \\
G\text{–}P: &\quad \cos(kL) = 0 \\
C\text{–}F: &\quad \cos(kL)\cosh(kL) + 1 = 0 \\
C\text{–}G: &\quad \tan(kL) + \tanh(kL) = 0 \\
C\text{–}P: &\quad \tan(kL) = \tanh(kL) \\
C\text{–}C: &\quad \cos(kL)\cosh(kL) - 1 = 0
\end{aligned}
\tag{1.12}
$$

The solutions of these transcendental equations are obtainable either analytically or numerically. For the beam pinned at both ends the non-trivial solution of Eq. (1.12) reads $k_j = j\pi$ $(j = 1, 2, \ldots)$. For the guided–pinned beam we get from Eq. (1.12) $kj = (2j - 1)\pi/2$. For the other combinations of boundary conditions the numerical solutions of the transcendental equations (1.12) are available, and the first three frequency coefficients k_j are listed below:

$$
\begin{array}{llll}
C\text{–}F: & k_1L = 1.875; & k_2L = 4.694; & k_3L = 7.854 \\
C\text{–}G: & k_1L = 2.365; & k_2L = 5.498; & k_3L = 8.639 \\
C\text{–}P: & k_1L = 3.927; & k_2L = 7.069; & k_3L = 10.210 \\
C\text{–}P: & k_1L = 4.73; & k_2L = 7.853; & k_3L = 10.996
\end{array}
\tag{1.13}
$$

For higher frequencies ($j \gg 1$), accurate asymptotic expressions can be given for the natural frequency coefficients k_j:

$$
\begin{array}{ll}
C\text{–}F: & k_j \approx (j - \tfrac{1}{2})\pi \\
C\text{–}G: & k_j \approx (j - \tfrac{1}{4})\pi \\
C\text{–}P: & k_j \approx (j + \tfrac{1}{4})\pi \\
C\text{–}C: & k_j \approx (j + \tfrac{1}{2})\pi
\end{array}
\tag{1.14}
$$

Exact solutions for various vibrating uniform beams are given in numerous references (see, e.g., Gorman, 1974).

1.3   Buckling of Uniform Homogeneous Columns

Buckling of uniform homogeneous columns subjected to compressive load at the ends is governed by the familiar differential equation

$$
D\frac{d^4 W}{dx^4} + P\frac{d^2 W}{dx^2} = 0
\tag{1.15}
$$

One, again, introduces the eigenvalue parameter

$$
\lambda^4 = \frac{P}{D}
\tag{1.16}
$$

leading to the following expression of the buckling mode $W(x)$:

$$
W = C_1 + C_2 x + C_3 \sin(\lambda x) + C_4 \cos(\lambda x)
\tag{1.17}
$$

The boundary conditions for the pinned or clamped ends coincide with their counterparts in the vibration case, yet for the free end the boundary condition (1.9) is replaced by

$$
D\frac{d^3 W}{dx^3} + P\frac{dW}{dx} = 0
\tag{1.18}
$$

The appropriate characteristic equations are:

$$
\begin{aligned}
P\text{--}P: & \quad \sin(\lambda L) = 0 \\
G\text{--}P: & \quad \cos(\lambda L) = 0 \\
C\text{--}F: & \quad \cos(\lambda L) = 0 \\
C\text{--}G: & \quad \sin(\lambda L) = 0 \\
C\text{--}P: & \quad \tan(\lambda L) = \lambda L \\
C\text{--}C: & \quad \cos(\lambda L) = 1
\end{aligned}
\tag{1.19}
$$

In the case of the $P\text{--}P, G\text{--}P, C\text{--}G, C\text{--}C$ and $C\text{--}F$ columns analytical solutions are obtained:

$$
\begin{aligned}
P\text{--}P: & \quad \lambda_j = j\pi \\
G\text{--}P: & \quad \lambda_j = \pi(2j-1)/2 \\
C\text{--}F: & \quad \lambda_j = \pi(2j-1)/2 \\
C\text{--}G: & \quad \lambda_j = j\pi \\
C\text{--}C: & \quad \lambda_j = 2j\pi
\end{aligned}
\tag{1.20}
$$

For $j = 1$, we obtain the buckling loads, denoted by P_{cr}:

$$
\begin{aligned}
P\text{--}P: & \quad P_{\mathrm{cr}} = \pi^2 D/L^2 \\
G\text{--}P: & \quad P_{\mathrm{cr}} = \pi^2 D/4L^2 \\
C\text{--}F: & \quad P_{\mathrm{cr}} = \pi^2 D/4L^2 \\
C\text{--}G: & \quad P_{\mathrm{cr}} = \pi^2 D/L^2 \\
C\text{--}C: & \quad P_{\mathrm{cr}} = 4\pi^2 D/L^2
\end{aligned}
\tag{1.21}
$$

For a $P\text{--}C$ column, the numerical solution of the transcendental equation (1.19.5) is available. It reads

$$
\lambda_1 L \approx 4.49
\tag{1.22}
$$

with the associated buckling load

$$
P_{\mathrm{cr}} \approx 20.2 \frac{D}{L^2}
\tag{1.23}
$$

Exact solutions are also available for *specific* non-uniform beams and columns, in both the vibration and the buckling contexts.

The governing equations describing buckling of non-uniform columns are differential equations with variable coefficients. Such equations may arise even in uniform columns. One such case is the initially vertical uniform column under its own weight with intensity q. We first consider the column that is clamped at $x = 0$ and free at $x = L$. The equation of bending of the column

$$D(x)\frac{d^2 W}{dx^2} = M(x) \tag{1.24}$$

where $M(x)$, the bending moment, is determined as the sum of elementary moments of weight intensity, acting on all elements of the part of the column until the cross-section x:

$$M = \int_x^L q[V(u) - W(x)]\,du = q\int_x^L V(u)\,du - qW(x)(L - x) \tag{1.25}$$

Substituting Eq. (1.25) into Eq. (1.24) and differentiating with respect to x we get

$$\frac{d}{dx}\left[D(x)\frac{d^2 W}{dx^2}\right] + q(L - x)\frac{dW}{dx} = 0 \tag{1.26}$$

If the flexural rigidity $D(x)$ is constant, we get an ordinary differential equation with variable coefficients

$$D\frac{d^3 W}{dx^3} + q(L - x)\frac{dW}{dx} = 0 \tag{1.27}$$

We introduce a new variable

$$z = \frac{2}{3}\sqrt{\frac{q(L - x)^3}{D}} \tag{1.28}$$

From this expression we find x and the derivatives of W with respect to x:

$$x = L - \sqrt[3]{\frac{9Dz^2}{4q}}$$

$$\frac{dW}{dx} = -\sqrt[3]{\frac{3qz}{2D}}\frac{dW}{dz}$$

$$\frac{d^2 W}{dx^2} = \sqrt[3]{\frac{9q^2}{4D^2}}\left(\frac{1}{3z^3}\frac{dW}{dz} + z^{2/3}\frac{d^2 W}{dz^2}\right)$$

$$\frac{d^2 W}{dx^3} = \frac{3q}{2D}\left(\frac{1}{z}\frac{dW}{dz} - \frac{d^2 W}{dz^2} - z\frac{d^3 W}{dz^3}\right)$$

(1.29)

Substitution of Eqs. (1.29) into Eq. (1.27) yields

$$\frac{d^3 W}{dz^3} + \frac{1}{z}\frac{d^2 W}{dz^2} + \left(1 - \frac{1}{9z^2}\right)\frac{dW}{dz} = 0 \qquad (1.30)$$

which is the Bessel equation with respect to the function dW/dz, with the solution

$$\frac{dW}{dz} = C_1 J_{1/3}(z) + C_2 J_{-1/3}(z) \qquad (1.31)$$

We will not determine the function W at this stage since it is sufficient to know its derivatives. Since at $x = 0$ the column is clamped, we have the following boundary conditions:

$$x = 0 \quad W = 0 \qquad \frac{dW}{dx} = 0 \quad \text{or} \quad \frac{dW}{dz} = 0 \qquad (1.32)$$

The upper end is free, i.e., there

$$x = L \quad z = L \qquad \frac{d^2 W}{dx^2} = 0 \quad \text{or} \quad z^{-1/3}\frac{dW}{dz} + 3z^{2/3}\frac{d^2 W}{dz^2} = 0 \qquad (1.33)$$

Following Dinnik (1912, 1929, 1955a,b) we start from the latter condition. For extremely small z, neglecting the terms of higher order, we write

$$\frac{dW}{dz} = D_1 z^{1/3} + D_2 z^{1/3} \qquad (1.34)$$

where D_1 and D_2 are new constants, proportional to C_1 and C_2, respectively. Substituting this expression into the boundary condition at the upper end, we get

$$D_1 = C_1 = 0 \qquad (1.35)$$

The condition at the lower end yields

$$C_2 J_{-1/3}\left[\frac{2}{3}\sqrt{\frac{qL^3}{D}}\right] = 0 \qquad (1.36)$$

i.e., either $C_2 = 0$, corresponding to straight equilibrium, or

$$\frac{2}{3}\sqrt{\frac{qL^3}{D}} = \alpha_n \qquad (1.37)$$

where α_n are roots of the transcendental equation $J_{-1/3}(\alpha) = 0$. The first root is 1.87. This value results in the buckling load

$$q_{cr} = 7.87\frac{D}{L^3} \tag{1.38}$$

Consider, now, the case in which the column is clamped at $x = 0$, the upper end can move but the slope is zero (guided case). At $x = 0$ the boundary conditions specified in Eq. (1.32) hold. At $x = L$

$$z = 0 \qquad \frac{dW}{dx} = 0 \quad \text{or} \quad z^{1/3}\frac{dW}{dz} = 0 \tag{1.39}$$

Consider the latter condition. We take dW/dz in Eq. (1.31), express the Bessel function as a series, and then multiply by $z^{1/3}$:

$$z^{1/3}\frac{dW}{dz} = \frac{C_1 z^{2/3}}{2^{1/3}\Gamma(4/3)}\left(1 - \frac{3z^2}{16} + \cdots\right) + \frac{C_2 2^{1/3}}{\Gamma(2/3)}\left(1 - \frac{3z^2}{8} + \cdots\right) \tag{1.40}$$

It is clear that in order for the right-hand side to vanish at $z = 0$, it is necessary to put $C_2 = 0$. The first boundary condition leads to either an uninteresting case of $C_2 = 0$, signifying the straight form of the equilibrium, or $C_2 \neq 0$, with

$$J_{1/3}\left[\frac{2}{3}\sqrt{\frac{qL^3}{D}}\right] = 0 \tag{1.41}$$

with first root equal to 2.90. Accordingly, the buckling intensity equals

$$q_{cr} = 18.9\frac{D}{L^3} \tag{1.42}$$

Now, consider now the case in which the lower end is clamped but the upper end carries the concentrated compressive load P. The bending moment reads

$$M = \int_x^L q[V(u) - W(x)]\,du + P(f - W) \tag{1.43}$$

where f is the displacement at the upper end. The differential equation reads

$$D\frac{d^3 W}{dx^3} + q(L^* - x)\frac{dW}{dx} = 0 \tag{1.44}$$

Note that Eq. (1.44) coincides with Eq. (1.27), except that L is replaced by L^*, where

$$L^* = \frac{(qL + P)}{q} \qquad (1.45)$$

has the dimension of length; L^* can be referred to as an *effective length* of the column. In perfect analogy with the case of the clamped–free column without the concentrated force, we obtain

$$\frac{dW}{dz} = C_1 J_{1/3}(z) + C_2 J_{-1/3}(z)$$

$$z = \frac{2}{3}\sqrt{\frac{q(L^* - x)^3}{D}} \qquad (1.46)$$

The boundary conditions at the lower end read

$$x = 0 \quad z = \frac{2}{3}\sqrt{\frac{q(L^*)^3}{D}} = z_{\mathrm{L}}$$

$$W = 0 \quad \frac{dW}{dx} = 0 \quad \text{or} \quad \frac{dW}{dz} = 0 \qquad (1.47)$$

At the upper end

$$x = L \quad z = \frac{2}{3}\sqrt{\frac{q(L^* - L)^3}{D}} = z_{\mathrm{U}} \quad \frac{d^2 W}{dx^2} = 0 \quad \frac{dW}{dz} + 3z\frac{d^2 W}{dz^2} = 0 \qquad (1.48)$$

where z_{L} and z_{U} represent the values of z taken at the lower and upper ends, respectively. The boundary conditions lead to the following equations:

$$C_1 J_{1/3}(z_{\mathrm{L}}) + C_2 J_{-1/3}(z_{\mathrm{U}}) = 0$$

$$C_1 J_{-2/3}(z_{\mathrm{L}}) - C_2 J_{2/3}(z_{\mathrm{U}}) = 0 \qquad (1.49)$$

The non-triviality requirement $C_1^2 + C_2^2 \neq 0$ leads to

$$J_{1/3}(z_{\mathrm{L}})J_{2/3}(z_{\mathrm{U}}) + J_{-1/3}(z_{\mathrm{L}})J_{-2/3}(z_{\mathrm{U}}) = 0 \qquad (1.50)$$

In the previous case we dealt with a homogeneous differential equation of the third order. If the column is pinned at $x = L$, then the horizontal force N, the reaction, should be taken into account. Instead of Eq. (1.27) we get

$$D(x)\frac{d^3 W}{dx^3} + q(L - x)\frac{dW}{dx} = N \qquad (1.51)$$

or with the new variable as in Eq. (1.27),

$$\frac{d^3W}{dz^3} + \frac{1}{z}\frac{d^2W}{dz^2} + \left(1 - \frac{1}{9z^2}\right)\frac{dW}{dz} = -\frac{2N}{3qz} = -bz \tag{1.52}$$

Let us first determine dW/dz. The complementary solution is given by Eq. (1.31). The particular solution is represented as

$$\frac{dW}{dz} = A_0 + A_1 z + A_2 z^2 + \cdots + A_n z^n + \cdots \tag{1.53}$$

Substituting it into Eq. (1.52) and equating the coefficients with equal powers of the variable z we get

$$A_0 = 0 \quad A_1(9-1) = -b \quad A_2(9\cdot4-1) = -9A_0$$
$$A_3(9\cdot9-1) = -9A_1, \ldots, A_n(9^2n-1) = -9A_{n-2} \tag{1.54}$$

This means that all the coefficients with even index vanish

$$A_0 = A_2 = A_4 = \cdots = A_{2n} = 0 \tag{1.55}$$

while the odd coefficients equal

$$A_1 = -\frac{b}{9\cdot1^2-1} \quad A_3 = \frac{9b}{(9\cdot1^2-1)(9\cdot3^2-1)}$$

$$A_5 = -\frac{9^2b}{(9\cdot1^2-1)(9\cdot3^2-1)(9\cdot5^2-1)}, \ldots, A_{2n+1} = -\frac{9A_{2n-1}}{9(2n+1)^2-1} \tag{1.56}$$

leading to

$$\frac{dW}{dz} = C_1 J_{1/3}(z) + C_2 J_{-1/3}(z) - \left(\frac{6N}{q}\right)C(z) \tag{1.57}$$

where

$$C(z) = z\left\{\frac{1}{9\cdot1^2-1} - \frac{(3z)^2}{(9\cdot1^2-1)(9\cdot3^2-1)} + \frac{(3z)^4}{(9\cdot1^2-1)(9\cdot3^2-1)(9\cdot5^2-1)}\right.$$

$$\left. - \cdots \pm \frac{(3z)^{2n}}{(9\cdot1^2-1)(9\cdot3^2-1)(9\cdot5^2-1)\cdots[9(2n+1)^2-1]}\right\} \tag{1.58}$$

This series is uniformly convergent for all finite values of the variable z. Integrating Eq. (1.57) we get

$$W = C_1 A(z) + C_2 B(z) - (6N/q)D(z) + C_3 \tag{1.59}$$

where

$$A(z) = \int J_{1/3}(z)\,dz \quad B(z) = \int J_{-1/3}(z)\,dz \quad D(z) = \int C(z)\,dz$$

$$\Gamma\left(\frac{4}{3}\right)A(z) = 3\left(\frac{z}{2}\right)^{4/3}\left[\frac{1}{2} - \frac{3}{5\cdot1\cdot4}\left(\frac{z}{2}\right)^2 + \frac{3^2}{8\cdot1\cdot2\cdot4\cdot7}\left(\frac{z}{2}\right)^4\right.$$

$$- \frac{3^3}{11\cdot1\cdot2\cdot4\cdot7\cdot10}\left(\frac{z}{2}\right)^6 + \cdots$$

$$\left. + (-1)^n \frac{3^n}{(3n+2)\cdot1\cdot2\cdot3\cdots n\cdot1\cdot4\cdot7\cdots(3n+1)}\left(\frac{z}{2}\right)^{2n} - \cdots\right]$$

$$\left(\frac{2}{3}\right)\Gamma\left(\frac{5}{3}\right)B(z) = 3\left(\frac{z}{2}\right)^{2/3}\left[1 - \frac{3}{4\cdot1\cdot2}\left(\frac{z}{2}\right)^2 + \frac{3^2}{7\cdot1\cdot2\cdot2\cdot5}\left(\frac{z}{2}\right)^4\right.$$

$$- \frac{3^3}{10\cdot1\cdot2\cdot3\cdot2\cdot5\cdot8}\left(\frac{z}{2}\right)^6 + \cdots$$

$$\left. + (-1)^n \frac{3^n}{(3n+1)\cdot1\cdot2\cdot3\cdots n\cdot2\cdot5\cdot8\cdots(3n-1)}\left(\frac{z}{2}\right)^{2n} - \cdots\right]$$

$$D(z) = \frac{z^2}{2}\left[\frac{1}{9\cdot1^2-1} - \frac{(3z)^2}{2(9\cdot1^2-1)(9\cdot3^2-1)}\right.$$

$$\left. + \frac{(3z)^4}{3(9\cdot1^2-1)(9\cdot3^2-1)(9\cdot5^2-1)} - \cdots\right]$$

$$\tag{1.60}$$

with Gamma function $\Gamma(x)$ evaluated at $x = 4/3$:

$$\Gamma\left(\frac{4}{3}\right) = 0.8910 \quad \frac{2}{3}\Gamma\left(\frac{5}{3}\right) = 1.3541 \tag{1.61}$$

Consider, now, the boundary conditions. Let the column be clamped at both ends. At the bottom

$$x = 0 \quad z = \frac{2}{3}\sqrt{\frac{qL^3}{D}} = z_L \quad W = \frac{dW}{dx} = \frac{dW}{dz} = 0 \tag{1.62}$$

while at the upper end

$$x = L \quad z = W = \frac{dW}{dx} = \frac{dW}{dz} = 0 \tag{1.63}$$

In order to satisfy the latter boundary conditions, it is necessary to require that

$$C_2 = C_3 = 0 \tag{1.64}$$

The conditions at $x = 0$ yield

$$C_1 A(z_L) - (6N/q)D(z_L) = 0$$
$$C_1 J_{1/3}(z_L) - (6N/q)C(z_L) = 0 \tag{1.65}$$

resulting in the equation for the determination of the buckling load

$$A(z_L)C(z_L) - D(z_L)J_{1/3}(z_L) = 0 \tag{1.66}$$

Dinnik reports 5.72 to be the minimum root of this equation, yielding the critical, buckling load

$$q_{cr} = 317.15D/L^3 \tag{1.67}$$

Likewise, for the clamped–pinned column

$$z_L = 4.83 \quad q_{cr} = 52.49D/L^3 \tag{1.68}$$

while for the pinned column

$$z_L = 2.87 \quad q_{cr} = 18.53D/L^3 \tag{1.69}$$

It should be noted that the buckling of uniform columns under their own weight was revisited by Willers (1941) and Engelhardt (1954). They demonstrated that the general solution for the column's slope can be written in terms of Bessel and Lommel functions:

$$\frac{dW}{dz} = C_1 J_{1/3}(z) + C_2 J_{-1/3}(z) + C_3 s_{0,1/3}(z) \tag{1.70}$$

Equation (1.70) contains the Lommel function

$$S_{\mu,\nu}(Z) = \sum_{m=0}^{\infty} \frac{(-1)^m z^{\mu+1+2m}}{\left[(\mu+1)^2 - \nu^2\right] \cdot \left[(\mu+3)^2 - \nu^2\right] \cdots \left[(\mu+2m+1)^2 - \nu^2\right]} \tag{1.71}$$

with parameters $\mu = 0$, $\nu = \frac{1}{3}$ (Erdelyi, 1953). Lommel functions have also been utilized in axisymmetric vibrations of a piezoelectric cylinder by Adelman and Stavsky (1975), and in the impact buckling of a bar in a compressive testing machine by Elishakoff (1980). Note that without going through the clever transformations of variables, one can solve the problems discussed in this section straightforwardly by the Frobenius (power series) method for differential equations having algebraic coefficients (Leissa, 2000b), as is shown in Section 1.4.

1.4 Some Exact Solutions for the Vibration of Non-uniform Beams

The literature usually deals with beams of variable cross-sectional area. The simplest case of this kind is a beam of constant width and linearly varying thickness h:

$$h = h_1 + (h_0 - h_1)(x/L) \tag{1.72}$$

where h_1 is the thickness at the cross-section $x = 0$ and h_0 is the thickness attained at the cross-section $x = L$, where L is the length of the beam. For the tapered beam, the moment of inertia and the cross-sectional area are expressed as

$$
\begin{aligned}
I &= bh^3/12 = (b/12)[h_1 + (h_0 - h_1)x/L]^3 \\
A &= bh = b[h_1 + (h_0 - h_1)x/L]
\end{aligned}
\tag{1.73}
$$

The governing differential equation is

$$E\frac{d^2}{dx^2}\left\{\frac{b}{12}\left[h_1 + (h_0 - h_1)\frac{x}{L}\right]^3 \frac{d^2W}{dx^2}\right\} - \rho b\omega^2\left[h_1 + (h_0 - h_1)\frac{x}{L}\right]W = 0 \tag{1.74}$$

We introduce a new coordinate

$$X = h_1 + (h_0 - h_1)x/L \tag{1.75}$$

Equation (1.74) becomes

$$\frac{d^2}{dX^2}\left(X^3\frac{d^2W}{dx^2}\right) = k^4 XW \tag{1.76}$$

where

$$k^4 = 12\rho\omega^2 L^4 / E(h_0 - h_1)^4 \qquad (1.77)$$

Introducing the linear operator M such that

$$M = \frac{1}{X}\frac{d}{dX}\left(X^2\frac{d}{dX}\right) \qquad (1.78)$$

Equation (1.76) can be put in the form

$$(M + k^2)(M - k^2)W = 0 \qquad (1.79)$$

which has a solution obtained from the equations

$$\begin{aligned}(M + k^2)W &= 0 \\ (M - k^2)W &= 0\end{aligned} \qquad (1.80)$$

or

$$\frac{d}{dX}\left(X^2\frac{dW}{dX}\right) \pm k^2 XW = 0 \qquad (1.81)$$

The general solution of Eq. (1.81) is then written as

$$W(X) = \left[A_1 J_1\left(2k\sqrt{X}\right) + A_2 Y_1\left(2k\sqrt{X}\right) + A_3 I_1\left(2k\sqrt{X}\right) + A_4 K_1\left(2k\sqrt{X}\right)\right]/\sqrt{X} \qquad (1.82)$$

where J_1 and Y_1 are first-order Bessel functions of the first and second kind, and I_1 and K_1 are first-order modified Bessel functions of the first and second kind, respectively. The above solution is due to Mabie and Rogers (1964). Satisfying boundary conditions, one gets four algebraic equations for the coefficients A_1, A_2, A_3 and A_4. Requirement of non-triviality of the coefficients yields a transcendental equation that should be solved numerically. The solution in terms of Bessel functions for inhomogeneous beams was pioneered by Kirchhoff (1879, 1882), who studied the free vibrations of a wedge or cone beams. Further contributions were provided by Ward (1913), Nicholson (1920) and Wrinch (1922). Considerable simplification in derivations occurred for the "complete" beams due to the absence of Bessel functions of the second kind in the solution. As of now there is an extensive literature, including the book by Gorman (1975). The results of numerical evaluation are usually presented in terms of tables of figures, in various references. These were presented for the variable cross-section beam that is clamped at one end and pinned at the other, by Mabie and Rogers (1968). Cantilever beams with constant width and linearly variable thickness, or beams with constant thickness and linearly variable width, with an end mass, have been investigated by Mabie and Rogers (1964). The same authors (Mabie and Rogers, 1974) also studied transverse vibrations of double tapered beams. With the solution based on Bessel functions, Conway and Dubil (1965) tackled the truncated wedge and cone beams, and presented tables of frequencies for combinations of clamped, pinned and free boundary conditions. Lee (1976) dealt with a cantilever with a mass at one end, while Goel (1976) extended the method to a wedge and a cone beam with resilient supports at both ends. Sanger (1968) considered a special class of non-uniform beams, with the geometry enabling the expression of the solution in terms of Bessel functions. Solution in terms of Bessel functions for a beam, part of which is tapered while the other part is uniform, was studied by Auciello and Ercolano (1997).

For the cone and the wedge beams, an exact solution obtained using the Frobenius method was proposed by Naguleswaran (1990, 1992, 1994a,b) and the results were tabulated for different constraint conditions. Wang (1967) also utilized the method of Frobenius, and considered the following variations of the cross-sectional area and the moment of inertia, respectively:

$$A = A_0(x/L)^n \quad I = I_0(x/L)^m \tag{1.83}$$

where the constants m and n are any two positive numbers, and A_0 and I_0 are the cross-sectional area and moment of inertia, respectively, of the beam at the end $x = L$.

1.4.1 The Governing Differential Equation

The differential equation that governs the free vibrations of the beams that are inhomogeneous and/or have a variable cross-sectional area reads:

$$\frac{d^2}{dx^2}\left(D(\xi)\frac{d^2 W}{dx^2}\right) - \rho A \omega^2 W = 0 \tag{1.84}$$

It can be expressed as

$$\xi^m \frac{d^4 W}{d\xi^4} + 2m\xi^{m-1}\frac{d^3 W}{d\xi^3} + m(m-1)\xi^{m-2}\frac{d^2 W}{d\xi^2} - \Omega^2 \xi^n W = 0 \tag{1.85}$$

$$\xi = x/L \quad \Omega^2 = \omega^2 L^4 \rho A/EI_0$$

We multiply Eq. (1.85) by ξ^{4-m} and let $\theta = 4 - m + n$, yielding

$$\xi^4 \frac{d^4 W}{d\xi^4} + 2m\xi^3 \frac{d^3 W}{d\xi^3} + m(m-1)\xi^2 \frac{d^2 W}{d\xi^2} - \Omega^2 \xi^\theta W = 0 \tag{1.86}$$

It has a general solution in the form of a linear combination of convergent series of an infinite form about the origin (Ince, 1956). In this case, the general solution can be obtained by the method of Frobenius. We introduce the differential operator

$$\delta = \xi \frac{d}{d\xi} \tag{1.87}$$

which gives the general relations

$$\xi^r \frac{d^r}{d\xi^r} = \delta(\delta - 1)(\delta - 2) \cdots (\delta - r + 1) \tag{1.88}$$

Moreover,

$$\delta(\delta - 1)(\delta - 2) \cdots (\delta - r + 1)x^s = s(s - 1)(s - 2) \cdots (s - r + 1)x^s \tag{1.89}$$

Equation (1.86) may be rewritten in the form

$$\delta(\delta - 1)(\delta + m - 2)(\delta + m - 3)W - \Omega^2 \xi^\theta W = 0 \tag{1.90}$$

We introduce the following notations

$$u = \frac{\Omega^2 \xi^\theta}{\theta^4} \quad \delta_u = u\left(\frac{d}{du}\right) \tag{1.91}$$

Then, Eq. (1.89) becomes

$$\delta_u \left(\delta_u - \frac{1}{\theta}\right)\left(\delta_u - \frac{2-m}{\theta}\right)\left(\delta_u - \frac{3-m}{\theta}\right)W - uW = 0 \tag{1.92}$$

which is due to Wang (1967). Equation (1.92) is a type of generalized hypergeometric equation (Erdelyi, 1953). Its general solution is a linear combination of the following independent generalized hypergeometric series:

$$
\begin{aligned}
W_1 &= {}_0F_3 \left(-; b_1, b_2, b_3; u\right) \\
W_2 &= u^{1-b_1} {}_0F_3 \left(-; 2 - b_1, b_2 - b_1 + 1, b_3 - b_1 + 1; u\right) \\
W_3 &= u^{1-b_2} {}_0F_3 \left(-; b_1 - b_2 + 1, 2 - b_2, b_3 - b_2 + 1; u\right) \\
W_4 &= u^{1-b_3} {}_0F_3 \left(-; b_1 - b_3 + 1, b_2 - b_3 + 1, 2 - b_3; u\right)
\end{aligned}
\tag{1.93}
$$

where

$$b_1 = (3 - m + n)/\theta \quad b_2 = (2 + n)/\theta \quad b_3 = (1 + n)/\theta \tag{1.94}$$

The generalized hypergeometric function is defined as

$$_0F_3(a_1, a_2, \ldots, a_p; b_1, b_2, \ldots, b_q; u) = 1 + \sum_{n=1}^{\infty} \left[\prod_{i=1}^{p} (a_i)_n \, u^n \right] \left[\prod_{j=1}^{q} (b_j)_n \, n! \right]^{-1}$$

(1.95)

The series in Eq. (1.93) are undefined or not linearly independent when either b_1, b_2 or b_3 is an integer or two of them differ by an integer. For these cases the logarithmic terms appear in the general solutions. The detailed derivations of the logarithmic solutions by the method of Frobenius were presented by Wang (1967) and are not recapitulated here. Wang (1967) considered the following four cases: (a) when two b coefficients are equal (this occurs if $m = 1$ or $m = 2$ or $m = 4$, which yields the coincident values for b_1 and b_3); (b) when one b value equals unity (indeed, when the parameter m equals 2 or 3, the value of b_2 or b_3 turns out to be unity); (c) when b_1 is a negative integer or zero (this happens when θ is a reciprocal of a positive integer); (d) when two b's differ by an integer (this case occurs for certain combinations of parameters m and n). For example, the combination $m = 3$ and $n = 1$ yields $b_1 = \frac{1}{2}$ and $b_3 = \frac{3}{2}$, which have a difference of unity. Some special cases follow from Wang's (1967) general formulation. For uniform beams, i.e., $m = n = 0$, $\theta = 4$ and $b_1 = \frac{3}{4}$, $b_2 = \frac{2}{4}$, $b_3 = \frac{1}{4}$ the hypergeometric functions reduce to the familiar solution in Eq. (1.6).

Another special combination of parameters m and n occurs when $m - n = 2$ or $\theta = 2$, which includes wedge-shaped and cone-shaped beams. The solution reduces to the Bessel functions on utilizing the relationship between the Bessel functions and the generalized hypergeometric functions (Rainville, 1960):

$$J_v(z) = \frac{(z/2)^v}{\Gamma(1+v)} {}_0F_1(-; 1+v; -z^2/4)$$

$$I_v(z) = \frac{(z/2)^v}{\Gamma(1+v)} {}_0F_1(-; 1+v; z^2/4)$$

(1.96)

The last special case is for beams with constant thickness and linearly tapered width. Then, $m = n = 1$, yielding $b_1 = \frac{3}{4}$, $b_2 = \frac{3}{4}$, $b_3 = \frac{1}{2}$. Since the values of b_1 and b_2 are equal, the series solutions for W_1 and W_2 in Eq. (1.93) coincide. The fundamental solution W_2 is replaced by

$$W_2 = W_1 \ln u - 4u^{1/4} \left[\frac{4^4}{5 \cdot 4^2 \cdot 3} \left(\frac{1}{5} + \frac{1}{3} + \frac{1}{2} \right) u \right.$$

$$\left. + \frac{4^4}{9 \cdot 8^2 \cdot 7} \frac{4^4}{5 \cdot 4^2 \cdot 3} \left(\frac{1}{9} + \frac{1}{7} + \frac{1}{4} + \frac{1}{5} + \frac{1}{3} + \frac{1}{2} \right) u^2 + \cdots \right]$$

(1.97)

This form of the solution for the beam with constant thickness and linearly tapered width was given by Ono (1925). Wang (1967) presented numerous solutions and attendant numerical evaluations for the fundamental and second frequencies.

1.5   Exact Solution for Buckling of Non-uniform Columns

The study of buckling of non-uniform columns was pioneered by Euler (1759). He considered the columns with flexural rigidity given as a polynomial $(a + bx/L)^m$. Actually, Euler represented the flexural rigidity as a product of the modulus of elasticity and the positive quantity k^2 (denoted by Euler as kk), unbeknown to him that the latter quantity was a moment of inertia. Euler referred to the product Ek^2 as a "moment of stiffness" (*moment du ressort or moment de roideur*). He hit upon the cases in which the buckling modes are given by the elementary functions. The analogous flexural rigidity variation,

$$D(x) = EI(x/b)^m \tag{1.98}$$

in the non-uniform columns was studied by Dinnik (1912). He demonstrated that for any m, equal to any, positive or negative, integer or decimal number, except 2, the equation

$$EI(x/b)^m W'' + PW = 0 \tag{1.99}$$

is integrable in terms of the Bessel functions of order

$$n = 1/(m - 2) \tag{1.100}$$

In these formulas, letting $m = 0$ and using the known relations between the Bessel functions of order $\frac{1}{2}$ and trigonometric functions

$$I_{1/2}(x) = \sin(x)\sqrt{2/\pi x} \quad I_{-1/2}(x) = \cos(x)\sqrt{2/\pi x} \tag{1.101}$$

we recover Euler's solution for the uniform column. At $m = 4$, we again get the Bessel functions of order $\pm\frac{1}{2}$, but for an argument other than at $m = 0$. Thus, for $m = 4$ the solution can be obtained both by Bessel and trigonometric functions. For $m = 4$, Eq. (1.97) is rewritten as

$$x^4 W'' + U^2 W = 0 \quad U^2 = Pb^4/EI \tag{1.102}$$

with the buckling mode

$$W(x) = x[A_i \cos(U/x) + A_2 \sin(U/x)] \tag{1.103}$$

whereas for $m = 2$, the mode shape reads

$$W(x) = \sqrt{x}\{A_1 \cos[\mu \ln(x/b)] + A_2 \sin[\mu \ln(x/b)]\} \tag{1.104}$$

with

$$\mu^2 = \frac{Pb^2}{EI} - \frac{1}{4}. \tag{1.105}$$

For the buckling of the inhomogeneous column with the elastic modulus variation

$$E(x) = E_0[1 \mp k(x/L)]^2 \tag{1.106}$$

Freudenthal (1966) introduced a new variable

$$v^2 = [1 \mp k(x/L)]^2 \tag{1.107}$$

and reduced the buckling equation to

$$v^2 \frac{d^2 W}{dv^2} + \lambda_0 \left(\frac{L}{k}\right) W = 0 \quad \lambda_0^2 = \frac{P}{E_0 I} \tag{1.108}$$

For $(2\lambda L/k)^2 > 1$ he derived a solution analogous to Eq. (1.104):

$$W = A_1 v \cos(a \ln v) + A_2 v \sin(a \ln v)$$

$$v = \frac{1}{2}\sqrt{|1 - (2\lambda L/k)^2|} \tag{1.109}$$

From the boundary conditions, $W = 0$ at $x = 0$ and $x = L$ the following critical value follows

$$P_{\text{cr}} = \frac{\pi^2 EI}{L^2} \left[\left(\frac{k}{\ln(1 \mp k)}\right)^2 - \left(\frac{k}{2\pi}\right)^2\right] \tag{1.110}$$

where the expression in brackets tends to unity as $k \to 0$. For

$$D(x) = D_0(x/L)^\alpha \tag{1.111}$$

Freudenthal (1966) gave a solution in terms of Bessel functions.

Columns with variable cross-section were extensively studied by Dinnik (1912, 1929, 1955a,b). He considered the columns with the following flexural rigidity distribution:

$$D(x) = b(L - x)^m \tag{1.112}$$

The columns were under the axial distributed loading

$$q = c(L - x)^n \qquad (1.113)$$

where b, c, m and n are constants. In the cross-section x

$$M(x) = \int_x^L c(L - u)^n [V(u) - W(x)] \, du$$

$$= c \int_x^L (L - u)^n V(u) \, du - \frac{c}{n + 1} (L - x)^{n+1} W(x) \qquad (1.114)$$

The differential equation reads

$$W''' - \frac{1 - m}{2} W'' + \frac{c}{b(n + 1)} (L - x)^{n-m+1} W' = 0 \qquad (1.115)$$

leading to

$$\frac{dW}{dx} = (L - x)^{(1-m)/2} \left[C_1 J_{-\frac{1-m}{n-m+3}}(y) + C_2 J_{-\frac{1-m}{n-m+3}}(y) \right], \qquad (1.116)$$

where

$$y^2 = 4c(L - x)^{n-m+3} / b(n + 1)(n - m + 3)^2 \qquad (1.117)$$

Letting $m = n = 0$, $b = B$ and $c = q$ results in the particular case of the uniform column:

$$\frac{dW}{dx} = (L - x)^{1/2} \left\{ C_1 J_{1/3} \left[\frac{2}{3} \sqrt{\frac{q(L - x)^3}{D}} \right] + C_2 J_{-1/3} \left[\frac{2}{3} \sqrt{\frac{q(L - x)^3}{D}} \right] \right\} \qquad (1.118)$$

For the cone, whose radius at the base equals R and whose height equals L, we get

$$r = R(L - x)/L \quad q = \pi R^2 k (L - x)^2 / L^2 \quad D = EI = \pi R^4 E (L - x)^4 / 4L^4 \qquad (1.119)$$

where r is the radius of the cone at distance x from the origin. Comparison of Eqs. (1.119) with Eqs. (1.112) and (1.113) results in

$$m = 4 \quad n = 2 \quad b = \pi R^4 E / 4L^4 \quad c = \pi R^2 k / L^2$$

$$\frac{dW}{dx} = (L - x)^{-3/2} [C_1 J_3(z) + C_2 Y_3(z)] \qquad (1.120)$$

$$z^2 = 16kL^2 (L - x)/3ER^2$$

For the clamped–free column, the buckling load is found from the equation

$$Y_3 \left(\sqrt{16kL^3/3ER^2} \right) = 0 \tag{1.121}$$

The first root, calculated by Dinnik (1955b) is 4.43. Hence, the critical length of the column equals

$$L_{cr} = 1.5453 \sqrt[3]{ER^2/K} \tag{1.122}$$

The optimal design problem of columns naturally involved variable cross-sections. The problem was first posed and solved by Lagrange (1770–1773). He attempted to determine the shape of a column of given length and volume so that the attendant buckling load attains a maximum. Further contributions were made by Clausen (1851), Bairstow and Stedman (1914), Blasius (1914), Darnley (1918), Nikolai (1955), Keller (1960), Tadjbakhsh and Keller (1962), Keller and Niordson (1966), Adali (1979, 1981), Cox (1992), Cox and Overton (1992), Banichuk (1974), Seyranian (1983, 1995), McCarthy (1999) and others. The finite element method in buckling optimization was employed by Simitses et al. (1973), Manickarajah et al. (2000) and others. Xie and Steven (1993) proposed a simple evolutionary method for the topology, shape and layout optimization of structures. In this method, a structure or the design domain is divided into a fine mesh of elements, and inefficient material is gradually removed according to the design criteria, and the residual structure evolves towards the desired optimum. Keller (1960) proved that the optimal solid convex cross-section against buckling failure has the form of an equivalent triangle; moreover, the dimension of that triangle should change along the axis according to the law found earlier for the optimal shape of a circular column. Szyszkowski and Watson (1988) proposed that the optimum shape of a structure with respect to buckling should have the configuration for which the specific bending energy due to the fundamental buckling mode is uniform. Gajewski and Życzkowski (1988) write: "... Tadjbakhsh and Keller (1962) derived the optimal solutions for columns clamped at one end and pinned at the other, and for clamped–clamped columns. The solution in the latter case, however is incorrect; it was obtained with respect to the first buckling mode, whereas a bimodal solution is here necessary (Olhoff and Rasmussen, 1977)." This sentiment seems to be shared by Cox (1992) and Cox and Overton (1992). Yet, Spillers and Meyers (1997) argued that the "work of Cox and Overton (1992), which claims to correct errors in the classic solution of Tadjbakhsh and Keller (1962) (for the optimal design of a column), itself contains an error and actually only serves to further establish the validity of the Tadjbakhsh–Keller solution." They also note that "it was... a surprise to many to see a rather robust discussion of the validity of the Tadjbakhsh–Keller solution ... running from the late 1970's into the mid-1980's." It appears that a detailed and careful literature review of the calculations and/or miscalculations in the optimum design of columns is called for, for the benefit of the general audience. The reader may consult reviews on selected topics by Ashley (1982), Banichuk (1982) and others. It is worth noting that the optimum design of vibrating structures was conducted by Niordson (1965, 1970), Karihaloo and Niordson (1973), Olhoff (1970, 1976a,b, 1977) and others (see also Section 1.9).

Exact methods of calculation of buckling loads and natural frequencies were extensively reviewed by Williams and Wittrick (1983).

1.6   Other Direct Methods (FDM, FEM, DQM)

In the earlier sections we considered exact solutions of differential equations describing buckling or vibration of structures. As Fried (1979) writes: "the theoretical limiting process in the mathematical formulation creates *differential* equations to describe physical phenomena or engineering processes. A symbolic solution to these equations in terms of some elementary functions, even if existing, is, as a rule, very hard to come by and the programming necessary to obtain such a symbolic solution impossible. To overcome this difficulty, numerical techniques of great generality have been invented, based on *discretization* of the problem, on the division of the continuous flow of events or continuous change of state into a series of discrete states formulated algebraically, with the limiting process on convergence deferred to the numerical stage of the solution."

Many symbolic algebraic or numerical methods have been developed in the past few decades that make it possible to carry out the vibration and buckling analysis of beams with arbitrarily varying cross-sectional areas or inhomogeneities. The class of approximate methods can be compared, in the terminology of Leissa (2000a) to the limitless Pandora's box. In these circumstances, we confine ourselves to presenting a superficial, yet hopefully still meaningful, overview of them.

The method of successive approximations, discussed by Engesser (1893) and Vianello (1898), was extensively employed in the "pre-FEM" literature. Sekhniashvili (1966) utilizes this method, as discussed by Bernshtein (1941) with attendant theoretical justification provided by Sushenkov (see Papkovich, 1933), who showed that in the limit the method leads to the fundamental frequency and the corresponding mode shape. Later, the method was utilized by Ratzensdorfer (1943), Popovich (1962), Mitelman (1970), Hodges (1997) and others. The numerical integration method was utilized by Newmark (1943), Lu et al. (1983) and Sakiyama (1986).

Special approximate methods for columns of varying flexural rigidity, including lower bound approximations, were developed by Miesse (1949),

Silver (1951), Abbassi (1958), Appl and Zorowski (1959), Fadle (1962, 1963), Mazurkiewicz (1964, 1965), Pnueli (1972a–c, 1999) and others.

Some authors (Thomson, 1949; Ram and Rao, 1951; Fadle, 1962; Ram, 1963; Bert, 1984a,b, 1987a,b) dealt with the so-called "*n*-section column" or "stepped column" which is a column consisting of *n* sections, each having a different but constant flexural rigidity. The exact solutions were obtained by solving simultaneous differential equations.

The basic idea of the finite-difference method (FDM) is the approximation of a derivative of a function at a point by an algebraic expression containing the value of the function at that point and at several nearby points. Thus, the governing differential equation is replaced by an algebraic equation. The replacement of a continuous function by an algebraic expression composed of the values of the function at several discrete points is equivalent to replacing an original distributed system by one with several lumped masses, i.e., reduction of the continuous system to a *n*-section column. The use of the FDM for columns with varying flexural rigidity and pinned ends was used by Salvadori (1951), Srinivasan (1964), Szidarovski (1964), Girijavallabhan (1969) and Iremonger (1980).

The differential quadrature method (DQM) was proposed Bellman and Casti (1971). In the DQM the discretization is accomplished by expressing at each grid point the calculus operator value of a function with respect to a coordinate direction at any discrete point as the weighted linear sum of the values of the function at *all* the discrete points chosen in that direction. The *r*th partial derivative of a function ψ is expressed as

$$\frac{\partial^r \psi}{\partial x^r}\bigg|_{x=x_i} - \sum_{k=1}^{N_x} A_{ik}^r \psi_{ki} \qquad i = 1, 2, \ldots, N_x \qquad (1.123)$$

Likewise,

$$\frac{\partial^s \psi}{\partial y^s}\bigg|_{y=y_j} = \sum_{l=1}^{N_y} B_{jl}^s \psi_{jl} \qquad j = 1, 2, \ldots, N_y \qquad (1.124)$$

where $A_{ik}^{(r)}$ and $B_{jl}^{(s)}$ are the respective weighting coefficients. Also $\psi_{ij} = \psi(x_i\, y_j)$. Bellman et al. (1972) presented an idea of using the polynomial test functions for the determination of the weighting coefficients. DQM was utilized for vibration problems by Bert et al. (1988, 1994), whereas Jang et al. (1989), and Sherbourne and Pandey (1991) employed it for buckling analysis. An extensive review of this method was provided by Bert and Malik (1996).

The finite element method (FEM) is a numerical technique that realizes the system as an assembly connected to one another at points called *nodes*. Each element is associated with generalized displacements and generalized forces, that are internal forces as far as overall structure is concerned, but they represent external forces when individual elements are involved. The structure's stiffness matrix is composed of the stiffness matrix of individual elements. This is usually done by applying at each node the conditions of equilibrium and deformation compatibility. The FEM to eigenvalue problem was applied by Karabalis and Beskos (1983), Yatram and Awadalla (1967), Tebede and Tall (1973) and Gallagher and Padlog (1963). The Convergence study of the finite element method to the exact solution was performed by Tong and Pian (1967), Strang and Fix (1969), Lindberg and Olson (1970), Fried (1971) and Tong et al. (1971), in the vibration context. Extensive convergence study in the buckling context was undertaken by Seide (1975). The Finite element method appears to be the most universal method that is uniformly available for direct computer applications. For extensive discussions of various numerical methods the readers must consult the books by Meirovitch (1980, 1997), Shabana (1994), Schuëller (1991), Krätzig and Niemann (1996) and others.

1.7   Eisenberger's Exact Finite Element Method

Often, it makes sense to represent the unknown buckling or vibration mode by the following infinite series:

$$w(x) = \sum_{i=0}^{\infty} a_i \psi_i(x) \tag{1.125}$$

For example, the buckling mode of the non-uniform column that is pinned at its ends, can be represented as the series in terms of the eigenfunctions of the uniform column:

$$w(x) = p_1 \sin \frac{\pi x}{L} + p_2 \sin \frac{2\pi x}{L} + p_3 \sin \frac{3\pi x}{L} + \cdots + p_n \sin \frac{n\pi x}{L} + \cdots \tag{1.126}$$

Since the functions themselves can be written as powers in terms of x, we can rewrite Eq. (1.126) as follows:

$$w(x) = a_0 + a_1 x + a_2 x^2 + a_3 x^3 + \cdots + a_n x^n + \cdots \tag{1.127}$$

Let us first demonstrate this method for the uniform column pinned at its ends. We substitute into the governing differential equation

$$\frac{d^2 W}{dx^2} + \lambda^2 W = 0 \quad \lambda^2 = \frac{P}{EI} \tag{1.128}$$

with boundary conditions

$$W(0) = W(L) = 0 \tag{1.129}$$

Substituting Eq. (1.127) into Eq. (1.128) leads to

$$1 \cdot 2a_2 x^0 + 2 \cdot 3a_3 x^1 + 3 \cdot 4a_4 x^2 + \cdots + \lambda^2 a_0 x^0 + \lambda^2 a_1 x^1 + \lambda^2 a_2 x^2 + \cdots = 0 \tag{1.130}$$

We order the terms as

$$x^0 \left(1 \cdot 2a_2 + \lambda^2 a_0\right) + x^1 \left(2 \cdot 3a_3 + \lambda^2 a_1\right) + x^2 \left(3 \cdot 4a_4 + \lambda^2 a_2\right) + \cdots = 0 \tag{1.131}$$

Since this expression must vanish for any x, we get

$$a_2 = -\frac{\lambda^2 a_0}{1 \cdot 2} \quad a_3 = -\frac{\lambda^2 a_1}{1 \cdot 2 \cdot 3} \quad a_4 = \frac{\lambda^4 a_0}{1 \cdot 2 \cdot 3 \cdot 4} \quad a_5 = \frac{\lambda^4 a_1}{5!} \quad a_6 = -\frac{\lambda^6 a_0}{6!} \tag{1.132}$$

and so on. Thus, the displacement becomes

$$W(x) = a_0 + a_1 x - \frac{\lambda^2 a_0}{2!} x^2 - \frac{\lambda^2 a_1}{3!} x^3 - \frac{\lambda^2 a_0}{4!} x^4 + \frac{\lambda^4 a_1}{5!} x^5 - \cdots \tag{1.133}$$

Due to the boundary condition at $x = 0$, $a_0 = 0$. Due to the boundary condition at $x = L$, we get

$$a_1 \left[L - \frac{\lambda^2 L^3}{3!} + \frac{\lambda^4 L^5}{5!} - \frac{\lambda^6 L^7}{7!} + \cdots + (-1)^n \frac{\lambda^{2n} L^{2n+1}}{(2n+1)!} \right] = 0 \tag{1.134}$$

For the solution to be non-trivial we set $a_1 \neq 0$, and the expression in parentheses must vanish. This leads, once we substitute $\lambda L = \alpha$, to

$$1 - \alpha^2/3! + \alpha^4/5! - \alpha^6/7! + \alpha^8/9! + \cdots = 0 \tag{1.135}$$

Taking into account that

$$\sin \alpha = \alpha - \alpha^3/3! + \alpha^5/5! - \alpha^7/7! + \cdots$$
$$\sin \pi n = \pi - \pi^3/3! + \pi^5 5! - \pi^7/7! + \cdots \tag{1.136}$$

we conclude that Eq. (1.134) is equivalent to Eq. (1.19.1) with $\alpha = n\pi$, and thus we obtain the minimal eigenvalue given in Eq. (1.20.1).

By taking only two terms we get the following approximation

$$\lambda^2 = 3!/L^2 \quad P_{cr} = 6EI/L^2 \tag{1.137}$$

which constitutes a 39% underestimation of the buckling load; but, with three terms

$$\lambda^2 = 9.48/L^2 \quad P_{cr} = 9.48EI/L^2 \tag{1.138}$$

and this yields a 3.94% error. With more terms the error diminishes rapidly. We followed Bürgermeister and Steup (1957) here in exposing the direct series representation method.

The method was greatly expanded and re-interpreted by Eisenberger (1991a–d). He considered bars, columns and beams of variable properties. To illustrate his approach, consider the behavior of the beam in the presence of variable axial forces

$$\frac{d^2}{d^{x}2}\left[D(x)\frac{d^2W}{dx^2}\right] - \frac{d}{dx}\left[N(x)\frac{dW}{dx}\right] = P(x) \tag{1.139}$$

where W is the transverse displacement, $N(x)$ the axial force, $P(x)$ the distributed transverse load along the member and $D(x) = EI(x)$ is the flexural rigidity. As Eisenberger (1991) notes,

> The solution for the general case of polynomial variation of $I(x)$, $N(x)$ and $P(x)$ along the beam is not generally available. Using the finite element technique, it is possible to derive the terms in the stiffness matrix. We assume that the shape functions for the element are polynomials and we have to find the appropriate coefficients. It is widely known that exact terms will result, and if one uses the solution of the differential equation as the shape functions, for the derivation of the terms in the stiffness matrix. In this work "exact" shape functions are used, to derive the exact flexural rigidity coefficients. These shape functions are "exact" up to the accuracy of the computer, or up to a preset value set by the analyst.

Following Eisenberger, the functions $D(x)$, $N(x)$ and $P(x)$ are written as

$$D(x) = \sum_{i=0}^{j} D_i x^i \quad N(x) = \sum_{i=0}^{l} N_i x^i \quad P(x) = \sum_{i=0}^{m} P_i x^i \tag{1.140}$$

where j, l and m are integers representing the number of terms in each series. Introducing a non-dimensional variable $\xi = x/L$, where L is the length of the beam Eq. (1.139) is rewritten as

$$\frac{d^2}{d\xi^2}\left[d(\xi)\frac{d^2W}{d\xi^2}\right] - \frac{d}{d\xi}\left[n(\xi)\frac{dW}{d\xi}\right] = p(\xi) \tag{1.141}$$

where

$$d(\xi) = \sum_{i=0}^{j} D_i L^i \xi^i = \sum_{i=0}^{j} d_i \xi^i$$

$$n(\xi) = \sum_{i=0}^{l} N_i L^{i+2} \xi^i = \sum_{i=0}^{l} n_i \xi^i \tag{1.142}$$

$$p(\xi) = \sum_{i=0}^{m} P_i L^{i+4} \xi^i = \sum_{i=0}^{m} p_i \xi^i$$

The solution $W(\xi)$ is chosen as an infinite series, as in Eq. (1.127);

$$W(\xi) = \sum_{i=0}^{\infty} w_i \xi^i \tag{1.143}$$

Substitution of Eq. (1.143) into Eq. (1.141) bearing in mind Eq. (1.142) results in

$$-\sum_{i=0}^{\infty}\sum_{k=0}^{i}(k+1)(i-k+1)n_{k+1}w_{i-k+1}\xi^i$$

$$-\sum_{i=0}^{\infty}\sum_{k=0}^{i}(i-k+1)(i-k+2)n_k w_{i-k+2}\xi^i$$

$$+\sum_{i=0}^{\infty}\sum_{k=0}^{i}(k+1)(k+2)(i-k+1)d_{k+2}w_{i-k+2}w_{i-k+2}\xi^i$$

$$+\sum_{i=0}^{\infty}\sum_{k=0}^{i}2(k+1)(i-k+1)(i-k+2)(i-k+3)d_{k+1}w_{i-k+3}\xi^i$$

$$+\sum_{i=0}^{\infty}\sum_{k=0}^{i}(i-k+1)(i-k+2)(i-k+3)(i-k+4)d_k w_{i-k+4}\xi^i$$

$$=\sum_{i=0}^{m} p_i \xi^i \tag{1.144}$$

This equation must be satisfied for every value of ξ. Hence, the following equality must hold

$$-\sum_{k=0}^{i}(k+1)(i-k+1)n_{k+1}w_{i-k+1} - \sum_{k=0}^{i}(i-k+1)(i-k+2)n_k w_{i-k+2}$$

$$+\sum_{k=0}^{i}(k+1)(k+2)(i-k+1)d_{k+2}w_{i-k+2}$$

$$+\sum_{k=0}^{i}2(k+1)(i-k+1)(i-k+2)(i-k+3)d_{k+1}w_{i-k+3}$$

$$+\sum_{k=1}^{i}(i-k+1)(i-k+2)(i-k+3)(i-k+4)d_k w_{i-k+4} = p_i \qquad (1.145)$$

allowing one to express w_{i+4} as follows:

$$w_{i+4} = \frac{1}{d_0(i+1)(i+2)(i+3)(i+4)}\left[p_i + \sum_{k=0}^{i}(k+1)(i-k+1)n_{k+1}w_{i-k+1} \right.$$

$$+\sum_{k=0}^{i}(i-k+1)(i-k+2)n_k w_{i-k+2}$$

$$-\sum_{k=0}^{i}(k+1)(k+2)(i-k+1)d_{k+2}w_{i-k+2}$$

$$-\sum_{k=0}^{i}2(k+1)(i-k+1)(i-k+2)(i-k+3)d_{k+1}w_{i-k+3}\xi$$

$$\left. -\sum_{k=0}^{i}(i-k+1)(i-k+2)(i-k+3)(i-k+4)d_k w_{i-k+4} \right] \qquad (1.146)$$

The missing first four coefficients are found by imposing boundary conditions; the next step consists in solving the problem numerically. For details one should consult the study by Eisenberger (1991a–d). The important contribution by Eisenberger (1991a–d) lies in his re-interpretation of the above series method. He proposed to use the above solution to form the stiffness matrix S in the context of the finite element method:

$$S = \int_0^1 [F''(\xi)]EI(\xi)F''(\xi)\,d\xi \qquad (1.147)$$

where $F''(\xi)$ are the second derivatives of the basic functions. Alternatively, "applying unit displacements at four degrees of freedom at the element ends one will find the exact shape functions as in the direct stiffness method. Then, since the shapes are exact, the flexural rigidity terms which are the shear forces and bending moments at the ends of the member, are found directly as the bending flexural rigidity $EI(x)$ multiplied by the third and second derivatives of the shape functions at $x = 0$ and $x = L$" (Eisenberger, 2000b). He studied longitudinal vibrations of a variable cross-section bar (1991), torsional analysis and vibrations of open and variable cross-section bars (Eisenberger, 1995a,b, 1997a,b), vibration frequencies of beams on variable one- and two-parameter elastic foundations, and apparently was amongst the first one to coin the term "an exact finite element method" (Eisenberger, 1990). One should also mention work by Banerjee and Williams (1985a,b, 1986) who gave solutions for a few cases of tapered members, as well as recent work by Eisenberger et al. (1995) and Abramovich et al. (1996), who applied the exact finite element method to vibration and buckling of composite beams and to Bernouilli–Euler and Bresse–Timoshenko beams (Eisenberger, 1995a,b).

A natural question arises: How accurate is the exact finite element method? In his papers, Eisenberger (1991a–d) performed extensive numerical analyses and compared his results with solutions provided by other investigators. Extremely good accuracy was recorded. As he notes: "The application of different sets of boundary conditions is straightforward as in the standard stiffness method of analysis ... Comparing this method to the finite element method or the finite difference method points out the ... advantage of the method: only one element is needed for the solution. Thus, the results are computed much faster." As he mentions, in his work "exact buckling loads (up to the accuracy of the computer) for variable cross section members with variable axial loads are given (implying that the method yields approximate buckling loads that can be made arbitrarily close to the exact values)." Thus, Eisenberger's exact FEM is a very attractive tool for the eigenvalue analysis of inhomogeneous structures. The exact finite element method was applied to buckling problems by Pieczara (1987), Waszczysyn and Pieczara (1990) and Janus and Waszczysyn (1991). The definitive works by Åkesson (1976), Åkesson and Tägnfors (1978) and Åkesson et al. (1972) in vibration theory, providing a unified computer program PFVIBAT, should be mentioned. Currently, the exact FEM method is being developed for variable thickness membranes and plates (Eisenberger, 2000b).

1.8 Semi-inverse or Semi-direct Methods

The principle behind of these methods lies in approximating the mode shape in vibration or buckling by the analytic expression containing a finite number of parameters c_i that are treated to be arbitrary in the outset of the solution process:

$$W(x) = \sum_{i=0}^{n} c_i \psi_i \qquad (1.148)$$

where $\psi_i(x)$ are coordinate functions that satisfy all or part of the boundary conditions pertinent to the problem. The functions $\psi_i(x)$ are chosen so as to represent in an appropriate manner the anticipated mode shape $W(x)$. Most convenient are coordinate functions drawn from the complete set of functions. The parameters c_i are determined from the condition of the best approximation of $W(x)$ by the series (1.148). These methods are referred to as *semi-inverse* ones by Grigoliuk and Selezov (1969) and appears to best represent their essence. Indeed, the *unknown* solution is postulated to be represented by *known* functions $\psi_i(x)$, whereas the coefficients c_i are determined by some condition of best approximation; it is as if we guess the solutions but not completely. The coefficients of the guessed solutions should still be evaluated. These methods can also be referred as *semi-direct* problems.

The first method of this kind is the method of Rayleigh (1873) suggested for the determination of the fundamental frequency of the string. He then generalized it for higher modes (Rayleigh, 1899a,b, 1911) [the impact of Lord Rayleigh on engineering vibration theory is described in an interesting paper by Crandall (1995)]. Ritz (1908) provided the mathematical foundation of the method. This method in its single-term application is usually called *Rayleigh's method*, while in the multi-term form is usually referred to as the *Rayleigh–Ritz* method (or sometimes simply as the Ritz method). Usually, the polynomial or trigonometric expressions are utilized as the coordinate functions, which should satisfy the geometric boundary conditions. Rayleigh's method is universally used, in almost all textbooks on both vibration and buckling, due to its simplicity. In the buckling context recent applications include those by Manicka Selvam (1997, 1998). Numerous applications of both the Rayleigh and the Rayleigh–Ritz method were given by Laura and Cortinez (1985, 1986a,b, 1987) and others. They published numerous studies in which the coordinate functions were polynomial functions, for beams, circular plates and rectangular plates for a wide range of practical problems. Thirty-six terms in conjunction with the Rayleigh–Ritz method were employed by Leissa (1973) to study the vibrations of plates under various boundary conditions, when the exact solution is unavailable. As Timoshenko (1953) wrote:

> The idea of calculating frequencies directly from an energy consideration, without solving differential equations, was ... elaborated by Walter Ritz (1909) and the Rayleigh–Ritz method is now widely used not only in studying vibration but in solving problems in elasticity, theory of structures, nonlinear mechanics, and other branches of physics. Perhaps no other single mathematical tool has led to as much research in the strength of materials and theory of elasticity.

The first application of Rayleigh's method to the buckling problem was performed by Bryan (1888). In 1910 Timoshenko utilized the Rayleigh and Rayleigh–Ritz methods extensively for buckling problems (Timoshenko, 1910). His book was submitted for the Zhuravsky prize. Boobnov (1913) provided one of the recommendation letters. In his overly positive review, Boobnov (1913), however, made an extremely constructive criticism of Timoshenko's work (may all criticisms turn out to be as efficient as Boobnov's!). He stated that the final equations can be obtained, more straight-forwardly, by substituting into the differential equation the series of type (1.148) with each coordinate function satisfying all boundary conditions. The result of such a substitution should be made orthogonal to each of the coordinate functions. Galerkin (1915) wrote an elegant paper in which he provided several evaluated examples of the above idea. Since then the method has been intensively utilized in the East. In the Russian literature it is mostly referred to as the Boobnov–Galerkin or the Galerkin method, although Grigoliuk (1975, 1996) maintains that it should be referred to as the Boobnov method. One can visualize that Galerkin's (1915a,b) work was a wonderful vehicle through which the idea exposed by Boobnov (1913) was not lost. "Well written, supplemented with many examples, executed in detail, this paper [by Galerkin (1915)] should have attracted an interest," according to Grigoliuk (1975). Indeed, clear exposition of the 1915 paper attracted Biezeno (1924), Henky (1927), Duncan (1937), and other Western scientists to this method. In the West it is most often called *the Galerkin method*. To digress, note that Crandall (1956), when describing the Ritz method wrote: "The same ideas had earlier been applied to eigenvalue problems by Lord Rayleigh. . .We call it the Ritz method when it is applied to equilibrium problems and the Rayleigh–Ritz method when applied to eigenvalue problems." Presently, the uniformly accepted term is the *Rayleigh–Ritz method* for *both* equilibrium and eigenvalue problems. Returning to the method advanced by Boobnov (1913) and Galerkin (1915a,b), it appears, as in the case of the Rayleigh–Ritz method, that the most appropriate name of the method would be the *Boobnov–Galerkin method*, as it is called in the extensive review of it, in the volume dedicated to Galerkin's 100th anniversary of birth, by Vorovich (1975). He stressed that in his paper Galerkin (1915a,b), in addition to utilizing Boobnov's idea, or proposing it independently, also provided "a mechanical treatment of the method as a procedure of choosing parameters of approximating the solution, during which on the widening classes of the possible displacements the total work of internal and external forces of the system turns out to be zero." Another widely used name for this technique is the *weighted residual method* (Finlayson, 1972; Meirovitch, 1997). As Schmidt (1990) wrote: "According to Finlayson (1972), the (Boobnov–) Galërkin method may be regarded as a special case of the method of weighted residuals." (see also Pomraning, 1966).

A general method for the analytical solution via the Boobnov–Galerkin method for the problem of buckling of a column with varying flexural rigidity and non-ideal boundary conditions was introduced by Durban and Baruch (1972), whereas the non-uniform beams of variable mass density were treated by Sekhniashvili (1950a,b, 1966), Lashenkov (1961) and others. It should be noted that coordinate functions utilized in the Boobnov–Galerkin method must satisfy all boundary conditions (geometric and force ones), whereas in the Rayleigh or Rayleigh–Ritz methods it is required that only the geometric boundary conditions be satisfied. In the former case the coordinate functions are referred to as *comparison functions*, whereas in the latter case they are called *trial functions* (Meirovitch, 1986).

It is remarkable that if the same set of functions is utilized, both the Rayleigh–Ritz and the Boobnov–Galerkin methods coincide (see, e.g., Duncan, 1938; Singer, 1962; Kodnar, 1964). Perhaps this fact led Rzhanit-syn (1955, p. 306) to write, after describing the Boobnov–Galerkin method: "These methods [by Ritz and Timoshenko] are inferior by their efficiency to the Boobnov–Galerkin method; hence, we will not touch upon them." It may appear, therefore, that the Boobnov–Galerkin method, in the words of Schmidt (1990) "loses its raison d'être." However, Schmidt (1990) "assumed that the Boobnov–Galerkin method amounts to replacing the dependent variable in a differential equation of any suitable form by a reasonable approximate function η (or a series of admissible functions), and then integrating the η-weighted approximate equation over the given interval. This assumption renders the method a general one that is quite different from the Rayleigh–Ritz method." According to Duncan (1938), ". . . Rayleigh's principle and the Galerkin method lead to identical results in all cases where both are applicable. Hence the choice of method will be governed entirely by convenience."

The requirement that all the conditions should be satisfied in the Boobnov–Galerkin method can be relaxed, when the so-called physical Boobnov–Galerkin method is applied (Batdorf, 1969; Leipholz, 1967a,b). In the latter case too the Boobnov–Galerkin and the Rayleigh–Ritz methods, with trial functions used as coordinate functions, lead to identical results. As Batdorf (1969) wrote: "The convergence of the Galerkin method and its relation to the Rayleigh–Ritz method are treated" by Kantorovich and Krylov (1958). "Briefly, when the differential equations are the Euler equations obtained by minimizing an energy integral, the series (truncated or not) is the same in both cases, and the functions appearing in the series are employed as the weighting functions referred to above, then the results obtained from the Galerkin method are identical to those obtained from the Rayleigh–Ritz method. Important differences develop, however, when the Galerkin method is applied to a differential equation obtained by reducing two or more simultaneous differential equations to a single one of higher order. The most familiar example of this is probably Donnell's eighth-order differential equation for cylinder buckling (Donnell, 1933). A great computational advantage arises when the Galerkin method is applied to the single equation of escalated order. Typically — an N-terms expansion leads to an Nth order determinantal equation which results in the same accuracy as is obtained from an (nN)th order determinantal equation derived from the original set of n Euler equations. Unfortunately, however, Galerkin's method applied to an escalated equation may lead to a completely incorrect answer, even in the limit of $N \to \infty$" (see, for instance Hoff and Soong, 1965; Yu and Lai, 1967). Pitfalls of such a situation are remedied in the papers by Batdorf (1969) and Leipholz (1967a,b); see also Volmir (1967).

Equivalence of the Boobnov–Galerkin method with the Fourier series method was established by Elishakoff and Lee (1986). Convergence of the Boobnov–Galerkin method was studied by Mikhlin (1950), Polskii (1949), Leipholz (1976) and others.

For buckling of non-uniform columns the Boobnov–Galerkin method was employed by Dimitrov (1953), Bürgermeister and Steup (1957) and others. Within the Boobnov–Galerkin method the coordinate functions must satisfy both geometric and force boundary conditions. This requirement, however, can be relaxed, as demonstrated by Leipholz (1967a,b), who also wrote an extensive review of the Boobnov–Galerkin method as applied to vibration problems (1976).

Rayleigh (1873) also suggested an interesting version of his method. Since Rayleigh's quotient provides an upper bound for the estimated natural frequency or the buckling load, he introduced an unknown power into the coordinate function, with subsequent minimization of the eigenvalue with respect to the power parameter. This provides the minimum value of the upper bound of the eigenvalue within the assumed class of functions. In particular, the original idea was to calculate an approximate value of the fundamental frequency of a string by assuming the vibration mode to be $W = 1 - (2x/L)^n$ with an adjustable power parameter n. This method was then utilized by Stodola (1927), Moskalenko (1973) and others, but remained largely unknown to the investigators. The method was reinvigorated by Schmidt (1985). As Schmidt (1985, p. 69) wrote,

> the most likely reason for [Rayleigh's unknown power parameter method not enjoying a wide acceptance] . . . seems to be the difficulties inherent in the minimization of the expression for the frequency (or some other eigenvalue) as a function of adjustable parameter . . . Nowadays, this need not be an insurmountable obstacle to the use of Rayleigh's original method, for the approximate minimization can be easily carried out on a hand-held (programmable) calculator.

One should add that utilization of computerized symbolic algebra is most convenient in the context of the original Rayleigh's method (Elishakoff and Tang, 1988; Elishakoff and Bert, 1988; Hodges, 1997). In this context an interesting paper by Laura (1985) on the interrelation between the computers and usefulness of old ideas appears to be very instructive. After Schmidt's (1985) studies, many papers followed devoted to buckling or vibration of structures, carefully summarized by Bert (1987a,b). See also the historical note by Gottlieb (1986).

Earlier, Biezeno and Grammel (1953) and Makushin (1963, 1964) suggested a variation of Rayleigh–Ritz's undetermined power method, introducing the undetermined multiplicative coefficient, rather than the unknown power. They approximated the coordinate functions as the sum where

$$W = \psi_1 + n\psi_2 \tag{1.149}$$

where ψ_1 and ψ_2 are basic component functions. For example, Biezeno and Grammel (1953, p. 126) utilized the approximation (1.148) with

$$\psi_1(\xi) = 3\xi^2 - \xi^3 \quad \psi_2(\xi) = 6\xi^2 - \xi^4 \tag{1.150}$$

for estimating the fundamental natural frequency of the clamped–free beam, via the Rayleigh method; they also employed the functions

$$\psi_1(\xi) = 6\xi^2 - 4\xi^3 + \xi^4 \quad \psi_2(\xi) = 20\xi^2 - 10\xi^3 + \xi^5 \tag{1.151}$$

whereas Makushin (1963, 1964) utilized a variety of functions ψ_j. The basic functions were static deflections due to different loads. Thus, the single-term coordinate function was utilized in Rayleigh's method, with attendant minimization of buckling loads with respect to the parameter n. Elishakoff (1987a–c) demonstrated that this method is equivalent to a two-term Rayleigh–Ritz method. Thus, a single-term Biezeno–Grammel–Makuskin method accomplishes the accuracy of the two-term Rayleigh–Ritz method. Indeed, undetermined multiplier representation is one form of the Rayleigh–Ritz method, for one could write Eq. (1.148) as $W(x) = c_1\psi_1 + c_2\psi_2 = c_1[\psi_1 + (c_2/c_1)\psi_2] = c_1[\psi_1 + n\psi_2]$, use the Ritz minimizing equations, and get the same results. As Leissa (2000b) notes, "a lot of researchers used the two-term form [of Eq. (1.148)] to solve the beam problems a few decades ago, although many did not call it the Ritz method."

Later on, this method was utilized by Elishakoff (1987a,b) in various contexts. Most recently, apparently independently, it was suggested by Hodges (1997). Bert's (1987b) extensive review describes Rayleigh's undetermined power method and its numerous applications. Laura and Cortinez (1988) were apparently the first investigators who introduced a parameter in an exponential function and studied the efficacy of this approximation. Also, Laura and Cortinez (1986b) first suggested and performed optimization of higher eigenvalues (Laura, 2000).

Rayleigh (1894, p. 112) mentions that "the period calculated from any hypothetical type [of the mode shape] cannot exceed that belonging to the gravest normal type." In other words, the estimate of the natural frequency evaluated via the assumed trial function cannot be less than the exact value of the fundamental natural frequency. There are instances, however, in the literature, when the Rayleigh or Rayleigh–Ritz method did not produce results that were in excess of the reported exact eigenvalues. Such a discrepancy must be attributed to the mistake made in either the realization of the approximate method (such as lack of satisfaction of boundary conditions) or evaluation of the exact solutions (Bhat, 1996, 2000). Grossi and Bhat (1991) studied free vibrations of tapered beams using the characteristic orthogonal polynomials method, developed earlier by Bhat (1985). They compared their results with those reported by Goel (1976) obtained by the use of Bessel functions. Since Goel's (1976) derivations are associated with the exact solution, while Grossi and Bhat's (1991) results are based on Rayleigh's method, the former results must be less than the latter ones. As Grossi and Bhat (1991) reported, for the tapered beams in question, "both the Rayleigh-Schmidt and Rayleigh–Ritz methods yield upper bounds." They added, that "unfortunately, several values reported by Goel (1976) are higher than the values obtained by the two mentioned methods." Auciello (1995) studied in detail this particular disagreement of the numerical results with general theorems. It was established that Goel's (1976) exact solution contained an error. It is remarkable, that as an interesting byproduct of the approximate analysis, the revision of the exact solution became necessary. An analogous situation took place also in regard to the fundamental frequency of the pinned circular plate. Laura et al. (1975) showed that for the Poisson's ratio $v = 0.3$, the non-dimensional fundamental frequency was $\Omega = 4.947$. It was found by the single polynomial comparison function via the Galerkin method. It turned to be *lower* than the value of the exact frequency, which was reported in the literature to equal 4.977. Several years later it was shown that their approximate value was "almost perfect" (Laura, 2000). Indeed, as Leissa (2000b) writes, "because of the inaccuracy and lack in detail in previously appearing literature [about the exact values] Narita and Leissa (1980) published a paper devoted solely to the pinned circular plate. In Table 4 of that section it gives the exact fundamental frequency to six figures as 4.93515, when Poisson's ratio is 0.3." Thus, an approximate result reported by Laura et al. (1975) is 0.24% greater than the exact value. Due to these two examples of the positive interplay between exact and appropriate analyses, namely when the updating of the exact results turned out to be necessary, it appears worthwhile to corroborate exact solutions with some approximate techniques, to distance oneself from possible inaccuracies.

In this monograph it is instructive to reproduce a problem (No. 113) from Feodosiev's (1996) famous problems and questions book on the strength of materials. The question is as follows:

Main approximate method of determining of critical loads is energy method. The sought form of equilibrium is given approximately with a view to satisfy the boundary conditions and the employed function to represent possibly closely the true form of equilibrium, that is not known to us, but intuitively chosen by the physical considerations of the problem. The question is, is there a danger, that when the approximating function becomes arbitrarily close to the exact displacement, that we will not get the exact value of the critical force.

Feodosiev (1996) mentions, in the response to the question that by utilizing "a formal approach" towards the choice of the approximating function, one can get a result that is quite far from reality. He suggested using the following function:

$$\psi(x) = \sin(\pi x/L) + (L/m)\sin(m\pi x/L) \tag{1.152}$$

for the column that is pinned at both ends. When $m \to \infty$, the approximating function tends to the exact mode shape

$$\lim_{m \to \infty} \psi(x) = \sin(\pi x/L) \tag{1.153}$$

The buckling load P, however, evaluated via the Rayleigh quotient equals

$$P_{cl} = \frac{EI \int_0^L (\psi'')^2 dx}{\int_0^L (\psi')^2 dx} = \frac{1+m^2}{2} \frac{\pi^2 EI}{L^2} \tag{1.154}$$

When $m \to \infty$, one gets a result that is infinitely far from the exact solution.

According to Feodosiev (1996), the specificity of the function (1.152) lies in the fact that while it approximates well the form of the displacement, its second derivative is far removed from the second derivative of the true displacement. According to Feodosiev (1996), this somewhat surprising example is instructive in the sense that while approximating the displacement function, one should try to care about a proper approximation of the highest derivative entering the expression of the energy.

This fact was well understood by Niedenfuhr (1952) who suggested selecting an approximating function for the beam curvature (second derivative of the displacement) rather than for the deflection itself. Bert (1987a,b) represented the beam curvature as follows:

$$Y'' = C(1 - X'') \tag{1.155}$$

where

$$X = 2x/L \tag{1.156}$$

with x varying in the interval $(-L/2, L/2)$, i.e., the origin of coordinates was chosen in the middle cross-section. Integrating twice, we obtain

$$Y = C\left[\frac{1}{2}x^2 - (2/L)'' \frac{x^{n+2}}{(n+1)(n+2)}\right] + C_1 x + C_2 \tag{1.157}$$

The condition of symmetry about the middle cross-section $x = 0$ leads to $C_1 = 0$. The boundary condition $Y(L/2) = 0$ is utilized to determine C_2.

Thus,

$$Y = C \left[\frac{X^2 - 1}{2} + \frac{1 - X^{n+2}}{(n+1)(n+2)} \right] \left(\frac{L}{2} \right)^2 \tag{1.158}$$

The Rayleigh's quotient yields

$$\omega^2 = \frac{\int_0^L EI(Y'')^2 \, dx}{\int_0^L \rho A Y^2 \, dx} = \frac{\int_0^1 EI(Y'')^2 \, dX}{\int_0^1 \rho A Y^2 \, dX} \tag{1.159}$$

leading, for the uniform beam, to the expression

$$\overline{\omega}^2 = \frac{32n^2}{2n+1} \left\{ \frac{2}{15}(n+1) + \frac{1}{n+2} \left[\frac{1}{n+3} - \frac{1}{n+5} - \frac{2}{3} \right. \right.$$

$$\left. \left. + \frac{1}{(n+1)(n+2)} \left(1 - \frac{2}{n+3} + \frac{1}{2n+5} \right) \right] \right\}^{-1} \tag{1.160}$$

The minimum value of $\overline{\omega} = 9.8702$ at $n = 1.70$. As Bert (1987a,b) noted, this is only 0.006% higher than the exact value. Eigenvalue analysis of prismatic members by the wavelet-based Boobnov–Galerkin method was performed by Jin and Ye (1999).

1.9 Inverse Eigenvalue Problems

Groetsch (1993) wrote:

> Classical applied mathematics is dominated by the Laplacian paradigm of known causes evolving continuously into uniquely determined effects. The classical *direct* problem is then to find the unique effect of a given cause by using the appropriate law of evolution. It is therefore no surprise that traditional teaching in mathematics and the natural sciences emphasizes the point of view that problems have a solution, this solution is unique, and the solution is insensitive to small changes in the problem. Such problems are called *well-posed* and typically arise from the so-called direct problems of natural science. The demands of science and technology have recently brought to the fore many problems that are inverse to the classical direct problems, that is, problems which may be interpolated as finding the cause of a given effect or finding a law of evolution given the cause and effect. Included among such problems are many questions of remote sensing or indirect measurement such as the determination of internal characteristics of an inaccessible region from measurements on its boundary, the determination of system parameters from input–output measurements, and the reconstruction of past events from measurement of the present

state . . . This study of inverse problems is very new and very old. The latest high-tech medical imaging devices are essentially inverse problem solvers; they reconstruct two- and three-dimensional object from projections. More than two thousand years ago, in the book VII of his *Republic*, Plato posed essentially the same problem in his allegory of the cave, namely, he considered the philosophical implications of reconstructing "reality" from observations of structures cast upon the wall.

Bolt (1980) illustrated the inverse problems in the following manner:

Suppose that we ski down mountain trail. As long as we neglect friction and know the undulations in the trail, we can calculate exactly the time it will take to travel from a point on the mountain side to the valley below. This is a direct problem. It was difficult for Galileo's contemporaries. It is, however, now old hat, and what is of current deeper interest is the following problem. If we start skiing from different places on the slope, and on each occasion we time our arrival at a fixed place on the valley floor, how can we calculate the topographic profile of undulations of the trail? This is the inverse problem. It is certainly a practical problem. It is also challenging and difficult and, indeed, in the general sense, it has no unique solution.

According to Tarantola (1987), To solve the forward problem is to predict the values of the observable parameters, given arbitrary values of the model parameters. To solve the inverse problem is to infer the values of the model parameters from given observed values of the observable parameters. Tanaka and Bui (1993) wrote:

There are many kinds of inverse problems across a wide variety of fields. In general, the inverse problem can be defined as the problem where one should estimate the cause from the result, while the direct problem is concerned with how to obtain the result from the cause. At present in engineering fields, CT scan, ultrasonic techniques, and so on can be successfully applied to some of the inverse problems including nondestructive evaluating or testing. On the other hand, different attempts have also been made recently in such a way that the computational software available for the direct problem is applied to corresponding inverse problem analysis. In most of these computational approaches, the inverse problem is formulated into a parameter identification problem in which the set of parameters corresponding to the lacking data to be estimated should be found by minimizing a suitable cost function. Two main difficulties encountered in the inverse analysis are non-uniqueness and ill-posedness of the inverse solution. The former difficulty should be overcome by selecting our useful solution from the engineering point of view. The latter difficulty could be circumvented or overcome from theoretical or mathematical considerations.

Who were the first investigators to tackle the inverse problems? According to Anger (2000), Lord Rayleigh was apparently the first investigator to pose an inverse vibration problem:

> one of the first inverse problems was explicitly solved by N.H. Abel ... in 1823. This problem consists of finding curve X in a vertical plane where X is assumed to be graph of an increasing function $x = \Phi(y)$ such that the falling time of a body (falling only under the influence of gravity) is a known function $t(y)$ of the height from which the body falls. This problem reduced to an Abel integral equation has a unique solution. If Φ is multi-valued then infinitely many solutions exist.

> In 1867 G.G. Stokes for the first time raised the question of the relationship between the Newtonian potential at the exterior of a closed surface containing an attracting body and the distribution of the body's density p producing the assigned potential. In his lectures on Celestial Mechanics at the University of Pavia in 1875/76 the astronomer G.V. Schiaparelli showed how the internal density of a body can be altered without affecting the outer potential ...

> In 1877 Lord Rayleigh in his book *"The Theory of Sound"* briefly discussed the possibility of inferring the density distribution from the frequencies of vibration.

As mentioned above, the interest in the inverse problems in vibration starts from the work of Rayleigh. Presently there is a single monograph, by Gladwell (1986) devoted entirely to the inverse problems in vibrations. He also wrote two extensive reviews on this subject (Gladwell, 1986b, 1996). The first paper on inverse vibrations is attributed to Ambarzumian (1929). He defined it as "a problem of unique determination of the mechanical system through the eigenvalue spectrum of corresponding linear differential equations". Further contributions belong to Krein (1933, 1934), Borg (1946), Gantmakher and Krein (1950), Gel'fand and Levitan (1951), Krein (1952), Levitan (1964) and others. Gladwell and his followers made this subject an indispensable part of modern vibration theory. Specifically, his book (1986) studies both discrete, as well as continuous second-order, and fourth-order systems. For extensive references, the reader must consult his two reviews (Gladwell, 1986, 1996), which address the following cardinal questions: "Is there a system of some specified type which exhibits the given behavior? Is there a more than one system of the specified type which exhibits the given behavior? If there is a system and it is unique, then how can it be found?"

According to Gladwell (1986), "in the context of ... classical theory, inverse problems are concerned with the construction of a model of a given type; e.g., a mass-spring system, a string, etc., which has given eigenvalues and/or eigenvectors; i.e., given *spectral* (emphasis by Gladwell) data." Gladwell (1986) explains his interest in inverse problems as follows:

> My interest in inverse problems was sparked by acquiring a copy of the translation of Gantmakher and Krein's beautiful book *Oscillation Matrices*

and Kernels and Small Oscillations of Mechanical Systems. During the first ten years that I owned the book I made a number of attempts to master it, without much success. One thing that I did understand and enjoy was their reconstruction of the positions and masses of a set of beads on a stretched string from a knowledge of the fixed–fixed and fixed–free spectra. In their reconstruction, the unknown quantities appear as the coefficients in a continued fraction representation of the ratio of two polynomials constructed from the given spectra. As a mathematician I was thrilled that a concept so esoteric and apparently useless as a continued fraction should appear (naturally) in the solution of a problem in mechanics.

One of the properties that may arise in inverse vibration problems is *isospectrality*. Two vibration systems which share the same natural frequencies are called isospectral. Gottlieb (1991) uncovered a family of *inhomogeneous* clamped circular plates having the same natural frequencies as a *homogeneous* plate. Abrate (1995) showed for some non-uniform rods and beams the equation of motion can be transformed into the equation of motion for a uniform rod or beam. It turns out that when ends are completely fixed, the natural frequencies of the non-uniform structures are the same as those of uniform rods and beams. Horgan and Chan (1999a) obtained closed-form solutions for different variations in mass density of longitudinally vibrating rods, namely with

$$m(x) = m_0(1 + \alpha x/L)^{-1} \tag{1.161}$$

$$m(x) = m_0(1 + \alpha x/L) \tag{1.162}$$

$$m(x) = m_0(1 + \alpha x/L)^2 \tag{1.163}$$

$$m(x) = m_0 \exp(-\alpha x/L) \tag{1.164}$$

where α is a measure of inhomogeneity. They also demonstrated an interesting phenomenon that certain inhomogeneous strings and rods have the same fundamental frequencies as their homogeneous counterparts.

Gladwell (1995b) studied a discrete system of in-line concentrated masses connected with each other and to the end pinned by ideal massless springs. Four methods were given for constructing a system which is isospectral to a given one. Since the discrete systems can be viewed as finite-difference or finite-element approximations of continuous systems, it is understood that isospectrality may occur in the distributed systems. Gladwell and Morassi (1995) demonstrated how to construct families of bars, in longitudinal or torsional vibration, which exhibited the isospectral property. One, simplest way is immediately detectable. As they mention, (p. 535), "The simplest, almost trivial pair of isospectral rods is obtained by physically turning the rod and restraints around." For the general case, from which the above simplest example turned out as a particular case, the Darboux (1882, 1915) lemma was utilized. Following Gladwell and Morassi (1995) let us expose the essence of this lemma. Let μ be a real number, and $g(x)$ be a non-trivial solution of the Sturm–Liouville equation:

$$-g''(x) + p(x)g = \mu g \tag{1.165}$$

where $p(x)$ is the potential function. If $f(x)$ is a non-trivial solution of

$$-f''(x) + p(x)f = \lambda f \tag{1.166}$$

and $\lambda \neq \mu$, then a new function $y(x)$ defined as

$$y(x) = \frac{1}{g(x)}[g(x)f'(x) - g'(x)f(x)] \tag{1.167}$$

is a non-trivial solution of the Sturm–Liouville equation

$$-y'' + q(x)y = \lambda y \tag{1.168}$$

where

$$q(x) = p(x) - 2\frac{d^2}{dx^2}\ln[g(x)] \tag{1.169}$$

The Darboux lemma enables one to determine a solution of a new Eq. (1.168), if we know two solutions $g(x)$ and $f(x)$ of another Eq. (1.165), corresponding to two different values λ, μ of a parameter. Gladwell and Morassi (1995) show that if $f(x)$ is non-trivial then $y(x)$ is likewise non-trivial. The second part of the Darboux lemma states that the general solution of the equation

$$-y'' + q(x)y = \mu y \tag{1.170}$$

reads

$$y = \frac{1}{g}\left(1 + c\int_0^x g^2(s)ds\right) \quad c = \text{const} \tag{1.171}$$

The Darboux lemma is utilized as follows. Suppose the rod is given with the cross-section $A(x)$ and spectrum $\{\lambda_n\}_0^\infty$ corresponding to end conditions

$$A(0)u'(0) - Ku(0) = 0 = A(1)u'(1) + Ku(1) \tag{1.172}$$

where u is the longitudinal displacement, and the coefficients k and K are positive. The Darboux lemma allows us to obtain any rod corresponding to a new potential $q(x)$ if one knows the spectrum for the rod with potential function $p(x)$, under some general conditions. Gladwell and Morassi (1995) also addressed an interesting problem of how by a finite or infinite sequence of Darboux transformations one can "flow" from one rod to another, such that they are isospectral, i.e., have the same spectrum. They exemplified their derivations by a single or double application of the Darboux lemma for rods. They showed that if one starts with one rod and a set of end conditions, one may construct many families of rods that have the same spectrum as the original rod for specific sets of boundary conditions. Gladwell (2000b) writes,

> The essential point is that if you have one system, either continuous or FEM, then one can construct an infinite family of other isospectral systems. The analysis also gives the modes of vibration of the new systems, so that on solving one problem you also solve many problems. My papers on inverse finite element vibration problems do the same thing: They allow one to form a family of isospectral systems from one system ... No one has yet obtained a family of isospectral beams in flexure, not finite element or continuous membranes.

Thus, having a solution of the inverse problem may be utilized for obtaining a solution of a direct problem, if it belongs to the isospectral family of structures.

A closely related paper is due to Kac (1966), who posed a provocative question "*Can one hear the shape of a drum?*" in the title of his study. He was referring to the problem of whether the Laplacian operator with Dirichlet boundary conditions could have identical spectra on two distinct planar regions. Gordon et al. (1992b) showed that the answer to the question by Kac (1966) was negative by constructing a counterexample; this research justifiably attracted a great deal of attention (Cipra, 1992; Chapman, 1995; Peterson, 1994). The simplest form of their example is a pair of regions bounded by eight-sided polygons. Buser et al. (1994) found other examples of this type. Thus, one cannot "hear the shape of a drum." As Driscoll (1997) wrote, while the isospectrality can be proven mathematically, analytical techniques are unable to produce eigenvalues themselves.

He described an algorithm due to Descloux and Tolley (1983) that combines together singular finite elements with domain decomposition, achieving results, accurate to 12 digits, for the most famous pair of isospectral drums. Some explicit solutions were derived by Gottlieb and McManus (1998).

As Gladwell (1986) wrote, "the first paper dealing with inverse problems for equations of order higher than two appears to have been by Niordson (1967). He used an iterative method to construct a cantilever beam with some specified n natural frequencies, by seeking a solution in a class of functions depending on n parameters" (see also Ainola, 1971). Barcilon (1974) showed that three spectra are needed to reconstruct the cross-sectional area and second moment of the area of the beam.

Niordson (1970) proposed a method of determining the thickness of an all-round pinned plate, by the given finite series of positive numbers, serving as eigenvalues of the problem. Niordson (1970) arrived at a conclusion that the solution was not unique, since only part of the spectrum was given.

For the square plate that has a constant density and thickness the natural frequencies have a form $\omega = \omega_1(m^2 + n^2)$, where $m, n = 1, 2, 3, \ldots$, and ω_1 is the nominal natural frequency. For such a plate the ratio of the three first eigenvalues is $2:5:5$, with the second natural frequency being a double one. Niordson (1970) posed a problem in the following form: "What should the thickness of the square plate be so that the first three natural frequencies relate as $2:6:6$ or $2:5:6$?" The thickness distributions obtained were given in graphical form.

Ram (1994) constructed discrete models for a non-uniform cantilever beam using eigenvector data. He showed that, for a given mesh a finite difference model may be constructed from two eigenvectors, one eigenvalue and the total mass of the beam. Yet, if the mesh is unknown, then for the determination of the model, including the grid points, one may need three eigenvectors, one eigenvalue and total mass and length of the beam. The author concluded that his results may be applied to the evaluation of discrete models for the non-uniform beam by using experimental modal analysis data. Ram's (1994) results are complementary to those of Barcilon (1979, 1982) and Gladwell (1984, 1986a) who showed that the discrete model may be reconstructed from the eigenvalues of the beam subject to three different end configurations.

One of the central problems of the inverse theory is the identification of elastic properties. Mota Soares et al. (1993) have suggested a scheme to identify the elastic parameters of laminated plates using the frequencies of free vibration. Yao and Qu (1998) used the optimization methods for the same purpose. The objective function of the optimization was defined as the difference between the measured frequencies and computed frequencies of the limited plates. The sensitivity of the structural eigenvalue with respect to the material parameters was analyzed.

A characteristic static test for the identification of elastic moduli can be described in the following way (Constantinescu, 1998): "Elastic moduli of a fixed plate are to be identified from displacement and forces measured simultaneously on a certain number of interior and boundary points." This type of test has traditionally been used in order to identify a homogeneous plate (Grediac and Vautrin, 1990, 1996; Grediac 1996). Changing the distributions of applied forces gives rise to a series of deflection-force measurements, representing a partial knowledge of the Dirichlet-to-Neumann data map characterizing the boundary response of a body. This map generally gives the correspondence among displacement, the Dirichlet boundary condition, and forces in the Neumann boundary conditions. A series of results (Nakamura and Uhlmann, 1995) showed that the knowledge of the Dirichlet-to-Neumann data map permits the identification of the *inhomogeneous* distribution of elastic moduli.

Gladwell and Morassi (1999) studied the effect of damage on the nodes of free vibration modes of a thin rod in longitudinal vibration. The damage (notch) was modeled as a spring. It was demonstrated that the nodes move toward the damaged region. Qualitative measures of such a movement may allow one to estimate the position and the damage. This investigation was a

combined theoretical/experimental analysis, with attendant good agreement.
A neural network solution of the inverse vibration problem was facilitated by
Mahmoud and Abu Kiefa (1999).

The problem of reconstructing structures while the material density and
modulus of elasticity are treated as constant constitutes the principal subject of classical inverse analysis in vibrations. Several methods are reported
in the definitive monograph by Gladwell (1986a). Recently, inhomogeneous
structures attracted more interest. As Wang and Wang (1994) mentioned, in
classical works: the condition and method for constructing of cross-sectional
area of a rod with constant density and modulus of elasticity E from frequency
data are given. The analysis and constructed parameters are very sensitive to
changes in the frequency data, so the method is inconvenient in engineering
problems. Likewise, Ram (1994) wrote: ... in the classical inverse problem it is
assumed that the cross-sectional area of the rod is variable while the Young's
modulus of elasticity and the rod density are constants.

In the papers by Ram (1994b) and Wang and Wang (1994) methods were
proposed to treat the cross-sectional area, the modulus of elasticity and the
material density as functions of the axial coordinate. It appears that this is a
more realistic situation than limiting oneself with structures of constant elastic
moduli.

1.10   Connection to the Work by Życzkowski and Gajewski

It is necessary to note that after this study was completed we uncovered
works by Życzkowski (1954–1991) which resonates well with the theme of
the present monograph, it is with great pleasure, that I am able to add this
section. Życzkowski (1991) writes:

> The stability analysis of non-prismatic and inhomogeneous bars makes
> it possible to apply a convenient inverse method. Thus, if we assume
> a certain equation of deflection line $w = w(x)$, we can find ... such a
> distribution of rigidity $EI(x)$ of the bar, for which the function $w(x)$ is the
> exact solution
>
> $$EI(x) = -\frac{M[w(x), x, P]}{w''(x)} \qquad (1.173)$$
>
> By introducing into the assumed equation of deflection line a certain
> number of free parameters, $w = w(x; a_1, a_2, \ldots, a_n)$, we can subsequently
> select their value so as to obtain minimum deviations of the resultant
> rigidity ... from the rigidity of the bar under consideration. The applic-
> ation of this method to elastic-plastic analysis of non-prismatic bars has
> been discussed by Życzkowski (1954).

Życzkowski (1955, 1956a) further developed his method. The generalization of this method to the buckling of circular plates was performed by

Gajewski and Życzkowski (1965a,b, 1966). In their 1966 paper they write the differential equation for the buckling of the circular plate in terms of the slope $\varphi = -dw/dr$,

$$z^2 D\varphi'' + (z^2 D' + zD)\varphi' + (vzD' - D + NR^2 z^2)\varphi = 0 \tag{1.174}$$

where $z = r/R$ is a non-dimensional polar coordinate and N is the required eigenvalue.

Then, the exact solution $\varphi = \varphi(z)$ is assumed to be known and we obtain a differential equation for the flexural rigidit D:

$$\left(\varphi' + \frac{v\varphi}{z}\right) D' + \frac{d}{dz}\left(\varphi' + \frac{\varphi}{z}\right) D = -NR^2\varphi \tag{1.175}$$

the general solution of which is written as

$$D(z) = \exp\left[-\int \left(\frac{(\varphi' + \varphi/z)'}{\varphi' + v\varphi/z}\right) dz\right]$$
$$\times \left\{C - NR^2 \int \frac{\varphi}{\varphi' + v\varphi(z)} \exp\left[\int \frac{(\varphi' + \varphi(z))'}{\varphi' + v\varphi/z} dz\right] dz\right\} \tag{1.176}$$

The constant C is chosen so as to eliminate a singularity of the function $D(z)$. A simplification is obtained by assuming that

$$\int \frac{[\varphi' + \varphi(z)]'}{\varphi' + v\varphi/z} dz = -\ln z^n \tag{1.177}$$

where n is a real parameter to be determined. This leads to the Euler-type equation for φ:

$$z^2\varphi'' + (n+1)z\varphi' - (1 - nv)\varphi = 0 \tag{1.178}$$

with solution

$$\varphi(z) = C_1 z^{b-n/2} + C_2 z^{-b-n/2} \tag{1.179}$$

where

$$2b = \sqrt{h^2 - 4nv} = 4 \tag{1.180}$$

These considerations lead to the following expression for the flexural rigidity

$$D(z) = \frac{NR^2}{(b - n/2 + v)(2 - n)}(z^n - z^2) \tag{1.181}$$

The critical force N_{cr} is determined under the assumption that the flexural rigidity is known at some point of the plate; the value of the flexural rigidity at $r = 0.5R$ is chosen as a known quantity. This rigidity is denoted as by $D_{0.5}$. This results in the following expression for the critical load:

$$N_{cr} = \frac{(b - n/2 + \nu)(n - 2)2^{n+2}}{2^n - 4} \frac{D_{0.5}}{R^2} \tag{1.182}$$

with an attendant expression for the flexural rigidity,

$$D(z) = \frac{2^{n+2}}{4 - 2^n}(z^n - z^2)D_{0.5} \tag{1.183}$$

As can be seen, a polynomial expression was obtained for $D(z)$. For other cases the readers can consult Gajewski and Życzkowski (1966).

In this monograph we postulate from the very outset that the flexural rigidity represents a polynomial function. One can pose a relevant question: is it possible that the vibration mode or the buckling mode of an inhomogeneous structure constitutes a trigonometric function? For uniform beams and columns, depending on the boundary conditions, the eigenfunction may be a trigonometric function. Calió and Elishakoff (2002, 2004a,b) demonstrated that the reply to the above question is in the affirmative.

1.11   Connection to Functionally Graded Materials

Functionally graded materials (FGMs) are characterized by continuous changes in their microstructure as distinguished from the piece-wise variation of the properties in conventional composite materials. An intensive study has been conducted recently on the behavior, vibration and buckling of structures made of FGMs. A definitive monograph on this topic is the one by Suresh and Mortensen (1998). For a discussion of the concept readers can consult the paper by Erdogan (1995); extensive activities associated with the FGM in Japan are described in the papers by Koizumi (1993, 1997). Stavsky's (1964, 1965) pioneering works, some three decades before the FGMs were developed, deal with continuously varying elastic moduli in the thickness direction. Naturally, inhomogeneous structures, i.e., those with variable moduli of elasticity, have long attracted investigators. The readers can consult the books by Olszak (1959), Kolchin (1971), Nesterenko (2001), as well as works by Mikhlin (1934), Glushkov (1939), Golecki (1959), Sherman (1959), Plotnikov (1959–1967a,b), Babich (1961a,b), Birger (1961), Du (1961), Ter-Mkrtchian (1961), Bufler (1963), Lekhnitskii (1964), Schile (1963), Schile and Sierakowski (1964), Kardomateas (1990) and many others.

The idea of the FGM stems from the desire to create new materials that enable engineers to fit a certain purpose. Actually, this idea is already in use in the strength of materials courses and in the concept of the economic use of materials, which, for example, prompts the use of an I-beam instead of one with a rectangular cross-section. A further development of this concept is the sandwich structure, i.e., a cross-section composed of strong facings and a soft core. However, in sandwich or composite structures the material properties in the thickness direction vary discontinuously. To avoid problems associated with discontinuous variation of the elastic properties the use of FGMs was suggested. In these the microstructure varies continuously in the thickness direction so as to result in the smoothly varying function $E(z)$.

The static behavior of structures made of FGMs was studied by Rooney and Ferrari (1995), Abid Mian and Spencer (1998), Horgan and Chan (1999b), Cheng and Batra (2000), Reddy (2000) and Venkataraman and Sankar (2003). Higher-order theory for FGM was developed by Aboudi et al. (1999). Vibration of structures made of FGMs, was investigated by Pradhan et al. (2000), Ng et al. (2001), and Yang and Shen (2003). Buckling problems attracted several investigators, namely Birman (1995, 1997), Feldman and Aboudi (1997), Najafizadeh and Eslami (2002a,b), and Javerheri and Eslami (2002). It should be stressed that the references on FGMs, listed here are only representative ones — they by no means exhaust this rapidly developing field.

In this monograph we consider structures made of materials that can be dubbed as *axially graded materials* (AGMs). Whereas we are unaware of experimental realization of AGMs, it is anticipated that such materials will be developed in the near future, for they will allow a *tailored* fit to a special purpose, i.e., with the static deflection not exceeding a specific level, or the buckling load not being less than a pre-specified level, or the natural frequency either exceeding or to being less than a pre-specified frequency. It is anticipated that the best functional grading will combine that both in axial and thickness directions.

1.12   Scope of the Present Monograph

This monograph is entirely dedicated to the buckling and vibrations of inhomogeneous structures. Whereas our main focus is on obtaining the closed-form solutions, the posing of the problem as an inverse one turns out to be effective. Yet, the formulation is very different from the classical inverse problems, where the frequency spectrum is postulated to be given. Here, we abstain from postulating the knowledge of the frequency spectrum. We pose a modest requirement that the fundamental mode of vibration is specified. Under these circumstances, we pose a problem of reconstructing the flexural rigidity of the structure when the variation of the material density is specified.

The essential feature of research reported in this monograph is a representation of postulated mode shape as a *polynomial* function. Specifically, the mode shapes are postulated as the *static* displacements of the associated uniform and homogeneous structures. This immediately begs the following question: Why is it at all possible for the structure to possess a mode shape that coincides with the static displacement of another structure? Indeed, as a well-known politician remarked, our formulation could have turned out to be as impossible an action as "trying to make eggs out of an omelette." Fortunately, as will be shown in subsequent chapters, having the (self-directed) *chutzpha* (audacity) to pose a seemingly strange question paid off. Numerous closed-form solutions will be derived for rods, columns, beams and plates. Along with the polynomial representation of the mode shape with attendant polynomial material density and flexural rigidity representation, some other alternatives are pursued, leading to a rational representation of the sought for flexural rigidity function. It is worth noting that trigonometric closed-form solutions that are characteristic to *uniform* structures found their utmost relevance to *inhomogeneous* structures too, as recent work by Calió and Elishakoff (2002) demonstrates. Forthcoming articles (2004a, 2004b, 2004c) of the authors clearly show numerous possibilities in which trigonometric closed-form solutions may be associated with inhomogeneous structures.

2

Unusual Closed-Form Solutions in Column Buckling

This chapter deals with obtaining closed-form solutions for inhomogeneous columns under various load conditions and boundary conditions. Among others, Engesser's (1893) and Duncan's (1937) closed-form solutions are generalized to a wide class of problems, including the generalized Euler problem, that of a column under its own weight, or under polynomially varying distributed axial loading. In these problems we postulate the buckling mode as the fourth-order polynomial.

2.1 New Closed-Form Solutions for Buckling of a Variable Flexural Rigidity Column

2.1.1 Introductory Remarks

The buckling of uniform columns under various loading and boundary conditions is a well-studied topic. As far as the columns with variable cross section are concerned, several exact solutions are available, in terms of logarithmic and trigonometric (Bleich, 1952; Dinnik, 1955a,b), Bessel (Dinnik, 1928, 1932) and Lommel (Willers, 1941; Engelhardt, 1954) functions. An exact solution in terms of series for buckling load, for variable cross-section columns with variable axial forces was furnished by Eisenberger (1991a–d). The closed-form solutions are extremely rare. Two cases will be described. For the column (Engesser, 1893) that is pinned at both its ends and possesses the following flexural rigidity:

$$D(x) = 4x(L - x)D_0/L^2 \tag{2.1}$$

where L is the length and x the axial coordinate, the governing differential equation reads

$$\left[4x(L - x)D_0/L^2\right]\frac{d^2w}{dx^2} + Pw = 0 \tag{2.2}$$

where $w(x)$ is a displacement. Substitution of the mode

$$w(x) = A[4x(L - x)/L^2] \tag{2.3}$$

where A is a constant, in Eq. (2.2) results in

$$4x(L - x)D_0/L^2(-8/L^2) + P4x(L - x)/L^2 = 0 \tag{2.4}$$

Since for buckling $A \neq 0$, one obtains the buckling load

$$P_{cl} = 8D_0/L^2 \tag{2.5}$$

(see also Rzhanitsyn, 1955). A second example belongs to Duncan (1937). Here, the flexural rigidity varies as

$$D(x) = \left[1 - \tfrac{3}{7}(x/L)^2\right] D_0 \tag{2.6}$$

so that the governing differential equation is

$$D_0\left[1 - \frac{3}{7}(x/L)^2\right]\frac{d^2w}{dx^2} + Pw = 0 \tag{2.7}$$

The buckling mode was guessed by Duncan (1937) as

$$w(x) = A[7(x/L) - 10(x/L)^3 + 3(x/L)^5] \tag{2.8}$$

Substituting Eq. (2.8) into Eq. (2.7), the classical buckling load becomes

$$P_{cl} = \tfrac{60}{7} D_0/L^2 \tag{2.9}$$

The author is unaware of other closed-form solutions for columns with variable flexural rigidity. Obtaining such solutions is worthwhile, since closed-form solutions could serve as benchmark solutions for the purpose of contrasting various approximate solutions with them. The results derived by Elishakoff and Rollot (1999) are reported. In what follows, one generalizes the two closed-form solutions discussed above.

2.1.2   Formulation of the Problem

The column buckling is governed by the differential equation

$$D(x)\frac{d^2w}{dx^2} + Pw = 0 \tag{2.10}$$

where $D(x)$ is defined as

$$D = D_0 r(z) \tag{2.11}$$

where D_0 is a constant and z is a non-dimensional coordinate defined as

$$z = x/L \tag{2.12}$$

The governing differential equation, Eq. (2.10), can be rewritten as

$$r\frac{d^2w}{dz^2} + k^2 w = 0 \tag{2.13}$$

where k^2 is a constant defined as

$$k^2 = PL^2/D_0 \tag{2.14}$$

One deduces the buckling load from Eq. (2.14):

$$P = k^2 D_0 / L^2 \tag{2.15}$$

In this study, $r(z)$ is assumed to be a polynomial of second degree. Three different variations for $r(z)$ are discussed that lead to new closed-form solutions for the buckling load:

$$r = \beta z - \gamma z^2 \quad r = 1 + \beta z - \gamma z^2 \quad r = 1 - \gamma z^2 \tag{2.16}$$

In this section, the displacement w is assumed to be a polynomial function that satisfies the differential equation and all boundary conditions. One finds new closed-form solutions for some particular choices of β and γ.

2.1.3 Uncovered Closed-Form Solutions

Case 1: $r = \beta z - \gamma z^2$
The variation of $D(x)$ is given by

$$D = D_0(\beta z - \gamma z^2) \tag{2.17}$$

and the displacement is a polynomial of degree 2:

$$w = az + bz^2 \tag{2.18}$$

The boundary conditions for a pinned column are:

$$w(0) = 0 \quad Dw''(0) = 0 \quad w(1) = 0 \quad Dw''(1) = 0 \tag{2.19a–d}$$

Equations (2.19a) and (2.19b) are always satisfied; Eqs. (2.19c) and (2.19d) lead to

$$b = -a \tag{2.20}$$

and

$$\beta = \gamma \tag{2.21}$$

Taking into account the boundary conditions in Eq. (2.19), one defines

$$r = \gamma(z - z^2) \tag{2.22}$$

and

$$w = a(z - z^2) \tag{2.23}$$

w has to satisfy the differential Eq. (2.13) for any z. This problem is solvable with the aid of the Mathematica® command *SolveAways* (Wolfram, 1996). Although for solving the problem the use of symbolic algebra is not absolutely necessary, it is an extremely convenient tool. *SolveAways* yields parameter values for which the given equation or system of equations which depend on a set of parameters is valid for all variable values. The result of *SolveAways* is given in the form of a list of all possible sets of values. *SolveAways* works primarily with linear and polynomial equations.

For this case two sets are obtained. The first one leads to a trivial solution with $a = 0$. The second set leads to

$$k^2 = 2\gamma \tag{2.24}$$

Finally, using Eq. (2.15) one deduces the buckling load,

$$P = 2\gamma D_0/L^2 \tag{2.25}$$

The buckling mode reads as

$$w = a(z - z^2) \tag{2.26}$$

This corresponds to the following definition of the flexural rigidity:

$$D = D_0\gamma(z - z^2) \tag{2.27}$$

One can now relate to the first example described in the introduction for $\gamma = 4$ (see Eq. (2.1)). One obtains the same buckling load,

$$P = 8D_0/L^2 \tag{2.28}$$

For $a = 4$, the same buckling mode

$$w = 4(z - z^2) \tag{2.29}$$

is found as in Eq. (2.3). Equation (2.25) allows one to optimize the column in the presence of the buckling constraint, demanding that the column does not buckle prior to a load level $\hat{P}$:

$$P_{cl} \geq \hat{P} \tag{2.30}$$

This yields the admissible region of variation of the parameter γ,

$$\gamma \geq \hat{P}L^2/2D_0 \tag{2.31}$$

so that the buckling load of the column will satisfy the inequality (2.30).

If one assumes that the displacement is a polynomial of higher degree, one can find higher buckling loads. The method proposed below is base of the logarithm of the Mathematica® function *SolveAways*. One assumes that the displacement is of the form

$$w = w_0 \sum_{j=1}^{N} a_j z^j \tag{2.32}$$

This displacement equals zero at $z = 0$ so it satisfies the first boundary condition. The second derivative of w reads as

$$w'' = w_0 \sum_{j=2}^{N} a_j j(j - 1)z^{j-2} \tag{2.33}$$

One defines the flexural rigidity as

$$D = D_0 \gamma(z - z^2) \tag{2.34}$$

This definition is very interesting because the flexural rigidity equals zero at both ends. So the bending moment Dw'' equals zero at the ends identically, irrespective of the definition of the displacement. The differential equation becomes

$$\gamma(z - z^2)w'' + k^2 w = 0 \tag{2.35}$$

Taking into account Eqs. (2.32) and (2.33), one obtains

$$\gamma(z - z^2) \sum_{j=2}^{N} a_j j(j - 1)z^{j-2} + k^2 \sum_{j=1}^{N} a_j z^j = 0 \tag{2.36}$$

Now, the equation is modified to have only z^j terms:

$$\sum_{j=1}^{N-1} \gamma a_{j+1} j(j+1) z^j - \sum_{j=2}^{N-1} \gamma a_j j(j-1) z^j - \gamma a_N N(N-1) z^N$$

$$+ k^2 a_1 z + \sum_{j=1}^{N} k^2 a_j z^j + k^2 a_N z^N = 0 \qquad (2.37)$$

Regrouping the terms of the same degree yields

$$(k^2 a_1 + 2\gamma a_2) z + \sum_{j=2}^{N-1} [\gamma a_{j+1} j(j+1) - \gamma a_j j(j-1)] z^j$$

$$\times [k^2 - \gamma N(N-1)] a_N z^N = 0 \qquad (2.38)$$

This equation must equal zero for any z. For z^N,

$$k^2 = \gamma N(N-1) \qquad (2.39)$$

and one deduces that

$$P = \gamma N(N-1) D_0 / L^2 \qquad (2.40)$$

The first buckling load is for $N = 2$:

$$P_{cl} = 2\gamma D_0 / L^2 \qquad (2.41)$$

As a result, a polynomial of degree N leads to mth buckling load with

$$N = m + 1 \qquad (2.42)$$

The other terms of the polynomial lead to the global definition of the coefficients. For the first power of z one obtains

$$a_2 = -m(m+1) a_1 / 2 \qquad (2.43)$$

whereas for jth power of z

$$a_{j+1} = \{[j(j-1) - m(m+1)] / j(j+1)\} a_j \qquad (2.44)$$

One notices that for $j = 1$, Eq. (2.44) reduces to Eq. (2.43). Hence, Eq. (2.44) is valid for any a_j. One chooses $a_1 = 1$. One can verify via Mathematica® that the sum of the coefficients a_j equals zero, and thus the boundary condition of the displacement at the end $z = 1$, is satisfied.

Now, one can consider that one has a column clamped beam at $z = 0$ and it has arbitrary boundary conditions at $z = 1$. If one assumes that w is a polynomial, as in Eq. (2.32), then the coefficient a_1 must vanish to have a slope equal to zero at $z = 0$. One has to take into account Eqs. (2.43) and (2.44). These relations imply that all a_j must equal zero. Finally, one arrives at a trivial solution. One concludes that the present method is in applicable for the column that has a clamped end at $z = 0$.

Case 2: $r = 1 + \beta z - \gamma z^2$

In this case, the flexural rigidity is given by

$$D = D_0(1 + \beta z - \gamma z^2) \tag{2.45}$$

The displacement is a polynomial of degree 5,

$$w = az + bz^2 + cz^3 + dz^4 + ez^5 \tag{2.46}$$

After simplifications, the boundary conditions (2.19) lead to the conditions

$$b = 0 \quad a + c + d + e = 0 \quad d = \beta^3 a \quad c = -2\beta^2 a \quad e = 0 \tag{2.47}$$

Mathematica®, leads to three different sets of solutions which respect the differential equation. The second and third one yield the same buckling mode. Hence, one has to consider two sets which correspond to different buckling loads and modes. The first set is defined as

$$k^2 = 12\beta^2 \quad \gamma = \beta^2 \quad d = \beta^3 a \quad c = -2\beta^2 a \quad e = 0 \tag{2.48}$$

This set of solutions implies the following definition of the flexural rigidity

$$D = D_0(1 + \beta z - \gamma z^2) \tag{2.49}$$

with attendant buckling load

$$P = 12\beta^2 D_0/L^2 \tag{2.50}$$

and buckling mode

$$w = a(z - 2\beta^2 z^3 + \beta^3 z^4) \tag{2.51}$$

These coefficients have to satisfy the boundary conditions (2.19). From this one finds three values of β that satisfy the above conditions:

$$\beta_1 = 1 \quad \beta_2 = (1 - \sqrt{5})/2 \quad \text{and} \quad \beta_3 = (1 + \sqrt{5})/2 \tag{2.52}$$

For the particular case $\beta_1 = 1$, the flexural rigidity is defined by

$$D = D_0(1 + z - z^2) \tag{2.53}$$

so that the buckling load equals

$$w = a(z - 2z^3 + z^4) \tag{2.54}$$

One notices that this function is a Duncan polynomial (1937). For the particular case $\beta_2 = (1 - \sqrt{5})/2$, the flexural rigidity is defined by

$$D = D_0[1 + (1 - \sqrt{5})z/2 - (1 - \sqrt{5})^2 z^2/4] \tag{2.55}$$

The buckling load equals

$$P = 3(\sqrt{5} - 1)^2 D_0/L^2 \tag{2.56}$$

The buckling mode is

$$w = a\{z - [(1 - \sqrt{5})^2/2]z^3 + [(1 - \sqrt{5})^3/8]z^4\} \tag{2.57}$$

The second set is defined as

$$k^2 = 20\gamma \quad \beta = -\sqrt{3\gamma} \quad d = -[5\gamma^{3/2}/\sqrt{3}]a \quad c = -(10\gamma/3)a \quad e = (2\gamma^2/3)a \tag{2.58}$$

This set of solutions implies the definition of the flexural rigidity

$$D = D_0(1 - \sqrt{3\gamma}z - \gamma z^2) \tag{2.59}$$

with the buckling load

$$P = 20\gamma D_0/L^2 \tag{2.60}$$

and the buckling mode

$$w = d[z - (10\gamma/3)z^3 - (5\gamma^{3/2}/\sqrt{3})z^4 + (2\gamma^2/3)z^5] \tag{2.61}$$

These coefficients have to satisfy the boundary conditions (2.19). From there one finds a single value of γ that satisfies these equations: $\gamma = (5 - \sqrt{21})/2$. Then, the flexural rigidity becomes

$$D = D_0\left\{1 - z\sqrt{[3(5 - \sqrt{21})/2]} - [5 - \sqrt{21})/2]z^2\right\} \tag{2.62}$$

The buckling mode becomes

$$w = a\left\{z - [5(5 - \sqrt{21})/3]z^3 - [5(5 - \sqrt{21})^{3/2}/2\sqrt{6}]z^4 + [(5 - \sqrt{21})^2/6]z^5\right\}$$

(2.63)

Case 3: $r = 1 - \gamma z^2$

In this last part, one has

$$D = D_0(1 - \gamma z^2)$$

(2.64)

We look for the buckling mode in the form of a fifth-order polynomial:

$$w = az + bz^2 + cz^3 + dz^4 + ez^5$$

(2.65)

After simplification, the boundary conditions (2.19) lead to the conditions

$$b = 0 \quad a + c + d + e = 0 \quad (3c + 6d + 10e)(1 - \gamma) = 0$$

(2.66)

The solution of the governing differential equation, Eq. (2.13), by Mathematica® leads to two different sets of solutions, which correspond to different buckling loads and modes. The first set is given by

$$k^2 = 6\gamma \quad c = -\gamma a \quad d = e = 0$$

(2.67)

This set of solutions implies the following definition of the buckling load:

$$P = 6\gamma D_0/L^2$$

(2.68)

The buckling mode is a cubic polynomial,

$$w = a(z - \gamma z^3)$$

(2.69)

These coefficients have to satisfy the boundary conditions (2.39). Thus, one finds a single value of γ that satisfies these equations: $\gamma = 1$. Therefore, the flexural rigidity is defined by

$$D = D_0(1 - z^2)$$

(2.70)

The buckling load is given by

$$P = 6D_0/L^2$$

(2.71)

The buckling mode reads as

$$w = a(z - z^3) \tag{2.72}$$

The second set is given by

$$k^2 = 20\gamma \quad c = -(10\gamma/3)a \quad e = -(7\gamma^2/3)a \quad d = 0 \tag{2.73}$$

This set of solutions implies the following definition of the buckling load:

$$P = 20\gamma D_0/L^2 \tag{2.74}$$

whereas the buckling mode is

$$w = a[z - (10\gamma/3)z^3 + (7\gamma^2/3)z^5] \tag{2.75}$$

These coefficients have to satisfy the boundary conditions (2.19). This allows us to find two values of γ that satisfy these equations: $\gamma = \frac{3}{7}$ or $\gamma = 1$. For the particular case $\gamma = 1$, the flexural rigidity is defined by

$$D = D_0(1 - z^2) \tag{2.76}$$

so the buckling load equals

$$P = 20D_0/L^2 \tag{2.77}$$

The buckling mode is

$$w = a\left[z - \left(\tfrac{10}{3}\right)z^3 + \left(\tfrac{7}{3}\right)z^5\right] \tag{2.78}$$

It is noticeable that this flexural rigidity is the same as that for the first set of the third case (Eqs. (2.78)–(2.80)), but a greater buckling load has been found implying that one determines the higher buckling loads by increasing the degree of freedom of the displacement in this case. For the particular case $\gamma = 3/7$, the flexural rigidity is defined by Eq. (2.6), so that the expression for the buckling load in Eq. (2.82) reduces to that in Eq. (2.9), and the buckling mode is

$$w = a\left[z - \left(\tfrac{10}{7}\right)z^3 + \left(\tfrac{3}{7}\right)z^5\right] \tag{2.79}$$

By choosing $a = 7A$, the displacement reduces to Eq. (2.8). This result and the buckling load match those of Duncan (1937).

2.1.4 Concluding Remarks

In this section Engesser's (1893) example is first generalized to a family of columns with a variable moment of inertia. Using the proposed approach, one also determines new closed-form solutions of columns with variable flexural rigidity, including the generalization of Duncan's (1937) solution. Then, a design criterion is discussed, so that the buckling load is required to exceed any prescribed value. Note that if one uses polynomials of higher degree, more degrees of freedom are allowed for the displacement. In such circumstances, the method leads to higher buckling loads. It appears remarkable that the closed-form solutions obtained are simpler than the exact solutions for many problems involving uniform columns. This unusual property will re-appear again and again throughout the monograph.

2.2 Inverse Buckling Problem for Inhomogeneous Columns

2.2.1 Introductory Remarks

The exact solutions for the buckling load of uniform beams are treated in almost any textbook on the mechanics of solids. Exact solutions for non-uniform columns is the subject of several works (see, e.g., the classical textbook by Timoshenko and Gere, 1961). These solutions are derived in terms of Bessel or Lommel functions, or some other, elementary or transcendental functions. As far as the closed-form solutions are concerned the results are much more restricted. All existing closed-form solutions are apparently listed, along with new solutions, in the recent study by Elishakoff and Rollot (1999) and in Section 2.1.

This section is devoted to obtaining additional closed-form solutions, which are posed as inverse problems. The formulation of the problem is as follows: Find the polynomial distribution of the Young modulus $E(x)$ of an inhomogeneous column of uniform cross-section with specified boundary conditions, so that the buckling mode will be a pre-selected polynomial function. It turns out that this seemingly simple formulation allows us to derive new closed-form solutions. The solutions obtained appear to be of much importance, since once the technology to construct columns with given variation of the modulus of elasticity is available, any pre-selected buckling load can be achieved for the appropriate design of the structure.

2.2.2 Formulation of the Problem

The differential equation that governs the buckling of the column under an axial load P, reads

$$\frac{d^2}{dx^2}\left[D(x)\frac{d^2w}{dx^2}\right] + P\frac{d^2w}{dx^2} = 0 \tag{2.80}$$

where $w(x)$ is the transverse dispacement, $D(x) = E(x)I(x)$ is the flexural rigidity, $E(x)$ is the modulus of elasticity, $I(x)$ the moment of inertia, and x the axial coordinate. We consider four sets of boundary conditions. For the column that is pinned at both its ends the simplest polynomial that satisfies the boundary conditions at $x = 0$ and $x = L$, with L being the length of the column, reads

$$\psi(\xi) = \xi - 2\xi^3 + \xi^4 \quad \xi = x/L \tag{2.81}$$

We pose the following question: Is there a column with polynomial variation of $E(\xi)$ that possesses the function in Eq. (2.81) as its fundamental buckling mode? Indeed, if the sought for solution exists, it corresponds to the fundamental buckling load, since $\psi(\xi)$ in Eq. (2.81) does not have internal nodes. This problem differs from the *direct* buckling problem, which presupposes the knowledge of the flexural rigidity D and requires the determination of the mode $w(x)$ and the buckling load P. Here, we are looking for the *cause*, i.e., the distribution of the flexural rigidity, knowing the *effect*, i.e., the buckling mode.

We are looking for the flexural rigidity $D(\xi)$ represented as follows:

$$D(\xi) = b_0 + b_1\xi + b_2\xi^2 \tag{2.82}$$

where b_0, b_1 and b_2 are the constants to be determined. The inverse problems may have no solution, multiple solutions or a unique solution. It turns out that in the case under study, an infinite number of solutions exists for reconstructing the column, which possess the function in Eq. (2.81) as their buckling mode. We now consider the different sets of boundary conditions.

2.2.3   Column Pinned at Both Ends

Using the non-dimensional axial coordinate ξ, defined in Eq. (2.81), the governing equation, Eq. (2.80), reduces to

$$\frac{d^2}{d\xi^2}\left[D(\xi)\frac{d^2\psi}{d\xi^2}\right] + PL^2\frac{d^2\psi}{d\xi^2} = 0 \tag{2.83}$$

With the buckling mode postulated in Eq. (2.81) we have for the term in Eq. (2.83)

$$PL^2\frac{d^2\psi}{d\xi^2} = qL^2(-12\xi + 12\xi^2) \tag{2.84}$$

whereas the first differential expression in Eq. (2.84) reads as

$$\frac{d^2}{d\xi^2}\left[D(\xi)\frac{d^2\psi}{d\xi^2}\right] = -12[2(b_1 - b_0) + 6\xi(b_2 - b_1) - 12b_2\xi^2] \qquad (2.85)$$

The sum of the expressions in the right-hand sides in Eqs. (2.84) and (2.85) must vanish, due to Eq. (2.83). Since the above sum must equal zero identically, for any value of ξ, we get the following expressions:

$$2(b_1 - b_0) = 0 \qquad (2.86)$$

$$-72(b_2 - b_1) - 12PL^2 = 0 \qquad (2.87)$$

$$144b_2 + 12PL^2 = 0 \qquad (2.88)$$

The solution of Eq. (2.88) yields

$$P = -12b_2/L^2 \qquad (2.89)$$

In order for the load P to remain compressive, it must be positive. We conclude, therefore, that the coefficient b_2 must be negative. Equations (2.86) and (2.87) lead, together with Eq. (2.89), to

$$b_0 = b_1 = -b_2 \qquad (2.90)$$

Thus, the flexural rigidity is defined up to the coefficient b_2:

$$D(\xi) = (-1 - \xi + \xi^2)b_2 \qquad (2.91)$$

We already established that b_2 must take a negative value. Hence Eq. (2.91) can be rewritten as

$$D(\xi) = (1 + \xi - \xi^2)|b_2| \qquad (2.92)$$

This function is in agreement with the physical realizability condition, namely, with the requirement of non-negativity of the the function $D(\xi)$ in the interval $[0; 1]$. It is depicted in Figure 2.1.

We have thus found the function $D(\xi)$ that corresponds to the postulated buckling mode in Eq. (2.90). Since $D(\xi)$ depends on the arbitrary negative constant b_2, we conclude that the solution of the posed problem is infinity in the class of polynomially varying flexural rigidities. Note that Eq. (2.97) coincides with Eq. (2.56) in the paper by Elishakoff and Rollot (1999). It pertains to the column that is pinned at both its ends. We will show that Eq. (2.97) is valid for two other sets of boundary conditions.

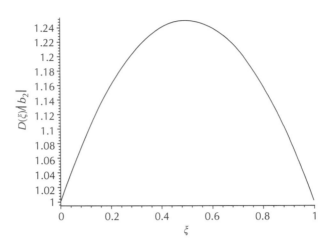

FIGURE 2.1
Variation of $D(\xi)/|b_2|$

2.2.4 Column Clamped at Both Ends

The boundary conditions

$$w = 0 \quad dw/d\xi = 0 \qquad \text{at} \quad \xi = 0 \quad \xi = 1 \tag{2.93}$$

are satisfied for the following polynomial function:

$$\psi(\xi) = \xi^2 - 2\xi^3 + \xi^4 \tag{2.94}$$

We are interested in establishing if this polynomial function may serve as a buckling shape of any inhomogeneous column. The expression for $PL^2\psi''$ reads, with primes denoting differentiation with respect to ξ, as follows:

$$PL^2\psi'' = qL^2(2 - 12\xi + 12\xi^2) \tag{2.95}$$

whereas the expression for $(D\psi'')''$ is

$$(D\psi'')'' = 2(2b_2 - 12b_1 + 12b_0) + 6(-12b_2 + 12b_1)\xi + 144b_2\xi^2 \tag{2.96}$$

We demand that the sum of the expressions in Eqs. (2.95) and (2.96) vanish for any ξ. This requirement leads to the following three equations:

$$2(2b_2 - 12b_1 + 12b_0) + 2PL^2 = 0 \tag{2.97}$$

$$6(-12b_2 + 12b_1) - 12PL^2 = 0 \tag{2.98}$$

$$144b_2 + 12PL^2 = 0 \tag{2.99}$$

From Eq. (2.99) we get the buckling load

$$P = -12b_2/L_2 \qquad (2.100)$$

which, remarkably, coincides with Eq. (2.89). Equations (2.97) and (2.98) lead to

$$b_0 = -b_2/6 \quad b_1 = -b_2 \qquad (2.101)$$

Hence, we obtain the sought for variation of the flexural rigidity

$$D(\xi) = \left(\tfrac{1}{6} + \xi - \xi^2 \right) |b_2| \qquad (2.102)$$

which takes a positive value throughout the column's axis, $\xi \in [0, 1]$, as seen from Figure 2.2.

2.2.5 Column Clamped at One End and Pinned at the Other

The boundary conditions read

$$w = 0 \quad D(\xi)\frac{d^2w}{d\xi^2} = 0 \qquad \text{at } \xi = 0$$

$$(2.103)$$

$$w = 0 \quad \frac{dw}{d\xi} = 0 \qquad \text{at } \xi = 1$$

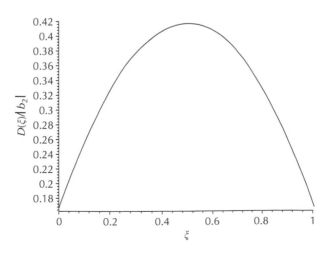

FIGURE 2.2
Variation of $D(\xi)/|b_2|$

The boundary conditions are satisfied by the following polynomial function:

$$\psi(\xi) = 3\xi^2 - 5\xi^3 + 2\xi^4 \tag{2.104}$$

Substitution of this expression into the governing differential equation, in conjunction with the postulated expression for the flexural rigidity results in

$$[2(6b_2 - 30b_1 + 24b_0) + 6(-30b_2 + 24b_1)\xi + 288b_2\xi^2]$$
$$+ PL^2(6 - 30\xi + 24\xi^2) = 0 \tag{2.105}$$

Since Eq. (2.105) is valid for any ξ, we get the following three equations:

$$\text{from } \xi^0: \quad 6b_2 - 18b_1 + 24b_0 = 0 \tag{2.106}$$

$$\text{from } \xi^1: \quad 6(-30b_2 + 24b_1) - 30PL^2 = 0 \tag{2.107}$$

$$\text{from } \xi^2: \quad 288b_2 + 24PL^2 = 0 \tag{2.108}$$

We arrive at three equations for the four unknowns: b_0, b_1, b_2 and P. We choose one of the parameters to be arbitrary, namely b_2. Then, Eq. (2.108) yields the same buckling load as in Eqs. (2.89) and (2.100):

$$P = -12b_2/L^2 \tag{2.109}$$

Equations (2.106)–(2.108) yield the following interrelation between the coefficients describing the flexural rigidity variation:

$$b_0 = -\tfrac{5}{16}b_2 \quad b_1 = -\tfrac{5}{4}b_2 \tag{2.110}$$

Substituting into Eq. (2.82) results in the variation of the flexural rigidity:

$$D(\xi) = \left(\tfrac{5}{16} + \tfrac{5}{4}\xi - \xi^2\right)|b_2| \tag{2.111}$$

which is a positive function within the length of the column. The function $D(\xi)$ for this case is depicted in Figure 2.3. Note that two other sets of boundary conditions were studied by Elishakoff (2000a). These are not reproduced here, to save space.

2.2.6   Concluding Remarks

The following conclusions appear to be relevant:

1. Inverse buckling problems with postulated polynomial buckling modes, as given in Eqs. (2.81), (2.94) or (2.104) for corresponding boundary conditions have closed-form solutions; namely, the variations of flexural rigidity

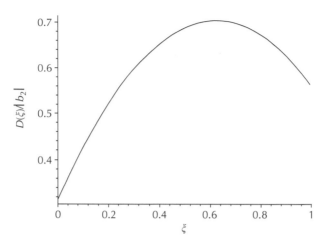

FIGURE 2.3
Variation of $D(\xi)/|b_2|$

corresponding to the above mode shapes are given in Eqs. (2.92), (2.102) and (2.111), respectively.

2. For three sets of boundary conditions, the fundamental buckling load is given by the same expression (see Eqs. (2.89), (2.100) and (2.109)). This conclusion may appear to be a paradoxical one at the first glance. To resolve it, let us consider the corresponding case of homogeneous and uniform columns. The fundamental buckling loads are

$$P_{P-P} = \frac{\pi^2 D_1}{L^2} \quad P_{C-C} = \frac{4\pi^2 D_2}{L^2} \quad P_{P-C} = \frac{4.493^2 D_3}{L^2} \approx \frac{2\pi^2 D_3}{L^2} \qquad (2.112)$$

where D_1 is the flexural rigidity of the column that is pinned at both its ends, D_2 corresponds to the column that is clamped at both ends and D_3 is associated with the column that is pinned at $x = 0$ and clamped at $x = L$. These three columns possess the same cross-sections and have the same lengths. Now, if

$$D_1 = 4D_2 \approx 2D_3 \qquad (2.113)$$

then all three columns have coalescing buckling loads. This implies that *only* those columns with different flexural rigidities — if they are under different boundary conditions, but have the same lengths and cross-sections — *may* share the same fundamental buckling load. The same phenomenon takes place in the case of our study: the columns with three different sets of boundary conditions share the same fundamental buckling load, since their moduli of elasticity are different.

3. Still, it appears to be intriguing that the search for the solution of the inverse buckling problem in the class of polynomial functions leads to the coincidence of buckling loads for pinned–pinned, clamped–clamped and

pinned–clamped boundary conditions. Here, we reported the case when the fundamental buckling loads are shared by columns under different boundary conditions [see also Gottlieb (1989) Gladwell and Morassi (1995) for vibration problems].

To compare the results for the buckling loads, let us calculate the average flexural rigidity in each of the three cases. The average flexural rigidity is defined as

$$D_{av} = \int_0^1 D(\xi)\, d\xi \tag{2.114}$$

Thus, for the pinned column

$$D_{av,S-S} = \tfrac{7}{6}|b_2| \tag{2.115}$$

For the clamped column

$$D_{av,C-C} = \tfrac{1}{3}|b_2| \tag{2.116}$$

For the column that is pinned at one end and clamped at the other

$$D_{av,S-C} = \tfrac{29}{48}|b_2| \tag{2.117}$$

Thus, the buckling loads can be put in the following forms, by first expressing $|b_2|$ via D_{av} in Eqs. (2.115)–(2.117):

$$P_{P-P} = 12|b_2|/L^2 = \tfrac{72}{7}D_{av}/L^2 \tag{2.118}$$

$$P_{C-C} = 12|b_2|/L^2 = 36D_{av}/L^2 \tag{2.119}$$

$$P_{P-C} = 12|b_2|/L^2 = \tfrac{576}{29}D_{av}/L^2 \tag{2.120}$$

If the average flexural rigidities of these columns are chosen to be the same, then, the buckling loads of the inhomogeneous columns are in the proportion

$$\tfrac{72}{7} : 36 : \tfrac{576}{29} \tag{2.121}$$

or $1:3.5:1.93$, vs. the corresponding proportion $1:4:(\approx 2)$ for the uniform columns.

4. The calculations of the buckling loads have been performed for columns possessing the flexural rigidity variations given in Eqs. (2.92), (2.102) and (2.111), respectively. The flexural rigidity in Eq. (2.92) is denoted as $D_{P-P}(\xi)$. The flexural rigidity function in Eq. (2.102) is designated as $D_{C-C}(\xi)$, whereas the flexural rigidity variation in Eq. (2.111) is denoted as $D_{C-P}(\xi)$. For each flexural rigidity function three different sets of boundary conditions have been examined. For the pinned column with the flexural rigidity

$D_{P-P}(\xi)$, the closed-form solution is reported in Eq. (2.89) as $12|b_2|/L^2$. For the column with flexural rigidity D_{P-P}, but under boundary conditions differing from both the pinned ends, the approximate values of the buckling loads are furnished by many approximate methods. Due to its simplicity, we employed the Boobnov–Galerkin method here. For consistency, the buckling load of the pinned–pinned inhomogeneous column was also evaluated by the Boobnov–Galerkin method, with the comparison function $\sin(\pi\xi)$, yielding 12.0145, the value that is just 0.12% above the closed-form expression. For the column with flexural rigidity $D_{P-P}(\xi)$ but with clamped–pinned boundary conditions, the buckling mode of the associated uniform column was utilized as the comparison function. The buckling load equals $23.7181|b_2|/L^2$, whereas for the clamped–clamped beam the buckling load derived by using the function $\sin^2(\pi\xi)$ as the comparison function, equals $45.5582|b_2|/L^2$. As is seen, for the columns with *identical* flexural rigidity, but under *different* boundary conditions, the buckling loads *differ*, as it should be. An analogous conclusion is reached for the columns with flexural rigidity equal to either $D_{C-P}(\xi)$ or $D_{C-C}(\xi)$. The results are summarized in Table 2.1.

5. For uniform columns the polynomial expressions of the buckling mode are usually utilized to facilitate the approximate solutions, via the Boobnov–Galerkin or the Rayleigh–Ritz methods. For example, Chajes (1974) uses the function $w(x) = xL^3 - 3x^3L + 2x^4$, as a *comparison function* in the Boobnov–Galerkin method for the column that is pinned at $x = 0$ and clamped at $x = L$. It is interesting that the same polynomial function (coincident with Eq. (2.104) once ξ is replaced by $1 - \xi$) turned out to be an *exact* buckling mode of the *inhomogeneous* column.

TABLE 2.1

Comparison of Exact and Approximate Solutions

Flexural rigidity	Boundary conditions	Comparison function in the Boobnov–Galerkin method	$P_{cr}L^2/|b_2|$ Exact solution	$P_{cr}L^2/|b_2|$ Approximate solution	Error (%)
$D_{P-P}(\xi) =$ $(1 + \xi - \xi^2)|b_2|$	P–P	$\sin(\pi\xi)$	12	12.0145	0.12
	C–P	$-1 + \xi + \cos(4.493\xi)$ $- \sin(4.493\xi)/4.493$		23.7181	–
	C–C	$\sin^2(\pi\xi)$		45.5582	–
$D_{C-P}(\xi) =$ $(\frac{5}{16} + \frac{5}{4}\xi - \xi^2)|b_2|$	P–P	$\sin(\pi\xi)$		6.46289	–
	C–P	$-1 + \xi + \cos(4.493\xi)$ $- \sin(4.493\xi)/4.493$	12	12.363	3.03
	C–C	$\sin^2(\pi\xi)$		23.3515	–
$D_{C-C}(\xi) =$ $(\frac{1}{6} + \xi - \xi^2)|b_2|$	P–P	$\sin(\pi\xi)$		3.78987	–
	C–P	$-1 + \xi + \cos(4.493\xi)$ $- \sin(4.493\xi)/4.493$		6.89555	–
	C–C	$\sin^2(\pi\xi)$	12	12.6595	5.50

Likewise, in his collection of problems, Volmir (1984) posed the question as to whether the Rayleigh's method as well as the Boobnov–Galerkin method could be used for the *approximate* estimation of the buckling load of a uniform column, pinned at both its ends, by utilizing the trial function $w(x) = xL^3 - 2x^3L + x^4$. In our case, the same buckling mode (coincident with Eq. (2.81)) serves as an *exact* expression for the inhomogeneous column.

6. The question arises as to the generality of the proposed method. In order for the method to be acceptable it should lead to a positive buckling load, corresponding to the case of a compressive force, and a modulus of elasticity that is a positive function. The formulation of some general conditions upon fulfillment of which the problem is amenable to a solution, is of interest.

7. The buckling loads turn out to depend on only a single coefficient b_2, once the function of modulus variation is obtained. By a suitable choice of b_2, the buckling load can be made arbitrarily large. This conclusion is valid within the context of elastic buckling that was presupposed in this study.

8. The solution of the buckling problems of uniform columns leads to irrational values of the buckling loads; for example, the buckling load of the pinned *uniform* column is written in terms of π^2 (see Eq. (2.112)). So is the buckling load, found by Euler (1759) in the case of the variable flexural rigidity column, where the buckling load is $\pi^2(D_0/L^2)a^2 \times (a+b)^2$ for the column with the flexural rigidity $D(\xi) = D_0(a+b\xi)^4$. Here, in the case of *inhomogeneous* and/or *non-uniform* columns, the solution turns out to be in terms of rational numbers. Other solutions of this kind have been reported by Elishakoff and Rollot (1999) (see Section 2.1).

9. Note that the flexural rigidity variation of the clamped–clamped column in Eq. (2.102) can be obtained from that of the pinned–pinned column in Eq. (2.92), by simple translation (Calió, 2000) of the curve.

2.3   Closed-Form Solution for the Generalized Euler Problem

2.3.1   Introductory Remarks

The motivation for this section is as follows. In recent decades numerous studies have been published on the so-called stochastic finite element method, which deals with stochastically inhomogeneous structures by the finite element method. Among various methods, the authors utilize the perturbation techniques of various orders, spectral finite element method, Neumann expansions, etc. For a review of some of these studies the reader can consult the review papers by Shinozuka and Yamazaki (1988) and Elishakoff et al. (1995). One of the techniques discussed by them is the finite element method combined with the Monte Carlo method. In the context of this method, this stochastic variables or fields are first simulated. Thus, a stochastic ensemble of a large amount of deterministic realizations is created. Each of the structures is solved by a deterministic FEM, and results are statistically analyzed. In order to elucidate the efficacy of such a technique, the closed-form deterministic problems will be of great help. However, even in the deterministic setting, the number of exact solutions, let alone the closed-form ones, is extremely limited [for exact solution in the stochastic setting one can consult the paper by Ellishakoff et al. (1995)]. It is important, therefore, to invest effort to generate closed-form solutions in the deterministic setting. Obviously, such solutions are interesting in their own right, for the community of researchers and engineers who occupy themselves with deterministic theoretical and applied mechanics.

The buckling of columns under their own weight is a problem that goes back to Leonhard Euler (1778), who looked for its solution in terms of power series with respect to the axial coordinate. For the clamped–free column of uniform flexural rigidity, the exact solution is given in terms of Bessel functions (Timoshenko and Gere, 1961) with the attendant critical load intensity q_{cr} equal to

$$q_{C-F} = 7.837D/L^3 \tag{2.122}$$

where D is the flexural rigidity, and L the length.

The column under its own weight was considered by Euler in three famous papers in 1778. For the fascinating and instructive story of these three papers the reader is directed to a most instructive book by Panovko and Gubanova (1965) on numerous fallacies and mistakes in applied mechanics.

In the first paper, Euler (1778a) arrived at a conclusion that a pinned column under its own weight does not buckle. In the second paper (1778b) he expressed a suspicion about the above conclusion, which he called a paradoxical one. In the third paper, Euler (1778c) found a mistake made in the first paper, and concluded that such a column is susceptible to buckling. Unfortunately, however, he made a numerical mistake and instead of the first buckling load calculated the second one. A correct numerical value was calculated by Dinnick (1912) some 134 years after Euler:

$$q_{P-P} = 18.6D/L^3 \tag{2.123}$$

Engelhardt (1954) reported results for the column that is clamped at one end and pinned at the other,

$$q_{C-S} = 52.50D/L^3 \tag{2.124}$$

as well as for the column that is clamped at both ends,

$$q_{C-C} = 74.6D/L^3 \tag{2.125}$$

Other exact solutions were furnished by Greenhill (1881), Willers (1941) and Frisch-Fay (1966). Dinnik (1912), Engelhardt (1954) and Frisch-Fay (1966) utilized the power series method, whereas Willers (1941) presented a solution in terms of Bessel and Lommel functions. The solution in terms of Bessel and Lommel functions was communicated to this writer by Pflüger (1975).

The exact finite element method in the context of the buckling was developed by Eisenberger (1991a–d). Both the column's, flexural rigidity and the axial load were considered to vary as polynomial functions. The buckling load was determined as the load that makes the determinant of the stiffness matrix zero (Eisenberger, 1991a–d). The latter mentions that in his work "exact buckling loads (up to the accuracy of the computer)" are given and that "only one element is need for the solution". Waszczyszyn and Pieczara (1990) also reported several results via the exact finite element method.

In the present section we obtain the closed-form solutions for the generalized Euler's problem, namely, columns with variable flexural rigidity under their own weight, following Elishakoff (2000a). It appears to be remarkable that for the column with *uniform* flexural rigidity transcendental equations are obtained for the buckling load determination, whereas for the more complicated case of the column with *variable* flexural rigidity, the closed-form solution turns out to be available. This seemingly curious fact is perhaps explainable by the observation that variability in flexural rigidity introduces additional degrees of freedom that enable us to obtain solutions in terms of polynomial functions.

2.3.2   Formulation of the Problem

The differential equation that governs the buckling of the non-uniform column under its own weight reads as

$$\frac{d^2}{dx^2}\left[D(x)\frac{d^2w}{dx^2}\right] + \frac{d}{dx}\left[q(L-x)\frac{dw}{dx}\right] = 0 \qquad (2.126)$$

where $w(x)$ is the buckling mode, $D(x) = E(x)I(x)$ is the flexural rigidity, E the modulus of elasticity, I the moment of inertia, q the intensity of the buckling load, L the length and x the axial coordinate.

The differential equation (2.134) is accompanied by appropriate boundary conditions. For the column that is pinned at its ends, the boundary conditions are

$$w = 0 \qquad \frac{d^2w}{dx^2} = 0 \qquad \text{at } x = 0 \quad x = L \qquad (2.127)$$

For simplicity, we introduce the non-dimensional coordinate $\xi = x/L$. The differential equation (2.126) then becomes

$$\frac{d^2}{d\xi^2}\left[D(\xi)\frac{d^2w}{d\xi^2}\right] + \frac{d}{d\xi}\left[qL^3(1-\xi)\frac{dw}{d\xi}\right] = 0 \tag{2.128}$$

We are looking for a polynomial representation of the flexural rigidity $D(\xi)$ in Eq. (2.134). Since the buckling mode $w(\xi)$ is also required sought as a polynomial, the left-hand side of Eq. (2.134) turns out to be a polynomial in terms of the non-dimensional axial coordinate ξ. To achieve the same order of the polynomial expressions in both the first and the second term in Eq. (2.134), the flexural rigidity must be a third-order polynomial:

$$D(\xi) = b_0 + b_1\xi + b_2\xi^2 + b_3\xi^3 \tag{2.129}$$

The buckling mode, denoted by $\psi(\xi)$, is chosen as a polynomial of fourth degree that satisfies all boundary conditions and can be written as

$$\psi(\xi) = \xi - 2\xi^3 + \xi^4 \tag{2.130}$$

The first term in the differential equation (2.128) becomes upon substitution of Eq. (2.129)

$$\frac{d^2}{d\xi^2}\left[D(\xi)\frac{d^2\psi}{d\xi^2}\right] = -12[2(b_1 - b_0) + 6\xi(b_2 - b_1) + 12\xi^2(b_3 - b_2) - 20b_3\xi^3] \tag{2.131}$$

whereas the second term in Eq. (2.128) becomes

$$\frac{d}{d\xi}\left[qL^3(1-\xi)\frac{d\psi}{d\xi}\right] = qL^3(-1 - 12\xi + 30\xi^2 - 16\xi^3) \tag{2.132}$$

The sum of the expressions in Eqs. (2.131) and (2.132) must vanish for every ξ. This leads to the following conditions:

$$\begin{aligned}
-24(b_1 - b_0) - qL^3 &= 0 \\
-72(b_2 - b_1) - 12qL^3 &= 0 \\
-144(b_3 - b_2) + 30qL^3 &= 0 \\
240b_3 - 16qL^3 &= 0
\end{aligned} \tag{2.133}$$

We arrive at four equations with five unknowns, which is solvable up to an arbitrary constant, which is taken here as the coefficient b_3. The buckling intensity equals

$$q_{cr} = 15b_3/L^3 \tag{2.134}$$

The other coefficients are:

$$b_0 = b_3$$
$$b_1 = \tfrac{3}{8}b_3 \tag{2.135}$$
$$b_2 = -\tfrac{17}{8}b_3$$

so that the attendant flexural rigidity reads as

$$D(\xi) = \left(1 + \tfrac{3}{8}\xi - \tfrac{17}{8}\xi^2 + \xi^3\right)b_3 \tag{2.136}$$

Figure 2.4 depicts $D(\xi)/b_3$. It is instructive to calculate the average flexural rigidity

$$D_{av} = \int_0^1 D(\xi)\,d\xi = \frac{35}{48}b_3 \tag{2.137}$$

The buckling intensity of the column with the average flexural rigidity is obtained by substituting D_{av} into Eq. (2.123), once it is expressed in terms of b_3 in Eq. (2.137):

$$q_{cr} = 18.6D_{av}/L^3 = 13.56b_3/L^3 \tag{2.138}$$

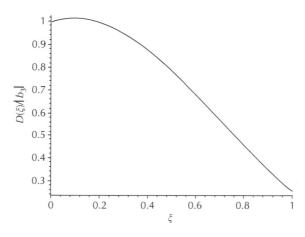

FIGURE 2.4
Variation of $D(\xi)/b_3$ for the column pinned at both its ends

which turns out to be 1.10 times smaller than its counterpart in Eq. (2.134) for the variable flexural rigidity column.

2.3.3 Column Clamped at Both Ends

The postulated buckling mode reads

$$\psi(\xi) = \xi^2 - 2\xi^3 + \xi^4 \tag{2.139}$$

The equations that are counterparts of Eq. (2.133) are

$$
\begin{aligned}
2(12b_0 - 12b_1 + 2b_2) + 2qL^3 &= 0 \\
6(12b_1 - 12b_2 + 2b_3) - 16qL^3 &= 0 \\
12(12b_2 - 12b_3) + 30qL^3 &= 0 \\
240b_3 - 16qL^3 &= 0
\end{aligned}
\tag{2.140}
$$

resulting in the buckling load

$$q_{cr} = 15b_3/L^3 \tag{2.141}$$

and the flexural rigidity

$$D(\xi) = \left(\tfrac{7}{48} + \tfrac{25}{24}\xi - \tfrac{17}{8}\xi^2 + \xi^3 \right) b_3 \tag{2.142}$$

Note that Eqs. (2.141) and (2.134) coincide. Figure 2.5 portrays the dependence of $D(\xi)/b_3$ on ξ. As in the previous case we calculate the average flexural rigidity, which equals

$$D_{av} = \tfrac{5}{24}b_3 \tag{2.143}$$

The column with constant flexural rigidity equal to D_{av} has a buckling load obtained from Eq. (2.125):

$$q_{cr} = 74.6D_{av}/L^3 = 15.54b_3/L^3 \tag{2.144}$$

which is about 4% larger than the column with variable flexural rigidity.

2.3.4 Column Pinned at One End and Clamped at the Other

The buckling mode is stipulated to be as follows:

$$\psi(\xi) = \xi - 3\xi^3 + 2\xi^4 \tag{2.145}$$

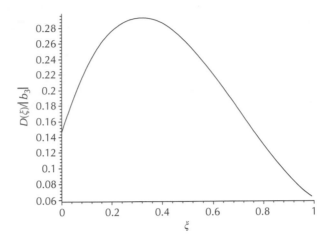

FIGURE 2.5
Variation of $D(\xi)/b_3$ for the column clamped at both its ends

The set of equations in terms of b_0, b_1, b_2, b_3 and q reads as follows:

$$2(24b_0 - 18b_1) - qL^3 = 0$$
$$6(24b_1 - 18b_2) - 18qL^3 = 0$$
$$12(24b_2 - 18b_3) + 51qL^3 = 0 \tag{2.146}$$
$$480b_3 - 32qL^3 = 0$$

The buckling intensity again equals

$$q_{cr} = 15b_3/L^3 \tag{2.147}$$

whereas the flexural rigidity variation associated with the buckling mode (2.145) and buckling intensity (2.147) is obtained as follows:

$$D(\xi) = \left(\tfrac{331}{512} + \tfrac{57}{128}\xi - \tfrac{61}{32}\xi^2 + \xi^3\right)b_3 \tag{2.148}$$

Figure 2.6 shows a plot of the ratio $D(\xi)/b_3$.

In order to contrast the results for the equivalent column with average flexural rigidity,

$$D_{av} = \tfrac{743}{1536}b_3 \tag{2.149}$$

we substitute D_{av} into Eq. (2.124) to arrive at

$$q_{cr} = 52.50D_{av}/L^3 = 25.39b_3/L^3 \tag{2.150}$$

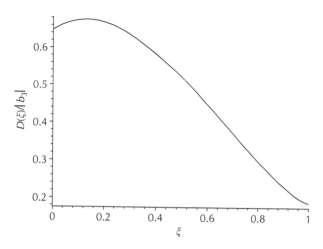

FIGURE 2.6
Variation of $D(\xi)/b_3$ for the pinned−clamped column

which is about 70% larger than its counterpart for the variable flexural rigidity.

2.3.5 Column Clamped at One End and Free at the Other

The buckling mode postulated for this case reads

$$\psi(\xi) = 6\xi^2 - 4\xi^3 + \xi^4 \qquad (2.151)$$

Substitution of this expression into the differential equation (2.126) yields

$$2(12b_0 - 24b_1 + 12b_2) + 6(12b_3 - 24b_2 + 12b_1)\xi + 12(12b_2 - 24b_3)\xi^2$$
$$+ 360b_3\xi^3 + qL^3(12 - 48\xi + 48\xi^2 - 16\xi^3) = 0 \qquad (2.152)$$

Since this equation must constitute an indentity for any value of ξ, we get the following equations:

$$2b_0 - 4b_1 + 2b_2 + qL^3 = 0$$
$$3b_3 - 6b_2 + 3b_1 - 2qL^3 = 0 \qquad (2.153)$$
$$3b_2 - 6b_3 + qL^3 = 0$$
$$240b_3 - 16pL^3 = 0$$

The last equation yields the expression for the critical parameter

$$q_{cr} = 15b_3/L^3 \tag{2.154}$$

where b_3 is an arbitrary constant. We also obtain

$$\begin{aligned} b_0 &= \tfrac{3}{2}b_3 \\ b_1 &= 3b_3 \\ b_2 &= -3b_3 \end{aligned} \tag{2.155}$$

leading to the following expression for the column's flexural rigidity:

$$D(\xi) = \left(\tfrac{3}{2} + 3\xi - 3\xi^2 + \xi^3\right) b_3 \tag{2.156}$$

The average flexural rigidity is

$$D_{av} = \tfrac{9}{4}b_3 \tag{2.157}$$

Substituting into Eq. (2.122) we get

$$q_{cr} = 7.837 D_{av}/L^3 = 17.63 b_3/L^3 \tag{2.158}$$

which is about 17% larger than its counterpart for the variable flexural rigidity.

Figure 2.7 depicts the variation of $D(\xi)/b_3$ in terms of ξ. The value of $D(\xi)$ at the origin and at the column's end are $1.5b_3$ and $2.5b_3$, respectively. The derivative of the function $D(\xi)$ vanishes at $\xi = 1$, implying

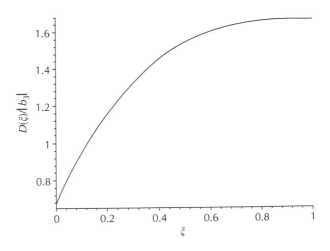

FIGURE 2.7
Variation of $D(\xi)/b_3$ for the clamped–free column

that the flexural rigidity is a positive-valued function, so long as b_3 is positive. Positiveness of b_3 stems also from the expression for the buckling intensity in Eq. (2.154). Thus, these two reasons for positivity of the flexural rigidity are compatible with each other. Note that two other sets of boundary conditions were studied by Elishakoff (2000f) and are not included here.

2.3.6   Concluding Remarks

Bulson (1970) noted in the preface to his monograph: "In many technical libraries the supports of bookshelves containing works on the stability of structure are in danger of buckling under the weight of literature. Can another work on this subject be justified?". It appears that the answer to this question is in the affirmative. Indeed, in the present work, we present the closed-form solutions for buckling of columns under their own weight, since the topic was initiated by Euler in 1778. In contrast to Euler and subsequent authors, we considered the column of variable flexural rigidity and derived the closed-form solutions for four sets of boundary conditions. The most significant conclusion derived in this section is the fact that, irrespective of boundary conditions, the buckling load in all four cases is given by the same analytical expression. Naturally, in these cases, the flexural rigidity variations along the column's axis are different.

In order to compare the buckling loads associated with the four uncovered columns of specific flexural rigidity, given respectively in Eqs. (2.136), (2.142), (2.148) and (2.156) we undertook the following calculations, to elucidate the results obtained. For example, the column with the flexural rigidity given in Eq. (2.136), under pinned–pinned boundary conditions, has the buckling load whose exact expression is $q_{cr} = 15b_3/L^3$ as given in Eq. (2.134). The question that arises naturally is: What are the buckling loads of the same column under other boundary conditions? At present there seem to be no exact solutions in these cases. Hence, an evaluation of approximate estimates was conducted, based on the Boobnov–Galerkin method. The associated comparison functions are listed in Tables 2.2–2.5. As is seen, the single-term approximation is extremely good. For the P–P column, the closed-form solution is $q_{cr}L^3/b_3 = 15$, whereas the approximation gives 15.02, with an attendant error of 0.133%.

As is clearly seen from Tables 2.2–2.5, for each of the four flexural rigidity distributions the buckling loads of columns under different boundary conditions differ from each other, as expected. The uncovered buckling load that is represented by a simple rational expression can be used as a benchmark solution for various numerical evaluation techniques. It is hoped that other researchers will be inspired to derive new closed-form solutions for more complicated and challenging problems, since even over 220 years after Leonhard Euler the field of buckling turns out to be inexhaustible.

TABLE 2.2

Comparison of Exact and Approximate Solutions

Flexural rigidity	Boundary conditions	Comparison function	$q_{cr}L^3/b_3$
$D_{P-P}(\xi) =$ $\left(1 + \frac{3}{8}\xi - \frac{17}{8}\xi^2 + \xi^3\right)b_3$	P–P	$\xi - 2\xi^3 + \xi^4$	15
		$\sin(\pi\xi)$	$5(7\pi^2 + 3)/24 \approx 15.02$
		$\psi_1(\xi) = \sin(\pi\xi),$ $\psi_2(\xi) = \sin(2\pi\xi)$	15.017
		$\psi_1(\xi) = \sin(\pi\xi),$ $\psi_2(\xi) = \sin(2\pi\xi)$	15.001
	P–C	$\frac{3}{2}\xi^2 - \frac{5}{2}\xi^3 + \xi^4$	$\frac{523}{12} = 43.58$
		$\psi_1(\xi) = \sin(\sqrt{2}\pi\zeta)$ $-\sqrt{2}\pi\cos(\sqrt{2}\pi\zeta)$ $+\sqrt{2}\pi(1-\xi)$	41.83
		$\psi_1(\xi)$ as above, $\psi_2(\xi) = \sin(7.7252\xi)$ $-7.7252\cos(7.7252\xi)$ $+7.7252(1-\xi)$	41.69
		$\psi_1(\xi)$ and $\psi_2(\xi)$ as above, $\psi_3(\xi) =$ homogeneous column's third buckling mode	41.65
	C–C	$\xi^2 - 2\xi^3 + \xi^4$	$\frac{235}{4} = 58.75$
		$\psi_1 =$ homogeneous column's first buckling mode	$5(28\pi^2 - 3)/24 \approx 56.49$
		$\psi_1 =$ as above $\psi_2 =$ homogeneous column's second buckling mode	56.77
		ψ_1 and $\psi_2 =$ as above $\psi_3 =$ homogeneous column's third buckling mode	56.75
	C–F	$6\xi^2 - 4\xi^3 + \xi^4$	$\frac{47}{6} = 7.83$

2.4   Some Closed-Form Solutions for the Buckling of Inhomogeneous Columns under Distributed Variable Loading

2.4.1   Introductory Remarks

The study of buckling of columns under their own weight was pioneered by Euler (1778a–c). Subsequent contributions were made by Heim (1838), Greenhill (1881), Jasinskii (1894) and others. Often it is referred to as the Greenhill's problem rather than Euler's problem (Love, 1944). The case when the axial distributed load is constant was studied, in addition to the above authors, by Dinnik (1912), Engelhardt (1954) and others. The notion of the variable intensity of the axial load was apparently introduced by Hauger

TABLE 2.3

Comparison of Exact and Approximate Solutions

Flexural rigidity	Boundary conditions	Comparison function	$q_{cr}L^3/b_3$
$D_{P-C}(\xi) =$ $\left(\frac{105}{512} + \frac{153}{128}\xi - \frac{72}{32}\xi^2 + \xi^3\right)b_3$	P–P	$\xi - 2\xi^3 + \xi^4$	$\frac{3351}{544} = 6.16$
		$\sin(\pi\xi)$	$\frac{139\pi^2+216}{256} = 6.20$
		$\psi_1(\xi) = \sin(\pi\xi),$ $\psi_2(\xi) = \sin(2\pi\xi)$	6.188
		$\psi_1(\xi) = \sin(\pi\xi),$ $\psi_2(\xi) = \sin(2\pi\xi)$	6.114
	P–C	$\frac{3}{2}\xi^2 - \frac{5}{2}\xi^3 + \xi^4$	15
		$\psi_n =$ homogeneous compressed column's nth mode shape, $\psi = \psi_1$	15.384
		Two terms	15.112
	C–C	$\xi^2 - 2\xi^3 + \xi^4$	$\frac{2478}{128} = 19.36$
		$\psi_1(\xi) = \sin^2(\pi\xi)$	$\frac{139\pi^2-54}{64} \approx 20.59$
		$\psi_1(\xi) = \sin^2(\pi\xi), \psi_2(\xi) = \sin^2(2\pi\xi)$	19.73
		$\psi_i(\xi) = \sin^2(\pi\xi), i = 1,2,3$	19.54
	C–F	$6\xi^2 - 4\xi^3 + \xi^4$	$\frac{159}{64} = 2.48$

TABLE 2.4

Comparison with the Boobnov–Galerkin Method

Flexural rigidity	Boundary conditions	Comparison function	$q_{cr}L^3/b_3$
$D_{C-C}(\xi) =$ $\left(\frac{7}{48} + \frac{25}{24}\xi - \frac{17}{8}\xi^2 + \xi^3\right)b_3$	P–P	$\xi - 2\xi^3 + \xi^4$	$\frac{80}{17} = 4.71$
		$\psi_1(\xi) = \sin(\pi\xi)$	$\frac{5(2\pi^2+3)}{24} = 4.74$[2pt]
		$\psi_1(\xi) = \sin(\pi\xi),$ $\psi_2(\xi) = \sin(2\pi\xi)$	4.73
		$\psi_1(\xi) = \sin(\pi\xi),$ $\psi_2(\xi) = \sin(2\pi\xi)$	4.68
	P–C	$\frac{3}{2}\xi^2 - \frac{5}{2}\xi^3 + \xi^4$	$\frac{407}{36} = 11.31$
		$\psi_n =$ compressed column's nth buckling mode (one term approximation)	11.63
		Two terms	11.43
		Three terms	11.30
	C–C	$\xi^2 - 2\xi^3 + \xi^4$	$\frac{235}{4} = 58.75$
		$\psi_1(\xi) = \sin^2(\pi\xi)$	15
		$\psi_1(\xi) = \sin^2(\pi\xi),$ $\psi_2(\xi) = \sin^2(2\pi\xi)$	15.21
		$\psi_i(\xi) = \sin^2(i\pi\xi), i = 1,2,3$	15.078
	C–F	$6\xi^2 - 4\xi^3 + \xi^4$	$\frac{17}{9} = 1.89$

TABLE 2.5

Comparison with the Boobnov–Galerkin Method

Flexural rigidity	Boundary conditions	Comparison function	$q_{cr}L^3/b_3$
$D_{C-F}(\xi) =$	P–P	$\xi - 2\xi^3 + \xi^4$	$\frac{780}{17} = 45.88$
$\left(\frac{3}{2} + 3\xi - 3\xi^2 + \xi^3\right)b_3$		$\psi_1(\xi) = \sin(\pi\xi)$	$\frac{3(3\pi^2+1)}{2} = 45.9$
		$\psi_1(\xi) = \sin(\pi\xi),$ $\psi_2(\xi) = \sin(2\pi\xi)$	41.53
		$\psi_1(\xi) = \sin(\pi\xi),$ $\psi_2(\xi) = \sin(2\pi\xi)$	41.29
	P–C	$\frac{3}{2}\xi^2 - \frac{5}{2}\xi^3 + \xi^4$	120
		$\psi_n =$ homogeneous compressed column's nth buckling mode (one term approximation)	131.59
		Two terms	109.76
		Three terms	109.00
	C–C	$\xi^2 - 2\xi^3 + \xi^4$	183
		$\psi_1(\xi) = \sin^2(\pi\xi)$	$\frac{3(12\pi^2-1)}{2} = 176.153$
		$\psi_1(\xi) = \sin^2(\pi\xi),$ $\psi_2(\xi) = \sin^2(2\pi\xi)$	175.82
		$\psi_i(\xi) = \sin^2(i\pi\xi), \quad i = 1, 2, 3$	175.79
	C–F	$6\xi^2 - 4\xi^3 + \xi^4$	15

(1966), who considered the non-conservative case as an analytical extension of the Beck's (1952) column that is acted upon the so-called *follower forces* (see also Leipholz and Bhalla, 1977). However, the very notion of the follower forces was rightfully criticized by Koiter (1996) due to the fact that no direct and pure experimental verification has been provided up to now of Beck's (1952) tangentially loaded column, although this is disputed by Sugiyama et al. (1999). We will limit ourselves to the consideration of the *conservative* case of the polynomially varying intensity of the axial loading.

We consider, in this study, the intensity of loading to be a polynomial function of the non-dimensional axial coordinate

$$\xi = x/L \tag{2.159}$$

where x is the axial coordinate and L the length. The variation of the axial intensity is taken as follows:

$$q(\xi) = q_0 + q_1\xi + \cdots + q_m\xi^m \tag{2.160}$$

where $q_j (j = 0, 1, \ldots, m)$ are given coefficients. The case with only q_0 present, and the rest of the coefficients vanishing, i.e., $q_j = 0$ $(j = 1, 2, \ldots, m)$ was studied by Elishakoff (2000a).

We are interested in closed-form solutions for the buckling load of the column. To this end, it is more convenient to pose the inverse problem of a special kind. We will postulate the prior knowledge of the buckling mode of the structure. We pose the problem as follows: find a special variation of the flexural rigidity, such that the buckling mode coincides with the preselected polynomial function that satisfies all boundary conditions.

2.4.2 Basic Equations

The differential equation that governs the buckling of the inhomogeneous column reads (Brush and Almroth, 1975) as follows:

$$\frac{d^2}{dx^2}\left[D(x)\frac{d^2w}{dx^2}\right] - \frac{d}{dx}\left[N(x)\frac{dw}{dx}\right] = 0 \tag{2.161}$$

where $D(x)$ is the flexural rigidity, $N(x)$ the axial force, $w(x)$ the buckling mode and x the axial coordinate. With the non-dimensional coordinate ξ, Eq. (2.169) becomes

$$\frac{d^2}{d\xi^2}\left[D(\xi)\frac{d^2w}{d\xi^2}\right] - L^2\frac{d}{d\xi}\left[N(\xi)\frac{dw}{d\xi}\right] = 0 \tag{2.162}$$

Consider first the column that is clamped at $\xi = 0$ and free at $\xi = 1$. The axial load reads

$$N(\xi) = -\int_x^L q(t)\, dt = -L\int_\xi^1 q(t)\, dt \tag{2.163}$$

Here

$$q(\xi) = \sum_{i=0}^m q_i \xi^i \tag{2.164}$$

leading to

$$N(\xi) = -L\sum_{i=0}^m \frac{q_i}{i+1}(1 - \xi^{i+1}) \tag{2.165}$$

We are interested in finding the critical value of q_0, when the ratios between the load parameters

$$g_i = q_i/q_0 \qquad i = 1, 2, \ldots, m \tag{2.166}$$

are specified. We restrict our considerations to the case of positive values of the ratios g_i.

The differential equation (2.162) reads, in conjunction with Eqs. (2.165) and (2.166), as follows:

$$\frac{d^2}{d\xi^2}\left[D(\xi)\frac{d^2w}{d\xi^2}\right] + q_0 L^3 \frac{d}{d\xi}\left[\sum_{i=0}^{m}\frac{g_i}{i+1}(1-\xi^{i+1})\frac{dw}{d\xi}\right] = 0 \tag{2.167}$$

Moreover, $g_0 = 1$. The simplest polynomial function that satisfies the boundary conditions of the clamped – free column is given by

$$\psi(\xi) = 6\xi^2 - 4\xi^3 + \xi^4 \tag{2.168}$$

We substitute $\psi(\xi)$ into Eq. (2.167) and demand that the resulting equation is satisfied identically for every value of ξ. This procedure leads to the set of algebraic equations for the coefficients b_i representing the flexural rigidity:

$$D(\xi) = \sum_{i=0}^{m+3} b_i \xi^i \tag{2.169}$$

Since the case $m = 0$ was considered in Section 2.3, it will not be recapitulated here. We first consider the particular case $m = 1$ prior to turning to the general case.

For $m = 1$, the result of substitution of Eqs. (2.165) and (2.166) into Eq. (2.167) is the fourth-order polynomial, whose coefficients must vanish. This condition results in the following set of equations:

$$24b_2 - 48b_1 + 24b_0 + q_0 L^3(12 + 6g_1) = 0$$
$$72b_3 - 144b_2 + 72b_1 + q_0 L^3(-48 - 12g_1) = 0$$
$$144b_4 - 288b_3 + 144b_2 + q_0 L^3(48 - 12g_1) = 0 \tag{2.170}$$
$$-280b_4 + 240b_3 + q_0 L^3(-16 + 24g_1) = 0$$
$$360b_4 - 10q_0 L^3 g_1 = 0$$

In Eqs. (2.170) we have five equations for six unknowns: b_0, b_1, b_2, b_3, b_4 and q_0. Thus, one of the coefficients can be chosen arbitrarily. We choose b_4 as an arbitrary constant, and the last equation leads to the buckling load $q_{0,\text{cr}}$:

$$q_{0,\text{cr}} = 36b_4/L^3 g_1 \tag{2.171}$$

The coefficients defining the flexural rigidity are

$$b_0 = (13g_1 + 18)b_4/5g_1 \quad b_1 = 2(13g_1 + 18)b_4/5g_1$$
$$b_2 = -6(g_1 + 6)b_4/5g_1 \quad b_3 = -4(2g_1 - 3)b_4/5g_1 \tag{2.172}$$

Hence, the sought for flexural rigidity $D(\xi)$ reads as follows:

$$D(\xi) = [(13g_1 + 18)b_4 + 2(13g_1 + 18)b_4\xi - 6(g_1 + 6)b_4\xi^2$$
$$- 4(2g_1 - 3)b_4\xi^3 + 5g_1\xi^4]/5g_1 \tag{2.173}$$

In order for the obtained expression for the flexural rigidity to be physically realizable, it should be non-negative for any positive value of the parameter g_1. To check the non-negativity of $D(\xi)$, let us consider the following function $f_1(\xi, g_1) = 5g_1 D(\xi)/b_4$,

$$f_1(\xi, g_1) = (13g_1 + 18) + 2(13g_1 + 18)\xi - 6(g_1 + 6)\xi^2$$
$$- 4(2g_1 - 3)\xi^3 + 5g_1\xi^4 \tag{2.174}$$

It is convenient to treat $f_1(\xi, g_1)$ as a function of the independent variable g_1, while ξ will be treated as a parameter. We get, therefore, the linear equation whose slope $\partial f_1/\partial g_1$ depends on ξ:

$$\partial f_1/\partial g_1 = 13 + 26\xi - 6\xi^2 - 8\xi^3 + 5\xi^4 \tag{2.175}$$

and is depicted in Figure 2.8. We conclude that the slope remains positive for $\xi \in [0; 1]$. We also note that at $g_1 = 0$, $f_1(\xi, 0) = 18 + 36\xi - 36\xi^2 + 12\xi^3$, which takes larger values than $18 + 12\xi^3$, for $\xi \in [0; 1]$. Since $18 + 12\xi^2$ is positive in the same interval, we conclude that for any positive g_1, $f_1(\xi, g_1)$ is positive. Hence the flexural rigidity $D(\xi)$ is positive. The dependence $D(\xi)$ is depicted in Figure 2.9 as a function of ξ and g_1.

We now turn to the general case of the mth order polynomial for the clamped–free column. We substitute Eqs. (2.176) and (2.177) into (2.175). The result is the set of $m + 4$ linear algebraic equations for $m + 5$ unknowns, namely $m + 4$ coefficients in Eq. (2.169) for $D(\xi)$ and the critical value of q_0 in

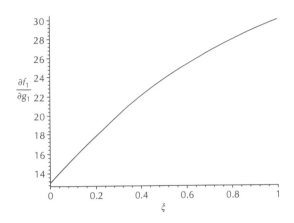

FIGURE 2.8
Variation of $\partial f_1/\partial g_1$ for the clamped–free column

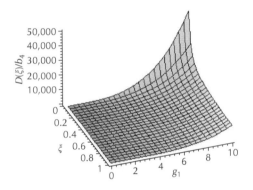

FIGURE 2.9
Variation of $D(\xi)/b_4$

Eq. (2.167):

from ξ^0: $24b_2 - 48b_1 + 24b_0 + 12q_0L^3G = 0$ (2.176)

from ξ^1: $72b_3 - 144b_2 + 72b_1 - 2q_0L^3(12G + 12) = 0$ (2.177)

from ξ^2: $144b_4 - 288b_3 + 144b_2 + 3q_0L^3(4G + 12 - 6g_1) = 0$ (2.178)

from $\xi^i, 3 \le i \le m + 1$:

$$12(i + 2)(b_{i+2} - 2b_{i+1} + b_i) - q_0L^3\left(12\frac{g_{i-1}}{i} - 12\frac{g_{i-2}}{i - 1} + 4\frac{g_{i-3}}{i - 2}\right) = 0$$
(2.179)

from ξ^{m+2}: $12(m + 4)(-2b_{m+3} + b_{m+2}) + q_0L^3\left(12\frac{g_m}{m + 1} - 4\frac{g_{m-1}}{m}\right) = 0$
(2.180)

from ξ^{m+3}: $12(m + 5)b_{m+3} - 4q_0L^3\frac{g_m}{m + 1} = 0$ (2.181)

where

$$g_0 = 1 \quad G = \sum_{i=0}^{m} \frac{g_i}{i+1} \tag{2.182}$$

It should be noted that the case $m = 1$ does not follow from Eqs. (2.176)–(2.181). In fact, Eq. (2.176) does not hold, for it is valid for $3 \leq i \leq m+1$, and when $m = 1$ the upper value of variation of i turns out to be smaller than the lower level, set at 3. On the other hand, the case $m = 2$ is included in the general formulation, for in this limiting case the upper value of variation of i, namely $m+1$ turns out to be equal to the lower value of variation of i, namely 3. Thus, we have only one special case, associated with $m = 1$, while all other cases are covered by the general formulation in Eqs. (2.176)–(2.181).

Note that the set (2.176)–(2.181) is a triangular system of equations, allowing the determination of $m+4$ unknowns once one of the unknowns is taken as an arbitrary constant. It is convenient to choose b_{m+3} as such a constant, since it appears in the last equation in the set (2.176)–(2.181).

The critical value of the load intensity q_0 is obtained as follows:

$$q_{0,\mathrm{cr}} = 3(m+1)(m+5)b_{m+3}/g_m L^3 \tag{2.183}$$

Note that although the case $m = 1$ is a special one, the buckling intensity associated with it follows from Eq. (2.25) by formally substituting in it $m = 1$. The coefficients defining $D(\xi)$ read as follows:

$$b_{m+2} = \left[\frac{(m+1)(m+5)g_{m-1}}{(m+4)m g_m} - \frac{m+7}{m+4} \right] b_{m+3} \tag{2.184}$$

$$b_i = (m+1)(m+5) \left[\frac{g_{i-3}}{i^2-4} - \frac{3g_{i-2}}{(i-1)(i+2)} + \frac{3g_{i-1}}{i(i+2)} \right] \frac{b_{m+3}}{g_m}$$

$$+ 2b_{i+1} - b_{i+2} \quad \text{for } 3 \leq i \leq m+1 \tag{2.185}$$

$$b_2 = (m+1)(m+5)(3g_1 - 2G - 6)b_{m+3}/8g_m - b_4 + 2b_3 \tag{2.186}$$

$$b_1 = (m+1)(m+5)(3g_1 + 2G - 2)b_{m+3}/4g_m - 2b_4 + 3b_3 \tag{2.187}$$

$$b_0 = (m+1)(m+5)(9g_1 - 2G - 2)b_{m+3}/8g_m - 3b_4 + 4b_3 \tag{2.188}$$

We observe that the critical buckling load depends on the coefficient g_m only. This does not mean that the other g_i coefficients do not affect the column's behavior, for all values of g_i define the flexural rigidity of the column. This conclusion may appear paradoxical, at first glance, since g_i are the ratios of the load coefficients, and it may seem that they should not influence the flexural rigidity. The resolution of this predicament is that the closed-form solutions are available only for such specific cases where the flexural rigidity and the other parameters defining the system, are intimately related. Since the coefficients g_i were fixed at the outset of the solution of the inverse problem, naturally, the sought for variation of $D(\xi)$ may, in some manner, depend on the input parameters g_i.

2.4.3   Column Pinned at Both Ends

The mode shape is taken as

$$\psi(\xi) = \xi - 2\xi^3 + \xi^4 \tag{2.189}$$

The flexural rigidity is again represented by Eq. (2.169). For the particular case $m = 1$, the system of equations (2.170) is replaced by the following:

$$-24b_1 + 24b_0 - q_0 L^3 = 0$$

$$-72b_2 + 72b_1 - q_0 L^3(72 + 7g_1) = 0$$

$$-144b_3 + 144b_2 + q_0 L^3(30 + 6g_1) = 0 \tag{2.190}$$

$$-240b_4 + 240b_3 + q_0 L^3(-16 + 12g_1) = 0$$

$$360b_4 - 10q_0 L^3 g_1 = 0$$

We have the same six unknowns as in the clamped–free case. We choose b_4 as an arbitrary constant to obtain $q_{0,\mathrm{cr}}$:

$$q_{0,\mathrm{cr}} = 36b_4/L^3 g_1 \tag{2.191}$$

Note that remarkably Eq. (2.191) coincides with Eq. (2.171). The coefficients determining the boundary flexural rigidity $D(\xi)$ are:

$$b_0 = 6(g_1 + 2)b_4/5g_1 \quad b_1 = 3(4g_1 + 3)b_4/10g_1,$$

$$b_2 = -(23g_1 + 51)b_4/10g_1 \quad b_3 = -4(g_1 - 3)b_4/5g_1. \tag{2.192}$$

Hence, the flexural rigidity $D(\xi)$ reads:

$$D(\xi) = [12(g_1 + 2)b_4 + 3(4g_1 + 3)b_4\xi - (23g_1 + 51)b_4\xi^2$$

$$- 8(g_1 - 3)b_4\xi^3 + 10g_1\xi^4]/10g_1 \tag{2.193}$$

The conditions of physical realizability are positivity of the flexural rigidity as well as the critical load intensity $q_{0,\mathrm{cr}}$. The positivity of $q_{0,\mathrm{cr}}$ is obvious when both b_4 and g_1 are positive. Concerning the flexural rigidity, let us look at the function $f_2(\xi, g_1) = 10g_1 D(\xi)/b_4$, which is proportional to the flexural rigidity,

$$f_2(\xi, g_1) = (12g_1 + 24) + 3(4g_1 + 3)\xi - (23g_1 + 51)\xi^2 - 8(g_1 - 3)\xi^3 + 10g_1\xi^4 \tag{2.194}$$

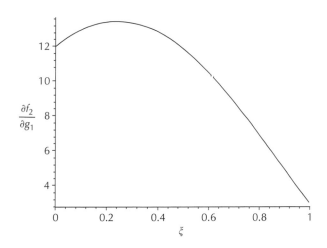

FIGURE 2.10
Variation of $\partial f_2/\partial g_1$ for the column pinned at both ends

As for the clamped–free column we treat $f_2(\xi, g_1)$ as a function of the independent variable g_1, while ξ will be treated as a parameter. We get, therefore, the linear equation whose slope $\partial f_2/\partial g_1$ depends on ξ:

$$\partial f_2/\partial g_1 = 12 + 12\xi - 23\xi^2 - 8\xi^3 + 10\xi^4 \qquad (2.195)$$

The variation of $\partial f_2/\partial g_1$ is portrayed in Figure 2.10. We observe that the slope remains positive, for $\xi \in [0; 1]$. The function $f_2(\xi, g_1)$ is increasing with g_1. The value of the function $f_2(\xi, g_1)$ at $g_1 = 0$ is $f_2(\xi, 0) = 24 + 9\xi - 51\xi^2 + 24\xi^3$, which is shown in Figure 2.11. Hence, for any positive g_1, $f_2(\xi, g_1)$ is positive and so is the flexural rigidity $D(\xi)$. The variation of $D(\xi)$ as a function of ξ is given as a function of variables ξ and g_1 in Figure 2.12.

For the case $m = 2$, the set (2.179) is replaced by

$$-24b_1 + 24b_0 - q_0 L^3 = 0$$

$$-72b_2 + 72b_1 - q_0 L^3(72 + 7g_1 + 4g_2) = 0$$

$$-144b_3 + 144b_2 + q_0 L^3(30 + 6g_1 + 3g_2) = 0$$

$$-240b_4 + 240b_3 + q_0 L^3(-16 + 12g_1) = 0 \qquad (2.196)$$

$$-360b_5 + 360b_4 + q_0 L^3(-10g_1 + 10g_2) = 0$$

$$504b_5 - 8q_0 L^3 g_2 = 0$$

leading to the critical value of $q_{0,\mathrm{cr}}$

$$q_{0,\mathrm{cr}} = 63b_5/L^3 g_2 \qquad (2.197)$$

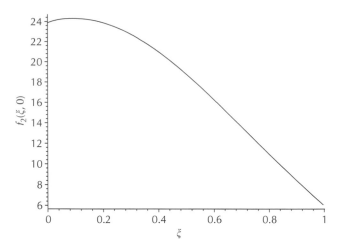

FIGURE 2.11
Variation of $f_2(\xi, 0)$ for the column pinned at both ends

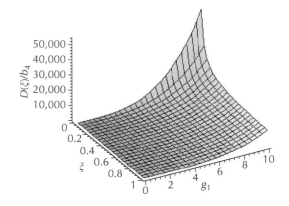

FIGURE 2.12
Variation of $D(\xi)/b_4$ for the column pinned at both ends

and the coefficients in $D(\xi)$:

$$b_0 = (115g_2 + 168g_1 + 336)b_5/80g_2 \quad b_1 = (115g_2 + 168g_1 + 126)b_5/80g_2$$
$$b_2 = -(165g_2 + 322g_1 + 714)b_5/80g_2 \quad b_3 = -(15g_2 + 28g_1 - 84)b_5/20g_2$$
$$b_4 = -(-3g_2 + 7g_1)b_5/4g_2$$

$$(2.198)$$

In order that the expression for the flexural rigidity is physically acceptable, we should check the positivity of $D(\xi)$ in the interval $[0; 1]$
for any positive value of the parameter g_1. We consider the function

$f_3(\xi, g_1, g_2) = 80g_2 D(\xi)/b_4$, which is proportional to $D(\xi)$:

$$f_3(\xi, g_1, g_2) = (115g_2 + 168g_1 + 336) + (115g_2 + 168g_1 + 126)\xi$$
$$- (165g_2 + 322g_1 + 714)\xi^2 - 4(15g_2 + 28g_1 - 84)\xi^3$$
$$+ 20(-3g_2 + 7g_1)\xi^4 + 80g_2\xi^5 \tag{2.199}$$

We will describe $f_3(\xi, g_1, g_2)$ as a function of g_1 and g_2, and take ξ as a parameter. Hence, the function can be expressed as follows:

$$f_3(\xi, g_1, g_2) = a(\xi)g_2 + b(\xi)g_1 + c(\xi) \tag{2.200}$$

where

$$a(\xi) = 115 + 115\xi - 165\xi^2 - 60\xi^3 - 60\xi^4 + 80\xi^5$$
$$b(\xi) = 168 + 168\xi - 322\xi^2 - 112\xi^3 + 140\xi^4 \tag{2.201}$$
$$c(\xi) = 336 + 126\xi - 714\xi^2 + 336\xi^3$$

Figure 2.13 shows the dependence of $a(\xi), b(\xi)$ and $c(\xi)$ in terms of ξ. We observe that they are all positive for $\xi \in [0; 1]$. This means that $f_3(\xi, g_1, g_2)$ is positive when g_1 and g_2 are both positive. We conclude, therefore, that the flexural rigidity is also positive when g_1 and g_2 are positive.

The general case for the pinned column is available for $m \geq 3$. As in the general case of the clamped–free column, we arrive at a triangular system of

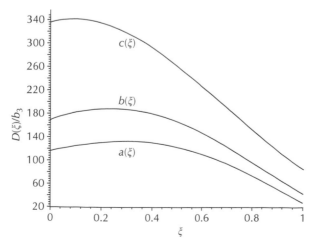

FIGURE 2.13
Variation of $a(\xi), b(\xi)$ and $c(\xi)$

$m + 4$ equations for $m + 5$ unknowns, that are the coefficients in $D(\xi)$ and q_0.

from ξ^0: $-24b_1 + 24b_0 - q_0L^3 = 0$ (2.202)

from ξ^1: $-72b_2 + 72b_1 - q_0L^3(12G + g_1) = 0$ (2.203)

from ξ^2: $-144b_3 + 144b_2 + q_0L^3(12G + 18 - g_2) = 0$ (2.204)

from $\xi^i, 3 \leq i \leq m$:

$$12(i + 2)(-b_{i+1} + b_i) - q_0L^3 \left(\frac{g_i}{i+1} - 6\frac{g_{i-2}}{i-1} + 4\frac{g_{i-3}}{i-2} \right) = 0 \qquad (2.205)$$

from ξ^{m+1}: $12(m + 3)(-b_{m+2} + b_{m+1}) + q_0L^3 \left(6\frac{g_{m-1}}{m} - 4\frac{g_{m-2}}{m-1} \right) = 0$

(2.206)

from ξ^{m+2}: $12(m + 4)\left(-b_{m+3} + b_{m+2}\right) + q_0L^3 \left(6\frac{g_m}{m+1} - 4\frac{g_{m-1}}{m} \right) = 0$

(2.207)

from ξ^{m+3}: $12(m + 5)b_{m+3} - 4q_0L^3\frac{g_m}{m+1} = 0$ (2.208)

We observe from Eq. (2.206) that the term $m - 1$ appears in the denominator. Therefore, it is necessary to separately consider the case $m = 1$. Moreover, in Eq. (2.205) $m \geq 3$. This implies that additional particular cases, associated respectively with $m = 1$ and $m = 2$, exist, as they were considered above. We take b_{m+3} to be an arbitrary constant. Equation (2.208) then yields the value of the critical load intensity:

$$q_{0,cr} = 3\frac{(m + 1)(m + 5)b_{m+3}}{g_m L^3} \qquad (2.209)$$

Note that this expression coincides with Eq. (2.183) for the clamped–free column. The remaining equations yield the coefficients in $D(\xi)$:

$$b_{m+2} = \frac{1}{2} \left[\frac{(2m^2 + 12m + 10)g_{m-1}}{(m + 4)mg_m} - \frac{m + 7}{m + 4} \right] b_{m+3} \qquad (2.210)$$

$$b_{m+1} = -\frac{1}{2} \left[\frac{(m + 1)(m + 5)(m + 6)g_{m-1}}{(m + 4)(m + 3)mg_m} - \frac{(m + 1)(m + 5)g_{m-2}}{(m - 1)(m + 3)g_m} + \frac{m + 7}{m + 4} \right] b_{m+3}$$

(2.211)

$$b_i = \frac{1}{4}(m+1)(m+5)\left[\frac{4g_{i-3}}{i^2-4} - \frac{6g_{i-2}}{(i-1)(i+2)} + \frac{g_i}{(i-1)(i+2)}\right]\frac{b_{m+3}}{g_m}$$

$$+ \frac{1}{4}b_{i+1} \quad \text{for } 3 \le i \le m \tag{2.212}$$

$$b_2 = (m+1)(m+5)(g_2 - 12G - 18)b_{m+3}/48g_m + b_3 \tag{2.213}$$

$$b_1 = (m+1)(m+5)(g_2 + 2g_1 + 12G - 18)b_{m+3}/48g_m + b_3 \tag{2.214}$$

$$b_0 = (m+1)(m+5)(g_2 + 2g_1 + 12G - 12)b_{m+3}/48g_m + b_3 \tag{2.215}$$

2.4.4 Column Clamped at Both Ends

The preselected buckling mode that fulfills the boundary conditions is

$$\psi(\xi) = \xi^2 - 2\xi^3 + \xi^4 \tag{2.216}$$

For this column we should first consider the particular case $m = 1$. For this value, the governing differential equation (2.12) yields the following equations:

$$4b_2 - 24b_1 + 24b_0 + q_0 L^3(2 + g_1) = 0$$

$$12b_3 - 72b_2 + 72b_1 - q_0 L^3(16 + 6g_1) = 0$$

$$24b_4 - 144b_3 + 144b_2 + q_0 L^3(30 + 3g_1) = 0 \tag{2.217}$$

$$-240b_4 + 240b_3 + q_0 L^3(-16 + 12g_1) = 0$$

$$360b_4 - 10q_0 L^3 g_1 = 0$$

In perfect analogy with the two previous cases, we obtain the critical value of q_0,

$$q_{0,\text{cr}} = 36b_4 L^3 g_1 \tag{2.218}$$

and the coefficients in $D(\xi)$,

$$b_0 = (73g_1 + 126)b_4/360g_1 \quad b_1 = (17g_1 + 30)b_4/12g_1$$
$$b_2 = -(103g_1 + 306)b_4/60g_1 \quad b_3 = -4(g_1 - 3)b_4/5g_1 \tag{2.219}$$

The validity of the obtained flexural rigidity imposes the condition $D(\xi) \ge 0$, for $\xi \in [0;1]$. We introduce again $f_4(\xi, g_1) = 360g_1 D(\xi)/b_4$, which we treat as a function of the independent variable g_1, while ξ will be treated as a parameter. We get, therefore, the linear equation whose slope $\partial f_4 / \partial g_1$ depends on ξ:

$$\partial f_4 / \partial g_1 = 73 + 510\xi - 618\xi^2 - 288\xi^3 + 360\xi^4 \qquad (2.220)$$

The slope $\partial f_4 / \partial g_1$ remains positive for $\xi \in [0;1]$ (see Figure 2.14, which depicts the variation of the slope as a function of ξ). The function $f_4(\xi, g_1)$ is an increasing function of g_1. Let us calculate $f_4(\xi, g_1) = 126 + 900\xi - 1836\xi^2 + 864\xi^3$, which is positive in the interval $[0;1]$ (see Figure 2.15). Hence, for any positive g_1, the flexural rigidity remains positive for $\xi \in [0;1]$; the variation of $D(\xi)$ in terms of ξ and g_1 is shown in Figure 2.16.

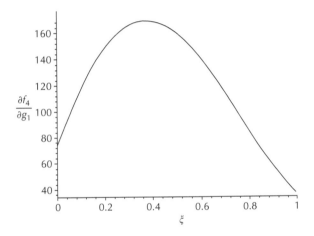

FIGURE 2.14
Variation of $\partial f_4 / \partial g_1$ for the column clamped at both ends

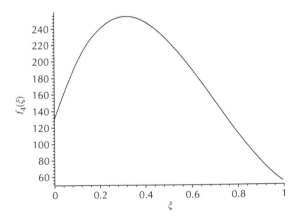

FIGURE 2.15
Variation of $f_4(\xi)$ for the column clamped at both ends

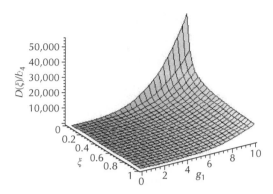

FIGURE 2.16
Variation of $D(\xi)/b_4$ for the column clamped at both ends

When $m \geq 2$, we have to satisfy the following set of equations:

from ξ^0: $4b_2 - 24b_1 + 24b_0 + 2q_0 L^3 G = 0$ $\qquad$ (2.221)

from ξ^1: $12b_3 - 72b_2 + 72b_1 - 2q_0 L^3 (6G + 2) = 0$ $\qquad$ (2.222)

from ξ^2: $24b_4 - 144b_3 + 144b_2 + 3q_0 L^3 (4G + 6 - g_1) = 0$ $\qquad$ (2.223)

from ξ^i $3 \leq i \leq m+1$:

$$(i + 2)(2b_{i+2} - 12b_{i+1} + 12b_i) - q_0 L^3 \left(2\frac{g_{i-1}}{i} - 6\frac{g_{i-2}}{i-1} + 4\frac{g_{i-3}}{i-2} \right) = 0$$
$$(2.224)$$

from ξ^{m+2}: $12(m + 4)(-b_{m+3} + b_{m+2}) + q_0 L^3 \left(6\frac{g_m}{m+1} - 4\frac{g_{m-1}}{m} \right) = 0$
$$(2.225)$$

from ξ^{m+3}: $12(m + 5)b_{m+3} - 4q_0 L^3 g_m /(m + 1) = 0$ $\qquad$ (2.226)

Since this is again a triangular system of $m + 4$ equations for $m + 5$ unknowns (the coefficients in $D(\xi)$ and q_0), we take one unknown as an arbitrary constant to determine the others. The judicious choice for this one is again b_{m+3} and Eq. (2.226) leads to the critical value of q_0:

$$q_{0,cr} = 3(m + 1)(m + 5)b_{m+3}/g_m L^3 \qquad (2.227)$$

Each of the remaining equations (2.221)–(2.225) lead, respectively, to the coefficients in $D(\xi)$:

$$b_{m+2} = \left[\frac{(m + 5)(m + 1)g_{m-1}}{(m + 4)m g_m} - \frac{1}{2}\frac{m + 7}{m + 4} \right] b_{m+3} \qquad (2.228)$$

$$b_i = (m+1)(m+5)\left[\frac{6g_{i-3}}{i^2-4} - \frac{9g_{i-2}}{(i-1)(i+2)} + \frac{3g_{i-1}}{i(i+2)}\right]\frac{b_{m+3}}{g_m}$$

$$+ b_{i+1} - \tfrac{1}{6}b_{i+2} \quad \text{for } 3 \leq i \leq m+1 \tag{2.229}$$

$$b_2 = (m+1)(m+5)(3g_1 - 12G - 18)b_{m+3}/48g_m - \tfrac{1}{6}b_4 + b_3 \tag{2.230}$$

$$b_1 = (m+1)(m+5)(3g_1 + 12G - 10)b_{m+3}/48g_m - \tfrac{1}{6}b_4 + \tfrac{5}{6}b_3 \tag{2.231}$$

$$b_0 = (m+1)(m+5)(15g_1 + 12G - 42)b_{m+3}/288g_m - \tfrac{5}{36}b_4 + \tfrac{2}{3}b_3 \tag{2.232}$$

2.4.5   Column that is Pinned at One End and Clamped at the Other

The polynomial function that satisfies the boundary conditions is chosen as

$$\psi(\xi) = \xi - 3\xi^3 + 2\xi^4 \tag{2.233}$$

In this case, we have to consider first of all the particular cases $m = 1$ and $m = 2$.

For $m = 1$, the governing differential equation (2.162) imposes the following equations:

$$-36b_1 + 48b_0 - q_0 L^3 = 0$$

$$-108b_2 + 144b_1 - q_0 L^3(18 + 10g_1) = 0$$

$$-216b_3 + 288b_2 + q_0 L^3(51 + 12g_1) = 0 \tag{2.234}$$

$$-360b_4 + 480b_3 + q_0 L^3(-32 + 18g_1) = 0$$

$$720b_4 - 20q_0 L^3 g_1 = 0$$

The critical value of q_0 is derived from the last one:

$$q_{0,cr} = 36b_4/L^3 g_1 \tag{2.235}$$

We obtain for the coefficients in $D(\xi)$:

$$b_0 = 3(166g_1 + 331)b_4/640g_1 \quad b_1 = (166g_1 + 171)b_4/160g_1$$
$$b_2 = -3(26g_1 + 61)b_4/40g_1 \quad b_3 = -3(g_1 - 4)b_4/5g_1 \tag{2.236}$$

We will now verify the positivity of $D(\xi)$ in the interval $[0; 1]$ for any positive g_1, in order to validate the solution. We define, $f_5(\xi, g_1) = 640g_1 D(\xi)/b_4$:

$$f_5(\xi, g_1) = 498g_1 + 993 + 4(166g_1 + 171)\xi - 48(26g_1 + 61)\xi^2$$
$$- 384(g_1 - 4)\xi^3 + 640g_1\xi^4 \tag{2.237}$$

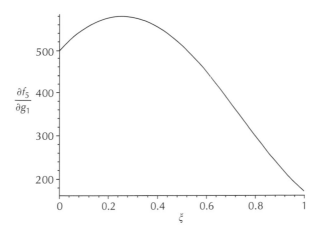

FIGURE 2.17
Variation of $\partial f_5/\partial g_1$ for the pinned–clamped column

Described as a function of the independent variable g_1, $f_5(\xi, g_1)$ is a linear function whose slope is $\partial f_5/\partial g_1$ is

$$\partial f_5/\partial g_1 = 498 + 664\xi - 1248\xi^2 - 384\xi^3 + 640\xi^4 \qquad (2.238)$$

The dependence of $\partial f_5/\partial g_1$ is shown in Figure 2.17. We observe that it is positive, for $\xi \in [0; 1]$, the function $f_5(\xi, g_1)$ is an increasing function of g_1. Let us calculate, $f_5(\xi, 0) = 993 + 684\xi - 2928\xi^2 + 1536\xi^3 > 0$ (see Figure 2.18). The flexural rigidity is thus positive for positive values of g_1. The dependence of $D(\xi)$ on ξ and g_1 is presented in Figure 2.19.

Concerning the case $m = 2$, we have the following set of equations,

$$
\begin{aligned}
&-36b_1 + 48b_0 - q_0 L^3 = 0 \\
&-108b_2 + 144b_1 - q_0 L^3(18 + 10g_1 + 6g_2) = 0 \\
&-216b_3 + 288b_2 + q_0 L^3(51 + 12g_1 + 7g_2) = 0 \\
&-360b_4 + 480b_3 + q_0 L^3(-32 + 18g_1) = 0 \\
&-540b_5 + 720b_4 + q_0 L^3(-20g_1 + 15g_2) = 0 \\
&1008b_5 - 16q_0 L^3 g_2 = 0
\end{aligned}
\qquad (2.239)
$$

with the last one yielding the critical value of q_0,

$$q_{0,\mathrm{cr}} = 63b_5/L^3 g_2 \qquad (2.240)$$

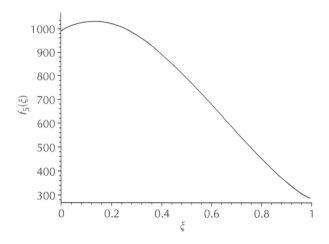

FIGURE 2.18
Variation of $f_5(\xi)$ for the pinned–clamped column

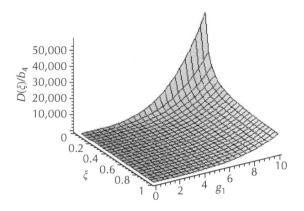

FIGURE 2.19
Variation of $D(\xi)/b_4$ for the pinned–clamped column

The rest of the equations result in the coefficients in $D(\xi)$:

$$b_0 = \frac{3(6,345g_2 + 9,296g_1 + 18\,536)b_5}{20,480g_2} \qquad b_1 = \frac{(6,345g_2 + 9,296g_1 + 9,576)b_5}{5,120g_2}$$

$$b_2 = -\frac{(2,365g_2 + 4,368g_1 + 10,248)b_5}{320g_2} \qquad b_3 = -\frac{3(45g_2 + 112g_1 - 448)b_5}{320g_2}$$

$$b_4 = (-9g_2 + 28g_1)b_5/16g_2$$

$$(2.241)$$

To validate this solution, the flexural rigidity $D(\xi)$ must be positive in the interval [0; 1] for any positive value of g_1 and g_2. As for the pinned column

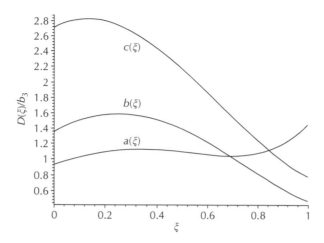

FIGURE 2.20
Variation of $a(\xi)$, $b(\xi)$ and $c(\xi)$

we consider the function $f_6(\xi, g_1, g_2) = g_2 D(\xi)/b_4$, which is a linear function of the independent variables g_1 and g_2, when ξ is taken as a parameter. We, therefore, describe the function f_6 as follows:

$$f_6(\xi, g_1, g_2) = a(\xi)g_2 + b(\xi)g_1 + c(\xi) \tag{2.242}$$

where

$$a(\xi) = \frac{3807}{4096} + \frac{1269}{1024}\xi - \frac{473}{256}\xi^2 - \frac{27}{64}\xi^3 + \frac{9}{16}\xi^4 + \xi^5$$

$$b(\xi) = \frac{1743}{1280} + \frac{581}{320}\xi - \frac{273}{80}\xi^2 - \frac{21}{20}\xi^3 + \frac{7}{4}\xi^4 \tag{2.243}$$

$$c(\xi) = \frac{6951}{2560} + \frac{1197}{640}\xi - \frac{1281}{160}\xi^2 + \frac{21}{5}\xi^3$$

Figure 2.20 shows the dependence of $a(\xi)$, $b(\xi)$ and $c(\xi)$. We observe that the function $b(\xi)$ takes negative values, especially $b(1) < 0$, while the other function $a(\xi)$ and $c(\xi)$ remain positive. Then, we will calculate $f_6(\xi = 1, g_1, g_2) = a(1)g_2 + b(1)g_1 + c(1)$.

Since $b(1) < 0$, for any g_2, we can find a positive g_1 that will give $f_6(\xi = 1, g_1, g_2) < 0$. We only have to take

$$g_1 > -[a(1)g_2 + c(1)]/b(1) > 0 \tag{2.244}$$

We prove, thus, that the flexural rigidity can take negative values for some particular positive value of the parameters g_1 and g_2, which implies that this solution is not physically acceptable for any positive g_1 and g_2.

We now consider the general case for $m \geq 3$. As above we get a triangular set of $m + 4$ equations for $m + 5$ unknowns:

from ξ^0: $-36b_1 + 48b_0 - q_0 L^3 = 0$ (2.245)

from ξ^1: $-108b_2 + 144b_1 - q_0 L^3 (18G + g_1) = 0$ (2.246)

from ξ^2: $-216b_3 + 288b_2 + q_0 L^3 (24G + 27 - g_2) = 0$ (2.247)

from ξ^i $3 \leq i \leq m$:

$$6(i + 2)(-3b_{i+1} + 4b_i) - q_0 L^3 \left(\frac{g_i}{i+1} - 9\frac{g_{i-2}}{i-1} + 8\frac{g_{i-3}}{i-2} \right) = 0 \qquad (2.248)$$

from ξ^{m+1}: $6(m + 3)(-3b_{m+2} + 4b_{m+1}) + q_0 L^3 \left(9\frac{g_{m-1}}{m} - 8\frac{g_{m-2}}{m-1} \right) = 0$

(2.249)

from ξ^{m+2}: $6(m + 4)(-3b_{m+3} + 4b_{m+2}) + q_0 L^3 \left(9\frac{g_m}{m+1} - 8\frac{g_{m-1}}{m} \right) = 0$

(2.250)

from ξ^{m+3}: $24(m + 5)b_{m+3} - 8q_0 L^3 g_m/(m + 1) = 0$ (2.251)

and we choose again b_{m+3} as an arbitrary constant. The critical value of q_0 reads

$$q_{0,cr} = 3(m + 1)(m + 5)b_{m+3}/g_m L^3 \qquad (2.252)$$

The coefficients in $D(\xi)$ are

$$b_{m+2} = \left[\frac{(m + 1)(m + 5)g_{m-1}}{(m + 4)m g_m} - \frac{3}{8}\frac{m + 7}{m + 4} \right] b_{m+3} \qquad (2.253)$$

$$b_{m+1} = \left[-\frac{3}{8} \frac{(m + 1)(m + 5)(m + 6)g_{m-1}}{(m + 4)(m + 3)m g_m} + \frac{(m^3 + 10m^2 + 29m - 20)g_{m-2}}{(m - 1)(m + 3)(m + 4)g_m} \right.$$

$$\left. -\frac{9}{32} \frac{m + 7}{m + 4} \right] b_{m+3} \qquad (2.254)$$

$$b_i = (m + 1)(m + 5)\left[\frac{g_{i-3}}{i^2 - 4} - \frac{9}{8}\frac{g_{i-2}}{(i - 1)(i + 2)} + \frac{1}{8}\frac{g_i}{(i + 1)(i + 2)} \right] \frac{b_{m+3}}{g_m}$$

$$+ \frac{3}{4}b_{i+1} \quad \text{for } 3 \leq i \leq m \qquad (2.255)$$

$$b_2 = (m + 1)(m + 5)(g_2 - 24G - 27)b_{m+3}/96g_m + \tfrac{3}{4}b_3 \qquad (2.256)$$

$$b_1 = (m + 1)(m + 5)(3g_2 + 8g_1 + 72G - 81)b_{m+3}/384g_m + \tfrac{9}{16}b_3 \qquad (2.257)$$

$$b_0 = (m + 1)(m + 5)(3g_2 + 8g_1 + 72G - 49)b_{m+3}/512g_m + \tfrac{27}{64}b_3 \qquad (2.258)$$

2.4.6 Concluding Remarks

As in the particular case $m = 0$, studied by Elishakoff (2000a), here too, for the different sets of boundary conditions the buckling intensities coincide. This pertains to the fundamental buckling load, since in all cases the postulated polynomial expressions of the buckling mode do not possess the internal nodes. It should be noted that whereas for the case $m = 0$ the physical realizability condition, namely that the flexural rigidity must be positive for $0 \leq \xi \leq 1$, can be easily verified, it is not so for the variable intensity. In this case the coefficients in the flexural rigidity depend upon the ratios g_i. Only such ratios are acceptable that lead to a positive flexural rigidity.

Since m can be any positive integer, we conclude that the number of closed-form solutions that are derived in this section, is a countable infinity.

Moreover, the buckling load depends solely on a single coefficient in the flexural rigidity variation and does not depend on other coefficients. This implies that an infinite number of columns, with different coefficients $b_0, b_1, \ldots, b_{m+2}$, can have the same buckling load, if only the shared coefficient b_{m+3} is fixed.

3

Unusual Closed-Form Solutions for Rod Vibrations

This chapter is devoted to vibration of inhomogeneous rods. While for the column buckling case the postulated buckling mode was a fourth-order polynomial, here a second-order polynomial suffices. The problem is posed as that of constructing a rod that possesses the pre-selected fundamental mode shape. The problem of constructing a rod with pre-selected second mode is studied as well.

3.1 Reconstructing the Axial Rigidity of a Longitudinally Vibrating Rod by its Fundamental Mode Shape

3.1.1 Introductory Remarks

Free longitudinal vibration of uniform, homogeneous rods is covered in nearly all vibration texts. The reader may consult, for example, books by Weaver et al. (1990) and Rao (1995). Vibration frequencies of tapered rods were studied by Conway et al. (1964). Graf (1975) pointed out that for rods of conical cross-sections, the equation of motion can be put in the form of the wave equation by an appropriate change of variable. This idea was further developed by Abrate (1995). Graf (1975) considered rods with various profiles, namely those designated as follows:

$$A(\xi) = A_0\xi \quad \text{(linear)} \tag{3.1}$$

$$A(\xi) = A_0\xi^2 \quad \text{(conical)} \tag{3.2}$$

$$A(\xi) = A_0 e^{\alpha\xi} \quad \text{(exponential)} \tag{3.3}$$

and

$$A(\xi) = A_0 \cosh^2(\beta\xi) \quad \text{(catenoidal)} \tag{3.4}$$

where A_0 is the cross-sectional area at the origin of the coordinates, $\xi = x/L$ is the non-dimensional axial coordinate, L the length of the rod, and α are β positive constants.

For the rod with of the linear area variation, Eq. (3.1), Graf (1975) obtained an exact solution in terms of the zeroth-order Bessel function, whereas for the conical rod, Eq. (3.2), the exact solution is obtainable via the application of the half-order Bessel functions. Series solutions for non-uniform bars were constructed by Eisenberger (1991). In another paper of Eisenberger (1990),

> the exact solution is obtained using one element for each segment with continuously varying properties, and the displacements and stresses are exact all along this rod.

The present chapter deals with closed-form solutions for inhomogeneous rods of uniform cross-section. Both the inertial coefficient and the longitudinal rigidity are treated as functions of the axial coordinate. Apparently, for the first time, closed-form solutions are presented for the inhomogeneous rod under two sets of boundary conditions. Closed-form expressions for the natural frequency can serve as benchmark solutions.

3.1.2   Formulation of the Problem

The differential equation governing the free longitudinal vibration of rods reads

$$\frac{\partial}{\partial x}\left[D(x)\frac{\partial u}{\partial x}\right] = R(x)\frac{\partial^2 u}{\partial t^2} \tag{3.5}$$

$$D(x) = E(x)A(x) \quad R(x) = \rho(x)A(x)$$

Here $u(x,t)$ is an axial displacement, which is a function of the axial coordinate x and time t, $D(x)$ is the longitudinal rigidity, $E(x)$ the modulus of elasticity that is varying along the axis, $R(x)$ the inertial coefficient, $\rho(x)$ and $A(x)$ are the inertial coefficient and the cross-sectional area, respectively, which are both functions of the axial coordinate x.

To find the natural frequency, we represent the displacement in the form

$$u(x,t) = U(x)e^{\omega x} \tag{3.6}$$

where $U(x)$ is the mode shape and ω is the natural frequency. Substituting Eq. (3.6) into Eq. (3.5) leads to

$$\frac{\partial}{\partial x}\left[E(x)A(x)\frac{\partial U}{\partial x}\right] + \rho(x)A(x)\omega^2 U = 0 \tag{3.7}$$

Consider rods whose ends are either clamped or free. At the clamped end the boundary condition is

$$U = 0 \tag{3.8}$$

whereas at the free end the boundary condition is

$$\frac{dU}{dx} = 0 \tag{3.9}$$

For the uniform and homogeneous rod, $A(x) = $ constant, $E(x) = $ constant and $\rho(x) = $ constant. We first find, as an auxiliary problem, the static displacements of such rods due to uniform axial loading. The static displacement for the clamped–clamped uniform rod reads (from the LHS of Eq. (3.5)):

$$U^{(\xi)}_{C-C} = \xi - \xi^2 \tag{3.10}$$

where ξ is the non-dimensional axial coordinate

$$\xi = x/L \tag{3.11}$$

For the clamped–free uniform rod the associated displacement is

$$U^{(\xi)}_{C-F} = \xi - \tfrac{1}{2}\xi^2 \tag{3.12}$$

We pose the following question: Are there non-uniform and/or inhomogeneous unloaded rods such that expressions given in either Eq. (3.10) or (3.12) constitute the exact solution for the natural frequency? This question may appear to be an artificial one in the first place. Indeed, why should static displacement of the uniform and homogeneous rod serve as a mode shape of either a non-uniform or a inhomogeneous rod?

In posing this question, we are guided here by previous experience in the derivation for the inhomogeneous beams (Elishakoff and Candan, 2000) where, indeed, the above phenomenon took place for four different sets of boundary conditions. We must immediately remark that if there are rods whose mode shapes are given by Eq. (3.10) or Eq. (3.12), then these mode shapes correspond to the fundamental frequencies since these mode shapes have no internal nodes.

3.1.3 Inhomogeneous Rods with Uniform Density

Consider first a clamped–clamped rod that has a constant inertial coefficient $R(x) = a_0 > 0$. We ask the following question: What are the coefficients b_0, b_1

and b_2 in the expression for the longitudinal rigidity

$$D(\xi) = b_0 + b_1\xi + b_2\xi^2 \tag{3.13}$$

so that the rod possesses a vibration mode given by Eq. (3.10)?

To this end, we substitute Eqs. (3.10) and (3.13) in the differential equation (3.7). The result is:

$$b_1 - 2b_0 + (-4b_1 + 2b_2 + kL^2)\xi - (6b_2 + kL^2)\xi^2 = 0 \tag{3.14}$$

where

$$k = \omega^2 L^2 \tag{3.15}$$

Since this expression must be valid for each ξ we conclude that

$$b_1 - 2b_0 = 0 \tag{3.16}$$
$$-4b_1 + 2b_2 + a_0 k = 0 \tag{3.17}$$
$$6b_2 + a_0 k = 0 \tag{3.18}$$

We get two expressions for the natural frequency coefficient k, from Eq. (3.17) and Eq. (3.18), respectively,

$$k = 2(2b_1 - b_2)/a_0 \tag{3.19}$$
$$k = -6b_2/a_0 \tag{3.20}$$

Since the expressions on the right-hand sides of Eqs. (3.19) and (3.20) coincide, we obtain, in view of Eq. (3.16).

$$b_1 = -b_2 \quad b_0 = -\tfrac{1}{2}b_2 \tag{3.21}$$

Thus, if the modulus of elasticity of the inhomogeneous rod reads

$$D(\xi) = (-\tfrac{1}{2} - \xi + \xi^2)b_2 \tag{3.22}$$

then, the natural frequency squared is given by

$$\omega^2_{C-C} = -6b_2/a_0 L^2 \tag{3.23}$$

with the mode shape given in Eq. (3.10). Since a_0 is the density of the rod, it has to be positive. Therefore, in order for the resulting expression for ω^2 to be positive, b_2 must be negative. Note that $D(\xi)$ must be positive throughout the interval $[0, 1]$. It is easy to confirm that $D(\xi)$ in Eq. (3.22) is positive.

Consider now the inhomogeneous clamped–free rod of uniform density and modulus of elasticity given in Eq. (3.13). The mode shape under these new circumstances is given in Eq. (3.12). We substitute Eqs. (3.12) and (3.13) into the governing differential equation (3.7). The result is a polynomial equation

$$(b_1 - b_0) + (-2b_1 + 2b_2 + kL^2 a_0)\xi - (3b_2 - \tfrac{1}{2}ka_0)\xi^2 = 0 \tag{3.24}$$

Hence, the expressions in parentheses must vanish. The resulting equations can be put in the following convenient form:

$$k_0 = k_1 + b_1 - b_0 \quad k_1 = 2(b_1 - b_2)/a_0 \quad k_2 = -6b_2/a_0 \tag{3.25}$$

Since

$$k_0 = k_1 = k_2 = k \tag{3.26}$$

we obtain

$$b_1 = -2b_2 \quad b_0 = -2b_2 \tag{3.27}$$

Note that the condition analogous to Eq. (3.26) was realized in the study by Elishakoff and Rollot (1999) using computerized symbolic algebra.

We arrive at the following conclusion: if the longitudinal rigidity is given by

$$D(\xi) = (-2 - 2\xi + \xi^2)b_2 \tag{3.28}$$

then, the mode shape is given by Eq. (3.12) and the natural frequency squared is given by the following closed-form expression

$$\omega_{C-F}^2 = -6b_2/a_0 L^2 \tag{3.29}$$

Here too, the value b_2 is negative; $D(\xi)$ in Eq. (3.28) is positive throughout the interval $0 \leq \xi \leq 1$.

We also observe that the fundamental frequency expressions for the clamped–clamped (C–C) rod in Eq. (3.23) and for the clamped–free (C–F) rod in Eq. (3.29) coincide if the coefficient of ξ^2 in $D(\xi)$ is the same in both cases. At first glance this may appear to be a paradoxical situation, since these two rods possess *different* mode shapes and different elastic moduli. To explain this result, let us list the fundamental frequencies of the associated uniform rods of the same length L:

$$\omega_{C-C} = (\pi/L)\sqrt{E_1/\rho_1} \quad \omega_{C-F} = (\pi/2L)\sqrt{E_2/\rho_2} \tag{3.30}$$

where ρ_1 and E_1 are the inertial coefficient and modulus of elasticity of the clamped–clamped rod, whereas ρ_2 and E_2 are the inertial coefficient and modulus of elasticity of the clamped–free rod, respectively. If the material densities are equal, $\rho_1 = \rho_2 = \rho_0$, but $E_2 = 4E_1$, then the fundamental natural frequencies coincide. This implies that two rods can *share* the same fundamental frequencies if their material characteristics *differ*. As discussed above when the elastic moduli of C–C and C–F rods differ they may share the same natural frequency. In the case of the inhomogeneous rods, considered above, the expressions for the elastic moduli in Eqs. (3.22) and (3.28) differ. Thus, the coincidence of frequencies should not be totally unexpected. Yet, let us review the conditions that led to the coincidence of the natural frequencies. We *postulated* the vibration modes and looked for rods with a *polynomial* variation of the density and longitudinal rigidity that possess the given mode shape. Such a search, remarkably, led to the same fundamental frequency of rods for two sets of boundary conditions. It will be shown later on that this interesting phenomenon repeats itself for rods with non-constant densities along the rod's axis.

3.1.4 Inhomogeneous Rods with Linearly Varying Density

Consider rods whose inertial coefficient is represented by the following function:

$$R(\xi) = a_0 + a_1\xi \tag{3.31}$$

We are looking for a rod whose longitudinal rigidity variation is represented by a cubic polynomial

$$D(\xi) = b_0 + b_1\xi + b_2\xi^2 + b_3\xi^3 \tag{3.32}$$

Note that the highest order in the polynomial expression for $E(\xi)$ is three, whereas in $\rho(\xi)$ it is unity. This is due to the fact that two differentiations are involved in the first term of the differential equation (3.7).

Substitution of Eqs. (3.31) and (3.32) into the governing equation, in view of the mode shape for the clamped–clamped rod (3.10), results in the polynomial expression as the left-hand side of the equation. Since it is valid for any ξ, we get the following set of equations:

$$b_1 - 2b_0 = 0 \quad -4b_1 + 2b_2 + a_0k = 0$$
$$-6b_2 + 3b_3 - a_0k + a_1k = 0 \quad -8b_3 - a_1k = 0 \tag{3.33}$$

These equations can conveniently be cast as four different expressions for the natural frequency coefficient, with subscripts added so as to distinguish them:

$$k_0 = k_1 + b_1 - 2b_0 \quad k_1 = 2(2b_1 - b_2)/a_0$$
$$k_2 = 3(-2b_2 + b_3)/kL^2(a_0 - a_1) \quad k_3 = -8b_3/a_1 \tag{3.34}$$

Since all k_α must coalesce, we get the following interconnection between the coefficients:

$$b_2 = (-5a_1 + 8a_0)b_3/6a_1 \quad b_1 = -(5a_1 - 16a_0)b_3/12a_1$$
$$b_0 = (5a_2 + 16a_1)b_3/24a_1 \tag{3.35}$$

We arrive at the following result: if the inertial coefficient is given by Eq. (3.31) and the longitudinal rigidity variation is

$$b_2 = \frac{(5a_2 + 16a_1)b_3}{24a_1} - \frac{(5a_1 - 16a_0)b_3}{12a_1}\xi + \frac{(-5a_1 + 8a_0)b_3}{6a_1}\xi^3 \tag{3.36}$$

then, the natural frequency of the clamped–clamped rod is given by the last expression in Eq. (3.34):

$$\omega^2 = -8b_3/a_1 L^2 \tag{3.37}$$

For the clamped–free rod expressions (3.31) and (3.32) should be utilized in conjunction with the mode shape in Eq. (3.12), to be substituted into the differential equation (3.7). It is valid if the following conditions are met:

$$b_1 - 2b_0 = 0$$
$$-4b_1 + 2b_2 + ka_0 = 0$$
$$-6b + 3b_3 - ka_0 + ka_1 = 0 \tag{3.38}$$
$$-8b_3 + 4b_4 - ka_1 + ka_2 = 0$$
$$-10b_4 - ka_2 = 0$$

The expressions are equivalent to the following set.

$$k_0 = k_1 + b_1 - 2b_0$$
$$k_1 = -2(-2b_1 + b_2)/a_0$$
$$k_2 = -3(2b_2 - b_3)/(a_0 - a_1) \tag{3.39}$$
$$k_3 = -4(2b_3 - b_4)/(a_1 - a_2)$$
$$k_4 = -10b_4/a_2$$

Solution of the above equations leads to the coefficients b_0, b_1, b_2 and b_3. The final expression for the modulus of elasticity is

$$96a_2 E(\xi) = -(9a_2 + 25a_1 + 80a_0)b_4 - 2(9a_2 + 25a_1 + 80a_0)b_4\xi$$
$$+ (-9a_2 - 25a_1 + 40a_0)b_4\xi^2 + 24(-3a_2 + 5a_1) + 96b_4\xi^4 \tag{3.40}$$

Thus, the inhomogeneous clamped–free rod with linearly varying inertial coefficient and variable modulus of elasticity in Eq. (3.40) has the fundamental natural frequency arising from the last expression in Eq. (3.39):

$$\omega^2 = -10b_4/L^2 a_2 \tag{3.41}$$

3.1.5   Inhomogeneous Rods with Parabolically Varying Inertial Coefficient

We study the free vibrations of the rod with the following inertial coefficient:

$$R(\xi) = a_0 + a_1\xi + a_2\xi^2 \tag{3.42}$$

with associated variable longitudinal rigidity

$$D(\xi) = b_0 + b_1\xi + b_2\xi^2 + b_3\xi^3 + b_4\xi^4 \tag{3.43}$$

Substitution of Eqs. (3.42) and (3.43) in conjunction with the mode shape of the C–C rod in Eq. (3.10) leads to the polynomial equation. We require it to be valid for any ξ. This leads to the following conditions:

$$b_1 - 2b_0 = 0 \quad -4b_1 + 2b_2 + a_0k = 0$$
$$-6b_2 + 3b_3 - a_0k + a_1k = 0 \quad -8b_3 - a_1k + ka_2 = 0 \tag{3.44}$$
$$-10b_4 - ka_2 = 0$$

Equation (3.44) leads to four expressions of natural frequency. We introduce, formally, the fifth one, namely k_0, as follows:

$$k_0 = k_1 + b_1 - 2b_0 = 0 \quad k_1 = 2(2b_1 - b_2)/a_0$$
$$k_2 = 3(-2b_2 + b_3)/kL^2(a_0 - a_1) \quad k_3 = -4(2b_3 - b_4)/(a_1 - a_2) \tag{3.45}$$
$$k_4 = -10b_4/a_2$$

Requiring $k_j = k_0$ ($j = 0, 1, 2, 3, 4$) we get

$$b_0 = -(9a_2 + 25a_1 + 80a_0)b_4/96a_2 \quad b_1 = -(9a_2 + 25a_1 + 80a_0)b_4/48a_2$$
$$b_2 = -(9a_2 + 25a_1 + 40a_0)b_4/24a_2 \quad b_3 = -(3a_2 - 5a_1)b_4/4a_2 \tag{3.46}$$

The results for the clamped–clamped rod can be summarized as follows: the longitudinal rigidity variation consistent with the variable inertial coefficient in Eq. (3.42) is defined by Eq. (3.43) with coefficients given in Eq. (3.46). The appropriate closed-form expression for the natural frequency is obtained from the last expression in Eq. (3.45):

$$\omega^2 = -10b_4/a_2L^2 \tag{3.47}$$

As before, the coefficients b_4 and a_2 should have opposite signs.

Let us turn now to the clamped–free rod. The mode shape is given in Eq. (3.12). Under new circumstances, Eqs. (3.44) must be replaced by the following ones:

$$b_1 - b_0 = 0 \quad -2b_1 + 2b_2 + ka_0 = 0 \quad -3b_2 + 3b_3 - \tfrac{1}{2}ka_0 + ka_1 = 0$$
$$-4b_3 + 4b_4 - \tfrac{1}{2}ka_1 + ka_2 = 0 \quad -5b_4 - \tfrac{1}{2}ka_2 = 0 \tag{3.48}$$

The coefficients b_j consistent with Eq. (3.48) read

$$b_0 = -(18a_2 + 25a_1 + 40a_0)b_4/12a_2$$
$$b_1 = -(18a_2 + 25a_1 + 40a_0)b_4/12a_2$$
$$b_2 = -(18a_2 + 25a_1 + 20a_0)b_4/12a_2 \tag{3.49}$$
$$b_3 = -(6a_2 - 5a_1)b_4/4a_2$$

with natural frequency arising from the last entry in Eq. (3.48):

$$\omega^2 = -10b_4/a_2L^2 \tag{3.50}$$

As is seen, Eqs. (3.47) and (3.50) coincide with each other.

3.1.6   Rod with General Variation of Inertial Coefficient ($m > 2$)

Consider now the general case, $m > 2$. We are looking for the following variations in the inertial coefficient and longitudinal rigidity variations:

$$R(\xi) = \sum_{i=0}^{m} a_i\xi^i$$
$$D(\xi) = \sum_{i=0}^{m+2} b_i\xi^i \tag{3.51}$$

For the clamped–clamped rod we are looking for the rods that possess the mode shape given in Eq. (3.10). Substitution into Eq. (3.5) yields

$$\sum_{i=0}^{m+2}(i+1)b_{i+1}\xi^i - 2\sum_{i=1}^{m+2}b_i\xi^i(i+1) - 2b_0 + \omega^2 L^2\sum_{i=1}^{m+1}a_{i-1}\xi^i$$

$$-\omega^2 L^2\sum_{i=2}^{m+2}a_{i-2}\xi^i = 0 \tag{3.52}$$

$$b_1 - 2b_0 = 0 \tag{3.53}$$

$$2b_2 - 4b_1 + \omega^2 L^2 a_0 = 0 \tag{3.54}$$

$$\cdots$$

$$(i+1)b_{i+1} - 2b_1(i+1) + \omega^2 L^2(a_{i-1} - a_{i-2}) = 0 \qquad 2 \le i \le m+1 \tag{3.55}$$

$$\cdots$$

$$-2(m+3)b_{m+2} - \omega^2 L^2 a_m = 0 \tag{3.56}$$

We obtain

$$b_0 = \tfrac{1}{2}b_1$$

$$\omega^2 = (4b_1 - 2b_2)/a_0 L^2$$

$$\cdots$$

$$\omega^2 = (i+1)(2b_i - b_{i+1})/(a_{i-1} - a_{i-2})L^2 \tag{3.57}$$

$$\cdots$$

$$\omega^2 = -2(m+3)b_{m+2}/a_m L^2$$

For the coefficients b_i we obtain

$$b_{m+1} = \frac{[-(2m+8)a_m + 2(m+3)a_{m-1}]b_{m+2}}{2a_m(m+2)}$$

$$\cdots$$

$$b_i = \frac{[i(a_i - a_{i-2} - 2a_{i-1}) + a_1 + 3a_{i-1} - 4a_{i-2}]b_{i+1} + (i+2)(a_{i-2} - a_{i-1})b_{i+2}}{2(i+1)(a_i - a_{i-1})}$$

$$\cdots$$

$$b_1 = [2(2a_0 + a_1)b_2 + -3a_0 b_3]/4(a_1 - a_0)$$

$$b_0 = \tfrac{1}{2}b_1$$

$$\tag{3.58}$$

As for the clamped–free rod, the result of substitution of Eqs. (3.51) into Eq. (3.5) in conjunction with Eq. (3.12) leads to

$$\sum_{i=0}^{m+2}(i+1)b_{i+1}\xi^i - \sum_{i=1}^{m+2}b_i\xi^i(i+1) - b_0 + \omega^2 L^2 \sum_{i=1}^{m+1}a_{i-1}\xi^i$$

$$-\omega^2 L^2 \sum_{i=2}^{m+2}a_{i-2}\xi^i = 0 \qquad (3.59)$$

Since this equation must be valid for any ξ we arrive at the following recurrent equations:

$$b_1 - b_0 = 0 \qquad (3.60)$$

$$2b_2 - 2b_1 + \omega^2 L^2 a_0 = 0 \qquad (3.61)$$

$$\cdots$$

$$(i+1)b_{i+1} - b_1(i+1) + \omega^2 L^2 (a_{i-1} - \tfrac{1}{2}a_{i-2}) = 0 \qquad 2 \le i \le m+1 \quad (3.62)$$

$$\cdots$$

$$-(m+3)b_{m+2} - \tfrac{1}{2}\omega^2 L^2 a_m = 0 \qquad (3.63)$$

These equations result in

$$b_0 = b_1$$
$$\omega^2 = 2(b_1 - b_2)/a_0 L^2$$
$$\cdots$$
$$\omega^2 = (i+1)(b_i - b_{i+1})/(a_{i-1} - a_{i-2})L^2 \qquad (3.64)$$
$$\cdots$$
$$\omega^2 = -2(m+3)b_{m+2}/a_m L^2$$
$$b_0 = b_1$$
$$\omega^2 = 2(b_1 - b_0)/L^2 a_0$$
$$\cdots$$
$$\omega^2 = \frac{(i+1)(b_i - b_{i+1})}{L^2(a_{i-1} - a_{i-2}/L)} \qquad (3.65)$$
$$\cdots$$
$$\omega^2 = -2(m+3)b_{m+2}/L^2 a_m$$

Note that the expression for ω^2 in Eq. (3.65) is the same as in Eq. (3.57). Compatibility of equation (3.65) yields

$$b_{m+1} = \frac{(m+3)a_{m-2} - (m+4)a_m}{(m+2)a_m} b_{m+2}$$

$$\cdots$$

$$b_i = \frac{1}{(i+1)(2a_i - a_{i-1})}\{[3a_{i-1} - 2a_{i-2} + 2a_i]b_{m+1} \\ + [i(a_{i-1} - 2a_{i-2}) + 2(a_{i-2} - 2a_{i-3})]b_{i+2}\}$$

$$(3.66)$$

$$\cdots$$

$$b_1 = [2(a_1 + a_0)b_2 - 3a_0b_3]/(2a_0 - 2a_1)$$
$$b_0 = b_1$$

The final expression for the natural frequency squared is

$$\omega^2 = -2(m+3)b_{m+2}/L^2 a_m \tag{3.67}$$

In order for the natural frequency to be a positive quantity, it is necessary that a_m and b_{m+2} have opposite signs.

3.1.7   Concluding Remarks

It should be emphasized that all previous expressions for the natural frequency can be put in the form (3.67) with proper choice of m. However, the expressions for b_i are derivable separately, for $m = 0, 1, 2$. To the best of the author's knowledge, most of the previous studies considered non-uniform rods in which the cross-section varied but the material properties remained constant. It can then be argued that one can actually manufacture such a rod so that the problem being studied can have practical applications. The present section considers cases in which both the longitudinal rigidity and/or the inertial coefficient vary continuously with position. This raises the most important question about this study: is it addressing a problem of practical interest, or is this an exercise in finding solutions to a differential equation? How does one make a rod with properties varying as prescribed in this section? To answer these somewhat provocative questions, we first note that early studies in inverse problems did not exclude variable material properties. For example, Krein (1952) considered a string with variable inertial coefficient. More recently, Ram and Elhay (1998) distinguished between the cases where A and E are constant while ρ varied with ξ, or A, E and ρ all varied with ξ. Important variables are, of course, the products EA and ρA. As far as the manufacturing of the rods with given axial variation of $E(\xi)$ and $\rho(\xi)$ is concerned, even if such a procedure does not exist today, its development in the future cannot be *a priori* excluded.

The coincidence of the *fundamental* natural frequencies of rods with different boundary conditions may appear to be a surprising fact, at the first glance. We have addressed this question above, albeit briefly. It is remarkable that there exist structures that have the same *complete* spectrum. For example, Gottlieb and McManus (1998) illustrated how two different polygonal membranes may have the same *entire* spectrum, thus forming so called *isospectral* structures. Readers may also consult work by Gottlieb (1989), Gladwell and Morassi (1995), Chapman (1995), and Sridhar and Kudrolli (1994) who experimentally verified the isospectral property of membranes of different shapes.

It is worth noting the similarities and differences of the present study with the general topic of inverse problems, covered, for example, in the definitive monographs by Gladwell (1986a) and Tarantola (1987). As Tarantola (1987) wrote, "to solve the *forward problem* is to predict the values of the observable parameters, given arbitrary values of the model parameters. To solve the *inverse problem* is to infer the values of the model parameters from given observed values of the observable parameters" (italics by Tarantola). In the vibration context, the inverse problem consists in reconstructing the structure by its observable vibration spectrum. The reconstruction of the continuous variations in axial rigidity and the inertial coefficient of a longitudinally vibrating rod was studied apparently independently by Ram (1994) and Wang and Wang (1994). In particular, Ram (1994b) proved that the density and axial rigidity functions are uniquely determined by two natural frequencies, their corresponding mode shapes and the total mass of the rod, when specially derived necessary and sufficient conditions are met for the construction of the physically realizable rod, i.e. with positive parameters. Wang and Wang (1994) demonstrated that for the rod's reconstruction one needs the knowledge of two positive square frequencies, two associated mode functions with piecewise continuous second-order derivatives, satisfying some necessary conditions. The objective of the present work is *different* from those of Ram (1994) or Wang and Wang (1994). Here, we are looking for closed-form solutions for natural frequencies with a specified fundamental vibration *mode* alone. In these circumstances we uncovered an *infinite* number of closed-form solutions, corresponding to the degree of variation in the mass density, with $m = 0, 1, 2, \ldots$.

In Ram's (1994) terminology, "in the classical inverse problem it is assumed that the cross-sectional area of the rod is variable while the Young's modulus of elasticity and the rod density are constants." Studies by Ram (1994) and Wang and Wang (1994) allow the cross-sectional area, the modulus of elasticity and the inertial coefficient to vary along the rod's axial coordinate. The present section is devoted to rods whose cross-sectional area is a constant. Even with this seeming restriction, an infinite number of rods are uncovered that possess a given polynomial mode shape. Once technology exists that allows construction of inhomogeneous rods with polynomially varying modulus of elasticity, it is an easy task to demand that the rod have any pre-selected fundamental natural frequency. Indeed, as Eq. (3.67) indicates, the fundamental frequency depends solely on a_m and b_{m+2}. If a technology allows for manufacturing rods with arbitrary a_m and b_{m+2}, one can get any desirable fundamental natural frequency. This leads both to an avoiding resonance condition for forced deterministic vibration, and the first frequency to lie outside the range of excitation of a rod under cutoff white noise with two cutoff frequencies $\omega_{c,1}$ and $\omega_{c,2}$ in the random vibration environment (Elishakoff, 1999). If the fundamental frequency ω_1 is less than $\omega_{C,1}$ the response level can be significantly reduced. Thus, a new passive vibration control mechanism may be obtained.

3.2 The Natural Frequency of an Inhomogeneous Rod may be Independent of Nodal Parameters

3.2.1 Introductory Remarks

The forward and inverse problems associated with vibration nodes have attracted several investigators. The motivation for the knowledge of nodes is described by Cha et al. (1998), who note: "Experimentalists and designers consider it important to know where the nodes of a structure are located. For example, the usefulness of accelerometers and actuators during testing and control depends on their placement. Dynamical measurements could easily be misleading or wrong if an accelerometer is inadvertently placed on or near a node. In another engineering application, shrouds are usually added to bladed disc assemblies to increase rigidity of the system. Thus, a shroud that is accidentally located at a node cannot achieve the objectives for which it was designed." While solving forward vibration problems, the structure is known and the problem consists in determining the vibration spectrum, i.e., vibration frequencies and vibration modes. Once the vibration mode is determined, the location of the node is also uniquely determined. Thus, with analytical or numerical methods, the information on the nodes is a byproduct of the analysis. One may wonder whether there exists a direct evaluation method of nodes and anti-nodes. Such an efficient method was proposed by Cha et al. (1998) and Cha and Pierre (1999). In the former paper, for example, the authors attach a virtual element in the form of either a virtual lumped mass or a virtual grounded spring to the system under consideration, and analyze the free vibration of the assembled system as a function of the location of the virtual element; the authors then plot the frequency of the assembled system against the constraint location of the virtual element; the nodes and anti-nodes of the structure under study are extracted by examining the maxima or minima of such curves. For the virtual lumped mass, for example, the nodes correspond to the maximum of the frequency curves, while for the virtual spring, they are associated instead with the minima of the frequency curves.

The inverse problem of reconstructing the structure from the nodal information was studied by McLaughlin (1998), Shen (1993), Lee and McLaughlin (1997) and Yang (1997). A review of these results was given by Gladwell (1996) and McLaughlin (1998). One should note also papers by Hald and

McLaughlin (1989, 1998) in which the density or Young's modulus are determined as piecewise constant functions from the nodal positions of mode shapes. Further, it is known that the density and Young's modulus can be uniquely determined from a dense subset of the nodes, as was demonstrated by Hald and McLaughlin (1998a,b), the pioneers in inverse nodal problems.

We deal with a question that is somewhat connected with the inverse vibration problem. Namely, we are concerned with the closed-form polynomial solutions for the inhomogeneous rods whose second mode shape may have a node at any preselected location on the axis of the rod. At first glance it is not clear that such a problem may have a solution. Yet, it is demonstrated that indeed, one can find rods with constant or continuously varying mass density, such that the nodal point of the second mode is located anywhere, except on the ends of the rod. The central point of the procedure that closely follows study of Neuringer and Elishakoff (2000) is to look for a polynomial variation of the modulus of elasticity such that three polynomial quantities: mode shape with node at any location, mass density, and modulus of elasticity, all satisfy the governing differential equation and boundary conditions. We define a *node* to be a point on the rod at which the displacement varies, *excluding* the boundaries.

3.2.2 The Nodal Parameters

The spatial dependence of the displacement $u(x)$ of a uniform (i.e. constant cross-section), inhomogeneous rod in longitudinal vibration is given by

$$\frac{d}{dx}\left[D(x)\frac{du}{dx}\right] + R(x)\omega^2 u = 0 \qquad (3.68)$$

where $D(x)$ is the longitudinal rigidity, $R(x)$ the inertial coefficient and ω the natural frequency. Introducing the dimensionless length $\xi = x/L$, where L is the rod length, the natural frequency coefficient,

$$k = L^2\omega^2 \qquad (3.69)$$

and carrying out the differentiation in Eq. (3.68), we obtain

$$\frac{dD}{d\xi}\frac{du}{d\xi} + D(\xi)\frac{d^2u}{d\xi^2} + kR(\xi)u = 0 \qquad (3.70)$$

The polynomial solutions sought for, $u(\xi)$, must satisfy the appropriate boundary conditions. For the first harmonic (i.e. the mode with one node) we take

$$u(\xi) = C_0 + C_1\xi + C_2\xi^2 + C_3\xi^3 \qquad (3.71)$$

For a C–C rod with a node at α_1, the relation to be satisfied is

$$u(0) = u(1) = u(\alpha_1) = 0 \tag{3.72}$$

Thus,

$$u(\xi) = C_1(\xi + \gamma_2\xi^2 + \gamma_3\xi^3) \tag{3.73}$$

where

$$\gamma_2 = -\frac{1 + \alpha_1}{\alpha_1} \quad \gamma_3 = \frac{1}{\alpha_1} \tag{3.74}$$

For a C–F rod with a node at α_1, the relation to be satisfied is

$$u(0) = u'(1) = u(\alpha_1) = 0 \tag{3.75}$$

Thus,

$$u(\xi) = C_1(\xi + \gamma_2\xi^2 + \gamma_3\xi^3)$$
$$\gamma_2 = \frac{3 - \alpha_1^2}{2\alpha_1^2 - 3\alpha_1} \quad \gamma_3 = \frac{\alpha_1 - 2}{2\alpha_1^2 - 3\alpha_1} \tag{3.76}$$

and C_1 is an arbitrary constant.

For the second harmonic (i.e. the mode with two nodes, at α_1 and α_2, respectively), we take for $u(\xi)$ the polynomial

$$u(\xi) = C_0 + C_1\xi + C_2\xi^2 + C_3\xi^3 + C_4\xi^4 \tag{3.77}$$

For a C–C rod with nodes at α_1 and α_2, we must satisfy

$$u(0) = u(1) = u(\alpha_1) = u(\alpha_2) = 0 \tag{3.78}$$

Thus,

$$u(\xi) = C_1(\xi + \gamma_2\xi^2 + \gamma_3\xi^3 + \gamma_4\xi^4) \tag{3.79}$$

where

$$\gamma_2 = -\frac{\alpha_1\alpha_2 + \alpha_1 + \alpha_2}{\alpha_1\alpha_2}$$

$$\gamma_3 = -\frac{\alpha_1 + \alpha_2 + 1}{\alpha_1\alpha_2} \tag{3.80}$$

$$\gamma_4 = \frac{1}{\alpha_1\alpha_2}$$

For the C–F case the satisfaction of the boundary conditions

$$u(0) = u'(1) = u(\alpha_1) = u(\alpha_2) = 0 \tag{3.81}$$

yields

$$\gamma_2 = \frac{-\alpha_1^2\alpha_2^2 + 3\alpha_1^2 + 3\alpha_2^2 + 3\alpha_1\alpha_2 - 4\alpha_1 - 4\alpha_2}{\alpha_1\alpha_2(2\alpha_1\alpha_2 - 3\alpha_2 - 3\alpha_1 + 4)}$$

$$\gamma_3 = \frac{\alpha_1\alpha_2^2 + \alpha_2\alpha_1^2 - 2\alpha_1^2 - 2\alpha_2^2 + 2\alpha_1\alpha_2 + 4}{\alpha_1\alpha_2(2\alpha_1\alpha_2 - 3\alpha_2 - 3\alpha_1 + 4)} \tag{3.82}$$

$$\gamma_2 = 3$$

If we are given the inertial coefficient in the polynomial form

$$R(\xi) = \sum_{i=0}^{m} a_i\xi^i \tag{3.83}$$

with upper limit m, then it is easily seen that, since the largest derivative in Eq. (3.70) is of second order, the polynomial for $E(\xi)$ must be of degree $m+2$. Thus,

$$D(\xi) = \sum_{i=0}^{m+2} b_i\xi^i \tag{3.84}$$

In what follows we investigate, both for the C–C and C–F conditions, the following three cases:

Case 1: The mode with single node and with inertial coefficient $R(\xi) = \alpha_0 = $ constant (i.e. $m = 0$).

Case 2: The mode with two nodes and with inertial coefficient constant ($m = 0$).

Case 3: The mode with a single arbitrarily prescribed nodal position and with inertial coefficient varying linearly with ξ (i.e. $m = 1$).

3.2.3   Mode with One Node: Constant Inertial Coefficient

With $m = 0$, $D(\xi)$ takes the polynomial form

$$D(\xi) = b_0 + b_1\xi + b_2\xi^2 \tag{3.85}$$

Now

$$u(\xi) = \xi + \gamma_2\xi^2 + \gamma_3\xi^3 \tag{3.86}$$

and hence,

$$u'(\xi) = 1 + 2\gamma_2\xi + 3\gamma_3\xi^2 \tag{3.87}$$

$$u''(\xi) = 2\gamma_2 + 6\gamma_3\xi \tag{3.88}$$

$$\frac{dD}{d\xi} = b_1 + 2b_2\xi \tag{3.89}$$

Substituting into Eq. (3.70), yields

$$(b_1 + 2b_2\xi)(1 + 2\gamma_2\xi + 3\gamma_3\xi^2) + (b_0 + b_1\xi + b_2\xi^2)(2\gamma_2 + 6\gamma_3\xi)$$
$$+ ka_0(\xi + \gamma_2\xi^2 + \gamma_3\xi^3) = 0 \tag{3.90}$$

Setting the net coefficient of each power of ξ equal to zero, yields

$$\text{from } \xi^0: \quad b_1 + 2\gamma_2b_0 = 0 \tag{3.91}$$

$$\text{from } \xi^1: \quad 2b_0 + 6\gamma_3b_0 + 4\gamma_2b_1 + ka_0 = 0 \tag{3.92}$$

$$\text{from } \xi^2: \quad 6\gamma_2b_2 + 9\gamma_3b_1 + \gamma_2ka_0 = 0 \tag{3.93}$$

$$\text{from } \xi^3: \quad 12\gamma_3b_2 + \gamma_3ka_0 = 0 \tag{3.94}$$

From Eq. (3.94),

$$k = -12b_2/a_0 \tag{3.95}$$

Substituting for k into Eq. (3.93) and solving for b_1, yields

$$b_1 = (2\gamma_2/3\gamma_3)b_2 \tag{3.96}$$

Substituting for b_0 from Eq. (3.92) into Eq. (3.93) and solving for b_1, we obtain

$$b_1 = \frac{10\gamma_2}{4\gamma_2^2 - 3\gamma_3} b_2 \qquad (3.97)$$

Setting Eq. (3.96) and (3.97) equal to each other, yields

$$\gamma_3 = \tfrac{2}{9}\gamma_2^2 \qquad (3.98)$$

Thus, the nodal position α_1 is determined for the C–C case as that value of α_1 lying between $0 < \alpha < 1$ which satisfies the equation

$$\frac{1}{\alpha_1} = \frac{2}{9} \frac{(1+\alpha_1)^2}{\alpha_1^2} \qquad (3.99)$$

Equation (3.99) is reducible to the polynomial equation

$$2\alpha_1^2 - 5\alpha_1 + 2 = (2\alpha_1 - 1)(\alpha_1 - 2)(\alpha_1 - 1)^2 = 0 \qquad (3.100)$$

For the C–F case, α_1 must satisfy

$$\frac{\alpha_1 - 2}{2\alpha_1^2 - 3\alpha_1} = \frac{2}{9}\left(\frac{3 - \alpha_1^2}{2\alpha_1^2 - 3\alpha_1}\right)^2 \qquad (3.101)$$

or

$$2\alpha_1^4 - 18\alpha_1^3 + 51\alpha_1^2 - 54\alpha_1 + 18 = (\alpha_1^2 - 6\alpha_1 + 6)(2\alpha_1^2 - 6\alpha_1 + 3) = 0 \qquad (3.102)$$

Once a solution for $\alpha_1 = (3 - \sqrt{3})/2 \cong 0.63$ is obtained, the solution for $E(\xi)$ is

$$D(\xi) = (-1 + 2\gamma_2\xi + 3\gamma_3\xi^2)b_2/3\gamma_3 \qquad (3.103)$$

where the nodal parameters are evaluated at that α_1 which satisfies Eq. (3.29) for the C–C case and Eq. (3.102) for the C–F case.

It can be immediately verified that $\alpha = \tfrac{1}{2}$ is a root of Eq. (3.100) and that to two significant figures $\alpha_1 = (3 - \sqrt{3})/2$ is a root of Eq. (3.102). It is surprising that the frequency $k = -12b_2/a_0$, where $b_2 < 0$ does not depend on the nodal parameters and depends only on the choice of the parameter b_2. If the same b_2 and a_0 are used, then the C–C and C–F cases have the same frequency. Note that since

$$\omega^2 = -12b_2/a_0 L^2 \qquad (3.104)$$

then, with b_2 negative a_0 must be positive as required.

With $\alpha_1 = \frac{1}{2}$, we obtain for the C–C case $\gamma_2 = -3$ and $\gamma_3 = 2$. Substituting into Eq. (3.103), yields

$$D(\xi) = (-\tfrac{1}{6} - \xi + \xi^2)b_2 \tag{3.105}$$

and the displacement becomes

$$u(\xi) = (\xi - 3\xi^2 + 2\xi^3)C_1 \tag{3.106}$$

With $\alpha_1 = (3 - \sqrt{3})/2$, we obtain for the C–F case $\gamma_2 = -(3 + \sqrt{3})/2$ and $\gamma_3 = (2 + \sqrt{3})/3$. Hence, using Eq. (3.103)

$$D(\xi) = [\sqrt{3} - 2 - (3 - \sqrt{3})\xi + \xi^2]b_2 \tag{3.107}$$

$$u(\xi) = \left[\xi - \tfrac{1}{2}(3 + \sqrt{3})\xi^2 + \tfrac{1}{3}(2 + \sqrt{3})\xi^3\right]C_1 \tag{3.108}$$

Three observations can now be made: (1) The maximum value of $D(\xi)$ occurs at the same place as the node! (2) For the C–C case $D(\xi)$ is symmetric about the nodal point but $u(\xi)$ is not antisymmetric about the nodal point. (3) Since $b_2 < 0$ ($k = -12b_2/\alpha_0$) it is seen from Eqs. (3.105) and (3.107) that $D(\xi)$ is positive for all $0 \le \xi \le 1$ as it should be. For the C–F case $D(\xi)$ is not symmetric and $u(\xi)$ is not antisymmetric about the nodal point.

The very interesting result of the correspondence of the nodal position and the position of the maximum of $D(\xi)$ is not fortuitous but is a consequence of the formalism. A proof of the coincidence follows: from Eq. (3.104), b_2 is required to be negative. In other words, $D(\xi) = (1 - 2\gamma_2\xi - 3\gamma_3\xi^2)|b_2|/3\gamma_3$.

The position of the maximum of $D(\xi)$, say ξ_1 occurs when $dE/d\xi = -2\gamma_2/3\gamma_3 - 2\xi = 0$, and corresponds to the negative second derivative. Hence,

$$\xi_1 = -\gamma_2/3\gamma_3 \tag{3.109}$$

The position of the node α_1 must satisfy

$$u(\alpha_1) = \alpha_1 + \gamma_2\alpha_1^2 + \gamma_3\alpha_1^3 = 0 \tag{3.110}$$

or

$$1 + \gamma_2\alpha_1 + \gamma_3\alpha_1^2 = 0 \qquad \alpha_1 \ne 0 \tag{3.111}$$

Hence,

$$\alpha_1 = \left[-\gamma_2 \pm \sqrt{\gamma_2^2 - 4\gamma_3}\right]\Big/2\gamma_3 \tag{3.112}$$

Now, the compatibility condition, Eq. (3.98), requires $\gamma_3 = (2/9)\gamma_2^2$. Substituting inside the radical for γ_3, we obtain

$$\alpha_1 = \frac{-\gamma_2 \pm \sqrt{\gamma_2^2 - (8/9)\gamma_2^2}}{2\gamma_3} = \frac{-\gamma_2 \pm \gamma_2/3}{2\gamma_3} \qquad (3.113)$$

Using the negative sign makes α_1 to be outside the interval $(0, 1)$. Hence, we take the positive sign, to get

$$\alpha_1 = -\gamma_2/3\gamma_3 \qquad (3.114)$$

Comparing Eqs. (3.103) and (3.104), we obtain $\xi_1 = \alpha_1$.

Summarizing, we conclude that when the material density is constant, we can find a mode shape with an interior node in the form of a third-degree polynomial (unique up to an arbitrary multiplicative constant). The nodal location is fixed by the boundary conditions and hence cannot be arbitrarily assigned. The corresponding elastic modulus is a second-degree polynomial uniquely determined up to an arbitrary (negative) multiplicative constant. This summation corrects some imprecise statements made in Section 3 of the study by Neuringer and Elishakoff (2000) (Storch, 2001b).

3.2.4 Mode with Two Nodes: Constant Density

Because of the introduction of an additional node, the satisfaction of the boundary conditions requires $u(\xi)$ to be of the fourth degree. Thus,

$$u(\xi) = \xi + \gamma_2\xi^2 + \gamma_3\xi^3 + \gamma_4\xi^4$$
$$u'(\xi) = 1 + 2\gamma_2\xi + 3\gamma_3\xi^2 + 4\gamma_4\xi^3$$
$$u''(\xi) = 2\gamma_2 + 6\gamma_3\xi + 12\gamma_4\xi^2$$
$$D(\xi) = b_0 + b_1\xi + b_2\xi^2$$
$$D'(\xi) = b_1 + 2b_2\xi$$

Substituting into Eq. (3.70), we get

$$(b_1 + 2b_2\xi)(1 + 2\gamma_2\xi + 3\gamma_3\xi^2 + 4\gamma_4\xi^3) + (b_0 + b_1\xi + b_2\xi^2)(2\gamma_2 + 6\gamma_3\xi + 12\gamma_4\xi^2)$$
$$+ ka_0(\xi + \gamma_2\xi^2 + \gamma_3\xi^3 + \gamma_4\xi^4) = 0 \qquad (3.115)$$

Equating like powers of ξ', yields

$$\text{from } \xi^0: \quad b_1 + 2\gamma_2 b_0 = 0 \tag{3.116}$$

$$\text{from } \xi^1: \quad 2b_2 + 4\gamma_2 b_1 + 6\gamma_3 b_0 + ka_0 = 0 \tag{3.117}$$

$$\text{from } \xi^2: \quad 6\gamma_2 b_2 + 9\gamma_3 b_1 + \gamma_2 ka_0 = 0 \tag{3.118}$$

$$\text{from } \xi^3: \quad 12\gamma_3 b_2 + 16\gamma_4 b_1 + \gamma_3 ka_0 = 0 \tag{3.119}$$

$$\text{from } \xi^4: \quad 20\gamma_4 b_2 + \gamma_4 ka_0 = 0 \tag{3.120}$$

From Eq. (3.120)

$$k = -20b_2/a_0 \tag{3.121}$$

Substituting for k from Eq. (3.121) into Eq. (3.119) and solving for b_1, we get

$$b_1 = \gamma_3/2\gamma_4 b_2 \tag{3.122}$$

Substituting for b_0 from Eq. (3.116) into Eq. (3.118) and solving for b_1, we get

$$b_1 = \frac{16\gamma_2}{9\gamma_3 - 6\gamma_4/\gamma_2} b_2 \tag{3.123}$$

Substituting for b_0 from Eq. (3.116) into Eq. (3.117) and solving for b_1, we get

$$b_1 = \frac{18}{4\gamma_2 - 3\gamma_3/\gamma_2} b_2 \tag{3.124}$$

From Eqs. (3.123)–(3.124), we get the two equations

$$\frac{1}{2}\frac{\gamma_3}{\gamma_4} = \frac{16\gamma_2}{3\gamma_3 - 6\gamma_4/\gamma_2} = \frac{18}{4\gamma_2 - 3\gamma_3/\gamma_2} \tag{3.125}$$

These coupled equations can be solved analytically for the case of the C–C rod (Storch, 2001b). The desired roots $0 < \alpha_1 < \alpha_2 < 1$ are given by

$$\alpha_1 = \tfrac{1}{2} + \sqrt{\tfrac{2}{7}} - \tfrac{1}{2}\sqrt{\tfrac{15}{7}} \cong 0.303$$
$$\alpha_2 = \tfrac{1}{14}(7 - 2\sqrt{14} + \sqrt{105}) \cong 0.697 \tag{3.126}$$

For the C–C case

$$E(\xi) = \left(\xi^2 - \xi + \sqrt{\tfrac{5}{6}} - 1 \right) b_2$$

$$u(\xi) = C_1 \frac{\xi(\xi - 1)[7\xi(\xi - 1) + \sqrt{30} - 4]}{\sqrt{30} - 4} \tag{3.127}$$

where C_1 and b_2 are arbitrary constants with $b_2 < 0$. For the C–F case we have the numerical solution (Storch, 2001b): [$\alpha_1 \cong 0.344$ and $\alpha_2 \cong 0.792$]

$$E(\xi) = (\xi^2 - 1.136\xi - 0.112)b_2$$
$$u(\xi) = C_1(-3.23\xi^4 + 7.34\xi^3 - 5.05\xi^2 + \xi) \tag{3.128}$$

3.2.5 Mode with One Node: Linearly Varying Material Coefficient

In this case the material coefficient is given by

$$R(\xi) = a_0 + a_1\xi = a_1(\beta + \xi) \quad \beta = \frac{a_0}{a_1} \tag{3.129}$$

because $m = 1$, $R(\xi)$ takes the polynomial form $E(\xi) = b_0 + b_1\xi + b_2\xi^2 + b_3\xi^3$ and the form for $u(\xi)$ remains the same, i.e. $u(\xi) = \xi + \gamma_2\xi^2 + \gamma_3\xi^3$. Letting $k = L^2\omega^2 a_1$ and making the usual substitutions into Eq. (3.70), we obtain

$$(b_1 + 2b_2\xi + 3b_3\xi^2)(1 + 2\gamma_2\xi + 3\gamma_3\xi^2)$$
$$+ (b_0 + b_1\xi + b_2\xi^2 + b_3\xi^3)(2\gamma_2 + 6\gamma_3\xi + 12\gamma_4\xi^2) \tag{3.130}$$
$$+ ka_1(\beta + \xi)(\xi + \gamma_2\xi^2 + \gamma_3\xi^3 + \gamma_4\xi^4) = 0$$

leading to the following equations:

$$\text{from } \xi^0: \quad b_1 + 2\gamma_2 b_0 = 0 \tag{3.131}$$
$$\text{from } \xi^1: \quad 2b_2 + 4\gamma_2 b_1 + 6\gamma_3 b_0 + ka_1\beta = 0 \tag{3.132}$$
$$\text{from } \xi^2: \quad 3b_3 + 6\gamma_2 b_2 + 9\gamma_3 b_1 + ka_1(\gamma_2\beta + 1) = 0 \tag{3.133}$$
$$\text{from } \xi^3: \quad 8\gamma_2 b_3 + 12\gamma_3 b_2 + ka_1(\gamma_3\beta + \gamma_2) = 0 \tag{3.134}$$
$$\text{from } \xi^4: \quad 15\gamma_3 b_3 + \gamma_3 ka_1 = 0 \tag{3.135}$$

From Eq. (3.135)

$$k = -15b_3/a_1 \tag{3.136}$$

Substituting for k from Eq. (3.136) into Eq. (3.134) and solving for b_2, we obtain

$$b_2 = (15\gamma_3\beta - 7\gamma_2)b_3/12\gamma_3 \tag{3.137}$$

For b_1, we get

$$b_1 = \frac{1}{9\gamma_3}\left\{-\frac{\gamma_2}{2\gamma_3}(15\gamma_3\beta - 7\gamma_2) + 15\gamma_2\beta + 12\right\}b_3 \tag{3.138}$$

Substituting for b_0 using Eq. (3.131) and b_2 using Eq. (3.137) into Eq. (3.125) and solving for b_1, we get

$$b_1 = \frac{[15\beta - (15\gamma_3\beta - 7\gamma_2)/6\gamma_3]}{(4\gamma_2 - 3\gamma_3/\gamma_2)}b_3 \tag{3.139}$$

Setting Eq. (3.135) equal to Eq. (3.136) and solving for β, we get

$$\beta = \frac{((7\gamma_2^2 + 24\gamma_3)/18\gamma_3^2) - (7\gamma_2^2/\gamma_3(4\gamma_2^2 - 3\gamma_3))}{25(4\gamma_2 - 3\gamma_3)/2\gamma_2 - (5/6)\gamma_2} \tag{3.140}$$

Note that a plot of the parameter β as defined in Eq. (3.137), shows that it is always positive for $0 < \alpha_1 < 1$ for both the C–C and C–F boundary conditions (Storch, 2001b).

The procedure now is as follows: using the arbitrarily chosen nodal position α_1, we evaluate the nodal parameters γ_2 and γ_3 using Eq. (3.74) or Eq. (3.76). Substituting into Eq. (3.140) we calculate β. The required linear inertial coefficient distribution is then given by:

$$R(\xi) = a_1(\beta + \xi) \tag{3.141}$$

Since we require $\omega^2 = -15b_3/L^2\alpha_1$; and for all $0 \le \xi \le 1$ we demand $D(\xi, \beta)$ and $R(\xi)$ to be both positive, there are three cases to consider.

Case 1: if $\beta > 0$, then $a_1 > 0$ and $b_3 < 0$.

Case 2: if $\beta < -1$, then $a_1 < 0$ and $b_3 > 0$.

Case 3: if $-1 < \beta < 0$, then the procedure fails.

For the moment, let $D(\xi, \beta)$ be the polynomial which is multiplied by the parameter b_3. If $D(\xi, \beta) > 0$, then $b_3 > 0$ and $D(\xi, \beta)$ is generated by $\beta < -1$. If, however, $D(\xi, \beta) < 0$, then $b_3 < 0$ and $D(\xi, \beta)$ is generated by $\beta > 0$.

It is now fairly clear how the method can be extended to the cases of $2, 3, \ldots, n$ prescribed nodal positions. For example, for the case of a mode with two arbitrarily chosen nodal positions take

$$u(\xi) = \sum_{i=0}^{4} C_i \xi^i \tag{3.142}$$

$$D(\xi) = \sum_{i=0}^{4} b_i \xi^i \tag{3.143}$$

and

$$R(\xi) = a_0 + a_1\xi + a_2\xi^2 = a_2(\beta_1 + \beta_2\xi + \xi^2) \tag{3.144}$$

where

$$\beta_1 = a_0/a_2 \quad \beta_2 = a_1/a_2 \tag{3.145}$$

and substitute into Eq. (3.70). In the process of evaluating the b_i, we arrive at two coupled linear equations for β_1 and β_2 whose solution yields the required quadratic density distribution.

3.3 Concluding Remarks

The longitudinal rigidity in Eq. (3.22) is denoted by $D_{C-C}(\xi)$

$$D_{C-C}(\xi) = (\tfrac{1}{2} + \xi - \xi^2)|b_2| \tag{3.146}$$

with subscript C–C indicating that for the clamped–clamped rod of longitudinal rigidity the closed-form natural frequency, given in Eq. (3.23) is obtained:

$$\omega_{C-C}^2 = 6|b_2|/a_0 L^2 \tag{3.147}$$

In perfect analogy, the longitudinal rigidity in Eq. (3.23) is designated $D_{C-F}(\xi)$:

$$D_{C-F}(\xi) = (2 + 2\xi - \xi^2)|b_2| \tag{3.148}$$

with the expression for ω_{C-F}^2 formally coinciding with Eq. (3.144). We are interested in calculating the natural frequencies of the rods with longitudinal rigidity equal to $D_{C-C}(\xi)$ under two sets of boundary conditions. Note again, that if the rod is clamped at both sides and has a longitudinal rigidity $D_{C-C}(\xi)$ the natural frequency is obtainable in the closed-form and is given in Eq. (3.147). Use of the Boobnov–Galerkin method with a single-term approximation with the comparison function

$$\varphi_{C-C}(\xi) = \sin(\pi\xi) \tag{3.149}$$

which constitutes the exact mode shape of the uniform clamped–clamped rod yields

$$\omega_{C-C}^2 = \left(\tfrac{3}{2}\pi^2 - \tfrac{1}{2}\right)|b_2|/a_0 L^2 \approx 6.08|b_2|/a_0 L^2 \tag{3.150}$$

and yields an error of 1.333% in comparison with the exact solution. If a clamped–free rod has a longitudinal rigidity $D_{C-C}(\xi)$ we can use, amongst others, either of the following comparison functions:

$$\varphi^{(1)}_{C-F}(\xi) = \sin(\pi\xi/2) \tag{3.151}$$

$$\varphi^{(2)}_{C-F}(\xi) = \xi - \tfrac{1}{2}\xi^2 \tag{3.152}$$

The expression in Eq. (3.151) constitutes an exact mode shape of the uniform clamped–free rod, whereas the expression in Eq. (3.152) represents a polynomial comparison function that satisfies all boundary conditions. The estimates of the natural frequency are

$$[\omega^{(1)}_{C-F}]^2 = (\pi^2/6)|b_2|/a_0 L^2 \approx 1.64|b_2|/a_0 L^2 \tag{3.153}$$

$$[\omega^{(2)}_{C-F}]^2 = \tfrac{13}{8}|b_2|/a_0 L^2 \approx 1.625|b_2|/a_0 L^2 \tag{3.154}$$

As can be seen, for the rods with the same longitudinal rigidity, but different boundary conditions, the expressions for the natural frequency are different: Eqs. (3.153) and (3.154) *differ* from Eq. (3.150) and Eq. (3.147).

Likewise, for the clamped–free rod, with longitudinal rigidity $D_{C-F}(\xi)$ we choose the comparison function that coincides with the exact mode shape of its homogeneous counterpart:

$$\varphi_{C-F}(\xi) = \sin(\tfrac{1}{2}\pi\xi) \tag{3.155}$$

The natural frequency squared reads

$$\omega^2_{C-F} = (\tfrac{2}{3}\pi^2 - \tfrac{1}{2})|b_2|/a_0 L^2 \approx 6.08|b_2|/a_0 L^2 \tag{3.156}$$

which is 1.33% in excess of the exact value given in Eq. (3.147). If, however, a clamped–clamped rod possesses the stiffness that formally coincides with $D_{C-F}(\xi)$, we choose the comparison functions

$$\varphi^{(1)}_{C-C}(\xi) = \sin(\pi\xi) \tag{3.157}$$

$$\varphi^{(2)}_{C-C}(\xi) = \xi - \xi^2 \tag{3.158}$$

The approximate natural frequencies squared read as

$$[\omega^{(1)}_{C-C}]^2 = \left(\tfrac{8}{3}\pi^2 - \tfrac{1}{2}\right)|b_2|/a_0 L^2 \approx 25.82|b_2|/a_0 L^2 \tag{3.159}$$

$$[\omega^{(2)}_{C-C}]^2 = 26|b_2|/a_0 L^2 \tag{3.160}$$

The numerial results are summarized in Table 3.1.

Naturally, the fundamental frequencies in Eqs. (3.156) and (3.157) are close to each other. Both differ from the value in Eq. (3.144) and (Eq. (3.153) as it should be.

TABLE 3.1

Comparison of Closed-Form Solutions with the Boobnov–Galerkin Method

Longitudinal rigidity	Boundary conditions	Comparison functions	Natural frequency $\lvert b_2\rvert/a_0 l^2$		
			Boobnov–Galerkin method	Exact solution	Percentage difference
$D_{C-C}(\xi) = (\frac{1}{2} + \xi - \xi^2)\lvert b_2\rvert$	C–F	$\sin(\frac{1}{2}\pi\xi)$	$\frac{1}{6}\pi^2 \approx 1.64$	–	–
	C–F	$\xi - \frac{1}{2}\xi^2$	$\frac{13}{8} \approx 1.625$	–	–
	C–C	$\sin(\pi\xi)$	$\frac{2}{3}\pi^2 - \frac{1}{2} \approx 6.08$	6	1.33
	C–F	$\sin(\frac{1}{2}\pi\xi)$	$\frac{2}{3}\pi^2 - \frac{1}{2} \approx 6.08$	6	1.33
$D_{C-F}(\xi) = (2 + 2\xi - \xi^2)\lvert b_2\rvert$	C–C	$\sin(\pi\xi)$	$\frac{8}{3}\pi^2 - \frac{1}{2} \approx 25.82$	–	–
	C–C	$\xi - \xi^2$	26.00	–	–

Thus, the conclusion obtained in this chapter that irrespective of the boundary conditions, the analytical expressions for the natural frequencies coincide, ought to be interpreted as follows: the proposed method "screens out," as it were, the *different* inhomogeneous rods that have the same natural frequencies at different boundary conditions. The main result of this chapter is the closed-form solution for the inhomogeneous rods. To the best of the author's knowledge no other closed-form solutions have been reported in the literature, since Bernoulli and Euler established the governing equation of the vibrating rods.

4

Unusual Closed-Form Solutions for Beam Vibrations

This chapter discusses the vibration of inhomogeneous beams under various boundary conditions. Associated aspects of stochastic analysis are also elucidated. Specifically, an infinite number of closed-form solutions is reported for both the free vibrations and reliabilities of vibrating beams. It is demonstrated that the stochastic beams may share deterministic vibration frequencies.

4.1 Apparently First Closed-Form Solutions for Frequencies of Deterministically and/or Stochastically Inhomogeneous Beams (Pinned–Pinned Boundary Conditions)

4.1.1 Introductory Remarks

The aim of this section is to find some closed-form solutions to the dynamic equation of a beam in which both the Young's modulus and the density are polynomial functions, with both the deterministic and stochastic inhomogeneities included. We look for the exact mode shape also as a polynomial function, with an attendant closed-form expression for the natural frequencies. The case considered is that of the beam pinned at both ends. For the bibliography of investigations on vibration and buckling of inhomogeneous beams, one may consult the papers by Eisenberger (1997a,b), and Rollot and Elishakoff (1999).

The importance of the solutions found lies in the possibility of their use as benchmark solutions against which the efficacy of various approximate methods could be ascertained. Additionally, presently there is a considerable literature on the so-called stochastic finite element method (SFEM), which deals with inhomogeneous structures involving random fields. The latter random functions can be represented as mean functions superimposed with deviation functions. The solution of the problem with properly chosen mean functions often constitutes an important part of the analysis (see e.g. Köylüoğlu et al., 1994; Elishakoff et al., 1995). Thus, the closed-form solutions, both in deterministic and stochastic settings possess attractive analytical advantages over approximate solutions, where inherent approximations of various natures are needed. For alternative formulations of the random eigenvalue problem, the reader may consult papers by Shinozuka and Astill (1972) and Zhu and Wu (1991).

4.1.2 Formulation of the Problem

The dynamic behavior of a beam, is described by the following equation

$$\frac{d^2}{dx^2}\left[D(x)\frac{d^2w(x)}{dx^2}\right] - R(x)\omega^2 w(x) = 0 \tag{4.1}$$

where $D(x) = EI$ is the flexural rigidity, E the Young's modulus, ρ the density, I the moment of inertia of the cross section, $R(x) = \rho A$ is the inertial coefficient, A the area of the cross section, $w(x)$ the displacement and ω the natural frequency.

In this section, it is assumed that the cross-sectional area is constant, but both Δ and P are specified as polynomial functions, given by

$$R(\xi) = \sum_{i=0}^{m} a_i \xi^i \tag{4.2}$$

$$D(\xi) = \sum_{i=0}^{n} b_i \xi^i \tag{4.3}$$

where $\xi = x/L$ is a non-dimensional axial coordinate.

We assume that $w(\xi)$ is also polynomial:

$$w(\xi) = \sum_{i=0}^{p} w_i \xi^i \tag{4.4}$$

where w_i are the sought for coefficients. In these expressions, m, n and p are, respectively, the degree of the polynomials for $R(\xi)$, $D(\xi)$ and $w(\xi)$.

Equation (4.1) can be re-written as

$$\frac{d^2}{d\xi^2}\left[D(\xi)\frac{d^2w(\xi)}{d\xi^2}\right] - kL^4 R(\xi)w(\xi) = 0 \tag{4.5}$$

where

$$k = \omega^2 \tag{4.6}$$

As the involved functions are assumed to be polynomial ones, the degrees of each polynomial function must be linked, namely

$$n + (p - 2) - 2 = m + p \tag{4.7}$$

or, simply

$$n - m = 4 \tag{4.8}$$

We observe that Eq. (4.8) does not depend on the degree p of the displacement $w(\xi)$. We arrive at the seemingly unexpected conclusion that any polynomial function for the displacement may be used in Eq. (4.5) if it also satisfies the boundary conditions. This fact will be used at a later stage. In view of Eq. (4.8) the expression for $D(\xi)$ can be written as follows:

$$D(\xi) = \sum_{i=0}^{m+4} b_i \xi^i \tag{4.9}$$

4.1.3 Boundary Conditions

The case of the pinned beam is associated with the following boundary conditions:

$$w(0) = 0 \tag{4.10}$$
$$D(0)w''(0) = 0 \tag{4.11}$$
$$w(1) = 0 \tag{4.12}$$
$$D(1)w''(1) = 0 \tag{4.13}$$

The solution to Eq. (4.11) can be found with either $D(0) = 0$ or $w''(0) = 0$. However, the Young's modulus being zero at one point has no physical sense; thus, Eq. (4.11) is equivalent to $w''(0) = 0$. Hence, we postulate that $b_0 > 0$. The same reasoning can be applied to Eq. (4.13). So, the displacement has to satisfy the following conditions:

$$w(0) = 0 \tag{4.14}$$
$$w''(0) = 0 \tag{4.15}$$
$$w(1) = 0 \tag{4.16}$$
$$w''(1) = 0 \tag{4.17}$$

Satisfaction of the boundary conditions (4.14–4.17) requires that the degree of the displacement polynomial must be at least 4. Assuming that $w(\xi)$ is a fourth-order polynomial

$$w(\xi) = w_0 + w_1\xi + w_2\xi^2 + w_3\xi^3 + w_4\xi^4 \tag{4.18}$$

The satisfaction of the boundary conditions yields

$$w(\xi) = w_1(\xi - 2\xi^3 + \xi^4) \tag{4.19}$$

4.1.4   Expansion of the Differential Equation

By substituting the different expressions for $D(\xi)$, $R(\xi)$ and $w(\xi)$ in Eq. (4.5), we obtain

$$w_1\left[\sum_{i=2}^{m+4} i(i-1)b_i\xi^{i-2}(-12\xi + 12\xi^2) + \sum_{i=0}^{m+4} 24b_i\xi^i + 2\sum_{i=1}^{m+4} ib_i\xi^{i-1}(-12 + 24\xi)\right.$$

$$\left. -kL^4\sum_{i=0}^{m} a_i\xi^i(\xi - 2\xi^3 + \xi^4)\right] = 0 \tag{4.20}$$

The latter expression can be re-written as follows:

$$-12\sum_{i=1}^{m+3} i(i+1)b_{i+1}\xi^i + 12\sum_{i=2}^{m+4} i(i-1)b_i\xi^i + 24\sum_{i=0}^{m+4} b_i\xi^i - 24\sum_{i=0}^{m+3}(i+1)b_{i+1}\xi^i$$

$$+48\sum_{i=1}^{m+4} ib_i\xi^i - kL^4\sum_{i=1}^{m+1} a_{i-1}\xi^i + 2kL^4\sum_{i=3}^{m+3} a_{i-3}\xi^i - kL^4\sum_{i=4}^{m+4} a_{i-4}\xi^i = 0 \tag{4.21}$$

Equation (4.21) has to be satisfied for any ξ. This requirement yields the following relations:

$$-24(b_1 - b_0) = 0 \qquad\qquad\qquad \text{for } i = 0 \tag{4.22}$$

$$-kL^4a_0 + 72(b_1 - b_2) = 0 \qquad\qquad \text{for } i = 1 \tag{4.23}$$

$$-kL^4a_1 + 144(b_2 - b_3) = 0 \qquad\qquad \text{for } i = 2 \tag{4.24}$$

$$L^4(2ka_0 - ka_2) + 240(b_3 - b_4) = 0 \quad \text{for } i = 3 \tag{4.25}$$

$$\cdots$$

$$L^4(2ka_{i-3} - ka_{i-4} - ka_{i-1}) + 12(i+1)(i+2)(b_i - b_{i+1}) = 0 \quad \text{for } 4 \le i \le m+1$$

$$(4.26)$$

$$\cdots$$

$$L^4(2ka_{m-1} - ka_{m-2}) + 12(m+3)(m+4)(b_{m+2} - b_{m+3}) = 0 \quad \text{for } i = m+2$$

$$(4.27)$$

$$L^4(2ka_m - ka_{m-1}) + 12(m+4)(m+5)(b_{m+3} - b_{m+4}) = 0 \quad \text{for } i = m+3$$

$$(4.28)$$

$$- kL^4 a_m + 12(m^2 + 11m + 30)b_{m+4} = 0 \quad \text{for } i = m+4$$

$$(4.29)$$

Note that Eqs. (4.22)–(4.29) are valid only if $m \ge 3$. For cases that satisfy the inequality $m < 3$, the reader is referred to Appendix A. Note also that Eqs. (4.22)–(4.29) have a recursive form.

The sole unknown in Eqs. (4.22)–(4.29) is the natural frequency coefficient k; yet, we observe that we have $m+5$ equations. We conclude that the parameters b_i and a_i have to satisfy some auxiliary conditions so that Eqs. (4.22)–(4.29) are compatible.

4.1.5   Compatibility Conditions

A first compatibility condition is given by Eq. (4.22), leading to $b_0 = b_1$. From the other equations, several expressions for k can be found. Its values determined from Eqs. (4.22)–(4.29) respectively, are listed below:

$$k = 72(b_1 - b_2)/L^4 a_0 \qquad (4.30)$$

$$k = 144(b_2 - b_3)/L^4 a_1 \qquad (4.31)$$

$$k = 240(a_2 - 2a_0)^{-1}(b_3 - b_4)L^4 \qquad (4.32)$$

$$\cdots$$

$$k = 12(i+1)(i+2)(a_{i-1} + a_{i-4} - 2a_{i-3})^{-1}[b_i - b_{i+1}]/L^4$$
$$\text{for } 4 \le i \le m+1 \qquad (4.33)$$

$$\cdots$$

$$k = 12(m+3)(m+4)(a_{m-2} - 2a_{m-1})^{-1}[b_{m+2} - b_{m+3}]/L^4 \qquad (4.34)$$

$$k = 12(m+4)(m+5)(a_{m-1} - 2a_m)^{-1}[b_{m+3} - b_{m+4}]/L^4 \qquad (4.35)$$

$$k = 12(m^2 + 11m + 30)b_{m+4}/L^4 a_m \qquad (4.36)$$

To check the compatibility of these expressions, all expressions for k have to be equal to each other. We consider two separate problems: (i) the material density coefficients a_i are specified; find coefficients b_i so that the closed-form solution holds; (ii) the elastic modulus coefficients b_i are specified; find coefficients a_i so that the closed-form solution is obtainable.

4.1.6   Specified Inertial Coefficient Function

Let us assume that the function $R(\xi)$, of the inertial coefficient, and hence all a_i ($i = 0, 2, \ldots, m$) are given. Let us observe that if b_{m+4} is specified then the expression given in Eq. (4.36) is the final formula for the natural frequency coefficient k. Then, Eqs. (4.30)–(4.36) allow an evaluation of the remaining parameters b_i. Note that b_{m+4} and a_m have to have the same sign due to the positivity of k.

From Eq. (4.35) we get

$$b_{m+3} = \left\{ \left[\frac{m^2 + 11m + 30}{(m+4)(m+5)} \right] \left(\frac{a_{m-1}}{a_m} - 1 \right) + 1 \right\} b_{m+4} \qquad (4.37)$$

Equation (4.34) yields

$$b_{m+2} = \left[\left(\frac{m+5}{m+3} \right) \frac{a_{m-2} - 2a_{m-1}}{a_{m-1} - 2a_m} + 1 \right] b_{m+3} - \left(\frac{m+5}{m+3} \right) \frac{a_{m-2} - 2a_{m-1}}{a_{m-1} - 2a_m} b_{m+4} \qquad (4.38)$$

Equation (4.33) results in

$$b_i = \left[\left(\frac{i+3}{i+1} \right) \frac{a_{i-1} - a_{i-4} - 2a_{i-3}}{a_i - a_{i-3} - 2a_{i-2}} + 1 \right] b_{i+1} - \left(\frac{i+3}{i+1} \right) \frac{a_{i-1} - a_{i-4} - 2a_{i-3}}{a_i - a_{i-3} - 2a_{i-2}} b_{i+2} \qquad (4.39)$$

where i belongs to the set $\{4, 5, \ldots, m+1\}$.

From Eq. (4.32) we obtain

$$b_3 = \left[\left(\frac{3}{2} \right) \frac{a_2 - 2a_0}{a_3 - a_0 - 2a_1} + 1 \right] b_4 - \left(\frac{3}{2} \right) \frac{a_2 - 2a_0}{a_3 - a_0 - 2a_1} b_5 \qquad (4.40)$$

Equation (4.31) leads to

$$b_2 = \left[\left(\frac{5}{3} \right) \frac{a_1}{a_2 - 2a_0} + 1 \right] b_3 - \left(\frac{5}{3} \right) \frac{a_1}{a_2 - 2a_0} b_4 \qquad (4.41)$$

From Eq. (4.29) we get

$$b_1 = \left(2\frac{a_0}{a_1} + 1 \right) b_2 - 2\frac{a_0}{a_1} b_3 \qquad (4.42)$$

And, finally, Eq. (4.22) yields

$$b_0 = b_1 \qquad (4.43)$$

Thus, for specified coefficients $a_0, a_1, \ldots, a_m$ and b_{m+4}, Eqs. (4.37)–(4.43) lead to the set of coefficients of the elastic modulus such that the beam possesses the mode shape given in Eq. (4.19). Note that if $a_i = a$, then the coefficients b_i do not depend on the parameter a.

To sum up, if

$$R(\xi) = \sum_{i=0}^{m} a_i \xi^i \quad D(\xi) = \sum_{i=0}^{m+4} b_i \xi^i \qquad (4.44)$$

where b_i are computed via Eqs. (4.37)–(4.43), the fundamental mode shape of a beam is

$$w(\xi) = w_1(\xi - 2\xi^3 + \xi^4) \qquad (4.45)$$

and the fundamental natural frequency squared reads

$$\omega^2 = 12(m^2 + 11m + 30)b_{m+4}/a_m L^4 \qquad (4.46)$$

As we have seen, in order to obtain the closed-from solution it is sufficient that (1) all coefficients a_i and (2) the coefficient b_{m+4} be specified. Yet, the requirements are not necessary ones. Indeed, one can assume that the coefficients a_i are given and, instead of b_{m+4}, any coefficient b_j ($j \neq m+4$) is specified. If this is the case, then from Eq. (4.33) one expresses b_{i+1} via b_i and k; substitution into subsequent equations allows us to express b_{m+2}, b_{m+3}, b_{m+4} via b_i; analogously, substitution of b_i into Eqs. (4.30)–(4.32) yields the sought for exact solutions.

In Tables 4.1 and 4.2, some sample specified function $D(x)$ and the attendant fundamental natural frequency coefficients are given. The polynomial functions $R_j(\xi)$ were specified as

$$R_1(\xi) = \sum_{i=0}^{m} \xi^i \quad R_2(\xi) = \sum_{i=0}^{m} (i+1)\xi^i \qquad (4.47)$$

respectively, in Tables 4.1 and 4.2.

4.1.7 Specified Flexural Rigidity Function

Consider, now, the case when the flexural rigidity function is specified, implying that all b_i ($i = 0, \ldots, m+4$) are given. The following question arises: is it possible to determine the material density coefficients a_i ($i = 0, \ldots, m$), such that the equations corresponding to Eqs. (4.22)–(4.29) are compatible? One

TABLE 4.1

Flexural Rigidities and Eigenvalues Corresponding to Different Values of m ($R_1(\xi)$ Given in Eq. (4.47.1))

m	$D_j(\xi)$	KL^4
0	$b(3 + 3\xi - 2\xi^2 - 2\xi^3 + \xi^4)$	$360b$
1	$(b/10)(59 + 59\xi - 11\xi^2 - 46\xi^3 - 4\xi^4 + 10\xi^5)$	$504b$
2	$(b/15)(135 + 135\xi - 5\xi^2 - 75\xi^3 - 33\xi^4 - 5\xi^5 + 15\xi^6)$	$672b$
3	$(b/35)(434 + 434\xi + 14\xi^2 - 196\xi^3 - 70\xi^4 - 70\xi^5 - 10\xi^6 + 35\xi^7)$	$864b$
4	$(b/28)(452 + 452\xi + 32\xi^2 - 178\xi^3 - 52\xi^4 - 52\xi^5 - 52\xi^6 - 7\xi^7 + 28\xi^8)$	$1{,}080b$
5	$(b/36)(648 + 648\xi + 69\xi^2 - 224\xi^3 - 56\xi^4 - 56\xi^5 - 56\xi^6 - 56\xi^7 - 8\xi^8 + 36\xi^9)$	$1{,}320b$
6	$(b/15)(371 + 371\xi + 41\xi^2 - 124\xi^3 - 25\xi^4 - 25\xi^5 - 25\xi^6 - 25\xi^7 - 25\xi^8 - 3\xi^9 + 15\xi^{10})$	$1{,}584b$
7	$(b/55)(1{,}628 + 1{,}628\xi + 198\xi^2 - 517\xi^3 - 88\xi^4 - 88\xi^5 - 88\xi^6 - 88\xi^7 - 88\xi^8 - 88\xi^9 - 10\xi^{10} + 55\xi^{11})$	$1{,}872b$
8	$(b/330)(5{,}715 + 57{,}151\xi + 1{,}492\xi^2 - 3{,}513\xi^3 - 510\xi^4 - 510\xi^5 - 510\xi^6 - 510\xi^7 - 510\xi^8 - 510\xi^9 - 510\xi^{10} - 55\xi^{11} + 330\xi^{12})$	$2{,}184b$
9	$(b/26)(1{,}053 + 1{,}053\xi + 143\xi^2 - 312\xi^3 - 39\xi^4 - 39\xi^5 - 39\xi^6 - 39\xi^7 - 39\xi^8 - 39\xi^9 - 39\xi^{10} - 39\xi^{11} - 4\xi^{12} + 26\xi^{13})$	$2{,}520b$

TABLE 4.2

Flexural Rigidities and Eigenvalues Corresponding to Different Values of m ($R_1(\xi)$ Given in Eq. (4.47.2))

m	$D_j(\xi)$	KL^4
0	$b(3 + 3\xi - 2\xi^2 - 2\xi^3 + \xi^4)$	$360b$
1	$(b/10)(38 + 38\xi + 3\xi^2 - 32\xi^3 - 11\xi^4 + 10\xi^5)$	$252b$
2	$(b/45)(203 + 203\xi + 63\xi^2 - 77\xi^3 - 119\xi^4 - 35\xi^5 + 45\xi^6)$	$224b$
3	$(b/140)(725 + 725\xi + 305\xi^2 - 115\xi^3 - 241\xi^4 - 325\xi^5 - 85\xi^6 + 140\xi^7)$	$216b$
4	$(b/140)(815 + 815\xi + 395\xi^2 - 25\xi^3 - 151\xi^4 - 235\xi^5 - 295\xi^6 - 70\xi^7 + 140\xi^8)$	$216b$
5	$(b/1{,}512)(9{,}751 + 9{,}751\xi + 5{,}131\xi^2 + 511\xi^3 - 875\xi^4 - 1{,}799\xi^5 - 2{,}549\xi^6 - 2{,}954\xi^7 - 644\xi^8 + 1{,}512\xi^9)$	$220b$
6	$(b/1{,}470)(10{,}388 + 10{,}388\xi + 5{,}768\xi^2 + 1{,}148\xi^3 - 238\xi^4 - 1{,}162\xi^5 - 1{,}822\xi^6 - 1{,}617\xi^7 - 2{,}702\xi^8 - 546\xi^9 + 1{,}470\xi^{10})$	$\dfrac{1584}{7}b$

immediately observes that there are $(m + 5)$ equations (4.22)–(4.29), while one has only $m + 1$ unknowns, $a_0, a_1, \ldots, a_m$. In actuality, however, one has only m unknowns. In order for the process of determining the coefficient a_i to proceed, one of the coefficients a_j should be specified. The most convenient assumption is to fix either a_0 or a_1 or a_m, since in these cases only one equation, respectively, Eq. (4.30) or Eq. (4.31) or Eq. (4.36) will be sufficient to determine the sought for expression for the natural frequency coefficient. Let us assume that the coefficient a_0 is given; thus, to check the compatibility of Eqs. (4.22)–(4.29), four b_i coefficients cannot be chosen arbitrarily.

Note that the natural frequency coefficient k has to be positive; thus, the difference $b_1 - b_2$ and the coefficient a_0 have to have the same sign. Moreover, as the coefficient a_0 is positive, the difference $b_1 - b_2$ should be positive. So, for $b_1 > b_2$, one substitutes the value of k determined from Eq. (4.30) into Eq. (4.31); this allows as to determine the coefficient a_1 so that the frequency coefficient k in Eq. (4.31) is positive, and so on.

First, Eq. (4.22) leads to

$$b_0 = b_1 \tag{4.48}$$

From Eq. (4.31), we get

$$a_1 = 2a_0 \frac{b_3 - b_2}{b_2 - b_1} \tag{4.49}$$

Then, Eq. (4.32) yields

$$a_2 = \frac{5a_1(b_4 - b_3) + 6a_0(b_3 - b_2)}{3(b_3 - b_2)} \tag{4.50}$$

Equation (4.33), where $i = 4$, results in

$$a_3 = \frac{a_0(4b_4 - 6b_5 + 2b_3) + a_1(4b_4 - 4b_3) + a_2(3b_5 - 3b_4)}{2(b_4 - b_3)} \tag{4.51}$$

From Eq. (4.33), where $4 \leq i \leq m$, we obtain

$$a_i = \frac{1}{(i+1)(b_{i+1} - b_i)} \{ a_{i-1}(i+3)(b_{i+2} - b_{i+1}) + 2a_{i-2}(i+1)(b_{i+1} - b_i)$$
$$+ a_{i-3}[b_{i+1}(i+5) - 2b_{i+2}(i+3) + b_i(i+1)] + a_{i-4}(i+3)(b_{i+2} - b_{i+1}) \} \tag{4.52}$$

Then, from Eqs. (4.36) and (4.30), one can find an expression for $b_m + 4$, so that, the compatibility of Eqs. (4.22)–(4.29) can be checked:

$$b_{m+4} = \frac{6a_m(b_1 - b_2)}{a_0(m^2 + 11m + 30)} \tag{4.53}$$

From Eqs. (4.35) and (4.36), a relation for b_{m+3} can be found

$$b_{m+3} = \frac{a_0 b_{m+4}(m^2 + 9m + 20) + 6a_{m-1}(b_1 - b_2) + 12a_m(b_2 - b_1)}{a_0(m+4)(m+5)} \tag{4.54}$$

Finally, Eqs. (4.34) and (4.35) yield an evaluation of b_{m+2}

$$b_{m+2} = \frac{a_0 b_{m+3}(m+3)(m+4) + 6a_{m-2}(b_1 - b_2) + 12a_{m-1}(b_2 - b_1)}{a_0(m+3)(m+4)} \tag{4.55}$$

To summarize, while specifying the elastic modulus function, only $m + 1$ coefficients b_i can be chosen arbitrarily; the remaining four coefficients are connected with the arbitrary ones via Eq. (4.48) and Eqs. (4.53)–(4.55).

Thus, if

$$R(\xi) = \sum_{i=0}^{m} a_i \xi^i \quad D(\xi) = \sum_{i=0}^{m+4} b_i \xi^i \tag{4.56}$$

where a_i and four of the coefficients b_i are computed via Eqs. (4.48)–(4.55), the fundamental mode shape of a beam is

$$w(\xi) = w_1(\xi - 2\xi^3 + \xi^4) \tag{4.57}$$

The fundamental natural frequency squared reads as

$$\omega^2 = 72(b_1 - b_2)/a_0 L^4 \tag{4.58}$$

The closed-form solutions could be utilized for comparison with approximate techniques. For example, utilization of the single term Boobnov–Galerkin method for the case

$$R(\xi) = 1 + 2\xi + 3\xi^2 + 4\xi^3 + 5\xi^4 \tag{4.59}$$

$$D(\xi) = b_8 L^8 \left(\tfrac{163}{28} + \tfrac{163}{28}\xi + \tfrac{79}{28}\xi^2 - \tfrac{5}{28}\xi^3 - \tfrac{151}{140}\xi^4 - \tfrac{47}{28}\xi^5 - \tfrac{59}{28}\xi^6 - \tfrac{1}{2}\xi^7 + \xi^8 \right) \tag{4.60}$$

yields, with $\sin(\pi\xi)$ taken as a comparison function, the following expression

$$k = \frac{b_8 L^4}{22{,}050} \left(\frac{6{,}945{,}750 - 409{,}185\pi^4 + 391{,}612\pi^8 - 2{,}716{,}875\pi^2 + 17{,}356\pi^6}{10\pi^4 - 19\pi^2 + 15} \right) \tag{4.61}$$

or numerically $k = 216.29697 b_8 L^4$, which differs from the exact solution $k = 216 b_8 L^4$ by 0.13%.

4.1.8 Stochastic Analysis

The preceding formulation allows one to perform a stochastic analysis to account for the possible randomness in the material density and elastic modulus.

4.1.8.1 *Probabilistically specified inertial coefficient function*

Assume that the coefficients a_i form a random vector with a joint probability density $f_A(a_1, a_2, \ldots, a_m)$, where $A^T = (A_1, A_2, \ldots, A_m)$ and capital letters denote a random variable A_i whose possible values are denoted by the lowercase letters a_i. As Eq. (4.36) suggests, the natural frequency squared, Ω^2, is also a random variable denoted by

$$\Omega^2 = \alpha^2 B_{m+4}/A_m$$
$$\alpha^2 = 12(m^2 + 11m + 30) \tag{4.62}$$

where the coefficient B_{m+4} constitutes either a deterministic or a random variable. Several cases allow a closed-form evaluation of the reliability r, defined in the present circumstances as the probability that the natural frequency squared Ω^2 does not exceed a pre-selected deterministic value ω_0^2:

$$r = \text{Prob}(\Omega^2 \leq \omega_0^2) = \text{Prob}(\alpha^2 B_{m+4} \leq \omega_0^2 A_m) \tag{4.63}$$

Let $B_{m+4} \equiv B$ be an exponentially distributed random variable with density

$$f_B(b) = \frac{1}{E(B)} \exp\left[-\frac{b}{E(B)}\right] \qquad b \geq 0 \tag{4.64}$$

and zero otherwise, $E(B)$ being the mean value of B. Likewise, the coefficient $A_m \equiv A$ has an exponential density

$$f_A(a) = \frac{1}{E(A)} \exp\left[-\frac{a}{E(A)}\right] \qquad a \geq 0 \tag{4.65}$$

and vanishes if $a < 0$, with $E(A)$ indicating the mean value of A. Since A is exponentially distributed, the random variable $\omega_0^2 A$ is also exponentially distributed with mean $\omega_0^2 E(A)$. Likewise $\alpha^2 B$ is an exponential random variable with mean $\alpha^2 E(B)$. The reliability is obtained as

$$r = \frac{\omega_0^2 E(A)}{\omega_0^2 E(A) + \alpha^2 E(B)} \tag{4.66}$$

It is remarkable that although all coefficients A_j $(j = 1, \ldots, m)$ are random, the probabilistic characterization of only a single coefficient A_m turns out to be needed, in addition to that of B_{m+4} for the reliability evaluation.

4.1.8.2 *Specified flexural rigidity function*

Assume now that the $m + 1$ coefficients b_i form a random vector with a joint probability density $f_B(b_1, \ldots, b_{m+1})$. The remaining four coefficients b_0, b_{m+2},

b_{m+3}, b_{m+4} are related to the above coefficients via Eqs. (4.48) and (4.53)–(4.55). Due to the randomness of coefficients $B_1, \ldots, B_{m+1}$, we conclude that the natural frequency squared is itself a random variable

$$\Omega^2 = \sigma^2(B_1 - B_2) \qquad (4.67)$$

where σ^2 is a coefficient

$$\sigma^2 = 72/L^4 a_0 \qquad (4.68)$$

The coefficient a_0 in Eq. (4.68) can be treated either as a deterministic or a random variable. For the sake of illustration, a particular case will be considered now, namely, the case when a_0 is a deterministic variable. The mean natural frequency squared equals

$$E[\Omega^2] = \sigma^2[E(B_1) - E(B_2)] \qquad (4.69)$$

whereas its variance reads as

$$\text{Var}[\Omega^2] = \sigma^4[\text{Var}(B_1) + \text{Var}(B_2)] \qquad (4.70)$$

As in Eq. (4.63) reliability is defined as a probability that the natural frequency squared, Ω^2, does not exceed a pre-selected value ω_0^2. Once the joint probability density of the coefficients b_1, b_2 and a_0 are specified, the reliability function r can be derived directly. The reliability is cast as

$$r = \text{Prob}[\sigma^2(B_1 - B_2) < \omega_0^2] \qquad (4.71)$$

A new random variable is introduced:

$$Z = \sigma^2(B_1 - B_2) - \omega_0^2 \qquad (4.72)$$

The reliability is re-written as

$$r = \text{Prob}(Z < 0) = \int_{-\infty}^{\infty} db_2 \left[\int_{-\infty}^{b_2 + \omega_0^2/\sigma^2} f_{B_2 B_1}(b_2, b_1)\, db_1 \right.$$

$$\left. - \int_{-\infty}^{b_2} f_{B_2 B_1}(b_2, b_1)\, db_1 \right] \qquad (4.73)$$

Let B_1 and B_2 be independent random variables; then, Eq (4.73) becomes

$$r = \int_{-\infty}^{\infty} f_{B_2}(b_2) \left[F_{B_1}\left(b_2 + \frac{\omega_0^2}{\sigma^2}\right) - F_{B_1}(b_2) \right] db_2 \qquad (4.74)$$

$$F_{B_1}\left(b_2 + \frac{\omega_0^2}{\sigma^2}\right) = \int_{-\infty}^{b_2 + \omega_0^2/\sigma^2} f_{B_1}(b_1) \, db_1 \qquad (4.75)$$

$$F_{B_1}(b_2) = \int_{-\infty}^{b_2} f_{B_1}(b_1) \, db_1$$

Let B_1 be a uniformly distributed random variable with density

$$f_{B_1}(b_1) = (\beta - \alpha)^{-1} \qquad \text{if } b_1 \in [\alpha, \beta] \qquad (4.76)$$

and zero otherwise; likewise, the coefficient B_2 has a uniform density

$$f_{B_2}(b_2) = (\delta - \gamma)^{-1} \qquad \text{if } b_2 \in [\gamma, \delta] \qquad (4.77)$$

Let us assume that $\alpha > \delta$. Thus, the positivity of Ω^2 is always assured since $B_1 > B_2$. We first calculate the expressions in Eq. (4.75):

$$F_{B_1}\left(b_2 + \frac{\omega_0^2}{\sigma^2}\right) = \begin{cases} 0 & \text{if } b_2 + \dfrac{\omega_0^2}{\sigma^2} \leq \alpha \\[2mm] \dfrac{b_2 - \left(\alpha - \dfrac{\omega_0^2}{\sigma^2}\right)}{\beta - \alpha} & \text{if } \alpha < b_2 + \dfrac{\omega_0^2}{\sigma^2} < \beta \\[2mm] 1 & \text{if } b_2 + \dfrac{\omega_0^2}{\sigma^2} \geq \beta \end{cases} \qquad (4.78)$$

$$F_{B_1}(b_2) = 0$$

From Eq. (4.74) it follows that for the reliability evaluation we need to find a region in which both f_{B_2} and F_{B_1} are non-zero. As Eq. (4.77) suggests f_{B_2} is non-zero if $\gamma \leq b_2 \leq \delta$. The function F_{B_1} differs from both zero and unity if $\alpha - \omega_0^2/\sigma^2 < b_2 < \beta - \omega_0^2/\sigma^2$; F_{B_1} equals unity if $b_2 > \beta - \omega_0^2/\sigma^2$. Thus, in order for the product $f_{B_2} F_{B_1}$ to be non-zero it is necessary and sufficient that b_2 belongs to the two following intervals:

$$I_1 = [\gamma, \delta] \qquad I_2 = [\alpha - \omega_0^2/\sigma^2, \infty] \qquad (4.79)$$

It is natural to inquire when these two intervals have no intersection. This takes place when the lower end of interval I_2 exceeds the upper-end of interval I_1, i.e., when $\alpha - \omega_0^2/\sigma^2 \geq \delta$, or in terms of ω_0^2, when $\omega_0^2 \leq \sigma^2(\alpha - \delta)$ (see Figure 4.1). Under these circumstances, the integrand in Eq. (4.69), and hence the reliability r both vanish identically.

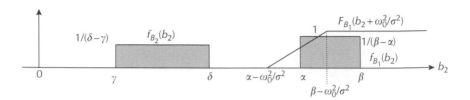

FIGURE 4.1
Probability density functions of variables B_1 and B_2, and the probability distribution function F_{B_1} for $\omega_0^2 \leq \sigma^2(\alpha - \delta)$, leading to zero reliability

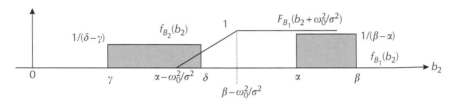

FIGURE 4.2
Probability density functions of variables B_1 and B_2, and the probability distribution function F_{B_1} for $\beta - \omega_0^2/\sigma^2 \geq \delta$; the reliability is given by Eq. (4.80)

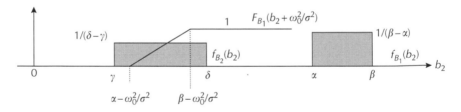

FIGURE 4.3
Probability density functions of variables B_1 and B_2, and the probability distribution function F_{B_1} for $\alpha - \omega_0^2/\sigma^2 \geq \gamma$ and $\gamma \leq \beta - \omega_0^2/\sigma^2 \leq \delta$; the reliability is given by Eq. (4.81)

Assume now that the lower end of the interval I_2 belongs to the interval I_1, i.e. $\gamma \leq \alpha - \omega_0^2/\sigma^2 \leq \delta$, and, moreover, $\beta - \omega_0^2/\sigma^2 \geq \delta$ (Figure 4.2).

This implies that $\sigma^2(\alpha - \delta) < \omega_0^2 \leq \sigma^2(\beta - \delta)$. The sought for region for b_2 is the interval $[\alpha - \omega_0^2/\sigma^2, \delta]$. The reliability is obtained as

$$r = \int_{\alpha - \omega_0^2/\sigma^2}^{\delta} \frac{b_2 - (\alpha - (\omega_0^2/\sigma^2))}{(\delta - \gamma)(\beta - \alpha)} db_2 = \frac{[\sigma^2(\delta - \alpha) + \omega_0^2]}{2\sigma^4(\delta - \gamma)(\beta - \alpha)} \qquad (4.80)$$

Now, consider the following case: $\alpha - \omega_0^2/\sigma^2 \geq \gamma$ and $\gamma \leq \beta - \omega_0^2/\sigma^2 \leq \delta$ (Figure 4.3) implying $\sigma^2(\alpha - \gamma) < \omega_0^2 \leq \sigma^2(\beta - \gamma)$.

The b_2 region is the interval $[\gamma, \delta]$. But to evaluate the reliability, this region has to be split into the union of two regions $[\gamma, \beta - \omega_0^2/\sigma^2]$, $[\beta - \omega_0^2/\sigma^2, \delta]$ since

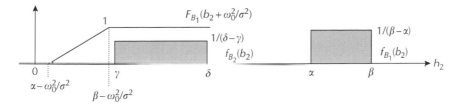

FIGURE 4.4
Probability density functions of variables B_1 and B_2, and the probability distribution function F_{B_1} for $\beta - \omega_0^2/\sigma^2 \le \gamma$, leading to unit reliability

the function $F_{B_1}(b_2 + \omega_0^2/\sigma^2)$ takes the value unity at $b_2 = \beta - \omega_0^2/\sigma^2$. Hence, the reliability reads as

$$
\begin{aligned}
r &= \int_r^{\beta-\omega_0^2/\sigma^2} \frac{b_2 - (\alpha - \omega_0^2/\sigma^2)}{(\delta - \gamma)(\beta - \alpha)} db_2 + \int_{\beta-\omega_0^2/\sigma^2}^{\delta} \frac{1}{\delta - \gamma} db_2 \\
&= \frac{(\sigma^2\beta - \omega_0^2) + 2\sigma^4\delta(\alpha - \beta) + \sigma^4\gamma(\sigma^2\gamma - 2\alpha\sigma^2 + 2\omega_0^2)}{2\sigma^4(\delta - \gamma)(\alpha - \beta)}
\end{aligned}
\tag{4.81}
$$

We now consider the case $\beta - \omega_0^2/\sigma^2 \le \gamma$ (Figure 4.4), meaning $\sigma^2(\beta - \gamma) < \omega_0^2$. The integration domain is the interval $[\gamma, \delta]$:

$$
r = \int_\gamma^\delta 1 \frac{1}{\delta - \gamma} db_2 = 1
\tag{4.82}
$$

There are two intermediate situations between the ones depicted in Figures 4.2 and 4.3, depending on the lengths of the intervals $\beta - \alpha$ and $\delta - \gamma$. If the length $\beta - \alpha$ of the interval b_1 is smaller than the length $\gamma - \delta$ of the interval b_2, then the two quantities, $\alpha - \omega_0^2/\sigma^2$ and $\beta - \omega_0^2/\sigma^2$ belong to the interval b_2 as shown in Figure 4.5. The reliability is given by

$$
\begin{aligned}
r &= \int_{\alpha-\omega_0^2/\sigma^2}^{\beta-\omega_0^2/\sigma^2} \frac{b_2 - (\alpha - \omega_0^2/\sigma^2)}{(\delta - \gamma)(\beta - \alpha)} db_2 + \int_{\beta-\omega_0^2/\sigma^2}^{\delta} \frac{1}{\delta - \gamma} db_2 \\
&= \frac{\sigma^2(2\delta - \alpha - \beta) + 2\omega_0^2}{2\sigma^2(\delta - \gamma)}
\end{aligned}
\tag{4.83}
$$

If, however, the length $\beta - \alpha$ of the interval b_1 is larger than the length $\gamma - \delta$ of the interval b_2, the sought for region is as illustrated in Figure 4.6, and the reliability is expressed as follows:

$$
r = \int_r^\delta \frac{b_2 - (\alpha - \omega_0^2/\sigma^2)}{(\delta - \gamma)(\beta - \alpha)} db_2 = \frac{\delta^2 - \gamma^2}{2(\delta - \gamma)(\beta - \alpha)} + \frac{\omega_0^2 - \sigma^2\alpha}{\sigma^2(\beta - \alpha)}
\tag{4.84}
$$

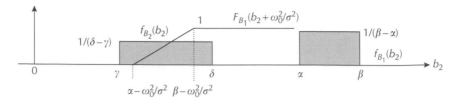

FIGURE 4.5
Case when the length $\beta - \alpha$ of the interval B_1 is smaller than the length of the interval B_2; the reliability is given in Eq. (4.83)

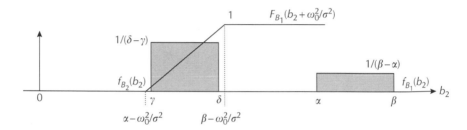

FIGURE 4.6
Case when the length $\beta - \alpha$ of the interval B_1 is larger than the length of the interval B_2; the reliability is given in Eq. (4.84)

The coefficients of variations c_1 and c_2 of the random variables B_1 and B_2, respectively:

$$c_1 = \frac{\sqrt{\text{Var}(B_1)}}{E(B_1)} \qquad c_2 = \frac{\sqrt{\text{Var}(B_2)}}{E(B_2)} \qquad (4.85)$$

are chosen to be equal — $c_1 = c_2 = c$. This implies that for a specified coefficient of variation c, the upper bounds of the interval are related to the lower bounds, as follows

$$\beta = \alpha\frac{1 + \sqrt{3}c}{1 - \sqrt{3}c} \qquad \delta = \gamma\frac{1 + \sqrt{3}c}{1 - \sqrt{3}c} \qquad (4.86)$$

α is fixed at 14, γ is fixed at 1. For the coefficient of variation 0.3, the reliability of unity is manifested for $y \geq 43.286$. Values associated with transition to unit reliability for the coefficients of variation of 0.4 and 0.5 are, respectively, $y \geq 76.151$ and $y \geq 193.994$.

It should be remarked that the transitional values of y from non-unit to unit reliability can be predicted for the uniformly distributed B_1 and B_2, *without* resort to the reliability calculations. As the natural frequency squared is proportional to the difference between B_1 and B_2 in Eq. (4.67), the largest value of the natural frequency is obtained when B_1 takes on the value of the upper bound of the interval, β, whereas B_2 takes on the value of the lower bound of the interval, γ. Thus, if the maximum natural frequency squared $\omega_{max}^2 = \sigma^2(\beta - \gamma)$ is smaller than $y = \omega_0^2$, then the inequality in Eq. (4.71) is satisfied automatically; hence, the reliability is identically unity. Likewise, the minimum natural frequency squared $\omega_{min}^2 = \sigma^2(\alpha - \delta)$. If this value exceeds $y = \omega_0^2$, then obviously for any pairs of values in the intervals B_1 and B_2, the inequality (4.71) will be violated, with resulting vanishing reliability.

On the other hand, for exponentially distributed variables, unit reliability is never achieved, as Eq. (4.66) suggests.

4.1.9   Nature of Imposed Restrictions

In this chapter, in order to obtain the closed-form solutions for natural frequencies deterministically and/or stochastically for an inhomogeneous pinned beam, the flexural rigidity and the inertial coefficient were assumed to be polynomial functions whose powers differ by four. One should stress that the a and b coefficients in Eqs. (4.2) and (4.3) cannot be specified independently in order for a closed-form solution to exist.

One should stress that the importance of the derived closed-form solution is not diminished by the fact that certain conditions must be met. The appearance of conditions is natural too. Indeed, it can be expected that the solution of the *inverse* problem would depend upon either part of or the entire given data. Thus, if the inhomogeneous beam has a polynomial inertial coefficient with given coefficients, it must be no surprise that the sought for flexural rigidity of the beam possesses the pre-selected mode shape, which is directly related to the specified inertial coefficients, in order to derive the closed-form solutions.

The following question arises: is there any resemblance in earlier literature to the type of thinking adopted in this chapter? The connection with previous work was found via Saint-Venant's *semi-inverse* method.

4.1.10   Concluding Remarks

The class of deterministic and stochastic solutions described contains an infinite number of closed-form solutions. Indeed, the degree m of the polynomial in the expression for the inertial coefficient in Eq. (4.44) can be chosen arbitrarily. Likewise, the coefficients a_i can be prescribed at will subject to a condition of positivity of both $P(\xi)$ and $\Delta(\xi)$.

It should be noted that there is a connection between the present work and the subject of "inverse problems" of vibration (Gladwell, 1986a–c, 1996). Indeed, whereas mathematical "direct problems" consist of finding solutions to equations with known *input* parameters, mathematical "inverse problems" deal with the reconstruction of the parameters of the governing equations when the *output* quantities are known. According to Gladwell (1986), "inverse problems are concerned with the construction of a model of a given type; e.g. a mass-spring system, a string, etc; which has given eigenvalues and/or eigenvectors or eigenfunctions, i.e. given *spectral* data. In general, if some such spectral data are given, there can be no system, a unique system, or many systems, having these properties." It is remarkable as the present section demonstrates, that there exist infinite beams, corresponding to $m = 0, 1, 2, \ldots$, that possess the fundamental mode given in Eq. (4.19).

The natural question arises: Is it possible to formulate the problem so as to obtain a unique solution? The reply to this question is in the affirmative. Indeed, one can pre-select not only the fundamental mode shape, but also the fundamental natural frequency denoted by ω_1. Then, Eq. (4.46) yields the coefficient b_{m+4} that accomplishes this goal:

$$b_{m+4} = \omega_1^2 a_m L^4 / 12(m^2 + 11m + 30) \qquad (4.87)$$

The polynomial expressions have been used prior to this study in deterministic analyses; yet, to the best of the authors' knowledge, this is the first collection of closed-form results in either the deterministic or the probabilistic setting for the natural frequencies and associated reliabilities of inhomogeneous beams.

It is also notable that whereas in the usual so called stochastic finite element method, only small coefficients of variation can be allowed, the present formulation is not bound to small coefficients of variation. Therefore, the deterministic and probabilistic closed-form solutions that were uncovered in this chapter, can be utilized as benchmark solutions.

4.2　Apparently First Closed-Form Solutions for Inhomogeneous Beams (Other Boundary Conditions)

4.2.1　Introductory Remarks

There are several articles that deal with the vibrations of beams that are non-uniform along their axes. Usually, the variation is attributed to the cross-sectional area. Then, for specific analytic expressions of such deterministic variations, exact solutions are given in terms of special functions. The first solution for the natural frequency of a tapered beam, that of the wedge, was pioneered by Kirchhoff (1882). The solution was given in terms of Bessel functions. Several other solutions of Kirchhoff's type followed. The appropriate bibliography of problems solved in terms of Bessel functions was given by Naguleswaran (1994a,b). The case where hypergeometric functions arise was discussed by other authors. In some special cases, for beams that are clamped at both ends, a transformation of the dependent variable is possible so that the tapered beam shares a natural frequency of the uniform one (Abrate,

1995). There are fewer studies that deal, in various approximate settings, with vibrations of beams with random inhomogeneities. For example, Shinozuka and Astill (1972), and Manohar and Keane (1993) considered several random eigenvalue problems via analytical methods. The finite element method in the stochastic setting was applied, amongst others, by Hart and Collins (1970), Nakagari et al. (1987), and Zhu and Wu (1991). For extensive bibliography devoted to vibrations of discrete and continuous structures with random parameters one may consult the extensive review by Ibrahim (1987) and its recent up-date by Manohar and Ibrahim (1997).

The present section deals with beams which exhibit inhomogeneity both in the material density and in the elastic modulus. These inhomogeneities are described in terms of polynomial functions. Simple closed-form expressions for both mode shapes and fundamental natural frequencies are uncovered for a special class of problems. Then, the problem is considered in the probabilistic setting with an attendant, seemingly paradoxical conclusion: beams with random properties may possess the deterministic fundamental natural frequencies. An extensive numerical study is conducted to substantiate this finding.

4.2.2 Formulation of the Problem

Consider a non-uniform beam of length L, with cross-sectional area A, moments of inertia, I, that are constant, and variable material density $\rho(x)$ and modulus of elasticity $E(x)$. The beam's vibrations are governed by the Bernouilli–Euler equation:

$$\frac{\partial^2}{\partial x^2}\left[E(x)I\frac{\partial^2 w(x,t)}{\partial x^2}\right] + \rho(x)A\frac{\partial^2 w(x,t)}{\partial x^2} = 0 \qquad (4.88)$$

where $w(x,t)$ is the displacement, x the axial coordinate and t the time.

We introduce the non-dimensional coordinate $\xi = x/L$, as well as consider harmonic vibration, so that the displacement $w(\xi, t)$ is represented as follows:

$$w(\xi, t) = W(\xi)e^{i\omega t} \qquad (4.89)$$

where $W(\xi)$ is the mode shape and ω is the sought for natural frequency. Thus, Eq. (4.82) becomes

$$\frac{d^2}{d\xi^2}\left[D(\xi)\frac{d^2 W(\xi)}{d\xi^2}\right] - kL^4 R(\xi)W(\xi) = 0 \qquad (4.90)$$

$$k = \omega^2 \quad D = EI \quad R = \rho A$$

The material density and elastic modulus are represented as polynomial functions

$$R(\xi) = \sum_{i=0}^{m} a_i \xi^i \quad D(\xi) = \sum_{i=0}^{n} b_i \xi^i \tag{4.91}$$

where m and n are positive integers. We restrict our consideration to the case $n = m + 4$, since the first term in Eq. (4.88) involves four spatial derivatives. We are looking for that special class of problems in which the mode shape $W(\xi)$ is represented by the simplest polynomial functions that satisfy a given set of boundary conditions.

4.2.3   Cantilever Beam

The beam is clamped at $\xi = 0$ and free at $\xi = 1$. The boundary conditions read as follows:

$$W(\xi) = 0 \quad \frac{dW(\xi)}{d\xi} = 0 \qquad \text{at } \xi = 0$$

$$E(\xi)I\frac{d^2 W(\xi)}{d\xi^2} = 0 \quad \frac{d}{d\xi}\left[E(\xi)I\frac{d^2 W(\xi)}{d\xi^2}\right] = 0 \qquad \text{at } \xi = 1 \tag{4.92}$$

A polynomial function that satisfies the boundary conditions in Eq. (4.92) is given by

$$W(\xi) = w_1(6\xi^2 - 4\xi^3 + \xi^4) \tag{4.93}$$

It coincides with the expression of the first comparison function in the set of polynomial functions introduced by Duncan (1937) for studying beam vibration in the context of the Boobnov–Galerkin method; w_1 is an indeterminate coefficient. We also note that the expression in parentheses is proportional to the static displacement of the uniform cantilever under constant loading.

We pose the following question: what should the coefficients a_i and b_i be so that the beam's vibration mode coincides with Eq. (4.93)?

By substituting the expressions for $D(\xi)$, $R(\xi)$, $W(\xi)$ into Eq. (4.90), we obtain

$$w_1\left[\sum_{i=2}^{m+4} i(i-1)b_i\xi^{i-2}(12\xi^2 - 24\xi + 12) + \sum_{i=0}^{m+4} 24b_i\xi^i + 2\sum_{i=1}^{m+4} ib_i\xi^{i-1}(24\xi - 24)\right.$$

$$\left. - kL^4\sum_{i=0}^{m} a_i\xi^i(\xi^4 - 4\xi^3 + 6\xi^2)\right] = 0 \tag{4.94}$$

The latter expression can be re-written as follows, in a more convenient form:

$$-24 \sum_{i=1}^{m+3} i(i+1)b_{i+1}\xi^i + 12 \sum_{i=2}^{m+4} i(i-1)b_i\xi^i + 12 \sum_{i=0}^{m+2} (i+2)(i+1)b_{i+2}\xi^i$$

$$+ 24 \sum_{i=0}^{m+4} b_i\xi^i - 48 \sum_{i=0}^{m+3} (i+1)b_{i+1}\xi^i + 48 \sum_{i=1}^{m+4} ib_i\xi^i - 6kL^4 \sum_{i=2}^{m+2} a_{i-2}\xi^i$$

$$+ 4kL^4 \sum_{i=3}^{m+3} a_{i-3}\xi^i - kL^4 \sum_{i=4}^{m+4} a_{i-4}\xi^i = 0 \tag{4.95}$$

Equation (4.95) has to be satisfied for any ξ. It will be shown later that one has to distinguish between two special sub-cases: (a) $m \leq 3$ and (b) $m > 3$. It appears instructive to first treat the particular cases $m = 0$, $m = 1$, $m = 2$ and $m = 3$.

4.2.3.1   Cantilever with uniform mass density (m = 0)

In this case, the expressions for $D(\xi)$ and $R(\xi)$ read as

$$R(\xi) = a_0 \quad D(\xi) = \sum_{i=0}^{4} b_i\xi^i \tag{4.96}$$

By substituting the latter expressions in Eq. (4.90), we obtain

$$-24 \sum_{i=1}^{3} i(i+1)b_{i+1}\xi^i + 12 \sum_{i=2}^{4} i(i-1)b_i\xi^i + 12 \sum_{i=0}^{2} (i+1)(i+2)b_{i+2}\xi^i$$

$$+ 24 \sum_{i=0}^{4} b_i\xi^i - 48 \sum_{i=0}^{3} (i+1)b_{i+1}\xi^i$$

$$+ 48 \sum_{i=1}^{4} ib_i\xi^i - kL^4 a_0(6\xi^2 - 5\xi^3 + 2\xi^4) = 0 \tag{4.97}$$

Equation (4.97) has to be satisfied for any ξ. This requirement yields

$$24(b_0 + b_2) - 48b_1 = 0 \tag{4.98}$$

$$72(b_1 + b_3) - 144b_2 = 0 \tag{4.99}$$

$$144(b_2 + b_4) - 288b_3 - 6kL^4 a_0 = 0 \tag{4.100}$$

$$240b_3 - 480b_4 + 4L^4 ka_0 = 0 \tag{4.101}$$

$$-kL^4 a_0 + 360b_4 = 0 \tag{4.102}$$

The sole unknown in Eqs. (4.13)–(4.15) is the natural frequency coefficient k, yet we have five equations. We conclude that the parameters b_i and a_i have to satisfy some auxiliary conditions so that Eqs. (4.13)–(4.15) are compatible.

Two compatibility conditions are given by Eq. (4.98) and Eq. (4.99), leading to

$$b_0 = 2b_1 - b_2 \tag{4.103}$$

$$b_1 = 2b_2 - b_3 \tag{4.104}$$

From Eqs. (4.100)–(4.102), three expressions for k can be found. These are listed below:

$$k = [144(b_2 + b_4) - 288b_3]/6L^4 a_0 \tag{4.105}$$

$$k = [-240b_3 + 480b_4]/L^4 4a_0 \tag{4.106}$$

$$k = 360b_4/L^4 a_0 \tag{4.107}$$

To satisfy the compatibility requirement, all expressions for k have to be equal to each other. We consider the case when the material density coefficients a_i are specified. Then, the problem is reduced to determining the coefficients b_i so that Eqs. (4.105)–(4.107) are compatible.

Since the function $R(\xi)$ of the inertial coefficient is given, so is the coefficient a_0. Let us observe that if b_4 is specified, then, the expression given in Eq. (4.107) is the final formula for the natural frequency coefficient k. Then Eqs. (4.105)–(4.107) allow an evaluation of the remaining parameters b_i. Note that b_4 and a_0 have to have the same sign since the natural frequency parameter k must be positive. From Eq. (4.106), we obtain

$$b_3 = -4b_4 \tag{4.108}$$

Equation (4.103) leads to

$$b_2 = -\tfrac{1}{2}b_2 + 4b_4 \tag{4.109}$$

The b_i where $i = \{0, 1, 2, 3\}$, can be re-written as follows:

$$b_3 = -4b_4 \tag{4.110}$$

$$b_2 = 6b_4 \tag{4.111}$$

$$b_1 = 16b_4 \tag{4.112}$$

$$b_0 = 26b_4 \tag{4.113}$$

To summarize if conditions (4.96) are satisfied, where b_i are given by Eqs. (4.110)–(4.117), then, the fundamental mode shape is expressed by Eq. (4.93). The fundamental natural frequency reads as

$$\omega^2 = 360b_4/a_0L^4 \tag{4.114}$$

4.2.3.2 Cantilever with linearly varying inertial coefficient ($m = 1$)

In this case, the expressions for $R(\xi)$ and $D(\xi)$ read as

$$R(\xi) = a_0 + a_1\xi \quad D(\xi) = \sum_{i=0}^{5} b_i\xi^i \tag{4.115}$$

By substituting the latter expressions in to Eq. (4.100), we obtain

$$-24\sum_{i=1}^{4} i(i+1)b_{i+1}\xi^i + 12\sum_{i=2}^{5} i(i-1)b_i\xi^i + 12\sum_{i=0}^{3}(i+1)(i+2)b_{i+2}\xi^i$$

$$+24\sum_{i=0}^{5} b_i\xi^i - 48\sum_{i=0}^{4}(i+1)b_{i+1}\xi^i$$

$$+48\sum_{i=1}^{5} ib_i\xi^i - kL^4(a_0 + a_1\xi)(6\xi^2 - 5\xi^3 + 2\xi^4) = 0 \tag{4.116}$$

Equation (4.116) has to be satisfied for any ξ. This requirement yields

$$24(b_0 + b_2) - 48b_1 = 0 \tag{4.117}$$

$$72(b_1 + b_3) - 144b_2 = 0 \tag{4.118}$$

$$144(b_2 + b_4) - 288b_3 - 6kL^4a_0 = 0 \tag{4.119}$$

$$240(b_3 + b_5) - 480b_4 + L^4(4ka_0 - 6ka_1) = 0 \tag{4.120}$$

$$360b_4 - 720b_5 + L^4(4ka_1 - ka_0) = 0 \tag{4.121}$$

$$-kL^4a_1 + 504b_5 = 0 \tag{4.122}$$

The coefficients b_i, where $i = \{0, 1, 2, 3, 4\}$, can be evaluated so that the compatibility of Eqs. (4.117)–(4.121) is checked:

$$b_4 = (7a_0 - 18a_1)b_5/5a_1 \tag{4.123}$$

$$b_3 = 2(11a_1 - 14a_0)b_5/5a_1 \tag{4.124}$$

$$b_2 = 2(31a_1 + 21a_0)b_5/5a_1 \tag{4.125}$$

$$b_1 = 2(51a_1 + 56a_0)b_5/5a_1 \tag{4.126}$$

$$b_0 = 2(71a_1 + 91a_0)b_5/5a_1 \tag{4.127}$$

We arrive at the following conclusion: if conditions (4.115) are satisfied, where b_i are given by Eqs. (4.123)–(4.127), then the fundamental mode shape is expressed by Eq. (4.93), where the fundamental natural frequency reads as

$$\omega^2 = 504b_5/a_1 L^4 \tag{4.128}$$

4.2.3.3   *Cantilever with parabolically varying inertial coefficient ($m = 2$)*

In this case, the expressions for $R(\xi)$ and $D(\xi)$ read as

$$R(\xi) = a_0 + a_1\xi + a_2\xi^2 \quad D(\xi) = \sum_{i=0}^{6} b_i\xi^i \tag{4.129}$$

By substituting the latter expressions into Eq. (4. 90), we obtain

$$-24\sum_{i=1}^{5} i(i+1)b_{i+1}\xi^i + 12\sum_{i=2}^{6} i(i-1)b_i\xi^i + 12\sum_{i=0}^{4}(i+1)(i+2)b_{i+2}\xi^i + 24\sum_{i=0}^{6} b_i\xi^i$$

$$-48\sum_{i=0}^{5}(i+1)b_{i+1}\xi^i + 48\sum_{i=1}^{6} ib_i\xi^i - kL^4(a_0 + a_1\xi + a_2\xi^2)(6\xi^2 - 5\xi^3 + 2\xi^4) = 0$$

$$\tag{4.130}$$

Equation (4.130) has to be satisfied for any ξ. This requirement is equivalent to

$$24(b_0 + b_2) - 48b_1 = 0 \tag{4.131}$$

$$72(b_1 + b_3) - 144b_2 = 0 \tag{4.132}$$

$$144(b_2 + b_4) - 288b_3 - 6kL^4 a_0 = 0 \tag{4.133}$$

$$240(b_3 + b_5) - 480b_4 + L^4(4ka_0 - 6ka_1) = 0 \tag{4.134}$$

$$360(b_4 + b_6) - 720b_5 + L^4(4ka_1 - ka_0 - 6ka_2) = 0 \tag{4.135}$$

$$504b_5 - 1008b_6 + L^4(4ka_2 - ka_1) = 0 \tag{4.136}$$

$$-kL^4 a_2 + 504b_6 = 0 \tag{4.137}$$

To satisfy the compatibility equations, b_i, where $i = \{0, 1, 2, 3, 4, 5\}$, have to be

$$b_5 = 2(2a_1 - 5a_2)b_6/3a \tag{4.138}$$

$$b_4 = (53a_2 - 72a_1 + 28a_0)b_6/15a_2 \tag{4.139}$$

$$b_3 = 4(39a_2 + 22a_1 - 28a_0)b_6/15a_2 \tag{4.140}$$

$$b_2 = (259a_2 + 248a_1 + 168a_0)b_6/15a_2 \tag{4.141}$$

$$b_1 = 2(181a_2 + 204a_1 + 224a_0)b_6/15a_2 \tag{4.142}$$

$$b_0 = (465a_2 + 568a_1 + 728a_0)b_6/15a_2 \tag{4.143}$$

Thus, if conditions (4.129) are satisfied, where b_i are given by Eqs. (4.138)–(4.143), the fundamental mode shape is given by Eq. (4.93), where the fundamental natural frequency is

$$\omega^2 = 672b_6/a_2L^4 \tag{4.144}$$

4.2.3.4 Cantilever with material inertial coeffiicent represented as a cubic polynomial ($m = 3$)

In this case, the expressions for $R(\xi)$ and $D(\xi)$ read

$$R(\xi) = a_0 + a_1\xi + a_2\xi^2 + a_3\xi^3 \quad D(\xi) = \sum_{i=0}^{7} b_i\xi^i \tag{4.145}$$

By substituting the latter expressions in Eq. (4. 90), we obtain

$$-24\sum_{i=1}^{6} i(i+1)b_{i+1}\xi^i + 12\sum_{i=2}^{7} i(i-1)b_i\xi^i + 12\sum_{i=0}^{5}(i+1)(i+2)b_{i+2}\xi^i$$

$$+ 24\sum_{i=0}^{7} b_i\xi^i - 48\sum_{i=0}^{6}(i+1)b_{i+1}\xi^i \tag{4.146}$$

$$+ 48\sum_{i=1}^{7} ib_i\xi^i - kL^4(a_0 + a_1\xi + a_2\xi^2 + a_3\xi^3)(6\xi^2 - 5\xi^3 + 2\xi^4) = 0$$

Equation (4.146) has to be satisfied for any ξ. This requirement yields

$$24(b_0 + b_2) - 48b_1 = 0 \tag{4.147}$$

$$72(b_1 + b_3) - 144b_2 = 0 \tag{4.148}$$

$$144(b_2 + b_4) - 288b_3 - 6kL^4a_0 = 0 \tag{4.149}$$

$$240(b_3 + b_5) - 480b_4 + L^4(4ka_0 - 6ka_1) = 0 \tag{4.150}$$

$$360(b_4 + b_6) - 720b_5 + L^4(4ka_1 - ka_0 - 6ka_2) = 0 \tag{4.151}$$

$$504(b_5 + b_7) - 1008b_6 + L^4(4ka_2 - ka_1 - 6ka_3) = 0 \tag{4.152}$$

$$672b_6 - 1344b_7 + L^4(4ka_3 - ka_2) = 0 \tag{4.153}$$

$$- kL^4a_3 + 864b_7 = 0 \tag{4.154}$$

To satisfy the compatibility equations, b_i, where $i = \{0,1,2,3,4,5,6\}$, have to be

$$b_6 = (9a_2 - 22a_3)b_7/7a_3 \tag{4.155}$$

$$b_5 = 3(7a_3 - 10a_2 + 4a_1)b_7/7a_3 \tag{4.156}$$

$$b_4 = (320a_3 + 195a_2 - 216a_1 + 84a_0)b_7/35a_3 \tag{4.157}$$

$$b_3 = (535a_3 + 468a_2 + 264a_1 - 336a_0)b_7/35a_3 \tag{4.158}$$

$$b_2 = 3(250a_3 + 259a_2 + 248a_1 + 168a_0)b_7/35a_3 \tag{4.159}$$

$$b_1 = (965a_3 + 1086a_2 + 1225a_1 + 1334a_0)b_7/35a_3 \tag{4.160}$$

$$b_0 = (1180a_3 + 1395a_2 + 1704a_1 + 2184a_0)b_7/35a_3 \tag{4.161}$$

In summary, if conditions (4.145) are satisfied, where b_i are given by Eqs. (4.155)–(4.161), then the fundamental mode shape is expressed by Eq. (4.97), where the fundamental natural frequency reads as

$$\omega^2 = 864b_7/a_3L^4 \tag{4.162}$$

4.2.3.5 Cantilever with material inertial coefficient represented as a higher order polynomial ($m > 3$)

Since Eq. (4.99) is valid for any ξ, we conclude that

$$24(b_0 + b_2) - 48b_1 = 0 \qquad \text{for } i = 0 \tag{4.163}$$

$$72(b_1 + b_3) - 144b_2 = 0 \qquad \text{for } i = 1 \tag{4.164}$$

$$144(b_2 + b_4) - 288b_3 - 6kL^4a_0 = 0 \qquad \text{for } i = 2 \tag{4.165}$$

$$240(b_3 + b_5) - 480b_4 + L^4(4ka_0 - 6ka_1) = 0 \qquad \text{for } i = 3 \tag{4.166}$$

$$\cdots$$

$$12(i+1)(i+2)(b_i + b_{i+2}) - 24(i+1)(i+2)b_{i+1}$$
$$+ L^4(4ka_{i-3} - 6ka_{i-2} - ka_{i-4}) = 0 \qquad \text{for } 4 \le i \le m+2 \tag{4.167}$$

$$\cdots$$

$$12(m^2 + 9m + 42)b_{m+3} - 24(m+4)(m+5)b_{m+4} + L^4(4ka_m - ka_{m-1}) = 0$$
$$\text{for } i = m+3 \tag{4.168}$$

$$-kL^4a_m + 12(m^2 + 11m + 30)b_{m+4} = 0 \qquad \text{for } i = m+4 \tag{4.169}$$

The only unknown in Eqs. (4.163)–(4.169) is the natural frequency coefficient k; yet, we have $m + 4$ equations. We conclude that the parameters b_i and a_i have to satisfy some auxiliary conditions so that Eqs. (4.163)–(4.169) are compatible.

Two compatibility conditions are given by Eqs. (4.163) and (4.164), leading to

$$b_0 = 2b_1 - b_2 \tag{4.170}$$

$$b_1 = 2b_2 - b_3 \tag{4.171}$$

From the other equations, several expressions for k can be found. These are determined from Eqs. (4.163)–(4.169), respectively, and are listed below

$$k = [144(b_2 + b_4) - 288b_3]/6L^4 a_0 \tag{4.172}$$

$$k = \frac{240(b_3 + b_5) - 480b_4}{L^4(6a_1 - 4a_0)} \tag{4.173}$$

$$\ldots \ldots$$

$$k = \frac{12(i + 1)(i + 2)(b_i + b_{i+2}) - 24(i + 1)(i + 2)b_{i+1}}{L^4(a_{i-4} + 6a_{i-2} - 4a_{i-3})} \tag{4.174}$$

$$\ldots \ldots$$

$$k = \frac{12(m^2 + 9m + 42)b_{m+3} - 24(m + 4)(m + 5)b_{m+4}}{L^4(a_{m-1} - 4a_m)} \tag{4.175}$$

$$k = 12(m^2 + 11m + 30)b_{m+4}/L^4 a_m \tag{4.176}$$

To meet the compatibility requirement, all expressions for k have to be equal to each other. We consider the case when the material density coefficients a_i are specified. Then, the problem is reduced to determining the coefficients b_i so that Eqs. (4.172)–(4.176) are compatible.

Let us assume that the function $R(\xi)$ of the inertial coeffcient, and hence all $a_i (i = 0, 1, \ldots, m)$ are given. Let us observe that if b_{m+4} is specified, then, the expression given in Eq. (4.176) is the final formula for the natural frequency coefficient k. Then Eqs. (4.172)–(4.176) allow an evaluation of the remaining parameters b_i. Note that b_{m+4} and a_m have to have the same sign since the natural frequency parameter k must be positive.

From Eq. (4.175), we obtain

$$b_{m+3} = \frac{b_{m+4}}{a_m(m^2 + 9m + 42)}[(m^2 + 11m + 30)a_{m-1} - 2(m^2 + 13m + 40)a_m] \tag{4.177}$$

Equation (4.174) leads to

$$b_i = \frac{1}{(i+1)(4a_{i-2} - 6a_{1-1} - a_{i-3})}$$
$$\times \{[-(i+3)a_{i-4} + 2(i+5)a_{i-3} + (2i-10)a_{i-2} - 12(i+1)a_{i-1}]b_{i+1}$$
$$+ [2(i+3)a_{i-4} - (7i-23)a_{i-3} + 8(i+4)a_{i-2} + 6(i+1)a_{i-1}]b_{i+2}$$
$$+ [-(i+3)a_{i-4} + 4(i+3)a_{i-3} + (2i-10)a_{i-2}]b_{i+3}\} \qquad (4.178)$$

where i takes values $\{4, 5, \ldots, m+2\}$.
Equation (4.173) yields

$$b_3 = \frac{(-4a_o + 12a_2 + a_1)b_4 + (11a_0 - 6a_2 - 14a_1)b_5 + (9a_1 - 6a_0)b_6}{a_0 + 6a_2 - 4a_1} \qquad (4.179)$$

From Eq. (4.177) we deduce that

$$b_2 = [(6a_1 + a_0)b_3 - (3a_1 + 8a_0)b_4 - 5a_0b_5]/(3a_1 - 2a_0) \qquad (4.180)$$

We conclude that for specified coefficients $a_0, \ldots, a_m$ and b_{m+4}, Eqs. (4.170), (4.171), (4.177) and (4.180) result in the set of coefficients for the elastic modulus such that the beam has a mode shape given in Eq. (4.6). It is remarkable that if $a_i = a$, then the coefficients b_i do not depend on the parameter a.

To summarize, if $R(\xi)$ and $D(\xi)$ vary as in Eq. (4.95) with b_i computed via Eqs. (4.170), (4.171) and (4.177)–(4.180), the fundamental mode shape of the beam is given by Eq. (4.97) and the fundamental natural frequency squared reads

$$\omega^2 = 12(m^2 + 11m + 30)b_{m+4}/a_m L^4 \qquad (4.181)$$

As we have established, in order for the closed-form solution to be obtainable, it is sufficient that (1) all coefficients a_i and (2) the coefficient b_{m+4} be specified. These requirements are not necessary: one can assume that all coefficients a_i are given and instead of the coefficient b_{m+4} any other coefficient b_j ($j \neq m+4$) is specified. If this is the case, then from Eq. (4.78) one expresses b_{j+1} via b_j and k; substitution into subsequent equations allows us to express b_{m+2}, b_{m+3} and b_{m+4} via b_i; analogously, substitution of b_i into Eqs. (4.77)–(4.79) yields sought for exact solutions.

Although the natural frequency expressions for uniform density in Eq. (4.100), for linearly varying density in Eq. (4.119), for parabolically varying density in Eq. (4.133) and cubically varying density in Eq. (4.149), are derived separately from the case $m > 3$, all these equations follow from Eq. (4.185) by substituting appropriate values for m. Hence, Eq. (4.185) is the final formula for any integer value m.

4.2.4   Beam that is Clamped at Both Ends

The beam is clamped at $\xi = 0$ and $\xi = 1$. The boundary conditions are

$$W(\xi) = 0 \quad \frac{dW(\xi)}{d\xi} = 0 \quad \text{at } \xi = 0 \quad \text{and} \quad \xi = 1 \qquad (4.182)$$

The simplest polynomial function that satisfies the boundary conditions in Eq. (4.182) is given by

$$W(\xi) = w_1(\xi^2 - 2\xi^3 + \xi^4) \qquad (4.183)$$

By substituting the expressions for $D(\xi)$, $R(\xi)$ and $W(\xi)$ in Eq. (4.90), we obtain

$$w_1 \left[\sum_{i=2}^{m+4} i(i-1)b_i\xi^{i-2}(12\xi^2 - 12\xi + 2) + \sum_{i=0}^{m+4} 24b_i\xi^i \right.$$

$$\left. + 2\sum_{i=1}^{m+4} ib_i\xi^{i-1}(24\xi - 12) - kL^4 \sum_{i=0}^{m} a_i\xi^i(\xi^4 - 2\xi^3 + \xi^2) \right] = 0 \qquad (4.184)$$

The latter expression can be cast in the following form

$$-12\sum_{i=1}^{m+3} i(i+1)b_{i+1}\xi^i + 12\sum_{i=2}^{m+4} i(i-1)b_i\xi^i + 2\sum_{i=0}^{m+2}(i+2)(i+1)b_{i+2}\xi^i$$

$$+ 24\sum_{i=0}^{m+4} b_i\xi^i - 24\sum_{i=0}^{m+3}(i+1)b_{i+1}\xi^i + 48\sum_{i=1}^{m+4} ib_i\xi^i - kL^4\sum_{i=2}^{m+2} a_{i-2}\xi^i$$

$$+ 2kL^4\sum_{i=3}^{m+3} a_{i-3}\xi^i - kL^4\sum_{i=4}^{m+4} a_{i-4}\xi^i = 0 \qquad (4.185)$$

Since Eq. (4.185) has to be satisfied for any ξ, we arrive at the following relations

$$24b_0 + 4b_2 - 24b_1 = 0 \quad \text{for } i = 0 \qquad (4.186)$$

$$72b_1 + 12b_3 - 72b_2 = 0 \quad \text{for } i = 1 \qquad (4.187)$$

$$144b_2 + 24b_4 - 144b_3 - kL^4a_0 = 0 \quad \text{for } i = 2 \qquad (4.188)$$

$$240b_3 + 40b_5 - 240b_4 + L^4(2ka_0 - ka_1) = 0 \quad \text{for } i = 3 \qquad (4.189)$$

$$\cdots$$

$$12(i+1)(i+2)b_i + 2(i+1)(i+2)b_{i+2} - 12(i+1)(i+2)b_{i+1}$$

$$+ L^4(2ka_{i-3} - ka_{i-2} - ka_{i-4}) = 0 \qquad \text{for } 4 \leq i \leq m+2 \tag{4.190}$$

$$\cdots$$

$$12(m+4)(m+5)b_{m+3} - 12(m+4)(m+5)b_{m+4} + L^4(2ka_m - ka_{m-1}) = 0$$

$$\text{for } i = m+3 \tag{4.191}$$

$$-kL^4 a_m + 12(m^2 + 11m + 30)b_{m+4} = 0 \qquad \text{for } i = m+4 \tag{4.192}$$

It should be borne in mind that Eqs. (4.186)–(4.192) are valid only if $m > 3$. For cases that satisfy the inequality $m \leq 4$, the reader is referred to Appendix A. Note also that Eqs. (4.186)–(4.192) have a recursive form, as do Eqs. (4.163)–(4.169) for the cantilever.

Two compatibility conditions are immediately detected for Eq. (4.186) and Eq. (4.187), resulting in

$$b_0 = b_1 - \tfrac{1}{6}b_2 \tag{4.193}$$

$$b_1 = b_2 - \tfrac{1}{6}b_3 \tag{4.194}$$

From the other equations, several expressions for k can be found. These are determined from Eqs. (4.188)–(4.192), respectively. The alternative analytical formulas for k are:

$$k = [144(b_2 - b_3) - 24b_4]/L^4 a_0 \tag{4.195}$$

$$k = [240(b_3 - b_4) - 40b_5]/L^4(a_1 - 2a_0) \tag{4.196}$$

$$\cdots$$

$$k = \frac{12(i+1)(i+2)(b_i - b_{i+1}) - 2(i+1)(i+2)b_{i+2}}{L^4(a_{i-4} + a_{i-2} - 2a_{i-3})} \tag{4.197}$$

$$\cdots$$

$$k = \frac{12(m+4)(m+5)(b_{m+3} - b_{m+4})}{L^4(a_{m-1} - 2a_m)} \tag{4.198}$$

$$k = 12(m^2 + 11m + 30)b_{m+4}/L^4 a_m \tag{4.199}$$

Equations (4.195)–(4.199) then allow an evaluation of the remaining parameters b_i. It is worth noting that b_{m+4} and a_m have to have the same sign due to the positivity of k. From Eq. (4.198), we get

$$b_{m+3} = \frac{b_{m+4}}{a_m(m+4)(m+5)}[(m^2 + 11m + 30)a_{m-1} - (m^2 + 13m + 40)a_m] \tag{4.200}$$

Equation (4.197) yields

$$
b_i = \frac{1}{6(i+1)(2a_{i-2} - a_{i-1} - a_{i-3})}
$$
$$
\times \{[-6(i+3)a_{i-4} + 6(i+5)a_{i-3} + 6(i-1)a_{i-2} - 6(i+1)a_{i-1}]b_{i+1}
$$
$$
+ [6(i+3)a_{i-4} - (37i+13)a_{i-3} + 4(2i+5)a_{i-2} - (i+1)a_{i-1}]b_{i+2}
$$
$$
+ [(i+3)a_{i-4} - 2(i+3)a_{i-3} + (i+3)a_{i-2}]b_{i+3}\} \tag{4.201}
$$

where i belongs to the set $\{4, 5, \ldots, m+2\}$. Equation (4.196) leads to

$$
b_3 = \frac{6(4a_0 + a_1 - 2a_2)b_4 - (38a_0 - 2a_2 - 22a_1)b_5 - 3(2a_0 - a_1)b_6}{12(2a_1 - a_0 - a_2)} \tag{4.202}
$$

From Eq. (4.195) we obtain

$$
b_2 = \frac{(18a_1 - 6a_0)b_3 + (3a_1 - 36a_0)b_4 - 5a_0 b_5}{18(a_1 - 2a_0)} \tag{4.203}
$$

Therefore, for specified coefficients $a_0, a_1, \ldots, a_m$ and b_{m+4}, Eqs. (4.193), (4.194) and (4.200)–(4.203) lead to the set of coefficients for the elastic modulus such that the beam possesses the mode shape given in Eq. (4.183). In perfect analogy with the cantilever, if $a_i = a$, then the coefficients b_i do not depend on the parameter a. We conclude that, if $R(\xi)$ and $D(\xi)$ vary as in Eq. (4.91), with attendant b_i computed via Eqs. (4.193), (4.194) and (4.200)–(4.203), the fundamental mode shape of the beam is governed by Eq. (4.183) and the fundamental natural frequency squared reads as

$$
\omega^2 = 12(m^2 + 11m + 30)b_{m+4}/L^4 a_m \tag{4.204}
$$

4.2.5 Beam Clamped at One End and Pinned at the Other

Consider now the beam that is clamped at $\xi = 0$ and pinned at $\xi = 1$. The boundary conditions are

$$
W(\xi) = 0 \quad \frac{dW(\xi)}{d\xi} = 0 \qquad \text{at } \xi = 0
$$
$$
W(\xi) = 0 \quad EI\frac{d^2 W(\xi)}{d\xi^2} = 0 \qquad \text{at } \xi = 1 \tag{4.205}
$$

The simplest polynomial function that satisfies the boundary conditions in Eq. (4.205), and does not have a node point in the interval $(0, 1)$ is given by

$$
W(\xi) = w_1\left(3\xi^2 - 5\xi^3 + 2\xi^4\right) \tag{4.206}
$$

By substituting the expressions for $D(\xi)$, $R(\xi)$ and $W(\xi)$ in Eq. (4.90), we obtain

$$
w_1 \left[\sum_{i=0}^{m+4} i(i-1)b_i\xi^{i-2}(24\xi^2 - 30\xi + 6) + \sum_{i=0}^{m+4} 48b_i\xi^i + 2\sum_{i=1}^{m+4} ib_i\xi^{i-1}(48\xi - 30) \right.
$$

$$
\left. - kL^4 \sum_{i=0}^{m} a_i\xi^i(2\xi^4 - 5\xi^3 + 3\xi^2) \right] = 0 \tag{4.207}
$$

Equation (4.207) can be re-written as follows:

$$
-30 \sum_{i=1}^{m+3} i(i+1)b_{i+1}\xi^i + 24 \sum_{i=2}^{m+4} i(i-1)b_i\xi^i + 6 \sum_{i=0}^{m+2} (i+2)(i+1)b_{i+2}\xi^i
$$

$$
+ 48 \sum_{i=0}^{m+4} b_i\xi^i - 60 \sum_{i=0}^{m+3} (i+1)b_{i+1}\xi^i + 96 \sum_{i=1}^{m+4} ib_i\xi^i
$$

$$
- 3kL^4 \sum_{i=2}^{m+2} a_{i-2}\xi^i + 5kL^4 \sum_{i=3}^{m+3} a_{i-3}\xi^i - 2kL^4 \sum_{i=4}^{m+4} a_{i-4}\xi^i = 0 \tag{4.208}
$$

Since Eq. (4.208) has to be satisfied for any ξ, we conclude that

$$48b_0 - 60b_1 + 12b_2 = 0 \qquad \text{for } i = 0 \tag{4.209}$$

$$144b_1 - 180b_2 + 36b_3 = 0 \qquad \text{for } i = 1 \tag{4.210}$$

$$288b_2 - 360b_3 + 72b_4 - 3kL^4a_0 = 0 \qquad \text{for } i = 2 \tag{4.211}$$

$$480b_3 - 600b_4 + 120b_5 + L^4(5ka_0 - 3ka_1) = 0 \qquad \text{for } i = 3 \tag{4.212}$$

$$\cdots$$

$$24(i+1)(i+2)b_i - 30(i+1)(i+2)b_{i+1} + 6(i+1)(i+2)b_{i+2}$$

$$+ L^4(5ka_{i-3} - 3ka_{i-2} - 2ka_{i-4}) = 0 \qquad \text{for } 4 \leq i \leq m+2 \tag{4.213}$$

$$24(m+4)(m+5)b_{m+3} - 30(m+4)(m+5)b_{m+4} + L^4(5ka_m - 2ka_{m-1}) = 0$$

$$\text{for } i = m+3 \tag{4.214}$$

$$- kL^4a_m + 12(m^2 + 11m + 30)b_{m+4} = 0 \qquad \text{for } i = m+4 \tag{4.215}$$

It should be borne in mind that Eqs. (4.209)–(4.215) are valid only if $m > 3$. For cases that satisfy the inequality $m \leq 3$, one can derive a solution in perfect analogy with Appendix A, which deals with a pinned beam. Note also that Eqs. (4.209)–(4.215) have a recursive form.

Two compatibility conditions are given by Eqs. (4.209) and (4.210), with

$$b_0 = \tfrac{1}{4}(5b_1 - b_2) \tag{4.216}$$

$$b_1 = \tfrac{1}{4}(5b_2 - b_3) \tag{4.217}$$

From the other equations, several expressions for k can be found. These are determined from Eqs. (4.211)–(4.215), respectively, and are listed below:

$$k = (96b_2 - 120b_3 + 24b_4)/L^4 a_0 \tag{4.218}$$

$$k = (480b_3 - 600b_4 + 120b_5)/L^4(3a_1 - 5a_0) \tag{4.219}$$

$$\cdots$$

$$k = \frac{6(i+1)(i+2)(4b_i - 5b_{i+1} + b_{i+2})}{L^4(2a_{i-4} + 3a_{i-2} - 5a_{i-3})} \tag{4.220}$$

$$\cdots$$

$$k = \frac{6(m+4)(m+5)(4b_{m+3} - 5b_{m+4})}{L^4(2a_{m-1} - 5a_m)} \tag{4.221}$$

$$k = 12(m^2 + 11m + 30)b_{m+4}/L^4 a_m \tag{4.222}$$

Then, Eqs. (4.218)–(4.222) allow an evaluation of the remaining parameters b_i. Note that b_{m+4} and a_m have to have the same sign due to the positivity of k. From Eq. (4.221), we obtain

$$b_{m+3} = \frac{b_{m+4}}{4a_m(m+4)}[4(m+6)a_{m-1} - 5(m+8)a_m] \tag{4.223}$$

Equation (4.220) yields

$$
\begin{aligned}
b_i = {} & \frac{1}{4(i+1)(5a_{i-2} - 3a_{i-1} - 2a_{i-3})} \\
& \times \{[-8(i+3)a_{i-4} + 10(i+5)a_{i-3} + (13i - 11)a_{i-2} - 15(i+1)a_{i-1}]b_{i+1} \\
& + [10(i+3)a_{i-4} - (23i + 73)a_{i-3} + 10(i+4)a_{i-2} + 3(i+1)a_{i-1}]b_{i+2} \\
& + [-2(i+3)a_{i-4} + 5(i+3)a_{i-3} - 3(i+3)a_{i-2}]b_{i+3}\}
\end{aligned}
\tag{4.224}
$$

where i belongs to the set $\{4, 5, \ldots, m+2\}$. Equation (4.219) results in

$$b_3 = \frac{(40a_o - 30a_2 + 14a_1)b_4 + (-71a_0 + 6a_2 + 35a_1)b_5 + (-9a_1 + 15a_0)b_6}{8(5a_1 - 3a_2 - 2a_0)} \tag{4.225}$$

From Eq. (4.219) we obtain

$$b_2 = \frac{5(5a_1 - a_0)b_3 - (3a_1 + 20a_0)b_4 + 5a_0b_5}{4(3a_1 - 5a_0)} \qquad (4.226)$$

Thus, for specified coefficients $a_0, a_1, \ldots, a_m$ and b_{m+4}, Eqs. (4.216), (4.217) and (4.221)–(4.226) lead to the set of coefficients for the elastic modulus such that the beam possesses the mode shape given in Eq. (4.206). Note that if $a_i = a$, then the coefficients b_i do not depend on the parameter a.

The results can be summarized as follows: if $R(\xi)$ and $D(\xi)$ vary as in Eq. (4.90) with b_i, computed via Eqs. (4.216), (4.217) and (4.224)–(4.226), the fundamental mode shape of the beam is given by Eq. (4.206) and the fundamental natural frequency squared is

$$\omega^2 = 12(m^2 + 11m + 30)b_{m+4}/L^4 a_m \qquad (4.227)$$

4.2.6   Random Beams with Deterministic Frequencies

As seen in Eqs. (4.181), (4.204) and (4.227) the fundamental natural frequency depends only on the terminal coefficients a_m and b_{m+4}. If either of these coefficients is random, so is the natural frequency. In the latter case one can ask whether the evaluation is reliable, the reliability being defined as the probability that the natural frequency does not exceed any pre-selected value. Such an analysis, for the beams that were pinned at their ends was conducted by Candan and Elishakoff (2000).

Yet, Eqs. (4.181), (4.204) and (4.227) for beams with three different boundary conditions lead to a remarkable conclusion: If the material density coefficients $a_0, a_1, \ldots, a_{m-1}$ and the elastic modulus coefficients $b_0, b_1, \ldots, b_{m+3}$ are *random*, but the quantities a_m and b_{m+4} are *deterministic*, the natural frequency is a *deterministic* variable too. Thus, although the beam is random, its fundamental frequency is *deterministic*. The present writers do not know of any other study that reports an analogous occurrence. The closest one is the study by Fraser and Budiansky (1969), which dealt with buckling of elastic beams on a non-linear elastic foundation; the beam possessed random initial imperfection — deviation from the straight line — with a given autocorrelation function. The beams were treated as having infinite length and the initial imperfections constituted an ergodic random field. Fraser and Budiansky (1969) used an approximate technique which resulted in deterministic buckling loads. The buckling load was defined as the maximum axial load the beam could support. Several other studies followed (see, for bibliography; Amazigo, 1976) which utilized different approximate analyses. Amazigo (1976) ascribed this seemingly paradoxical behavior to the property of ergodicity that was postulated for the random initial imperfection.

Elishakoff (1979) re-examined the Fraser–Budiansky problem for finite beams on non-linear elastic foundations. The Monte Carlo method was utilized in conjunction with developing a special procedure for solving a nonlinear boundary-value problem for each realization of the random beam. In *finite* beams, the buckling loads did not turn out to be deterministic. Yet, the coefficient of variation of the buckling load was shown to be a *decreasing* function with increase in length.

How can one detect that a complex structure may possess deterministic eigenvalues? Such structures obviously are analyzed by approximate analytical and/or numerical techniques. In order to answer this question we simulate the realistic situation and apply the Monte Carlo method to check the validity of the main conclusion of this study, namely, that random beams may have deterministic frequencies.

The particular case considered hereinafter is $m = 2$. The coefficients a_0 and a_1 were taken to be exponentially distributed independent random variables and 1089 realizations of beams were simulated. For each realization of coefficients a_0 and a_1 the appropriate coefficients $b_0, \ldots, b_5$ were evaluated. For simplicity the coefficients $a_m = a_2$ and $b_{m+4} = b_6$ were fixed at unity. For each realization of the beam the finite element method was applied. For one of the simulated beams, with material density coefficients $a_0 = 0.557602, a_1 = 0.387297$ and $a_2 = 1$, the associated elastic modulus coefficients are given in Tables 4.5, 4.6, 4.7 and 4.8, depending on the boundary condition. The convergence of the finite element method is illustrated in Table 4.3 with the exact solution for the natural frequency coefficient being $k = 672$. The percentagewise errors from the exact solution are listed in Table 4.4. For the subsequent calculations, the number of elements was taken to be equal to four, with maximum error, which for the clamped–clamped beam, was only 0.41%. This is in agreement with the observation by Gupta and Rao (1978) who concluded that "the finite element procedure developed for the eigenvalue analysis of doubly tapered and twisted Bresse–Timoshenko beams has been found to give reasonably accurate results even with four finite

TABLE 4.3

Convergence of the Natural Frequency Squared for Different Boundary Conditions ($a_0 = 0.557602$, $a_1 = 0.387297$, $a_2 = 1$)

Number of elements	Pinned	Clamped–pinned	Clamped–clamped	Clamped–free
1	791.789673	1146.29273	–	696.96113
2	678.667018	691.102962	717.006005	673.551345
3	673.321172	675.750735	680.733099	672.307085
4	672.418974	673.188072	674.761629	672.097261
5	672.171821	672.487131	673.131839	672.039859
6	672.082919	672.235081	672.546128	672.019228
7	672.044777	672.126948	672.294903	672.01038
8	672.026255	672.074438	672.172916	672.006086
9	672.016394	672.046481	672.107973	672.0038
10	672.010758	672.030501	672.070852	672.002493

TABLE 4.4

Percentagewise Error

Number of Elements	Pinned	Clamped–pinned	Clamped–clamped	Clamped–free
1	17.8258442	70.5792753	–	3.714453869
2	0.992115774	2.842702679	6.697322173	0.230854911
3	0.196602976	0.558145089	1.299568304	0.045697173
4	0.062347321	0.176796429	0.410956696	0.014473363
5	0.025568601	0.072489732	0.168428423	0.005931399
6	0.012339137	0.034982292	0.081269048	0.00286131
7	0.006663244	0.018891071	0.043884375	0.001544643
8	0.003906994	0.011077083	0.025731548	0.000905655
9	0.002439583	0.006916815	0.016067411	0.000565476
10	0.001600893	0.004538839	0.010543452	0.000370982

elements." Sample calculations for ten realizations of the random beam are listed in Tables 4.5–4.8 for various boundary conditions. For the case of the beams pinned at both ends, Table 4.5 also lists the results obtainable from the single-term Boobnov–Galerkin method with a sinusoidal comparison function, $\sin \pi\xi$. It can be observed from Table 4.5 that the sample frequencies are concentrated around the exact solution $k = 672$ none exceeding the value 673. The same occurs for the C–F beams (Table 4.6). For clamped pinned beams none of the random frequencies in Table 4.7 exceeds 674, whereas for the clamped-clamped beam all frequencies in Table 4.8 are below 675.

Results of the Monte Carlo simulation for 1089 sample beams are given in Table 4.9. At this size of the sample, according to Massey (1951), at the level of significance of 0.01, the maximum absolute difference between exact and empirical reliabilities is smaller than $1.63/\sqrt{1089} = 0.049$. It lists mean values, standard deviation and the coefficient of variation for the natural frequencies. It is seen that for all four sets of boundary conditions, the standard deviation is much smaller than the mean natural frequency. The resulting coefficients of variation all are less than 10^{-5}. Extreme smallness of the coefficient of variation supports our theoretical finding that the natural frequency constitutes a deterministic quantity.

Likewise, it appears that if the results of the Monte Carlo simulation of a complex structure exhibit small coefficients of variation for the *eigenvalues* irrespective of the moderate or large coefficients of variation of the *input* stochastic quantities, one is facing the phenomenon discovered in this study, namely the random structures possessing the deterministic eigenvalues. Soize (1999, 2000a,b) also found that some stochastic systems may possess a deterministic fundamental frequency.

The difference from the paper by Fraser and Budiansky (1969) lies in the fact that the deterministic property of the buckling loads was possibly due to the ergodicity of the input random fields (Amazigo, 1976) expanding from minus infinity to plus infinity, combined with approximate analysis. Here,

TABLE 4.5

Sample Calculations for Fundamental and Second Natural Frequencies of Pinned Beams

Coefficients $a[j]$	Approximation: $\sum_{i=1}^{6} w_i \sin(i\pi\xi)$ Coefficients $b[j]$	First natural frequency coefficient k		Second natural frequency coefficient k	
		Boobnov–Galerkin method	FEM	Boobnov–Galerkin method	FEM
$a[0] = 0.557602$	$b[0] = 5.133779$				
	$b[1] = 5.133779$				
$a[1] = 0.387297$	$b[2] = -0.070508$	672.006907	672.000673	10,565.5	10,565.4
$a[2] = 1.000000$	$b[3] = -1.877895$				
	$b[4] = -1.555322$				
	$b[5] = -1.150270$				
	$b[6] = 1.000000$				
$a[0] = 0.421035$	$b[0] = 4.948464$				
	$b[1] = 4.948464$				
$a[1] = 0.642941$	$b[2] = 1.018801$	672.006899	672.000673	10,801.0	10,800.8
$a[2] = 1.000000$	$b[3] = -1.981591$				
	$b[4] = -2.423793$				
	$b[5] = -0.809412$				
	$b[6] = 1.000000$				
$a[0] = 0.817484$	$b[0] = 8.824267$				
	$b[1] = 8.824267$				
$a[1] = 1.373393$	$b[2] = 1.194419$	672.006882	672.000672	10,703.8	10,703.7
$a[2] = 1.000000$	$b[3] = -5.214750$				
	$b[4] = -3.436841$				
	$b[5] = 0.164524$				
	$b[6] = 1.000000$				

TABLE 4.5
Continued

Approximation: $\sum_{i=1}^{6} w_i \sin(i\pi\xi)$

Coefficients $a[j]$	Coefficients $b[j]$	First natural frequency coefficient k		Second natural frequency coefficient k	
		Boobnov–Galerkin Method	FEM	Boobnov–Galerkin Method	FEM
	$b[0] = 4.225417$				
$a[0] = 0.032587$	$b[1] = 4.225417$				
$a[1] = 1.283644$	$b[2] = 3.921267$	672.006868	672.000673	11,829.6	11,829.4
$a[2] = 1.000000$	$b[3] = -2.069073$				
	$b[4] = -4.686583$				
	$b[5] = 0.044859$				
	$b[6] = 1.000000$				
	$b[0] = 9.427710$				
$a[0] = 0.927113$	$b[1] = 9.427710$				
$a[1] = 1.368769$	$b[2] = 0.774654$	672.006882	672.000671	10,653.4	10,653.3
$a[2] = 1.000000$	$b[3] = -5.612935$				
	$b[4] = -3.221101$				
	$b[5] = 0.158359$				
	$b[6] = 1.000000$				
	$b[0] = 3.659425$				
$a[0] = 0.300851$	$b[1] = 3.659425$				
$a[1] = 0.371173$	$b[2] = 0.851482$	672.006913	672.000673	10,872.9	10,872.7
$a[2] = 1.000000$	$b[3] = -0.880659$				
	$b[4] = -1.995894$				
	$b[5] = -1.171769$				
	$b[6] = 1.000000$				

a[0] = 1.595434	b[0] = 10.158333	b[4] = 1.215579	672.006896	672.000671
a[1] = 0.039957	b[1] = 10.158333	b[5] = −1.613390			10,274.3	10,274.2
a[2] = 1.000000	b[2] = −4.732383	b[6] = 1.000000		
	b[3] = −4.918851			

a[0] = 1.595434   b[0] = 10.158333   b[4] = 1.215579
a[1] = 0.039957   b[1] = 10.158333   b[5] = −1.613390
a[2] = 1.000000   b[2] = −4.732383   b[6] = 1.000000
                  b[3] = −4.918851
672.006896   672.000671   10,274.3   10,274.2

a[0] = 0.092588   b[0] = 8.976387   b[4] = −9.249257
a[1] = 3.231425   b[1] = 8.976387   b[5] = 2.641900
a[2] = 1.000000   b[2] = 8.112237   b[6] = 1.000000
                  b[3] = −6.967747
672.006847   672.000671   11,639.2   11,639.0

a[0] = 1.676221   b[0] = 11.890765   b[4] = 0.011065
a[1] = 0.604673   b[1] = 11.890765   b[5] = −0.860436
a[2] = 1.000000   b[2] = −3.753968   b[6] = 1.000000
                  b[3] = −6.575775
672.006888   672.000671   10,363.8   10,363.7

a[0] = 1.032913   b[0] = 12.660406   b[4] = −5.819134
a[1] = 2.533571   b[1] = 12.660406   b[5] = 1.711428
a[2] = 1.000000   b[2] = 3.019887   b[6] = 1.000000
                  b[3] = −8.803445
672.006870   672.000671   10,802.4   10,802.3

TABLE 4.6

Sample Calculations for Natural Frequencies of Clamped–Free Beams

Coefficients $a[j]$	Coefficients $b[j]$		Natural frequency coefficient k by the FEM
$a[0] = 0.557602$	$b[0] = 72.727949$ $b[1] = 51.321537$ $b[2] = 29.915125$ $b[3] = 8.508713$	$b[4] = 2.715164$ $b[5] = -2.816937$ $b[6] = 1.000000$	672.097261
$a[1] = 0.387297$			
$a[2] = 1.000000$			
$a[0] = 0.421035$	$b[0] = 75.780285$ $b[1] = 54.196254$ $b[2] = 32.612222$ $b[3] = 11.028191$	$b[4] = 1.233148$ $b[5] = -2.476078$ $b[6] = 1.000000$	672.097963
$a[1] = 0.642941$			
$a[2] = 1.000000$			
$a[0] = 0.817484$	$b[0] = 122.681037$ $b[1] = 85.905145$ $b[2] = 49.129253$ $b[3] = 12.353362$	$b[4] = -1.532985$ $b[5] = -1.502142$ $b[6] = 1.000000$	672.096366
$a[1] = 1.373393$			
$a[2] = 1.000000$			
$a[0] = 0.032587$	$b[0] = 81.188912$ $b[1] = 60.021739$ $b[2] = 38.854566$ $b[3] = 17.687394$	$b[4] = -2.567330$ $b[5] = -1.621808$ $b[6] = 1.000000$	672.099769
$a[1] = 1.283644$			
$a[2] = 1.000000$			
$a[0] = 0.927113$	$b[0] = 127.826609$ $b[1] = 89.053628$ $b[2] = 50.280647$ $b[3] = 11.507667$	$b[4] = -1.306147$ $b[5] = -1.508308$ $b[6] = 1.000000$	672.096056
$a[1] = 1.368769$			
$a[2] = 1.000000$			
$a[0] = 0.300851$	$b[0] = 59.656390$ $b[1] = 43.214658$ $b[2] = 26.772926$ $b[3] = 10.331195$	$b[4] = 2.313291$ $b[5] = -2.838436$ $b[6] = 1.000000$	672.098978
$a[1] = 0.371173$			
$a[2] = 1.000000$			

TABLE 4.6

Continued

Coefficients $a[j]$	Coefficients $b[j]$		Natural frequency coefficient k by the FEM
	$b[0] = 109.944777$		
$a[0] = 1.595434$	$b[1] = 72.870466$	$b[4] = 6.319681$	
$a[1] = 0.039957$	$b[2] = 35.796155$	$b[5] = -3.280057$	672.093444
$a[2] = 1.000000$	$b[3] = -1.278156$	$b[6] = 1.000000$	
	$b[0] = 157.856878$		
$a[0] = 0.092588$	$b[1] = 114.793376$	$b[4] = -11.804677$	
$a[1] = 3.231425$	$b[2] = 71.729875$	$b[5] = 0.975233$	672.098338
$a[2] = 1.000000$	$b[3] = 28.666374$	$b[6] = 1.000000$	
	$b[0] = 135.249566$		
$a[0] = 1.676221$	$b[1] = 90.643587$	$b[4] = 3.759849$	
$a[1] = 0.604673$	$b[2] = 46.037608$	$b[5] = -2.527103$	672.093899
$a[2] = 1.000000$	$b[3] = 1.431629$	$b[6] = 1.000000$	
	$b[0] = 177.068593$		
$a[0] = 1.032913$	$b[1] = 123.896129$	$b[4] = -6.699704$	
$a[1] = 2.533571$	$b[2] = 70.723666$	$b[5] = 0.044762$	672.096116
$a[2] = 1.000000$	$b[3] = 17.551202$	$b[6] = 1.000000$	

we do not use the ergodicity assumption, the beams have finite length and the results are obtained in closed form.

4.3 Inhomogeneous Beams that may Possess a Prescribed Polynomial Second Mode

4.3.1 Introductory Remarks

The exact solutions for the vibration frequencies of non-uniform rods and beams were considered, respectively, by Ward (1913) and Kirchhoff (1882), apparently for the first time, in terms of Bessel functions. General-purpose

TABLE 4.7

Sample Calculations for Natural Frequencies of Clamped–Pinned Beams

Coefficients $a[j]$	Coefficients $b[j]$	Natural frequency coefficient k by the FEM
$a[0] = 0.557602$ $a[1] = 0.387297$ $a[2] = 1.000000$	$b[0] = 2.386304$ $b[1] = 2.172488$ $b[2] = 1.317223$ $b[3] = -2.103834$ $b[4] = -0.175201$ $b[5] = -1.566937$ $b[6] = 1.000000$	673.188072
$a[0] = 0.421035$ $a[1] = 0.642941$ $a[2] = 1.000000$	$b[0] = 2.352934$ $b[1] = 2.172475$ $b[2] = 1.450641$ $b[3] = -1.436697$ $b[4] = -1.197058$ $b[5] = -1.226078$ $b[6] = 1.000000$	673.190547
$a[0] = 0.817484$ $a[1] = 1.373393$ $a[2] = 1.000000$	$b[0] = 4.127546$ $b[1] = 3.778540$ $b[2] = 2.382518$ $b[3] = -3.201570$ $b[4] = -2.648377$ $b[5] = -0.252142$ $b[6] = 1.000000$	673.18739
$a[0] = 0.032587$ $a[1] = 1.283644$ $a[2] = 1.000000$	$b[0] = 2.159529$ $b[1] = 2.078161$ $b[2] = 1.752691$ $b[3] = 0.450809$ $b[4] = -3.844270$ $b[5] = -0.371808$ $b[6] = 1.000000$	673.197488
$a[0] = 0.927113$ $a[1] = 1.368769$ $a[2] = 1.000000$	$b[0] = 4.382875$ $b[1] = 3.997323$ $b[2] = 2.455116$ $b[3] = -3.713713$ $b[4] = -2.429862$ $b[5] = -0.258308$ $b[6] = 1.000000$	673.186588[6pt]
$a[0] = 0.300851$ $a[1] = 0.371173$ $a[2] = 1.000000$	$b[0] = 1.757614$ $b[1] = 1.630702$ $b[2] = 1.123053$ $b[3] = -0.907543$ $b[4] = -0.606097$ $b[5] = -1.588436$ $b[6] = 1.000000$	673.192617

TABLE 4.7

Continued

Coefficients $a[j]$	Coefficients $b[j]$	Natural frequency coefficient k by the FEM
$a[0] = 1.595434$	$b[0] = 4.457584$ $b[4] = 2.804104$	
$a[1] = 0.039957$	$b[1] = 3.912576$ $b[5] = -2.030057$	673.179294
$a[2] = 1.000000$	$b[2] = 1.732547$ $b[6] = 1.000000$	
	$b[3] = -6.987571$	
$a[0] = 0.092588$	$b[0] = 4.521305$ $b[4] = -9.575612$	
$a[1] = 3.231425$	$b[1] = 4.324959$ $b[5] = 2.225233$	673.194604
$a[2] = 1.000000$	$b[2] = 3.539577$ $b[6] = 1.000000$	
	$b[3] = 0.398050$	
$a[0] = 1.676221$	$b[0] = 5.293016$ $b[4] = 1.260761$	
$a[1] = 0.604673$	$b[1] = 4.693411$ $b[5] = 1.277103$	673.180924
$a[2] = 1.000000$	$b[2] = 2.294989$ $b[6] = 1.000000$	
	$b[3] = -7.298696$	
$a[0] = 1.032913$	$b[0] = 5.961455$ $b[4] = -5.726776$	
$a[1] = 2.533571$	$b[1] = 5.483693$ $b[5] = 1.294762$	673.187544
$a[2] = 1.000000$	$b[2] = 3.572644$ $b[6] = 1.000000$	
	$b[3] = -4.071551$	

finite element codes can handle this problem successfully. Since then numerous studies have been performed to obtain both exact and approximate solutions. Presently, the attraction to exact solutions appears to be disappearing due to the finite element method's versality and its fast convergence.

The natural question arises: Does it makes sense to still look for closed form solutions? One argument for obtaining exact solutions is that they can serve as benchmark solutions. The second argument is less articulated: if the exact solution is extremely simple and reveals some interesting underpinnings of the problem, it appears to be worth pursuing. Closed-form solutions will be obtained in the present section for the natural frequency and mode shape of inhomogeneous beams that possess uniform cross-section along the beam's

TABLE 4.8

Sample Calculations for Natural Frequencies of Clamped–Clamped Beams

Coefficients $a[j]$	Coefficients $b[j]$	Natural frequency coefficient k by the FEM
$a[0] = 0.557602$ $a[1] = 0.387297$ $a[2] = 1.000000$	$b[0] = 1.013858$ $b[1] = 1.159905$ $b[2] = 0.876280$ $b[3] = -1.701751$ $b[4] = 0.144678$ $b[5] = -1.150270$ $b[6] = 1.000000$	674.761629
$a[0] = 0.421035$ $a[1] = 0.642941$ $a[2] = 1.000000$	$b[0] = 0.973584$ $b[1] = 1.130086$ $b[2] = 0.939010$ $b[3] = -1.146453$ $b[4] = -0.723793$ $b[5] = -0.809412$ $b[6] = 1.000000$	674.76113
$a[0] = 0.817484$ $a[1] = 1.373393$ $a[2] = 1.000000$	$b[0] = 1.755885$ $b[1] = 2.023840$ $b[2] = 1.607728$ $b[3] = -2.496669$ $b[4] = -1.736841$ $b[5] = 0.164524$ $b[6] = 1.000000$	674.764592
$a[0] = 0.032587$ $a[1] = 1.283644$ $a[2] = 1.000000$	$b[0] = 0.819969$ $b[1] = 0.997884$ $b[2] = 1.067493$ $b[3] = 0.417655$ $b[4] = -2.986583$ $b[5] = 0.044859$ $b[6] = 1.000000$	674.758823
$a[0] = 0.927113$ $a[1] = 1.368769$ $a[2] = 1.000000$	$b[0] = 1.878854$ $b[1] = 2.157733$ $b[2] = 1.673270$ $b[3] = -2.906774$ $b[4] = -1.521101$ $b[5] = 0.158359$ $b[6] = 1.000000$	674.764933
$a[0] = 0.300851$ $a[1] = 0.371173$ $a[2] = 1.000000$	$b[0] = 0.713686$ $b[1] = 0.831554$ $b[2] = 0.707207$ $b[3] = -0.746080$ $b[4] = -0.295894$ $b[5] = -1.171769$ $b[6] = 1.000000$	674.758412

TABLE 4.8

Continued

Coefficients $a[j]$	Coefficients $b[j]$	Natural frequency coefficient k by the FEM
$a[0] = 1.595434$ $a[1] = 0.039957$ $a[2] = 1.000000$	$b[0] = 2.040809$ $b[1] = 2.261035$ $b[2] = 1.321356$ $b[3] = -5.638072$ $b[4] = 2.915579$ $b[5] = -1.613390$ $b[6] = 1.000000$	674.765655
$a[0] = 0.092588$ $a[1] = 3.231425$ $a[2] = 1.000000$	$b[0] = 1.768522$ $b[1] = 2.140223$ $b[2] = 2.230211$ $b[3] = 0.539927$ $b[4] = -7.549257$ $b[5] = 2.641900$ $b[6] = 1.000000$	674.763433
$a[0] = 1.676221$ $a[1] = 0.604673$ $a[2] = 1.000000$	$b[0] = 2.388134$ $b[1] = 2.671119$ $b[2] = 1.697904$ $b[3] = -5.839285$ $b[4] = 1.711065$ $b[5] = -0.860436$ $b[6] = 1.000000$	674.766146
$a[0] = 1.032913$ $a[1] = 2.533571$ $a[2] = 1.000000$	$b[0] = 2.525862$ $b[1] = 2.927879$ $b[2] = 2.412098$ $b[3] = -3.094684$ $b[4] = -4.119134$ $b[5] = 1.711428$ $b[6] = 1.000000$	674.767408

TABLE 4.9

Statistical Properties of Natural Frequency for Different Boundary Conditions (1089 Samples)

Boundary conditions	Pinned–pinned	Clamped–free	Clamped–clamped	Clamped–pinned
Mean	672.4186337	672.0966348	674.762564	673.1879431
Standard deviation	0.000767083	0.002587103	0.0046755	0.006558834
Coefficient of variation	1.14078×10^{-6}	3.8493×10^{-6}	6.9291×10^{-6}	9.74295×10^{-6}

axis. Usually, once the properties of the structure are known one determines
the location of nodes and anti-nodes (Gürgoze, 1999). We ask a question about
the type of inverse problems that exist: find the distribution of the elastic
modulus and the material density that yield a pre-selected, polynomial mode
shape. We pose the additional question: Is it possible to determine the mode
shape that has a nodal point at a pre-selected location on the beam's axis? A
somewhat related problem, for spring-mass systems, was treated by Cha and
Peirre (1999). They studied a case of a one-dimensional arbitrarily pinned
chain of oscillators and showed that the desired node position can either
coincide with the oscillator chain location or differ from it. In the former
case, it is always possible to get a node at any *pre-selected* location on the
structure whereas in the latter case one can induce the node for certain normal
modes only. Earlier, McLaughlin (1988) and Hald and McLaughlin (1998b)
studied the inverse problem of the special type, namely, of reconstructing the
structure from the set of nodal points or nodal lines. The motivation for this
section stems from the work by Candan and Elishakoff (2000) who focused
on the *fundamental* natural frequency of inhomogenous beams. Here, we are
concerned with the *second* frequency. This section closely follows the study
by Neuringer and Elishakoff (2001).

4.3.2   Basic Equation

The dynamical behavior of a uniform, inhomogeneous beam is described by
the following equation

$$\frac{d^2}{dx^2}\left[E(x)I\frac{d^2W}{dx^2} \right] - \rho(x)\omega^2 AW(x) = 0 \qquad (4.228)$$

where E is the modulus of elasticity, $\rho(x)$ the density, I the moment of inertia,
A the cross-sectional area, $W(x)$ the modal displacement and ω the natural
frequency. Letting $\xi = x/L$ represent the non-dimensional axial coordinate, where L is the length of the beam, introducing the natural frequency
coefficient

$$k = \omega^2 A/I \qquad (4.229)$$

and carrying out the differentiation in Eq. (4.228), we obtain

$$\frac{d^2E}{d\xi^2}\frac{d^2W}{d\xi^2} + 2\frac{dE}{d\xi}\frac{d^3W}{d\xi^3} + E\frac{d^4W}{d\xi^4} - kL^4\rho(\xi)W = 0 \qquad (4.230)$$

The objective of this section is to determine the polynomial function $W(\xi)$ that
satisfies the boundary conditions of simple support $W = W'' = 0$ at $\xi = 0$ and
$\xi = 1$, and has an arbitrarily prescribed node at $\xi = \alpha$ $(0 < \alpha < 1)$, and serves

as the mode shape satisfying Eq. (4.230) with polynomial variations of $E(\xi)$ and $\rho(\xi)$. The mode shape satisfying the conditions

$$W(0) = W''(0) = W(1) = W''(1) = W(\alpha) = 0 \qquad (4.231)$$

will be sought for the fifth-order polynomial in terms of ξ:

$$W(\xi) = \gamma_0 + \gamma_1 \xi + \gamma_2 \xi^2 + \gamma_3 \xi^3 + \gamma_4 \xi^4 + \gamma_5 \xi^5 \qquad (4.232)$$

Satisfying the boundary conditions (4.232) yields the following set of equations:

$$
\begin{aligned}
\gamma_0 &= 0 \\
\gamma_2 &= 0 \\
\gamma_0 + \gamma_1 + \gamma_2 + \gamma_3 + \gamma_4 + \gamma_5 &= 0 \\
2\gamma_2 + 6\gamma_3 + 12\gamma_4 + 20\gamma_5 &= 0 \\
\gamma_0 + \gamma_1\alpha + \gamma_2\alpha^2 + \gamma_3\alpha^3 + \gamma_4\alpha^4 + \gamma_5\alpha^5 &= 0
\end{aligned}
\qquad (4.233)
$$

We have five equations with six unknowns. One of the unknowns, namely γ_1, is set equal to unity, resulting in five inhomogeneous equations in five unknowns. The determinant of the system equals

$$
\begin{vmatrix}
1 & 0 & 0 & 0 & 0 \\
0 & 1 & 0 & 0 & 0 \\
1 & 1 & 1 & 1 & 1 \\
0 & 2 & 6 & 12 & 20 \\
1 & \alpha^2 & \alpha^3 & \alpha^4 & \alpha^5
\end{vmatrix}
= 6\alpha^5 - 14\alpha^4 + 8\alpha^3 = 2\alpha^3(\alpha - 1)(3\alpha - 4) \qquad (4.234)
$$

and vanishes at $\alpha = 0$, $\alpha = 1$ and $\alpha = \frac{4}{3}$. We exclude values $\alpha = 0$ and $\alpha = 1$ from consideration, since these cases do not correspond to internal nodes. Likewise $\alpha = \frac{4}{3}$ is not included in the study, since it corresponds to a node outside the beam. Hence, for $\alpha \neq 0$, $\alpha \neq 1$ and $\alpha \neq \frac{4}{3}$ the above inhomogeneous set has a unique solution, which leads to the following mode shape:

$$W(\xi) = \xi + \gamma_3 \xi^3 + \gamma_4 \xi^4 + \gamma_5 \xi^5 \qquad (4.235)$$

where

$$\gamma_3 = -2\left(3\alpha^3 - 2\alpha^2 - 2\alpha - 2\right)/\gamma$$

$$\gamma_4 = \left(3\alpha^3 + 2\alpha^2 - 7\alpha - 7\right)/\gamma$$

$$\gamma_5 = -3\left(\alpha^2 - \alpha - 1\right)/\gamma \tag{4.236}$$

$$\gamma = \alpha^2\left(3\alpha - 4\right)$$

If we are given the material density in the polynomial form

$$\rho(\xi) = \sum_{i=0}^{m} a_i \xi^i \tag{4.237}$$

with upper limit m, then it is easily seen that the largest derivative in Eq. (4.230) is of the fourth order, and the polynomial for $E(\xi)$, to be used in Eq. (4.234) must have the upper limit $m + 4$:

$$E(\xi) = \sum_{i=0}^{m+4} b_i \xi^i \tag{4.238}$$

The required expressions in Eq. (4.230) are

$$W(\xi) = \xi + \gamma_3 \xi^3 + \gamma_4 \xi^4 + \gamma_5 \xi^5$$

$$W''(\xi) = 6\gamma_3 \xi + 12\gamma_4 \xi^2 + 20\gamma_5 \xi^3$$

$$W'''(\xi) = 6\gamma_3 + 24\gamma_4 \xi + 60\gamma_5 \xi^2$$

$$W^{IV}(\xi) = 24\gamma_4 + 120\gamma_5 \xi$$

$$E'(\xi) = \sum_{i=0}^{m+4} i b_i \xi^{i-1} \tag{4.239}$$

$$E''(\xi) = \sum_{i=0}^{m+4} i(i-1) b_i \xi^{i-2}$$

We will start the analysis with particular cases.

4.3.3   A Beam with Constant Mass Density

Let us first take the case where the density is uniform, i.e. $m = 0$ so that

$$a_0 > 0 \qquad a_1 = a_2 = \cdots = a_m = 0 \tag{4.240}$$

Substituting Eq. (4.238) and (4.239) into Eq. (4.230) we get the following expressions for each term of the governing equation

$$\frac{d^2 E}{d\xi^2}\frac{d^2 W}{d\xi^2} = \left(2b_2 + 6b_3 + 12b_4\xi^2\right)\left(6\gamma_3\xi + 12\gamma_4\xi^2 + 20\gamma_5\xi^3\right) \tag{4.241}$$

$$2\frac{dE}{d\xi}\frac{d^3 W}{d\xi^3} = 2\left(b_1 + 2b_2\xi + 3b_3\xi^2 + 4b_4\xi^3\right)\left(6\gamma_3 + 24\gamma_4\xi + 60\gamma_5\xi^2\right) \tag{4.242}$$

$$E\frac{d^4 W}{d\xi^4} = \left(b_0 + b_1\xi + b_2\xi^2 + b_3\xi^3 + b_4\xi^4\right)\left(24\gamma_4 + 120\gamma_5\xi^2\right) \tag{4.243}$$

$$-kL^4\rho\left(\xi\right)W\left(\xi\right) = -kL^4 a_0\left(\xi + \gamma_3\xi^3 + \gamma_4\xi^4 + \gamma_5\xi^5\right) \tag{4.244}$$

Setting the sum of Eqs. (4.241)–(4.244) to zero and setting the coefficient of the various power of ξ equal to zero, we get a set of six linear algebraic homogeneous equations for the six unknowns $b_{i,i=0...4}$ and k:

$$\begin{aligned}
2\gamma_4 b_0 + \gamma_3 b_1 &= 0 \\
120\gamma_5 b_0 + 72\gamma_4 b_1 + 36\gamma_3 b_2 - kL^4 a_0 &= 0 \\
240\gamma_5 b_1 + 144\gamma_4 b_2 + 72\gamma_3 b_3 &= 0 \\
400\gamma_5 b_2 + 240\gamma_4 b_3 + 120\gamma_3 b_4 - kL^4\gamma_3 a_0 &= 0 \\
600\gamma_5 b_3 + 360\gamma_4 b_4 - kL^4\gamma_4 a_0 &= 0 \\
840\gamma_5 b_4 - kL^4\gamma_5 a_0 &= 0
\end{aligned} \tag{4.245}$$

In order to have a non-trivial solution the determinant of the matrix of the set (4.245) must vanish, yielding the following equation:

$$265{,}420{,}800 L^4\gamma_4\gamma_5 a_0\left(108\gamma_4^4 - 675\gamma_3\gamma_4^2\gamma_5 - 4{,}375\gamma_5^3 + 750\gamma_3^2\gamma_5^2\right) = 0 \tag{4.246}$$

Substituting the expressions for γ_3, γ_4 and γ_5 in terms of α into Eq. (4.246), we obtain

$$-21{,}499{,}084{,}800 L^4 a_0 (972\alpha^{17} - 8{,}262\alpha^{16} + 15{,}498\alpha^{15} + 49{,}005\alpha^{14} - 188{,}595\alpha^{13}$$
$$- 18{,}603\alpha^{12} + 688{,}293\alpha^{11} - 370{,}557\alpha^{10} - 1{,}214{,}845\alpha^9 + 890{,}775\alpha^8$$
$$+ 1{,}261{,}667\alpha^7 - 839{,}737\alpha^6 - 846{,}457\alpha^5 + 317{,}555\alpha^4 + 333{,}335\alpha^3 - 8{,}668\alpha^2$$
$$- 46{,}032\alpha - 7{,}672)/\left[\alpha^{12}(3\alpha - 4)^6\right] = 0 \tag{4.247}$$

The solution of this equation for α within the interval $[0; 1]$ is $\alpha = \frac{1}{2}$. Hence, for this value of α, $\gamma_3 = -10$, $\gamma_4 = 15$ and $\gamma_5 = -6$. We conclude that for

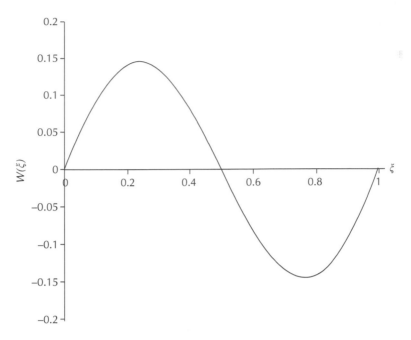

FIGURE 4.7
Second mode shape of the pinned beam with constant mass density

the case $m = 0$, there exists only a single fifth-degree polynomial solution for $w(\xi)$, namely

$$W(\xi) = \xi - 10\xi^3 + 15\xi^4 - 6\xi^5 \tag{4.248}$$

which is depicted in Figure 4.7.

Hence, with the mode shape (4.248), we obtain the following expression for k from the last equation in the set (4.245):

$$k = \frac{840b_4}{L^4 a_0} \tag{4.249}$$

Note that the fundamental frequency coefficient is (Candan and Elishakoff, 2000):

$$k = \frac{360b_4}{L^4 a_0} \tag{4.250}$$

Solving for the b_i $(i = 0, 1, \ldots, 4)$ in terms of b_4, Eq. (4.238) for $E(\xi)$ becomes

$$E(\xi) = \left(\tfrac{1}{3} + \xi - 2\xi^3 + \xi^4\right) b_4 \tag{4.251}$$

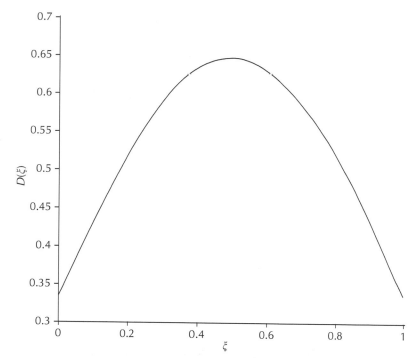

FIGURE 4.8
Variation of the stiffness of a pinned beam with constant mass density

The variation of the obtained flexural rigidity is portrayed in Figure 4.8.

Thus, for the case $m = 0$, there exists only a single fifth-degree polynomial solution for $W(\xi)$ satisfying certain prescribed boundary conditions and having a definite nodal position, which is generated by a unique elastic modulus distribution of the fourth degree. It is remarkable that in the case of the uniform material density ($m = 0$) the variation of the elastic modulus does not depend on a_0.

4.3.4 A Beam with Linearly Varying Mass Density

In the case of a linear variation in the material density $\rho(\xi)$, namely

$$\rho(\xi) = a_0 + a_1\xi \tag{4.252}$$

consider first the case when a_0 differs from zero. We rewrite the mass density in Eq. (4.252) as follows:

$$\rho(\xi) = a_0(1 + \beta\xi) \qquad \text{where } \beta = a_1/a_0 \neq 0 \tag{4.253}$$

The modulus of elasticity, according to Eq. (4.238), takes the form

$$E(\xi) = \sum_{i=0}^{5} b_i \xi^i \tag{4.254}$$

Substitution of Eqs. (4.253) and (4.229) into Eq. (4.230) along with Eq. (4.235) yields the polynomial equation

$$C_6 \xi^6 + C_5 \xi^5 + C_4 \xi^4 + C_3 \xi^3 + C_2 \xi^2 + C_1 \xi + C_0 = 0 \tag{4.255}$$

where

$$
\begin{aligned}
C_0 &= 24\gamma_4 b_0 + 12\gamma_3 b_1 \\
C_1 &= 120\gamma_5 b_0 + 72\gamma_4 b_2 + 36\gamma_3 b_3 - kL^4 a_0 \\
C_2 &= 240\gamma_5 b_1 + 144\gamma_4 b_2 + 72\gamma_3 b_3 - kL^4 a_0 \beta \\
C_3 &= 400\gamma_5 b_2 + 240\gamma_4 b_3 + 120\gamma_3 b_4 - kL^4 a_0 \gamma_3 \\
C_4 &= 600\gamma_5 b_3 + 360\gamma_4 b_4 + 180\gamma_3 b_5 - kL^4 a_0 (\gamma_4 - \beta\gamma_3) \\
C_5 &= 840\gamma_5 b_4 + 504\gamma_4 b_5 - kL^4 a_0 (\gamma_5 + \beta\gamma_4) \\
C_6 &= 1120\gamma_5 b_5 - kL^4 a_0 \beta\gamma_5
\end{aligned}
\tag{4.256}
$$

In order for Eq. (4.256) to be valid for any ξ, we demand all coefficients $C_i (i = 0, 1, \ldots, 6)$ to vanish, leading to seven equations for the seven unknowns, namely b_i $(i = 0, 1, \ldots, 5)$ and k. Since this set is homogeneous, for non-triviality of the solution, its determinant

$$
\Delta = \begin{vmatrix}
24\gamma_4 & 12\gamma_3 & 0 & 0 & 0 & 0 & 0 \\
120\gamma_5 & 72\gamma_4 & 36\gamma_3 & 0 & 0 & 0 & -L^4 a_0 \\
0 & 240\gamma_5 & 144\gamma_4 & 72\gamma L^3 & 0 & 0 & -L^4 a_0 \beta \\
0 & 0 & 400\gamma_5 & 240\gamma_4 & 120\gamma_3 & 0 & -L^4 a_0 \gamma_3 \\
0 & 0 & 0 & 600\gamma_5 & 360\gamma_4 & 180\gamma_3 & -L^4 a_0 (\gamma_4 + \beta\gamma_3) \\
0 & 0 & 0 & 0 & 840\gamma_5 & 504\gamma_4 & -L^4 a_0 (\gamma_5 + \beta\gamma_4) \\
0 & 0 & 0 & 0 & 0 & 1120\gamma_5 & -L^4 a_0 \beta\gamma_5
\end{vmatrix}
\tag{4.257}
$$

must vanish. This leads to the following equation

$$L^4 a_0 \gamma_5 (G_1 + G_2 \beta) = 0 \tag{4.258}$$

where

$$\lambda_1 = 1{,}857{,}945{,}600$$

$$G_1 = \lambda_1 \left(700{,}000\gamma_4\gamma_5^4 - 120{,}000\gamma_5^3\gamma_4\gamma_3^2 + 108{,}000\gamma_5^2\gamma_4^3\gamma_3 - 17{,}280\gamma_5\gamma_4^5\right)$$

$$G_2 = \lambda_1 \left(175{,}000\gamma_5^4\gamma_3 - 210{,}000\gamma_5^3\gamma_4^2 - 17{,}625\gamma_5^3\gamma_3^3 + 78{,}200\gamma_5^2\gamma_4^2\gamma_3^2 \right.$$

$$\left. -49{,}140\gamma_5\gamma_4^4\gamma_3 + 7{,}128\gamma_4^6\right) \tag{4.259}$$

Since $\gamma_5 \neq 0$ (i.e. the roots of $\alpha^2 - \alpha - 1$ lie outside $[0; 1]$ according to Eq. (4.236)) and it was assumed that a_0 differs from zero, the expression in parentheses in (4.258) must vanish, leading to the following expression for β:

$$\beta = -G_1/G_2 \tag{4.260}$$

The latter expression allows us to pose the following question: What is the value of the ratio β of the mass density coefficients corresponding to a vibration node placed at an arbitrary location α?

The expressions for G_1 and G_2 in terms of α read as follows:

$$G_1 = 20(972\alpha^{17} - 8{,}262\alpha^{16} - 15{,}498\alpha^{15} + 49{,}005\alpha^{14} - 188{,}595\alpha^{13} - 18{,}603\alpha^{12}$$

$$+ 688{,}293\alpha^{11} - 370{,}557\alpha^{10} - 1{,}214{,}845\alpha^9 + 890{,}775\alpha^8 + 1{,}261{,}667\alpha^7$$

$$- 839{,}737\alpha^6 - 846{,}457\alpha^5 + 31{,}755\alpha^4 + 333{,}335\alpha^3 - 8{,}668\alpha^2$$

$$- 46{,}032\alpha - 7{,}672)$$

$$G_2 = 8{,}016\alpha^{18} - 62{,}451\alpha^{17} + 102{,}384\alpha^{16} + 310{,}824\alpha^{15} - 987{,}795\alpha^{14}$$

$$+ 6{,}129\alpha^{13} + 1{,}026{,}444\alpha^{12} + 3{,}368{,}349\alpha^{11} - 4{,}463{,}541\alpha^{10} - 8{,}101{,}030\alpha^9$$

$$+ 9{,}604{,}699\alpha^8 + 9{,}339{,}499\alpha^7 - 8{,}927{,}306\alpha^6 - 7{,}068{,}461\alpha^5 + 3{,}583{,}555\alpha^4$$

$$+ 3{,}150{,}739\alpha^3 - 225{,}591\alpha^2 - 511{,}026\alpha - 85{,}171 \tag{4.261}$$

When G_2 equals zero, the value of β tends to infinity, implying that either a_1 increases without limit or that a_0 tends to zero. The approximate value of α for this limiting case is

$$\alpha = 0.5571986621 \tag{4.262}$$

The following question arises: Can α take on other values besides $\alpha = \alpha_1$ for finite values of β? The reply must include the consideration of physical realizability of the problem, which demands that $\rho(\xi)$ must be positive when $\xi \in [0; 1]$, i.e.

$$1 + \beta\xi \geq 0 \quad \text{or} \quad \beta \geq -1 \tag{4.263}$$

Therefore, the allowable region of variation of α is

$$\alpha \in [\alpha_0; \alpha_1] \tag{4.264}$$

where α_0 denotes the value of α, corresponding to $\beta = -1$, namely

$$\alpha_0 = 0.4428013379 \tag{4.265}$$

Various mode shapes corresponding to α in the allowable region are presented in Figure 4.9.

Physical realizability also imposes the requirement that the elastic modulus $E(\xi)$ be positive. This latter property can be checked straightforwardly; in fact, the flexural rigidity coefficient b_i that involve β and γ_i are functions of α, since β and γ_i depend on α.

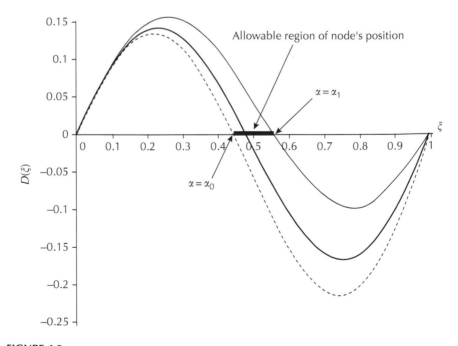

FIGURE 4.9
Various second mode shapes of the inhomogeneous pinned beam with linearly varying mass density $(\rho(\xi) = a_0(1 + \beta\xi))$

Consider α in the allowable region. The system (4.256) has a non-trivial solution. The second frequency reads as

$$k = 1120b_5/L^4 a_1 \tag{4.266}$$

The coefficients b_i in the elastic modulus $E(\xi)$ are derived from the set (4.256):

$$b_0 = \frac{b_5}{37{,}500\gamma_5^5 \beta}\left[\left(-21{,}600\gamma_3\gamma_4^3\gamma_5 + 23{,}625\gamma_3^2\gamma_4\gamma_5^2 + 3{,}564\gamma_4^5 - 105{,}000\gamma_4\gamma_5^3\right)\beta \right.$$
$$\left. + \left(46{,}800\gamma_3\gamma_4^2\gamma_5^2 - 27{,}000\gamma_3^2\gamma_5^3 - 8{,}640\gamma_4^4\gamma_5 + 350{,}000\gamma_5\right)\right]$$

$$b_1 = -\frac{b_5}{7{,}500\gamma_5^4 \beta}\left[\left(1{,}188\gamma_4^4 - 6{,}210\gamma_3\gamma_4^2\gamma_5 + 3{,}525\gamma_3^2\gamma_5^2 - 35{,}000\gamma_5\right)\beta \right.$$
$$-2{,}880\gamma_4^3\gamma_5 + 13{,}200\gamma_3\gamma$$

$$b_2 = -\frac{b_5}{125\gamma_5^3 \beta}\left[\left(-33\gamma_4^3 + 145\gamma_3\gamma_4\gamma_5\right)\beta + 80\gamma_4^2\gamma_5 - 300\gamma_3\gamma_5^2\right]$$

$$b_3 = \frac{b_5}{150\gamma_5^2 \beta}\left[\left(-66\gamma_4^2 + 235\gamma_3\gamma_5\right)\beta + 160\gamma_4\gamma_5\right]$$

$$b_4 = \frac{b_5}{15\gamma_5 \beta}\left(11\gamma_4\beta + 20\gamma_5\right)$$

$$\tag{4.267}$$

Consider separately the case $a_0 = 0$, $\rho(\xi) = a_1\xi$. We again obtain seven equations for seven unknowns. The requirement of its determinant to vanish reads as

$$\lambda_2 a_1 L^4 G_3/G_4 = 0 \tag{4.268}$$

where

$$\lambda_2 = -3{,}611{,}846{,}246{,}40$$
$$G_3 = 8{,}019\alpha^{20} - 70{,}470\alpha^{19} + 15{,}681\alpha^{18} + 270{,}891\alpha^{17} - 14{,}011{,}003\alpha^{16}$$
$$\quad + 6{,}831{,}003\alpha^{15} + 2{,}008{,}110\alpha^{14} + 2{,}335{,}577\alpha^{13} - 8{,}858{,}334\alpha^{12}$$
$$\quad - 7{,}005{,}838\alpha^{11} + 22{,}169{,}270\alpha^{10} + 78{,}355{,}830\alpha^9 - 27{,}871{,}504\alpha^8$$
$$\quad - 7{,}480{,}654\alpha^7 + 19{,}579{,}322\alpha^6 + 6{,}635{,}645\alpha^5 - 6{,}959{,}885\alpha^4 - 3{,}436{,}174\alpha^3$$
$$\quad + 651{,}446\alpha^2 + 596{,}197\alpha + 85{,}171$$
$$G_4 = \alpha^{14}(3\alpha - 4)$$

$$\tag{4.269}$$

The solution of this equation yields $\alpha = \alpha_1 = 0.55719$, which matches the root of G_2 in Eq. (4.262).

The frequency coefficient k is given by

$$k = 1120b_5/L^4 a_1 \tag{4.270}$$

We denote $\tilde{\gamma}_3$, $\tilde{\gamma}_4$, $\tilde{\gamma}_5$ the values of the mode shape coefficients for $\alpha = \alpha_1$. The coefficients in the expression for the flexural rigidity read

$$b_0 = \frac{\tilde{\gamma}_4 b_5}{12{,}500\tilde{\gamma}_5^5} \left(-7{,}200\tilde{\gamma}_3\tilde{\gamma}_4^2\tilde{\gamma}_5 + 7{,}875\tilde{\gamma}_3^2\tilde{\gamma}_5^2 + 1{,}188\tilde{\gamma}_4^4 - 35{,}000\tilde{\gamma}_5^3\right)$$

$$= 0.1169104775b_5$$

$$b_1 = -\frac{b_5}{7{,}500\tilde{\gamma}_5^4} \left(1{,}188\tilde{\gamma}_4^4 - 6{,}210\tilde{\gamma}_3\tilde{\gamma}_4^2\tilde{\gamma}_5 + 3{,}525\tilde{\gamma}_3^2\tilde{\gamma}_5^2 - 35{,}000\tilde{\gamma}_5^3\right)$$

$$= 0.3434953999b_5 \tag{4.271}$$

$$b_2 = -\frac{\tilde{\gamma}_4 b_5}{125\tilde{\gamma}_5^3} \left(-33\tilde{\gamma}_4^3 + 145\tilde{\gamma}_3\tilde{\gamma}_5\right) = 0.7826410468b_5$$

$$b_3 = \frac{b_5}{150\tilde{\gamma}_5^2} \left(-66\tilde{\gamma}_4^2 + 235\tilde{\gamma}_3\tilde{\gamma}_5\right) = -0.114362033b_5$$

$$b_4 = \frac{11\tilde{\gamma}_4 b_5}{15\tilde{\gamma}_5} = -1.852849279b_5$$

Figure 4.10 depicts the variation of the flexural rigidity in this case.

4.3.5   A Beam with Parabolically Varying Mass Density

Let us study now the parabolically varying material density case, namely

$$\rho(x) = a_0 + a_1\xi + a_2\xi^2 \tag{4.272}$$

We study first the case when a_0 differs from zero. We rewrite the mass density in Eq. (4.272) as follows:

$$\rho(x) = a_0 \left(1 + \beta_1\xi + \beta_2\xi^2\right) \quad \beta_1 = a_1/a_0 \quad \beta_2 = a_2/a_0 \neq 0 \tag{4.273}$$

The elastic flexural rigidity takes the form

$$E(\xi) = \sum_{i=0}^{6} b_i\xi^i \tag{4.274}$$

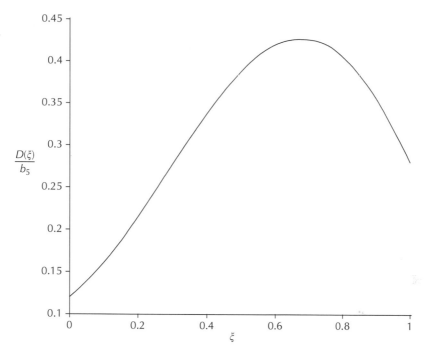

FIGURE 4.10
Variation of the flexural rigidity of a pinned beam with linearly varying mass density $\rho(\xi) = \alpha_1 \xi$

The governing differential equation (4.230) yields the following set of eight equations for eight unknowns:

$$24\gamma_4 b_0 + 12\gamma_3 b_1 = 0$$

$$120\gamma_5 b_0 + 72\gamma_4 b_1 + 36\gamma_3 b_2 + kL^4 a_0 = 0$$

$$240\gamma_5 b_1 + 144\gamma_4 b_2 + 72\gamma_3 b_3 - kL^4 a_0 \beta_1 = 0$$

$$400\gamma_5 b_2 + 240\gamma_4 b_3 + 120\gamma_3 b_2 - kL^4 a_0 (\gamma_3 + \beta_2) = 0$$

$$600\gamma_5 b_3 + 360\gamma_4 b_4 + 180\gamma_3 b_5 - kL^4 a_0 (\gamma_4 + \gamma_3 \beta_1) = 0 \qquad (4.275)$$

$$840\gamma_5 b_4 + 504\gamma_4 b_5 + 252\gamma_3 b_6 - kL^4 a_0 (\gamma_5 + \gamma_4 \beta_1 + \gamma_3 \beta_2) = 0$$

$$1120\gamma_5 b_5 + 672\gamma_4 b_6 - kL^4 a_0 (\gamma_5 \beta_1 + \gamma_4 \beta_2) = 0$$

$$1440\gamma_5 b_6 - kL^4 a_0 \gamma_5 \beta_2 = 0$$

In order to find a non-trivial solution, the determinant of the matrix of the set (4.275) must vanish. This leads to the following determinantal equation:

$$L^4 a_0 \gamma_5 (A\beta_2 + B\beta_1 + C) = 0 \qquad (4.276)$$

where

$$A = \lambda_3 \left(-15{,}552\gamma_4^7 + 118{,}260\gamma_5\gamma_4^5\gamma_3 + 378{,}000\gamma_5^3\gamma_4^3 - 243{,}000\gamma_5^2\gamma_4^3\gamma_3^2 \right.$$

$$\left. -630{,}000\gamma_5^4\gamma_4\gamma_3 + 120{,}375\gamma_5\gamma_4\gamma_3^2 \right)$$

$$B = \lambda_3 \left(35{,}640\gamma_5\gamma_4^6 - 245{,}700\gamma_5^2\gamma_4^4\gamma_3 - 1{,}050{,}000\gamma_5^4\gamma_4^2 + 391{,}500\gamma_5^2\gamma_4^3\gamma_3^2 \right.$$

$$\left. +875{,}000\gamma_5^5\gamma_3 - 88{,}125\gamma_5^4\gamma_3^3 \right)$$

$$C = \lambda_3 \left(-86{,}400\gamma_5^2\gamma_4^5 + 540{,}000\gamma_5^3\gamma_4^3\gamma_3 + 3{,}500{,}000\gamma_5^5\gamma_4 - 600{,}000\gamma_5^4\gamma_4\gamma_3^2 \right)$$

$$\lambda_3 = -535{,}088{,}332{,}800 \tag{4.277}$$

Substituting Eq. (4.256) into Eq. (4.277), we get the following expression for A, B and C in terms of α:

$$A = G_5 G_6 \quad B = G_7 G_8 \quad C = G_8 G_9 \tag{4.278}$$

where

$$G_5 = 5{,}832\alpha^{18} - 53{,}703\alpha^{17} + 163{,}377\alpha^{16} - 102{,}978\alpha^{15} - 382{,}085\alpha^{14}$$

$$+ 711{,}387\alpha^{13} - 634{,}768\alpha^{12} + 1{,}165{,}347\alpha^{11} - 94{,}623\alpha^{10} - 4{,}158{,}540\alpha^9$$

$$+ 2{,}937{,}847\alpha^8 + 4{,}518{,}597\alpha^7 - 3{,}352{,}918\alpha^6 - 3{,}061{,}283\alpha^5 + 1{,}432{,}465\alpha^4$$

$$+ 1{,}292{,}917\alpha^3 - 110{,}598\alpha^2 - 219{,}228\alpha - 36{,}538 \tag{4.279}$$

$$G_6 = 3\alpha^3 + 3\alpha^2 7\alpha - 7 \tag{4.280}$$

$$G_7 = 5(8{,}019\alpha^{18} - 62{,}451\alpha^{17} + 102{,}384\alpha^{16} + 310{,}824\alpha^{15} - 987{,}795\alpha^{14}$$

$$+ 6{,}129\alpha^{13} + 1{,}026{,}444 + 3{,}368{,}349\alpha^{11} - 4{,}463{,}541\alpha^{10} - 8{,}101{,}030\alpha^9$$

$$+ 9{,}604{,}699\alpha^8 + 9{,}339{,}499\alpha^7 - 8{,}907{,}306\alpha^6 - 7{,}068{,}461\alpha^5 + 3{,}583{,}555\alpha^4$$

$$+ 3{,}150{,}739\alpha^3 - 225{,}591\alpha^2 - 511{,}026\alpha - 85{,}171) \tag{4.281}$$

$$G_8 = \alpha^2 - Ga - 1 \tag{4.282}$$

$$G_9 = 5(19{,}440\alpha^{17} - 165{,}240\alpha^{16} + 309{,}960\alpha^{15} + 980{,}100\alpha^{14} + 3{,}771{,}900\alpha^{13}$$

$$- 372{,}060\alpha^{12} + 13{,}765{,}860\alpha^{11} - 7{,}411{,}140\alpha^{10} - 24{,}296{,}900\alpha^9$$

$$+ 17{,}815{,}500\alpha^8 + 25{,}233{,}340\alpha^7 + 16{,}794{,}740\alpha^6 - 16{,}929{,}140\alpha^5$$

$$+ 6{,}351{,}100\alpha^4 + 6{,}666{,}700\alpha^3 - 173{,}360\alpha^2 - 920{,}640\alpha - 153{,}440)$$

$$(4.283)$$

Since $\gamma_5 \neq 0$ and a_0 was assumed to be non-zero, Eq. (4.276) becomes

$$A\beta_2 + B\beta_1 + C = 0 \qquad (4.284)$$

Solving Eq. (4.284) for β_2 we get

$$\beta_2 = -(B\beta_1 + C)/A \qquad (4.285)$$

When $A = 0$ the value of β_2 grows without bound, implying that α_2 tends to infinity or that a_0 tends to zero. Consider α_3 the value of α that satisfies $A = 0$; α_3 equals

$$\alpha_3 = 0.6020659819 \qquad (4.286)$$

An approximation for the right limit α_4 in the allowed interval (α_3, α_4) is obtained as follows:

We know that the system (4.275) has a non-trivial solution when the relationship (4.285) between β_1, β_2 and α holds. Figure 4.11 shows the variation of β_2 in terms of β_1 and α, as given in Eq. (4.285) by the gray surface; the interval of variation for β_1 is taken as $[-20; 20]$.

Along with the relationship (4.285) one has also to take into account the requirement of physical realizability of the problem, which imposes the condition that both mass density and flexural rigidity have to be non-negative for any ξ within the interval $[0; 1]$. At this stage the non-negativity of the mass density can be checked, whereas this property for the flexural rigidity can be checked by direct numerical evaluation.

Consider the expression for the mass density given in Eq. (4.273). Since a_0 is the value taken by $\rho(\xi)$ at $\xi = 0$, it must be positive. Thus, in order for the mass density to be positive the following trinomial

$$\rho(\xi)/a_0 = 1 + \beta_1\xi + \beta_2\xi^2 \qquad (4.287)$$

has to be positive within the interval $[0; 1]$. Note that a trinomial can either have real roots or not, and in each case it does not have the same sign. When it has no real root (i.e. its discriminant is negative), it keeps the same sign for all real numbers and its sign is that of the coefficient of the term ξ^2. When it has real roots (i.e. its discriminant is positive) it shares the same sign as the coefficient of the term ξ^2 outside the interval defined by its roots.

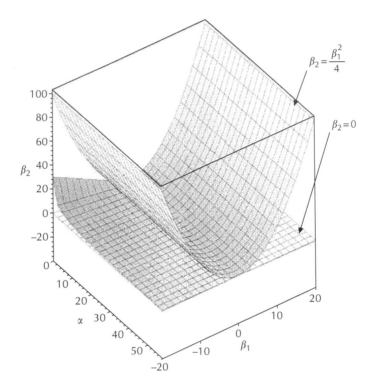

FIGURE 4.11
Variation of the parameter β_2 given in Eq. (4.285) in terms of $\alpha \in [0.61; 1]$ and β_1

We demand the trinomial (4.287) to be positive for all real numbers, leading to the following requirement

$$\delta = \beta_1^2 - 4\beta_2 < 0 \quad \beta_2 > 0 \tag{4.288}$$

where δ is the discriminant of the trinomial (4.287). In addition to the surface representing Eq. (4.285), Figure 4.12 also depicts the surface on which the discriminant vanishes ($\delta = 0$).

The two surfaces intersect on the curve that is shown in bold. The observable part of the surface that satisfies Eq. (4.285) is above the bold line; in this region the discriminant is negative, while $\beta_2 > 0$, implying that the mass density is non-negative. The maximum value of α on the bold curve is $\alpha_4 \approx 0.66$. Thus, we arrive at the interval $(\alpha_3, \alpha_4) = (0.602; 0.66)$ in which the node can be placed so as to lead to a physically realizable problem.

Consider α in an allowable region. The system (4.275) has a non-trivial solution. The second frequency coefficient is derived from the last equation of the set (4.275):

$$k = 1440b_6/L^4\alpha_0\beta_2 \tag{4.289}$$

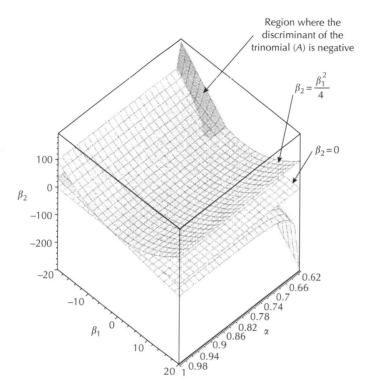

FIGURE 4.12
Variation of the parameter β_2 given in Eq. (4.309) in terms of $\alpha \in [0, 0.58]$ and β_1

The coefficient b_i in the elastic modulus $E(\xi)$ reads as

$$b_0 = -\frac{3b_6}{875,000\gamma_5^6\beta_2}\left[\left(315,000\gamma_5^4\gamma_4 - 105,300\gamma_5\gamma_4^4\gamma_3 + 166,050\gamma_5^2\gamma_4^2\gamma_3^2 - 37,125\gamma_5^3\gamma_3^3\right.\right.$$

$$-378,000\gamma_5^3\gamma_4^3 + 15,552\gamma_4^6\left) \beta_2 + \left(216,000\gamma_5^2\gamma_4^3 - 236,250\gamma_5^3\gamma_4\gamma_3^2 - 35,640\gamma_5\gamma_4^5\right.$$

$$+1,050,000\gamma_5^4\gamma_4\left) \beta_1 - 46,800\gamma_3\gamma_4^2\gamma_5^2 - 27,000\gamma_5^2\gamma_3^3 - 8,640\gamma_4^4\gamma_5 + 350,000\gamma_5\right]$$

$$b_1 = \frac{3b_6}{875,000\gamma_5^5\beta_2}\left[\left(-630,000\gamma_5^3\gamma_4 + 2,592\gamma_4^5 - 15,390\gamma_5\gamma_4^3\gamma_3 + 16,650\gamma_5^2\gamma_4\gamma_3^2\right)\beta_2\right.$$

$$+ \left(31,050\gamma_5^2\gamma_4^4\gamma_3 - 5,940\gamma_5\gamma_4^4 - 17,625\gamma_5^3\gamma_3^2 + 175,000\gamma_5^4\right)\beta_1 + 14,400\gamma_5^2\gamma_4^3$$

$$-66,000\gamma_5^3\gamma_4\gamma_3\right]$$

$$b_2 = -\frac{9b_6}{17,500\gamma_5^4\beta_2}\left[\left(-7,000\gamma_5^3 + 288\gamma_4^4 - 1,470\gamma_5\gamma_4^2\gamma_3 + 825\gamma_5^2\gamma_3^2\right) \beta_2\right.$$

$$+ \left(2,900\gamma_5^2\gamma_4\gamma_3 - 660\gamma_5\gamma_4^3\right) \beta_1 + 1,600\gamma_5^2\gamma_4^2 - 6,000\gamma_5^3\gamma_3\right]$$

$$b_3 = \frac{3b_6}{1{,}750\gamma_5^3\beta_2}\left[\left(144\gamma_4^3 - 615\gamma_4\gamma_5\gamma_3\right)\beta_2 + \left(1{,}175\gamma_5^2\gamma_3 - 330\gamma_5\gamma_4^2\right)\beta_1 + 800\gamma_5^2\gamma_4\right]$$

$$b_4 = -\frac{3b_6}{350\gamma_5^2\beta_2}\left[\left(48\gamma_4^2 - 165\gamma_5\gamma_3\right)\beta_2 - 110\gamma_5\gamma_4\beta_1 - 200\gamma_5^2\right]$$

$$b_5 = \frac{3b_6}{35\gamma_5\beta_2}(8\gamma_4\beta_2 - 15\gamma_5\beta_1)$$

$$\text{(4.290)}$$

Now, consider the case $a_0 = 0$, with $\rho(\xi) = a_1\xi + a_2\xi^2$. We rewrite the mass density as follows:

$$\rho(\xi) = a_1\xi(1 + \theta\xi) \qquad \theta = a_2/a_1 \neq 0 \qquad \text{(4.291)}$$

The parameter a_1 is assumed to be non-zero. We obtain eight equations for eight unknowns:

$$24\gamma_4 b_0 + 12\gamma_3 b_1 = 0$$
$$120\gamma_5 b_0 + 72\gamma_4 b_1 + 36\gamma_3 b_2 = 0$$
$$240\gamma_5 b_1 + 144\gamma_4 b_2 + 72\gamma_3 b_3 - kL^4 a_0 = 0$$
$$400\gamma_5 b_2 + 240\gamma_4 b_3 + 120\gamma_3 b_2 - kL^4 a_1\theta = 0$$
$$600\gamma_5 b_3 + 360\gamma_4 b_4 + 180\gamma_3 b_5 - kL^4 a_1\gamma_3 = 0 \qquad \text{(4.292)}$$
$$840\gamma_5 b_4 + 504\gamma_4 b_5 + 252\gamma_3 b_6 - kL^4 a_1(\gamma_4 + \gamma_3\theta) = 0$$
$$1120\gamma_5 b_5 + 672\gamma_4 b_6 - kL^4 a_1(\gamma_5 + \gamma_4\theta) = 0$$
$$1440\gamma_5 b_6 - kL^4 a_1\gamma_5\theta = 0$$

The requirement that the determinant of the system (4.292) should vanish reads as

$$A\theta + B = 0 \qquad \text{(4.293)}$$

with A and B defined in Eq. (4.277). The solution of this equation for θ yields the following expression

$$\theta = -B/A \qquad \text{(4.294)}$$

Note that when the denominator in Eq. (4.294) equals zero, the value of θ tends to infinity, implying that a_1 tends to zero, or that a_2 grows without bound. The corresponding node location α, which is a solution of the equation $A = 0$, is α_3 given in Eq. (4.286).

Moreover, the requirement of positivity of the mass density, demands

$$1 + \theta\xi \geq 0 \quad \text{or} \quad \theta \geq -1 \tag{4.295}$$

Hence, the allowable region of variation of α is

$$\alpha \in [\alpha_2, \alpha_3] \tag{4.296}$$

where α_2 corresponding to the value of α when $\theta = -1$, reads as

$$\alpha_2 = 0.5 \tag{4.297}$$

Various mode shapes corresponding to α in the allowable region are presented in Figure 4.13.

The requirement of physical realizability also imposes the condition that the elastic modulus $E(\xi)$ be positive. This latter property can be checked straightforwardly; in fact, the flexural rigidity coefficients b_i that involve β and γ_i are functions of α, since θ and γ_i depend upon α.

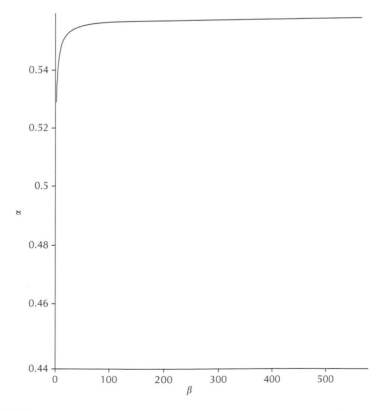

FIGURE 4.13
Variation of the node position α in terms of the parameter β

Consider α in the allowable region. The system (4.292) has a non-trivial solution. The second frequency is derived from the last equation of the set (4.292):

$$k = 1440b_6/L^4a_1\theta \tag{4.298}$$

The coefficients b_i in the elastic modulus $E(\xi)$ are given in Eq. (4.290) with $\beta_2 = \theta$.

Consider, finally, the case $a_0 = 0$ and $a_1 = 0$. The mass density, then, equals

$$\rho(\xi) = a_2\xi^2 \tag{4.299}$$

Again we obtain eight equations for eight unknowns:

$$\begin{aligned}
&24\gamma_4 b_0 + 12\gamma_3 b_1 = 0 \\
&120\gamma_5 b_0 + 72\gamma_4 b_1 + 36\gamma_3 b_2 = 0 \\
&240\gamma_5 b_1 + 144\gamma_4 b_2 + 72\gamma_3 b_3 = 0 \\
&400\gamma_5 b_2 + 240\gamma_4 b_3 + 120\gamma_3 b_2 - kL^4 a_2 = 0 \\
&600\gamma_5 b_3 + 360\gamma_4 b_4 + 180\gamma_3 b_5 = 0 \\
&840\gamma_5 b_4 + 504\gamma_4 b_5 + 252\gamma_3 b_6 - kL^4 a_2\gamma_3 = 0 \\
&1120\gamma_5 b_5 + 672\gamma_4 b_6 - kL^4 a_2\gamma_4 = 0 \\
&1440\gamma_5 b_6 - kL^4 a_2\gamma_5 = 0
\end{aligned} \tag{4.300}$$

The requirement for the determinant of (4.300) to vanish reads

$$\lambda_3 L^2 a_2 G_{10}/G_{11} = 0 \tag{4.301}$$

where

$$\lambda_3 = -3{,}120{,}635{,}156{,}889{,}600$$

$$\begin{aligned}
G_{10} = {}&17{,}496\alpha^{23} - 161{,}109\alpha^{22} + 414{,}315\alpha^{21} + 371{,}709\alpha^{20} - 3{,}027{,}699\alpha^{19} \\
&+ 227{,}126\alpha^{18} + 5{,}284{,}392\alpha^{17} - 4{,}903{,}488\alpha^{16} - 237{,}482\alpha^{15} \\
&- 18{,}436{,}704\alpha^{14} + 2{,}640{,}556\alpha^{13} + 79{,}775{,}162\alpha^{12} - 28{,}942{,}438\alpha^{11} \\
&- 135{,}634{,}072\alpha^{10} + 46{,}346{,}616\alpha^9 + 13{,}757{,}347\alpha^8 - 25{,}073{,}663\alpha^7 \\
&- 88{,}091{,}388\alpha^6 - 3{,}925{,}062\alpha^5 + 3{,}130{,}985\alpha^4 + 8{,}634{,}725\alpha^3 \\
&+ 3{,}733{,}764\alpha^2 - 2{,}046{,}128\alpha - 255{,}766
\end{aligned}$$

$$G_{11} = \alpha^{16}(3\alpha - 4)^8$$

$$\tag{4.302}$$

The denominator of the determinant is zero when $\alpha = 0$, which is one end of the beam or $\alpha = \frac{4}{3}$, which is a node outside the beam. The solution of the determinantal equation (4.301) is $\alpha = \alpha_3$ given in Eq. (4.286). Denoting by $\hat{\gamma}_3$, $\hat{\gamma}_4$ and $\hat{\gamma}_5$ the value taken by the coefficients γ_i in this case, the mode shape becomes

$$W(\xi) = \xi + \hat{\gamma}_3 \xi^3 + \hat{\gamma}_4 \xi^4 + \hat{\gamma}_5 \xi^5 \tag{4.303}$$

This yields the following solutions for the second frequency and the coefficients b_i in the flexural rigidity:

$$k = 1440 b_6 / L^4 a_2 \tag{4.304}$$

$$b_0 = \frac{27 b_6}{875{,}000 \hat{\gamma}_5^6} \left(11{,}700 \hat{\gamma}_3 \hat{\gamma}_4^4 \hat{\gamma}_5 - 18{,}450 \hat{\gamma}_3^2 \hat{\gamma}_4^2 \hat{\gamma}_5^2 + 4{,}125 \hat{\gamma}_3^3 \hat{\gamma}_5^3 - 35{,}000 \hat{\gamma}_3 \hat{\gamma}_5^4 \right.$$

$$\left. -1{,}728 \hat{\gamma}_4^6 + 42{,}000 \hat{\gamma}_4^2 \right)$$

$$\approx 0.4899 b_6$$

$$b_1 = \frac{27 \hat{\gamma}_4 b_6}{43{,}750 \hat{\gamma}_5^5} \left(144 \hat{\gamma}_4^4 - 855 \hat{\gamma}_3 \hat{\gamma}_4^2 \hat{\gamma}_5 + 925 \hat{\gamma}_3^2 \hat{\gamma}_5^2 - 3{,}500 \hat{\gamma}_5^3 \right) \approx 0.1417 b_6$$

$$b_2 = -\frac{9 b_6}{17{,}500 \hat{\gamma}_5^4} \left(288 \hat{\gamma}_4^4 - 1{,}470 \hat{\gamma}_3 \hat{\gamma}_4^2 \hat{\gamma}_5 + 825 \hat{\gamma}_3^2 \hat{\gamma}_5^2 - 7{,}000 \hat{\gamma}_5^3 \right) \approx 0.3172 b_6$$

$$b_3 = -\frac{9 \hat{\gamma}_4 b_6}{1750 \hat{\gamma}_5^3} \left(205 \hat{\gamma}_3 \hat{\gamma}_5 - 48 \hat{\gamma}_4^2 \right) \approx 0.6495 b_6$$

$$b_4 = \frac{9 b_6}{350 \hat{\gamma}_5^2} \left(55 \hat{\gamma}_3 \hat{\gamma}_5 - 16 \hat{\gamma}_4^2 \right) \approx -0.1788 b_6$$

$$b_5 = \frac{24 \hat{\gamma}_4 b_6}{35 \hat{\gamma}_5} \approx -1.7466 b_6$$

$$\tag{4.305}$$

Figure 4.14 depicts the flexural rigidity obtained in this case. The case of the general variation of the mass density is considered in Appendix B.

4.4   Concluding Remarks

In this section for three sets of boundary conditions the closed-form solutions have been derived for the mode shapes and natural frequencies of inhomogeneous beams. The following conclusions have been reached.

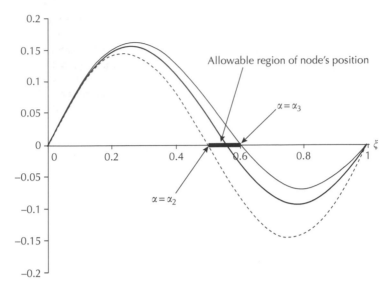

FIGURE 4.14
Various second mode shapes of the inhomogeneous pinned beam with parabolically varying mass density ($\rho(\xi) = a_1\xi(1 + \theta\xi)$)

Inhomogeneous beams may possess the natural mode that is coincident with the static deflection of the associated uniform beam under a uniformly distributed load.

The fundamental frequencies in all three cases coincide with each other, as a comparison of Eqs. (4.181), (4.204) and (4.227) reveals. The remaining case of the inhomogeneous beam that is pinned at both its ends was studied by Candan and Elishakoff (2000). There too, the beam turned out to possess the fundamental frequency given in Eqs. (4.181), (4.204) and (4.227).

Although the expressions for fundamental frequencies of inhomogeneous beams with four different boundary conditions coalesce, the beam's characteristics in each case are *different*. Namely, although they share the same material density variation as in Eq. (4.2), the b_i coefficients in the elastic modulus variation *differ*. This leads to the interesting conclusion that the beams with different elastic modulus variation may have the same natural frequency, although the beams are under differing boundary conditions. This conclusion may at first glance appear to be counterintuitive. Indeed, if one anticipates that the fundamental frequency of the clamped–clamped beam must be greater than its counterpart for the beam that is pinned at both its ends. Yet, it must be borne in mind that in the cases that we have considered the beams' characteristics are different: The clamped–clamped beam and the pinned beam have different expressions for the elastic modulus.

The intricate connection of the subject of this study with the inverse problems should be mentioned. As Gladwell (1996) stressed,

> classical direct problems have involved the analysis and derivation of the behavior of the system (e.g., forced response, natural frequencies, current flow, stresses, etc) from its properties such as density or mass, conductivity, elastic constant, crack lengths, etc. Inverse problem are concerned with the determination or estimation of such properties from behavior.

It turns out that although each of the beams has different boundary conditions, and, moreover, each of them has different $D(\xi)$ these beams have the same frequency. Two vibrating system which have the same natural frequencies are called *isospectral*. In our particular case, beams of different boundary conditions share the *first* natural frequency. Gottlieb (1991), Driscoll (1997) and others have constructed examples of isospectral structures. In particular, Gottlieb (1991) showed that clamped *inhomogeneous* circular plates have the same vibration spectrum as their homogeneous counterparts. In our cases, the second and other frequencies do not coincide. For example, the second natural frequency squared of the pinned beam is $10881.18bI/(AL^4)$, while the clamped-clamped beam has a second natural frequency squared $5607.68I/(AL^4)$. Clamped-free and clamped-pinned beams' second natural frequencies squared are respectively $42727.97I/(AL^4)$ and $8013.24I/(AL^4)$. These values are obtained by the finite element method. The difference between the present work and those associated with the inverse vibration problem lies in our desire to obtain closed-form solutions to find *any* beam that has a polynomial mode shape.

The expressions for the squared fundamental frequency depend solely upon two coefficients a_m and b_{m+4}. If, by any procedure, these coefficients could be fixed by the designer, one can have a beam that has a *pre-selected* fundamental natural frequency so that the unwanted resonance condition can be avoided. Whereas we are unaware of a procedure with such a derivable feature at present, its possible development in the future cannot be *a priori* ruled out.

If the coefficients a_m and b_{m+4} are deterministic but the remaining coefficients are random, the natural fundamental frequency squared is a deterministic quantity. In complex structures that must be analyzed by approximate methods this remarkable phenomenon could be validated if the coefficient of variation of the output quantity turns out to be much smaller than its counterparts for the input parameter.

5

Beams and Columns with Higher-Order Polynomial Eigenfunctions

In Chapters 2 and 4 the mode shape that was forced as the beam or column was a fourth-order polynomial. The natural question that arises is this: can these structures possess eigenfunctions that are represented by polynomials of degree greater than four? The reply to this question is shown to be in the affirmative, if some conditions are met.

5.1 Family of Analytical Polynomial Solutions for Pinned Inhomogeneous Beams. Part 1: Buckling

5.1.1 Introductory Remarks

The closed-form solutions for buckling of columns were reported by Engesser (1893), Tukerman (1929), Duncan (1937) and Elishakoff and Rollot (1999). The latter studies concentrated on pinned columns under axial compression. The second-order differential equation was employed:

$$D(x)\frac{d^2W}{dx^2} + Pw = 0 \tag{5.1}$$

where $W(x)$ is the buckling mode, P the compressive force, $D(x)$ the flexural rigidity and x the axial coordinate. For the flexural rigidity variation, given as

$$D(x) = \left[1 - \tfrac{3}{7}(x/L)^2\right]D_0 \tag{5.2}$$

where L is the length and D_0 is the flexural rigidity at the origin of the coordinates, Duncan (1937) suggested the buckling mode

$$W(x) = A[7(x/L) - 10(x/L)^3 + 3(x/L)^5] \tag{5.3}$$

leading, upon substitution of Eqs. (5.2) and (5.3) into Eq. (5.1) the buckling load

$$P_{cr} = \frac{60}{7} D_0/L^2 \tag{5.4}$$

Elishakoff and Rollot (1999) studied several additional cases in which the closed-form solution is available. It is instructive to report here one of the cases. For the flexural rigidity variation

$$D(x) = \left[1 + \left(1 - \sqrt{5}\right) x/L - \tfrac{1}{4} \left(1 - \sqrt{5}\right)^2 (x/L)^2\right] \tag{5.5}$$

They derived the buckling mode

$$W(x) = A[x/L - 2(x/L)^3 + (x/L)^4] \tag{5.6}$$

resulting in the buckling load

$$P_{cr} = 3\left(\sqrt{5} - 1\right)^2 D_0/L^2 \tag{5.7}$$

We note that in the two cases studied by Duncan (1937) and Elishakoff and Rollot (1999), the buckling mode is represented as Duncan polynomials, proposed by Duncan (1937) in the context of the Boobnov–Galerkin method. Namely, Eq. (5.3) represents the fifth-order polynomial, whereas Eq. (5.6) constitutes a fourth-order polynomial.

Naturally, the following question arises: are there buckling problems in which the buckling modes will be represented by polynomials of higher degree? The present section attempts to answer this question. We pose the inverse problem of *reconstructing* the flexural rigidity of the column in order that a pre-selected polynomial function serves as an *exact* mode shape of the buckling of inhomogeneous columns. We consider, here, two kinds of buckling problems, namely the buckling of an inhomogeneous beam under a compressive axial load P, formerly studied by Elishakoff (2000a), for the specialized case, and the buckling of an inhomogeneous beam under an axially distributed load $q(x)$, earlier treated by Elishakoff and Guédé (2001), with the aid of the fourth-degree polynomials. Here, we study the feasibility of having *higher* order polynomials as exact mode shapes.

5.1.2   Choosing a Pre-selected Mode Shape

We treat, here, the problem of reconstructing the flexural rigidity of an inhomogeneous column subjected to a buckling load, when its mode shape is known as a pre-selected polynomial function that satisfies the boundary conditions. The pre-selected mode shape is identified here as a deflection of a uniform beam under the various distributed loads. Consider a uniform beam, pinned at both its ends, and subjected to the load $q(x) = q_0 x^n$. The governing differential equation reads:

$$D_0 \frac{d^4 w}{dx^4} = q_0 x^n \qquad (5.8)$$

where $w(x)$ is the static displacement due to the load $q_0 x^n$, D_0 is the uniform flexural rigidity and x the axial coordinate. We introduce the non-dimensional coordinate

$$\xi = x/L \qquad (5.9)$$

where L is the length. Equation (5.8) becomes

$$EI \frac{d^4 w}{d\xi^4} = \alpha \xi^n \qquad (5.10)$$

The solution of this equation that satisfies all boundary conditions of the pinned–pinned column reads:

$$w(\xi) = \alpha L^{n+4} \psi(\xi) / D_0 (n+1)(n+2)(n+3)(n+4)$$
$$\psi(\xi) = \xi^{n+4} - \tfrac{1}{6}(n^2 + 7n + 12)\xi^3 + \tfrac{1}{6}(n^2 + 7n + 6)\xi \qquad (5.11)$$

Note that such an approach to construct the mode shape is in line with an observation that the buckling mode $\sin(\pi \xi)$ of the uniform pinned–pinned column, can be viewed as the proportional response of the uniform beam without an axial force but subjected to the transverse load $\sin(\pi \xi)$; whereas this observation is not needed to solve the problem of the uniform column, it is useful to solve the problem of the inhomogeneous column, in order to formulate the *postulated* mode shape.

5.1.3   Buckling of the Inhomogeneous Column under an Axial Load

Consider the polynomial function $\psi(\xi)$ given in Eq. (5.11). We address the following problem: can the function $\psi(\xi)$ serve as an exact mode shape for the inhomogeneous beam under a compression load P and under an axially distributed load? The answer to this question was shown to be in the affirmative for $n = 0$ (see Elishakoff, 1999). We deal here with cases $n > 0$. The differential equation that governs the buckling of the column under the axial load P reads:

$$\frac{d^2}{dx^2}\left[D(x) \frac{d^2 W}{dx^2} \right] + P \frac{d^2 W}{dx^2} = 0 \qquad (5.12)$$

where now we consider an inhomogeneous column with flexural rigidity $D(\xi)$; W_G is the buckling mode in contrast to $w(x)$ in Eq. (5.8), where the latter is a static displacement and P is the buckling load.

With the non-dimensional coordinate $\xi = x/L$, and for $W(\xi) \equiv \psi(\xi)$, the differential equation (5.12) becomes

$$\frac{d^2}{d\xi^2}\left[D(\xi)\frac{d^2\psi}{d\xi^2}\right] + PL^2\frac{d^2\psi}{d\xi^2} = 0 \qquad (5.13)$$

We emphasize again that we postulate the buckling mode $W(x)$ to *coalesce* with the function $\psi(\xi)$ that is proportional to the *static* displacement $w(\xi)$ in Eq. (11). At first glance, this postulate may appear to be an artificial one. Yet, it will be shown, that it leads to uncovering of a series of closed-form solutions.

In order to achieve the same order in both terms in the governing differential equation (5.13), the flexural rigidity is taken as a second-order polynomial:

$$D(\xi) = b_0 + b_1\xi + b_2\xi^2 \qquad (5.14)$$

Substitution of Eqs. (5.11) and (5.14) into both terms of the left-hand side in Eq. (5.13) yields

$$\frac{d^2}{d\xi^2}\left[D(\xi)\frac{d^2\psi}{d\xi^2}\right] = (n+3)(n+4)\left[-2b_1 - 6b_2\xi + (n+1)(n+2)b_0\xi^n\right.$$

$$\left. +(n+2)(n+3)b_1\xi^{n+1} + (n+3)(n+4)b_2\xi^{n+2}\right] \qquad (5.15)$$

$$PL^2\frac{d^2\psi}{d\xi^2} = PL^2(n+3)(n+4)(\xi^{n+2} - \xi) \qquad (5.16)$$

Note that the left-hand side of the differential equation (5.13) becomes a polynomial, which must vanish for any ξ within $[0; 1]$. This forces all coefficients of the polynomial equation to be zero. Therefore, we have to collect the terms with the same power of ξ on the left-hand side of Eq. (5.13), and require that their coefficients vanish. In view of Eqs. (5.15) and (5.16), we consider two main cases: $n < 2$ and $n \geq 2$.

Let us first treat the simplest case $n = 0$; the governing equation yields the following set of algebraic equations

$$b_0 - b_1 = 0$$

$$6b_1 - 6b_2 - PL^2 = 0 \qquad (5.17)$$

$$144b_2 + 12PL^2 = 0$$

From the last equation of the system (5.17), we get

$$P_{cr} = -12b_2/L^2 \tag{5.18}$$

Since the critical compressive load is positive, b_2 must be negative. The rest of the equations in (5.17) yield

$$b_1 = -b_2 \quad b_0 = b_2 \tag{5.19}$$

The flexural rigidity is, then,

$$D(\xi) = (1 - \xi + \xi^2)b_2 = (-1 + \xi - \xi^2)|b_2| \tag{5.20}$$

We note that Eq. (5.18) agrees with the result of Elishakoff (2001d). Figure 5.1 portrays the variation of $D(\xi)/|b_2|$ for ξ within $[0; 1]$.

When $n = 1$, instead of (5.17) we arrive at

$$\begin{aligned}
b_1 &= 0 \\
6b_0 - 6b_2 - PL^2 &= 0 \\
b_1 &= 0 \\
20b_2 + PL^2 &= 0
\end{aligned} \tag{5.21}$$

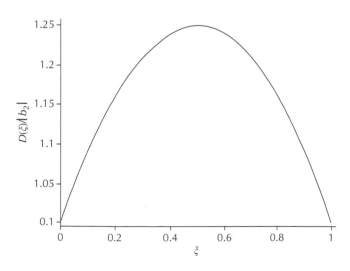

FIGURE 5.1
Variation of $D(\xi)/|b2|$ with ξ for the inhomogeneous beam under an axial load P, when $n = 0$

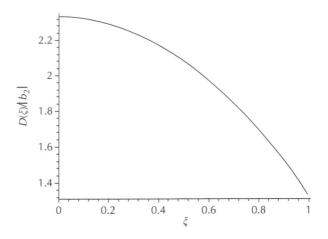

FIGURE 5.2
Variation of $D(\xi)/|b2|$ with ξ for the inhomogeneous beam under an axial load P, when $n = 1$

whose solution is

$$P_{cr} = -20b_2/L^2 \tag{5.22}$$

$$b_0 = -\tfrac{7}{3}b_2 \quad b_1 = 0 \tag{5.23}$$

The flexural rigidity is then

$$D(\xi) = \left(-\tfrac{7}{3} + \xi^2\right)b_2 = \left(\tfrac{7}{3} - \xi^2\right)|b_2| = \left(1 - \tfrac{3}{7}\xi^2\right)\tfrac{7}{3}|b_2| \tag{5.24}$$

Note that Eq. (5.24) essentially coincides with Duncan's (1937) solution given in Eq. (5.2) with $D_0 \equiv 7|b_2|/3$. We obtained it not by guessing, but by using the polynomial representations. Figure 5.2 depicts the variation of $D(\xi)/|b_2|$ when ξ belongs to $[0; 1]$.

Concerning the general case $n \geq 2$, we obtain from Eqs. (5.15) and (5.16), the following homogeneous set of algebraic equations for four unknowns, namely b_0, b_1, b_2 and P:

$$b_1 = 0$$
$$-b_2 - PL^2 = 0$$
$$b_0 = 0 \tag{5.25}$$
$$b_1 = 0$$
$$(n+3)(n+4)b_2 + PL^2 = 0$$

The solution of the system (5.25) is trivial with $b_0 = b_1 = b_2 = P = 0$.

As is seen, by postulating the buckling mode shapes given in Eq. (5.11) with $n = 0$ and $n = 1$ we obtain, respectively, the solutions given by Elishakoff (2001d) and Duncan (1937).

5.1.4   Buckling of Columns under an Axially Distributed Load

Here, we are interested in the buckling of an inhomogeneous beam under a distributed axial load. The problem is posed as follows: find the flexural rigidity $D(x)$ of the inhomogeneous column so that the polynomial function $\psi(\xi)$ given in Eq. (5.11) serves as an exact mode shape of its buckling under an axially distributed load. The differential equation that governs the buckling of the column under the axial distributed load $q(x)$, reads

$$\frac{d^2}{dx^2}\left[D(x)\frac{d^2 W}{dx^2}\right] - \frac{d}{dx}\left[N(x)\frac{dW}{dx}\right] = 0 \tag{5.26}$$

where

$$N(x) = -\int_x^L q(u)\,du, \quad q(u) = \sum_{i=0}^m q_i u^i \tag{5.27}$$

With the non-dimensional coordinate $\xi = x/L$, and for $W(\xi) \equiv \psi(\xi)$, Eq. (5.26) becomes

$$\frac{d^2}{d\xi^2}\left[D(\xi)\frac{d^2\psi}{d\xi^2}\right] - L^2\frac{d}{d\xi}\left[N(\xi)\frac{d\psi}{d\xi}\right] = 0 \tag{5.28}$$

The distributed axial load is rewritten as

$$q(\xi) = \sum_{i=0}^m q_i L^i \xi^i = q_0 \sum_{i=0}^m g_i L^i \xi^i \tag{5.29}$$

We introduce the parameters

$$g_i = q_i/q_0 \quad \text{for } i = 0, 1, \ldots, m \tag{5.30}$$

The axial load N reads

$$N(\xi) = \int_\xi^1 q(\gamma)\,d\gamma = -q_0 L \sum_{i=0}^m \frac{g_i}{i+1}(1 - \xi^{i+1}) \tag{5.31}$$

Therefore, the differential equation (5.28) is rewritten as

$$\frac{d^2}{d\xi^2}\left[D(\xi)\frac{d^2\psi}{d\xi^2}\right] + q_0L^3\frac{d}{d\xi}\left[\sum_{i=0}^{m}\frac{g_i}{i+1}(1-\xi^{i+1})\frac{d\psi}{d\xi}\right] = 0 \qquad (5.32)$$

In order to achieve the same polynomial order both terms of Eq. (5.32), the flexural rigidity is taken as a polynomial of degree $m+3$,

$$D(\xi) = \sum_{i=0}^{m+3} b_i\xi^i \qquad (5.33)$$

We consider now special cases.

5.1.4.1   *Uniformly distributed axial load (m = 0)*

We consider various possible values of n. Consider first $n = 0$; the governing equation (5.32) yields the following homogeneous set of four equations for five unknowns, namely for b_0, b_1, b_2, b_3 and q_0:

$$-24(b_1 - b_0) - q_0L^3 = 0$$
$$-72(b_2 - b_1) - 12q_0L^3 = 0 \qquad (5.34)$$
$$-144(b_3 - b_2) + 30q_0L^3 = 0$$
$$240b_3 - 16q_0L^3 = 0$$

which is solvable up to an arbitrary constant, which is taken here as the coefficient b_3. The buckling intensity equals

$$q_{0,\mathrm{cr}} = 15b_3/L^3 \qquad (5.35)$$

The other coefficients are:

$$b_0 = b_3 \quad b_1 = \tfrac{3}{8}b_3 \quad b_2 = -\tfrac{17}{8}b_3 \qquad (5.36)$$

so that the attendant flexural rigidity reads:

$$D(\xi) = \left(1 + \tfrac{3}{8}\xi - \tfrac{17}{8}\xi^2 + \xi^3\right)b_3 \qquad (5.37)$$

Note that Eq. (5.35) coincides with Elishakoff and Guédé's (2001) solution. Figure 5.3 represents the variation of the flexural rigidity in this case.

It is shown in Appendix C that only trivial solutions are deduced when $n \geq 1$, by this method.

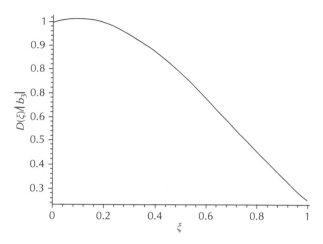

FIGURE 5.3
Variation of $D(\xi)/|b_2|$ with ξ for the column under axially distributed load P, for $m = 0$

5.1.4.2   *Linearly varying distributed axial load (m = 1)*

When $n = 0$, instead of (5.34), we get

$$-24b_1 + 24b_0 - q_0L^3 = 0$$

$$-72b_2 + 72b_1 - q_0L^3(72 + 7g_1) = 0$$

$$-144b_3 + 144b_2 + q_0L^3(30 + 6g_1) = 0 \tag{5.38}$$

$$-240b_4 + 240b_3 + q_0L^3(-16 + 12g_1) = 0$$

$$360b_4 - 10q_0L^3g_1 = 0$$

We have six unknowns for five equations. We choose b_4 as an arbitrary constant to obtain $q_{0,cr}$:

$$q_{0,cr} = 36b_4/g_1L^3 \tag{5.39}$$

The coefficients determining the boundary flexural rigidity $D(\xi)$ read

$$b_0 = 6(g_1 + 2)b_4/5g_1 \quad b_1 = 3(4g_1 + 3)b_4/10g_1$$
$$b_2 = -(23g_1 + 51)b_4/10g_1 \quad b_3 = -4(g_1 - 3)b_4/5g_1 \tag{5.40}$$

For $n \geq 1$, no non-trivial solution is derivable, as shown in Appendix D.

5.1.4.3   *Parabolically distributed axial load (m = 2)*

Consider the case $n = 0$; the differential equation (5.32) produces a homogeneous linear algebraic system of six equations for seven unknowns, specifically the coefficients b_i in the flexural rigidity and the critical load q_0:

$$-24b_1 + 24b_0 - q_0 L^3 = 0$$
$$-72b_2 + 72b_1 - q_0 L^3 (72 + 7g_1 + 4g_2) = 0$$
$$-144b_3 + 144b_2 + q_0 L^3 (30 + 6g_1 + 3g_2) = 0$$
$$-240b_4 + 240b_3 + q_0 L^3 (-16 + 12g_1) = 0 \tag{5.41}$$
$$-360b_5 + 360b_4 + q_0 L^3 (-10g_1 + 10g_2) = 0$$
$$504b_5 - 8q_0 L^3 g_2 = 0$$

leading to the critical value of $q_{0,\mathrm{cr}}$,

$$q_{0,\mathrm{cr}} = 63b_5 / g_2 L^3 \tag{5.42}$$

and the coefficients in $D(\xi)$,

$$b_0 = (115g_2 + 168g_1 + 336)b_5/80g_2 \quad b_1 = (115g_2 + 168g_1 + 126)b_5/80g_2$$
$$b_2 = -(165g_2 + 322g_1 + 714)b_5/80g_2 \quad b_3 = -(15g_2 + 28g_1 - 84)b_5/20g_2$$
$$b_4 = -(-3g_2 + 7g_1)b_5/4g_2$$

$$\tag{5.43}$$

The critical load in (5.42) is identical to the result obtained previously by Elishakoff and Guédé (2001) by other means. When $n = 1$, we have

$$-40b_1 - \tfrac{7}{3}q_0 L^3 = 0$$
$$-120b_2 + 120b_0 - q_0 L^3 \left(20 + \tfrac{37}{3}g_1 + \tfrac{20}{3}g_2\right) = 0$$
$$-240b_3 + 240b_1 + q_0 L^3 \left(30 - \tfrac{7}{3}g_2\right) = 0$$
$$-400b_4 + 400b_2 + q_0 L^3 \left(20 + 30g_1 + \tfrac{20}{3}g_2\right) = 0 \tag{5.44}$$
$$-600b_5 + 600b_3 + q_0 L^3 \left(-25 + \tfrac{50}{3}g_2\right) = 0$$
$$840b_4 - 15q_0 L^3 g_1 = 0$$
$$1120b_5 - \tfrac{35}{3}q_0 L^3 g_2 = 0$$

which represents a homogeneous set of seven algebraic equations for seven unknowns. In order to find a non-trivial solution the determinant of the system must be zero. This condition is expressed by the simple linear equation $11g_2 + 36 = 0$, resulting in

$$g_2 = -36/11 \tag{5.45}$$

Due to the negative sign of the parameter g_2, we check the sign of the load $N(\xi)$. We substitute Eq. (5.40) into the expression for the load $N(\xi)$ given in Eq. (5.31), for $m = 2$. We obtain

$$N(\xi) = -q_0 L \left[(11g_1 - 2) - 22\xi - 11g_1\xi^2 + 24\xi^3 \right] / 22 \tag{5.46}$$

which is not negative for arbitrary values of the parameter g_1. In the following, we consider the parameter g_1 so that the load is compressive. Upon substitution of Eq. (5.39) into equations of the set (5.38), we get for the critical load

$$q_{0,cr} = -88b_5/3L^3 \tag{5.47}$$

Since the critical load is positive, the coefficient b_5 must be negative. The coefficients in the flexural rigidity read

$$b_0 = \left(-\tfrac{253}{189}g_1 + \tfrac{14}{45} \right) b_5 \quad b_1 = 77b_5/45 \quad b_2 = \left(\tfrac{176}{105}g_1 - \tfrac{2}{15} \right) b_5$$
$$b_3 = -26b_5/3 \quad b_4 = -11g_1b_5/21 \tag{5.48}$$

For the flexural rigidity to agree with a realistic problem, it must be non-negative for any ξ in $[0; 1]$. A necessary condition to fulfill is the non-negativity of the coefficient b_0. Since the coefficient b_5 is negative this condition is expressed as follows:

$$-\frac{253}{189}g_1 + \frac{14}{45} \leq 0 \quad \text{or} \quad g_1 \geq \frac{294}{1265} \approx 0.2324 \tag{5.49}$$

Figure 5.4 depicts the flexural rigidity for various values of the parameter g_1, whereas Figure 5.5 portrays the corresponding load $N(\xi)$.

For $n = 2$, instead of (5.38) we get the following eight equations for seven unknowns:

$$-60b_1 - 4q_0L^3 = 0$$
$$-180b_2 - q_0L^3(30 + 19g_1 + 10g_2) = 0$$
$$-360b_3 + 360b_0 + q_0L^3(45 - 4g_2) = 0$$
$$-600b_4 + 600b_1 + 30q_0L^3g_1 = 0$$
$$-900b_5 + 900b_2 + q_0L^3(30 + 15g_1 + 35g_2) = 0 \tag{5.50}$$
$$1260b_3 - 36q_0L^3 = 0$$
$$1680b_4 - 21q_0L^3g_1 = 0$$
$$2160b_5 - 16q_0L^3g_2 = 0$$

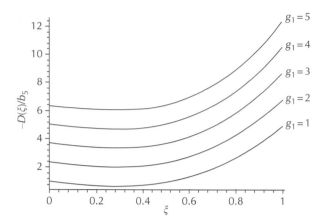

FIGURE 5.4
Variation of the flexural rigidity

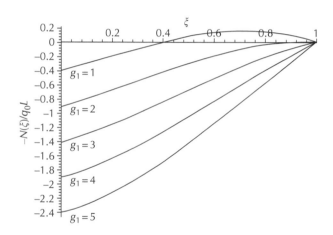

FIGURE 5.5
Variation of $N(\xi)$

To obtain a non-trivial solution the matrix of the system (5.50) must have a rank less than 7. According to Uspensky (1948), "a matrix is of rank p if it contains minors of order p different from 0, while all minors of order $p + 1$ (if there are such) are zero." Therefore, the matrix of (5.50) admits a rank less than 7, if all minors of order 7 are zero. This condition yields a set of eight linear algebraic equations among which two are identically zero, and the rest reduce to the following two equations with two unknowns, namely the parameters g_1 and g_2:

$$38{,}516{,}954{,}112 g_2 + 142{,}216{,}445{,}952 g_1 = -213{,}324{,}668{,}928$$
$$19{,}999{,}187{,}712 g_1 = -35{,}554{,}111{,}488$$

(5.51)

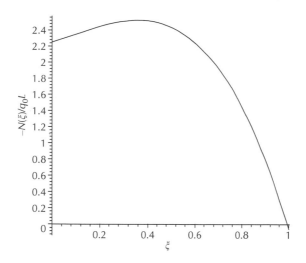

FIGURE 5.6
Variation of $N(\xi)$

whose solution is

$$g_1 = \tfrac{16}{9} \quad g_2 = -\tfrac{472}{39} \tag{5.52}$$

Note that the load $N(\xi)$ is entirely determined by (5.52). We check whether it is compressive. Substitution of (5.52) into (5.31), for $m = 2$, results in the load $N(\xi)$:

$$N(\xi) = -q_0 L \left(-\tfrac{251}{117} - \xi - \tfrac{8}{9}\xi^2 + \tfrac{472}{117}\xi^3 \right) \tag{5.53}$$

which is positive (see Figure 5.6) and, therefore, it does not correspond to a compressive load. We conclude that for $m = 2$, there can be a non-trivial solution, yet the system is not realistic. It is shown in Appendix E that when $n \geq 3$, only trivial solutions are produced.

5.1.4.4 Cubically distributed axial load (m = 3)

Consider $n = 0$; the governing differential equation again yields a homogeneous set of seven equations for eight unknowns:

$$24b_0 - 24b_1 - q_0 L^3 = 0$$

$$72b_1 - 72b_2 - q_0 L^3(12 + 7g_1 + 4g_2 + 3g_3) = 0$$

$$144b_2 - 144b_3 + q_0 L^3(30 + 6g_1 + 3g_2 + 3g_3) = 0$$

$$240b_3 - 240b_4 - q_0 L^3(16 - 12g_1 + g_3) = 0$$

$$360b_4 - 360b_5 - q_0 L^3 (1 + 10g_1 - 10g_2) = 0$$

$$504b_5 - 504b_6 - q_0 L^3 (8g_2 - 9g_3) = 0$$

$$672b_6 - 7q_0 L^3 g_3 = 0$$

$$(5.54)$$

which is solvable expressing all unknowns in terms of b_6, which is regarded as a parameter. We get thus, for the critical value of the buckling load,

$$q_{0,cr} = 96b_6/g_3 L^3 \qquad (5.55)$$

and the coefficients b_i in the flexural rigidity

$$b_0 = (177g_3 + 230g_2 + 336g_1 + 672)b_6/105g_3$$

$$b_1 = (177g_3 + 230g_2 + 336g_1 + 252)b_6/105g_3$$

$$b_2 = -(243g_3 + 330g_2 + 644g_1 + 1428)b_6/105g_3 \qquad (5.56)$$

$$b_3 = -(33g_3 + 120g_2 + 224g_1 + 672)b_6/105g_3$$

$$b_4 = (-15g_3 - 24g_2 + 56g_1)b_6/21g_3 \quad b_5 = (-15g_3 + 32g_2)b_6/21g_3$$

We note that Elishakoff and Guédé (2001) deduced the same result for the critical load in (5.55) as part of their general case.

Consider now $n = 1$; this time, we obtain a set of eight equations for eight unknowns,

$$-40b_1 - \tfrac{7}{3}q_0 L^3 = 0$$

$$-120b_2 + 120b_0 - q_0 L^3 \left(20 + \tfrac{37}{3}g_1 + \tfrac{20}{3}g_2 + 5g_3\right) = 0$$

$$-240b_3 + 240b_1 + q_0 L^3 \left(30 - \tfrac{7}{3}g_2\right) = 0$$

$$-400b_4 + 400b_2 + q_0 L^3 \left(20 + 30g_1 + \tfrac{7}{3}g_2 + \tfrac{8}{3}g_3\right) = 0 \qquad (5.57)$$

$$-600b_5 + 600b_3 + q_0 L^3 (-25 + \tfrac{50}{3}g_2) = 0$$

$$-840b_6 + 840b_4 + q_0 L^3 (-15g_1 + 15g_3)$$

$$1120b_5 - \tfrac{35}{3}q_0 L^3 g_2 = 0$$

$$1440b_5 - 10q_0 L^3 g_3 = 0$$

In order for this homogeneous linear algebraic system to have a non-trivial solution, its determinant must vanish, leading to the following determinantal equation:

$$2{,}861{,}236{,}244g_2 + 9{,}364{,}045{,}824 = 0 \qquad (5.58)$$

from which we get the following expression for the parameter g_2:

$$g_2 = -\frac{36}{11} \tag{5.59}$$

We remark that this expression, suprisingly, coincides with the parameter g_2 obtained for the parabolically varying axial load in Eq. (5.45). We have to verify whether the load obained for g_2 given in Eq. (5.59) corresponds to a compressive load. The load $N(\xi)$ is given by

$$N(\xi) = -q_0 L \left[(11g_3 + 22g_1 - 4) - 44\xi - 22g_1\xi^2 + 48\xi^3 - 11g_3\xi^4 \right]/44 \tag{5.60}$$

One of the parameters g_1 or g_3 can be treated as an arbitrary constant, while the other is chosen straightforwardly enough so that the load $N(\xi)$ remains negative for ξ within $[0; 1]$. Once the parameters g_1 and g_3 are chosen judiciously, the critical buckling load is given by

$$q_{0,cr} = 144b_6/g_3L^3 \tag{5.61}$$

The coefficients of the flexural rigidity are

$$b_0 = (6{,}677g_3 + 12{,}650g_1 - 2{,}940)b_6/1{,}925g_3 \quad b_1 = -42b_6/5g_3$$
$$b_2 = (4{,}873g_3 + 15{,}840g_1 + 1{,}260)b_6/1{,}925g_3 \quad b_3 = 156b_6/11g_3 \tag{5.62}$$
$$b_4 = -(11g_3 - 18g_1)b_6/7g_3 \quad b_5 = -54b_6/11g_3$$

Likewise, the non-negativity of the flexural rigidity, for the problem to be realistic, can be checked directly. For $n = 2$ instead of (5.57) we obtain

$$-60b_1 - 4q_0L^3 = 0$$

$$-180b_2 - q_0L^3\left(30 + 19g_1 + 10g_2 + \tfrac{15}{2}g_3\right) = 0$$

$$360b_3 + 360b_0 + q_0L^3(45 - 4g_2) = 0$$

$$-600b_4 + 400b_1 + q_0L^3(30g_1 - 4g_3) = 0$$

$$-900b_5 + 900b_2 + q_0L^3\left(30 + 15g_1 + 35g_2 + \tfrac{15}{2}g_3\right) = 0 \tag{5.63}$$

$$-1260b_6 + 1260b_3 + q_0L^3\left(-36 + \tfrac{45}{2}g_3\right) = 0$$

$$1680b_4 - 21q_0L^3g_1 = 0$$

$$2160b_5 - 16q_0L^3g_2 = 0$$

$$2700b_6 - \tfrac{27}{2}q_0L^3g_3 = 0$$

which represents a homogeneous set of nine algebraic equations always for eight unknowns. In order to find a non-trivial solution, the matrix of the system (5.63) must have a rank less than 7. According to Uspensky's (1948) definition of the rank, the matrix of (5.63) admits a rank less than 7, if all minors of order 8 are zero. This conditions yields a set of nine linear algebraic equations with three unknowns as the parameters g_i whose solution is

$$g_1 = \tfrac{8}{45}g_3 + \tfrac{16}{9} \quad g_2 = -\tfrac{398}{195}g_3 - \tfrac{472}{39} \tag{5.64}$$

where g_3 is an arbitrary parameter. The load obtainable from substitution of Eq. (5.64) into (5.31) for $m = 3$, is

$$N(\xi) = q_0L[(5020 + 799g_3) + 2340\xi + (2080 + 208g_3)\xi^2$$
$$- (1592g_3 + 9440)\xi^3 + 585g_3\xi^4]/2340 \tag{5.65}$$

A necessary condition for $N(\xi)$ to be compressive is

$$5020 + 799g_3 \leq 0 \tag{5.66}$$

or $g_3 \leq -5020/799 \approx 6.29$. From Eq. (5.66) we note that the parameter g_3 has to be negative. Hereinafter g_3 is taken to satisfy the condition (5.66). Hence, upon substitution of (5.54) into the set (5.53), we arrive at the following critical buckling load:

$$q_{0,cr} = 200b_6/g_3L^3 \tag{5.67}$$

Since the critical value of the buckling load is positive and g_3 is negative, the coefficient b_6 must be negative. The coefficients of the flexural rigidity read

$$b_0 = (17{,}462g_3 - 113{,}465)b_6/2{,}457g_3 \quad b_1 = -40b_6/3g_3$$
$$b_2 = -(11{,}153g_3 + 66{,}980)b_6/1{,}053g_3 \quad b_3 = -2(9g_3 - 20)b_6/7g_3 \tag{5.68}$$
$$b_4 = 4(g_3 + 10)b_6/9g_3 \quad b_5 = -16(199g_3 + 1{,}180)b_6/1{,}053g_3$$

We conclude that the coefficients in the flexural rigidity and the critical buckling load are given, respectively, in (5.62) and (5.61), provided the distributed buckling load equals

$$q(\xi) = 1 + \left(\tfrac{8}{45}g_3 + \tfrac{16}{9}\right)\xi - \left(\tfrac{398}{195}g_3 + \tfrac{472}{39}\right)\xi^2 + g_3\xi^3 \tag{5.69}$$

In order for the problem to be physically realizable, we demand that the flexural rigidity remains non-negative when ξ belongs to the interval $[0; 1]$. This requirement forces the value of the flexural rigidity at $\xi = 0$ to be positive, leading to the following inequality:

$$113{,}465 + 17{,}462g_3 \leq 0 \tag{5.70}$$

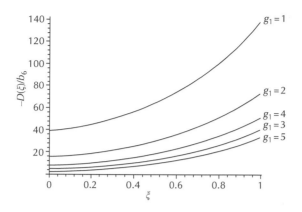

FIGURE 5.7
Variation of the flexural rigidity

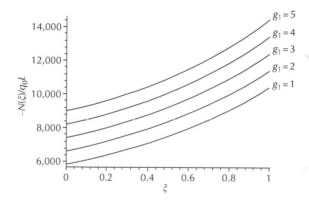

FIGURE 5.8
Variation of $N(\xi)$

for $g_3 \leq -113{,}465/17{,}462 \approx 6.5$. Figure 5.7 depicts the flexural rigidity for various values of the parameter g_3, while Figure 5.8 presents the corresponding distribution of $N(\xi)$, for the buckling value q_0.

5.1.4.5 Quartically distributed axial load (m = 4)

Let us take $n = 0$; the governing differential equation again yields a homogeneous set of eight equations for nine unknowns:

$$24b_0 - 24b_1 - q_0L^3 = 0$$

$$72b_1 - 72b_2 - q_0L^3\left(12 + 7g_1 + 4g_2 + 3g_3 + \tfrac{12}{5}g_4\right) = 0$$

$$144b_2 - 144b_3 + q_0L^3\left(30 + 6g_1 + 3g_2 + 3g_3 + \tfrac{12}{5}g_4\right) = 0$$

$$240b_3 - 240b_4 - q_0L^3(16 - 12g_1 + g_3) = 0$$

$$360b_4 - 360b_5 - q_0L^3(10g_1 - 10g_2 - g_4) = 0$$

$$504b_5 - 504b_6 - q_0L^3(8g_2 - 9g_3) = 0$$

$$672b_6 - 672b_7 - q_0L^3\left(-7g_3 + \tfrac{42}{5}g_4\right) = 0$$

$$864b_7 - \tfrac{32}{5}q_0L^3g_4 = 0$$

$$\text{(5.71)}$$

which is solvable by expressing all unknowns in terms of b_6, which is regarded as an arbitrary parameter. We get for the critical value of the buckling load,

$$q_{0,cr} = 135b_7/g_4L^3 \tag{5.72}$$

The coefficients b_i in the flexural rigidity read

$$b_0 = (434g_4 + 531g_3 + 690g_2 + 1008g_1 + 2016)b_7/224g_4$$

$$b_1 = (434g_4 + 531g_3 + 690g_2 + 1008g_1 + 756)b_7/224g_4$$

$$b_2 = -(574g_4 + 729g_3 + 990g_2 + 1932g_1 + 4284)b_7/224g_4$$

$$b_3 = -(70g_4 + 99g_3 + 360g_2 + 672g_1 - 2016)b_7/224g_4$$

$$b_4 = -5(14g_4 + 45g_3 + 72g_2 + 168g_1)b_7/224g_4$$

$$b_5 = -(154g_4 + 225g_3 - 480g_2)b_7/224g_4 \quad b_6 = (45g_3 - 22g_4)b_7/32g_4$$

$$\text{(5.73)}$$

We note that Elishakoff and Guédé (2001) deduced the same result for the critical load in (5.60) as a particular case analysis. They did not consider, however, the case for which $n \geq 1$.

Consider now $n = 1$; we obtain a set of nine equations for nine unknowns:

$$-40b_1 - \tfrac{7}{3}q_0L^3 = 0$$

$$-120b_2 + 120b_0 - q_0L^3\left(20 + \tfrac{37}{3}g_1 + \tfrac{20}{3}g_2 + 5g_3 + 4g_4\right) = 0$$

$$-240b_3 + 240b_1 + q_0L^3\left(30 - \tfrac{7}{3}g_2\right) = 0$$

$$-400b_4 + 400b_2 + q_0L^3\left(20 + 30g_1 + \tfrac{20}{3}g_2 + \tfrac{8}{3}g_3 + 4g_4\right) = 0$$

$$-600b_5 + 600b_3 + q_0L^3\left(-25 + \tfrac{50}{3}g_2 - \tfrac{7}{3}g_4\right) = 0 \qquad \text{(5.74)}$$

$$-840b_6 + 840b_4 + q_0L^3(-15g_1 + 15g_3)$$

$$-1120b_7 + 1120b_5 + q_0L^3\left(-\tfrac{35}{3}g_2 + 14g_4\right)$$

$$1440b_6 - 10q_0L^3g_3 = 0$$

$$1800b_7 - 9q_0L^3g_4 = 0$$

The determinantal equation reads

$$243{,}465{,}191{,}424g_4 + 51{,}502{,}252{,}032g_2 + 168{,}552{,}824{,}832 = 0 \qquad (5.75)$$

from which we get the following expression for the parameter g_2

$$g_2 = -\tfrac{26}{55}g_4 - \tfrac{36}{11} \qquad (5.76)$$

The critical buckling load is given as

$$q_{0,cr} = 200b_7/g_4 L^3 \qquad (5.77)$$

The coefficients of the flexural rigidity are, therefore,

$$b_0 = (6{,}677g_3 + 12{,}650g_1 + 1{,}372g_4 - 2{,}940)b_7/1{,}386g_4 \quad b_1 = -35b_7/3g_4$$
$$b_2 = -(4{,}873g_3 + 15{,}840g_1 + 588g_4 - 1{,}260)b_7/1{,}386g_4$$
$$b_3 = 13(150 + 7g_4)b_7/99g_4 \quad b_4 = -25(11g_3 - 18g_1)b_7/126g_4$$
$$b_5 = -(225 + 82g_4)b_7/33g_4 \quad b_6 = 25g_3b_7/18g_4$$

$$(5.78)$$

The physical realizability of the problem, which requires the flexural rigidity to be non-negative can be checked straightforwardly for any set of parameters g_1, g_3 and g_4.

For $n = 2$, instead of (5.74) we obtain

$$-60b_1 - 4q_0L^3 = 0$$
$$-180b_2 - q_0L^3\left(30 + 19g_1 + 10g_2 + \tfrac{15}{2}g_3 + 6g_4\right) = 0$$
$$-360b_3 + 360b_0 + q_0L^3(45 - 4g_2) = 0$$
$$-600b_4 + 600b_1 + q_0L^3(30g_1 - 4g_3) = 0$$
$$-900b_5 + 900b_2 + q_0L^3\left(30 + 15g_1 + 35g_2 + \tfrac{15}{2}g_3 + 2g_4\right) = 0 \qquad (5.79)$$
$$-1260b_6 + 1260b_3 + q_0L^3\left(-36 + \tfrac{45}{2}g_3\right) = 0$$
$$-1680b_7 + 1680b_4 + q_0L^3(-21g_1 + 21g_4) = 0$$
$$2160b_5 - 16q_0L^3g_2 = 0$$
$$2700b_6 - \tfrac{27}{2}q_0L^3g_3 = 0$$
$$3300b_7 - 12q_0L^3g_4 = 0$$

which represents a homogeneous set of ten algebraic equations always for nine unknowns. In order to find a non-trivial solution, the matrix of the system (5.78) must have a rank less than 9. The matrix of (5.78) admits a rank less than 8, if all minors of order 9 are zero. This condition yields a set of ten linear algebraic equations with four unknowns as the parameters g_i whose solution is

$$g_2 = -\frac{8961}{2860}g_4 - \frac{597}{52}g_1 + \frac{108}{13} \quad g_3 = \frac{117}{88}g_4 + \frac{45}{8}g_1 - 10 \tag{5.80}$$

where g_1 and g_4 are arbitrary parameters. Upon substitution of (5.80) into the set (5.79), we arrive at the following critical buckling load:

$$q_{0,cr} = 275b_7/g_4L^3 \tag{5.81}$$

The coefficients of the flexural rigidity are

$$b_0 = (124{,}703g_4 + 480{,}205g_1 - 298{,}980)b_7/8{,}736g_4 \quad b_1 = -55b_7/3g_4$$
$$b_2 = 55(11{,}153g_1 + 3{,}195g_4 - 7{,}920)b_7/7{,}488g_4$$
$$b_3 = -(1{,}053g_4 + 4{,}455g_1 - 9{,}680)b_7/224g_4$$
$$b_4 = -(39g_4 - 55g_1)b_7/16g_4 \tag{5.82}$$
$$b_5 = (7{,}920 - 10{,}945g_1 + 2{,}987g_4)b_7/468g_4$$
$$b_6 = (117g_4 + 495g_1 - 880)b_7/64g_4$$

We conclude that for the physically realizable problems the coefficients in the flexural rigidity and the critical buckling load are given, respectively, in (5.82) and (5.83), provided the distributed buckling load equals

$$q(\xi) = 1 + g_1\xi + \left(-\frac{8961}{2860}g_4 - \frac{597}{52}g_1 + \frac{108}{13}\right)\xi^2$$
$$+ \left(\frac{117}{88}g_4 + \frac{45}{8}g_1 - 10\right)\xi^3 + g_4\xi^4 \tag{5.83}$$

For $n = 3$, we have a homogeneous set of 12 linear algebraic equations for 9 unknowns:

$$-84b_1 - 6q_0L^3 = 0$$

$$-252b_2 - q_0L^3\left(42 + 27g_1 + 14g_2 + \tfrac{21}{2}g_3 + \tfrac{42}{5}g_4\right) = 0$$

$$-504b_3 + q_0L^3(63 - 6g_2) = 0$$

$$-840b_4 + 840b_0 + q_0L^3(42g_1 - 6g_3) = 0$$

$$-1260b_5 + 1260b_1 + q_0L^3(35g_2 - 6g_4) = 0$$

$$-1764b_6 + 1764b_2 + q_0L^3\left(42 + 21g_1 + 14g_2 + 42g_3 + \tfrac{42}{5}g_4\right) = 0$$

$$-1260b_6 + 1260b_3 + q_0L^3\left(-36 + \tfrac{42}{5}g_3\right) = 0$$

$$-2352b_7 + 2352b_3 + q_0L^3\left(-49 + \tfrac{147}{5}g_4\right) = 0$$

$$3024b_4 - 28q_0L^3g_1 = 0$$

$$3780b_5 - 21q_0L^3g_2 = 0$$

$$4620b_6 - \tfrac{35}{2}q_0L^3g_3 = 0$$

$$5544b_7 - \tfrac{77}{5}q_0L^3g_4 = 0$$

$$(5.84)$$

In order to find a non-trivial solution, the matrix of the system (5.84) must have a rank less than 9. The matrix of (5.57) admits a rank less than 8 if all minors of order 9 are zero. This condition yields a set of 55 linear algebraic equations with four unknowns as the parameters g_i whose solution is

$$g_2 = \frac{315}{253} \quad g_3 = \frac{321}{115} - \frac{22}{5}g_1 \quad g_4 = -\frac{2325}{253} \qquad (5.85)$$

where g_1 is an arbitrary parameter. Hence, the axial load equals

$$N(\xi) = -q_0L[1391 - 3060g_1 - 5060\xi - 2530g_1\xi^2 - 2100\xi^3$$
$$+ (5566g_1 - 3531)\xi^4 + 9300\xi^5]/5060$$

Again, we demand the load $N(\xi)$ to be negative for any ξ within the interval $[0; 1]$. This leads to the following necessary condition

$$1391 - 3060g_1 \geq 0$$

or $g_1 \leq 1391/3060 \approx 0.46$. Upon substitution of (5.85) into the set (5.84), we arrive at the following critical buckling load

$$q_{0,\mathrm{cr}} = -6072b_7/155L^3$$

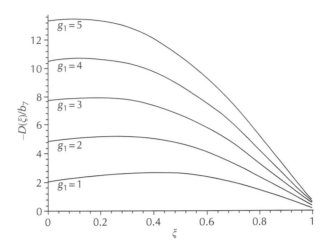

FIGURE 5.9
Variation of the flexural rigidity

Since the critical value of the buckling load must be positive, the coefficient b_7 must be negative. The coefficients of the flexural rigidity read

$$b_0 = \left(\frac{22{,}264}{7{,}875} g_1 - \frac{21{,}186}{27{,}125} \right) b_7 \quad b_1 = \frac{3{,}036}{1{,}085} b_7 \quad b_2 = \left(\frac{1{,}391}{775} - \frac{16{,}192}{5{,}425} g_1 \right) b_7$$

$$b_3 = -\frac{669}{155} b_7 \quad b_4 = -\frac{506}{1{,}395} g_1 b_7 \quad b_5 = -\frac{42}{155} b_7 \quad b_6 = \left(\frac{506}{775} g_1 - \frac{321}{775} \right) b_7$$

The physical realizability of the problem forces the flexural rigidity to be non-negative for any ξ in $[0; 1]$. A necessary condition is given by $D(0) \geq 0$, or

$$\frac{22{,}264}{7{,}875} g_1 - \frac{21{,}186}{27{,}125} \leq 0$$

or $g_1 \leq 8667/31372 \approx 0.28$. Figure 5.9 shows the variation of the flexural rigidity for different values of the parameter g_1, and Figure 5.10 depicts the corresponding load $N(\xi)$.

5.1.5   Concluding Remarks

The postulated buckling modes constitute deflections of uniform beams under distributed loading varying with the coordinate as a polynomial. Exact closed-form solutions are obtained for both the buckling loads and the associated flexural rigidity distributions. We also list, in order to obtain a complete picture, the cases in which the polynomial representation does not lead to a non-trivial solution.

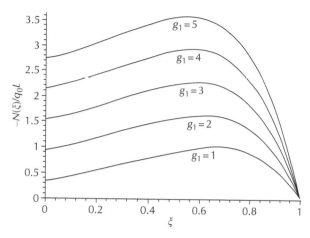

FIGURE 5.10
Variation of $N(\xi)$

The buckling loads constitute simple rational expressions, which are much simpler than the known exact solutions for uniform columns, the latter involving Bessel and/or Lommel functions. It is hoped that the closed-form solutions reported in this section will be utilized as benchmark solutions for designing tailored structures with specified buckling loads. In particular, it appears that having such closed-form solutions could be very valuable to computational code developers for verification. It is envisioned that such solutions can be *embedded* in larger purpose computational tools.

5.2 Family of Analytical Polynomial Solutions for Pinned Inhomogeneous Beams. Part 2: Vibration

5.2.1 Introductory Comments

As was mentioned already, Duncan (1937) constructed an example with a closed-form solution of the buckling load of a variable cross-section rod. The mode shape was taken as a polynomial of fifth degree. Elishakoff and Rollot (1999) found cases in which the simpler, fourth-order polynomial serves as an exact buckling mode of the flexural rigidity column. Later on, Elishakoff and Candan (2000) found numerous solutions for natural frequencies of vibrating beams with *variable* density and *variable* flexural rigidity, apparently for the first time in the literature. They postulated for the fourth-order polynomial to serve as an exact mode shape. In the present section we pose the following question: is the simplest fourth-order polynomial the only possible mode shape for the inhomogeneous beam, or are higher order polynomial mode shapes also acceptable? Note that the fourth-order polynomial represents an exact static deflection of the beam under the uniform load q_0. One can generate the higher order polynomials as the static deflections of the uniform beams to the loading represented as $q_0\xi^n$, where q_0 is a constant, ξ the non-dimensional coordinate and n a positive integer. The static deflections so obtained satisfy all the boundary conditions. The question is then re-formulated as follows: can such a static deflection serve as an exact mode shape of an inhomogeneous beam? For the case $n = 0$, the reply to this question was shown to be in the affirmative by Elishakoff and Candan (2000). In this section we address cases when n differs from zero, and obtain a series of new closed-form solutions.

The authors are unaware of any work in which the polynomial expressions serve as exact, closed-form mode shapes, other than this and the one by Candan and Elishakoff (2000).

5.2.2   Formulation of the Problem

Consider a uniform beam, pinned at both its ends. It is subjected to the load $q(\xi) = q_0\xi^n$. The governing differential equation reads:

$$D_0\frac{d^4w}{dx^4} = q_0x^n$$

where D_0 is the flexural rigidity of the beam, $w(x)$ the static displacement, x the axial coordinate and q_0 the load parameter. We introduce the non-dimensional coordinate

$$\xi = x/L \tag{5.86}$$

where L is the length. The equation becomes

$$D_0\frac{d^4w}{d\xi^4} = \alpha\xi^n \tag{5.87}$$

The solution of this equation, which satisfies the boundary conditions, reads:

$$w(\xi) = \frac{\alpha L^{n+4}}{(n+1)(n+2)(n+3)(n+4)D_0}\psi(\xi) \tag{5.88}$$

$$\psi(\xi) = \xi^{n+4} - \tfrac{1}{6}(n^2 + 7n + 12)\xi^3 + \tfrac{1}{6}(n^2 + 7n + 6)\xi$$

We address the following problem: can the function $\psi(\xi)$ serve as an exact mode shape for the inhomogeneous beam? We will explore cases of reconstruction of the flexural rigidity $D(\xi)$ for various axial variations of the inertial coefficient $R(\xi) = \rho(\xi)A$.

5.2.3 Basic Equations

The differential equation that governs the mode shape $W(\xi)$ of the inhomogeneous beam reads:

$$\frac{d^2}{dx^2}\left[D(x)\frac{d^2 W}{dx^2}\right] - \omega^2 R(x)W(x) = 0 \tag{5.89}$$

where $W(x)$ is the mode shape in contrast to the function $w(x)$ denoting the static displacement in an auxiliary problem. The function $D(x)$ constitutes the flexural rigidity that should be *reconstructed*, and $R(x)$ is the inertial coefficient.

With the non-dimensional coordinate ξ from Eq. (5.86), the governing equation becomes

$$\frac{d^2}{d\xi^2}\left[D(\xi)\frac{d^2 \psi}{d\xi^2}\right] - \Omega^2 R(\xi)\psi(\xi) = 0 \tag{5.90}$$

where the substitution

$$w(\xi) = \psi(\xi) \tag{5.91}$$

has been made. We are confronted with determining $R(\xi)$ and $D(\xi)$ such that $\psi(\xi)$ serves as an exact mode shape. The variation of the inertial coefficient is taken as follows:

$$R(\xi) = \sum_{i=0}^{m} a_i \xi^i \tag{5.92}$$

where a_i $(i = 0, 1, \ldots, m)$ are given coefficients. The second term of the governing equation (5.90) is a polynomial function whose degree is $n + m + 4$. In order to achieve the same polynomial order in the first term of this equation, the flexural rigidity $D(\xi)$ has to be taken as a polynomial of degree $m + 4$:

$$D(\xi) = \sum_{i=0}^{m+4} b_i \xi^i \tag{5.93}$$

The left-hand side of the governing equation (5.90) becomes a polynomial in terms of ξ; in order for it to vanish for every ξ in the interval $[0;1]$, we have to demand that the coefficients of ξ raised to any power, be zero. This leads to a set of $n + m + 5$ linear algebraic equations. The problem is posed as one of *reconstruction* of the flexural rigidity when the inertial coefficient and the mode shape are given. We have to determine the coefficients b_i of the flexural rigidity as well as the natural frequency ω, which represent $m + 6$ unknowns, namely, $m + 5$ coefficients of the flexural rigidity function and the $(n + 6)$th unknown natural frequency ω.

Note that when n is zero, we get a triangular system of $m + 5$ equations for $m + 6$ unknowns as studied by Elishakoff and Candan (2000). They succeeded in expressing the unknowns in terms of one of the coefficients b_i, namely b_{m+4}, treated as an arbitrary constant. Yet, when n is larger than zero, the number of equations is greater than the number of unknowns. It is shown that in some cases new closed-form solutions are obtained. In other cases, as has been shown in the Appendices G–J, only a trivial solution can be established. We will specify the order of the polynomial on the right-hand side of Eq. (5.92) and pursue the determination of the coefficients b_i in Eq. (5.93).

5.2.4   Constant Inertial Coefficient ($m = 0$)

Consider first the case $n = 0$; the pre-selected mode shape is given by

$$\psi(\xi) = \xi^4 - 2\xi^3 + \xi \tag{5.94}$$

The inertial coefficient reduces to $R(\xi) = a_0$. The governing differential equation leads to the following set of equations:

$$-24b_1 + 24b_0 = 0$$
$$-72b_2 + 72b_1 - a_0\Omega^2 = 0$$
$$-144b_3 + 144b_2 = 0 \tag{5.95}$$
$$-240b_4 + 240b_3 + 2a_0\Omega^2 = 0$$
$$360b_4 - a_0\Omega^2 = 0$$

We have five equations for six unknowns: b_0, b_1, b_2, b_3, b_4 and Ω^2. We declare b_4 as an arbitrary parameter. We get Ω^2 from the last equation of the set (5.95):

$$\Omega^2 = 360b_4/a_0 \tag{5.96}$$

The rest of the equations lead to

$$b_0 = 3b_4 \quad b_1 = 3b_4 \quad b_2 = -2b_4 \quad b_3 = -2b_4 \tag{5.97}$$

The flexural rigidity is then given by the following expression:

$$D(\xi) = (3 + 3\xi - 2\xi^2 - 2\xi^3 + \xi^4)b_4 \tag{5.98}$$

which is positive within the interval [0; 1] (see Figure 5.11).

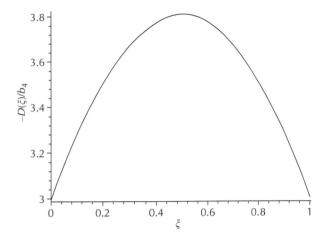

FIGURE 5.11
Variation of the flexural rigidity

This expression along with Eq. (5.12) was derived previously by Elishakoff and Candan (2000). Consider now the case $n = 1$. Instead of Eq. (5.95) we obtain

$$b_1 = 0$$

$$-120b_2 + 120b_0 - \tfrac{7}{3}a_0\Omega^2 = 0$$

$$-b_3 + b_1 = 0$$

$$-400b_4 + 400b_2 + \tfrac{10}{3}a_0\Omega^2 = 0 \quad (5.99)$$

$$b_3 = 0$$

$$840b_4 - a_0\Omega^2 = 0$$

We get a linear system of six equations for six unknowns. Since the determinant of the system (5.99) vanishes

$$\begin{vmatrix} 0 & 1 & 0 & 0 & 0 & 0 \\ 120 & 0 & -120 & 0 & 0 & -\tfrac{7}{3}a_0 \\ 0 & 1 & 0 & -1 & 0 & 0 \\ 0 & 0 & 400 & 0 & 400 & \tfrac{10}{3}a_0 \\ 0 & 0 & 0 & 1 & 0 & 0 \\ 0 & 0 & 0 & 0 & 840 & -a_0 \end{vmatrix} = 0 \quad (5.100)$$

we can solve the set, expressing the unknowns with b_4 taken as an arbitrary constant. Thus, we get

$$\Omega^2 = 840b_4/a_0 \quad (5.101)$$

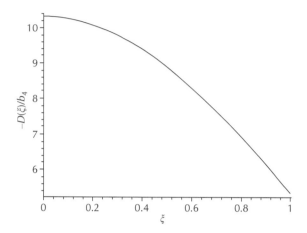

FIGURE 5.12
Variation of the flexural rigidity

and the coefficients b_i are

$$b_0 = \tfrac{31}{3}b_4 \quad b_1 = 0 \quad b_2 = -6b_4 \quad b_3 = 0 \tag{5.102}$$

The flexural rigidity,

$$D(\xi) = \left(\tfrac{31}{3} - 6\xi^2 + \xi^4 \right) b_4 \tag{5.103}$$

which is depicted in Figure 5.12, is positive within the interval $[0; 1]$. It is shown in Appendix G that, for n larger than unity, only trivial solutions are obtainable for pinned beams with constant mass density.

5.2.5 Linearly Varying Inertial Coefficient ($m = 1$)

Consider first the case $n = 0$; instead of (5.95) we have

$$-24b_1 + 24b_0 = 0$$

$$-72b_2 + 72b_1 - a_0\Omega^2 = 0$$

$$-144b_3 + 144b_2 - a_1\Omega^2 = 0$$

$$-240b_4 + 240b_3 + 2a_0\Omega^2 = 0 \tag{5.104}$$

$$-360b_5 + 360b_4 - (a_0 - 2a_1)\Omega^2 = 0$$

$$504b_5 - a_1\Omega^2 = 0$$

This set represents a linear algebraic system of six equations for the seven unknowns, $b_0, b_1, \ldots, b_5$ and Ω^2. Hence, we express them in terms of b_5, treated as an arbitrary constant. We get Ω^2 from the last equation of the set (5.104):

$$\Omega^2 = 504b_5/a_1 \tag{5.105}$$

while the remaining equations lead to

$$
\begin{aligned}
&b_0 = (1/10a_1)(17a_1 + 42a_0)b_5 \quad b_1 = (1/10a_1)(17a_1 + 42a_0)b_5 \\
&b_2 = (1/10a_1)(17a_1 - 28a_0)b_5 \quad b_3 = -(1/5a_1)(9a_1 + 14a_0)b_5 \quad \text{(5.106)} \\
&b_4 = -(1/5a_1)(9a_1 - 7a_0)b_5
\end{aligned}
$$

For pinned beams with linearly varying mass density no polynomial solutions can be derived for n larger than zero (see Appendix H).

5.2.6 Parabolically Varying Inertial Coefficient ($m = 2$)

We consider first the case $n = 0$. We obtain the following set of seven equations with eight unknowns:

$$
\begin{aligned}
&-b_1 + b_0 = 0 \\
&72b_2 + 72b_1 - a_0\Omega^2 = 0 \\
&-144b_3 + 144b_2 - a_1\Omega^2 = 0 \\
&-240b_4 + 240b_3 + 2a_0\Omega^2 - a_2\Omega^2 = 0 \quad \text{(5.107)} \\
&-360b_5 + 360b_4 - a_0\Omega^2 + 2a_1\Omega^2 = 0 \\
&-504b_6 + 504b_5 - a_1\Omega^2 + 2a_2\Omega^2 = 0 \\
&672b_6 - a_2\Omega^2 = 0
\end{aligned}
$$

We can, therefore, find a non-trivial solution when b_6 is regarded as a parameter. From the last equation of the set (5.107), we obtain the natural frequency squared:

$$\Omega^2 = 672b_6/a_2 \tag{5.108}$$

The remaining equations yield the coefficients in the flexural rigidity as follows:

$$
\begin{aligned}
&b_0 = (84a_0 + 34a_1 + 17a_2)b_6/15a_2 \quad b_1 = (84a_0 + 34a_1 + 17a_2)b_6/15a_2 \\
&b_2 = -(56a_0 + 34a_1 + 17a_2)b_6/15a_2 \quad b_3 = -(56a_0 + 36a_1 + 17a_2)b_6/15a_2 \\
&b_4 = (28a_0 - 36a_1 - 25a_2)b_6/15a_2 \quad b_5 = (4a_1 - 5a_2)b_6/3a_2
\end{aligned}
$$

$$\tag{5.109}$$

When $n = 1$, the governing differential equation yields a set of eight linear algebraic equations for eight unknowns, namely the coefficients b_i in the flexural rigidity and the natural frequency squared Ω^2:

$$b_1 = 0 \quad -120b_2 + 120b_0 - \tfrac{7}{3}r_0\Omega^2 = 0$$

$$-240b_3 + 240b_1 - \tfrac{7}{3}r_1\Omega^2 = 0$$

$$-400b_4 + 400b_2 + \tfrac{10}{3}r_0\Omega^2 - \tfrac{7}{3}r_2\Omega^2 = 0$$

$$-600b_5 + 600b_3 + \tfrac{10}{3}r_1\Omega^2 = 0 \tag{5.110}$$

$$-840b_6 + 840b_4 - r_0\Omega^2 + \tfrac{10}{3}r_2\Omega^2 = 0$$

$$1120b_5 - r_1\Omega^2 = 0 \quad 1440b_6 - r_2\Omega^2 = 0$$

The determinant of the system, evaluated by the aid of the computerized algebraic code MAPLE®, is $1{,}895{,}104{,}512{,}000{,}000{,}000a_1$. In order to find a non-trivial solution, the determinant must be zero, leading to the requirement $a_1 = 0$. We conclude that for an inertial coefficient equal to

$$R(\xi) = a_0 + a_2\xi^2 \tag{5.111}$$

where a_0 and a_2 are arbitrary, the set (5.110) yields

$$\Omega^2 = 1440b_6/r_2 \tag{5.112}$$

and

$$b_0 = (620a_0 + 129a_2)b_6/35a_2 \quad b_1 = 0 \quad b_2 = -3(120a_0 - 43a_2)b_6/35a_2$$

$$b_3 = 0 \quad b_4 = 3(4a_0 - 11a_2)b_6/7a_2 \quad b_5 = 0 \tag{5.113}$$

Concerning the case $n = 2$, the governing equation yields a set of nine equations for eight unknowns, namely the coefficients b_i in the flexural rigidity and the natural frequency squared, Ω^2:

$$b_1 = 0 \quad -180b_2 - 4r_0\Omega^2 = 0$$

$$-360b_3 + 360b_0 - 4r_1\Omega^2 = 0$$

$$-600b_4 + 400b_1 + 5r_0\Omega^2 - 4r_2\Omega^2 = 0$$

$$-900b_5 + 900b_2 + 5r_1\Omega^2 = 0$$

$$-1260b_6 + 900b_3 + 5r_2\Omega^2 = 0 \quad 1680b_4 - r_0\Omega^2 = 0$$

$$2160b_5 - r_1\Omega^2 = 0 \quad 2700b_6 - r_2\Omega^2 = 0$$

$$(5.114)$$

Since the number of equations is larger than the number of unknowns, in order to have a non-trivial solution, the rank of the set (5.114) must be less than 8. According to the definition of the rank, a matrix is of rank p if it contains minors of order p different from 0, while all minors of order $p+1$ (if there are such) are zero (Uspensky, 1948). Accordingly, the rank of the matrix of the system (5.114) is less than 8 if all minors of order 8 vanish. This last condition leads to a set of nine equations out of which three are identically zero; the remaining equations can be reduced to a set of two equations as follows:

$$- 219{,}991{,}064{,}832a_1 + 959{,}961{,}010{,}176a_2 = 0$$

$$- 111{,}424{,}045{,}824a_0 + 959{,}961{,}010{,}176a_2 = 0$$

$$(5.115)$$

The fulfillment of (5.115) yields

$$a_0 = \frac{56}{65}a_2 \quad a_1 = \frac{2688}{715}a_2 \qquad (5.116)$$

where r_2 is treated as an arbitrary constant. We deduced that when the inertial coefficient is

$$R(\xi) = \left(\frac{56}{65} + \frac{2688}{715}\xi + \xi^2\right) a_2 \qquad (5.117)$$

with an arbitrary parameter a_2, the set (5.114) yields

$$\Omega^2 = 2{,}700b_6/a_2 \qquad (5.118)$$

$$b_0 = \frac{103{,}172}{1{,}001}b_6 \quad b_1 = 0 \quad b_2 = -\frac{672}{13}b_6 \quad b_3 = -\frac{68}{7}b_6 \quad b_4 = \frac{18}{13}b_6 \quad b_5 = \frac{672}{143}b_6$$

$$(5.119)$$

The flexural rigidity is then

$$D(\xi) = \left(\frac{103{,}172}{1{,}001} - \frac{672}{13}\xi^2 - \frac{68}{7}\xi^3 + \frac{18}{13}\xi^4 + \frac{673}{143}\xi^5 + \xi^6\right) b_6 \qquad (5.120)$$

which is positive within the interval [0; 1] (see Figure 5.13).

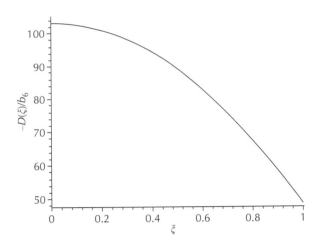

FIGURE 5.13
Variation of the flexural rigidity

For $n = 3$, we have to satisfy the following set of ten equations for eight unknowns

$$b_1 = 0 \quad - 252b_2 - 6a_0\Omega^2 = 0$$

$$-504b_3 - 6a_1\Omega^2 = 0$$

$$-840b_4 + 840b_0 + 7a_0\Omega^2 - 6a_2\Omega^2 = 0$$

$$-1260b_5 + 1260b_1 + 7a_1\Omega^2 = 0 \qquad (5.121)$$

$$-1764b_6 + 1764b_2 + 7a_2\Omega^2 = 0 \quad b_3 = 0$$

$$3024b_4 - a_0\Omega^2 = 0 \quad 3780b_5 - a_1\Omega^2 = 0$$

$$4620b_6 - a_2\Omega^2 = 0$$

A non-trivial solution is obtainable, if the rank of the set (5.121) is less than 8. This demands that 45 minors of order 8 of the matrix of the linear system (5.121) be zero. We then obtain 45 equations out of which 24 are identically zero, whereas 15 are proportional to a_1. Thus, the entire system reduces to

$$a_1 = 0$$

$$- 219{,}251{,}183{,}203{,}068{,}936{,}192a_0 + 1{,}391{,}401{,}739{,}557{,}937{,}479{,}680a_2 = 0$$
$$(5.122)$$

The solution of (5.122) is

$$a_0 = \tfrac{26}{165}a_2 \quad a_1 = 0 \qquad (5.123)$$

where a_0 represents an arbitrary constant. Therefore, when the inertial coefficient equals

$$R(\xi) = \left(\frac{26}{165} + \xi^2 \right) a_2 \tag{5.124}$$

the set (5.121) yields

$$\Omega^2 = 4620 b_6 / a_2 \tag{5.125}$$

and

$$b_0 = \frac{7337}{270} b_6 \quad b_1 = 0 \quad b_2 = -\frac{52}{3} b_6 \quad b_3 = 0 \quad b_4 = \frac{13}{54} b_6 \quad b_5 = 0 \tag{5.126}$$

The flexural rigidity is then

$$D(\xi) = \left(\frac{7337}{270} - \frac{52}{3} \xi^2 + \frac{13}{54} \xi^4 + \xi^6 \right) b_6 \tag{5.127}$$

Figure 5.14 depicts the variation of the flexural rigidity $D(\xi)$ in terms of ξ in $[0; 1]$. It can be shown that for n larger than 4 only trivial solutions are obtained (Appendix I).

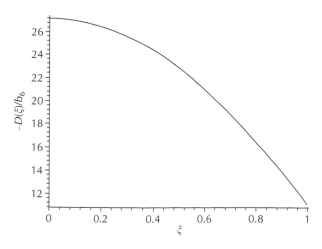

FIGURE 5.14
Variation of the flexural rigidity

5.2.7   Cubic Inertial Coefficient ($m = 3$)

When $n = 0$, instead of (5.104) we obtain

$$-b_1 + b_0 = 0$$
$$-72b_2 + 72b_1 - a_0\Omega^2 = 0$$
$$-144b_3 + 144b_2 - a_1\Omega^2 = 0$$
$$-240b_4 + 240b_3 + 2a_0\Omega^2 - a_2\Omega^2 = 0$$
$$-360b_6 + 360b_4 - a_0\Omega^2 + 2a_1\Omega^2 - a_3\Omega^2 = 0 \quad (5.128)$$
$$-504b_6 + 504b_5 - a_1\Omega^2 + 2a_2\Omega^2 = 0$$
$$-672b_7 + 672b_6 - a_2\Omega^2 + 2a_3\Omega^2 = 0$$
$$672b_7 - a_3\Omega^2 = 0$$

which constitutes eight equations for nine unknowns. We can, therefore, solve the system taking b_7 as a parameter. From the last equation of the set (5.128), we obtain the natural frequency squared:

$$\Omega^2 = 864b_7/a_3 \quad (5.129)$$

And the remaining equations yield the coefficients in the flexural rigidity as follows:

$$b_0 = (252a_0 + 102a_1 + 51a_2 + 29a_3)b_6/35a_3$$
$$b_1 = (252a_0 + 102a_1 + 51a_2 + 29a_3)b_6/35a_3$$
$$b_2 = -(168a_0 - 102a_1 - 51a_2 - 29a_3)b_7/35a_3$$
$$b_3 = -(168a_0 - 108a_1 - 51a_2 - 29a_3)b_7/35a_3 \quad (5.130)$$
$$b_4 = (84a_0 - 108a_1 - 75a_2 + 29a_3)b_7/35a_3$$
$$b_5 = (12a_1 - 15a_2 - 11a_3)b_7/7a_3$$
$$b_6 = (9a_2 - 11a_3)b_7/7a_3$$

Consider now the case $n = 1$; the governing differential equation gives a set of nine linear algebraic equations for nine unknowns:

$$b_1 = 0 \quad -120b_2 + 120b_0 - \tfrac{7}{3}a_0\Omega^2 = 0$$
$$-240b_3 + 240b_1 - \tfrac{7}{3}a_1\Omega^2 = 0$$
$$-400b_4 + 400b_2 + \tfrac{10}{3}a_0\Omega^2 - \tfrac{7}{3}a_2\Omega^2 = 0$$

$$-600b_5 + 600b_3 + \tfrac{10}{3}a_1\Omega^2 - \tfrac{7}{3}a_3\Omega^2 = 0$$

$$-840b_6 + 840b_4 - a_0\Omega^2 + \tfrac{10}{3}a_2\Omega^2 = 0$$

$$-1120b_7 + 1120b_5 - a_1\Omega^2 + \tfrac{10}{3}a_3\Omega^2 = 0$$

$$1440b_6 - a_2\Omega^2 = 0 \quad 1800b_7 - a_3\Omega^2 = 0$$

$$(5.131)$$

Again, we impose the condition that the determinant of the system vanishes, so that we get a non-trivial solution. It produces the following equation:

$$-98{,}991{,}341{,}568a_3 - 341{,}118{,}812{,}160a_1 = 0 \tag{5.132}$$

whose solution reads

$$a_1 = -\tfrac{74}{255}a_3 \tag{5.133}$$

Thus, provided the inertial coefficient equals

$$R(\xi) = a_0 - \tfrac{74}{255}a_3\xi + a_2\xi^2 + a_3\xi^3 \tag{5.134}$$

the set (5.131) yields

$$\Omega^2 = 1800b_7/a_3 \tag{5.135}$$

$$b_0 = (129a_2 + 620a_0)b_7/28a_3 \quad b_1 = 0$$

$$b_2 = -3(-43a_2 + 120a_0)b_7/28a_3 \quad b_3 = -259b_7/51 \tag{5.136}$$

$$b_4 = 15(4a_0 - 11a_2)b_7/28a_3 \quad b_5 = -82b_7/17 \quad b_6 = 5a_2b_6/4a_3$$

For the case $n = 2$, the governing differential equation yields a set of ten linear algebraic equations for nine unknowns:

$$b_1 = 0 \quad -180b_2 - 4a_0\Omega^2 = 0$$

$$-360b_3 + 360b_0 - 4a_1\Omega^2 = 0$$

$$-600b_4 + 600b_1 + 5a_0\Omega^2 - 4a_2\Omega^2 = 0$$

$$-900b_5 + 900b_2 + 5a_1\Omega^2 - 4a_3\Omega^2 = 0$$

$$-1260b_6 + 1260b_3 - a_0\Omega^2 + 5a_2\Omega^2 = 0$$

$$-1680b_7 + 1680b_4 - a_0\Omega^2 + 5a_3\Omega^2 = 0$$

$$2160b_5 - a_1\Omega^2 = 0 \quad 2700b_6 - a_2\Omega^2 = 0$$

$$3300b_7 - a_3\Omega^2 = 0$$

$$(5.137)$$

A non-trivial solution is obtainable, if the rank of the set (5.137) is less than 9. This requires that the ten minors of order 9 of the matrix of the system (5.137) are zero. We then get ten equations which reduce to the following two equations:

$$-316{,}787{,}133{,}358{,}080a_0 + 72{,}597{,}051{,}394{,}560a_1 - 63{,}357{,}426{,}671{,}616a_3 = 0$$

$$36{,}769{,}935{,}121{,}920a_0 - 31{,}678{,}713{,}335{,}808a_2 + 12{,}702{,}341{,}223{,}936a_3 = 0$$

$$(5.138)$$

The solution of (5.138) is then

$$a_1 = \tfrac{48}{55}a_3 + \tfrac{48}{11}a_0 \quad a_2 = \tfrac{247}{616}a_3 + \tfrac{65}{56}a_0 \tag{5.139}$$

where a_0 and a_3 are considered as arbitrary constants. We conclude that for an inertial coefficient taken as follows

$$R(\xi) = a_0 + \left(\tfrac{48}{55}a_3 + \tfrac{48}{11}a_0\right)\xi + \left(\tfrac{247}{616}a_3 + \tfrac{65}{56}a_0\right)\xi^2 + a_3\xi^3 \tag{5.140}$$

the set (5.137) yields

$$\Omega^2 = 3{,}300b_7/a_3 \tag{5.141}$$

$$b_0 = 5(25{,}793a_0 + 4{,}805a_3)b_7/882a_3 \quad b_1 = 0$$

$$b_2 = -220a_0b_7/3a_3 \quad b_3 = -221(55a_0 + 19a_3)b_7/882a_3$$

$$b_4 = (55a_0 - 247a_3)b_7/28a_3 \quad b_5 = 4(5a_0 + a_3)b_7/3a_3$$

$$b_6 = 13(55a_0 + 19a_3)b_7/504a_3$$

$$(5.142)$$

It is shown in Appendix J that for n larger than 3, the set of algebraic equations deduced from the governing equation does not possess a non-trivial solution.

5.2.8 Particular Case $m = 4$

We consider first the case $n = 0$. We get a set of nine equations for ten unknowns:

$$-b_1 + b_0 = 0$$

$$-72b_2 + 72b_1 - a_0\Omega^2 = 0$$

$$-144b_3 + 144b_2 - a_1\Omega^2 = 0$$

$$-240b_4 + 240b_3 + 2a_0\Omega^2 - a_2\Omega^2 = 0$$

$$-360b_5 + 360b_4 - a_0\Omega^2 + 2a_1\Omega^2 - a_3\Omega^2 = 0 \qquad (5.143)$$

$$-504b_6 + 504b_5 - a_1\Omega^2 + 2a_2\Omega^2 - a_4\Omega^2 = 0$$

$$-672b_7 + 504b_6 - a_2\Omega^2 + 2a_3\Omega^2 = 0$$

$$-864b_8 + 864b_7 - a_3\Omega^2 + 2a_4\Omega^2 = 0$$

$$1080b_8 - a_4\Omega^2 = 0$$

We can, therefore, solve the system by taking b_8 as a parameter. The last equation of the set (5.143) yields the natural frequency squared:

$$\Omega^2 = 1080b_8/a_4 \qquad (5.144)$$

and the remaining equations give the coefficients in the flexural rigidity as follows:

$$b_0 = (252a_0 + 102a_1 + 51a_2 + 29a_3 + 18a_4)b_8/28a_4$$

$$b_1 = (252a_0 + 102a_1 + 51a_2 + 29a_3 + 18a_4)b_8/28a_4$$

$$b_2 = (-168a_0 + 102a_1 + 51a_2 + 29a_3 + 18a_4)b_8/28a_4$$

$$b_3 = (-168a_0 - 108a_1 + 51a_2 + 29a_3 + 18a_4)b_8/28a_4 \qquad (5.145)$$

$$b_4 = (84a_0 - 108a_1 - 75a_2 + 29a_3 + 18a_4)b_8/28a_4$$

$$b_5 = (60a_1 - 75a_2 - 55a_3 + 18a_4)b_8/28a_4$$

$$b_6 = (45a_2 - 55a_3 - 42a_4)b_8/28a_4 \quad b_7 = (5a_3 - 6a_4)b_8/4a_4$$

Note that Eq. (5.140) coincides with the result of Elishakoff and Candan (2000), as part of a general case. Indeed, substituting $m = 4$ in their Eq. (5.145) for the general natural frequency squared we get the same expression. When $n = 1$, the governing differential equation yields a set of ten linear algebraic equations for ten unknowns, specifically the coefficients b_i in the flexural rigidity and the natural frequency squared Ω^2:

$$b_1 = 0 \quad -120b_2 + 120b_0 - \tfrac{7}{3}a_0\Omega^2 = 0$$

$$-240b_3 + 240b_1 - \tfrac{7}{3}a_1\Omega^2 = 0$$

$$-400b_4 + 400b_2 + \tfrac{10}{3}a_0\Omega^2 - \tfrac{7}{3}a_2\Omega^2 = 0$$

$$-600b_5 + 600b_3 + \tfrac{10}{3}a_1\Omega^2 = 0 \tag{5.146}$$

$$-840b_6 + 840b_4 - a_0\Omega^2 + \tfrac{10}{3}a_2\Omega^2 = 0$$

$$1120b_5 - a_1\Omega^2 = 0 \quad 1440b_6 - a_2\Omega^2 = 0$$

To find a non-trivial solution, the determinant must be zero. This leads to the following equation:

$$72{,}597{,}051{,}394{,}560a_1 - 63{,}357{,}426{,}671{,}616a_3 = 0 \tag{5.147}$$

which is solved as follows:

$$a_3 = -\tfrac{255}{74}a_1 \tag{5.148}$$

Hence, for the inertial coefficient

$$R(\xi) = a_0 + a_1\xi + a_2\xi^2 - \tfrac{255}{74}r_1\xi^3 + a_4\xi^4 \tag{5.149}$$

where a_0 and a_2 are arbitrary, the set (5.146) has the following non-trivial solution

$$\Omega^2 = 2{,}200b_8/a_4 \tag{5.150}$$

$$b_0 = (20{,}460a_0 + 4{,}257a_2 + 1{,}526a_4)b_8/756a_4 \quad b_1 = 0$$

$$b_2 = -(11{,}880a_0 - 4{,}257a_2 + 1{,}526a_4)b_8/756a_4 \quad b_3 = -385a_1b_8/18a_4$$

$$b_4 = (1{,}980a_0 - 5{,}445a_2 + 1{,}526a_4)b_8/756a_4 \quad b_5 = 2{,}255a_1b_8/111a_4$$

$$b_6 = (165a_2 - 442a_4)b_8/108a_4 \quad b_7 = -935a_1b_8/222a_4$$

$$\tag{5.151}$$

Concerning the case $n = 2$, the governing equation yields the following set

$$b_1 = 0 \quad -180b_2 - 4a_0\Omega^2 = 0$$

$$-360b_3 + 360b_0 - 4a_1\Omega^2 = 0$$

$$-600b_4 + 400b_1 + 5a_0\Omega^2 - 4a_2\Omega^2 = 0$$

$$-900b_5 + 900b_2 + 5a_1\Omega^2 - 4a_3\Omega^2 = 0$$

$$-1260b_6 + 1260b_3 + 5a_2\Omega^2 - 4a_4\Omega^2 = 0 \tag{5.152}$$

$$-1680b_7 + 1680b_4 - a_0\Omega^2 + 5a_3\Omega^2 = 0$$

$$-2160b_8 + 2160b_5 - a_1\Omega^2 + 5a_4\Omega^2 = 0$$

$$2700b_6 - a_2\Omega^2 = 0 \quad 3300b_7 - a_3\Omega^2 = 0$$

$$3960b_8 - a_4\Omega^2 = 0$$

We have a set of 11 equations for 10 unknowns, namely the coefficients b_i in the flexural rigidity and the natural frequency squared, Ω^2. In order to have a non-trivial solution, the rank of the set (5.152) must be less than 10, or all 11 minors of order 10 of the matrix of the set (5.152) should vanish. This last condition leads to a set of 11 equations whose solution reads

$$a_2 = -\frac{65}{77}a_0 + \frac{1{,}235}{2{,}688}a_1 + \frac{8{,}645}{46{,}464}a_4 \quad a_3 = -5a_0 + \frac{55}{48}a_1 + \frac{245}{528}a_4 \tag{5.153}$$

where a_0, a_1 and a_4 are treated as arbitrary constants. Hence, we observe that for an inertial coefficient equal to

$$R(\xi) = a_0 + a_1\xi + \left(-\frac{65}{77}a_0 + \frac{1{,}235}{2{,}688}a_1 + \frac{8{,}645}{46{,}464}a_4\right)\xi^2$$

$$+ \left(-5a_0 + \frac{55}{48}a_1 + \frac{245}{528}a_4\right)\xi^3 + a_4\xi^4 \tag{5.154}$$

the set (5.152) yields

$$\Omega^2 = 3{,}960b_8/a_4 \tag{5.155}$$

$$b_0 = (933{,}504a_0 + 2{,}907{,}025a_1 + 769{,}993a_4)b_8/77{,}616a_4 \quad b_1 = 0$$

$$b_2 = -88a_0b_8/a_4 \quad b_3 = (933{,}504a_0 - 508{,}079a_1 + 769{,}993a_4)b_8/77{,}616a_4$$

$$b_4 = (136{,}224a_0 - 29{,}887a_1 - 12{,}103a_4)b_8/2{,}464a_4 \quad b_5 = (11a_1 - 49a_4)b_8/6a_4$$

$$b_6 = -(4{,}224a_0 - 2{,}299a_1 - 931a_4)b_8/44{,}352a_4$$

$$b_7 = -(528a_0 - 121a_1 - 49a_4)b_8/88a_4$$

$$\tag{5.156}$$

For $n = 3$, we have to satisfy the following set of equations:

$$b_1 = 0 \quad -252b_2 - 6a_0\Omega^2 = 0$$

$$-504b_3 - 6a_1\Omega^2 = 0$$

$$-840b_4 + 840b_0 + 7a_0\Omega^2 - 6a_2\Omega^2 = 0$$

$$-1260b_5 + 1260b_1 + 7a_1\Omega^2 - 6a_3\Omega^2 = 0$$

$$-1764b_6 + 1764b_2 + 7a\Omega^2 - 6a_4\Omega^2 = 0 \qquad (5.157)$$

$$-2352b_7 + 2352b_3 + 7a_3\Omega^2 = 0$$

$$-3024b_8 + 3024b_4 - a_0\Omega^2 + 7a_4\Omega^2 = 0$$

$$3780b_5 - a_1\Omega^2 = 0 \quad 4620b_6 - a_2\Omega^2 = 0$$

$$5544b_7 - a_3\Omega^2 = 0 \quad 6552b_8 - a_4\Omega^2 = 0$$

We get a set of 12 equations for 10 unknowns. A non-trivial solution is obtainable, if the rank of the set (5.157) is less than 10. This forces all 66 minors of order 10 of the matrix of the linear system (5.157) to be zero. This condition is satisfied in this case, leading to the following inertial coefficient:

$$a_0 + \left(\tfrac{165}{26}a_0 + \tfrac{165}{182}a_4\right)\xi^2 + a_4\xi^4 \qquad (5.158)$$

where a_0 and a_4 are treated as arbitrary constants. The set (5.157) then yields

$$\Omega^2 = 6{,}552b_8/a_4 \qquad (5.159)$$

$$b_0 = (51{,}359a_0 + 5{,}935a_4)b_8/210a_4 \quad b_1 = 0$$

$$b_2 = -156a_0b_8/a_4 \quad b_3 = 0$$

$$\qquad\qquad\qquad\qquad\qquad\qquad\qquad\qquad (5.160)$$

$$b_4 = (13a_0 - 85a_4)b_8/6a_4 \quad b_5 = 0$$

$$b_6 = (63a_0 + 9a_4)b_8/7a_4 \quad b_7 = 0$$

5.2.9 Concluding Remarks

The procedure described in the body of this study can be applied for the larger values of m. We discuss some general patterns obtainable for various values of m, representing the *maximum* power in the polynomial expression of the beam's material density. The solution depends on the value of the integer n, with $n + 4$ defining the *maximum* degree of the polynomial in the postulated mode shape. We note that for $n = 0$, the number of equations is smaller than the number of unknowns and hence one of the coefficients b_i can be

TABLE 5.1

Pinned–Pinned Beam: $m = 4$

n	Ω^2	Inertial coefficient	Flexural rigidity $D(\xi)r_m/b_{m+4}$
1	$\dfrac{2{,}200b_8}{a_4}$	$a_0 + a_1\xi + a_2\xi^2 - \dfrac{255}{74}a_1\xi^3 + a_4\xi^4$	$\dfrac{20{,}460a_0 - 4{,}257a_2 + 1{,}526a_4}{756} - \dfrac{11{,}880a_0 - 4{,}257a_2 - 1{,}526a_4}{756}\xi^2$ $- \dfrac{385a_1}{18}\xi^3 + \dfrac{1{,}980a_0 - 5{,}445a_2 + 1{,}526a_4}{756}\xi^4 + \dfrac{2{,}255a_1}{111}\xi^5 + \dfrac{165a_2 - 442a_4}{108}\xi^6$ $- \dfrac{935a_1}{222}\xi^7 + a_4\xi^8$
2	$\dfrac{3{,}960b_8}{a_4}$	$a_0 + a_1\xi$ $+ \left(-\dfrac{65}{77}a_0 + \dfrac{1{,}235}{2{,}688}a_1 + \dfrac{8{,}645}{46{,}464}a_4\right)\xi^2$ $+ \left(-5a_0 + \dfrac{55}{48}a_1 + \dfrac{245}{528}a_4\right)\xi^3$ $+ a_4\xi^4$	$\dfrac{933{,}504a_0 + 2{,}907{,}025a_1 + 769{,}993a_4}{77616} - 88a_0\xi^2 + \dfrac{933{,}504a_0 - 508{,}079a_1 + 769{,}993a_4}{77{,}616}\xi^3$ $+ \dfrac{136{,}224a_0 - 29{,}887a_1 - 12{,}103a_4}{2{,}464}\xi^4 + \dfrac{11a_1 - 49a_4}{6}\xi^5$ $- \dfrac{4{,}224a_0 - 2{,}299a_1 - 931a_4}{44{,}352}\xi^6 - \dfrac{528a_0 - 121a_1 - 49a_4}{88}\xi^7 + a_4\xi^8$
3	$\dfrac{6{,}552b_8}{a_4}$	$a_0 + \left(\dfrac{165}{26}a_0 + \dfrac{165}{182}a_4\right)\xi^2 + a_4\xi^4$	$\dfrac{51{,}359a_0 + 5{,}935a_4}{210} - 156a_0\xi^2 + \dfrac{13a_0 - 85a_4}{6}\xi^4 + \dfrac{63a_0 + 9a_4}{7}\xi^6 + a_4\xi^8$

TABLE 5.2

Pinned–Pinned Beam: $m = 5$

n	Ω^2	Inertial coefficient	Flexural rigidity $D(\xi)r_5/b_9$
1	$\dfrac{2,640b_9}{a_5}$	$a_0 + a_1\xi + a_2\xi^2$ $- \left(\dfrac{255}{74}a_1 + \dfrac{1,015}{2,442}\right)\xi^3$ $+ a_4\xi^4 + a_5\xi^5$	$\dfrac{20,460a_0 + 4,257a_2 + 1,526a_4}{630} - \dfrac{11,880a_0 - 4,257a_2 - 1,526a_4}{630}\xi^2 - \dfrac{77a_1}{3}\xi^3$ $+ \dfrac{1,980a_0 - 5,445a_2 + 1,526a_4}{630}\xi^4 + \dfrac{8,118a_1 + 1,421a_5}{333}\xi^5$ $+ \dfrac{165a_2 - 442a_4}{90}\xi^6 - \dfrac{1,683a_1 + 1,498a_5}{333}\xi^7 + \dfrac{6a_4}{5}\xi^8 + r_5\xi^9$
2	$\dfrac{4,680b_9}{a_5}$	$a_0 + a_1\xi$ $+ \left(\dfrac{65}{56}a_0 + \dfrac{247}{616}a_3 - \dfrac{5}{14}a_5\right)\xi^2$ $+ a_3\xi^3 + \left(\dfrac{528}{49}a_0 - \dfrac{121}{49}a_1 + \dfrac{528}{245}a_3\right)\xi^4$ $+ a_5\xi^5$	$\dfrac{7,954,375a_0 - 866,580a_1 + 1,42,9987a_3 - 93,555a_5}{56,595} - 104a_0\xi^2$ $+ \dfrac{7,954,375a_0 - 2,076,360r_1 + 1,429,987a_3 - 93,555a_5}{56,595}\xi^3$ $+ \dfrac{39(55a_0 - 247a_3 + 220a_5)}{770}\xi^4 - \dfrac{26(20a_0 - 5a_1 + 4a_3)}{6}\xi^5 + \dfrac{9,295a_0 + 3,211a_3 - 38,280a_5}{44,352}\xi^6$ $+ \dfrac{78a_3}{55}\xi^7 + \dfrac{240a_0 - 55a_1 + 48a_3}{245}\xi^8 + a_5\xi^9$
3	$\dfrac{7,644b_9}{a_5}$	$r_0 - \dfrac{6,995}{39,949}r_5\xi + r_2\xi^2 + \dfrac{6,666}{39,949}r_5\xi^3$ $+ \left(-7r_0 + \dfrac{182}{165}r_2\right)\xi^4 + r_5\xi^5$	$\dfrac{23,133r_0 + 15,431r_2}{2,970} - 182r_0\xi^2 + \dfrac{6,995}{439}\xi^3 + \dfrac{7(10,032r_0 - 1,547r_2)}{594}\xi^4$ $- \dfrac{88,961}{6,585}\xi^5 + \dfrac{91r_2}{55}\xi^6 + \dfrac{101}{439}\xi^7 + \dfrac{49(165r_0 - 26r_2)}{990}\xi^8 + r_5\xi^9$

TABLE 5.3

Pinned–Pinned Beam: $m = 6$

n	Ω^2	Inertial coefficient	Flexural rigidity $D(\xi)r_5/b_9$
1	$\dfrac{3,120b_{10}}{a_6}$	$a_0 + a_1\xi + a_2\xi^2 + a_3\xi^3 + a_4\xi^4$ $-\left(\dfrac{1,683}{203}a_1 + \dfrac{2,442}{1,015}a_3\right)\xi^5 + a_6\xi^6$	$\dfrac{265,980a_0 + 55,341a_2 + 19,838a_4 + 9,205a_6}{6,930}$ $-\dfrac{154,440a_0 - 55,341a_2 - 19,838a_4 - 9,205a_6}{6,930}\xi^2 - \dfrac{91a_1}{3}\xi^3$ $+\dfrac{25,740a_0 - 70,785a_2 + 19,838a_4 + 9,205a_6}{6,930}\xi^4 - \dfrac{13(15a_1 + 14a_3)}{15}\xi^5$ $+\dfrac{2,145a_2 - 5,746a_4 + 1,315a_6}{990}\xi^6 + \dfrac{13(1275a_1 + 428a_3)}{435}\xi^7 + \dfrac{78a_4 - 205a_6}{55}\xi^8$ $-\dfrac{39(255a_1 + 74a_3)}{1015}\xi^9 + a_6\xi^{10}$
2	$\dfrac{5,460b_{10}}{a_6}$	$a_0 + a_1\xi + a_2\xi^2 + a_3\xi^3$ $+\left(\dfrac{528}{49}a_0 - \dfrac{121}{49}a_1 + \dfrac{528}{245}a_3 + \dfrac{44}{49}a_6\right)\xi^4$ $+\left(\dfrac{13}{4}a_0 - \dfrac{14}{5}a_2 + \dfrac{247}{220}a_3 + \dfrac{80}{143}a_6\right)\xi^5$ $+a_6\xi^6$	$\left(\dfrac{397,816,705a_0 + 45,062,160a_1 + 13,621,608a_2 +68,897,413a_3 + 26,639,680a_6}{2,522,520}\right) - \dfrac{364a_0}{3}\xi^2$ $\left(\dfrac{-397,816,705a_0 + 107,970,720a_1 - 13,621,608a_2 -68,897,413a_3 - 26,639,680a_6}{2,522,520}\right)\xi^3 - \dfrac{91(-5a_0 + 4a_2)}{10}\xi^4$ $+\dfrac{91(-20a_0 + 5a_1 - 4a_3)}{15}\xi^5 - \dfrac{213,785a_0 - 199,056a_2 + 73,853a_3 + 36,800a_6}{51,480}\xi^6$ $+\dfrac{91a_3 - 400a_6}{55}\xi^7 + \dfrac{13(240a_0 - 55a_1 + 48a_3 + 20a_6)}{210}\xi^8$ $+\dfrac{7(9,295a_0 - 8,008a_2 + 3,211a_3 + 1,600a_6)}{17,160}\xi^9 + a_6\xi^{10}$

TABLE 5.3
Continued

n	Ω^2	Inertial coefficient	Flexural rigidity $D(\xi)r_5/b_9$
3	$\dfrac{8{,}820b_{10}}{a_6}$	$a_0 - \dfrac{6{,}995}{39{,}949}a_3\xi + a_2\xi^2 + a_3\xi^3$ $+ \left(-7a_0 + \dfrac{182}{165}a_2 + \dfrac{68}{165}a_6\right)\xi^4$ $+ \dfrac{39{,}949}{6{,}666}a_3\xi^5 + a_6\xi^6$	$\dfrac{7(23{,}133a_0 + 15{,}431a_2 + 3{,}545a_6)}{2574} - 210r_0\xi^2 + \dfrac{244{,}825a_3}{2{,}222}\xi^3$ $+ \dfrac{35(10{,}032a_0 - 1{,}547a_2 + 709a_6)}{2{,}574}\xi^4 - \dfrac{622{,}727a_3}{6{,}666}\xi^5 + \dfrac{21a_2 - 136a_6}{11}\xi^6 + \dfrac{35a_3}{22}\xi^7$ $- \dfrac{7(1{,}155a_0 - 182a_2 - 68a_6)}{858}\xi^8 + \dfrac{15{,}365a_3}{2{,}222}\xi^9 + a_6\xi^{10}$

designated as an arbitrary one. For the particular case $n = 1$, the number of equations equals the number of unknowns. Since the set is a homogeneous one, the determinant must vanish for a non-trivial solution. This leads to a condition between the inertial coefficients, for which the solution is obtainable. For $n \geq 2$, the number of equations exceeds that of the unknowns. Hence, for the availability of the non-trivial solution the rank should be less than the number of unknowns. This results, generally, in a large number of determinantal equations which yield relationships between the inertial coefficients. Tables 5.1–5.3 report closed-form solutions for particular cases $4 \leq m \leq 6$, respectively.

6

Influence of Boundary Conditions on Eigenvalues

In this chapter, we discuss the effect of boundary conditions in detail. In particular, it is shown that for the same polynomial order of inertial coefficient and the power in the mode shape, the analytical expression for the natural frequencies is insensitive to the boundary conditions. The method, as it were, screens the different beams with coinciding frequencies.

6.1 The Remarkable Nature of Effect of Boundary Conditions on Closed-Form Solutions for Vibrating Inhomogeneous Bernoulli–Euler Beams

6.1.1 Introductory Remarks

The methods of direct or inverse vibration problems for homogeneous or inhomogeneous structures have been amply discussed in the literature. For an extensive review the reader may consult the paper by Elishakoff (2000h) or Chapter 1. In nearly all the cases the solution is not given in closed-form, even if it is exact.

In this chapter we derive closed-form solutions for the inhomogeneous Bernoulli–Euler beams under three sets of boundary conditions. It complements the case where the beam is pinned at both ends. As a particular case, it includes results obtained by Candan and Elishakoff (2000) and Elishakoff and Candan (2001).

The closed-form solution is made possible by a seemingly modest requirement posed in this study. We postulate the knowledge of only the fundamental mode shape of the beam. Whereas in most of the inverse problems a knowledge of the vibration frequency spectrum is assumed, here only the mode fundamental shape is specified. It may appear curious, at first glance, that we postulate static displacements of homogeneous beams in the capacity of the mode shape. Fortunately, the so formulated problem turns out to have solutions for the numerous cases that are reported in this study.

A remarkable feature accompanies the reported results: for the same order of inertial coefficient and the power in the mode shape, the analytical expression for the natural frequencies are insensitive to the boundary conditions, contrary to one's initial anticipation; this counterintuitive by-product is explained by the fact that the flexural rigidity distribution, associated with coinciding frequencies, do differ. Thus, we obtain different beams that have the same boundary conditions.

6.1.2   Construction of Postulated Mode Shapes

Consider a uniform beam, clamped at one end and free at the other. It is subjected to the load $q(\xi) = q_0\xi^n$. The governing differential equation reads:

$$EI\frac{d^4w}{dx^4} = q_0x^n \tag{6.1}$$

We introduce the non-dimensional coordinate

$$\xi = x/L \tag{6.2}$$

where L is the length. The equation becomes

$$\frac{d^4w}{d\xi^4} = \alpha\xi^n \quad \alpha = q_0L^{n+4}/EI \tag{6.3}$$

The solution of this equation, which satisfies the boundary conditions is given by

$$W(\xi) = \frac{\alpha}{(n+1)(n+2)(n+3)(n+4)}\psi(\xi)$$
$$\psi(\xi) = \xi^{n+4} - \tfrac{1}{6}(n+2)(n+3)(n+4)\xi^3 + \tfrac{1}{2}(n+1)(n+3)(n+4)\xi^2 \tag{6.4}$$

For the beam that is pinned at one end and clamped at the other, the functions $w(\xi)$ and $\psi(\xi)$ are, respectively,

$$W(\xi) = \frac{\alpha}{2(n+1)(n+2)(n+3)(n+4)}\psi(\xi)$$
$$\psi(\xi) = 2\xi^{n+4} - (n+3)\xi^3 + (n+1)\xi \tag{6.5}$$

Finally, for the beam that is clamped at both its ends, we get

$$W(\xi) = \frac{\alpha}{(n+1)(n+2)(n+3)(n+4)}\psi(\xi)$$
$$\psi(\xi) = \xi^{n+4} - (n+2)\xi^3 + (n+1)\xi^2 \tag{6.6}$$

We address the following problem: can the function $\psi(\xi)$ serve as an exact mode shape for the inhomogeneous Bernoulli–Euler beam? We will investigate various cases of reconstruction of the flexural rigidity for different variations of the mass density.

6.1.3 Formulation of the Problem

The differential equation that governs the mode shape $W(\xi)$ of the inhomogeneous Bernoulli–Euler beam reads:

$$\frac{d^2}{dx^2}\left[D(x)\frac{d^2W}{dx^2}\right] - \omega^2 R(x)W(x) = 0 \tag{6.7}$$

where $R(x) = \rho(x)A(x)$ is the inertial coefficient. Introducing the non-dimensional coordinate ξ from Eq. (6.2), the governing equation becomes

$$\frac{d^2}{d\xi^2}\left[D(\xi)\frac{d^2\psi}{d\xi^2}\right] - \Omega^2 R(\xi)\psi(\xi) = 0 \tag{6.8}$$

where the postulate

$$w(\xi) = \psi(\xi) \tag{6.9}$$

has been invoked. We are confronted with determining $R(\xi)$ and $D(\xi)$ such that $\psi(\xi)$ serves as an exact mode shape of the beam under suitable boundary conditions.

The variation of the inertial coefficient is taken as follows

$$R(\xi) = \sum_{i=0}^{m} a_i \xi^i \tag{6.10}$$

where a_i $(i = 0, 1, \ldots, m)$ are given coefficients. The first term of the governing equation (6.9) is a polynomial function whose degree is $n + m + 4$. In order to achieve the same polynomial order in the first term of this equation, the flexural rigidity $D(\xi)$ must be represented as

$$D(\xi) = \sum_{i=0}^{m+4} b_i \xi^i \tag{6.11}$$

The left-hand side of the governing equation (6.9) becomes a polynomial of ξ, in order for it to vanish for every ξ in the interval $[0; 1]$, we have to demand that its coefficients in front of ξ taken in any power, to be zero. This leads to a set of $n + m + 5$ linear algebraic equations. Since the problem was posed as the reconstruction of the flexural rigidity, when the inertial coefficient and the mode shape are given, we have to determine the coefficients b_i of the flexural rigidity as well as the natural frequency ω, which constitute $m + 6$ unknowns, namely $m + 5$ coefficients in the flexural rigidity and the natural frequency ω.

We note that when n is zero, we get a triangular system of $m + 5$ equations for $m + 6$ unknowns, as studied by Elishakoff and Candan (2001) who, therefore, succeeded in expressing the unknowns in terms of one of the coefficients b_i, namely b_{m+4}, taken as an arbitrary constant. Yet, when n is larger than zero, the number of equations is greater than the number of unknowns.

6.1.4 Closed-Form Solutions for the Clamped–Free Beam

6.1.4.1 Constant inertial coefficient ($m = 0$)

The inertial coefficient reduces to $R(\xi) = a_0$. The governing differential equation leads to the following set of equations:

$$24b_2 - 48b_1 + 24b_0 = 0$$

$$72b_3 - 144b_2 + 72b_1 = 0$$

$$144b_4 - 288b_3 + 144b_2 - 6a_0\Omega^2 = 0 \qquad (6.12)$$

$$-480b_4 + 240b_3 + 4a_0\Omega^2 = 0$$

$$360b_4 - a_0\Omega^2 = 0$$

We have five equations for six unknowns: b_0, b_1, b_2, b_3, b_4 and Ω^2. We declare b_4 to be a known parameter. We get Ω^2 from the last equation of the set (6.13):

$$\Omega^2 = 360b_4/a_0 \qquad (6.13)$$

The rest of the equations of the set (6.13) leads to

$$b_0 = 26b_4 \quad b_1 = 16b_4 \quad b_2 = 6b_4 \quad b_3 = -4b_4 \qquad (6.14)$$

The flexural rigidity is then given by

$$D(\xi) = (26 + 16\xi + 6\xi^2 - 4\xi^3 + \xi^4)b_4 = [26 + 16\xi + 2\xi^2 + (\xi - 2)^2\xi^2]b_4 \qquad (6.15)$$

This expression was first derived by Elishakoff and Candan (2001).

The requirement of physical realizability of the problem forces both the flexural rigidity and the inertial coefficient to be non-negative. One can see that when $a_0 > 0$ and $b_4 > 0$ the inertial coefficient as well as the flexural rigidity are positive (see Figure 6.1).

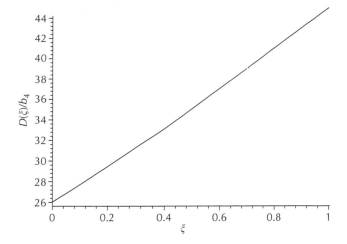

FIGURE 6.1
Variation of the flexural rigidity for the clamped–free beam

6.1.4.2 Linearly varying inertial coefficient ($m = 1$)

First, consider the case $n = 0$; instead of Eq. (6.13) we have

$$24b_2 - 48b_1 + 24b_0 = 0$$
$$72b_3 - 144b_2 + 72b_1 = 0$$
$$144b_4 - 288b_3 + 144b_2 - 6a_0\Omega^2 = 0$$
$$240b_5 - 480b_4 + 240b_3 + (4a_0 - 6a_1)\Omega^2 = 0 \qquad (6.16)$$
$$-720b_5 + 360b_4 - (a_0 - 4a_1)\Omega^2 = 0$$
$$504b_5 - a_1\Omega^2 = 0$$

This set represents a linear algebraic system of six equations for the seven unknowns, $b_0, b_1, \ldots, b_5$ and Ω^2. Hence, we express them in terms of b_5, treated as an arbitrary constant. We get Ω^2 from the last equation of the set (6.17):

$$\Omega^2 = 504b_5/a_1 \qquad (6.17)$$

while the remaining equations lead to

$$b_0 = 2b_5(17a_1 + 91a_0)/5a_1 \quad b_1 = 2b_5(51a_1 + 56a_0)/5a_1$$
$$b_2 = 2b_5(31a_1 + 21a_0)/5a_1 \quad b_3 = 2b_5(11a_1 - 14a_0)/5a_1 \qquad (6.18)$$
$$b_4 = -b_5(18a_1 - 7a_0)/5a_1$$

For the problem to be realistic we demand that both the flexural rigidity and the inertial coefficient be non-negative. One can show that this takes place when the parameters of the system, namely r_0, r_1 and b_5, are non-negative.

Now consider the case $n = 1$. The governing differential equation yields a set of six equations for six unknowns:

$$80b_2 - 120b_1 = 0$$

$$240b_3 - 360b_2 + 120b_0 = 0$$

$$480b_4 - 720b_3 + 240b_1 - 20a_0\Omega^2 = 0$$

$$800b_5 - 1200b_4 + 400b_2 + (10a_0 - 20a_1)\Omega^2 = 0 \qquad (6.19)$$

$$-1800b_5 + 600b_3 + 10a_1\Omega^2 = 0$$

$$840b_4 - a_0\Omega^2 = 0$$

$$1120b_5 - a_1\Omega^2 = 0$$

In order to have a non-trivial solution the determinant of the system (6.19) must vanish, leading to the following determinantal equation:

$$320a_0 - 249a_1 = 0 \qquad (6.20)$$

whose solution is

$$a_0 = \tfrac{249}{320}a_1 \qquad (6.21)$$

We can solve the set expressing the unknowns with b_5 taken as an arbitrary constant. Thus, we get

$$\Omega^2 = 1120b_5/a_1 \qquad (6.22)$$

and the coefficients b_i are

$$b_0 = \frac{16,477}{120}b_5 \quad b_1 = \frac{471}{20}b_5 \quad b_2 = \frac{1,413}{40}b_5 \quad b_3 = -\frac{47}{3}b_5 \quad b_4 = \frac{83}{80}b_5 \tag{6.23}$$

We conclude that the solution for the flexural rigidity is

$$D(\xi) = \left(\frac{16,477}{120} + \frac{471}{20}\xi + \frac{1,413}{40}\xi^2 - \frac{47}{3}\xi^3 + \frac{83}{80}\xi^4 + \xi^5 \right) b_5$$

$$= \left[\frac{16,477}{120} + \frac{471}{20}\xi + \frac{1,413}{40}\xi^2 \left(1 - \frac{1,880}{4,209}\xi \right) + \frac{83}{80}\xi^4 + \xi^5 \right] b_5 \qquad (6.24)$$

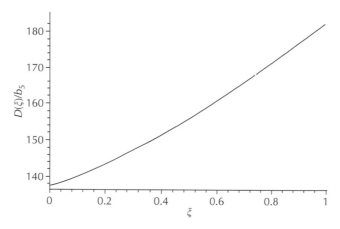

FIGURE 6.2
Variation of the flexural rigidity for the clamped–free beam

provided the inertial term reads

$$R(\xi) = \left(\tfrac{249}{320} + \xi\right) a_1 \tag{6.25}$$

One clearly sees that, when the parameters r_1 and b_5 are non-negative the problem is realizable. Figure 6.2 depicts the variation of the flexural rigidity for ξ within the interval $[0; 1]$.

6.1.4.3 Parabolically varying inertial coefficient ($m = 2$)

We first consider the case $n = 0$. We get the following set of equations

$$24b_2 - 48b_1 + 24b_0 = 0 \quad 72b_3 - 144b_2 + 72b_1 = 0$$

$$144b_4 - 288b_3 + 144b_2 - 6a_0\Omega^2 = 0$$

$$240b_5 - 480b_4 + 240b_3 + (4a_0 - 6a_1)\Omega^2 = 0$$

$$360b_6 - 720b_5 + 360b_4 - (a_0 - 4a_1 + 6a_2)\Omega^2 = 0$$

$$-1008b_6 + 504b_5 - (a_1 - 4a_2)\Omega^2 = 0 \quad 672b_6 - a_2\Omega^2 = 0 \tag{6.26}$$

which constitutes seven equations for eight unknowns. We can, therefore, solve the system taking b_6 as a parameter. From the last equation of the set (6.26), we obtain the natural frequency squared:

$$\Omega^2 = 672b_6/a_2 \tag{6.27}$$

The remaining equations yield the coefficients in the flexural rigidity as follows:

$$b_0 = b_6(728a_0 + 568a_1 + 465a_2)/15a_2 \quad b_1 = 2b_6(224a_0 + 204a_1 + 181a_2)/15a_2$$
$$b_2 = b_6(168a_0 + 248a_1 + 259a_2)/15a_2 \quad b_3 = -4b_6(28a_0 - 22a_1 - 39a_2)/15a_2$$
$$b_4 = b_6(28a_0 - 72a_1 + 53a_2)/15a_2 \quad b_5 = 2b_6(2a_1 - 5a_2)/3a_2$$

$$(6.28)$$

As in the previous cases for non-negative parameters the flexural rigidity and the inertial coefficient remain positive for ξ within the interval $[0; 1]$.

When $n = 1$, the governing differential equation yields a set of eight linear algebraic equations for eight unknowns, namely the coefficients b_i in the flexural rigidity and the natural frequency squared Ω^2:

$$80b_2 - 120b_1 = 0 \quad 240b_3 - 360b_2 + 120b_0 = 0$$
$$480b_4 - 720b_3 + 240b_1 - 20a_0\Omega^2 = 0$$
$$800b_5 - 1200b_4 + 400b_2 + (10a_0 - 20a_1)\Omega^2 = 0$$
$$1200b_6 - 1800b_5 + 600b_3 + (10a_1 - 20a_2)\Omega^2 = 0 \qquad (6.29)$$
$$-2520b_6 + 840b_4 - (a_0 - 10a_2)\Omega^2 = 0$$
$$1120b_5 - a_1\Omega^2 = 0 \quad 1440b_6 - a_2\Omega^2 = 0$$

In order to find a non-trivial solution, the determinant must be zero, which leads to the following determinantal equation:

$$320a_0 - 249a_1 + 454a_2 = 0 \qquad (6.30)$$

whose solution is

$$a_1 = (320a_0 + 454a_2)/249 \qquad (6.31)$$

where a_0 and a_2 are arbitrary. Upon substitution of the expression (6.31) into the set (6.29) we get the solution for the natural frequency squared:

$$\Omega^2 = 1{,}440b_6/a_2 \qquad (6.32)$$

and the coefficients in the flexural rigidity

$$b_0 = (13{,}1816a_0 + 135{,}915a_2)b_6/581a_2 \quad b_1 = 18(1{,}256a_0 + 1{,}811a_2)b_6/581a_2$$
$$b_2 = 27(1{,}256a_0 + 1{,}811a_2)b_6/581a_2b_3 = -4(3{,}760a_0 + 1{,}347a_2)b_6/581a_2$$
$$b_4 = 3(4a_0 - 33a_2)b_6/7a_2 \quad b_5 = 6(160a_0 + 227a_2)b_6/581a_2$$

$$(6.33)$$

The physical realizability of the system (i.e. positivity of both the flexural rigidity and the inertial coefficient) can be checked straightforwardly for non-negative parameters of the problem.

When $n = 2$, the governing differential equation yields a set of nine linear algebraic equations for eight unknowns:

$$180b_2 - 240b_1 = 0 \quad 540b_3 - 720b_2 = 0$$

$$1080b_4 - 1440b_3 + 360b_0 - 45a_0\Omega^2 = 0$$

$$1800b_5 - 2400b_4 + 600b_1 + (20a_0 - 45a_1)\Omega^2 = 0$$

$$2700b_6 - 3600b_5 + 900b_2 + (20a_1 - 45a_2)\Omega^2 = 0 \tag{6.34}$$

$$-5040b_6 + 1260b_3 + 20a_2\Omega^2 = 0 \quad 1680b_4 - a_0\Omega^2 = 0$$

$$2160b_5 - a_1\Omega^2 = 0 \quad 2700b_6 - a_2\Omega^2 = 0$$

In order to have a non-trivial solution the rank of the set (6.34) must be less than 8. As is well known the definition of the rank reads: "a matrix is of rank p if it contains minors of order p different from 0, while all minors of order $p+1$ (if there are such) are zero," Uspensky (1948). According to this definition, the rank of the matrix of the system is less than 8 if all minors of order 8 vanish. We produce the minors, extracting one row of the matrix among nine; therefore, we obtain a set of nine determinantal equations for three unknowns, namely the parameters r_0, r_1 and r_2. The solution of this set reads

$$a_0 = \frac{5,169}{715}a_2 \quad a_1 = \frac{1,128}{385}a_2 \tag{6.35}$$

where a_2 is an arbitrary constant. Substituting the expressions (6.35) into the set (6.34) we get the solution for the natural frequency squared:

$$\Omega^2 = 2,700b_6/a_2 \tag{6.36}$$

and the coefficients in the flexural rigidity,

$$b_0 = 9,007,511b_6/4,004 \quad b_1 = -153b_6/7 \quad b_2 = -204b_6/7$$

$$b_3 = -272b_6/7 \quad b_4 = 46,521b_6/4,004 \quad b_5 = 282b_6/77 \tag{6.37}$$

We have to verify the physical realizability of the problem, which requires both the flexural rigidity and the inertial term to be positive. Figures 6.3 and 6.4 portray, respectively, the variation of the flexural rigidity and the inertial coefficient in terms of ξ within the interval $[0; 1]$; they show that when the parameters of the system b_6 and a_2 are non-negative, both the flexural rigidity and the inertial coefficient remain positive.

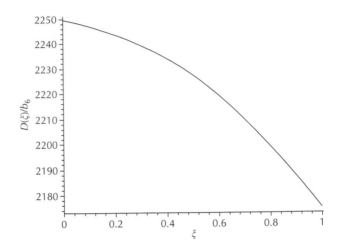

FIGURE 6.3
Variation of the flexural rigidity for the clamped–free beam

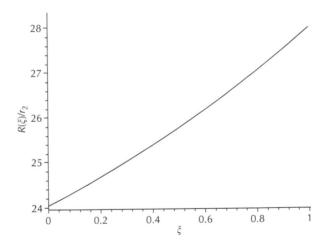

FIGURE 6.4
Variation of the inertial coefficient for the clamped–free beam

6.1.4.4 *Cubic inertial coefficient* ($m = 3$)

When $n = 0$, we obtain a homogeneous linear algebraic system of eight equations for nine unknowns

$$24b_2 - 48b_1 + 24b_0 = 0$$

$$72b_3 - 144b_2 + 72b_1 = 0$$

$$144b_4 - 288b_3 + 144b_2 - 6a_0\Omega^2 = 0$$

$$240b_5 - 480b_4 + 240b_3 + (4a_0 - 6a_1)\Omega^2 = 0$$

$$360b_6 - 720b_5 + 360b_4 - (a_0 - 4a_1 + 6a_2)\Omega^2 = 0$$
$$205b_7 - 1008b_6 + 504b_5 - (a_1 - 4a_2 + 6a_3)\Omega^2 = 0$$
$$-1344b_7 + 672b_6 - (a_2 - 4a_3)\Omega^2 = 0$$
$$864b_7 - a_3\Omega^2 = 0$$

$$(6.38)$$

Therefore, we solve the system taking, say, b_7 as a parameter. From the last equation of the set (6.38), we obtain the natural frequency squared:

$$\Omega^2 = 864b_7/a_3 \tag{6.39}$$

and the remaining equations yield the coefficients in the flexural rigidity as follows:

$$b_0 = (2184a_0 + 1704a_1 + 1395a_2 + 1180a_3)b_7/35a_3$$
$$b_1 = (1344a_0 + 1224a_1 + 1086a_2 + 965a_3)b_7/35a_3$$
$$b_2 = 3(168a_0 + 248a_1 + 259a_2 + 250a_3)b_7/35a_3$$
$$b_3 = -(336a_0 - 264a_1 - 468a_2 - 535a_3)b_7/35a_3 \tag{6.40}$$
$$b_4 = (84a_0 - 216a_1 + 159a_2 + 320a_3)b_7/35a_3$$
$$b_5 = 3(4a_1 - 10a_2 + 7a_3)b_7/7a_3 \quad b_6 = (9a_2 - 22a_3)b_7/7a_3$$

Again, the realizability of the problem can be checked straightforwardly for non-negative parameters of the problem.

Now consider $n = 1$; the governing differential equation yields a set of nine linear algebraic equations for nine unknowns:

$$80b_2 - 120b_1 = 0$$
$$240b_3 - 360b_2 + 120b_0 = 0$$
$$480b_4 - 720b_3 + 240b_1 - 20a_0\Omega^2 = 0$$
$$800b_5 - 1200b_4 + 400b_2 + (10a_0 - 20a_1)\Omega^2 = 0$$
$$1200b_6 - 1800b_5 + 600b_3 + (10a_1 - 20a_2)\Omega^2 = 0 \tag{6.41}$$
$$1680b_7 - 2520b_6 + 840b_4 - (a_0 - 10a_2 + 20a_3)\Omega^2 = 0$$
$$-3360b_7 + 1120b_5 - (a_1 - 10a_3)\Omega^2 = 0$$
$$1440b_6 - a_2\Omega^2 = 0$$
$$1800b_7 - a_3\Omega^2 = 0$$

Again, we require that the determinant of the system (6.42) vanish in order to get a non-trivial solution. This leads to the following equation

$$1600a_0 - 1245a_1 + 2270a_2 - 2786a_3 = 0 \tag{6.42}$$

or

$$r_1 = \frac{320}{249}a_0 + \frac{454}{249}a_2 - \frac{2786}{1245}a_3 \tag{6.43}$$

where r_0, r_2 and r_3 are arbitrary. We substitute the expression (6.43) into the set (6.41) and get the solution for the natural frequency squared:

$$\Omega^2 = 1{,}800b_7/a_3 \tag{6.44}$$

and the coefficients in the flexural rigidity

$$b_0 = (659{,}080a_0 + 679{,}575a_2 - 396{,}776a_3)b_7/2{,}324a_3$$
$$b_1 = 5(11{,}304a_0 + 16{,}299a_2 - 7{,}048a_3)b_7/1{,}162a_3$$
$$b_2 = 15(11{,}304a_0 + 16{,}299a_2 - 7{,}048a_3)b_7/2{,}324a_3$$
$$b_3 = -(18{,}800a_0 - 6{,}735a_2 - 9{,}952a_3)b_7/581a_3 \tag{6.45}$$
$$b_4 = (60a_0 - 495a_2 + 1{,}144a_3)b_7/28a_3$$
$$b_5 = 3(800a_0 + 1{,}135a_2 - 6{,}456a_3)b_7/1{,}162a_3 \quad b_6 = 5a_2b_7/4a_3$$

For the case $n = 2$, the governing differential equation produces a set of ten linear algebraic equations for nine unknowns, namely eight coefficients b_i in the flexural rigidity and the natural frequency squared Ω^2:

$$180b_2 - 240b_1 = 0 \quad 540b_3 - 720b_2 = 0$$
$$1080b_4 - 1440b_3 + 360b_0 - 45a_0\Omega^2 = 0$$
$$1800b_5 - 2400b_4 + 600b_1 + (20a_0 - 45a_1)\Omega^2 = 0$$
$$2700b_6 - 3600b_5 + 900b_2 + (20a_1 - 45a_2)\Omega^2 = 0$$
$$3780b_7 - 5040b_6 + 1260b_3 + 20a_2\Omega^2 = 0 \tag{6.46}$$
$$-6720b_7 + 1680b_4 - (a_0 - 20a_3)\Omega^2 = 0$$
$$2160b_5 - a_1\Omega^2 = 0 \quad 2700b_6 - a_2\Omega^2 = 0$$
$$3300b_7 - a_3\Omega^2 = 0$$

In order to have a non-trivial solution, the rank of the set (6.46) must be less than 9. The rank of the matrix of the system is less than 9 if all minors of order

9 vanish, leading to a set of ten equations for four unknowns, namely the parameters a_0, a_1, a_2 and a_3. The solution of this set reads

$$a_0 = \frac{5,169}{715}a_2 - \frac{39,812}{7,865}a_3 \quad a_1 = \frac{1,128}{385}a_2 - \frac{5,427}{4,235}a_3 \tag{6.47}$$

where a_2 and a_3 are arbitrary constants. Upon substitution of the expressions (6.47) into the set (6.46) we get the solution for the natural frequency squared:

$$\Omega^2 = 3,300b_7/a_3 \tag{6.48}$$

and the coefficients in the flexural rigidity

$$b_0 = (99,082,621a_2 - 53,799,120a_3)b_7/36,036a_3$$
$$b_1 = -(748r_2 - 1,809a_3)b_7/28a_3 \quad b_2 = -(748a_2 - 1,809a_3)b_7/21a_3$$
$$b_3 = -4(748a_2 - 1,809a_3)b_7/63a_3 \quad b_4 = (56,859a_2 - 181,096r_3)b_7/4,004a_3$$
$$b_5 = (4,136a_2 - 1,809a_3)b_7/924a_3 \quad b_6 = 11a_2b_7/9a_3$$

$$\tag{6.49}$$

For the case $n = 3$, the governing differential equation yields a set of 11 linear algebraic equations for 9 unknowns:

$$336b_2 - 420b_1 = 0 \quad 1,008b_3 - 1,260b_2 = 0$$
$$2,016b_4 - 2,520b_3 - 84a_0\Omega^2 = 0$$
$$3,360b_5 - 4,200b_4 + 840b_0 + (35a_0 - 84a_1)\Omega^2 = 0$$
$$5,040b_6 - 6,300b_5 + 1,260b_1 + (35a_1 - 84a_2)\Omega^2 = 0$$
$$7,056b_7 - 8,820b_6 + 1,764b_2 + (35a_2 - 84a_3)\Omega^2 = 0$$
$$-11,760b_7 + 2,352b_3 + 35a_3\Omega^2 = 0$$
$$3,024b_4 - a_0\Omega^2 = 0 \quad 3,780b_5 - a_1\Omega^2 = 0$$
$$4,620b_6 - a_2\Omega^2 = 0 \quad 5,544b_7 - a_3\Omega^2 = 0$$

$$\tag{6.50}$$

In order to have a non-trivial solution, the rank of the set (6.50) must be less than 9. The rank of the matrix of the system is less than 9 if all minors of order 9 vanish. To produce all minors of order 9 of the matrix of the set (6.50) we have to extract 2 rows among 11, leading to a set of $C_9^2 = 55$ equations for 4 unknowns, namely the parameters a_0, a_1, a_2 and a_3. The computerized symbolic algebraic package MAPLE® yields the following solution:

$$a_0 = \frac{93}{220}a_3 \quad a_1 = \frac{5748}{715}a_3 \quad a_2 = \frac{161}{52}a_3 \tag{6.51}$$

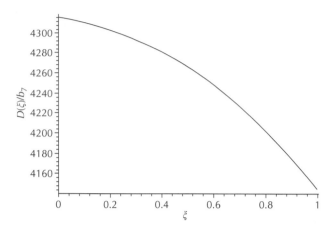

FIGURE 6.5
Variation of the flexural rigidity for the clamped–free beam

where a_3 is an arbitrary constant, by using the command SOLVE. In Eq. (6.51), by back substitution of Eq. (6.51) into the 55 equations, we verify that the expressions for a_i given in Eq. (6.51) indeed substitute a solution. Upon substitution of the expressions (6.51) into the set (6.49) we get the solution for the natural frequency squared:

$$\Omega^2 = 5{,}544b_7/a_3 \tag{6.52}$$

and the coefficients in the flexural rigidity

$$b_0 = 11{,}221{,}529b_7/2{,}600 \quad b_1 = -248b_7/5 \quad b_2 = -62b_7 \quad b_3 = -155b_7/2$$
$$b_4 = 31b_7/40 \quad b_5 = 3{,}832b_7/325 \quad b_6 = 483b_7/130$$

$$\tag{6.53}$$

Figure 6.5 depicts the flexural rigidity, which is positive for ξ within $[0; 1]$, while Figure 6.6 represents the variation of the inertial coefficient.

6.1.4.5 *Quartic inertial coefficient* ($m = 4$)

When $n = 0$, we obtain

$$24b_2 - 48b_1 + 24b_0 = 0$$
$$72b_3 - 144b_2 + 72b_1 = 0$$
$$144b_4 - 288b_3 + 144b_2 - 6a_0\Omega^2 = 0$$
$$240b_5 - 480b_4 + 240b_3 + (4a_0 - 6a_1)\Omega^2 = 0$$

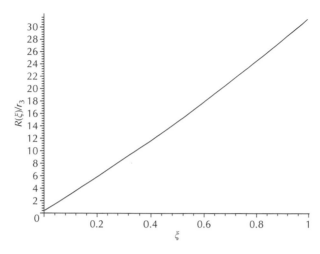

FIGURE 6.6
Variation of the inertial coefficient for the clamped–free beam

$$360b_6 - 720b_5 + 360b_4 - (a_0 - 4a_1 + 6a_2)\Omega^2 = 0$$

$$504b_7 - 1008b_6 + 504b_5 - (a_1 - 4a_2 + 6a_3)\Omega^2 = 0$$

$$672b_8 - 1344b_7 + 672b_6 - (a_2 - 4a_3 + 6a_4)\Omega^2 = 0$$

$$-1728b_8 + 864b_7 - (a_3 - 4a_4)\Omega^2 = 0$$

$$1080b_8 - a_4\Omega^2 = 0$$

$$(6.54)$$

which constitutes nine equations for ten unknowns. We can, therefore, solve the system taking b_8 as a parameter. From the last equation of the set (6.54), we obtain the natural frequency squared:

$$\Omega^2 = 1080b_8/a_4 \qquad (6.55)$$

The remaining equations yield the coefficients in the flexural rigidity as follows:

$$b_0 = (2184a_0 + 1704a_1 + 1395a_2 + 1180a_3 + 1022a_4)b_8/28a_4$$

$$b_1 = (1344a_0 + 1224a_1 + 1086a_2 + 965a_3 + 864a_4)b_8/28a_4$$

$$b_2 = (504a_0 + 744a_1 + 777a_2 + 750a_3 + 706a_4)b_8/28a_4$$

$$b_3 = -(336a_0 - 264a_1 - 468a_2 - 535a_3 - 548a_4)b_8/28a_4 \qquad (6.56)$$

$$b_4 = (84a_0 - 216a_1 + 159a_2 + 320a_3 + 390a_4)b_8/28a_4$$

$$b_5 = (60a_1 - 150a_2 + 105a_3 + 232a_4)b_8/28a_4$$

$$b_6 = (45a_2 - 110a_3 + 74a_4)b_8/28a_4 \quad b_7 = (5a_3 - 12a_4)b_8/4a_4$$

Now consider the case $n = 1$. The governing differential equation yields a set of ten linear algebraic equations for ten unknowns:

$$80b_2 - 120b_1 = 0$$

$$240b_3 - 360b_2 + 120b_0 = 0$$

$$480b_4 - 720b_3 + 240b_1 - 20a_0\Omega^2 = 0$$

$$800b_5 - 1200b_4 + 400b_2 + (10a_0 - 20a_1)\Omega^2 = 0$$

$$1200b_6 - 1800b_5 + 600b_3 + (10a_1 - 20a_2)\Omega^2 = 0$$

$$1680b_7 - 2520b_6 + 840b_4 - (a_0 - 10a_2 + 20a_3)\Omega^2 = 0 \qquad (6.57)$$

$$2240b_8 - 3360b_7 + 1120b_5 - (a_1 - 10a_3 + 20a_4)\Omega^2 = 0$$

$$-4320b_8 - 1440b_6 - (a_2 - 10a_4)\Omega^2 = 0$$

$$1800b_7 - a_3\Omega^2 = 0$$

$$2200b_8 - a_4\Omega^2 = 0$$

Again, we impose the condition that the determinant of the system (6.57) has to vanish in order to get a non-trivial solution. This leads to the following equation:

$$1,600a_0 - 1,245a_1 + 2,270a_2 - 2,786a_3 + (50,928/11)a_4 = 0 \qquad (6.58)$$

which yields the following solution:

$$a_1 = \frac{320}{249}a_0 + \frac{454}{249}a_2 - \frac{2,786}{245}a_3 + \frac{16,976}{4,565}a_4 \qquad (6.59)$$

where a_0, a_2, a_3 and a_4 are arbitrary. Upon substitution of the expression (6.59) into the set (6.57) we get the solution for the natural frequency squared:

$$\Omega^2 = 2,200b_8/a_4 \qquad (6.60)$$

and the coefficients in the flexural rigidity

$$b_0 = (7,249,880a_0 + 7,475,325a_2 - 4,364,536a_3 + 12,218,962a_4)b_8/20,916a_4$$

$$b_1 = (621,720a_0 + 896,445a_2 - 387,640a_3 + 1,459,794a_4)b_8/10,458a_4$$

$$b_2 = (621,720a_0 + 896,445a_2 - 387,640a_3 + 1,459,794a_4)b_8/6,972a_4$$

$$b_3 = (-206,800a_0 + 74,085a_2 + 109,472a_3 + 114,898a_4)b_8/5,229a_4$$

$$b_4 = (660a_0 - 5,445a_2 + 12,584a_3 - 9,282a_4)b_8/252a_4$$

$$b_5 = (8,800a_0 + 12,485a_2 - 71,016a_3 + 155,442a_4)b_8/3,486a_4$$

$$b_6 = (55a_2 - 442a_4)b_8/36a_4 \qquad b_7 = 11a_3b_8/9a_4$$

$$(6.61)$$

For the case $n = 2$, the governing differential equation yields a set of 11 linear algebraic equations for 10 unknowns:

$$180b_2 - 240b_1 = 0 \quad 540b_3 - 720b_2 = 0$$

$$1080b_4 - 1440b_3 + 360b_0 - 45a_0\Omega^2 = 0$$

$$1800b_5 - 2400b_4 + 600b_1 + (20a_0 - 45a_1)\Omega^2 = 0$$

$$2700b_6 - 3600b_5 + 900b_2 + (20a_1 - 45a_2)\Omega^2 = 0$$

$$3780b_7 - 5040b_6 + 1260b_3 + (20a_2 - 45a_3)\Omega^2 = 0 \qquad (6.62)$$

$$5040b_8 - 6720b_7 + 1680b_4 - (a_0 - 20a_3 + 45a_4)\Omega^2 = 0$$

$$-8640b_8 + 2160b_5 - (a_1 - 20a_4)\Omega^2 = 0$$

$$2700b_6 - a_2\Omega^2 = 0 \quad 3300b_7 - a_3\Omega^2 = 0$$

$$3960b_8 - a_4\Omega^2 = 0$$

In order to have a non-trivial solution, the rank of the set (6.62) must be less than 10. The rank of the matrix of the system is less than 10 if all minors of order 10 vanish, leading to a set of 11 equations for 5 unknowns. The solution of this set reads

$$a_0 = \frac{5,169}{715}a_2 - \frac{39,812}{7,865}a_3 + \frac{489}{1,573}a_4 \quad a_1 = \frac{1,128}{385}a_2 - \frac{5,427}{4,235}a_3 - \frac{196}{121}a_4$$

$$(6.63)$$

where a_2, a_3 and a_4 are arbitrary constants. Upon substitution of the expressions (6.63) into the set (6.62) we get the solution for the natural frequency squared:

$$\Omega^2 = 3,960b_8/a_4 \qquad (6.64)$$

and the coefficients in the flexural rigidity

$$b_0 = (99,082,621a_2 - 53,799,120a_3 - 4,730,670a_4)b_8/30,030a_4$$

$$b_1 = -3(748a_2 - 1,809a_3)b_8/70a_4 \quad b_2 = -2(748a_2 - 1,809a_3)b_8/35a_4$$

$$b_3 = -8(748a_2 - 1,809a_3)b_8/105a_4$$

$$b_4 = 3(56,859a_2 - 181,096a_3 + 346,360a_4)b_8/10,010a_4$$

$$b_5 = (41,36a_2 - 1,809a_3 - 27,440a_4)b_8/770a_4 \quad b_6 = 22a_2b_8/15a_4$$

$$b_7 = 6a_3b_8/5a_4$$

$$(6.65)$$

For the case $n = 3$, the governing differential equation yields a set of 12 linear algebraic equations for 10 unknowns:

$$336b_2 - 420b_1 = 0 \quad 1{,}008b_3 - 1{,}260b_2 = 0$$

$$2{,}016b_4 - 2{,}520b_3 - 84a_0\Omega^2 = 0$$

$$3{,}360b_5 - 4{,}200b_4 + 840b_0 + (35a_0 - 84a_1)\Omega^2 = 0$$

$$5{,}040b_6 - 6{,}300b_5 + 1{,}260b_1 + (35a_1 - 84a_2)\Omega^2 = 0$$

$$7{,}056b_7 - 8{,}820b_6 + 1{,}764b_2 + (35a_2 - 84a_3)\Omega^2 = 0 \qquad (6.66)$$

$$9{,}408b_8 - 11{,}760b_7 + 2{,}352b_3 + (35a_3 - 84a_4)\Omega^2 = 0$$

$$-15{,}120b_8 + 3{,}024b_4 - (a_0 - 35a_4)\Omega^2 = 0$$

$$3{,}780b_5 - a_1\Omega^2 = 0 \quad 4{,}620b_6 - a_2\Omega^2 = 0$$

$$5{,}544b_7 - a_3\Omega^2 = 0 \quad 6{,}552b_8 - a_4\Omega^2 = 0$$

In order to have a non-trivial solution, the rank of the set (6.66) must be less than 10. The rank of the matrix of the system is less than 10 if all minors of order 10 vanish, leading to a set of 66 equations for 5 unknowns. The solution of this set reads

$$a_1 = \frac{7{,}664}{403}a_0 + \frac{107{,}864}{5{,}239}a_4 \quad a_2 = \frac{8{,}855}{1{,}209}a_0 + \frac{128{,}777}{15{,}717}a_4$$

$$a_3 = \frac{220}{93}a_0 + \frac{3{,}784}{1{,}209}a_4 \qquad (6.67)$$

where a_0 and a_4 are arbitrary constants. Substituting the expressions (6.67) into the set (6.66) we get the solution for the natural frequency squared:

$$\Omega^2 = 6{,}552b_8/a_4 \qquad (6.68)$$

as well as the coefficients in the flexural rigidity,

$$b_0 = (145{,}879{,}877a_0 + 157{,}082{,}669a_4)b_8/12{,}090a_4$$

$$b_1 = -32(65a_0 + 17a_4)b_8/15a_4$$

$$b_2 = -8(65a_0 + 17a_4)b_8/3a_4 \quad b_3 = -10(65a_0 + 17a_4)b_8/3a_4$$

$$b_4 = (13a_0 - 425a_4)b_8/6a_4 \quad b_5 = 16(12{,}454a_0 + 13{,}483a_4)b_8/6{,}045a_4$$

$$b_6 = 46(455a_0 + 509a_4)b_8/2{,}015a_4 \quad b_7 = 4(65a_0 + 86a_4)b_8/93a_4$$

$$(6.69)$$

6.1.4.6 *Quintic inertial coefficient* ($m = 5$)

When $n = 0$, we obtain a set of 10 equations for 11 unknowns

$$24b_2 - 48b_1 + 24b_0 - 0$$

$$72b_3 - 144b_2 + 72b_1 = 0$$

$$144b_4 - 288b_3 + 144b_2 - 6a_0\Omega^2 = 0$$

$$240b_5 - 480b_4 + 240b_3 + (4a_0 - 6a_1)\Omega^2 = 0$$

$$360b_6 - 720b_5 + 360b_4 - (a_0 - 4a_1 + 6a_2)\Omega^2 = 0$$

$$504b_7 - 1008b_6 + 504b_5 - (a_1 - 4a_2 + 6a_3)\Omega^2 = 0 \qquad (6.70)$$

$$672b_8 - 1344b_7 + 672b_6 - (a_2 - 4a_3 + 6a_4)\Omega^2 = 0$$

$$864b_9 - 1728b_8 + 864b_7 - (a_3 - 4a_4 + 6a_5)\Omega^2 = 0$$

$$-2160b_9 + 1080b_8 - (a_4 - 4a_5)\Omega^2 = 0$$

$$1320b_9 - a_5\Omega^2 = 0$$

We can, therefore, solve the system taking b_9 as a parameter. From the last equation of the set (6.70), we obtain the natural frequency squared:

$$\Omega^2 = 1320b_9/a_5 \qquad (6.71)$$

The remaining equations yield the coefficients in the flexural rigidity as follows:

$$b_0 = (24{,}024a_0 + 18{,}744a_1 + 15{,}345a_2 + 12{,}980a_3 + 11{,}242a_4$$
$$+ 9{,}912a_5)b_9/252a_5$$

$$b_1 = (14{,}784a_0 + 13{,}464a_1 + 11{,}946a_2 + 10{,}615a_3 + 9{,}504a_4$$
$$+ 8{,}582a_5)b_9/252a_5$$

$$b_2 = (5{,}544a_0 + 8{,}184a_1 + 8{,}547a_2 + 8{,}250a_3 + 7{,}766a_4 + 7{,}252a_5)b_9/252a_5$$

$$b_3 = -(3{,}696a_0 - 2{,}904a_1 - 5{,}148a_2 - 5{,}885a_3 - 6{,}028a_4 - 5{,}922a_5)b_9/252a_5$$

$$b_4 = (924a_0 - 2{,}376a_1 + 1{,}749a_2 + 3{,}520a_3 + 4{,}290a_4 + 4{,}592a_5)b_9/252a_5$$

$$b_5 = (660a_1 - 1{,}650a_2 + 1{,}155a_3 + 2{,}552a_4 + 3{,}262a_5)b_9/252a_5$$

$$b_6 = (495a_2 - 1{,}210a_3 + 814a_4 + 1{,}932a_5)b_9/252a_5$$

$$b_7 = (55a_3 - 132a_4 + 86a_5)b_9/36a_5 \qquad b_8 = (11a_4 - 26a_5)b_9/9a_5$$

$$(6.72)$$

We note that Eq. (6.71) was derived by Elishakoff and Candan (2000), as part of their general case, when the substitution $m = 5$ is made in their equations. Note that the case $n \neq 0$ was not studied by Elishakoff and Candan (2000).

Now, consider the case $n = 1$. The governing differential equation yields a set of 11 linear algebraic equations for 11 unknowns:

$$80b_2 - 120b_1 = 0$$

$$240b_3 - 360b_2 + 120b_0 = 0$$

$$480b_4 - 720b_3 + 240b_1 - 20a_0\Omega^2 = 0$$

$$800b_5 - 1200b_4 + 400b_2 + (10a_0 - 20a_1)\Omega^2 = 0$$

$$1200b_6 - 1800b_5 + 600b_3 + (10a_1 - 20a_2)\Omega^2 = 0$$

$$1680b_7 - 2520b_6 + 840b_4 - (a_0 - 10a_2 + 20a_3)\Omega^2 = 0 \qquad (6.73)$$

$$2240b_8 - 3360b_7 + 1120b_5 - (a_1 - 10a_3 + 20a_4)\Omega^2 = 0$$

$$2880b_9 - 4320b_8 - 1440b_6 - (a_2 - 10a_4 + 20a_5)\Omega^2 = 0$$

$$-5400b_9 + 1800b_7 - (a_3 - 10a_5)\Omega^2 = 0$$

$$2200b_8 - a_4\Omega^2 = 0, \quad 2640b_9 - a_5\Omega^2 = 0$$

We impose the condition that the determinant of the system (6.73) should vanish so that we get a non-trivial solution. This leads to the following equation:

$$1{,}600a_0 - 1{,}245a_1 + 2{,}270a_2 - 2{,}786a_3 + \frac{50{,}928}{11}a_4 = 0 \qquad (6.74)$$

or,

$$a_0 = \frac{249}{320}a_1 - \frac{227}{160}a_2 + \frac{1{,}393}{800}a_3 - \frac{3{,}183}{1{,}100}a_4 + \frac{7{,}553}{1{,}760}a_5 \qquad (6.75)$$

where a_1, a_2, a_3, a_4 and a_5 are arbitrary. Upon substitution of the expression (6.75) into the set (6.73) we get the solution for the natural frequency squared:

$$\Omega^2 = 2{,}640b_9/a_5 \qquad (6.76)$$

and the coefficients in the flexural rigidity

$$b_0 = (2{,}718{,}705a_1 - 1{,}354{,}430a_2 + 3{,}980{,}394a_3 - 4{,}221{,}472a_4$$
$$+ 8{,}201{,}270a_5)b_9/8400a_5$$

$$b_1 = (233{,}145a_1 + 6{,}930a_2 + 334{,}906a_3 - 163{,}488a_4 + 585{,}830a_5)b_9/4{,}200a_5$$

$$b_2 = (233{,}145a_1 + 6{,}930a_2 + 334{,}906a_3 - 163{,}488a_4 + 585{,}830a_5)b_9/2{,}800a_5$$

$$b_3 = -(7{,}755a_1 - 17{,}710a_2 + 12{,}078a_3 - 34{,}376a_4 + 36{,}610a_5)b_9/210a_5$$

$$b_4 = (41{,}085a_1 - 510{,}510a_2 + 1{,}098{,}658a_3 - 895{,}344a_4 + 2{,}365{,}790a_5)b_9/16{,}800a_5$$

$$b_5 = (165a_1 - 1{,}342a_3 + 3{,}132a_4 - 2{,}450a_5)b_9/70a_5$$

$$b_6 = (55a_2 - 442a_4 + 1{,}040a_5)b_9/30a_5 \quad b_7 = (22a_3 - 175a_5)b_9/15a_5$$

$$b_8 = 6a_4b_9/5a_5$$

$$(6.77)$$

For the case $n = 2$, the governing differential equation yields a set of 12 linear algebraic equations for 11 unknowns:

$$180b_2 - 240b_1 = 0, \quad 540b_3 - 720b_2 = 0$$

$$1{,}080b_4 - 1{,}440b_3 + 360b_0 - 45a_0\Omega^2 = 0$$

$$1{,}800b_5 - 2{,}400b_4 + 600b_1 + (20a_0 - 45a_1)\Omega^2 = 0$$

$$2{,}700b_6 - 3{,}600b_5 + 900b_2 + (20a_1 - 45a_2)\Omega^2 = 0$$

$$3{,}780b_7 - 5{,}040b_6 + 1{,}260b_3 + (20a_2 - 45a_3)\Omega^2 = 0$$

$$5{,}040b_8 - 6{,}720b_7 + 1{,}680b_4 - (a_0 - 20a_3 + 45a_4)\Omega^2 = 0$$

$$6{,}480b_9 - 8{,}640b_8 + 2{,}160b_5 - (a_1 - 20a_4 + 45a_5)\Omega^2 = 0$$

$$-10{,}800b_9 + 2{,}700b_6 - (a_2 - 20a_5)\Omega^2 = 0$$

$$3{,}300b_7 - a_3\Omega^2 = 0 \quad 3{,}960b_8 - a_4\Omega^2 = 0$$

$$4{,}680b_9 - a_5\Omega^2 = 0$$

$$(6.78)$$

In order to have a non-trivial solution, the rank of the set must be less than 10. The rank of the matrix of the system is less than 10 if all minors of order 10 vanish, leading to a set of 11 equations for 5 unknowns. The solution of this set reads

$$a_0 = \frac{5{,}169}{715}a_2 - \frac{39{,}812}{7{,}865}a_3 + \frac{489}{1{,}573}a_4 + \frac{23{,}310}{1{,}859}a_5$$

$$a_1 = \frac{1{,}128}{385}a_2 - \frac{5{,}427}{4{,}235}a_3 - \frac{196}{121}a_4 + \frac{843}{143}a_5$$

$$(6.79)$$

where a_2, a_3, a_4 and a_5 are arbitrary constants. Upon substitution of the expressions (6.79) into the set (6.78) we get the solution for the natural frequency squared:

$$\Omega^2 = 4{,}680b_9/a_5 \qquad (6.80)$$

and the coefficients in the flexural rigidity,

$$b_0 = (1{,}288{,}074{,}073a_2 - 699{,}388{,}560a_3 - 61{,}498{,}710a_4$$

$$+ 2{,}226{,}377{,}230a_5)b_9/330{,}330a_5$$

$$b_1 = -3(9{,}724a_2 - 23{,}517a_3 + 17{,}710a_5)b_9/770a_5$$

$b_2 = -2(9{,}724a_2 - 23{,}517a_3 + 17{,}710a_5)b_9/385a_5$

$b_3 = -8(9{,}724a_2 - 23{,}517a_3 + 17{,}710a_5)b_9/1{,}155a_5$

$b_4 = 3(739{,}167a_2 - 2{,}354{,}248a_3 + 4{,}502{,}680a_4 + 1{,}282{,}050a_5)b_9/110{,}110a_5$

$b_5 = (53{,}768a_2 - 23{,}517a_3 - 356{,}720a_4 + 908{,}600a_5)b_9/8{,}470a_5$

$b_6 = 2(13a_2 - 230a_5)b_9/15a_5 \quad b_7 = 78a_3b_9/55a_5 \quad b_8 = 13a_4b_9/11a_5$

$$(6.81)$$

For the case $n = 3$, the governing differential equation yields a set of 13 linear algebraic equations for 11 unknowns:

$336b_2 - 420b_1 = 0 \quad 1{,}008b_3 - 1{,}260b_2 = 0$

$2{,}016b_4 - 2{,}520b_3 - 84a_0\Omega^2 = 0$

$3{,}360b_5 - 4{,}200b_4 + 840b_0 + (35a_0 - 84a_1)\Omega^2 = 0$

$5{,}040b_6 - 6{,}300b_5 + 1{,}260b_1 + (35a_1 - 84a_2)\Omega^2 = 0$

$7{,}056b_7 - 8{,}820b_6 + 1{,}764b_2 + (35a_2 - 84a_3)\Omega^2 = 0$

$9{,}408b_8 - 11{,}760b_7 + 2{,}352b_3 + (35a_3 - 84a_4)\Omega^2 = 0$

$12{,}096b_9 - 15{,}120b_8 + 3{,}024b_4 - (a_0 - 35a_4 + 84a_5)\Omega^2 = 0$

$-18{,}900b_9 + 3{,}780b_5 - (a_1 - 35a_5)\Omega^2 = 0$

$4{,}620b_6 - a_2\Omega^2 = 0 \quad 5{,}544b_7 - a_3\Omega^2 = 0$

$6{,}552b_8 - a_4\Omega^2 = 0 \quad 7{,}644b_9 - a_5\Omega^2 = 0$

$$(6.82)$$

In order to have a non-trivial solution, the rank of the set (6.82) must be less than 11. The rank of the matrix of the system is less than 11 if all minors of order 11 vanish, leading to a set of 78 equations for 5 unknowns. The solution of this set reads

$$a_0 = \frac{93}{220}a_3 - \frac{86}{65}a_4 + \frac{60}{91}a_5 \quad a_1 = \frac{5{,}748}{715}a_3 - \frac{3{,}864}{845}a_4 - \frac{148}{91}a_5$$

$$a_2 = \frac{161}{52}a_3 - \frac{253}{169}a_4$$

$$(6.83)$$

where a_0 and a_4 are arbitrary constants. Substituting the expressions (6.83) into the set (6.82) we get the solution for the natural frequency squared:

$$\Omega^2 = 7{,}644b_9/a_5 \tag{6.84}$$

as well as the coefficients in the flexural rigidity,

$$b_0 = 7(13{,}261{,}807a_3 - 7{,}726{,}372a_4 + 13{,}261{,}807a_5)b_9/15{,}600a_5$$

$$b_1 - -28(403a_3 - 1{,}012a_4)b_9/165a_5 \quad b_2 - -7(403a_3 - 1{,}012a_4)b_9/33a_5$$

$$b_3 = -32(403a_3 - 1{,}012a_4)b_9/132a_5$$

$$b_4 = 7(403a_3 - 32{,}428a_4 + 79{,}200a_5)b_9/2{,}640a_5$$

$$b_5 = 28(6{,}227a_3 - 3{,}542a_4 - 26{,}455a_5)b_9/10{,}725a_5$$

$$b_6 = 161(91a_3 - 44a_4)b_8/2{,}860a_5 \quad b_7 = 91a_3b_9/66a_5 \quad b_8 = 7a_4b_9/6a_5$$

$$(6.85)$$

6.1.5 Closed-Form Solutions for the Pinned–Clamped Beam

6.1.5.1 Constant inertial coefficient (m = 0)

First consider the case $n = 0$; the inertial coefficient reduces to $R(\xi) = a_0$. The governing differential equation leads to the following set of equations:

$$-36b_1 + 48b_0 = 0 \quad -108b_2 + 144b_1 - a_0\Omega^2 = 0$$

$$-216b_3 + 288b_2 = 0 \quad -360b_4 + 480b_3 + 3a_0\Omega^2 = 0 \qquad (6.86)$$

$$720b_4 - 2a_0\Omega^2 = 0$$

We have five equations for six unknowns: b_0, b_1, b_2, b_3, b_4 and Ω^2. We declare b_4 to be an arbitrary parameter. We get Ω^2 from the last equation of the set (6.86):

$$\Omega^2 = 360b_4/a_0 \qquad (6.87)$$

The remaining equations lead to

$$b_0 = \tfrac{159}{128}b_4 \quad b_1 = \tfrac{53}{32}b_4 \quad b_2 = -\tfrac{9}{8}b_4 \quad b_3 = -\tfrac{3}{2}b_4 \qquad (6.88)$$

The flexural rigidity is then given by the following expression:

$$D(\xi) = \left(\tfrac{159}{128} + \tfrac{53}{32}\xi - \tfrac{9}{8}\xi^2 - \tfrac{3}{2}\xi^3 + \xi^4 \right) b_4 \qquad (6.89)$$

which is positive within the interval [0; 1] (see Figure 6.7).

Now consider now the case $n = 1$. We obtain

$$b_1 = 0 \quad -144b_2 + 240b_0 - 2a_0\Omega^2 = 0$$

$$-b_3 + 2b_1 = 0 \quad -480b_4 + 800b_2 + 4a_0\Omega^2 = 0 \qquad (6.90)$$

$$b_3 = 0 \quad 1680b_4 - 2a_0\Omega^2 = 0$$

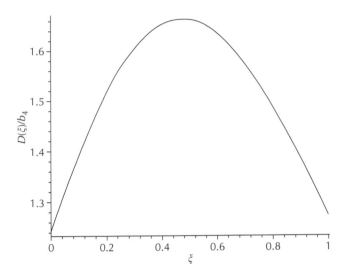

FIGURE 6.7
Variation of the flexural rigidity for the pinned–clamped beam

We get a linear system of six equations for six unknowns. Since the determinant of the system (6.91) vanishes, i.e.

$$\begin{vmatrix} 0 & 1 & 0 & 0 & 0 & 0 \\ 240 & 0 & -144 & 0 & 0 & -2a_0 \\ 0 & 2 & 0 & -1 & 0 & 0 \\ 0 & 0 & 800 & 0 & 480 & 4a_0 \\ 0 & 0 & 0 & 1 & 0 & 0 \\ 0 & 0 & 0 & 0 & 1680 & -2a_0 \end{vmatrix} = 0 \tag{6.91}$$

we can solve the set (6.91), expressing the unknowns with b_4 taken as an arbitrary constant. Thus, we get

$$\Omega^2 = 840b_4/a_0 \tag{6.92}$$

and the coefficients b_i are

$$b_0 = \tfrac{121}{25}b_4 \quad b_1 = 0 \quad b_2 = -\tfrac{18}{5}b_4 \quad b_3 = 0 \tag{6.93}$$

The flexural rigidity,

$$D(\xi) = \left(\tfrac{121}{25} - \tfrac{18}{5}\xi^2 + \xi^4 \right) b_4 \tag{6.94}$$

which is depicted in Figure 6.8, is positive within the interval [0; 1], when the parameter b_4 is non-negative.

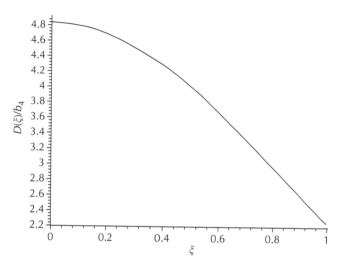

FIGURE 6.8
Variation of the flexural rigidity for the pinned–clamped beam

6.1.5.2 *Linearly varying inertial coefficient ($m = 1$)*

First consider the case $n = 0$. We have

$$-36b_1 + 48b_0 = 0 \quad -108b_2 + 144b_1 - a_0\Omega^2 = 0$$

$$-216b_3 + 288b_2 - a_1\Omega^2 = 0$$

$$-360b_4 + 480b_3 + 3a_0\Omega^2 = 0 \tag{6.95}$$

$$-540b_5 + 720b_4 - (2a_0 - 3a_1)\Omega^2 = 0$$

$$1008b_5 - 2a_1\Omega^2 = 0$$

This set represents a linear algebraic system of six equations for the seven unknowns, $b_0, b_1, \ldots, b_5$ and Ω^2. Hence, we express them in terms of b_5, which is treated as an arbitrary constant. We get Ω^2 from the last equation of the set (6.95):

$$\Omega^2 = 504b_5/a_1 \tag{6.96}$$

while the remaining equations lead to

$$b_0 = 3(2968a_0 + 951a_1)b_5/5120a_1 \quad b_1 = (2968a_0 + 951a_1)b_5/1280a_1$$

$$b_2 = -(504a_0 - 317a_1)b_5/320a_1 \quad b_3 = -3(56a_0 + 27a_1)b_5/80a_1 \tag{6.97}$$

$$b_4 = (28a_0 - 27a_1)b_5/20a_1$$

Comparison of Eqs. (6.96) and (6.17) reveals that the natural frequencies of the cantilever beam and the pinned–clamped beam coincide. It must be stressed that this remarkable phenomenon takes place for beams with different flexural rigidities. Indeed, the expressions in Eq. (6.97) differ from those in Eq. (6.18). An analogous phenomenon will be recorded also for the clamped–clamped beam, when the values of the integers "m" and "n" coincide for the beams with different boundary conditions.

6.1.5.3   Parabolically varying inertial coefficient ($m = 2$)

We first consider the case $n = 0$. We obtain the following set of seven equations with eight unknowns

$$-36b_1 + 48b_0 = 0$$

$$-108b_2 + 144b_1 - a_0\Omega^2 = 0$$

$$-216b_3 + 288b_2 - a_1\Omega^2 = 0$$

$$-360b_4 + 480b_3 + (3a_0 - a_2)\Omega^2 = 0 \qquad\qquad (6.98)$$

$$-540b_5 + 720b_4 - (2a_0 - 3a_1)\Omega^2 = 0$$

$$-756b_6 + 1008b_5 - (2a_1 - 3a_2)\Omega^2 = 0$$

$$1344b_6 - 2a_2\Omega^2 = 0$$

We can, therefore, find a non-trivial solution when b_6 is regarded as a parameter. From the last equation of the set (6.98), we obtain the natural frequency squared:

$$\Omega^2 = 672b_6/a_2 \qquad\qquad (6.99)$$

The remaining equations yield the coefficients in the flexural rigidity as follows:

$$b_0 = (47{,}488a_0 + 15{,}216a_1 + 6{,}021a_2)b_6/20{,}480a_2$$
$$b_1 = (47{,}488a_0 + 15{,}216a_1 + 6{,}021a_2)b_6/15{,}360a_2$$
$$b_2 = -(8{,}064a_0 - 5{,}072a_1 - 2{,}007a_2)b_6/3{,}840a_2 \qquad\qquad (6.100)$$
$$b_3 = -(896a_0 + 432a_1 - 223a_2)b_6/320a_2$$
$$b_4 = (448a_0 - 432a_1 - 225a_2)b_6/240a_2 \quad b_5 = (16a_1 - 15a_2)b_6/12a_2$$

When $n = 1$, the governing differential equation yields a set of eight linear algebraic equations for eight unknowns, namely the seven coefficients b_i in the flexural rigidity, and the natural frequency squared Ω^2:

$$b_1 = 0$$

$$-144b_2 + 240b_0 - 2a_0\Omega^2 = 0$$

$$-288b_3 + 480b_1 - 2a_1\Omega^2 = 0$$

$$-480b_4 + 800b_2 + (4a_0 - 2a_2)\Omega^2 = 0$$

$$-720b_5 + 1200b_3 + 4a_1\Omega^2 = 0$$

$$-1008b_6 + 1680b_4 - (2a_0 - 4a_2)\Omega^2 = 0$$

$$2240b_5 - 2a_1\Omega^2 = 0$$

$$2880b_6 - 2a_2\Omega^2 = 0$$

$$(6.101)$$

The determinant of the system (obtained by the MAPLE® program) equals 143,146,830,790,656,000,000a_1. In order to find a non-trivial solution, the determinant must be zero, leading to the requirement $a_1 = 0$. We conclude that for an inertial coefficient

$$R(\xi) = a_0 + a_2\xi^2 \tag{6.102}$$

where r_0 and r_2 are arbitrary, the set (6.102) yields

$$\Omega^2 = 1{,}440b_6/a_2 \tag{6.103}$$

and

$$b_0 = 3(2420a_0 + 333a_2)b_6/875a_2 \quad b_1 = 0 \quad b_2 = -9(120a_0 - 37a_2)b_6/175a_2$$
$$b_3 = 0 \quad b_4 = 3(20a_0 - 33a_2)b_6/35a_2 \quad b_5 = 0$$

$$(6.104)$$

Concerning the case $n = 2$, the governing equation yields a set of eight equations for nine unknowns, namely the coefficients b_i in the flexural rigidity and the natural frequency squared Ω^2:

$$b_1 = 0 \quad -180b_2 - 3a_0\Omega^2 = 0$$

$$-360b_3 + 720b_0 - 3a_1\Omega^2 = 0$$

$$-600b_4 + 1200b_1 + (5a_0 - 3a_2)\Omega^2 = 0$$

$$-900b_5 + 1800b_2 + 5a_1\Omega^2 = 0 \tag{6.105}$$

$$-1260b_6 + 2520b_3 + 5a_2\Omega^2 = 0$$

$$3360b_4 - 2a_0\Omega^2 = 0 \quad 4320b_5 - 2a_1\Omega^2 = 0$$

$$5400b_6 - 2r_2\Omega^2 = 0$$

Since the number of equations is larger than that of the unknowns, in order to have a non-trivial solution, the rank of the set (6.105) must be less than 8.

The rank of the matrix of the system (6.105) is less than 8 if all minors of order 8 vanish. This last condition leads to a set of nine equations, whose solution reads

$$a_0 = \frac{42}{65}a_2 \quad a_1 = \frac{3024}{715}a_2 \tag{6.106}$$

where a_2 is treated as an arbitrary constant. We deduce that when the inertial coefficient is

$$R(\xi) = \left(\frac{42}{65} + \frac{3024}{715}\xi + \xi^2\right)a_2 \tag{6.107}$$

where a_2 is an arbitrary parameter, the set (6.106) yields

$$\Omega^2 = 2,700b_6/a_2 \tag{6.108}$$

and

$$b_0 = \frac{45,197}{1,001}b_6 \quad b_1 = 0 \quad b_2 = -\frac{378}{13}b_6 \quad b_3 = -\frac{34}{7}b_6$$

$$b_4 = \frac{27}{26}b_6 \quad b_5 = \frac{756}{143}b_6 \tag{6.109}$$

The flexural rigidity is, then,

$$D(\xi) = \left(\frac{45,197}{1,001} - \frac{378}{13}\xi^2 - \frac{34}{7}\xi^3 + \frac{27}{26}\xi^4 + \frac{756}{143}\xi^5 + \xi^6\right)b_6 \tag{6.110}$$

Figures 6.9 and 6.10, that depict, respectively, the variation of the flexural rigidity and the inertial coefficient, show that both quantities are positive, unless the parameters b_6 and a_2 are non-negative.

For $n = 3$, we have to satisfy the following set of ten equations for eight unknowns:

$$b_1 = 0 \quad -216b_2 - 4a_0\Omega^2 = 0$$

$$-432b_3 - 4a_1\Omega^2 = 0$$

$$-720b_4 + 1680b_0 + (6a_0 - 4a_2)\Omega^2 = 0$$

$$-1080b_5 + 2520b_1 + 6a_1\Omega^2 = 0 \tag{6.111}$$

$$-1512b_6 + 3528b_2 + 6a_2\Omega^2 = 0$$

$$b_3 = 0, \quad 6048b_4 - 2a_0\Omega^2 = 0$$

$$7560b_5 - 2a_1\Omega^2 = 0 \quad 9240b_6 - 2a_2\Omega^2 = 0$$

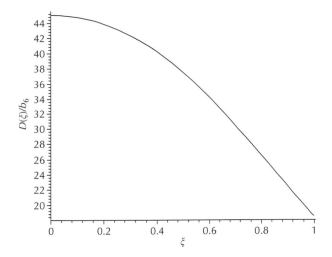

FIGURE 6.9
Variation of the flexural rigidity for the pinned–clamped beam

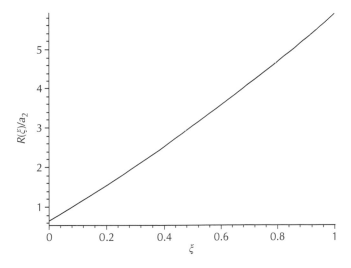

FIGURE 6.10
Variation of the inertial coefficient for the pinned–clamped beam

A non-trivial solution is obtainable, provided the rank of the set (6.111) is less than 8. This requires that the $C_{10}^8 = 45$ minors of order 8 of the matrix of the linear system (6.111) are zero. We, then, obtain 45 equations, whose solution is

$$a_0 = 234a_2/2695 \quad a_1 = 0 \tag{6.112}$$

where a_2 represents an arbitrary constant. Therefore, when the inertial coefficient equals

$$R(\xi) = \left(\frac{234}{2695} + \xi^2\right) a_2 \tag{6.113}$$

the set (6.111) yields

$$\Omega^2 = 4620 b_6 / a_2 \tag{6.114}$$

Moreover,

$$b_0 = \frac{33{,}011}{3{,}430} b_6 \quad b_1 = 0 \quad b_2 = -\frac{52}{7} b_6 \quad b_3 = 0 \quad b_4 = \frac{13}{98} b_6 \quad b_5 = 0 \tag{6.115}$$

The flexural rigidity is then

$$D(\xi) = \left(\frac{33{,}011}{3{,}430} - \frac{52}{7}\xi^2 + \frac{13}{98}\xi^4 + \xi^6\right) b_6 \tag{6.116}$$

Figure 6.11 depicts the variation of the flexural rigidity $D(\xi)$ in terms of ξ in [0; 1], whereas Figure 6.12 portrays the variation of the inertial coefficient.

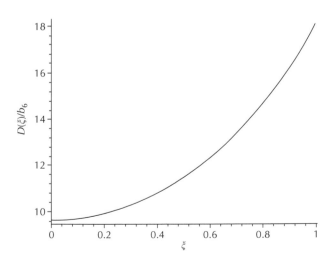

FIGURE 6.11
Variation of the flexural rigidity for the pinned–clamped beam

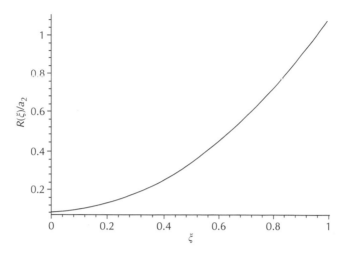

FIGURE 6.12
Variation of the inertial coefficient for the pinned–clamped beam

6.1.5.4 Cubic inertial coefficient ($m = 3$)

When $n = 0$, we get

$$-36b_1 + 48b_0 = 0$$
$$-108b_2 + 144b_1 - a_0\Omega^2 = 0$$
$$-216b_3 + 288b_2 - a_1\Omega^2 = 0$$
$$-360b_4 + 480b_3 + (3a_0 - a_2)\Omega^2 = 0$$
$$-540b_5 + 720b_4 - (2a_0\Omega - 3a_1 + a_3)\Omega^2 = 0 \qquad (6.117)$$
$$-756b_6 + 1008b_5 - (2a_1 - 3a_2)\Omega^2 = 0$$
$$-1008b_7 + 1344b_6 - (2a_2\Omega^2 - 3a_3)\Omega^2 = 0$$
$$1728b_7 - 2a_3\Omega^2 = 0$$

which constitutes eight equations for nine unknowns. We can, therefore, solve the system taking b_7 as a parameter. From the last equation of the set (6.117), we obtain the natural frequency squared:

$$\Omega^2 = 864b_7/a_3 \qquad (6.118)$$

while the remaining equations yield the coefficients in the flexural rigidity as follows:

$$b_0 = 9(189{,}952a_0 + 60{,}864a_1 + 24{,}084a_2 + 10{,}827a_3)b_6/573{,}440a_3$$
$$b_1 = 3(189{,}952a_0 + 60{,}864a_1 + 24{,}084a_2 + 10{,}827a_3)b_6/143{,}360a_3$$

$$b_2 = -3(32{,}256a_0 - 20{,}288a_1 - 8{,}028a_2 - 3{,}609a_3)b_7/35{,}840a_3$$

$$b_3 = -9(3{,}584a_0 + 1{,}728a_1 - 892a_2 - 401a_3)b_7/8{,}960a_3$$

$$b_4 = 3(1{,}792a_0 - 1{,}728a_1 - 900a_2 + 401a_3)b_7/2{,}240a_3$$

$$b_5 = 3(64a_1 - 60a_2 - 33a_3)b_7/112a_3 \quad b_6 = 3(12a_2 - 11a_3)b_7/28a_3$$

$$(6.119)$$

Now, consider the case $n = 1$; the governing differential equation gives a set of nine linear algebraic equations for nine unknowns:

$$b_1 = 0$$

$$-144b_2 + 240b_0 - 2a_0\Omega^2 = 0$$

$$-288b_3 + 480b_1 - 2a_1\Omega^2 = 0$$

$$-480b_4 + 800b_2 + (4a_0 - 2a_2)\Omega^2 = 0$$

$$-720b_5 + 1200b_3 + (4a_1 - 2a_3)\Omega^2 = 0 \qquad (6.120)$$

$$-1008b_6 + 1680b_4 - (2a_0 - 4a_2)\Omega^2 = 0$$

$$-1344b_7 + 2240b_5 - (2a_1 - 4a_3)\Omega^2 = 0$$

$$2880b_6 - 2a_2\Omega^2 = 0$$

$$3600b_7 - 2a_3\Omega^2 = 0$$

Again, we require the determinant of the system to vanish, so that we get a non-trivial solution. It produces the following equation:

$$1002a_3 + 5225a_1 = 0 \qquad (6.121)$$

which produces a relationship

$$a_1 = -1002a_3/5\,225 \qquad (6.122)$$

where a_3 is arbitrary. Thus, provided the inertial coefficient equals

$$R(\xi) = a_0 - (1002/5225)a_3\xi + a_2\xi^2 + a_3\xi^3 \qquad (6.123)$$

the set (6.120) yields

$$\Omega^2 = 1800b_7/a_3 \qquad (6.124)$$

$$b_0 = 3(333r_2 + 2420r_0)b_7/700a_3 \quad b_1 = 0 \quad b_2 = -9(-37a_2 + 120a_0)b_7/140a_3$$
$$b_3 = 501/209b_7 \quad b_4 = 3(20a_0 - 33a_2)b_7/28a_3$$
$$b_5 = -3054/1045b_7 \quad b_6 = \tfrac{5}{4}a_2b_6/a_3$$

$$(6.125)$$

For the case $n = 2$, the governing differential equation yields a set of ten linear algebraic equations for nine unknowns:

$$b_1 = 0$$

$$-180b_2 - 3a_0\Omega^2 = 0$$

$$-360b_3 + 720b_0 - 3a_1\Omega^2 = 0$$

$$-600b_4 + 1200b_1 + (5a_0 - 3a_2)\Omega^2 = 0$$

$$-900b_5 + 1800b_2 + (5a_1 - 3a_3)\Omega^2 = 0$$

$$-1260b_6 + 2520b_3 + 5a_2\Omega^2 = 0 \qquad\qquad (6.126)$$

$$-1680b_7 + 3360b_4 - (2a_0 - 5a_3)\Omega^2 = 0$$

$$4320b_5 - 2a_1\Omega^2 = 0$$

$$5400b_6 - 2a_2\Omega^2 = 0$$

$$6600b_7 - 2a_3\Omega^2 = 0$$

A non-trivial solution is obtainable if the rank of the set (6.126) is less than 9. This requires the $C_{10}^9 = C_{10}^1 = 10$ minors of order 9 of the matrix of the system (6.126) to be zero. We then get an algebraic set of 10 equations whose solution is

$$a_0 = \frac{42}{65}a_2 - \frac{19}{110}a_3 \quad a_1 = \frac{3024}{715}a_2 - \frac{288}{605}a_3 \qquad (6.127)$$

where a_2 and a_3 are considered to be arbitrary constants. We deduce that for an inertial coefficient

$$R(\xi) = \frac{42}{65}a_2 - \frac{19}{110}a_3 + \left(\frac{3024}{715}a_2 - \frac{288}{605}a_3\right)\xi + a_2\xi^2 + a_3\xi^3 \qquad (6.128)$$

the set (6.126) yields

$$\Omega^2 = 3300b_7/a_3 \qquad (6.129)$$

$$b_0 = (497167a_2 - 58968a_3)b_7/9009a_3 \quad b_1 = 0 \quad b_2 = -(924a_2 - 247a_3)b_7/26a_3$$
$$b_3 = -374a_2b_7/63a_3 \quad b_4 = (66a_2 - 247a_3)b_7/52a_3$$
$$b_5 = 4(231a_2 - 26a_3)b_7/143a_3 \quad b_6 = 11a_2b_7/9a_3$$

$$(6.130)$$

6.1.5.5 Quartic inertial coefficient (m = 4)

We first consider the case $n = 0$. We get a set of nine equations for ten unknowns

$$-36b_1 + 48b_0 = 0$$

$$-108b_2 + 144b_1 - a_0\Omega^2 = 0$$

$$-216b_3 + 288b_2 - a_1\Omega^2 = 0$$

$$-360b_4 + 480b_3 + (3a_0 - a_2)\Omega^2 = 0$$

$$-540b_5 + 720b_4 - (2a_0 - 3a_1 + a_3)\Omega^2 = 0 \qquad (6.131)$$

$$-756b_6 + 1008b_5 - (2a_1 - 3a_2 + a_4)\Omega^2 = 0$$

$$-1008b_7 + 1344b_6 - (2a_2 - 3a_3)\Omega^2 = 0$$

$$-1296b_8 + 1728b_7 - (2a_3 - 3a_4)\Omega^2 = 0$$

$$2160b_8 - 2a_4\Omega^2 = 0$$

We can, therefore, solve the system taking, say, b_8 as an arbitrary parameter. The last equation of the set (6.132) yields the natural frequency squared:

$$\Omega^2 = 1080b_8/a_4 \qquad (6.132)$$

The remaining equations give the coefficients in the flexural rigidity as follows:

$$b_0 = 9(379{,}904a_0 + 121{,}728a_1 + 48{,}168a_2 + 21{,}654a_3 + 10{,}611a_4)b_8/917{,}504a_4$$

$$b_1 = 3(379{,}904a_0 + 121{,}728a_1 + 48{,}168a_2 + 21{,}654a_3 + 10{,}611a_4)b_8/229{,}376a_4$$

$$b_2 = -3(64{,}512a_0 - 40{,}576a_1 - 16{,}056a_2 - 7{,}218a_3 - 3{,}537a_4)b_8/57{,}344a_4$$

$$b_3 = -9(7{,}168a_0 + 3{,}456a_1 - 1784a_2 - 802a_3 - 393a_4)b_8/14{,}336a_4$$

$$b_4 = 3(3{,}584a_0 - 3{,}456a_1 - 1{,}800a_2 + 802a_3 + 393a_4)b_8/3{,}584a_4$$

$$b_5 = 3(640a_1 - 600a_2 - 330a_3 + 131a_4)b_8/896a_4$$

$$b_6 = 3(120a_2 - 110a_3 - 63a_4)b_8/224a_4 \quad b_7 = (10a_3 - 9a_4)b_8/8a_4$$

$$(6.133)$$

Note that Eq. (6.132) coincides with the result by Elishakoff and Candan (2001), as a special case of their treatment.

For $n = 1$, the governing differential equation yields a set of ten linear algebraic equations with ten unknowns, specifically for nine coefficients b_i in the flexural rigidity and the natural frequency squared Ω^2, as follows:

$$b_1 = 0$$

$$-144b_2 + 240b_0 - 2a_0\Omega^2 = 0$$

$$-288b_3 + 480b_1 - 2a_1\Omega^2 = 0$$

$$-480b_4 + 800b_2 + (4a_0 - 2a_2)\Omega^2 = 0$$

$$-720b_5 + 1200b_3 + (4a_1 - 2a_3)\Omega^2 = 0$$

$$-1008b_6 + 1680b_4 - (2a_0 - 4a_2 + 2a_4)\Omega^2 = 0$$

$$-1344b_7 + 2240b_4 - (2a_1 - 4a_3)\Omega^2 = 0$$

$$-1728b_8 + 2880b_4 - (2a_2 - 4a_4)\Omega^2 = 0$$

$$3600b_7 - 2a_3\Omega^2 = 0$$

$$4400b_8 - 2a_4\Omega^2 = 0$$

$$(6.134)$$

To find a non-trivial solution, the determinant must be zero. This leads to the following equation

$$5225a_1 + 1002a_3 = 0 \qquad (6.135)$$

which results in

$$a_3 = -5225a_1/1002 \qquad (6.136)$$

Hence, for the inertial coefficient

$$R(\xi) = a_0 + a_1\xi + a_2\xi^2 - \frac{5225}{1002}a_1\xi^3 + a_4\xi^4 \qquad (6.137)$$

where a_0 and a_2 are arbitrary, the set (6.134) has the following non-trivial solution

$$\Omega^2 = 2{,}200b_8/r_4 \qquad (6.138)$$

$$b_0 = \frac{(665{,}500a_0 + 91{,}575a_2 + 21{,}654a_4)b_8}{52{,}500a_4} \qquad b_1 = 0$$

$$b_2 = -\frac{(33{,}000a_0 - 10{,}175a_2 - 2{,}406a_4)b_8}{3500a_4} \qquad b_3 = -\frac{275a_1b_8}{18a_4}$$

$$b_4 = \frac{(5{,}500a_0 - 9{,}075a_2 + 2{,}406a_4)b_8}{2{,}100a_4} \qquad b_5 = \frac{27{,}995a_1b_8}{1{,}503a_4}$$

$$b_6 = \frac{(275a_2 - 442a_4)b_8}{180a_4} \qquad b_7 = -\frac{57{,}475a_1b_8}{90{,}18a_4}$$

$$(6.139)$$

Concerning the case $n = 2$, the governing equation yields the following set

$$b_1 = 0 \quad -180b_2 - 3a_0\Omega^2 = 0$$

$$-360b_3 + 720b_0 - 3a_1\Omega^2 = 0$$

$$-600b_4 + 1200b_1 + (5a_0 - 3a_2)\Omega^2 = 0$$

$$-900b_5 + 1800b_2 + (5a_1 - 3a_3)\Omega^2 = 0$$

$$-1260b_6 + 2520b_3 + (5a_2 - 3a_4)\Omega^2 = 0$$

$$-1680b_7 + 3360b_4 - (2a_0 - 5a_3)\Omega^2 = 0$$

$$-2160b_8 + 4320b_5 - (2a_1 - 5a_4)\Omega^2 = 0$$

$$5400b_6 - 2a_2\Omega^2 = 0 \quad 6600b_7 - 2a_3\Omega^2 = 0$$

$$7920b_8 - 2r_4\Omega^2 = 0$$

$$(6.140)$$

We have a set of 11 equations for 10 unknowns, namely the coefficients b_i in the flexural rigidity and the natural frequency squared Ω^2. In order to have a non-trivial solution, the rank of the set (6.140) must be less than 10, or all $C_{11}^{10} = 11$ minors of order 10 of the matrix of the set (6.140) should vanish. This last condition leads to a set of 11 equations whose solution reads

$$r_0 = \frac{42}{65}a_2 - \frac{19}{110}a_3 \quad r_1 = \frac{3024}{715}a_2 - \frac{288}{605}a_3 - \frac{49}{242}a_4 \qquad (6.141)$$

where a_2, a_3 and a_4 are treated as arbitrary constants. Hence, we arrive at the following conclusion. For an inertial coefficient equal to

$$R(\xi) = \frac{42}{65}a_2 - \frac{19}{110}a_3 + \left(\frac{3{,}024}{715}a_2 - \frac{288}{605}a_3 - \frac{49}{242}a_4\right)\xi + a_2\xi^2 + a_3\xi^3 \qquad (6.142)$$

the set (6.140) yields

$$\Omega^2 = 3{,}960b_8/a_4 \qquad (6.143)$$

$$b_0 = (3{,}77{,}336a_2 - 471{,}744a_3 - 59{,}085a_4)b_8/60{,}060a_4 \quad b_1 = 0$$
$$b_2 = -3(924a_2 - 247a_3)b_8/65a_4 \quad b_3 = -11(68a_2 - 45a_4)b_8/105a_4$$
$$b_4 = 3(66a_2 - 247r_3)b_8/130a_4 \quad b_5 = (5{,}544a_2 - 624a_3 - 3{,}185a_4)b_8/715a_4$$
$$b_6 = 22a_2b_8/15a_4 \quad b_7 = 6a_3b_8/5a_4$$

$$(6.144)$$

For $n = 3$, we have to satisfy the following set of equations

$$b_1 = 0 \quad -216b_2 - 4a_0\Omega^2 = 0$$

$$-432b_3 - 4a_1\Omega^2 = 0$$

$$-720b_4 + 1{,}680b_0 + (6a_0 - 4a_2)\Omega^2 = 0$$

$$-1{,}080b_5 + 2{,}520b_1 + (6a_1 - 4a_3)\Omega^2 = 0$$

$$-1{,}512b_6 + 3{,}528b_2 + (6a_2 - 4a_4)\Omega^2 = 0$$

$$-2{,}016b_7 + 4{,}704b_3 + 6a_3\Omega^2 = 0$$

$$-2{,}592b_8 + 6{,}048b_4 - (2a_0 - 6a_4)\Omega^2 = 0$$

$$7{,}560b_5 - 2a_1\Omega^2 = 0 \quad 9{,}240b_6 - 2a_2\Omega^2 = 0$$

$$11{,}088b_7 - 2a_3\Omega^2 = 0 \quad 13{,}104b_8 - 2a_4\Omega^2 = 0$$

$$(6.145)$$

We get a set of 12 equations for 10 unknowns. A non-trivial solution is obtainable, if the rank of the set (6.145) is less than 10. This forces all 66 minors of order 10 of the matrix of the linear system (6.145) to be zero. This condition is satisfied in this case, leading to the following inertial coefficient:

$$R(\xi) = a_0 + \left(\frac{2696}{234} a_0 + \frac{55}{78} a_4 \right) \xi^2 + a_4 \xi^4 \qquad (6.146)$$

where r_0 and r_4 are treated as arbitrary constants. The set (6.145) then yields

$$\Omega^2 = 6{,}552b_8/a_4 \qquad (6.147)$$

$$b_0 = (231{,}077a_0 + 12{,}345a_4)b_8/1{,}470a_4 \quad b_1 = 0 \quad b_2 = -364a_0b_8/3a_4 \quad b_3 = 0$$
$$b_4 = (91a_0 - 255a_4)b_8/42a_4 \quad b_5 = 0 \quad b_6 = (49a_0 + 3a_4)b_8/3a_4 \quad b_7 = 0$$

$$(6.148)$$

6.1.5.6 Quintic inertial coefficient ($m = 5$)

When $n = 0$, we obtain a set of 10 equations for 11 unknowns

$$-36b_1 + 48b_0 = 0$$

$$-108b_2 + 144b_1 - a_0\Omega^2 = 0$$

$$-216b_3 + 288b_2 - a_1\Omega^2 = 0$$

$$-360b_4 + 480b_3 + (3a_0 - a_2)\Omega^2 = 0$$

$$-540b_5 + 720b_4 - (2a_0 - 3a_1 + a_3)\Omega^2 = 0$$

$$-756b_6 + 1008b_5 - (2a_1 - 3a_2 + a_4)\Omega^2 = 0$$

$$-1008b_7 + 1344b_6 - (2a_2 - 3a_3 + a_5)\Omega^2 = 0$$

$$-1296b_8 + 1728b_7 - (2a_3 - 3a_4)\Omega^2 = 0$$

$$-1620b_9 + 2160b_8 - (2a_4 - 3a_5)\Omega^2 = 0$$

$$2640b_9 - 2a_5\Omega^2 = 0$$

$$(6.149)$$

We can, therefore, solve the system taking b_9 as a parameter. From the last equation of the set (6.149), we obtain the natural frequency squared:

$$\Omega^2 = 1320 b_9 / a_5 \tag{6.150}$$

The remaining equations yield the coefficients in the flexural rigidity as follows:

$$
\begin{aligned}
b_0 &= (8{,}357{,}888 a_0 + 2{,}678{,}016 a_1 + 1{,}059{,}696 a_2 + 476{,}388 a_3 + 233{,}442 a_4 \\
&\quad + 121{,}743 a_5) b_9 / 1{,}835{,}008 a_5 \\
b_1 &= (8{,}357{,}888 a_0 + 2{,}678{,}016 a_1 + 1{,}059{,}696 a_2 + 476{,}388 a_3 + 233{,}442 a_4 \\
&\quad + 121{,}743 a_5) b_9 / 1{,}376{,}256 a_5 \\
b_2 &= (1{,}419{,}264 a_0 - 892{,}672 a_1 - 353{,}232 a_2 - 158{,}796 a_3 - 77{,}814 a_4 \\
&\quad - 40{,}581 a_5) - b_9 / 344{,}064 a_5 \\
b_3 &= (157{,}696 a_0 + 76{,}032 a_1 - 39{,}248 a_2 - 17{,}644 a_3 - 8{,}646 a_4 \\
&\quad - 4{,}509 a_5) b_9 / 28672 a_5 \\
b_4 &= (78{,}848 a_0 - 76{,}032 a_1 - 39{,}600 a_2 + 17{,}644 a_3 + 8{,}646 a_4 \\
&\quad + 4{,}509 a_5) b_9 / 1{,}835{,}008 a_5 \\
b_5 &= (14{,}080 a1 - 13{,}200 a_2 - 7{,}260 a_3 + 2{,}882 a_4 + 1{,}503 a_5) b_9 / 5{,}376 a_5 \\
b_6 &= (2{,}640 a_2 - 2{,}420 a_3 - 1{,}386 a_4 + 501 a_5) b_9 / 1344 a_5 \\
b_7 &= (220 a_3 - 198 a_4 - 117 a_5) b_9 / 144 a_5 \quad b_8 = (44 a_4 - 39 a_5) b_9 / 36 a_5
\end{aligned}
\tag{6.151}
$$

Now consider the case $n = 1$. The governing differential equation yields a set of 11 linear algebraic equations for 11 unknowns:

$$
\begin{aligned}
b_1 &= 0 \\
-144 b_2 + 240 b_0 - 2 a_0 \Omega^2 &= 0 \\
-288 b_3 + 480 b_1 - 2 a_1 \Omega^2 &= 0 \\
-480 b_4 + 800 b_2 + (4 a_0 - 2 a_2) \Omega^2 &= 0 \\
-720 b_5 + 1200 b_3 + (4 a_1 - 2 a_3) \Omega^2 &= 0 \\
-1008 b_6 + 1680 b_4 - (2 a_0 - 4 a_2 + 2 a_4) \Omega^2 &= 0 \\
-1344 b_7 + 2240 b_5 - (2 a_1 - 4 a_3 + 2 a_5) \Omega^2 &= 0 \\
-1728 b_8 + 2880 b_6 - (2 a_2 - 4 a_4) \Omega^2 &= 0 \\
-2160 b_9 + 3600 b_7 - (2 a_3 - 4 a_5) \Omega^2 &= 0 \\
4400 b_8 - 2 a_4 \Omega^2 &= 0 \\
5280 b_9 - 2 a_5 \Omega^2 &= 0
\end{aligned}
\tag{6.152}
$$

We impose the condition that the determinant of the system (6.152) should vanish, so that we get a non-trivial solution. This leads to the following equation:

$$57{,}475a_1 + \frac{5{,}010}{11}a_3 + 3{,}015a_5 = 0 \tag{6.153}$$

or,

$$a_1 = -\frac{1{,}002}{5{,}225}a_3 - \frac{603}{11{,}495}a_5 \tag{6.154}$$

where a_3 and a_5 are arbitrary. Upon substitution of the expression (6.154) into the set (6.152) we get the solution for the natural frequency squared:

$$\Omega^2 = 2{,}640b_9/a_5 \tag{6.155}$$

and the coefficients in the flexural rigidity,

$$b_0 = (665{,}500a_0 + 91{,}575a_2 + 21{,}654a_4)b_9/43{,}750a_5 \quad b_1 = 0$$
$$b_2 = -3(33{,}000a_0 - 10{,}175a_2 + 2{,}406a_4)b_9/8{,}750a_5$$
$$b_3 = -(3{,}674a_3 + 1{,}005a_5)b_9/1{,}045a_5$$
$$b_4 = (5{,}500a_0 - 9{,}075a_2 + 2{,}406a_4)b_9/1{,}750a_5$$
$$b_5 = -(22{,}396a_3 - 4{,}355a_5)b_9/5{,}225a_5 \quad b_6 = (275a_2 - 442a_4)b_9/150a_5$$
$$b_7 = (22a_3 - 35a_5)b_9/15a_5 \quad b_8 = 6a_4b_9/5a_5 \tag{6.156}$$

For the case $n = 2$, the governing differential equation yields a set of 12 linear algebraic equations for 11 unknowns:

$$b_1 = 0 \quad -180b_2 - 3a_0\Omega^2 = 0$$
$$-360b_3 + 720b_0 - 3a_1\Omega^2 = 0$$
$$-600b_4 + 1200b_1 + (5a_0 - 3a_2)\Omega^2 = 0$$
$$-900b_5 + 1800b_2 + (5a_1 - 3a_3)\Omega^2 = 0$$
$$-1260b_6 + 2520b_3 + (5a_2 - 3a_4)\Omega^2 = 0$$
$$-1680b_7 + 3360b_4 - (2a_0 - 5a_3 + 3a_5)\Omega^2 = 0 \tag{6.157}$$
$$-2160b_8 + 4320b_5 - (2a_1 - 5a_4)\Omega^2 = 0$$
$$-2700b_9 + 5400b_6 - (a_2 - 5a_5)\Omega^2 = 0$$
$$6600b_7 - 2a_3\Omega^2 = 0 \quad 7920b_8 - 2a_4\Omega^2 = 0$$
$$9360b_9 - 2a_5\Omega^2 = 0$$

In order to have a non-trivial solution, the rank of the set (6.157) must be less than 10. The rank of the matrix of the system is less than 10 if all minors of order 10 vanish, leading to a set of 11 equations for 5 unknowns. The solution of this set reads

$$a_1 = \tfrac{72}{11}a_0 + \tfrac{36}{55}a_3 - \tfrac{49}{242}a_4 \quad a_2 = \tfrac{65}{42}a_0 + \tfrac{247}{924}a_3 - \tfrac{5}{28}a_5 \tag{6.158}$$

where a_0, a_3, a_4 and a_2 are arbitrary. Upon substitution of the expression (6.158) into the set (6.157) we get the solution for the natural frequency squared:

$$\Omega^2 = 4{,}680/a_5 \tag{6.159}$$

and the coefficients in the flexural rigidity

$$b_0 = (258{,}526{,}840a_0 + 24{,}841{,}388a_3 - 2{,}481{,}570a_4 - 441{,}045a_5)b_9/2{,}134{,}440a_5$$

$$b_1 = 0 \quad b_2 = -78a_0b_9/a_5$$

$$b_3 = -(1{,}264{,}120a_0 + 218{,}348a_3 - 540{,}540a_4 + 40{,}095a_5)b_9/97{,}020a_5$$

$$b_4 = 39(110a_0 - 247a_3 + 165r_5)b_9/1{,}540a_5$$

$$b_5 = 13(660a_0 + 66a_3 - 545a_4)b_9/605a_5$$

$$b_6 = (18{,}590a_0 + 3{,}211a_3 - 28{,}710a_5)b_9/6{,}930a_5$$

$$b_7 = 78a_3b_9/55a_5 \quad b_8 = 13a_4b_9/11a_5$$

$$\tag{6.160}$$

For the case $n = 3$, the governing differential equation produces a set of 13 linear algebraic equations for 11 unknowns:

$$b_1 = 0 \quad -216b_2 - 4a_0\Omega^2 = 0$$

$$-432b_3 - 4a_1\Omega^2 = 0$$

$$-720b_4 + 1{,}680b_0 + (6r_0 - 4a_2)\Omega^2 = 0$$

$$-1{,}080b_5 + 2{,}520b_1 + (6a_1 - 4a_3)\Omega^2 = 0$$

$$-1{,}512b_6 + 3{,}528b_2 + (6a_2 - 4a_4)\Omega^2 = 0$$

$$-2{,}016b_7 + 4{,}704b_3 + (6a_3 - 4a_5)\Omega^2 = 0 \tag{6.161}$$

$$-2{,}592b_8 + 6{,}048b_4 - (2a_0 - 6a_4)\Omega^2 = 0$$

$$-3{,}240b_9 + 7{,}560b_5 - (2a_1 - 6a_5)\Omega^2 = 0$$

$$9{,}240b_6 - 2a_2\Omega^2 = 0 \quad 11{,}088b_7 - 2a_3\Omega^2 = 0$$

$$13{,}104b_8 - 2a_4\Omega^2 = 0 \quad 15{,}288b_9 - 2a_5\Omega^2 = 0$$

In order to have a non-trivial solution, the rank of the set (6.161) must be less than 11. The rank of the matrix of the system is less than 11 if all minors of order 11 vanish, leading to a set of 78 equations for 5 unknowns. The solution of this set reads

$$a_1 = -\frac{317,565}{3,918,187}a_5 \quad a_2 = \frac{2,695}{234}a_0 + \frac{55}{78}a_4 \quad a_3 = \frac{6,666}{79,963}a_5 \tag{6.162}$$

where a_0, a_4 and a_3 are arbitrary constants. Substituting of the expressions (6.162) into the set (6.161) we get the solution for the natural frequency squared:

$$\Omega^2 = 7,644b_9/a_5 \tag{6.163}$$

as well as the coefficients in the flexural rigidity,

$$b_0 = 7(231,077a_0 + 12,345a_4)b_9/1,260a_5 \quad b_1 = 0 \quad b_2 = -1274a_0b_9/9a_5$$
$$b_3 = 35,285b_9/6,151a_5 \quad b_4 = (91a_0 - 255r_4)b_9/36a_5$$
$$b_5 = -535,321b_9/92,265a_5 \quad b_6 = 7(49a_3 + 3a_4)b_8/18a_5$$
$$b_7 = 707b_9/615a_5 \quad b_8 = 7a_4b_9/6a_5$$

$$\tag{6.164}$$

6.1.6 Closed-Form Solutions for the Clamped–Clamped Beam

6.1.6.1 Constant inertial coefficient (m = 0)

The inertial coefficient reduces to $R(\xi) = a_0$. The governing differential equation leads to the following set of equations:

$$4b_2 - 24b_1 + 24b_0 = 0$$
$$12b_3 - 72b_2 + 72b_1 = 0$$
$$24b_4 - 144b_3 + 144b_2 - a_0\Omega^2 = 0 \tag{6.165}$$
$$-240b_4 + 240b_3 + 2a_0\Omega^2 = 0$$
$$360b_4 - a_0\Omega^2 = 0$$

We have five equations for six unknowns: b_0, b_1, b_2, b_3, b_4 and Ω^2. We declare b_4 to be a known parameter. We get Ω^2 from the last equation of the set (6.165):

$$\Omega^2 = 360b_4/a_0 \tag{6.166}$$

The rest of the equations lead to

$$b_0 = \tfrac{11}{18}b_4 \quad b_1 = \tfrac{2}{3}b_4 \quad b_2 = \tfrac{1}{3}b_4 \quad b_3 = -2b_4 \tag{6.167}$$

The flexural rigidity is then given by

$$D(\xi) = \left(\tfrac{11}{18} + \tfrac{2}{3}\xi + \tfrac{1}{3}\xi^2 - 2\xi^3 + \xi^4 \right) b_4 \qquad (6.168)$$

which is positive within the interval $[0; 1]$ (see Figure 6.13), unless parameters b_4 and a_0 are non-negative.

6.1.6.2   Linearly varying inertial coefficient ($m = 1$)

Consider first the case $n = 0$; instead of (6.166) we have

$$4b_2 - 24b_1 + 24b_0 = 0$$
$$12b_3 - 72b_2 + 72b_1 = 0$$
$$24b_4 - 144b_3 + 144b_2 - a_0\Omega^2 = 0$$
$$40b_5 - 240b_4 + 240b_3 + (2a_0 - a_1)\Omega^2 = 0 \qquad (6.169)$$
$$-360b_5 + 360b_4 - (a_0 - 2a_1)\Omega^2 = 0$$
$$504b_5 - a_1\Omega^2 = 0$$

This set represents a linear algebraic system of six equations for the seven unknowns, $b_0, b_1, \ldots, b_5$ and Ω^2. Hence we express them in terms of b_5, treated as an arbitrary constant. We get Ω^2 from the last equation of the set (6.169):

$$\Omega^2 = 504b_5/a_1 \qquad (6.170)$$

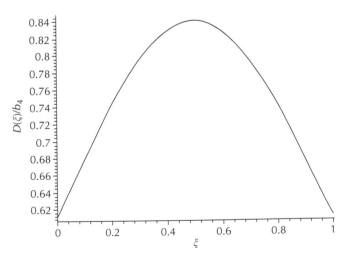

FIGURE 6.13
Variation of the flexural rigidity for the clamped–clamped beam

while the remaining equations lead to

$$b_0 = (61a_1 + 154a_0)b_5/180a_1 \quad b_1 = (37a_1 + 84a_0)b_5/90a_1$$

$$b_2 = (13a_1 + 14a_0)b_5/30a_1 \quad b_3 = 2(a_1 - 21a_0)b_5/15a_1 \qquad (6.171)$$

$$b_4 = -(9a_1 - 7a_0)b_5/5a_1$$

Equation (6.170) coincides with the Elishakoff and Candan (2000) solution. Now consider the case $n = 1$. Instead of (6.166) we obtain

$$8b_2 - 36b_1 = 0$$

$$24b_3 - 108b_2 + 120b_0 = 0$$

$$48b_4 - 216b_3 + 240b_1 - 2a_0\Omega^2 = 0$$

$$80b_5 - 360b_4 + 400b_2 + (3a_0 - 2a_1)\Omega^2 = 0 \qquad (6.172)$$

$$-540b_5 + 600b_3 + 3a_1\Omega^2 = 0$$

$$840b_4 - a_0\Omega^2 = 0$$

$$1120b_5 - a_1\Omega^2 = 0$$

We get a linear system of 6 equations for six unknowns. In order to have a non-trivial solution the determinant of the system (6.172) must vanish, leading to the following determinantal equation $3200a_0 - 1629a_1 = 0$, whose solution is

$$a_0 = \frac{1629}{3200}a_1 \qquad (6.173)$$

We can solve the set expressing the unknowns with b_5 taken as an arbitrary constant. Thus, we get

$$\Omega^2 = 1{,}120b_5/a_1 \qquad (6.174)$$

and the coefficients b_i are

$$b_0 = \frac{100{,}051}{40{,}000}b_5 \quad b_1 = \frac{771}{2{,}000}b_5 \quad b_2 = \frac{6{,}939}{4{,}000}b_5 \quad b_3 = -\frac{47}{10}b_5 \quad b_4 = \frac{543}{800}b_5$$

$$(6.175)$$

We conclude that the solution for the flexural rigidity is

$$D(\xi) = \left(\frac{100{,}051}{40{,}000} + \frac{771}{2{,}000}\xi + \frac{6{,}939}{4{,}000}\xi^2 - \frac{47}{10}\xi^3 + \frac{543}{800}\xi^4 + \xi^5 \right) b_5 \qquad (6.176)$$

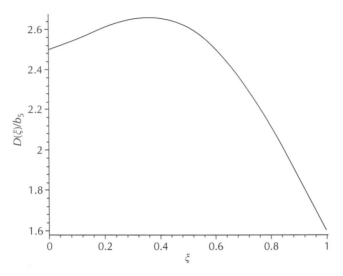

FIGURE 6.14
Variation of the flexural rigidity for the clamped–clamped beam

provided the inertial term reads

$$R(\xi) = \left(\frac{1629}{3200} + \xi\right) a_1 \qquad (6.177)$$

Figure 6.14 depicts the variation of the flexural rigidity, which is positive within the interval $[0; 1]$, for the non-negative parameters b_5 and a_1.

6.1.6.3 *Parabolically varying inertial coefficient (m = 2)*

We first consider the case $n = 0$. We get the following set of equations

$$4b_2 - 24b_1 + 24b_0 = 0$$
$$12b_3 - 72b_2 + 72b_1 = 0$$
$$24b_4 - 144b_3 + 144b_2 - a_0\Omega^2 = 0$$
$$40b_5 - 240b_4 + 240b_3 + (2a_0 - a_1)\Omega^2 = 0 \qquad (6.178)$$
$$60b_6 - 360b_5 + 360b_4 - (a_0 - 2a_1 + a_2)\Omega^2 = 0$$
$$-504b_6 + 504b_5 - (a_1 - 2a_2)\Omega^2 = 0$$
$$672b_6 - a_2\Omega^2 = 0$$

which has seven equations for eight unknowns. We can, therefore, solve the system taking b_6 as a parameter. From the last equation of the set (6.178), we obtain the natural frequency squared:

$$\Omega^2 = 672b_6/a_2 \qquad (6.179)$$

and the remaining equations yield the coefficients in the flexural rigidity as follows,

$$b_0 = (1232a_0 + 488a_1 + 219a_2)b_6/1080a_2$$

$$b_1 = (672a_0 + 296a_1 + 137a_2)b_6/540a_2$$

$$b_2 = (112a_0 + 104a_1 + 55a_2)b_6/180a_2 \quad b_3 = -2(84a_0 - 4a_1 - 7a_2)b_6/45a_2$$

$$b_4 = (56a_0 - 72a_1 + a_2)b_6/30a_2 \quad b_5 = (4a_1 - 5a_2)b_6/3a_2$$

$$(6.180)$$

When $n = 1$, the governing differential equation yields a set of eight linear algebraic equations for eight unknowns, namely the coefficients b_i in the flexural rigidity and the natural frequency squared Ω^2:

$$8b_2 - 36b_1 = 0$$

$$24b_3 - 108b_2 + 120b_0 = 0$$

$$48b_4 - 216b_3 + 240b_1 - 2a_0\Omega^2 = 0$$

$$80b_5 - 360b_4 + 400b_2 + (3a_0 - 2a_1)\Omega^2 = 0$$

$$120b_6 - 540b_5 + 600b_3 + (3a_1 - 2a_2)\Omega^2 = 0 \qquad (6.181)$$

$$-756b_6 + 840b_4 - (a_0 - 3a_2)\Omega^2 = 0$$

$$1120b_5 - a_1\Omega^2 = 0$$

$$1440b_6 - a_2\Omega^2 = 0$$

In order to find a non-trivial solution, the determinant must be zero, which leads to the following determinantal equation:

$$3200a_0 - 1629a_1 + 1362a_2 = 0 \qquad (6.182)$$

whose solution is

$$a_1 = \frac{3200}{1629}a_0 + \frac{454}{543}a_2 \qquad (6.183)$$

where a_0 and a_2 are arbitrary. Upon substitution of the expression (6.183) into the set (6.181) we get the solution for the natural frequency squared:

$$\Omega^2 = 1{,}440b_6/a_2 \qquad (6.184)$$

and the coefficients in the flexural rigidity,

$$b_0 = (1{,}143{,}440a_0 + 339{,}949a_2)b_6/181{,}000a_2$$

$$b_1 = 3(20{,}560a_0 + 9{,}321a_2)b_6/63{,}350a_2$$

$$b_2 = 27(20{,}560a_0 + 9{,}321a_2)b_6/126{,}700a_2$$

$$b_3 = -2(37{,}600a_0 + 1{,}433a_2)b_6/6{,}335a_2 \quad b_4 = 3(40a_0 - 99a_2)b_6/70a_2$$

$$b_5 = 2(1{,}600a_0 + 681a_2)b_6/1{,}267a_2$$

$$(6.185)$$

When $n = 2$, the governing differential equation yields a set of nine linear algebraic equations for eight unknowns, namely the coefficients b_i in the flexural rigidity and the natural frequency squared, Ω^2:

$$12b_2 - 48b_1 = 0$$

$$36b_3 - 144b_2 = 0$$

$$72b_4 - 288b_3 + 360b_0 - 3a_0\Omega^2 = 0$$

$$120b_5 - 480b_4 + 600b_1 + (4a_0 - 3a_1)\Omega^2 = 0$$

$$180b_6 - 720b_5 + 900b_2 + (4a_1 - 3a_2)\Omega^2 = 0 \qquad (6.186)$$

$$-1008b_6 + 1260b_3 + 4a_2\Omega^2 = 0$$

$$1680b_4 - a_0\Omega^2 = 0$$

$$2160b_5 - a_1\Omega^2 = 0$$

$$2700b_6 - a_2\Omega^2 = 0$$

In order to have a non-trivial solution, the rank of the set (6.186) must be less than 8. The rank of the matrix of the system is less than 8 if all minors of order 8 vanish, leading to a set of nine equations for three unknowns, namely the parameters a_0, a_1 and a_2. The solution of this set reads

$$a_0 = \frac{1{,}723}{2{,}145}a_2 \quad a_1 = \frac{376}{385}a_2 \qquad (6.187)$$

where a_2 is an arbitrary constant. Upon substitution of the expressions (6.187) into the set (6.186) we get the solution for the natural frequency squared:

$$\Omega^2 = 2{,}700b_6/a_2 \qquad (6.188)$$

and the coefficients in the flexural rigidity,

$$b_0 = \frac{1{,}160{,}969}{100{,}100}b_6 \quad b_1 = -\frac{17}{35}b_6 \quad b_2 = -\frac{68}{35}b_6 \quad b_3 = -\frac{272}{35}b_6$$

$$b_4 = \frac{5{,}169}{4{,}004}b_6 \quad b_5 = \frac{94}{77}b_6$$

$$(6.189)$$

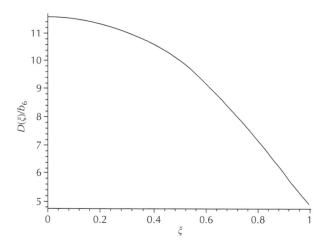

FIGURE 6.15
Variation of the flexural rigidity for the clamped–clamped beam

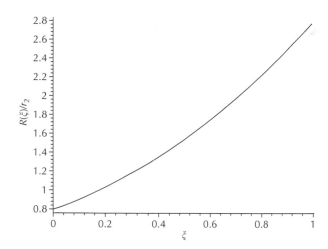

FIGURE 6.16
Variation of the inertial coefficient for the clamped–clamped beam

Figure 6.15 portrays the variation of the flexural rigidity in terms of ξ within the interval $[0; 1]$, and Figure 6.16 depicts the variation of the inertial coefficient.

6.1.6.4 Cubic inertial coefficient ($m = 3$)

When $n = 0$, instead of (6.165) we obtain

$$4b_2 - 24b_1 + 24b_0 = 0$$
$$12b_3 - 72b_2 + 72b_1 = 0$$

$$24b_4 - 144b_3 + 144b_2 - a_0\Omega^2 = 0$$

$$40b_5 - 240b_4 + 240b_3 + (2a_0 - a_1)\Omega^2 = 0$$

$$60b_6 - 360b_5 + 360b_4 - (a_0 - 2a_1 + a_2)\Omega^2 = 0$$

$$84b_7 - 504b_6 + 504b_5 - (a_1 - 2a_2 + a_3)\Omega^2 = 0$$

$$-672b_7 + 672b_6 - (a_2 - 2a_3)\Omega^2 = 0$$

$$864b_7 - a_3\Omega^2 = 0$$

$$(6.190)$$

which has eight equations for nine unknowns. We can, therefore, solve the system taking b_7 as a parameter. From the last equation of the set (6.190), we obtain the natural frequency squared:

$$\Omega^2 = 864b_7/a_3 \qquad (6.191)$$

and the remaining equations yield the coefficients in the flexural rigidity as follows,

$b_0 = (11{,}088a_0 + 4{,}392a_1 + 1{,}971a_2 + 970a_3)b_7/7{,}560a_3$

$b_1 = (12{,}096a_0 + 5{,}328a_1 + 2{,}466a_2 + 1{,}225a_3)b_7/7{,}560a_3$

$b_2 = (336a_0 + 312a_1 + 165a_2 + 85a_3)b_7/420a_3$

$b_3 = -(6{,}048a_0 - 288a_1 - 504a_2 - 305a_3)b_7/1{,}260a_3$

$b_4 = (504a_0 - 648a_1 + 9a_2 + 50a_3)b_7/210a_3 \quad b_5 = (72a_1 - 90a_2 - a_3)b_7/42a_3$

$b_6 = (9a_2 - 11a_3)b_7/7a_3$

$$(6.192)$$

Now consider $n = 1$. The governing differential equation yields a set of nine linear algebraic equations for nine unknowns, namely the coefficients b_i in the flexural rigidity and the natural frequency squared Ω^2:

$$8b_2 - 36b_1 = 0$$

$$24b_3 - 108b_2 + 120b_0 = 0$$

$$48b_4 - 216b_3 + 240b_1 - 2a_0\Omega^2 = 0$$

$$80b_5 - 360b_4 + 400b_2 + (3a_0 - 2a_1)\Omega^2 = 0$$

$$120b_6 - 540b_5 + 600b_3 + (3a_1 - 2a_2)\Omega^2 = 0 \qquad (6.193)$$

$$168b_7 - 756b_6 + 840b_4 - (a_0 - 3a_2 + 2a_3)\Omega^2 = 0$$

$$-1008b_7 + 1120b_5 - (a_1 - 3a_3)\Omega^2 = 0$$

$$1440b_6 - a_2\Omega^2 = 0$$

$$1800b_7 - a_3\Omega^2 = 0$$

Again, we impose the condition that determinant of the system should vanish, in order to get a non-trivial solution. This leads to the following equation:

$$40{,}725a_0 - 80{,}000a_1 + 34{,}050a_2 - 23{,}263a_3 = 0 \tag{6.194}$$

whose solution is

$$a_2 = -\frac{1{,}600}{681}a_0 + \frac{543}{454}a_1 + \frac{23{,}263}{34{,}050}a_3 \tag{6.195}$$

where a_0, a_2 and a_3 are arbitrary. Upon substitution of the expression (6.195) into the set (6.193) we get the solution for the natural frequency squared:

$$\Omega^2 = 1{,}800b_7/r_3a \tag{6.196}$$

and the coefficients in the flexural rigidity,

$$b_0 = (64{,}852{,}000a_0 + 76{,}488{,}525a_1 + 27{,}253{,}687a_3)b_7/27{,}240{,}000a_3$$
$$b_1 = -(252{,}000a_0 - 2{,}097{,}225a_1 - 847{,}723r_3)b_7/3{,}178{,}000a_3$$
$$b_2 = -9(252{,}000a_0 - 2{,}097{,}225a_1 - 847{,}723a_3)b_7/6{,}356{,}000a_3$$
$$b_3 = -(6{,}440{,}000a_0 + 322{,}425a_1 - 190{,}261a_3)b_7/476{,}700a_3 \tag{6.197}$$
$$b_4 = (9{,}282{,}000a_0 - 4{,}031{,}775a_1 + 293{,}843a_3)b_7/635{,}600a_3$$
$$b_5 = 9(25a_1 - 61a_3)b_7/140a_3$$
$$b_6 = -(80{,}000a_0 - 40{,}725a_1 - 23{,}263a_3)b_7/27{,}240a_3$$

For the case $n = 2$, the governing differential equation yields a set of ten linear algebraic equations for nine unknowns, namely the coefficients b_i in the flexural rigidity and the natural frequency squared Ω^2:

$$12b_2 - 48b_1 = 0 \quad 36b_3 - 144b_2 = 0$$
$$72b_4 - 288b_3 + 360b_0 - 3a_0\Omega^2 = 0$$
$$120b_5 - 480b_4 + 600b_1 + (4a_0 - 3a_1)\Omega^2 = 0$$
$$180b_6 - 720b_5 + 900b_2 + (4a_1 - 3a_2)\Omega^2 = 0$$
$$252b_7 - 1008b_6 + 1260b_3 + (4a_2 - 3a_3)\Omega^2 = 0 \tag{6.198}$$
$$-1344b_7 + 1680b_4 - (a_0 - 4a_3)\Omega^2 = 0$$
$$2160b_5 - a_1\Omega^2 = 0 \quad 2700b_6 - a_2\Omega^2 = 0$$
$$3300b_7 - a_3\Omega^2 = 0$$

In order to have a non-trivial solution, the rank of the set (6.198) must be less than 9. The rank of the matrix of the system is less than 9 if all minors of order

9 vanish, leading to a set of ten equations for four unknowns, namely the parameters a_0, a_1, a_2 and a_3. The solution of this set reads

$$a_1 = \frac{14{,}664}{12{,}061}a_0 + \frac{1{,}191{,}849}{3{,}316{,}775}a_3 \quad a_2 = \frac{2{,}145}{1{,}723}a_0 + \frac{48{,}684}{94{,}765}a_3 \tag{6.199}$$

where a_0 and a_3 are arbitrary constants. Upon substitution of the expressions (6.199) into the set (6.198) we get the solution for the natural frequency squared:

$$\Omega^2 = 3{,}300b_7/a_3 \tag{6.200}$$

and the coefficients in the flexural rigidity,

$$b_0 = (319{,}266{,}475a_0 + 65{,}733{,}964a_3)b_7/18{,}091{,}500a_3$$
$$b_1 = -121(22{,}100a_0 - 5{,}191a_3)b_7/3{,}618{,}300a_3$$
$$b_2 = -121(22{,}100a_0 - 5{,}191r_3)b_7/904{,}575a_3$$
$$b_3 = -484(22{,}100a_0 - 5{,}191a_3)b_7/904{,}575a_3$$
$$b_4 = (275r_0 - 988a_3)b_7/140a_3 \quad b_5 = (1{,}344{,}200a_0 + 397{,}283a_3)b_7/723{,}660a_3$$
$$b_6 = (39{,}325a_0 + 16{,}228a_3)b_7/25{,}845a_3$$
$$\tag{6.201}$$

For the case $n = 3$, the governing differential equation yields a set of 11 linear algebraic equations for 9 unknowns, namely the coefficients b_i in the flexural rigidity and the natural frequency squared Ω^2:

$$16b_2 - 60b_1 = 0 \quad 48b_3 - 180b_2 = 0$$
$$94b_4 - 360b_3 - 4a_0\Omega^2 = 0$$
$$160b_5 - 600b_4 + 840b_0 + (5a_0 - 4a_1)\Omega^2 = 0$$
$$240b_6 - 900b_5 + 1260b_1 + (5a_1 - 4a_2)\Omega^2 = 0$$
$$336b_7 - 1260b_6 + 1764b_2 + (5a_2 - 4a_3)\Omega^2 = 0 \tag{6.202}$$
$$-1680b_7 + 2352b_3 + 5a_3\Omega^2 = 0$$
$$3024b_4 - a_0\Omega^2 = 0 \quad 3780b_5 - a_1\Omega^2 = 0$$
$$4620b_6 - a_2\Omega^2 = 0 \quad 5544b_7 - a_3\Omega^2 = 0$$

In order to have a non-trivial solution, the rank of the set (6.202) must be less than 9. The rank of the matrix of the system is less than 9 if all minors of order 9 vanish, leading to a set of 55 equations for 4 unknowns, namely the parameters a_0, a_1, a_2 and a_3. The solution of this set reads

$$a_0 = \frac{279}{1540}a_3 \quad a_1 = \frac{1916}{2145}a_3 \quad a_2 = \frac{161}{156}a_3 \tag{6.203}$$

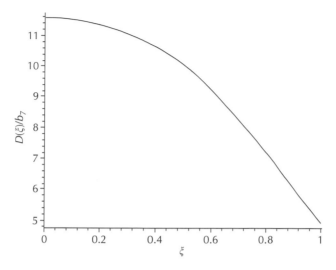

FIGURE 6.17
Variation of the flexural rigidity for the clamped–clamped beam

where r_3 is an arbitrary constant. Upon substitution of the expressions (6.203) into the set (6.202) we get the solution for the natural frequency squared:

$$\Omega^2 = 5{,}544 b_7 / a_3 \tag{6.204}$$

and the coefficients in the flexural rigidity,

$$b_0 = \frac{60{,}508{,}393}{3{,}439{,}800} b_7 \quad b_1 = -\frac{248}{315} b_7 \quad b_2 = -\frac{62}{21} b_7 \quad b_3 = -\frac{155}{14} b_7$$

$$b_4 = \frac{93}{280} b_7 \quad b_5 = \frac{3{,}832}{2{,}925} b_7 \quad b_6 = \frac{161}{130} b_7 \tag{6.205}$$

Figures 6.17 and 6.18 depict, respectively, the flexural rigidity and the inertial coefficient, which are positive for ξ within $[0; 1]$, when the parameters b_7 and r_3 are non-negative.

6.1.6.5   Quartic inertial coefficient ($m = 4$)

When $n = 0$, we obtain

$$4b_2 - 24b_1 + 24b_0 = 0 \quad 12b_3 - 72b_2 + 72b_1 = 0$$

$$24b_4 - 144b_3 + 144b_2 - a_0\Omega^2 = 0$$

$$40b_5 - 240b_4 + 240b_3 + (2a_0 - a_1)\Omega^2 = 0$$

$$60b_6 - 360b_5 + 360b_4 - (a_0 - 2a_1 + a_2)\Omega^2 = 0$$

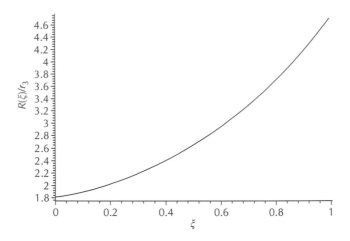

FIGURE 6.18
Variation of the inertial coefficient for the clamped–clamped beam

$$84b_7 - 504b_6 + 504b_5 - (a_1 - 2a_2 + a_3)\Omega^2 = 0$$

$$112b_8 - 672b_7 + 672b_6 - (a_2 - 2a_3 + a_4)\Omega^2 = 0$$

$$-864b_8 + 864b_7 - (a_3 - 2a_4)\Omega^2 = 0$$

$$1080b_8 - a_4\Omega^2 = 0$$

$$(6.206)$$

which has nine equations for ten unknowns. We can, therefore, solve the system taking b_8 as a parameter. From the last equation of the set (6.207), we obtain the natural frequency squared:

$$\Omega^2 = 1{,}080b_8/a_4 \tag{6.207}$$

and the remaining equations yield the coefficients in the flexural rigidity as follows:

$$b_0 = (33{,}264a_0 + 13{,}176a_1 + 5{,}913a_2 + 2{,}910a_3 + 1{,}535a_4)b_8/18{,}144a_4$$

$$b_1 = (12{,}096a_0 + 5{,}328a_1 + 2{,}466a_2 + 1{,}225a_3 + 648a_4)b_8/6{,}048a_4$$

$$b_2 = (3{,}024a_0 + 2{,}808a_1 + 1{,}485a_2 + 765a_3 + 409a_4)b_8/3{,}024a_4$$

$$b_3 = -(6{,}048a_0 - 288a_1 - 504a_2 - 305a_3 - 170a_4)b_8/1{,}008a_4$$

$$b_4 = (1{,}512a_0 - 1{,}944a_1 + 27a_2 + 150a_3 + 101a_4)b_8/504a_4$$

$$b_5 = (360a_1 - 450a_2 - 5a_3 + 32a_4)b_8/168a_4 \quad b_6 = 5(27a_2 - 33a_3 - r_4)b_8/84a_4$$

$$b_7 = (5a_3 - 6a_4)b_8/4a_4$$

$$(6.208)$$

Now consider $n = 1$; the governing differential equation yields a set of ten linear algebraic equations for ten unknowns, namely the coefficients b_i in the flexural rigidity and the natural frequency squared Ω^2:

$$8b_2 - 36b_1 = 0$$
$$24b_3 - 108b_2 + 120b_0 = 0$$
$$48b_4 - 216b_3 + 240b_1 - 2a_0\Omega^2 = 0$$
$$80b_5 - 360b_4 + 400b_2 + (3a_0 - 2a_1)\Omega^2 = 0$$
$$120b_6 - 540b_5 + 600b_3 + (3a_1 - 2a_2)\Omega^2 = 0$$
$$168b_7 - 756b_6 + 840b_4 - (a_0 - 3a_2 + 2a_3)\Omega^2 = 0 \qquad (6.209)$$
$$224b_8 - 1008b_7 + 1120b_5 - (a_1 - 3a_3 + 2a_4)\Omega^2 = 0$$
$$-1296b_8 - 1440b_6 - (a_2 - 3a_4)\Omega^2 = 0$$
$$1800b_7 - a_3\Omega^2 = 0$$
$$2200b_8 - a_4\Omega^2 = 0$$

Again, we impose the condition that the determinant of the system (6.209) should vanish in order to get a non-trivial solution. This leads to the following equation:

$$880,000a_0 - 447,975a_1 + 374,550a_2 - 255,893a_3 + 217,344a_4 = 0 \qquad (6.210)$$

which can be solved to give

$$a_3 = \frac{80,000}{23,263}a_0 - \frac{40,725}{23,263}a_1 + \frac{34,050}{23,263}a_2 + \frac{217,344}{255,893}a_4 \qquad (6.211)$$

where a_0, a_1, a_2 and a_4 are arbitrary. Upon substitution of the expression (6.211) into the set (6.209) we get the solution for the natural frequency squared:

$$\Omega^2 = 2,200b_8/a_4 \qquad (6.212)$$

and the coefficients in the flexural rigidity,

$$b_0 = (2,979,3258,000a_0 + 5,406,687,000a_1 + 7,494,763,925a_2 + 3,041,064,039a_4)b_8$$
$$/4,187,340,000a_4$$
$$b_1 = (136,466,000a_0 + 31,419,000a_1 + 63,579,225a_2 + 26,711,763a_4)11b_8$$
$$/1,465,569,000a_4$$

$b_2 = (136{,}466{,}000a_0 + 31{,}419{,}000a_1 + 63{,}579{,}225a_2 + 26{,}711{,}763a_4)11b_8$

$\quad /325{,}682{,}000r_4$

$b_3 = (1{,}087{,}020{,}000a_0 + 123{,}156{,}000a_1 - 52{,}321{,}775a_2 - 31{,}409{,}157a_4)b_8$

$\quad /73{,}278{,}450a_4$

$b_4 = (5{,}801{,}279{,}000a_0 - 2{,}562{,}417{,}000a_1 + 242{,}420{,}475a_2 + 271{,}535{,}433a_4)b_8$

$\quad /293{,}113{,}800a_4$

$b_5 = (26{,}840{,}000a_0 - 16{,}861{,}900a_1 + 11{,}423{,}775a_2 + 557{,}349a_4)b_8/162{,}841ar_4$

$b_6 = (275a_2 - 663a_4)b_8/180a_4$

$b_7 = (880{,}000a_0 - 447{,}975a_1 + 374{,}550a_2 + 217{,}344a_4)b_8/209{,}367a_4$

$$(6.213)$$

For the case $n = 2$, the governing differential equation yields a set of ten linear algebraic equations for nine unknowns, namely the coefficients b_i in the flexural rigidity and the natural frequency squared, Ω^2:

$$12b_2 - 48b_1 = 0 \quad 36b_3 - 144b_2 = 0$$

$$72b_4 - 288b_3 + 360b_0 - 3a_0\Omega^2 = 0$$

$$120b_5 - 480b_4 + 600b_1 + (4a_0 - 3a_1)\Omega^2 = 0$$

$$180b_6 - 720b_5 + 900b_2 + (4a_1 - 3a_2)\Omega^2 = 0$$

$$252b_7 - 1008b_6 + 1260b_3 + (4a_2 - 3a_3)\Omega^2 = 0$$

$$12b_2 - 48b_1 = 0 \quad 36b_3 - 144b_2 = 0 \qquad\qquad (6.214)$$

$$72b_4 - 288b_3 + 360b_0 - 3a_0\Omega^2 = 0$$

$$120b_5 - 480b_4 + 600b_1 + (4a_0 - 3a_1)\Omega^2 = 0$$

$$180b_6 - 720b_5 + 900b_2 + (4a_1 - 3a_2)\Omega^2 = 0$$

$$252b_7 - 1008b_6 + 1260b_3 + (4a_2 - 3a_3)\Omega^2 = 0$$

In order to have a non-trivial solution, the rank of the set (6.214) must be less than 10. The rank of the matrix of the system is less than 10 if all minors of order 10 vanish, leading to a set of 11 equations for 5 unknowns. The solution of this set reads

$$a_0 = \frac{113{,}596}{39{,}195}a_1 - \frac{397{,}283}{195{,}975}a_2 + \frac{2{,}068{,}751}{2{,}155{,}725}a_4$$

$$a_3 = -\frac{4{,}235}{603}a_1 + \frac{4{,}136}{603}a_2 - \frac{1{,}372}{603} \qquad\qquad (6.215)$$

where a_1, a_2 and a_4 are arbitrary constants. Upon substitution of the expressions (6.215) into the set (6.214) we get the solution for the natural frequency squared:

$$\Omega^2 = 3{,}960 b_8/a_4 \tag{6.216}$$

and the coefficients in the flexural rigidity,

$$b_0 = (20{,}089{,}520 a_1 - 8{,}507{,}763 a_2 + 6{,}050{,}974 a_4)b_8/653{,}250 a_4$$
$$b_1 = -(605 a_1 - 484 a_2 + 196 a_4)b_8/150 a_4$$
$$b_2 = -2(605 a_1 - 484 a_2 + 196 a_4)b_8/75 a_4$$
$$b_3 = -8(605 a_1 - 484 a_2 + 196 a_4)b_8/75 a_4 \tag{6.217}$$
$$b_4 = (2{,}887{,}720 a_1 - 2{,}737{,}757 a_2 + 1{,}236{,}904 a_4)b_8/43{,}550 a_4$$
$$b_5 = (55 a_1 - 196 a_4)b_8/30 a_4 \quad b_6 = 22 a_2 b_8/15 a_4$$
$$b_7 = -2(4{,}235 a_1 - 4{,}136 a_2 + 1{,}372 a_4)b_8/1{,}005 a_4$$

For the case $n = 3$, the governing differential equation yields a set of 11 linear algebraic equations for 9 unknowns, namely the coefficients b_i in the flexural rigidity and the natural frequency squared, Ω^2:

$$16 b_2 - 60 b_1 = 0 \quad 48 b_3 - 180 b_2 = 0$$
$$96 b_4 - 360 b_3 - 4 a_0 \Omega^2 = 0$$
$$160 b_5 - 600 b_4 + 840 b_0 + (5 a_0 - 4 a_1)\Omega^2 = 0$$
$$240 b_6 - 900 b_5 + 1260 b_1 + (5 a_1 - 4 a_2)\Omega^2 = 0$$
$$336 b_7 - 1260 b_6 + 1764 b_2 + (5 a_2 - 4 a_3)\Omega^2 = 0 \tag{6.218}$$
$$448 b_8 - 1680 b_7 + 2352 b_3 + (5 a_3 - 4 a_4)\Omega^2 = 0$$
$$-2160 b_8 + 3024 b_4 - (a_0 - 5 a_4)\Omega^2 = 0$$
$$3780 b_5 - a_1 \Omega^2 = 0 \quad 4620 b_6 - a_2 \Omega^2 = 0$$
$$5544 b_7 - a_3 \Omega^2 = 0 \quad 6552 b_8 - a_4 \Omega^2 = 0$$

In order to have a non-trivial solution, the rank of the set (6.218) must be less than 10. The rank of the matrix of the system is less than 10 if all minors of order 10 vanish, leading to a set of 66 equations for 5 unknowns. The solution of this set reads

$$a_0 = \frac{279}{1{,}540} a_3 - \frac{86}{455} a_4 \quad a_1 = \frac{1{,}916}{2{,}145} a_3 - \frac{1{,}288}{7{,}605} a_4 \quad a_2 = \frac{161}{156} a_3 - \frac{253}{1{,}521} a_4$$
$$\tag{6.219}$$

where a_3 and a_4 are arbitrary constants. Upon substitution of the expressions (6.219) into the set (6.218) we get the solution for the natural frequency squared:

$$\Omega^2 = 6{,}552b_8/a_4 \tag{6.220}$$

and the coefficients in the flexural rigidity

$$b_0 = (214{,}529{,}757a_3 - 55{,}489{,}124a_4)b_8/10{,}319{,}400a_4$$

$$b_1 = -8(1{,}209a_3 - 1{,}012a_4)b_8/10{,}395a_4$$

$$b_2 = -2(1{,}209a_3 - 1{,}012a_4)b_8/693a_4 \quad b_3 = -5(1{,}209a_3 - 1{,}012a_4)b_8/462a_4$$

$$b_4 = (1{,}209a_3 - 32{,}428a_4)b_8/3{,}080a_4 \quad b_5 = 8(18{,}681a_3 - 3{,}542a_4)b_8/96{,}525a_4$$

$$b_6 = 23(273a_3 - 44a_4)b_8/4{,}290a_4 \quad b_7 = 13a_3b_8/11a_4$$

$$\tag{6.221}$$

6.1.6.6   *Quintic inertial coefficient ($m = 5$)*

When $n = 0$, we obtain a set of 10 equations for 11 unknowns:

$$4b_2 - 21b_1 + 24b_0 = 0 \quad 12b_3 - 72b_2 + 72b_1 = 0$$

$$24b_4 - 144b_3 + 144b_2 - a_0\Omega^2 = 0$$

$$40b_5 - 240b_4 + 240b_3 + (2a_0 - a_1)\Omega^2 = 0$$

$$60b_6 - 360b_5 + 360b_4 - (a_0 - 2a_1 + a_2)\Omega^2 = 0$$

$$84b_7 - 504b_6 + 504b_5 - (a_1 - 2a_2 + a_3)\Omega^2 = 0 \tag{6.222}$$

$$112b_8 - 672b_7 + 672b_6 - (a_2 - 2a_3 + a_4)\Omega^2 = 0$$

$$144b_9 - 864b_8 + 864b_7 - (a_3 - 2a_4 + a_5)\Omega^2 = 0$$

$$-1080b_9 + 1080b_8 - (a_4 - 2a_5)\Omega^2 = 0$$

$$1320b_9 - a_5\Omega^2 = 0$$

We can, therefore, solve the system taking b_9 as a parameter. From the last equation of the set (6.222), we obtain the natural frequency squared:

$$\Omega^2 = 1320b_9/a_5 \tag{6.223}$$

The remaining equations yield the coefficients in the flexural rigidity as follows:

$$b_0 = 11(33{,}264a_0 + 13{,}176a_1 + 5{,}913a_2 + 2{,}910a_3 + 1{,}535a_4$$

$$+ 854a_5)b_9/163{,}296a_5$$

$$b_1 = (133{,}056a_0 + 58{,}608a_1 + 27{,}126a_2 + 13{,}475a_3 + 7{,}128a_4$$
$$+ 3{,}969a_5)b_9/54{,}432a_5$$

$$b_2 = (33{,}264a_0 + 30{,}888a_1 + 16{,}335a_2 + 8{,}415a_3 + 4{,}499a_4$$
$$+ 2{,}513a_5)b_9/27{,}216a_5$$

$$b_3 = -(66{,}528a_0 - 3{,}168a_1 - 5{,}544a_2 - 3{,}355a_3 - 1{,}870a_4$$
$$- 1{,}057a_5)b_9/9{,}072a_5$$

$$b_4 = (16{,}632a_0 - 21{,}384a_1 + 297a_2 + 1{,}650a_3 + 1{,}111a_4 + 658a_5)b_9/4{,}536a_5$$

$$b_5 = (3{,}960a_1 - 4{,}950a_2 - 55a_3 + 352a_4 + 259a_5)b_9/1{,}512a_5$$

$$b_6 = (1{,}485a_2 - 1{,}815a_3 - 55a_4 + 119a_5)b_9/756a_5$$

$$b_7 = (55a_3 - 66a_4 - 3a_5)b_9/36a_5 \quad b_8 = (11a_4 - 13a_5)b_9/9a_5$$

$$(6.224)$$

Now consider the case $n = 1$. The governing differential equation yields a set of 11 linear algebraic equations for 11 unknowns:

$$8b_2 - 36b_1 = 0 \quad 24b_3 - 108b_2 + 120b_0 = 0$$
$$48b_4 - 216b_3 + 240b_1 - 2a_0\Omega^2 = 0$$
$$80b_5 - 360b_4 + 400b_2 + (3a_0 - 2a_1)\Omega^2 = 0$$
$$120b_6 - 540b_5 + 600b_3 + (3a_1 - 2a_2)\Omega^2 = 0$$
$$168b_7 - 756b_6 + 840b_4 - (a_0 - 3a_2 + 2a_3)\Omega^2 = 0 \qquad (6.225)$$
$$224b_8 - 1008b_7 + 1120b_5 - (a_1 - 3a_3 + 2a_4)\Omega^2 = 0$$
$$288b_9 - 1296b_8 - 1440b_6 - (a_2 - 3a_4 + 2a_5)\Omega^2 = 0$$
$$-1620b_9 + 1800b_7 - (a_3 - 3a_5)\Omega^2 = 0$$
$$2200b_8 - a_4\Omega^2 = 0 \quad 2640b_9 - a_5\Omega^2 = 0$$

We impose the condition that the determinant of the system (6.225) should vanish so that we get a non-trivial solution. This leads to the following expression for the coefficient a_0:

$$a_0 = \frac{1{,}629}{3{,}200}a_1 - \frac{681}{1{,}600}a_2 + \frac{23{,}263}{80{,}000}a_3 - \frac{1{,}698}{6{,}875}a_4 + \frac{67{,}809}{352{,}000}a_5 \qquad (6.226)$$

where a_1, a_2, a_3, a_4 and a_5 are arbitrary. Upon substitution of the expression (6.226) into the set (6.225) we get the solution for the natural frequency squared:

$$\Omega^2 = 2,640 b_9/a_5 \tag{6.227}$$

and the coefficients in the flexural rigidity,

$$b_0 = (70,750,350 a_1 - 17,834,300 a_2 + 29,793,258 a_3 - 14,847,024 a_4$$
$$+ 17,093,285 a_5)b_9/12,000,000 a_5$$

$$b_1 = (3,816,450 a_1 + 207,900 a_2 + 1,501,126 a_3 - 264,528 a_4$$
$$+ 745,395 a_5)b_9/4,200,000 a_5$$

$$b_2 = 3(3,816,450 a_1 + 207,900 a_2 + 1,501,126 a_3 - 264,528 a_4$$
$$+ 745,395 a_5)b_9/2,800,000 a_5$$

$$b_3 = -(232,650 a_1 - 177,100 a_2 + 108,702 a_3 - 103,128 a_4 + 74,095 a_5)b_9/21,000 a_5$$

$$b_4 = (2,687,850 a_1 - 15,315,300 a_2 + 11,602,558 a_3 - 7,987,104 a_4$$
$$+ 7,434,735 a_5)b_9/1,680,000 a_5$$

$$b_5 = 3(550 a_1 - 1,342 a_3 + 1,044 a_4 - 735 a_5)b_9/700 a_5$$

$$b_6 = (275 a_2 - 663 a_4 + 520 a_5)b_9/150 a_5 \quad b_7 = (44 a_3 - 105 a_5)b_9/30 a_5$$

$$b_8 = 6 a_4 b_9/5 a_5$$

$$\tag{6.228}$$

For the case $n = 2$, the governing differential equation yields a set of 12 linear algebraic equations for 11 unknowns:

$$12 b_2 - 48 b_1 = 0 \quad 36 b_3 - 144 b_2 = 0$$

$$72 b_4 - 288 b_3 + 360 b_0 - 3 a_0 \Omega^2 = 0$$

$$120 b_5 - 480 b_4 + 600 b_1 + (4 a_0 - 3 a_1)\Omega^2 = 0$$

$$180 b_6 - 720 b_5 + 900 b_2 + (4 a_1 - 3 a_2)\Omega^2 = 0$$

$$252 b_7 - 1008 b_6 + 1260 b_3 + (4 a_2 - 3 a_3)\Omega^2 = 0$$

$$336 b_8 - 1344 b_7 + 1680 b_4 - (a_0 - 4 a_3 + 3 a_4)\Omega^2 = 0 \tag{6.229}$$

$$432 b_9 - 1728 b_8 + 2160 b_5 - (a_1 - 4 a_4 + 3 a_5)\Omega^2 = 0$$

$$-2160 b_9 + 2700 b_6 - (a_2 - 4 a_5)\Omega^2 = 0$$

$$3300 b_7 - a_3 \Omega^2 = 0 \quad 3960 b_8 - a_4 \Omega^2 = 0$$

$$4680 b_9 - a_5 \Omega^2 = 0$$

In order to have a non-trivial solution, the rank of the set (6.229) must be less than 10. The rank of the matrix of the system is lessthan 10 if all minors of order 10 vanish, leading to a set of 11 equations for 5 unknowns. The solution of this set reads

$$a_0 = \frac{1{,}723}{2{,}145}a_2 - \frac{16{,}228}{39{,}325}a_3 + \frac{163}{7{,}865}a_4 + \frac{518}{1{,}859}a_5$$

$$a_1 = \frac{376}{385}a_2 - \frac{603}{4{,}235}a_3 - \frac{196}{605}a_4 + \frac{281}{715}a_5 \tag{6.230}$$

where a_2, a_3, a_4 and a_5 are arbitrary constants. Upon substitution of the expressions (6.230) into the set (6.229) we get the solution for the natural frequency squared:

$$\Omega^2 = 4{,}680b_9/a_5 \tag{6.231}$$

and the coefficients in the flexural rigidity,

$$b_0 = (166{,}018{,}567a_2 - 42{,}735{,}888a_3 - 6{,}833{,}190a_4 + 56{,}045{,}066a_5)b_9$$

$$/8{,}258{,}250a_5$$

$$b_1 = -(9{,}724a_2 - 7{,}839a_3 + 3{,}542a_5)b_9/11{,}550a_5$$

$$b_2 = -2(9{,}724a_2 - 7{,}839a_3 + 3{,}542a_5)b_9/5{,}775a_5$$

$$b_3 = -8(9{,}724a_2 - 7{,}839a_3 + 3{,}542a_5)b_9/5{,}775a_5$$

$$b_4 = 3(1{,}231{,}945a_2 - 6{,}142{,}968a_3 + 4{,}502{,}680a_4 + 427{,}350a_5)b_9/550{,}550a_5$$

$$b_5 = (53{,}768a_2 - 7{,}839a_3 - 214{,}032a_4 + 181{,}720a_5)b_9/25{,}410a_5$$

$$b_6 = 2(13a_2 - 46a_5)b_9/15a_5 \quad b_7 = 78a_3b_9/55a_5 \quad b_8 = 13a_4b_9/11a_5$$

$$\tag{6.232}$$

For the case $n = 3$, the governing differential equation yields a set of 13 linear algebraic equations for 11 unknowns:

$$16b_2 - 60b_1 = 0$$

$$48b_3 - 180b_2 = 0$$

$$96b_4 - 360b_3 - 4a_0\Omega^2 = 0$$

$$160b_5 - 600b_4 + 840b_0 + (5a_0 - 4a_1)\Omega^2 = 0$$

$$240b_6 - 900b_5 + 1260b_1 + (5a_1 - 4a_2)\Omega^2 = 0$$

$$336b_7 - 1260b_6 + 1764b_2 + (5a_2 - 4a_3)\Omega^2 = 0 \tag{6.233}$$

$$448b_8 - 1680b_7 + 2352b_3 + (5a_3 - 4a_4)\Omega^2 = 0$$

$$576b_9 - 2160b_8 + 3024b_4 - (a_0 - 5a_4 + 4a_5)\Omega^2 = 0$$

$$-2700b_9 + 3780b_5 - (a_1 - 5a_5)\Omega^2 = 0$$

$$4620b_6 - a_2\Omega^2 = 0 \quad 5544b_7 - a_3\Omega^2 = 0$$

$$6552b_8 - a_4\Omega^2 = 0 \quad 7644b_9 - a_5\Omega^2 = 0$$

In order to have a non-trivial solution, the rank of the set (6.233) must be less than 11. The rank of the matrix of the system is less than 11 if all minors of order 11 vanish, leading to a set of 78 equations for 5 unknowns. The solution of this set reads

$$a_0 = \frac{279}{1540}a_3 - \frac{86}{455}a_4 + \frac{20}{637}a_5 \quad a_1 = \frac{1916}{2145}a_3 - \frac{1288}{7605}a_4 - \frac{148}{637}a_5$$

$$a_2 = \frac{161}{156}a_3 - \frac{253}{1521}a_4 \tag{6.234}$$

where a_3, a_4 and a_5 are arbitrary constants. Substituting the expressions (6.234) into the set (6.233) we get the solution for the natural frequency squared:

$$\Omega^2 = 7{,}644b_9/a_5 \tag{6.235}$$

as well as the coefficients in the flexural rigidity

$$b_0 = (214{,}529{,}757a_3 - 55{,}489{,}124a_4 - 7{,}637{,}760a_5)b_9/8{,}845{,}200a_5$$

$$b_1 = -4(1{,}209a_3 - 1{,}012a_4)b_9/4{,}455a_5 \quad b_2 = -(1{,}209a_3 - 1{,}012a_4)b_9/297a_5$$

$$b_3 = -5(1{,}209a_3 - 1{,}012a_4)b_9/396a_5$$

$$b_4 = (1{,}209a_3 - 32{,}428a_4 + 26{,}400a_5)b_9/2{,}640a_5$$

$$b_5 = 4(130{,}767a_3 - 24{,}794a_4 - 714{,}285a_5)b_9/289{,}575a_5$$

$$b_6 = 161(273a_3 - 44a_4)b_8/25{,}740a_5 \quad b_7 = 91a_3b_9/66a_5 \quad b_8 = 7a_4b_9/6a_5 \tag{6.236}$$

6.1.7   Concluding Remarks

In this section, a thorough investigation has been conducted with a single objective in mind: obtaining novel closed-form solutions for the beam vibrations. We consider many variations of the inertial coefficient, including constant or polynomially varying sub-cases. In the latter case, the maximum power that was considered was fixed at five. One can directly obtain solutions also for the inertial coefficients that are of higher degree than the quintic one. The derivations are unusual in the sense that, for the inhomogeneous beams, simple closed-form polynomial solutions have been derived, while even for uniform beams transcendental equations should be solved for obtaining the natural frequencies. Likewise, a remarkable nature of the effect of boundary conditions was reported. We are unaware of any parallel study in beam vibrations. One should anticipate that with this new "pipe" of thought being opened for closed-form solutions, many new solutions will be reported in the future, hopefully triggered by the present investigation.

7

Boundary Conditions Involving Guided Ends

In this chapter, we continue our discussion on the effect of boundary conditions on the vibration modes and natural frequencies; we concentrate, here, on the effect of the guided ends.

7.1 Closed-Form Solutions for the Natural Frequency for Inhomogeneous Beams with One Guided Support and One Pinned Support

7.1.1 Introductory Remarks

Closed-form solutions for inhomogeneous beams have been obtained by Elishakoff and Rollot (1999) and Elishakoff and Candan (2001). In particular, the former study dealt with the stability of inhomogeneous columns, whereas the latter was devoted to their vibration. It contained both deterministic and probabilistic formulations, with the deterministic relationship serving as a transfer function for the probabilistic calculations. In both cases, a polynomial representation of the mode shape was postulated, and a closed-form solution was obtained by formulating an inverse vibration problem. In this section, we deal with the vibrations of a beam that has a guided support on the left end and a pinned support on the right end. Here, we demand a function that satisfies all boundary conditions, to serve as a *mode shape* of the vibrating beam. We then construct an inhomogeneous beam that has the postulated function as the mode shape. It is shown, remarkably, that the expression for the natural frequency of the guided–pinned beam coincides with that of the pinned–pinned beam. Specific cases of variations of material density are given, for constant, linear, parabolic, cubic and quartic variations. These constitute particular cases. The general case is also treated, when the variation is at least quintic. The closed-form rational expressions for fundamental natural frequencies are derived for all the above cases.

7.1.2   Formulation of the Problem

The governing differential equation of the dynamic behavior of a beam (assuming that the cross-sectional area A of the beam is constant, as well as the moment of inertia I) is

$$\frac{d^2}{d\xi^2}\left[D(\xi)\frac{d^2w(\xi)}{d\xi^2}\right] - kL^4R(\xi)w(\xi) = 0 \qquad (7.1)$$

where $w(\xi)$ is the mode shape, ξ is the non-dimensional coordinate ($\xi = x/L$), L the length, $E(\xi)$ the Young's modulus and $R(\xi)$ the inertial coefficient. Moreover,

$$k = \omega^2 \qquad (7.2)$$

is the frequency coefficient where ω^2 is the sought for natural frequency.

In this study, we assume that $R(\xi)$, $D(\xi)$ and $w(\xi)$ are polynomial functions, given by

$$R(\xi) = \sum_{i=0}^{m} a_i\xi^i \qquad (7.3)$$

$$D(\xi) = \sum_{i=0}^{n} b_i\xi^i \qquad (7.4)$$

$$w(\xi) = \sum_{i=0}^{p} w_i\xi^i \qquad (7.5)$$

where m, n and p are, respectively, the degrees of the coefficients of $R(\xi)$, $D(\xi)$ and $w(\xi)$. These are linked by the orders of the derivatives Eq. (7.1), namely $n - m = 4$.

7.1.3   Boundary Conditions

The boundary conditions are as follows:

$$w'(0) = 0 \qquad (7.6)$$
$$w'''(0) = 0 \qquad (7.7)$$
$$w(1) = 0 \qquad (7.8)$$
$$w''(1) = 0 \qquad (7.9)$$

We have four boundary conditions, so we must choose at least $p = 4$. One can check that the following polynomial function agrees with the boundary conditions (7.6)–(7.9):

$$w(\xi) = 1 - \tfrac{6}{5}\xi^2 + \tfrac{1}{5}\xi^4 \tag{7.10}$$

7.1.4 Solution of the Differential Equation

Equation (7.1) with the polynomial functions $\rho(\xi)$, $E(\xi)$ and $w(\xi)$ yields

$$-\frac{12}{5}\sum_{i=0}^{m+2}(i+1)(i+2)b_{i+2}\xi^i + \frac{12}{5}\sum_{i=2}^{m+4}i(i-1)b_i\xi^i + \frac{48}{5}\sum_{i=1}^{m+4}ib_i\xi^i + \frac{24}{5}\sum_{i=0}^{m+4}b_i\xi^i$$

$$-kL^4\sum_{i=0}^{m}a_i\xi^i + \frac{6}{5}kL^4\sum_{i=2}^{m+2}a_{i-2}\xi^i - \frac{1}{5}kL^4\sum_{i=4}^{m+4}a_{i-4}\xi^i = 0 \tag{7.11}$$

Note that the second, third and fourth sums in Eq. (7.11) can be combined into the single sum $\frac{12}{5}\sum_{i=0}^{m+4}(i+1)(i+2)b_i\xi^i$ (Storch, 2001b). The equations above must be satisfied for any ξ, so we have for each ith power of ξ the following equations:

from ξ^0: $24(b_0 - b_2) - 5kL^4 a_0 = 0$ (7.12)

from ξ^1: $72(b_1 - b_3) - 5kL^4 a_1 = 0$ (7.13)

from ξ^2: $144(b_2 - b_4) + kL^4(6a_0 - 5a_2) = 0$ (7.14)

from ξ^3: $240(b_3 - b_5) + kL^4(6a_1 - 5a_3) = 0$ (7.15)

$\cdots$

from ξ^i: $12(i+1)(i+2)(b_i - b_{i+2}) + kL^4(6a_{i-2} - a_{i-4} - 5a_i) = 0$

       for $4 \le i \le m$ (7.16)

$\cdots$

from ξ^{m+1}: $12(m+2)(m+3)(b_{m+1} - b_{m+3}) + kL^4(6a_{m-1} - a_{m-3}) = 0$ (7.17)

from ξ^{m+2}: $12(m+3)(m+4)(b_{m+2} - b_{m+4}) + kL^4(6a_m - a_{m-2}) = 0$ (7.18)

from ξ^{m+3}: $12(m+4)(m+5)b_{m+3} - kL^4 a_{m-1} = 0$ (7.19)

from ξ^{m+4}: $12(m+5)(m+6)b_{m+4} - kL^4 a_m = 0$ (7.20)

Note that Eq. (7.16) is valid only for $4 \le i \le m$. From Eqs. (7.12)–(7.20), we have $m + 5$ relations between the coefficients a_i and b_i, these having a recursive form between b_i and b_{i+2}. The sole unknown is the natural frequency coefficient k. Thus, there must be other relations between a_i and b_i to assure the compatibility of the Eqs. (7.12)–(7.20). These relations will be formulated at a later stage.

We first treat the cases in which $m < 5$. The general case will be treated in Section 7.1.6.

7.1.5   The Degree of the Material Density is Less than Five

7.1.5.1   *Uniform inertial coefficient* ($m = 0$)

In this sub-case, $R(\xi)$ and $D(\xi)$ read

$$R(\xi) = a_0 \quad D(\xi) = \sum_{i=0}^{4} b_i \xi^i \tag{7.21}$$

By the substitution of Eq. (7.21) into Eq. (7.1), we obtain

$$24(b_0 - b_2) - 5kL^4 a_0 = 0 \tag{7.22}$$

$$b_1 = b_3 \tag{7.23}$$

$$144(b_2 - b_4) + 6kL^4 a_0 = 0 \tag{7.24}$$

$$b_3 = 0 \tag{7.25}$$

$$360b_4 - kL^4 a_0 = 0 \tag{7.26}$$

We obtain five equations for six unknowns: b_0, b_1, b_2, b_3, b_4 and k. We take b_4 to be an arbitrary constant. The coefficients b_i, then, become

$$b_0 = 61b_4 \tag{7.27}$$

$$b_1 = 0 \tag{7.28}$$

$$b_2 = -14b_4 \tag{7.29}$$

$$b_3 = 0 \tag{7.30}$$

$$k = 360b_4/L^4 a_0 \tag{7.31}$$

The fundamental natural frequency reads

$$\omega^2 = 360b_4/L^4 a_0 \tag{7.32}$$

Figure 7.1 depicts the variation of $D(\xi)/b_4$.

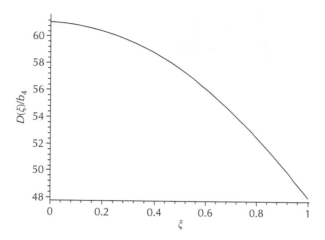

FIGURE 7.1
Variation of $D(\xi)/b_4$, $\xi \in [0;1]$, for constant density

7.1.5.2 Linearly varying inertial coefficient ($m = 1$)

Here, $R(\xi)$ and $D(\xi)$ are specified as

$$\rho(\xi) = a_0 + a_1\xi \quad E(\xi) = \sum_{i=0}^{5} b_i\xi^i \tag{7.33}$$

The substitution of Eq. (7.33) into Eq. (7.1) yields

$$24(b_0 - b_2) - 5kL^4 a_0 = 0 \tag{7.34}$$

$$72(b_1 - b_3) - 5kL^4 a_1 = 0 \tag{7.35}$$

$$144(b_2 - b_4) + 6kL^4 a_0 = 0 \tag{7.36}$$

$$240(b_3 - b_5) + 6kL^4 a_1 = 0 \tag{7.37}$$

$$360b_4 - kL^4 a_0 = 0 \tag{7.38}$$

$$504b_5 - kL^4 a_1 = 0 \tag{7.39}$$

We have six equations with seven unknowns, b_0, b_1, b_2, b_3, b_4, b_5 and k. The coefficient b_5 is taken to be an arbitrary constant. Then, the solution of the set (7.34)–(7.39) is expressed as follows:

$$b_0 = 427a_0b_5/5a_1 \tag{7.40}$$

$$b_1 = 117b_5/5 \tag{7.41}$$

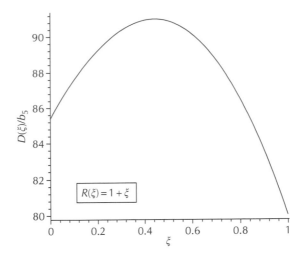

FIGURE 7.2
Variation of $D(\xi)/b_5$, $\xi \in [0;1]$, for a linear variation of the density

$$b_2 = -98a_0b_5/5a_1 \tag{7.42}$$

$$b_3 = -58b_5/5 \tag{7.43}$$

$$b_4 = 7a_0b_5/5a_1 \tag{7.44}$$

$$k = 504b_5/L^4a_1 \tag{7.45}$$

The fundamental natural frequency is expressed by the following formula:

$$\omega^2 = 504b_5/L^4a_1 \tag{7.46}$$

To illustrate this case, we portray in Figure 7.2 the function $D(\xi)/b_5$ for $a_0 = a_1 = 1$.

7.1.5.3 Parabolically varying inertial coefficient ($m = 2$)

The inertial coefficient and the elastic modulus are expressed as follows:

$$R(\xi) = \sum_{i=0}^{2} a_i\xi^i \quad D(\xi) = \sum_{i=0}^{6} b_i\xi^i \tag{7.47}$$

The substitution of Eq. (7.47) into Eq. (7.1) results in

$$24(b_0 - b_2) - 5kL^4a_0 = 0 \tag{7.48}$$

$$72(b_1 - b_3) - 5kL^4a_1 = 0 \tag{7.49}$$

$$144(b_2 - b_4) + kL^4(6a_0 - 5a_2) = 0 \tag{7.50}$$

$$240(b_3 - b_5) + 6kL^4 a_1 = 0 \tag{7.51}$$

$$360(b_4 - b_6) + kL^4(6a_2 \quad a_0) = 0 \tag{7.52}$$

$$504b_5 - kL^4 a_1 = 0 \tag{7.53}$$

$$672b_6 - kL^4 a_2 = 0 \tag{7.54}$$

We note that the above set contains seven equations for eight unknowns, one of which, namely b_6 is taken to be an arbitrary constant. Hence,

$$b_0 = (1708a_0 + 197a_2)b_6/15a_2 \tag{7.55}$$

$$b_1 = 156a_1 b_6/5a_2 \tag{7.56}$$

$$b_2 = -(392a_0 - 197a_2)b_6/15a_2 \tag{7.57}$$

$$b_3 = -232a_1 b_6/15a_2 \tag{7.58}$$

$$b_4 = (28a_0 - 153a_2)b_6/15a_2 \tag{7.59}$$

$$b_5 = 4a_1 b_6/3a_2 \tag{7.60}$$

$$k = 672b_6/L^4 a_2 \tag{7.61}$$

leading to the fundamental natural frequency

$$\omega^2 = 672b_6/L^4 a_2 \tag{7.62}$$

Figure 7.3 illustrates the dependence $D(\xi)/b_6$ for the specific case $a_0 = a_1 = a_2 = 1$.

7.1.5.4 *Material density as a cubic polynomial ($m = 3$)*

In this particular case, $R(\xi)$ and $D(\xi)$ are represented as the following polynomial functions:

$$R(\xi) = \sum_{i=0}^{3} a_i \xi^i \quad D(\xi) = \sum_{i=0}^{7} b_i \xi^i \tag{7.63}$$

The requirement that Eq. (7.1) is valid for every ξ imposes the following conditions:

$$24(b_0 - b_2) - 5kL^4 a_0 = 0 \tag{7.64}$$

$$72(b_1 - b_3) - 5kL^4 a_1 = 0 \tag{7.65}$$

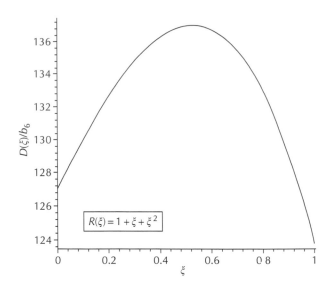

FIGURE 7.3
Variation of $D(\xi)/b_6$, $\xi \in [0; 1]$, for parabolic variation of the density

$$144(b_2 - b_4) + kL^4(6a_0 - 5a_2) = 0 \tag{7.66}$$

$$240(b_3 - b_5) + kL^4(6a_1 - 5a_3) = 0 \tag{7.67}$$

$$360(b_4 - b_6) + kL^4(6a_2 - a_0) = 0 \tag{7.68}$$

$$504(b_5 - b_7) + kL^4(6a_3 - a_1) = 0 \tag{7.69}$$

$$672b_6 - kL^4a_2 = 0 \tag{7.70}$$

$$864b_7 - kL^4a_3 = 0 \tag{7.71}$$

The coefficients b_i, to assure the compatibility of Eqs. (7.64)–(7.71), must satisfy the following relations, in terms of b_7:

$$b_0 = 3(1708a + 197a_2)b_7/35a_3 \tag{7.72}$$

$$b_1 = (1404a_1 + 305a_3)b_7/35a_3 \tag{7.73}$$

$$b_2 = -3(392a_0 - 197a_2)b_7/35a_3 \tag{7.74}$$

$$b_3 = (-696a_1 + 305a_3)b_7/35a_3 \tag{7.75}$$

$$b_4 = 3(28a_0 - 153a_2)b_7/35a_3 \tag{7.76}$$

$$b_5 = -(-12a_1 + 65a_3)b_7/7a_3 \tag{7.77}$$

$$b_6 = 9a_2b_7/7a_3 \tag{7.78}$$

$$k = 864b_7/L^4a_3 \tag{7.79}$$

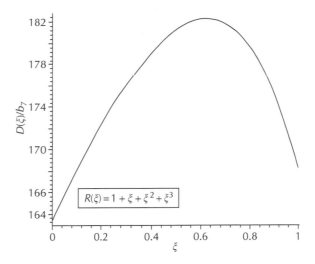

FIGURE 7.4
Variation of $D(\xi)/b_7$, $\xi \in [0;1]$, for cubic variation of the density

The fundamental natural frequency reads

$$\omega^2 = 864b_7/L^4 a_3 \tag{7.80}$$

The dependence $D(\xi)/b_7$ for the specific case $a_j = 1$ vs. ξ is shown in Figure 7.4.

7.1.5.5 Inertial coefficient as a quartic polynomial (*m = 4*)

In these circumstance, $R(\xi)$ and $D(\xi)$ are polynomial functions given by

$$R(\xi) = \sum_{i=0}^{4} a_i \xi^i \quad D(\xi) = \sum_{i=0}^{8} b_i \xi^i \tag{7.81}$$

Substitution of the above expressions into Eq. (7.1) results in

$$24(b_0 - b_2) - 5kL^4 a_0 = 0 \tag{7.82}$$

$$72(b_1 - b_3) - 5kL^4 a_1 = 0 \tag{7.83}$$

$$144(b_2 - b_4) + kL^4(6a_0 - 5a_2) = 0 \tag{7.84}$$

$$240(b_3 - b_5) + kL^4(6a_1 - 5a_3) = 0 \tag{7.85}$$

$$360(b_4 - b_6) + kL^4(6a_2 - a_0 - 5a_4) = 0 \tag{7.86}$$

$$504(b_5 - b_7) + kL^4(6a_3 - a_1) = 0 \tag{7.87}$$

$$672(b_6 - b_8) + kL^4(6a_4 - a_2) = 0 \tag{7.88}$$

$$864b_7 - kL^4a_3 = 0 \tag{7.89}$$

$$1080b_8 - kL^4a_4 = 0 \tag{7.90}$$

We have nine equations for ten unknowns, b_0, b_1, b_2, b_3, b_4, b_5, b_6, b_7, b_8 and k. We express these unknowns in terms of b_8. Thus,

$$b_0 = (5124a_0 + 591a_2 + 178a_4)b_8/28a_4 \tag{7.91}$$

$$b_1 = (1404a_1 + 305a_3)b_8/28a_4 \tag{7.92}$$

$$b_2 = -(1176a_0 - 591a_2 - 178a_4)b_8/28a_4 \tag{7.93}$$

$$b_3 = -(696a_1 - 305a_3)b_8/28a_4 \tag{7.94}$$

$$b_4 = (84a_0 - 459a_2 + 178a_4)b_8/28a_4 \tag{7.95}$$

$$b_5 = 5(12a_1 - 65a_3)b_8/28a_4 \tag{7.96}$$

$$b_6 = (45a_2 - 242a_4)b_8/28a_4 \tag{7.97}$$

$$b_7 = 5a_3b_8/4a_4 \tag{7.98}$$

$$k = 1080b_8/L^4a_4 \tag{7.99}$$

The fundamental natural frequency is given by

$$\omega^2 = 1080b_8/L^4a_4 \tag{7.100}$$

Figure 7.5 portrays the ratio $D(\xi)/b_8$ for $a_j = 1$.
    In what follows, we present the general case, with $5 \leq i \leq m$.

7.1.6   General Case: Compatibility Conditions

For the unknown k we have different expressions arising from Eqs. (7.12)–(7.20):

$$k = 24(b_0 - b_2)/5a_0L^4 \tag{7.101}$$

$$k = 72(b_1 - b_3)/5a_1L^4 \tag{7.102}$$

$$k = -144(b_2 - b_4)/(6a_0 - 5a_2)L^4 \tag{7.103}$$

$$k = -240(b_3 - b_5)/(6a_1 - 5a_3)L^4 \tag{7.104}$$

$$\cdots$$

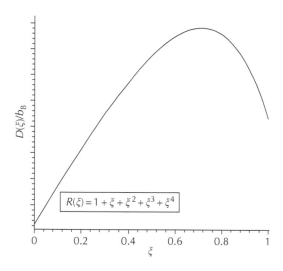

FIGURE 7.5
Variation of $D(\xi)/b_8$, $\xi \in [0;1]$, for quintic variation of the density

$$k = -12(i+1)(i+2)(b_i - b_{i+2})/(6a_{i-2} - a_{i-4} - 5a_i)L^4 \quad \text{for } 4 \le i \le m \tag{7.105}$$

$$\cdots$$

$$k = -12(m+2)(m+3)(b_{m+1} - b_{m+3})/(6a_{m-1} - a_{m-3})L^4 \tag{7.106}$$

$$k = -12(m+3)(m+4)(b_{m+2} - b_{m+4})/(6a_m - a_{m-2})L^4 \tag{7.107}$$

$$k = 12(m+4)(m+5)b_{m+3}/a_{m-1}L^4 \tag{7.108}$$

$$k = 12(m+5)(m+6)b_{m+4}/a_m L^4 \tag{7.109}$$

Compatibility conditions demand that all these expressions, as they represent the same natural frequency coefficient, should be equal.

Equations (7.101)–(7.109), in conjunction with the knowledge of the coefficients a_i permit us to obtain a closed-form solution for the natural frequency.

We assume the inertial coefficient coefficients ($a_i = 0, \ldots, m$) to be known. From the above equations (7.101)–(7.109), we can compute the coefficient b_i. First, let us observe Eq. (7.109). The knowledge of b_{m+4} leads to the natural frequency. Moreover, b_{m+4} and a_m have the same sign (due to the positivity of k). Second, we need only one coefficient b_i to determine all b_j, $j \ne i$. This is due to the recursive form of Eqs. (7.101)–(7.109). We assume that the coefficient b_{m+4} is known. Then, we calculate the other coefficients b_i, $i = 0, \ldots, m+3$. From Eq. (7.108) in conjunction with Eq. (7.109), we get

$$b_{m+3} = \frac{m+6}{m+4} \frac{a_{m-1}}{a_m} b_{m+4} \tag{7.110}$$

Then, Eqs. (7.107) and (7.109) lead to

$$b_{m+2} = -\frac{(m+7)(5m+24)a_m - (m+5)(m+6)a_{m-2}}{(m+3)(m+4)a_m}b_{m+4} \qquad (7.111)$$

Analogously, Eqs. (7.106) and (7.108) yield

$$b_{m+1} = -\frac{(m+6)\left((5m^2+49m+114)a_{m-1} - (m+4)(m+5)a_{m-3}\right)}{(m+2)(m+3)(m+4)a_m}b_{m+4}$$

$$(7.112)$$

Eq. (7.105), with $i = m$ and Eq. (7.107) result in

$$b_m = 8\frac{2m^3 + 30m^2 + 136m + 183}{(m+1)(m+2)(m+3)(m+4)}b_{m+4}$$
$$- \frac{(m+5)^2(m+6)(5m+14)}{(m+1)(m+2)(m+3)(m+4)}\frac{a_{m-2}}{a_m}b_{m+4}$$
$$+ \frac{(m+5)(m+6)}{(m+1)(m+2)}\frac{a_{m-4}}{a_m}b_{m+4} \qquad (7.113)$$

Equation (7.105), with $i = m-1$ and Eq. (7.106) becomes

$$b_{m-1} = 8\frac{(m+6)(2m^3 + 24m^2 + 82m + 75)}{m(m+1)(m+2)(m+3)(m+4)}\frac{a_{m-1}}{a_m}b_{m+4}$$
$$- \frac{(m+4)(m+5)(m+6)(5m+9)}{m(m+1)(m+2)(m+3)}\frac{a_{m-3}}{a_m}b_{m+4}$$
$$+ \frac{(m+5)(m+6)}{m(m+1)}\frac{a_{m-5}}{a_m}b_{m+4} \qquad (7.114)$$

We need to calculate b_m and b_{m-1} in order to use the general expression of b_i for $5 \le i \le m - 2$

$$b_i = \left[\frac{(i+3)(i+4)}{(i+1)(i+2)}\frac{6a_{i-2} - a_{i-4} - 5a_i}{6a_i - a_{i-2} - 5a_{i+2}} + 1\right]b_{i+2}$$
$$- \left[\frac{(i+3)(i+4)}{(i+1)(i+2)}\frac{6a_{i-2} - a_{i-4} - 5a_i}{6a_i - a_{i-2} - 5a_{i+2}}\right]b_{i+4} \qquad (7.115)$$

Note that Eq. (7.115) is only valid for $i \le m - 2$ because of the coefficient a_{i+2}. Indeed, $i + 2 < m + 1$. Just as $m - 1 > i > 3$ (due to the coefficient a_{i-4}), we must have $m > 4$ (this explains why cases $m \le 4$ are particular cases). Now, we calculate the coefficients b_3, b_2, b_1 and b_0. Equation (104) leads to

$$b_3 = \frac{21}{10}\left(\frac{6a_1 - 5a_3}{6a_3 - a_1 - 5a_5} + 1\right)b_5 - \frac{21}{10}\left(\frac{6a_1 - 5a_3}{6a_3 - a_1 - 5a_5}\right)b_7 \qquad (7.116)$$

Equation (7.103) results in

$$b_2 = \left(\frac{5}{2} \frac{6a_0 - 5a_2}{6a_2 - a_0 - 5a_4} + 1 \right) b_4 - \frac{5}{2} \left(\frac{6a_0 - 5a_2}{6a_2 - a_0 - 5a_4} \right) b_6 \qquad (7.117)$$

Note that in Elishakoff and Becquet (2000a) there is a typographical error in Eq. 177 (Storch, 2001b). From Eq. (7.102), we obtain

$$b_1 = \frac{1}{10} \left(\frac{234a_1 + 45a_3 + 50a_5}{-6a_3 + a_1 + 5a_5} \right) b_5 - \frac{1}{10} \left(\frac{224a_1 + 105a_3}{-6a_3 + a_1 + 5a_5} \right) b_7 \qquad (7.118)$$

Eq. (7.101) gives

$$b_0 = \frac{1}{2} \left(\frac{122a_0 + 13a_2 + 10a_4}{-6a_2 + a_0 + 5a_4} \right) b_4 - \frac{1}{2} \left(\frac{120a_0 + 25a_2}{-6a_2 + a_0 + 5a_4} \right) b_6 \qquad (7.119)$$

Figure 7.6 portrays the function, $D(\xi)/b_{16}$ for $m = 15$ with a_i specified as $16 - i$. From Eq. (7.109) in view of equation Eq. (7.2), we deduce the natural frequency squared

$$\omega^2 = 12(m + 5)(m + 6)b_{m+4}I/a_m AL^4 \qquad (7.120)$$

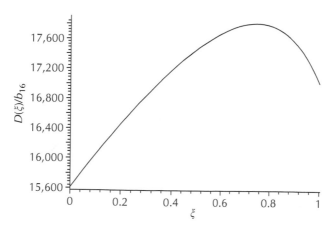

FIGURE 7.6
Variation of $D(\xi)/b_{16}$, $\xi \in [0;1]$, $R(\xi) = \sum_0^{15}(16 - i)\xi^i$

which, remarkably, coincides with its counterpart for the pinned–pinned beam. Still, the expression for the coefficients b_j differ for these two cases. It should also be noted that by formally substituting $m = 0, m = 1, m = 2, m = 3$ and $m = 4$, we get the expression derived in Eqs. (7.32), (7.46), (7.62), (7.80) and (7.100), respectively.

7.1.7 Concluding Comments

Apparently, for the first time in the literature, we obtained closed-form solutions for the natural frequencies of the inhomogeneous beams with one guided support and the other pinned. We hope that this investigation will be followed by an intensified search for additional closed-form solutions for inhomogeneous and/or non-uniform structures, and, most importantly, we hope that the interest in experimental implementation of the proposed procedures will lead to the designs tailored to the required fundamental natural frequency.

7.2 Closed-Form Solutions for the Natural Frequency for Inhomogeneous Beams with One Guided Support and One Clamped Support

7.2.1 Introductory Remarks

This section closely follows the recent studies by Elishakoff and Rollot (1999), Elishakoff and Candan (2001) and Becquet and Elishakoff (2001). It deals with closed-form solutions for beam eigenvalues; buckling is considered in the former study, while vibration is studied in the others. They treat, respectively, pinned–pinned beams and guided–pinned beams. Research by Elishakoff and Candan (2001) deals with three other boundary conditions: pinned–clamped, clamped–free and clamped–clamped beams. Here, we complete the investigation by studying the vibration of a clamped–guided beam. Like our previous studies, this investigation is posed as an inverse vibration problem. The first step is to postulate the mode shape of the vibrating beam, which is represented by a polynomial function satisfying all boundary conditions. We ought to note that in all the cases we obtain the same expression for the natural frequency. As we treat an inhomogeneous beam (the Young's modulus and density are given by polynomial functions), the problem for the engineer is to have an accurate knowledge of the material density in order to obtain the Young's modulus that corresponds to the selected mode shape.

In this section, a clamped–guided beam is studied. Moreover, we treat two specific cases, which are associated with constant and linear variations of the density. The fundamental natural frequency is given for all cases of variation of the material density.

7.2.2   Formulation of the Problem

The dynamic behavior of a beam, with a constant cross-sectional area A and a constant moment of inertia I, is given by

$$\frac{d^2}{d\xi^2}\left[D(\xi)\frac{d^2 w(\xi)}{d\xi^2}\right] - kL^4 R(\xi)w(\xi) = 0 \qquad (7.121)$$

where $D(\xi)$, $w(\xi)$ and $R(\xi)$ are, respectively, the longitudinal rigidity, the mode shape and the inertial coefficient; L is the length of the beam and k contains the fundamental natural frequency

$$k = \omega^2 \qquad (7.122)$$

In Eq. (7.121) the modulus of elasticity and the material density are taken as functions of the axial coordinate. The independent parameter of Eq. (7.121) is the non-dimensional axial coordinate $\xi = x/L$ (x being the dimensional coordinate). We assume that the properties of the inhomogeneous beam are as follows

$$R(\xi) = \sum_{i=0}^{m} a_i \xi^i \qquad (7.123)$$

$$D(\xi) = \sum_{i=0}^{n} b_i \xi^i \qquad (7.124)$$

$$w(\xi) = \sum_{i=0}^{p} w_i \xi^i \qquad (7.125)$$

Consequently, m, n and p (the respective degrees of the polynomial functions $D(\xi)$, $w(\xi)$ and $R(\xi)$) are linked by Eq. (7.121), i.e., $n - m = 4$.

7.2.3   Boundary Conditions

We treat the case of a clamped–guided beam. The boundary conditions read

$$w(0) = 0 \qquad (7.126)$$
$$w'(0) = 0 \qquad (7.127)$$
$$w'(1) = 0 \qquad (7.128)$$
$$w'''(1) = 0 \qquad (7.129)$$

Note that in the paper by Elishakoff and Becquet (2000a) the equivalent of Eq. (7.129) contains a misprint (Storch, 2002a). Since the degree of the mode shape $w(\xi)$ is a polynomial function, it must be at least 4 for it has to satisfy the boundary conditions. The mode shape reads

$$w(\xi) = \xi^2 - \xi^3 + \tfrac{1}{4}\xi^4 \tag{7.130}$$

7.2.4   Solution of the Differential Equation

Equation (7.121) is expanded by substituting Eqs. (7.123), (7.124) and (7.125) into it. We obtain

$$2 \sum_{i=0}^{m+2} (i+1)(i+2)b_{i+2}\xi^i - 6 \sum_{i=1}^{m+3} i(i+1)b_{i+1}\xi^i$$

$$+ 3 \sum_{i=2}^{m+4} i(i-1)b_i\xi^i - 12 \sum_{i=0}^{m+3} (i+1)b_{i+1}\xi^i$$

$$+ 12 \sum_{i=1}^{m+4} ib_i\xi^i + 6 \sum_{i=0}^{m+4} b_i\xi^i - kL^4 \sum_{i=2}^{m+2} a_{i-2}\xi^i$$

$$+ kL^4 \sum_{i=3}^{m+3} a_{i-3}\xi^i - \frac{1}{4}kL^4 \sum_{i=4}^{m+4} a_{i-4}\xi^i = 0 \tag{7.131}$$

Note that the second, third, fourth, fifth and sixth terms can be slightly simplified (Storch, 2002a) to $-6\sum_{i=0}^{m+3}(i+1)(i+2)b_{i+1}\xi^i + 3\sum_{i=0}^{m+4}(i+1)(i+2)b_i\xi^i$. Equation (7.131) is valid for any ξ, so for each coefficient in front of ξ^i, $0 \le i \le m+4$, we have a single equation. These read

from ξ^0: $4b_2 - 12b_1 + 6b_0 = 0$ $\qquad\qquad\qquad\qquad\qquad$ (7.132)

from ξ^1: $12b_3 - 36b_2 + 18b_1 = 0$ $\qquad\qquad\qquad\qquad\qquad$ (7.133)

from ξ^2: $24b_4 - 72b_3 + 36b_2 - kL^4 a_0 = 0$ $\qquad\qquad\qquad$ (7.134)

from ξ^3: $40b_5 - 120b_4 + 60b_3 + kL^4(a_0 - a_1) = 0$ $\qquad\qquad$ (7.135)

$$\cdots$$

from ξ^i: $(i+1)(i+2)(2b_{i+2} - 6b_{i+1} + 3b_i)$

$\qquad\qquad + kL^4(a_{i-3} - a_{i-2} - 1/4a_{i-4}) = 0 \qquad$ for $4 \le i \le m+2$ (7.136)

$$\cdots$$

from ξ^{m+3}: $(m+4)(m+5)(-6b_{m+4} + 3b_{m+3}) + kL^4(a_m - 1/4a_{m-1}) = 0$
$$\tag{7.137}$$

from ξ^{m+4}: $3(m+5)(m+6)b_{m+4} - 1/4kL^4 a_m = 0$ $\qquad\qquad$ (7.138)

We obtained a system of $m + 5$ equations $(0, \ldots i, \ldots, m + 4)$; Eq. (7.136) is a recursive equation. We are looking for the unknown k. Thus, the coefficients a_i and b_i are linked by other relations, given at a later stage. These relations are valid for the general case for $m \geq 2$. The case $m < 1$ is given in Section 7.2.5. The general case is treated in Section 7.2.6.

7.2.5   Cases of Uniform and Linear Densities

7.2.5.1   Uniform inertial coefficient

$R(\xi)$ and $D(\xi)$ are given by

$$R(\xi) = a_0 \quad D(\xi) = \sum_{i=0}^{4} b_i \xi^i \tag{7.139}$$

Equation (7.121), with the above expressions for $D(\xi)$ and $R(\xi)$, leads to

$$4b_2 - 12b_1 + 6b_0 = 0 \tag{7.140}$$

$$-36b_2 + 12b_3 + 18b_1 = 0 \tag{7.141}$$

$$36b_2 - 72b_3 + 24b_4 - kL^4 a_0 = 0 \tag{7.142}$$

$$60b_3 - 120b_4 + kL^4 a_0 = 0 \tag{7.143}$$

$$90b_4 - 1/4kL^4 a_0 = 0 \tag{7.144}$$

We have six unknowns, b_0, b_1, b_2, b_3, b_4 and k, and five equations, given by Eqs. (7.140)–(7.144). Thus, b_0, b_1, b_2, b_3 and k are calculated in terms of b_4. These coefficients and k are obtained as follows

$$b_0 = 88b_4/9 \tag{7.145}$$

$$b_1 = 16b_4/3 \tag{7.146}$$

$$b_2 = 4b_4/3 \tag{7.147}$$

$$b_3 = -4b_4 \tag{7.148}$$

$$k = 360b_4/L^4 a_0 \tag{7.149}$$

From Eq. (7.149), the fundamental natural frequency is derived

$$\omega^2 = 360b_4/L^4 a_0 \tag{7.150}$$

Figure 7.7 represents $D(\xi)/b_4$ for $a_0 = 1$.

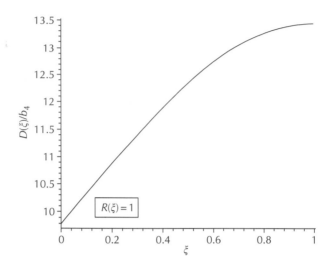

FIGURE 7.7
Variation of $D(\xi)/b_4$, $\xi \in [0; 1]$, for the constant density

7.2.5.2 Linear inertial coefficient

In this second case, $R(\xi)$ and $D(\xi)$ are as follows

$$R(\xi) = a_0 + a_1\xi \quad D(\xi) = \sum_{i=0}^{5} b_i\xi^i \qquad (7.151)$$

Equation (7.121), valid for every ξ, leads to the following

$$4b_2 - 12b_1 + 6b_0 = 0 \qquad (7.152)$$

$$-36b_2 + 12b_3 + 18b_1 = 0 \qquad (7.153)$$

$$36b_2 - 72b_3 + 24b_4 - kL^4 a_0 = 0 \qquad (7.154)$$

$$60b_3 - 120b_4 + 40b_5 + kL^4(a_0 - a_1) = 0 \qquad (7.155)$$

$$90b_4 - 180b_5 + kL^4\left(a_1 - \tfrac{1}{4}a_0\right) = 0 \qquad (7.156)$$

$$126b_5 - \tfrac{1}{4}kL^4 a_1 = 0 \qquad (7.157)$$

The six equations in the system (7.152)–(7.157) are a system of seven unknowns. Hence, b_5 is taken as an arbitrary coefficient. The unknowns b_0, b_1, b_2, b_3, b_4 and k read

$$b_0 = 8b_5(77a_0 + 61a_1)/45a_1 \qquad (7.158)$$

$$b_1 = 8b_5(42a_0 + 37a_1)/45a_1 \qquad (7.159)$$

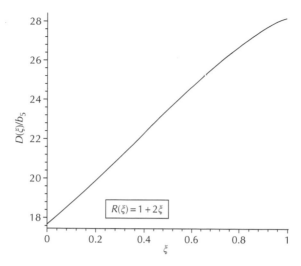

FIGURE 7.8
Variation of $D(\xi)/b_5$, $\xi \in [0; 1]$, for linear variation of the density

$$b_2 = 4b_5(7a_0 + 13a_1)/15a_1 \tag{7.160}$$

$$b_3 = -4b_5(21a_0 - 2a_1)/15a_1 \tag{7.161}$$

$$b_4 = b_5(7a_0 - 18a_1)/5a_1 \tag{7.162}$$

$$k = 504b_5/L^4 a_1 \tag{7.163}$$

leading to

$$\omega^2 = 504b_5/L^4 a_1 \tag{7.164}$$

The dependence $D(\xi)/b_5$ on ξ is shown in the Figure 7.8 for the particular case $R(\xi) = 1 + 2\xi$, $a_0 = 1, a_2 = 1$.

7.2.6 General Case: Compatibility Condition

We calculate k from Eqs. (7.132)–(7.138). Obviously, these different expressions for k are equal to each other. Thus,

$$k = 12(2b_4 - 6b_3 + 3b_2)/L^4 a_0 \tag{7.165}$$

$$k = 20(2b_5 - 6b_4 + 3b_3)/L^4(a_1 - a_0) \tag{7.166}$$

$$\cdots$$

$$k = 4(i + 1)(i + 2)(2b_{i+2} - 6b_{i+1} + 3b_i)/L^4(-4a_{i-3} + 4a_{i-2} + a_{i-4}) \tag{7.167}$$

$$\cdots$$

$$k = -12(m + 4)(m + 5)(2b_{m+4} - b_{m+3})/L^4(-4a_m + a_{m-1}) \qquad (7.168)$$

$$k = 12(m + 5)(m + 6)b_{m+4}/L^4a_m \qquad (7.169)$$

Let us assume that the material density coefficients a_i are known. The above expressions for k allow us to obtain the material Young's modulus coefficients b_i.

As the material density coefficients are known, Eqs. (7.165)–(7.169) lead to the determination of the coefficients $b_i(i = 2, \ldots, m + 4)$. The coefficients b_0 and b_1 are calculated using Eqs. (7.132) and (7.133).

Yet, k is also unknown: thus, we have $m + 5$ equations (Eqs. (7.132), (7.133) and (7.165)–(7.169)) with $m + 6$ unknowns ($b_i, i = 0, \ldots, m + 4$ and k). We need to fix one of the coefficients b_j in order to compute the other coefficients b_i, $i \neq j$ and k. Here, we choose b_{m+4} to be specified.

Equation (7.168) yields

$$b_{m+3} = \frac{(m + 6)a_{m-1} - (2m + 16)a_m}{(m + 4)a_m} b_{m+4} \qquad (7.170)$$

Then, Eq. (7.167), with $i = m + 2$ results in

$$b_{m+2} = \frac{(3m^2 + 33m + 90)a_{m-2} - (6m^2 + 78m + 252)a_{m-1} - (2m^2 + 14m - 48)a_m}{3(m + 3)(m + 4)a_m} b_{m+4}$$

$$(7.171)$$

With Eq. (7.167), we have the general expression for b_i, for $4 \leq i \leq m + 1$. Hence,

$$b_i = -\frac{[(6 + 2i)a_{i-4} + (24 + 8i)(a_{i-2} - a_{i-3})]b_{i+3}}{3(i + 1)(4a_{i-2} - 4a_{i-1} - a_{i-3})}$$
$$- \frac{[-(18 + 6i)a_{i-4} + (70 + 22i)a_{i-3} - (64 + 16i)a_{i-2} - 8(i + 1)a_{i-1}]b_{i+2}}{3(i + 1)(4a_{i-2} - 4a_{i-1} - a_{i-3})}$$
$$- \frac{[(9 + 3i)a_{i-4} - (30 + 6i)a_{i-3} - 12(i - 1)a_{i-2} + 24(i + 1)a_{i-1}]b_{i+1}}{3(i + 1)(4a_{i-2} - 4a_{i-1} - a_{i-3})}$$

$$(7.172)$$

Equation (7.166) yields

$$b_3 = 2\frac{6(a_0 - a_1)b_6 + (14a_1 + 4a_2 - 17a_0)b_5 + (3a_1 - 12a_2 + 6a_0)b_4}{3(-4a_1 + 4a_2 + a_0)} \qquad (7.173)$$

Equation (7.165) results in

$$b_2 = \frac{(24a_0 + 6a_1)b_4 + (3a_0 - 18a_1)b_3 - 10a_0b_5)}{9(a_0 - a_1)} \qquad (7.174)$$

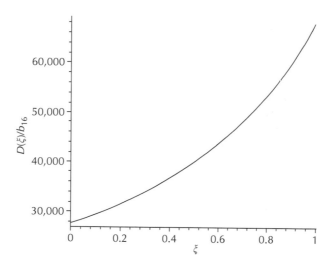

FIGURE 7.9
Variation of $D(\xi)/b_{16}$, $\xi \in [0;1]\rho(\xi) = \sum_{i=0, i\neq 4, i\neq 12}^{15}(16-i)\xi^i + 3\xi^4 + 8\xi^{12}$

From Eqs. (7.133) and (7.132), we obtain

$$b_1 = -\tfrac{2}{3}b_3 + 2b_2$$
$$b_0 = -\tfrac{2}{3}b_2 + 2b_1 \tag{7.175}$$

Equation (7.169) leads to the natural frequency ω

$$\omega^2 = 12(m+5)(m+6)b_{m+4}/L^4 a_m \tag{7.176}$$

In Figure 7.9, we present the function $D(\xi)/b_{16}$.

7.2.7 Concluding Remarks

It is remarkable that the expression for the natural frequency obtained within the present formulation for the clamped–guided beam coincides with its counterparts considered in previous chapters. Thus, the natural frequency expressions remain invariant, irrespective of the boundary conditions. It should not be forgotten, however, that the flexural rigidity coefficients *depend* on the boundary conditions. This implies that for a given expression for material density, in order that beams under *differing* boundary conditions have the same frequency, they must have *differing* expressions for flexural rigidity. This feature is in line with one's anticipation, since even uniform beams with different boundary conditions may possess the same natural frequency if the flexural rigidity is properly tailored. The present formulation screens, as it were, the beams that have the same frequencies under differing boundary conditions.

7.3 Class of Analytical Closed-Form Polynomial Solutions for Guided–Pinned Inhomogeneous Beams

7.3.1 Introductory Remarks

In the above material, we uncovered numerous closed-form solutions for inhomogeneous beams. We postulated that the mode shape was a fourth-order polynomial function. After representing the material density and the flexural rigidity too as polynomial functions, for different boundary conditions, the natural frequency of the beam was determined in a closed-form fashion. The following natural question arises: does a polynomial of general order, and not just the fourth-order one, exist to represent the true mode shape of an inhomogeneous beam? Elishakoff and Guédé (2001) worked out this problem for a beam that was pinned at both ends. They constructed a $(n + 4)$th order polynomial that satisfied boundary conditions, and then obtained a closed-form solution for the natural frequency.

In this section, we intend to determine closed-form solutions for the natural frequency when the beam has one guided end and one pinned end. The mode shape of the beam is represented by a $(n+4)$th order polynomial which satisfies all boundary conditions. We first treat special cases for which n is fixed at a specific integer value. Then, we deal with the general case for which n is arbitrary.

7.3.2 Formulation of the Problem

Let us begin with an auxiliary problem of a homogeneous and uniform $(D = \text{const.})$ beam, which has a guided left end and a pinned right end, and is subjected to the distributed load $q(\xi) = q_0\xi^n$. The governing differential equation of its shape $w(x)$ is as follows:

$$\frac{Dd^4w}{dx^4} = q_0x^n \tag{7.177}$$

With $\xi = x/L$ the dimensionless coordinate, where $0 \le \xi \le 1$ and L is its length, the above differential equation becomes

$$\frac{d^4w}{d\xi^4} = \beta\xi^n \quad \beta = q_0\frac{L^{n+4}}{D} \tag{7.178}$$

We solve the above differential equation in conjunction with the following boundary conditions

$$w'(0) = 0 \quad w'''(0) = 0 \quad w(1) = 0 \quad w''(1) = 0 \tag{7.179}$$

The solution of Eq. (7.178) that satisfies all boundary conditions may be expressed as

$$w(\xi) = \beta\psi(\xi)/(n+1)(n+2)(n+3)(n+4)$$
$$\psi(\xi) = \xi^{n+4} - (n^2 + 7n + 12)\xi^2/2 + (n^2 + 7n + 10)/2 \tag{7.180}$$

We now return to our original problem of a vibrating inhomogeneous beam. We will confine ourselves to *inhomogeneous* beams whose exact mode shape *coincides* with the expression for $\psi(\xi)$ in Eq. (7.180). The dynamic behavior of a beam is governed by the following equation

$$\frac{d^2\left(D(\xi)d^2w/d\xi^2\right)}{d\xi^2} - \omega^2 L^4 R(\xi)w(\xi) = 0 \tag{7.181}$$

where $D(\xi)$ is the flexural rigidity, $R(\xi) = \rho(\xi)A$ is the inertial coefficient (A is the cross-sectional area, $\rho(\xi)$ the material density) and ω the natural frequency of the beam. We demand that $w(\xi) = C\psi(\xi)$ in Eq. (7.180.2), with C a non-zero constant. Thus, Eq. (7.181) can be written as follows

$$\frac{d^2(D(\xi)d^2\psi/d\xi^2)}{d\xi^2} - \Omega^2 R(\xi)\psi(\xi) = 0 \quad \Omega^2 = \omega^2 L^4 \tag{7.182}$$

The functions $D(\xi)$ and $R(\xi)$ are taken to be polynomials. The problem is to construct the flexural rigidity $D(\xi)$ for various variations of the inertial coefficient $R(\xi)$ when the mode shape is given by Eq. (7.180). We assume that the inertial coefficient $R(\xi)$ is represented by

$$R(\xi) = \sum_{i=0}^{m} a_i\xi^i \tag{7.183}$$

where each coefficient a_i is specified. This implies that the second term in Eq. (7.182) is an $(m+n+4)$th order polynomial function. To have the same polynomial order in the first term of the governing equation, the degree of $D(\xi)$ must be $m+4$. Hence, we represent $D(\xi)$ as follows

$$D(\xi) = \sum_{i=0}^{m+4} b_i\xi^i \tag{7.184}$$

Equation (7.182) must be satisfied for every $\xi \in [0;1]$, this imposes the condition that each coefficient of ξ^i must vanish. We, thus, obtain a system of linear equations, in which each coefficient b_i, for $i = 0, \ldots, m + 4$, of the flexural rigidity and the natural frequency coefficient Ω, constitute the unknowns, hereinafter denoted collectively as the vector $\alpha^T = [b_0 \ldots b_i \ldots b_{m+4}\Omega^2]$. Therefore, there are $m+6$ unknowns. The degree n of the mode shape determines the number of equations of the system obtained by substituting $\psi(\xi)$ and Eqs. (7.183) and (7.184) into Eq. (7.185). As the degree of the polynomial function derived from the governing equation is $n + m + 4$, we obtain a system of $n + m+5$ equations by requiring that the coefficients of each degree of ξ must vanish. In the following, this system is written as $A\alpha = 0$, where A is a $(m + 6) \times (n + m + 5)$ matrix.

The special case $n = 0$ has been studied by Becquet and Elishakoff (2001). We immediately note that when $n = 1$, we obtain a square system, which describes a non-trivial system if the determinant of the system det(A) vanishes. The following sections deal with cases where the degree m of the inertial coefficient $R(\xi)$ is specified at either of four values 1, 2, 3 or 4. We treat the case m being any integer greater than 4 in Section 8.

7.3.3   Constant Inertial Coefficient ($m = 0$)

In this section, we treat the case of a constant inertial coefficient ($m = 0$), i.e.,

$$R(\xi) = a_0 \quad D(\xi) = b_0 + b_1 x + b_2 x^2 + b_3 x^3 + b_4 x^4 \qquad (7.185)$$

The first case is $n = 0$. The mode shape may be given by

$$\psi(\xi) = 1 - \tfrac{6}{5}\xi^2 + \tfrac{1}{5}\xi^4 \qquad (7.186)$$

The above expressions, in conjunction with Eq. (7.182), lead to

$$\begin{aligned}
&24(b_0 - b_2) - 5\Omega^2 a_0 = 0 \\
&b_1 = b_3 \\
&24(b_2 - b_4) + \Omega^2 a_0 = 0 \\
&b_3 = 0 \\
&360b_4 - \Omega^2 a_0 = 0
\end{aligned} \qquad (7.187)$$

This is a set of five equations with six unknowns, b_0, b_1, b_2, b_3, b_4 and Ω^2. One unknown, namely b_4, is taken to be an arbitrary parameter. Then, the solution of this system is

$$D(\xi) = (61 - 14\xi^2 + \xi^4)b_4 \qquad (7.188)$$

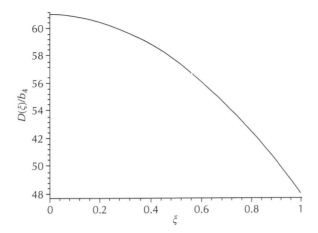

FIGURE 7.10
Variation of $D(\xi)/b_4$ for the beam of constant mass density $(m = 0, n = 0)$

From the last equation of the set (7.187), we get Ω^2

$$\Omega^2 = 360b_4/a_0 \tag{7.189}$$

For $\xi \in [0; 1]$, $D(\xi)$ is positive as Figure 7.10 shows.

In case $n = 1$, we have six equations with six unknowns, namely b_0, b_1, b_2, b_3, b_4 and Ω^2. To have a non-trivial solution, the determinant $\det(A)$ of the system must equal zero

$$\det(A) = \begin{vmatrix} 0 & 0 & 0 & 0 & 840 & -a_0 \\ 0 & 0 & 0 & 1 & 0 & 0 \\ 0 & 0 & 1 & 0 & 0 & 0 \\ 0 & 24 & 0 & 0 & -24 & a_0 \\ 1 & 0 & 0 & -1 & 0 & 0 \\ 0 & 0 & -40 & 0 & 0 & -9 \end{vmatrix} = -181{,}440a_0 = 0 \tag{7.190}$$

Thus, we demand that a_0 vanish. Yet, $a_0 = 0$ corresponds to the beam with zero density. We conclude that the case $n = 1$ does not result in an acceptable solution.

In Section 7.3.9, the remaining cases $n > 1$ are treated. It turns out that only trivial solutions are obtained. We conclude that, for the beam with a uniform density $m = 0$, the only non-trivial solution is obtained when the mode shape is represented by a fourth-order polynomial.

7.3.4 Linearly Varying Inertial Coefficient ($m = 1$)

We now treat the case of a beam with a linearly varying inertial coefficient $R(\xi) = a_0 + a_1\xi$. The governing equation, for $n = 0$, yields the following equations:

$$24(b_0 - b_2) - 5\Omega^2 a_0 = 0$$
$$72(b_1 - b_3) - 5\Omega^2 a_1 = 0$$
$$144(b_2 - b_4) + 6\Omega^2 a_0 = 0$$
$$240(b_3 - b_5) + 6\Omega^2 a_1 = 0 \qquad (7.191)$$
$$360 b_4 - \Omega^2 a_0 = 0$$
$$504 b_5 - \Omega^2 a_1 = 0$$

We get six equations with seven unknowns $\alpha^T = [b_0, b_1, b_2, b_3, b_4, b_5, \Omega^2]$. This requires that unknowns have to be expressed in terms of an arbitrary constant, taken here to be b_5. Through it, the coefficients of $D(\xi)$ and Ω^2 are obtained as follows

$$b_0 = 427 a_0 b_5 / 5 a_1$$
$$b_1 = 117 b_5 / 5$$
$$b_2 = -98 a_0 b_5 / 5 a_1$$
$$b_3 = -58 b_5 / 5 \qquad (7.192)$$
$$b_4 = 7 a_0 b_5 / 5 a_1$$
$$\Omega^2 = 504 b_5 / a_1$$

The flexural rigidity $D(\xi)$ reads

$$D(\xi) = \tfrac{1}{5}(427 a_0 / a_1 + 117\xi - 98 a_0 \xi^2 / a_1 - 58\xi^3 + 7 a_0 \xi^4 / a_1 + 5\xi^5) b_5 \quad (7.193)$$

The case $n = 1$ results in a system of seven equations with the seven unknowns $\alpha^T = [b_0, b_1, b_2, b_3, b_4, b_5, \Omega^2]$. In order to obtain a non-trivial solution, the determinant $\det(A)$ of the system must vanish

$$\det(A) = \begin{vmatrix} 0 & 0 & -40 & 0 & 0 & 0 & -9a_0 \\ 120 & 0 & 0 & -120 & 0 & 0 & -9a_1 \\ 0 & 240 & 0 & 0 & -240 & 0 & 10a_0 \\ 0 & 0 & 400 & 0 & 0 & -400 & 10a_1 \\ 0 & 0 & 0 & 1 & 0 & 0 & 0 \\ 0 & 0 & 0 & 0 & 840 & 0 & -a_0 \\ 0 & 0 & 0 & 0 & 0 & 1120 & -a_1 \end{vmatrix}$$

$$= 10,450,944(a_1 - 28a_0/3) \times 10^6 = 0 \qquad (7.194)$$

yielding a requirement $a_1 = 28a_0/3$. Once a_1 is set at this specific value, we can solve the system by taking b_5 as a parameter. We get

$$\Omega^2 = 1120 b_5 / a_0 \qquad (7.195)$$

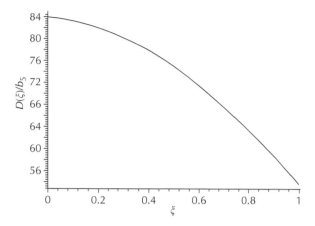

FIGURE 7.11
Variation of $D(\xi)/b_5$ for the beam of linearly varying mass density ($m = 1, n = 1$)

The functions $R(\xi)$ and $D(\xi)$ read, respectively,

$$R(\xi) = a_0(1 + 28\xi/3)$$
$$D(\xi) = (84 - 34\xi/7 - 27\xi^2 + \xi^4/7 + \xi^5)b_5 \qquad (7.196)$$

Figure 7.11 depicts the ratio $D(\xi)/b_5$, which is positive within the interval $[0;1]$. For the linearly varying inertial coefficients, the two cases $n = 0$ and $n = 1$ are the only ones that yield non-trivial solutions. For the case $n > 1$, one can consult the Section 7.3.9.

7.3.5 Parabolically Varying Inertial Coefficient ($m = 2$)

The inertial coefficient is represented by

$$R(\xi) = a_0 + a_1\xi + a_2\xi^2 \qquad (7.197)$$

We look for coefficients b_i of the flexural rigidity $D(\xi) = \sum_{i=0}^{6} b_i\xi^i$ and the natural frequency coefficient squared, Ω^2. In this case, for $n = 0$, Eq. (7.182) results in the following set of equations

$$
\begin{aligned}
24(b_0 - b_2) - 5\Omega^2 a_0 &= 0 \\
72(b_1 - b_3) - 5\Omega^2 a_1 &= 0 \\
144(b_2 - b_4) + \Omega^2(6a_0 - 5a_2) &= 0 \\
240(b_3 - b_5) + 6\Omega^2 a_1 &= 0 \\
360(b_4 - b_6) + \Omega^2(6a_2 - a_0) &= 0 \\
504b_5 - \Omega^2 a_1 &= 0 \\
672b_6 - \Omega^2 a_2 &= 0
\end{aligned}
\qquad (7.198)
$$

These are seven equations with eight unknowns $\alpha^T = [b_0, b_1, b_2, b_3, b_4, b_5, b_6, \Omega^2]$. We express all unknowns in terms of an arbitrary constant, b_6, for example. We obtain

$$b_0 = (1708a_0 + 197a_2)b_6/15a_2 \quad b_1 = 156a_1b_6/5a_2$$

$$b_2 = -(392a_0 - 197a_2)b_6/15a_2 \quad b_3 = -232a_1b_6/15r_2$$

$$b_4 = (28a_0 - 153a_2)b_6/15a_2 \quad b_5 = 4a_1b_6/3a_2 \tag{7.199}$$

$$\Omega^2 = 672b_6/a_2$$

For $n = 1$, we get from Eq. (7.182) a set of eight equations with eight unknowns $\alpha^T = [b_0, b_1, b_2, b_3, b_4, b_5, b_6, \Omega^2]$

$$A\alpha = \begin{bmatrix} 0 & 0 & -40 & 0 & 0 & 0 & 0 & -9a_0 \\ 120 & 0 & 0 & -120 & 0 & 0 & 0 & -9a_1 \\ 0 & 240 & 0 & 0 & -240 & 0 & 0 & 10a_0 - 9a_2 \\ 0 & 0 & 400 & 0 & 0 & -400 & 0 & 10a_1 \\ 0 & 0 & 0 & 600 & 0 & 0 & -600 & 10a_2 \\ 0 & 0 & 0 & 0 & 840 & 0 & 0 & -a_0 \\ 0 & 0 & 0 & 0 & 0 & 1120 & 0 & -a_1 \\ 0 & 0 & 0 & 0 & 0 & 0 & 1440 & -a_2 \end{bmatrix} \begin{bmatrix} b_0 \\ b_1 \\ b_2 \\ b_3 \\ b_4 \\ b_5 \\ b_6 \\ \Omega^2 \end{bmatrix} = \begin{bmatrix} 0 \\ 0 \\ 0 \\ 0 \\ 0 \\ 0 \\ 0 \\ 0 \end{bmatrix}$$
$$\tag{7.200}$$

The determinant, $\det(A)$, of the system must be zero to obtain a non-trivial solution:

$$\det(A) = 9,029,615,619 \times 10^9(-a_1 + 28a_0/3) \tag{7.201}$$

Equation (7.201) leads to $a_1 = 28a_0/3$. When this requirement is satisfied, the above set of equations yields

$$b_0 = (1008a_0/a_2 - 23)b_6$$

$$b_1 = (54 - 408a_0/7a_2)b_6$$

$$b_2 = -324a_0b_6/a_2$$

$$b_3 = -23b_6 \tag{7.202}$$

$$b_4 = 12a_0b_6/7a_2$$

$$b_5 = 12a_0b_6/a_2$$

$$\Omega^2 = 1440b_6/a_2$$

Other cases are treated in Section 7.3.9. For specific cases $n = 3$ or 4, only trivial solutions are reported. It is shown that for the general case $n > 4$ too, a trivial solution can be derived. To summarize, for the parabolically varying inertial coefficient, we obtain a closed-form solution for the natural frequency when the mode shape of the beam is taken as a fourth, fifth- or sixth-degree polynomial function.

7.3.6 Cubically Varying Inertial Coefficient ($m = 3$)

If the inertial coefficient is a third-order polynomial function $R(\xi) = a_0 + a_1\xi + a_2\xi^2 + a_3\xi^3$, Eq. (7.184) forces the flexural rigidity to be a seventh-order polynomial. We begin with the case $n = 0$. Equation (7.182) leads to a system of eight equations with nine unknowns $\alpha^T = [b_0, b_1, b_2, b_3, b_4, b_5, b_6, b_7, \Omega^2]$. In order to solve it, we have to take one unknown, namely b_7 to be an arbitrary constant. Thus, we are able to solve the following system $A\alpha = 0$ with

$$
A = \begin{bmatrix}
24 & 0 & -24 & 0 & 0 & 0 & 0 & 0 & -5a_0 \\
0 & 72 & 0 & -72 & 0 & 0 & 0 & 0 & -5a_1 \\
0 & 0 & 144 & 0 & -144 & 0 & 0 & 0 & -5a_2 \\
0 & 0 & 0 & 240 & 0 & -240 & 0 & 0 & 6a_1 - 5a_3 \\
0 & 0 & 0 & 0 & 360 & 0 & -360 & 0 & 6a_2 - a_0 \\
0 & 0 & 0 & 0 & 0 & 504 & 0 & -504 & 6a_3 - a_1 \\
0 & 0 & 0 & 0 & 0 & 0 & 672 & 0 & -a_2 \\
0 & 0 & 0 & 0 & 0 & 0 & 0 & 864 & -a_3
\end{bmatrix}
$$

$$(7.203)$$

We derive the following coefficients b_i and Ω^2, respectively:

$$
\begin{aligned}
b_0 &= 3(1708a_o + 197a_2)b_7/35a_3 \\
b_1 &= (1404a_1 + 305a_3)b_7/35a_3 \\
b_2 &= -3(392a_0 - 197a_2)b_7/35a_3 \\
b_3 &= (-696a_1 + 305a_3)b_7/35a_3 \\
b_4 &= 3(28a_0 - 153a_2)b_7/35a_3 \\
b_5 &= (12a_1 - 65a_3)b_7/7a_3 \\
b_6 &= 9a_2b_7/7a_3 \\
\Omega^2 &= 864b_7/a_3
\end{aligned}
$$

$$(7.204)$$

Equation (7.182), with $n = 1$ and the above expressions for the flexural rigidity and the inertial coefficients, results in a square system of nine equations with nine unknowns $b_0, b_1, b_2, b_3, b_4, b_5, b_6, b_7$ and Ω^2

$$
A\alpha = \begin{bmatrix}
0 & 0 & -40 & 0 & 0 & 0 & 0 & 0 & -9a_0 \\
120 & 0 & 0 & -120 & 0 & 0 & 0 & 0 & -9a_1 \\
0 & 240 & 0 & 0 & -240 & 0 & 0 & 0 & 10a_0 \\
0 & 0 & 400 & 0 & 0 & -400 & 0 & 0 & 10a_1 - 9a_3 \\
0 & 0 & 0 & 600 & 0 & 0 & 600 & 0 & 10a_2 \\
0 & 0 & 0 & 0 & 840 & 0 & 0 & -840 & -a_0 + 10a_3 \\
0 & 0 & 0 & 0 & 0 & 1120 & 0 & 0 & -a_1 \\
0 & 0 & 0 & 0 & 0 & 0 & 1440 & 0 & -a_2 \\
0 & 0 & 0 & 0 & 0 & 0 & 0 & 1800 & -a_3
\end{bmatrix}
\begin{bmatrix}
b_0 \\
b_1 \\
b_2 \\
b_3 \\
b_4 \\
b_5 \\
b_6 \\
b_7 \\
\Omega^2
\end{bmatrix}
=
\begin{bmatrix}
0 \\
0 \\
0 \\
0 \\
0 \\
0 \\
0 \\
0 \\
0
\end{bmatrix}
$$

$$(7.205)$$

The determinant must vanish for the non-triviality of the solution

$$\det(A) = 1{,}516{,}975{,}423{,}488 \times 10^{10}(15a_1/14 - a_3 - 10a_0) = 0 \qquad (7.206)$$

which is satisfied when $a_3 = 15a_1/14 - 10a_0$. For this case, a non-trivial solution is expressed in terms of an arbitrary constant b_7

$$
\begin{aligned}
b_0 &= -7(108a_1 - 23a_2)b_7/2(28a_0 - 3a_1) \\
b_1 &= -(2576a_0 - 429a_1 + 1323a_2)b_7/7(28a_0 - 3a_1) \\
b_2 &= 1134a_0 b_7/(28a_0 - 3a_1) \\
b_3 &= 161a_2 b_7/2(28a_0 - 3a_1) \\
b_4 &= -(4046a_0 - 429a_1)b_7/7(28a_0 - 3a_1) \\
b_5 &= -9a_1 b_7/2(28a_0 - 3a_1) \\
b_6 &= -7a_2 b_7/2(28a_0 - 3a_1) \\
\Omega^2 &= 1800b_7/a_3
\end{aligned}
\qquad (7.207)
$$

Section 8.3.9 treats the other cases for which $n > 1$. For the cubically varying inertial coefficient, we obtain a closed-form solution for the natural frequency when a fourth-, a fifth-, sixth- or seventh-degree polynomial represents the mode shape.

7.3.7 Coefficient Represented by a Quartic Polynomial ($m = 4$)

The inertial coefficient equals $R(\xi) = a_0 + a_1\xi + a_2\xi^2 + a_3\xi^3 + a_4\xi^4$ and, consequently, the flexural rigidity is given by $D(\xi) = \sum_{i=0}^{8} b_i\xi^i$. We begin with the case $n = 0$. Equation (7.182) results in a set of nine equations $A\alpha = 0$ with ten unknowns $\alpha^T = [b_0, b_1, b_2, b_3, b_4, b_5, b_6, b_7, b_8, \Omega^2]$:

$$
A =
\begin{bmatrix}
24 & 0 & -24 & 0 & 0 & 0 & 0 & 0 & 0 & -5a_0 \\
0 & 72 & 0 & -72 & 0 & 0 & 0 & 0 & 0 & -5a_1 \\
0 & 0 & 144 & 0 & -144 & 0 & 0 & 0 & 0 & 6a_0 - 5a_2 \\
0 & 0 & 0 & 240 & 0 & -240 & 0 & 0 & 0 & 6a_1 - 5a_3 \\
0 & 0 & 0 & 0 & 360 & 0 & -360 & 0 & 0 & -a_0 + 6a_2 - 5a_4 \\
0 & 0 & 0 & 0 & 0 & 504 & 0 & -504 & 0 & 6a_3 - a_1 \\
0 & 0 & 0 & 0 & 0 & 0 & 672 & 0 & -672 & 6a_4 - a_2 \\
0 & 0 & 0 & 0 & 0 & 0 & 0 & 864 & 0 & -a_3 \\
0 & 0 & 0 & 0 & 0 & 0 & 0 & 0 & 1080 & -a_4
\end{bmatrix}
$$

$$(7.208)$$

We solve this system with b_8 taken as a parameter. The coefficients of $D(\xi)$ and Ω^2 are given by

$$
\begin{aligned}
b_0 &= (5124a_o + 591a_2 + 178a_4)b_8/28a_4 \\
b_1 &= (1404a_1 + 305a_3)b_8/28a_4 \\
b_2 &= -(1176a_o - 591a_2 - 178a_4)b_8/28a_4 \\
b_3 &= -(696a_1 - 305a_3)b_8/28a_4 \\
b_4 &= (84a_0 - 459a_2 + 178a_4)b_8/28a_4 \\
b_5 &= 5(12a_1 - 65a_3)b_8/28a_4 \\
b_6 &= (45a_2 - 242a_4)b_8/28a_4 \\
b_7 &= 5a_3b_8/4a_4 \\
\Omega^2 &= 1080b_8/a_3
\end{aligned}
\tag{7.209}
$$

Consider the second case, $n = 1$. Equation (7.182) results in a set of ten equations with ten unknowns $\alpha^T = [b_0, b_1, b_2, b_3, b_4, b_5, b_6, b_7, b_8, \Omega^2]$. In order to obtain a non-trivial solution, the determinant of this system must vanish

$$
\det(A) =
\begin{vmatrix}
0 & 0 & -40 & 0 & 0 & 0 & 0 & 0 & 0 & -9a_0 \\
120 & 0 & 0 & -120 & 0 & 0 & 0 & 0 & 0 & -9a_1 \\
0 & 240 & 0 & 0 & -240 & 0 & 0 & 0 & 0 & 10a_0 - 9a_2 \\
0 & 0 & 400 & 0 & 0 & -400 & 0 & 0 & 0 & 10a_1 - 9a_3 \\
0 & 0 & 0 & 600 & 0 & 0 & -600 & 0 & 0 & 10a_2 - 9a_4 \\
0 & 0 & 0 & 0 & 840 & 0 & 0 & -840 & 0 & 10a_3 - a_0 \\
0 & 0 & 0 & 0 & 0 & 1120 & 0 & 0 & -1120 & 10a_4 - a_1 \\
0 & 0 & 0 & 0 & 0 & 0 & 1440 & 0 & 0 & -a_2 \\
0 & 0 & 0 & 0 & 0 & 0 & 0 & 1800 & 0 & -a_3 \\
0 & 0 & 0 & 0 & 0 & 0 & 0 & 0 & 2200 & -a_4
\end{vmatrix}
$$

$$
= -12{,}569{,}224{,}937{,}472 \times 10^{12} \left(-\tfrac{770}{29}a_0 + \tfrac{165}{58}a_1 - \tfrac{77}{29}a_3 + a_4 \right) = 0
\tag{7.210}
$$

The above relation yields $a_4 = \tfrac{770}{29}a_0 - \tfrac{165}{58}a_1 + \tfrac{77}{29}a_3$. For this case, we express all the unknowns in terms of an arbitrary constant chosen to be b_8

$$
b_0 = (83{,}160a_0 + 6{,}750a_1 - 3{,}335a_2 + 8{,}316a_3)b_8/18(140a_0 - 15a_1 + 14a_3)
$$

$$
b_1 = -29(1{,}020a_0 - 945a_2 + 286a_3)b_8/63(140a_0 - 15a_1 + 14a_3)
$$

$$
b_2 = -2{,}610a_0b_8/(140a_0 - 15a_1 + 14a_3)
$$

$$
b_3 = (83{,}160a_0 - 8{,}910a_1 - 3{,}335a_2 + 8{,}316a_3)b_8/18(140a_0 - 15a_1 + 14a_3)
$$

$$
b_4 = 58(15a_0 - 143a_3)b_8/63(140a_0 - 15a_1 + 14a_3)
$$

$$
b_5 = -29(90a_0 - 10a_1 + 9a_3)b_8/(140a_0 - 15a_1 + 14a_3)
$$

$$
b_6 = 145a_2b_8/18(140a_0 - 15a_1 + 14a_3)
$$

$$
b_7 = 58a_3b_8/9(140a_0 - 15a_1 + 14a_3)
$$

$$
\Omega^2 = 2{,}200b_8/a_4
$$

$$
\tag{7.211}
$$

Section 8.3.9 deals with cases $m = 4$ and $n > 1$. For the quartically varying inertial coefficient, one can represent the mode shape by a fourth-, fifth-, sixth-, seventh- or eighth-degree order polynomial to obtain a non-trivial solution.

Now, we treat the general case when m is an arbitrary constant.

7.3.8 General Case

In the above sections, we dealt with the case when m was fixed. Obviously, the study with particular cases where m is fixed could be carried on with higher values for m. The natural question arises: can we calculate coefficients b_i, for $i = 0, \ldots, m + 4$ when m is not specified in advance? This section presents a positive reply to this question. We explain, in the following, for which cases a non-trivial solution is derived and, moreover, what the general closed-form solution for the natural frequency is.

The first term in Eq. (7.182) becomes

$$
\frac{d^2}{d\xi^2}\left[D(\xi)\frac{d^2\psi}{d\xi^2}\right] = -\sum_{i=0}^{m+2}(n^2 + 7n + 12)(i + 1)(i + 2)b_{i+2}\xi^i
$$

$$
+ \sum_{i=0}^{m+4}(n + 4)(n + 3)(n + 2)(n + 1)b_i\xi^{i+n}
$$

$$
+ 2\sum_{i=1}^{m+4}(n + 4)(n + 3)(n + 2)ib_i\xi^{i+n}
$$

$$
+ \sum_{i=2}^{m+4}(n + 4)(n + 3)i(i - 1)b_i\xi^{i+n} \tag{7.212}
$$

The second term in Eq. (7.182) reads

$$
-\Omega^2 R(\xi)\psi(\xi) = -\Omega^2\sum_{i=0}^{m}\tfrac{1}{2}(n^2 + 7n + 10)a_i\xi^i
$$

$$
+ \Omega^2\sum_{i=0}^{m}\tfrac{1}{2}(n^2 + 7n + 12)a_i\xi^{i+2} - \Omega^2\sum_{i=0}^{m}a_i\xi^{i+4+n} \tag{7.213}
$$

Substituting these expressions into Eq. (7.182) we get a polynomial function that equals zero. Thus, to satisfy this equality, each coefficient of ξ^i must vanish. In order for the above expressions to be readable, we explain how one can rearrange the terms. Equation (7.121) contains four terms

$$\frac{d^2\left(D(\xi)d^2\psi/d\xi^2\right)}{d\xi^2} - -\sum_{i=0}^{m+2}\beta_i\xi^i + \sum_{i=0}^{m+4}\chi_i\xi^{i+n} + \sum_{i=1}^{m+4}\delta_i\xi^{i+n} \mid \sum_{i=2}^{m+4}\varepsilon_i\xi^{i+n}$$

(7.214)

or, in another form,

$$\frac{d^2\left(D(\xi)d^2\psi/d\xi^2\right)}{d\xi^2} = -\sum_{i=0}^{m+2}\beta_i\xi^i + \sum_{i=n}^{n+m+4}\chi_{i-n}\xi^i$$

$$+ \sum_{i=n+1}^{n+m+4}\delta_{i-n}\xi^i + \sum_{i=n+2}^{n+m+4}\varepsilon_{i-n}\xi^i$$

(7.215)

One can check, in accordance with the previous formula, that the polynomial function in Eq. (7.215) can be rewritten as follows

$$\frac{d^2\left(D(\xi)d^2\psi/d\xi^2\right)}{d\xi^2} = -(\beta_0 + \beta_1\xi + \cdots + \beta_{m+1}\xi^{m+1} + \beta_{m+2}\xi^{m+2})$$

$$+ \chi_0\xi^n + (\chi_1 + \delta_1)\xi^{n+1} + (\chi_2 + \delta_2 + \varepsilon_2)\xi^{n+2} + \cdots$$

$$+ (\chi_i + \delta_i + \varepsilon_i)\xi^{n+i} + \cdots$$

$$+ (\chi_{m+4} + \delta_{m+4} + \varepsilon_{m+4})\xi^{n+m+4}$$

(7.216)

We have a coefficient, namely $\beta_i = (n^2 + 7n + 12)(i + 1)(i + 2)b_{i+2}$, for ξ^i for $i = 0, \ldots, m + 2$. For $i = n, \ldots, n + m + 4$, we have other coefficients: $\chi_i = (n + 4)(n + 3)(n + 2)(n + 1)b_i$, $\delta_i = (n + 4)(n + 3)(n + 2)_i b_i$ and $\varepsilon_i = (n + 4)(n + 3)i(i - 1)b_i$. But for $i = m + 3, \ldots, n - 1$ there are no coefficients, or, in other words, the coefficients are zero. Thus, two cases result directly from Eq. (7.216): (1) the first is when the order n of the polynomial function that represents the mode shape is greater than $m + 2$. Equation (7.216) can, then, be rewritten as $\sum_{i=0}^{m+2} K_i\xi^i + \sum_{i=n}^{m+n+4} L_{i-n}\xi^i$. One could check that this case results in a system of $m + 3$ equations (for each coefficient K_i) plus $m + 5$ equations (corresponding to the coefficients L_i). We thus have a total of $2m + 8$ equations; (2) the second case pertains to $n \leq m + 2$; Eq. (7.216) is equivalent to $\sum_{i=0}^{n+m+4} M_i\xi^i$ and we obtain directly $n + m + 5$ equations.

Let us first assume that the inequality $n > m + 2$ holds. From Eq. (7.213), we observe that we have one coefficient of ξ^i for $i = 0, \ldots, m + 2$ and for $i = n + 4, \ldots, n + m + 4$. Hence, for $i = n, n + 1, n + 2, n + 3$, the coefficients of

b_i are zero. Finally, Eq. (7.182) is given by

$$\frac{d^2(D(\xi)d^2\psi/d\xi^2)}{d\xi^2} - \Omega^2 R(\xi)\psi(\xi)$$

$$= -\left((\beta_0 + \Omega^2 \tfrac{1}{2}(n^2 + 7n + 10)a_0) - \left(\beta_1 + \Omega^2 \tfrac{1}{2}(n^2 + 7n + 10)a_1\right)\xi\right.$$

$$- \left\{\beta_2 + \Omega^2 \left[\tfrac{1}{2}(n^2 + 7n + 10)a_2 - \tfrac{1}{2}(n^2 + 7n + 12)a_0\right]\right\}\xi^2$$

$$- \cdots$$

$$- \left\{\beta_i + \Omega^2 \left[\tfrac{1}{2}(n^2 + 7n + 10)a_i - \tfrac{1}{2}(n^2 + 7n + 12)a_{i-2}\right]\right\}\xi^i$$

$$- \cdots$$

$$- \left[\beta_{m+1} - \tfrac{1}{2}(n^2 + 7n + 12)a_{m-1}\right]\xi^{m+1}$$

$$- \left[\beta_{m+2} - \tfrac{1}{2}(n^2 + 7n + 12)a_m\right]\xi^{m+2}\right)$$

$$+ \chi_0\xi^n + (\chi_1 + \delta_1)\xi^{n+1} + (\chi_2 + \delta_2 + \varepsilon_2)\xi^{n+2}$$

$$+ (\chi_3 + \delta_3 + \varepsilon_3)\xi^{n+3}$$

$$+ (\chi_4 + \delta_4 + \varepsilon_4 - \Omega^2 a_0)\xi^{n+4}$$

$$+ \cdots$$

$$+ (\chi_i + \delta_i + \varepsilon_i - \Omega^2 a_{i-4})\xi^{n+i}$$

$$+ \cdots$$

$$+ (\chi_{m+4} + \delta_{m+4} + \varepsilon_{m+4} - \Omega^2 a_m)\xi^{n+m+4} \tag{7.217}$$

If the above expression has to be satisfied for every ξ within the interval $[0; 1]$, we have to demand that each coefficient of ξ^i for $i = 0, \ldots, m+2, n, \ldots, n+m+4$ must vanish. Thus, we obtain the following equations:

from ξ^0: $(n^2 + 7n + 12)2b_2 + \Omega^2 \tfrac{1}{2}(n^2 + 7n + 10)a_0 = 0$ $\qquad$ (7.218)

from ξ^1: $(n^2 + 7n + 12)6b_3 + \Omega^2 \tfrac{1}{2}(n^2 + 7n + 10)a_1 = 0$ $\qquad$ (7.219)

from ξ^2: $(n^2 + 7n + 12)12b_4 + \Omega^2 \left[\tfrac{1}{2}(n^2 + 7n + 10)a_2 - \tfrac{1}{2}(n^2 + 7n + 12)a_0\right] = 0$
$$\tag{7.220}$$

$$\cdots$$

from ξ^l: $(n^2 + 7n + 12)(i + 2)(i + 1)b_{i+2}$

$$+ \Omega^2 \left[\tfrac{1}{2}(n^2 + 7n + 10)a_i - \tfrac{1}{2}(n^2 + 7n + 12)a_{i-2}\right] = 0 \tag{7.221}$$

$$\cdots$$

from ξ^{m+1}: $-(n^2 + 7n + 12)(m + 3)(m + 2)b_{m+3} + \Omega^2 \tfrac{1}{2}(n^2 + 7n + 12)a_{m-1} = 0$
$$\tag{7.222}$$

from ξ^{m+2}: $-(n^2 + 7n + 12)(m + 4)(m + 3)b_{m+4} + \Omega^2 \tfrac{1}{2}(n^2 + 7n + 12)a_m = 0$
$$\tag{7.223}$$

fromξ^n: $b_0 = 0$ (7.224)

fromξ^{n+1}: $b_1 = 0$ (7.225)

from ξ^{n+2}: $b_2 = 0$ (7.226)

from ξ^{n+3}: $b_3 = 0$ (7.227)

fromξ^{n+4}: $(n+6)(n+5)(n+4)(n+3)b_4 - \Omega^2 a_0 = 0$ (7.228)

$$\cdots$$

from ξ^{i+n}: $(i+2+n)(i+1+n)(n+4)(n+3)b_i - \Omega^2 a_{i-4} = 0$ (7.229)

$$\cdots$$

fromξ^{m+n+4}: $(m+n+6)(m+n+5)(n+4)(n+3)b_{m+4} - \Omega^2 a_m = 0$ (7.230)

We get, in this case, two sets of equations: the first set given in Eqs. (7.218)–(7.223) constitutes $m+3$ equations; the second set, namely Eqs. (7.224)–(7.230), contains $m+5$ equations. The above system has $2m+8$ equations in total with $m+6$ unknowns, namely b_i, for $i = 0,\ldots,m+4$ and Ω^2. Let us check if this system can yield a non-trivial solution. From Eqs. (7.218) and (7.219), knowing $b_2 = b_3 = 0$ from Eqs. (7.226) and (7.227), we get

$$ar_0 = 0 \tag{7.231}$$

$$ar_1 = 0 \tag{7.232}$$

The substitution of Eqs. (7.231) and (7.232) into Eqs. (7.228) and (7.229) with $i = 5$ results in

$$b_4 = 0 \tag{7.233}$$

$$b_5 = 0 \tag{7.234}$$

From Eq. (7.220) we obtain

$$r_2 = 0 \tag{7.235}$$

Then, we rewrite Eqs. (7.221) and (7.229) in either of the following forms:

$$Ab_{i+2} + Ba_i + Ca_{i-2} = 0 \tag{7.236}$$

with three non-zero coefficients A, B and C, for $i = 3,\ldots,m$,

$$Db_i + Ea_{i-4} = 0 \tag{7.237}$$

with two non-zero coefficients D and E, for $i = 5,\ldots,m+3$.
   Equation (7.237) for $i = 6$ leads to

$$b_6 = 0 \tag{7.238}$$

Equation (7.236), for $i = 3$ results in

$$a_3 = 0 \qquad (7.239)$$

Now, we use the recursivity of Eqs. (7.237) and (7.236). Equations (7.238) and (7.239) are used here to initialize the following recursive equation: Eq. (7.237) for $6 < j < m + 1$, bearing in mind $a_{j-4} = 0$, leads to

$$b_j = 0 \qquad (7.240)$$

and Eq. (7.236), for $3 < i < m$, bearing in mind $b_{i+2} = 0$ and $a_{i-2} = 0$, yields

$$a_i = 0 \qquad (7.241)$$

We employ Eq. (7.237) for $i = m + 2$ and Eq. (7.236) for $i = m$:

$$b_{m+2} = 0 \quad a_m = 0 \qquad (7.242)$$

We, thus, demonstrated that the coefficient a_m is zero. This result is not compatible with our hypothesis of an mth order polynomial representing the inertial coefficient. Thus, we arrive at the conclusion that the case $n > m + 2$ yields a trivial solution.

The second case is $n = m + 2$. When we take an $(m + 2)$th order polynomial for the mode shape, and get

$$\frac{d^2(D(\xi)d^2\psi/d\xi^2)}{d\xi^2} = -(\beta_0 + \beta_1\xi + \cdots + \beta_{m+1}\xi^{m+1} + (\beta_{m+2} - \chi_0)\xi^{m+2})$$

$$+ (\chi_1 + \delta_1)\xi^{n+1} + (\chi_2 + \delta_2 + \varepsilon_2)\xi^{n+2}$$

$$+ \cdots + (\chi_i + \delta_i + \varepsilon_i)\xi^{n+i} + \cdots + (\chi_{m+4} + \delta_{m+4} + \varepsilon_{m+4})\xi^{n+m+4}$$

$$(7.243)$$

The natural question arises: What is the difference from the previous case? Equations (7.218)–(7.224) are the same in this case since we have the same coefficients in Eq. (7.216) and Eq. (7.243) for $i = 0, \ldots, m + 1$. Likewise, Eqs. (7.225)–(7.230) are identical in the both cases. We observe that Eq. (7.243) has a different coefficient for of ξ^{m+2} than Eq. (7.216). The consequence is that Eqs. (7.223) and (7.224) become the following single equation:

$$- (n^2 + 7n + 12)(m + 3)(m + 4)b_{m+4} + (n + 4)(n + 3)(n + 2)(n + 1)b_0$$

$$+ \tfrac{1}{2}\Omega^2(n^2 + 7n + 12)a_m = 0 \qquad (7.244)$$

Finally, the case $n = m + 2$ results in the following system of $n + m + 5$ equations:

from ξ^0: $\quad (n^2 + 7n + 12)2b_2 + \Omega^2 \frac{1}{2}(n^2 + 7n + 10)a_0 - 0$ $\hfill (7.245)$

from ξ^1: $\quad (n^2 + 7n + 12)6b_3 + \Omega^2 \frac{1}{2}(n^2 + 7n + 10)a_1 = 0$ $\hfill (7.246)$

from ξ^2: $\quad (n^2 + 7n + 12)12b_4 + \Omega^2 \left[\frac{1}{2}(n^2 + 7n + 10)a_2 - \frac{1}{2}(n^2 + 7n + 12)a_0 \right] = 0$ $\hfill (7.247)$

$$\cdots$$

from ξ^i: $\quad (n^2 + 7n + 12)(i+2)(i+1)b_{i+2}$
$$+ \Omega^2 \left[\frac{1}{2}(n^2 + 7n + 10)a_i - \frac{1}{2}(n^2 + 7n + 12)a_{i-2} \right] = 0 \hfill (7.248)$$

$$\cdots$$

from ξ^{m+1}: $\quad -(n^2 + 7n + 12)(m+3)(m+2)b_{m+3} + \Omega^2 \frac{1}{2}(n^2 + 7n + 12)a_{m-1} = 0$ $\hfill (7.249)$

from ξ^{m+2}: $\quad -(n^2 + 7n + 12)(m+4)(m+3)b_{m+4} + (n+4)(n+3)(n+2)(n+1)b_0$
$$+ \Omega^2 \frac{1}{2}(n^2 + 7n + 12)a_m = 0 \hfill (7.250)$$

from ξ^{n+1}: $\quad b_1 = 0$ $\hfill (7.251)$

from ξ^{n+2}: $\quad b_2 = 0$ $\hfill (7.252)$

from ξ^{n+3}: $\quad b_3 = 0$ $\hfill (7.253)$

from ξ^{n+4}: $\quad (n+6)(n+5)(n+4)(n+3)b_4 - \Omega^2 a_0 = 0$ $\hfill (7.254)$

$$\cdots$$

from ξ^{n+i}: $\quad (i+2+n)(i+1+n)(n+4)(n+3)b_i - \Omega^2 a_{i-4} = 0$ $\hfill (7.255)$

$$\cdots$$

from ξ^{n+m+4}: $\quad (m+n+6)(m+n+5)(n+4)(n+3)b_{m+4} - \Omega^2 a_m = 0$ $\hfill (7.256)$

One can show that, in this case, one arrives at a trivial solution. To do this we first observe that Eqs. (7.251)–(7.253) lead to

$$b_1 = 0 \quad b_2 = 0 \quad b_3 = 0 \hfill (7.257)$$

We repeat the same procedure as in the case $n > m + 2$. The recursive Eqs. (7.248) and (7.255) allow us to demonstrate that

$$a_i = 0 \qquad \text{for } 0 \le i \le m \hfill (7.258)$$

This is a trivial solution. Thus, a non-trivial, closed-form solution for the natural frequency is not obtained if one takes the mode shape as a polynomial of degree $m + 2$.

The third case to be considered is $n = m + 1$. The equations obtained are as follows:

from ξ^0: $(n^2 + 7n + 12)2b_2 + \Omega^2 \frac{1}{2}(n^2 + 7n + 10)a_0 = 0$ (7.259)

from ξ^1: $(n^2 + 7n + 12)6b_3 + \Omega^2 \frac{1}{2}(n^2 + 7n + 10)a_1 = 0$ (7.260)

from ξ^2: $(n^2 + 7n + 12)12b_4 + \Omega^2 \left[\frac{1}{2}(n^2 + 7n + 10)a_2 \right.$

$$\left. - \frac{1}{2}(n^2 + 7n + 12)a_0 \right] = 0 \tag{7.261}$$

$$\cdots$$

from ξ^i: $(n^2 + 7n + 12)(i + 2)(i + 1)b_{i+2} + \Omega^2 \left[\frac{1}{2}(n^2 + 7n + 10)a_i \right.$

$$\left. - \frac{1}{2}(n^2 + 7n + 12)a_{i-2} \right] = 0 \tag{7.262}$$

$$\cdots$$

from ξ^{m+1}: $- (n^2 + 7n + 12)(m + 3)(m + 2)b_{m+3} + (n + 4)(n + 3)(n + 2)(n + 1)b_0$

$$+ \Omega^2 \frac{1}{2}(n^2 + 7n + 12)a_{m-1} = 0 \tag{7.263}$$

from ξ^{m+2}: $- (n^2 + 7n + 12)(m + 4)(m + 3)b_{m+4} + (n + 4)(n + 3)(n + 3)(n + 2)b_1$

$$+ \Omega^2 \frac{1}{2}(n^2 + 7n + 12)a_m = 0 \tag{7.264}$$

from ξ^{m+3}: $b_2 = 0$ (7.265)

from ξ^{m+4}: $b_3 = 0$ (7.266)

from ξ^{n+4}: $(n + 6)(n + 5)(n + 4)(n + 3)b_4 - \Omega^2 a_0 = 0$ (7.267)

$$\cdots$$

from ξ^{i+n}: $(i + 2 + n)(i + 1 + n)(n + 4)(n + 3)b_i - \Omega^2 a_{i-4} = 0$ (7.268)

$$\cdots$$

from ξ^{m+n+4}: $(m + n + 6)(m + n + 5)(n + 4)(n + 3)b_{m+4} - \Omega^2 a_m = 0$ (7.269)

As in the case $n = m + 2$, Eq. (7.265) yields

$$b_2 = 0 \tag{7.270}$$

Thus, Eq. (7.259) reads

$$a_0 = 0 \tag{7.271}$$

From Eqs. (7.266) and (7.260), we obtain

$$b_3 = 0 \quad a_1 = 0 \tag{7.272}$$

Equations (7.267) and (7.261) result in

$$b_4 = 0 \quad a_2 = 0 \tag{7.273}$$

We use the property of recursivity of Eqs. (7.262) and (7.268) like in the previous cases and we obtain

$$a_i = 0 \qquad \text{for } 0 \le i \le m \tag{7.274}$$

All coefficients a_i, for $i = 0, \ldots, m$, must vanish, which contradicts the realistic situation. Thus, the case $n = m + 1$ does not allow us to obtain a non-trivial solution.

Now, we treat the particular case $n = m$. This case is different from the previous cases because we directly observe from Eq. (7.182) that only one coefficient vanishes, namely $b_3 = 0$. We obtain the following $n + m + 5$ equations with $m + 6$ unknowns, $b_0, b_1, \ldots, b_{m+4}$ and Ω^2:

from ξ^0: $(n^2 + 7n + 12)2b_2 + \Omega^2 \frac{1}{2}(n^2 + 7n + 10)a_0 = 0$ \hfill (7.275)

from ξ^1: $(n^2 + 7n + 12)6b_3 + \Omega^2 \frac{1}{2}(n^2 + 7n + 10)a_1 = 0$ \hfill (7.276)

from ξ^2: $(n^2 + 7n + 12)12b_4 + \Omega^2 \left[\frac{1}{2}(n^2 + 7n + 10)a_2 - \frac{1}{2}(n^2 + 7n + 12)a_0 \right] = 0$
\hfill (7.277)

$$\cdots$$

from ξ^i: $(n^2 + 7n + 12)(i + 2)(i + 1)b_{i+2} + \Omega^2 \left[\frac{1}{2}(n^2 + 7n + 10)a_i \right.$

$$\left. - \frac{1}{2}(n^2 + 7n + 12)a_{i-2} \right] = 0 \tag{7.278}$$

$$\cdots$$

from ξ^m: $- (n^2 + 7n + 12)(m + 2)(m + 1)b_{m+2} + (n + 4)(n + 3)(n + 2)(n + 1)b_0$

$$+ \Omega^2 \frac{1}{2}[(n^2 + 7n + 12)r_{m-2} - (n^2 + 7n + 10)a_m] = 0 \tag{7.279}$$

from ξ^{m+1}: $- (n^2 + 7n + 12)(m + 3)(m + 2)b_{m+3} + (n + 4)(n + 3)(n + 3)(n + 2)b_1$

$$+ \Omega^2 \frac{1}{2}(n^2 + 7n + 12)a_{m-1} = 0 \tag{7.280}$$

from ξ^{m+2}: $- (n^2 + 7n + 12)(m + 4)(m + 3)b_{m+4} + (n + 4)(n + 3)(n + 4)(n + 3)b_2$

$$+ \Omega^2 \frac{1}{2}(n^2 + 7n + 12)a_m = 0 \tag{7.281}$$

from ξ^{m+3}: $b_3 = 0$ \hfill (7.282)

from ξ^{m+4}: $(n + 6)(n + 5)(n + 4)(n + 3)b_4 - \Omega^2 a_0 = 0$ \hfill (7.283)

$$\cdots$$

from ξ^{m+i}: $(i + 2 + n)(i + 1 + n)(n + 4)(n + 3)b_i - \Omega^2 a_{i-4} = 0$ \hfill (7.284)

$$\cdots$$

from ξ^{2m+4}: $(m + n + 6)(m + n + 5)(n + 4)(n + 3)b_{m+4} - \Omega^2 a_m = 0$ \hfill (7.285)

Equations (7.276) and (7.282) lead to $r_1 = 0$. Consequently, Eq. (7.284) with $i = 5$ yields $b_5 = 0$.

Hence, we deduce from Eq. (7.278) with $i = 3$ that $r_3 = 0$. Equation (7.284), with $i = 7$ results in $b_7 = 0$. We note that in the above results only the coefficients b_{2i+1}, for $i = 1, 2, 3$, and r_{2i+1}, for $i = 0, 1$, equal zero. Equation (7.282) allows us to write

$$a_i = Aa_{i-2} + B_i b_{i+2} \qquad \text{for } 3 \leq i \leq m - 1$$
$$A = (n^2 + 7n + 12)/(n^2 + 7n + 10) \quad B_i = A(i + 1)(i + 2)/\Omega^2 \tag{7.286}$$

where A and B_i are two coefficients different from 0. Equation (7.284) reads

$$b_i = Ca_{i-4} \qquad \text{for } 5 \leq i \leq m + 4$$
$$C = \Omega^2/(i + 2 + n)(i + 1 + n)(n + 4)(n + 3) \tag{7.287}$$

with C a non-zero constant.

From Eq. (7.282), we get $b_3 = 0$. Then, Eq. (7.276) leads to $a_1 = 0$. Equation (7.284), evaluated for $i = 5$, results in $b_5 = 0$. The previous results will be utilized now with the recursive Eqs. (7.286) and (7.287). Equation (7.286) with $i = 3$ leads to $a_3 = 0$. Then, Eq. (7.287) with $i = 7$ results in $b_7 = 0$. One can check that the previous results, and Eqs. (7.286) and (7.287) impose the condition that b_j and a_j are zero when j is odd.

Now consider Eq. (7.286) with $i = m - 1$ and Eq. (7.287) with $i = m + 3$ and suppose m is even:

$$a_{m-1} = Aa_{m-3} + B_{m-1} b_{m+1}$$
$$b_{m+3} = Ca_{m-1} \tag{7.288}$$

As m is even, the numbers $m + 1$ and $m - 3$ are odd. Thus, Eq. (7.288) leads to $a_{m-1} = 0$ and $b_{m+3} = 0$. In this case, we do obtain a non-trivial solution because a_m does not have to be a zero constant. The second case is for odd m:

$$a_{m-2} = Aa_{m-2} + B_m b_m$$
$$b_{m+2} = Ca_{m-2} \tag{7.289}$$

Equation (7.289) leads to $a_{m-2} = 0$ and $b_{m+2} = 0$. In this case too, we cannot arrive at $r_m = 0$. We can illustrate this by the following example. If $m = 6$, the coefficients that vanish are: b_3, b_5, b_7, b_9 and a_1, a_3, a_5. If $m = 7$, the following coefficients are zero: b_3, b_5, b_7, b_9 and a_1, a_3, a_5. One can check that we never have $a_m = 0$ or $b_{m+4} = 0$, in the case $n = m$.

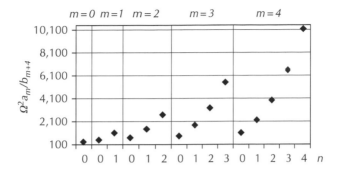

FIGURE 7.12
Variation of $\Omega^2 a_m/b_{m+4}$ for different values of m and n

The general expression for the natural frequency for $n = m$ reads

$$\omega^2 = (n + 3)(n + 4)(m + n + 5)(m + n + 6)b_{m+4}/a_m L^4 \qquad (7.290)$$

Figure 7.12 depicts the variation of $\Omega^2 a_m/b_{m+4}$ in terms of m and n.

Thus, in the general case, in order to obtain a non-trivial solution, a necessary condition is to choose a polynomial function of order $n < m + 1$ to represent the mode shape.

In the following section, we distinguish between two cases: (1) $n \le m + 2$ with the number of equations being less than $2m + 8$; (2) $n > m + 2$, resulting in $2m + 8$ equations.

7.3.9 Particular Cases Characterized by the Inequality $n \ge m + 2$

7.3.9.1 Constant inertial coefficient ($m = 0$)

This section presents the solution for the case $n \ge 2$, in which the number of equations is greater than the number of unknowns. We substitute the mode shape given by Eq. (7.180) into Eq. (7.182). The first term of Eq. (7.182) reads

$$\frac{d^2}{d\xi^2}\left[D(\xi)\frac{d^2\psi}{d\xi^2}\right] = (n + 3)(n + 4)\left[(n + 5)(n + 6)b_4\xi^{n+4} + (n + 4)(n + 5)b_3\xi^{n+3}\right.$$

$$+ (n + 3)(n + 4)b_2\xi^{n+2} + (n + 2)(n + 3)b_1\xi^{n+1}$$

$$\left. + (n + 1)(n + 2)b_0\xi^n - 12b_4\xi^2 - 6b_3\xi - 2b_2\right] \qquad (7.291)$$

The second term of Eq. (7.180) becomes

$$-\Omega^2 R(\xi)\psi(\xi) = -\Omega^2 a_0 \left[\xi^{n+4} - \tfrac{1}{2}(n^2 + 7n + 12)\xi^2 + \tfrac{1}{2}(n^2 + 7n + 10)\right]$$
(7.292)

The sum of Eqs. (7.291) and Eq. (7.292) is a polynomial function that must vanish. Each coefficient of ξ^i for $i = \{0, 1, 2, n, n + 1, n + 2, n + 3, n + 4\}$ is equated to zero. Here, we have to distinguish between two distinct cases: (1) $n = 2$, which results in a system of seven equations, and (2) $n > 2$, which gives a system of eight equations.

We start with the case $n = 2$. The following equations are obtained from Eq. (7.282):

$$60b_2 + 14\Omega^2 a_0 = 0 \tag{7.293a}$$

$$b_3 = 0 \tag{7.293b}$$

$$360(b_0 - b_4) + 15\Omega^2 a_0 = 0 \tag{7.293c}$$

$$b_1 = 0 \tag{7.293d}$$

$$b_2 = 0 \tag{7.293e}$$

$$b_3 = 0 \tag{7.293f}$$

$$1680b_4 - \Omega^2 a_0 = 0 \tag{7.293g}$$

constituting a system of seven equations with six unknowns. Since Eq. (7.293a) and Eq. (7.293e) result in $r_0 = 0$, we obtain a trivial solution.

Let us assume now that $n > 2$. Equation (7.182) results in eight equations:

$$\text{from } \xi^0: \quad 4(n + 3)(n + 4)b_2 + (n^2 + 7n + 10)\Omega^2 a_0 = 0 \tag{7.294a}$$

$$\text{from } \xi^1: \quad b_3 = 0 \tag{7.294b}$$

$$\text{from } \xi^2: \quad 24b_4 - \Omega^2 a_0 = 0 \tag{7.294c}$$

$$\text{from } \xi^n: \quad b_0 = 0 \tag{7.294d}$$

$$\text{from } \xi^{n+1}: \quad b_1 = 0 \tag{7.294e}$$

$$\text{from } \xi^{n+2}: \quad b_2 = 0 \tag{7.294f}$$

$$\text{from } \xi^{n+3}: \quad b_3 = 0 \tag{7.294g}$$

$$\text{from } \xi^{n+4}: \quad (n + 1)(n + 2)(n + 3)(n + 4)b_4 - \Omega^2 a_0 = 0 \tag{7.294h}$$

Since $b_2 = 0$, from Eq. (7.294a) we obtain $a_0 = 0$, leading to a trivial solution. We conclude that for $n > 1$, only a trivial solution is derived.

7.3.9.2 Linear inertial coefficient ($m = 1$)

Let us treat the case of a linearly by varying inertial coefficient when the mode shape is represented by an nth order polynomial, with $n > 1$. The first term of Eq. (7.182) is

$$\frac{d^2}{d\xi^2}\left[D(\xi)\frac{d^2\psi}{d\xi^2}\right] = (n+3)(n+4)[(n+6)(n+7)b_5\xi^{n+5}$$

$$+ (n+5)(n+6)b_4\xi^{n+4} + (n+4)(n+5)b_3\xi^{n+3}$$

$$+ (n+3)(n+4)b_2\xi^{n+2} + (n+2)(n+3)b_1\xi^{n+1}$$

$$+ (n+1)(n+2)b_0\xi^n - 20b_5\xi^3 - 12b_4\xi^2 - 6b_3\xi - 2b_2]$$

$$(7.295)$$

The second term in Eq. (182) reads

$$-\Omega^2 R(\xi)\psi(\xi) = -\Omega^2\left[a_0\xi^{n+4} - \tfrac{1}{2}a_0(n^2 + 7n + 12)\xi^2 + \tfrac{1}{2}a_0(n^2 + 7n + 10)\right.$$

$$\left. + a_1\xi^{n+5} - \tfrac{1}{2}a_1(n^2 + 7n + 12)\xi^3 + \tfrac{1}{2}a_1(n^2 + 7n + 10)\xi\right]$$

$$(7.296)$$

Each coefficient of ξ^i for $i = \{0, 1, 2, 3, n, n+1, n+2, n+3, n+4, n+5\}$ must vanish. We have to distinguish between the two cases: (1) $n \le 3$, with $n+6$ equations and (2) $n > 3$ with ten equations.

We start with the case $n = 2$. The following equations are derived from Eq. (7.182):

$$60b_2 + 14\Omega^2 a_0 = 0 \tag{7.297a}$$

$$180b_3 + 14\Omega^2 a_1 = 0 \tag{7.297b}$$

$$360(b_0 - b_4) + 15\Omega^2 a_0 = 0 \tag{7.297c}$$

$$600(b_1 - b_5) + 15\Omega^2 a_1 = 0 \tag{7.297d}$$

$$b_2 = 0 \tag{7.297e}$$

$$b_3 = 0 \tag{7.297f}$$

$$1680b_4 - \Omega^2 a_0 = 0 \tag{7.297g}$$

$$2160b_5 - \Omega^2 a_1 = 0 \tag{7.297h}$$

constituting a system of eight equations with seven unknowns. Equations (7.297b) and (7.297f) result in $a_1 = 0$. This is incompatible with our main premise that the inertial coefficient is a linearly varying function. The case $n = 3$ also results in a trivial solution; we obtain directly $b_3 = 0$ and this yields $a_1 = 0$.

Consider now the general case $n > 3$. From Eq. (7.182) we obtain the following ten equations:

from ξ^0: $4(n + 3)(n + 4)b_2 + (n^2 + 7n + 10)\Omega^2 a_0 = 0$ (7.298a)

from ξ^1: $12(n + 3)(n + 4)b_3 + (n^2 + 7n + 10)\Omega^2 a_1 = 0$ (7.298b)

from ξ^2: $-24(n + 3)(n + 4)b_4 + (n^2 + 7n + 12)\Omega^2 a_0 = 0$ (7.298c)

from ξ^3: $-40(n + 3)(n + 4)b_5 + (n^2 + 7n + 12)\Omega^2 a_1 = 0$ (7.298d)

from ξ^n: $b_0 = 0$ (7.298e)

from ξ^{n+1}: $b_1 = 0$ (7.298f)

from ξ^{n+2}: $b_2 = 0$ (7.298g)

from ξ^{n+3}: $b_3 = 0$ (7.298h)

from ξ^{n+4}: $(n + 3)(n + 4)(n + 5)(n + 6)b_4 - \Omega^2 a_0 = 0$ (7.298i)

from ξ^{n+5}: $(n + 3)(n + 4)(n + 6)(n + 7)b_5 - \Omega^2 a_1 = 0$ (7.298j)

From Eq. (7.298h) we have $b_3 = 0$. This when substituted into Eq. (7.298b) results in $a_1 = 0$, which again contradicts the assumed linearity of the material density.

We conclude that we have only trivial solutions with $n \geq 2$ for the linearly varying inertial coefficient.

7.3.9.3   Parabolic inertial coefficient ($m = 2$)

The first term of Eq. (7.182) read

$$\frac{d^2}{d\xi^2}\left[D(\xi)\frac{d^2\psi}{d\xi^2}\right] = (n + 3)(n + 4)[(n + 7)(n + 8)b_6\xi^{n+6} + (n + 6)(n + 7)b_5\xi^{n+5}$$

$$+ (n + 5)(n + 6)b_4\xi^{n+4} + (n + 4)(n + 5)b_3\xi^{n+3}$$

$$+ (n + 3)(n + 4)b_2\xi^{n+2} + (n + 2)(n + 3)b_1\xi^{n+1}$$

$$+ (n + 1)(n + 2)b_0\xi^n - 30b_6\xi^4 - 20b_5\xi^3$$

$$- 12b_4\xi^2 - 6b_3\xi - 2b_2]$$ (7.299)

The second term in Eq. (7.180) is

$$-\Omega^2 R(\xi)\psi(\xi) = -\Omega^2[a_0\xi^{n+4} - \tfrac{1}{2}a_0(n^2 + 7n + 12)\xi^2 + \tfrac{1}{2}a_0(n^2 + 7n + 10)$$
$$+ a_1\xi^{n+5} - \tfrac{1}{2}a_1(n^2 + 7n + 12)\xi^3 + \tfrac{1}{2}a_1(n^2 + 7n + 10)\xi$$
$$+ a_2\xi^{n+6} - \tfrac{1}{2}a_2(n^2 + 7n + 12)\xi^4 + \tfrac{1}{2}a_2(n^2 + 7n + 10)\xi^2]$$

$$(7.300)$$

Each coefficient of ξ^i for $i = \{0, 1, 2, 3, 4, n, n+1, n+2, n+3, n+4, n+5, n+6\}$ must be identically zero. The two separate cases that must be treated are (1) $n \le 4$ leading to a set of $n + 7$ equations and (2) $n > 4$ resulting in 12 equations.

For $n = 2$, Eqs. (7.299) and (7.300) lead to

$$60b_2 + 14\Omega^2 a_0 = 0 \qquad (7.301a)$$

$$180b_3 + 14\Omega^2 a_1 = 0 \qquad (7.301b)$$

$$360(b_0 - b_4) + 15\Omega^2 a_0 - 14\Omega^2 a_2 = 0 \qquad (7.301c)$$

$$600(b_1 - b_5) + 15\Omega^2 a_1 = 0 \qquad (7.301d)$$

$$900(b_2 - b_6) + 15\Omega^2 a_2 = 0 \qquad (7.301e)$$

$$b_3 = 0 \qquad (7.301f)$$

$$1680b_4 - \Omega^2 a_0 = 0 \qquad (7.301g)$$

$$2160b_5 - \Omega^2 r_1 = 0 \qquad (7.301h)$$

$$2700b_6 - \Omega^2 a_2 = 0 \qquad (7.301i)$$

Equations (7.301f) and (7.301b) yield $b_3 = 0$ and $a_1 = 0$. Equation (7.301h) results in $b_5 = 0$. Then, Eq. (7.301d) leads to $b_1 = 0$. From the remaining equations, we get

$$b_0 = 9531b_6/98$$
$$b_2 = -44b_6$$
$$b_4 = 11b_6/98$$
$$\Omega^2 = 2700b_6/a_2$$
$$a_2 = 315a_0/22$$

$$(7.302)$$

Thus, we derive

$$R(\xi) = a_0(1 + 315\xi^2/22)$$
$$D(\xi) = (9531/98 - 44\xi^2 + 11\xi^4/98 + \xi^6)b_6$$

$$(7.303)$$

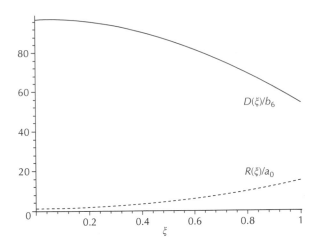

FIGURE 7.13
Variation of $D(\xi)/b_6$ (solid line) and $R(\xi)/r_0$ (dashed-dotted line), for the beam of parabolic mass density

Figure 7.13 portrays $R(\xi)$ and $D(\xi)$ for $\xi \in [0; 1]$. The cases $n = 3$ and $n = 4$ yield trivial solutions: $a_0 = a_1 = a_2 = 0$.

The general case is $n > 4$. We obtain from Eq. (7.182) the following 12 equations:

from ξ^0: $4(n+3)(n+4)b_2 + (n^2 + 7n + 10)\Omega^2 a_0 = 0$ (7.304a)

from ξ^1: $12(n+3)(n+4)b_3 + (n^2 + 7n + 10)\Omega^2 a_1 = 0$ (7.304b)

from ξ^2: $24(n+3)(n+4)b_4 + \Omega^2[(n^2 + 7n + 10)a_2 - (n^2 + 7n + 12)a_0] = 0$ (7.304c)

from ξ^3: $-40(n+3)(n+4)b_5 + (n^2 + 7n + 12)\Omega^2 a_1 = 0$ (7.304d)

from ξ^4: $-60(n+3)(n+4)b_6 + (n^2 + 7n + 12)\Omega^2 a_2 = 0$ (7.304e)

from ξ^n: $b_0 = 0$ (7.304f)

from ξ^{n+1}: $b_1 = 0$ (7.304g)

from ξ^{n+2}: $b_2 = 0$ (7.304h)

from ξ^{n+3}: $b_3 = 0$ (7.304i)

from ξ^{n+4}: $(n+3)(n+4)(n+5)(n+6)b_4 - \Omega^2 a_0 = 0$ (7.304j)

from ξ^{n+5}: $(n+3)(n+4)(n+6)(n+7)b_5 - \Omega^2 a_1 = 0$ (7.304k)

from ξ^{n+6}: $(n+3)(n+4)(n+7)(n+8)b_6 - \Omega^2 a_2 = 0$ (7.304l)

Equations (7.304a) and (7.304h) give $r_0 = 0$; Eq. (7.304j), therefore, yields $b_4 = 0$ and Eq. (7.304c) leads to $a_2 = 0$. This constitutes a trivial solution. Hence, only trivial solutions are derived for the case $n > 2$.

7.3.9.4 Cubic inertial coefficient ($m = 3$)

Cases associated with $n > 1$, for the cubically varying inertial coefficient, are treated as follows. The first term of Eq. (7.182) reads

$$\frac{d^2}{d\xi^2}\left[D(\xi)\frac{d^2\psi}{d\xi^2}\right] = (n+3)(n+4)[(n+8)(n+9)b_7\xi^{n+7} + (n+7)(n+8)b_6\xi^{n+6}$$

$$+ (n+6)(n+7)b_5\xi^{n+5} + (n+5)(n+6)b_4\xi^{n+4}$$

$$+ (n+4)(n+5)b_3\xi^{n+3} + (n+3)(n+4)b_2\xi^{n+2}$$

$$+ (n+2)(n+3)b_1\xi^{n+1} + (n+1)(n+2)b_0\xi^{n}$$

$$- 42b_7\xi^5 - 30b_6\xi^4 - 20b_5\xi^3 - 12b_4\xi^2 - 6b_3\xi - 2b_2]$$

$$(7.305)$$

The second term in Eq. (7.180) results in

$$-\Omega^2 R(\xi)\psi(\xi) = -\Omega^2[a_0\xi^{n+4} - a_0(n^2+7n+12)\xi^2/2 + a_0(n^2+7n+10)/2$$

$$+ a_1\xi^{n+5} - a_1(n^2+7n+12)\xi^3/2 + a_1(n^2+7n+10)\xi/2$$

$$+ a_2\xi^{n+6} - a_2(n^2+7n+12)\xi^4/2 + a_2(n^2+7n+10)\xi^2/2$$

$$+ a_3\xi^{n+7} - a_3(n^2+7n+12)\xi^5/2 + a_3(n^2+7n+10)\xi^3/2]$$

$$(7.306)$$

The degree n determines the number of equations. We have to consider, according to Eqs. (7.305) and (7.306) two separate cases: (1) $n \le 5$, which results in a situation with the number of equations being less than ten, and (2) $n > 5$, which invariably leads to a system of ten equations.

For $n = 2$, we obtain from Eqs. (7.305) and (7.306) a system of ten equations with nine unknowns:

$$60b_2 + 14\Omega^2 a_0 = 0 \tag{7.307a}$$

$$180b_3 + 14\Omega^2 a_1 = 0 \tag{7.307b}$$

$$360(b_0 - b_4) + 15\Omega^2 r_0 - 14\Omega^2 a_2 = 0 \tag{7.307c}$$

$$600(b_1 - b_5) + 15\Omega^2 r_1 - 14\Omega^2 a_3 = 0 \tag{7.307d}$$

$$900(b_2 - b_6) + 15\Omega^2 a_2 = 0 \tag{7.307e}$$

$$1260(b_3 - b_7) + 15\Omega^2 a_3 = 0 \tag{7.307f}$$

$$1680b_4 - \Omega^2 a_0 = 0 \tag{7.307g}$$

$$2160b_5 - \Omega^2 a_1 = 0 \tag{7.307h}$$

$$2700b_6 - \Omega^2 a_2 = 0 \tag{7.307i}$$

$$3300b_7 - \Omega^2 a_3 = 0 \tag{7.307j}$$

This system contains more equations than unknowns. In order to obtain a solution in terms of an arbitrary constant, the rank of this system must be less than the number of unknowns. According to the definition of the rank, a matrix is of rank p if it contains minors of order p different from 0, while all minors of order $p + 1$ (if there are such) are zero. So, all minors of order 9 must vanish to have a rank lower than 9. This leads to ten equations, four of which are identically zero, which can be reduced to the following two relations:

$$a_1 = \frac{402}{2695} a_3 \quad a_0 = \frac{22}{315} a_2 \tag{7.308}$$

Thus, we obtain the following expressions of b_i in terms of b_7:

$$b_0 = 11{,}649 a_2 b_7 / 98 a_3 \tag{7.309a}$$

$$b_1 = 19{,}087 b_7 / 294 \tag{7.309b}$$

$$b_2 = -484 a_2 b_7 / 9 a_3 \tag{7.309c}$$

$$b_3 = -268 b_7 / 7 \tag{7.309d}$$

$$b_4 = 121 a_2 b_7 / 882 a_3 \tag{7.309e}$$

$$b_5 = 67 b_7 / 294 \tag{7.309f}$$

$$b_6 = 11 a_2 b_7 / 9 a_3 \tag{7.309g}$$

$$\Omega^2 = 3300 b_7 / a_3 \tag{7.309h}$$

For $R(\xi)$ and $D(\xi)$ we finally get

$$R(\xi) = a_0 \left(1 + \frac{315}{22} \xi^2 \right) + a_1 \left(\xi + \frac{2{,}695}{402} \xi^3 \right)$$

$$D(\xi) = \left(\frac{11{,}649}{98} - \frac{484}{9} \xi^2 + \frac{121}{882} \xi^4 + \frac{11}{9} \xi^6 \right) a_2 / a_3$$

$$+ \frac{19{,}087}{294} \xi - \frac{268}{7} \xi^3 + \frac{67}{294} \xi^5 + \xi^7) b_7 \tag{7.310}$$

The above functions are valid for $\xi \in [0; 1]$. For $a_1 = a_0$, Figure 7.14 presents the variation of $D(\xi)/b_7$.

Now consider the case $n = 3$. We get a system of eleven equations with nine unknowns. In order to obtain a rank equal to 8, all minors of order 9 must

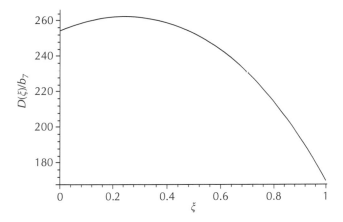

FIGURE 7.14
Variation of $D(\xi)/b_7$ for the beam of cubic mass density ($m = 3, n = 2, a_1 = a_0$)

vanish. This results in $C_{11}^2 = 11!/9!/2! = 55$ leading to the following relations between the coefficients r_i:

$$a_1 = 0 \quad a_2 = \frac{25}{24}a_0 \quad a_3 = \frac{264}{13}a_0 \tag{7.311}$$

We have to express the unknowns in terms of a parameter, namely b_7. We obtain the following expressions:

$$b_0 = 132b_7 \quad b_1 = -\frac{1{,}235}{264}b_7 \quad b_2 = -65b_7 \quad b_3 = 0$$

$$b_4 = \frac{13}{144}b_7 \quad b_5 = 0 \quad b_6 = \frac{65}{1{,}056}b_7 \quad \Omega^2 = \frac{5{,}544b_7}{a_3} \tag{7.312}$$

The following expressions for $D(\xi)$ and $R(\xi)$ are derived:

$$R(\xi) = a_0\left(1 + \frac{25}{24}\xi^2 + \frac{264}{13}\xi^3\right)$$

$$D(\xi) = \left(132 - \frac{1{,}235}{264}\xi - 65\xi^2 + \frac{13}{144}\xi^4 + \frac{65}{1{,}056}\xi^6 + \xi^7\right)b_7 \tag{7.313}$$

The functions $D(\xi)$ and $R(\xi)$ are presented in Figure 7.15. The cases $n = 4$ and $n = 5$ result in trivial solutions.

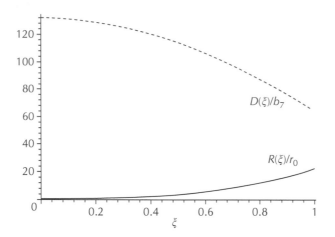

FIGURE 7.15
Variations of $D(\xi)/b_7$ (dashed-dotted line) and $R(\xi)/r_0$ (solid line) for the beam of cubic mass density

The general case consists of the requirement that $n > 5$. We obtain from Eq. (7.182) the following ten equations:

from ξ^0: $4(n+3)(n+4)b_2 + (n^2+7n+10)\Omega^2 a_0 = 0(7.16a)$ $\qquad$ (7.314a)

from ξ^1: $12(n+3)(n+4)b_3 + (n^2+7n+10)\Omega^2 a_1 = 0$ $\qquad$ (7.314b)

from ξ^2: $24(n+3)(n+4)b_4 + \Omega^2[(n^2+7n+10)a_2 - (n^2+7n+12)a_0] = 0$
$\qquad\qquad$ (7.314c)

from ξ^3: $40(n+3)(n+4)b_5 + \Omega^2[(n^2+7n+10)a_3 - (n^2+7n+12)a_1] = 0$
$\qquad\qquad$ (7.314d)

from ξ^4: $60(n+3)(n+4)b_6 - (n^2+7n+12)\Omega^2 a_2 = 0$ $\qquad$ (7.314e)

from ξ^5: $84(n+3)(n+4)b_7 - (n^2+7n+12)\Omega^2 a_3 = 0$ $\qquad$ (7.314f)

from ξ^n: $b_0 = 0$ $\qquad$ (7.314g)

from ξ^{n+1}: $b_1 = 0$ $\qquad$ (7.314h)

from ξ^{n+2}: $b_2 = 0$ $\qquad$ (7.314i)

from ξ^{n+3}: $b_3 = 0$ $\qquad$ (7.314j)

from ξ^{n+4}: $(n+3)(n+4)(n+5)(n+6)b_4 - \Omega^2 a_0 = 0$ $\qquad$ (7.314k)

from ξ^{n+5}: $(n+3)(n+4)(n+6)(n+7)b_5 - \Omega^2 a_1 = 0$ $\qquad$ (7.314l) from ξ^{n+6}: $(n+3)(n+4)(n+7)(n+8)b_6 - \Omega^2 a_2 = 0$ (7.314m)

from ξ^{n+7}: $(n+3)(n+4)(n+8)(n+9)b_7 - \Omega^2 a_3 = 0$ (7.314n)

Equations (7.314i) and (7.314j) result in $a_0 = 0$ and $a_1 = 0$; Eqs. (7.314k) and (7.314l) yield $b_4 = 0$ and $b_5 = 0$, and Eqs. (7.314c) and (7.314d) give $a_2 = 0$ and $a_3 = 0$. Thus, a trivial solution is derived. We get only trivial solutions when $n > 4$.

7.3.9.5   *Quartic polynomial inertial coefficient* ($m = 4$)

Here, we treat cases associated with $n > 1$. Equation (7.182) with $m = 4$ and the mode shape given by Eq. (7.180) yield

$$\frac{d^2}{d\xi^2}\left[D(\xi)\frac{d^2\psi}{d\xi^2}\right] = (n+3)(n+4)[(n+9)(n+10)b_8\xi^{n+8}$$

$$+ (n+8)(n+9)b_7\xi^{n+7} + (n+7)(n+8)b_6\xi^{n+6}$$

$$+ (n+6)(n+7)b_5\xi^{n+5} + (n+5)(n+6)b_4\xi^{n+4}$$

$$+ (n+4)(n+5)b_3\xi^{n+3} + (n+3)(n+4)b_2\xi^{n+2}$$

$$+ (n+2)(n+3)b_1\xi^{n+1} + (n+1)(n+2)b_0\xi^{n}$$

$$- 56b_8\xi^6 - 42b_7\xi^5 - 30b_6\xi^4 - 20b_5\xi^3$$

$$- 12b_4\xi^2 - 6b_3\xi - 2b_2]$$ (7.315)

The second term in Eq. (7.180) reads

$$-\Omega^2 R(\xi)\psi(\xi) = -\Omega^2\left[a_0\xi^{n+4} - \tfrac{1}{2}a_0(n^2+7n+12)\xi^2 + \tfrac{1}{2}a_0(n^2+7n+10)\right.$$

$$+ a_1\xi^{n+5} - \tfrac{1}{2}a_1(n^2+7n+12)\xi^3 + \tfrac{1}{2}a_1(n^2+7n+10)\xi$$

$$+ a_2\xi^{n+6} - \tfrac{1}{2}a_2(n^2+7n+12)\xi^4 + \tfrac{1}{2}a_2(n^2+7n+10)\xi^2$$

$$+ a_3\xi^{n+7} - \tfrac{1}{2}a_3(n^2+7n+12)\xi^5 + \tfrac{1}{2}a_3(n^2+7n+10)\xi^3$$

$$\left.+ a_4\xi^{n+8} - \tfrac{1}{2}a_4(n^2+7n+12)\xi^6 + \tfrac{1}{2}a_4(n^2+7n+10)\xi^4\right]$$ (7.316)

Analogously to the previous cases, we treat separately the case corresponding to a constant number of equations and the case for which the number of equations of the system depends on n. These two cases are, respectively: (1) $n \leq 6$ and (2) $n > 6$.

For the case $n = 2$, we obtain from Eqs. (7.315) and (7.316) a system of eleven equations with ten unknowns $\alpha^T = [b_0, b_1, b_2, b_3, b_4, b_5, b_6, b_7, b_8, \Omega^2]$, with the following matrix:

$$[A] = \begin{bmatrix}
0 & 0 & -60 & 0 & 0 & 0 & 0 & 0 & 0 & -14a_0 \\
0 & 0 & 0 & -180 & 0 & 0 & 0 & 0 & 0 & -14a_1 \\
360 & 0 & 0 & 0 & -360 & 0 & 0 & 0 & 0 & 15a_0 - 14a_2 \\
0 & 600 & 0 & 0 & 0 & -600 & 0 & 0 & 0 & 15a_1 - 14a_3 \\
0 & 0 & 900 & 0 & 0 & 0 & -900 & 0 & 0 & 15a_2 - 14a_4 \\
0 & 0 & 0 & 1260 & 0 & 0 & 0 & -1260 & 0 & 15a_3 \\
0 & 0 & 0 & 0 & 1680 & 0 & 0 & 0 & -1680 & 15a_4 - a_0 \\
0 & 0 & 0 & 0 & 0 & 2160 & 0 & 0 & 0 & -a_1 \\
0 & 0 & 0 & 0 & 0 & 0 & 2700 & 0 & 0 & -a_2 \\
0 & 0 & 0 & 0 & 0 & 0 & 0 & 3300 & 0 & -a_3 \\
0 & 0 & 0 & 0 & 0 & 0 & 0 & 0 & 3960 & -a_4
\end{bmatrix}$$

$$(7.317)$$

To obtain a non-trivial solution and in order to express all the unknowns in terms of an arbitrary constant, all the minors of order 10 of the matrix A must vanish. This conditions leads to eleven equations, which result in

$$a_1 = \frac{402}{2,695}a_3 \qquad a_0 = \frac{22}{315}a_2 - \frac{1}{15}a_4 \qquad (7.318)$$

We express the unknowns in terms of b_7:

$$b_0 = \frac{[4,938a_0 + (34,694/21)a_2]b_8}{14a_4} \qquad b_1 = \frac{209,957a_1b_8}{402a_4} \qquad b_2 = \frac{-924a_0b_8}{a_4}$$

$$b_3 = -308a_1b_8/a_4 \qquad b_4 = \frac{[7,248a_0 + (10,582/21)a_2]b_8}{14a_4}$$

$$b_5 = \frac{11}{6}a_1b_8/a_4 \qquad b_6 = \frac{22}{15}a_2b_8/a_4 \qquad b_7 = \frac{539}{67}a_1b_8/a_4 \qquad (7.319)$$

$$\Omega^2 = 3,960b_8/a_4$$

The second case is when the mode shape is represented by a seventh-order polynomial function; in this case, we obtain a system of $n + m + 5 = 12$ equations with ten unknowns $\alpha^T = [b_0, b_1, b_2, b_3, b_4, b_5, b_6, b_7, b_8, \Omega^2]$, namely

$A\alpha^T = 0$. The matrix A reads

$$[A] = \begin{bmatrix} 0 & 0 & -84 & 0 & 0 & 0 & 0 & 0 & 0 & -20a_0 \\ 0 & 0 & 0 & -252 & 0 & 0 & 0 & 0 & 0 & -20a_1 \\ 0 & 0 & 0 & 0 & -504 & 0 & 0 & 0 & 0 & 21a_0 - 20a_2 \\ 840 & 0 & 0 & 0 & 0 & -840 & 0 & 0 & 0 & 21a_1 - 20a_3 \\ 0 & 1260 & 0 & 0 & 0 & 0 & -1260 & 0 & 0 & 21a_2 - 20a_4 \\ 0 & 0 & 1764 & 0 & 0 & 0 & 0 & -1764 & 0 & 21a_3 \\ 0 & 0 & 0 & 2352 & 0 & 0 & 0 & 0 & -2352 & 21a_4 \\ 0 & 0 & 0 & 0 & 3024 & 0 & 0 & 0 & 0 & -a_0 \\ 0 & 0 & 0 & 0 & 0 & 3780 & 0 & 0 & 0 & -a_1 \\ 0 & 0 & 0 & 0 & 0 & 0 & 4620 & 0 & 0 & -a_2 \\ 0 & 0 & 0 & 0 & 0 & 0 & 0 & 5544 & 0 & -a_3 \\ 0 & 0 & 0 & 0 & 0 & 0 & 0 & 0 & 6552 & -a_4 \end{bmatrix}$$

$$(7.320)$$

The compatibility of this system requires that all minors of order 10 vanish. This condition yields $C_{12}^2 = 12!/10!/2! = 66$ equations, reduced to

$$a_2 = \frac{25}{24}a_0 \quad a_3 = \frac{264}{13}a_0 \quad a_4 = \frac{208}{23}a_1 \tag{7.321}$$

The flexural rigidity coefficients are

$$b_0 = \frac{23[(47{,}520/13)a_0 - 187a_1]b_8}{240a_1}$$

$$b_1 = \frac{-23[(475/8)a_0 - (11{,}440/23)a_1]b_8}{110a_1}$$

$$b_2 = -345a_0b_8/2a_1 \quad b_3 = -\frac{115}{2}b_8$$

$$(7.322)$$

$$b_4 = \frac{23a_0b_8}{96a_1} \quad b_5 = \frac{23}{120}b_8$$

$$b_6 = \frac{115}{704}a_0b_8/a_1 \quad b_7 = \frac{69}{26}a_0b_8/a_1$$

$$\Omega^2 = 6{,}552b_8/a_4$$

We deal now with the particular case $n = m = 4$. The following matrix system is given by Eq. (7.182):

$$
A = \begin{bmatrix}
0 & 0 & -112 & 0 & 0 & 0 & 0 & 0 & 0 & -27a_0 \\
0 & 0 & 0 & -336 & 0 & 0 & 0 & 0 & 0 & -27a_1 \\
0 & 0 & 0 & 0 & -672 & 0 & 0 & 0 & 0 & 28a_0 - 27a_2 \\
0 & 0 & 0 & 0 & 0 & -1120 & 0 & 0 & 0 & 28a_1 - 27a_3 \\
1680 & 0 & 0 & 0 & 0 & 0 & -1680 & 0 & 0 & 28a_2 - 27a_4 \\
0 & 2352 & 0 & 0 & 0 & 0 & 0 & -2352 & 0 & 28a_3 \\
0 & 0 & 3136 & 0 & 0 & 0 & 0 & 0 & -3136 & 28a_4 \\
0 & 0 & 0 & 1 & 0 & 0 & 0 & 0 & 0 & 0 \\
0 & 0 & 0 & 0 & 5040 & 0 & 0 & 0 & 0 & -a_0 \\
0 & 0 & 0 & 0 & 0 & 6160 & 0 & 0 & 0 & -a_1 \\
0 & 0 & 0 & 0 & 0 & 0 & 7392 & 0 & 0 & -a_2 \\
0 & 0 & 0 & 0 & 0 & 0 & 0 & 8736 & 0 & -a_3 \\
0 & 0 & 0 & 0 & 0 & 0 & 0 & 0 & 10192 & -a_4
\end{bmatrix}
$$

$$(7.323)$$

To obtain a solution, we demand that all minors of order 10 vanish. This condition results in $C_{13}^3 = 13!/10!/3! = 286$ equations, which reduces to 4 relations between the coefficients a_i

$$a_1 = 0 \quad a_2 = \frac{418}{405}a_0 \quad a_3 = 0 \quad a_4 = \frac{273}{10}a_0 \qquad (7.324)$$

Since the above relations must be satisfied, we can solve our system of 13 equations to yield

$$R(\xi)/a_0 = 1 + \frac{418}{405}\xi^2 + \frac{273}{10}\xi^4$$

$$D(\xi) = \left(\frac{573{,}833}{3{,}645} - 90\xi^2 + \frac{2}{27}\xi^4 + \frac{38}{729}\xi^6 + \xi^8 \right) b_8$$

$$(7.325)$$

The Figure 7.16 depicts $R(\xi)/a_0$ and $D(\xi)/b_8$.

The cases $n = 5$ and $n = 6$ yield trivial solutions. The demonstration to this effect is performed later on, with $n = m + 1$ and $n = m + 2$. In the general case, $n > 6$, we get

from ξ^0: $4(n+3)(n+4)b_2 + (n^2 + 7n + 10)\Omega^2 a_0 = 0$ (7.326a)

from ξ^1: $12(n+3)(n+4)b_3 + (n^2 + 7n + 10)\Omega^2 a_1 = 0$ (7.326b)

from ξ^2: $24(n+3)(n+4)b_4 + \Omega^2[(n^2 + 7n + 10)a_2 - (n^2 + 7n + 12)a_0] = 0$

$$(7.326c)$$

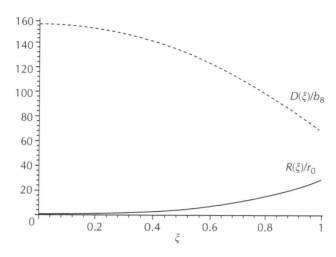

FIGURE 7.16
Variations of $D(\xi)/b_8$ (dashed-dotted line) and $R(\xi)/r_0$ (solid line) for the beam of quartic mass density

from ξ^3: $\quad 40(n+3)(n+4)b_5 + \Omega^2[(n^2+7n+10)a_3 - (n^2+7n+12)a_1] = 0$
$$\tag{7.326d}$$

from ξ^4: $\quad 60(n+3)(n+4)b_6 + \Omega^2[(n^2+7n+10)a_4 - (n^2+7n+12)a_2] = 0$
$$\tag{7.326e}$$

from ξ^5: $\quad 84(n+3)(n+4)b_7 - (n^2+7n+12)\Omega^2 a_3 = 0 \qquad\qquad \text{(7.326f)}$

from ξ^6: $\quad 112(n+3)(n+4)b_8 - (n^2+7n+12)\Omega^2 a_4 = 0 \qquad\qquad \text{(7.326g)}$

from ξ^n: $\quad b_0 = 0 \qquad\qquad\qquad\qquad\qquad\qquad\qquad\qquad\qquad \text{(7.326h)}$

from ξ^{n+1}: $\quad b_1 = 0 \qquad\qquad\qquad\qquad\qquad\qquad\qquad\qquad\qquad \text{(7.326i)}$

from ξ^{n+2}: $\quad b_2 = 0 \qquad\qquad\qquad\qquad\qquad\qquad\qquad\qquad\qquad \text{(7.326j)}$

from ξ^{n+3}: $\quad b_3 = 0 \qquad\qquad\qquad\qquad\qquad\qquad\qquad\qquad\qquad \text{(7.326k)}$

from ξ^{n+4}: $\quad (n+3)(n+4)(n+5)(n+6)b_4 - \Omega^2 a_0 = 0 \qquad\qquad \text{(7.326l)}$

from ξ^{n+5}: $\quad (n+3)(n+4)(n+6)(n+7)b_5 - \Omega^2 a_1 = 0 \qquad\qquad \text{(7.326m)}$

from ξ^{n+6}: $\quad (n+3)(n+4)(n+7)(n+8)b_6 - \Omega^2 a_2 = 0 \qquad\qquad \text{(7.326n)}$

from ξ^{n+7}: $\quad (n+3)(n+4)(n+8)(n+9)b_7 - \Omega^2 a_3 = 0 \qquad\qquad \text{(7.326o)}$

from ξ^{n+8}: $\quad (n+3)(n+4)(n+9)(n+10)b_8 - \Omega^2 a_4 = 0 \qquad\qquad \text{(7.326p)}$

It can be shown that the solution of this system is trivial.

We conclude that we find a closed-form solution for the natural frequency, for the inertial coefficient represented by a fourth-order polynomial, only for a mode shape taken as an $(n+4)$th order polynomial, with $n = 0, 1, 2, 3$ and 4. We note that these results agree with those obtained for the general case.

7.3.10   Concluding Remarks

Several closed-form solutions for the natural frequency of a beam that is guided at one end and pinned at the other end have been reported in this section. When the inertial coefficient is given by an mth order polynomial function, closed-form solutions are obtained for different mode shapes of the beam. These are represented by a polynomial function of order $n + 4$. These closed-form solutions have the same formulation for different values of m and n. Moreover, closed-form solution are derived for $n \leq m$. It appears that the present method cannot uncover closed-form solutions for $n > m$. This conjecture needs further investigation.

7.4   Class of Analytical Closed-Form Polynomial Solutions for Clamped–Guided Inhomogeneous Beams

7.4.1   Introductory Remarks

Becquet and Elishakoff (2001a) treated the vibration of a clamped–guided inhomogeneous beam and derived a new class of closed-form solutions. Their study yielded the fundamental natural frequency for a specific mode shape, postulated as a fourth-order polynomial. Elishakoff and Becquet (2001b) dealt with the case of a guided–pinned inhomogeneous beam where the mode shape was represented by an $(n+4)$th order polynomial function, where n is an arbitrary positive integer.

This section deals with an $(n+4)$th order polynomial function mode shape for a clamped–guided beam. We treat the general case where the material density is represented a polynomial of order m, where m is an arbitrary positive integer. We construct the flexural rigidity, also represented as a polynomial function, so that, in concert with the postulated mode shape, it is compatible with the material density representation. Closed-form solutions are derived for arbitrary values of m and n. Several particular cases are exemplified.

7.4.2   Formulation of the Problem

Let us first consider an auxiliary problem that will yield a function to be postulated as a mode shape in our subsequent analysis. We consider the *static* problem of a *uniform* clamped–guided beam subjected to the load $q(\xi) = q_0 \xi^n$. The governing differential equation is as follows, with D_0 representing a constant flexural rigidity:

$$D_0 \frac{d^4 w}{dx^4} = q_0 x^n \tag{7.327}$$

Here, the dimensionless coordinate ξ is introduced as x/L, where $0 \le \xi \le 1$ and L is the length of the beam:

$$\frac{d^4 w}{d\xi^4} = \lambda \xi^n \quad \lambda = \frac{q_0 L^{n+4}}{D} \tag{7.328}$$

Equation (7.327) is solved in conjunction with the boundary condition $w(0) = w'(0) = 0$ and $w'(1) = w'''(1) = 0$. We obtain an $(n+4)$th degree polynomial function:

$$w(\xi) = \lambda \psi(\xi)/(n+1)(n+2)(n+3)(n+4)$$
$$\psi(\xi) = \xi^{n+4} - \tfrac{1}{6}(n+2)(n+3)(n+4)\xi^3 \tag{7.329}$$
$$+ \tfrac{1}{2}[(n+2)(n+3)(n+4)/2 - (n+4)]\xi^2$$

Now, we return to the original dynamic problem of the vibration of the inhomogeneous beam. We postulate that its mode shape *coincides* with the function $\psi(\xi)$. Hence, the beam vibration is governed by the following ordinary differential equation with variable coefficients:

$$\frac{d^2(D(\xi)d^2 \psi/d\xi^2)}{d\xi^2} - \Omega^2 R(\xi)\psi(\xi) = 0 \tag{7.330}$$

where $D(\xi)$ is the flexural rigidity, $R(\xi) = \rho(\xi)A(\xi)$ is the inertial coefficient ($A(\xi)$ being the cross-sectional area and $\rho(\xi)$ the material density) and $\Omega^2 = \omega^2 L^4$, ω being the natural frequency. Both functions, $D(\xi)$ and $R(\xi)$, are represented by polynomials. Let us assume the degree of $R(\xi)$ to be m. This implies that the degree of the second term in Eq. (7.330) is $m + n + 4$. To have the same degree in the first term of Eq. (7.330), the degree of the function $D(\xi)$ must be $m + 4$:

$$R(\xi) = \sum_{i=0}^{m} a_i \xi^i \quad D(\xi) = \sum_{i=0}^{m+4} b_i \xi^i \tag{7.331}$$

In this study, we assume that $R(\xi)$ is specified. The unknowns are the coefficients defining the variable flexural rigidity $D(\xi)$, and the coefficient Ω^2.

There are $m + 5$ unknown coefficients in $D(\xi)$. Since Ω^2 is also an unknown, we have in total $m + 6$ unknowns. The substitution of $D(\xi)$ and $R(\xi)$ into Eq. (7.330) yields a polynomial expression. It is valid for every ξ within the interval $[0; 1]$; therefore, each coefficient of ξ^i for every i must vanish. We obtain a homogeneous system of equations. The number of equations is determined by the values of both n and m. Indeed, the degree of the polynomial function is $n + m + 4$. The particular case, which treats a fifth-order polynomial for the mode shape, with $n = 1$, results in a square system of $n + m + 5 = m + 6$ with $m + 6$ unknowns. For $n > 1$, we have more equations than unknowns. In the following, the unknowns are grouped as a vector $\alpha^T = [b_0 b_1 \cdots b_{m+4} \Omega^2]$. The matrix of the system of equations is denoted by A.

We first treat the general case. Then, we treat the exceptional cases of constantly, linearly, parabolically and quartically varying inertial coefficients, as particular cases.

7.4.3 General Case

We are looking for the coefficients of the flexural rigidity b_i for $i = 0, \ldots, m+4$ and the coefficient Ω^2. The substitution of Eq. (7.331) into Eq. (7.330) yields the following equation:

$$
\frac{1}{2} \sum_{i=0}^{m+2} (n^3 + 9n^2 + 24n + 16)(i + 1)(i + 2)b_{i+2}\xi^i
$$

$$
- \sum_{i=0}^{m+3} (2n^3 + 18n^2 + 52n + 48)(i + 1)b_{i+1}\xi^i
$$

$$
- \sum_{i=1}^{m+3} (n^3 + 9n^2 + 26n + 24)i(i + 1)b_{i+1}\xi^i
$$

$$
+ \sum_{i=0}^{m+4} (n^4 + 10n^3 + 35n^2 + 50n + 24)b_i\xi^{i+n}
$$

$$
+ \sum_{i=1}^{m+4} (2n^3 + 18n^2 + 52n + 48)ib_i\xi^{i+n} + \sum_{i=2}^{m+4} (n^2 + 7n + 12)(i - 1)ib_i\xi^{i+n}
$$

$$
- \Omega^2 \sum_{i=0}^{m} a_i\xi^{i+n+4} + \frac{1}{6}\Omega^2 \sum_{i=0}^{m} a_i(n + 2)(n + 3)(n + 4)\xi^{i+3}
$$

$$
- \frac{1}{2}\Omega^2 \sum_{i=0}^{m} a_i \left[\frac{1}{2}(n + 2)(n + 3)(n + 4) - (n + 4)\right]\xi^{i+2} = 0 \qquad (7.332)
$$

Let us assume that $n > m + 3$. We rewrite the above equation as follows:

$$
\beta_0 b_2 + \chi_0 b_1 + [\beta_1 b_3 + (\chi_1 + \delta_1)b_2]\xi + [\beta_4 b_2 + (\chi_1 + \delta_1)b_3 + \nu a_0]\xi^2
$$

$$
+ \sum_{i=3}^{m+2} [\beta_i b_{i+2} + (\chi_i + \delta_i)b_{i+1} + \mu a_{i-3} + \nu a_{i-2}]\xi^i
$$

$$
+ [(\chi_{m+3} + \delta_{m+4})b_{m+4} + \mu a_m]\xi^{m+3} + \varepsilon b_0 \xi^n + (\varepsilon + \phi_1)b_1 \xi^{n+1}
$$

$$
+ \sum_{i=2}^{m+4} (\varepsilon + \phi_i + \varphi_i)b_i \xi^{i+n} - \sum_{i=0}^{m} \Omega^2 a_i \xi^{i+n+4} = 0 \tag{7.333a}
$$

$$
\begin{aligned}
\beta_i &= (n^3 + 9n^2 + 24n + 16)(i+1)(i+2) \\
\chi_i &= -(2n^3 + 18n^2 + 52n + 48)(i+1) \\
\delta_i &= -(n^3 + 9n^2 + 26n + 24)i(i+1) \\
\mu &= (n+2)(n+3)(n+4)\Omega^2/6 \\
\varepsilon_i &= (n^4 + 10n^3 + 35n^2 + 50n + 24) \\
\phi_i &= (2n^3 + 18n^2 + 52n + 48)i \\
\varphi_i &= (n^2 + 7n + 12)(i-1)i \\
\nu &= -[(n+2)(n+3)(n+4)/2 - (n+4)]\Omega^2/2
\end{aligned} \tag{7.333b}
$$

with the coefficients in Eq. (7.333a) being defined in Eq. (7.333b). To have the same power of ξ in each line to follow, we rewrite Eq. (7.333a) as follows:

$$
\frac{d^2(D(\xi)d^2\psi/d\xi^2)}{d\xi^2} - \Omega^2 R(\xi)\psi(\xi)
$$

$$
= \beta_0 b_2 + \chi_0 b_1 + [\beta_1 b_3 + (\chi_1 + \delta_1)b_2]\xi
$$

$$
+ [\beta_4 b_2 + (\chi_1 + \delta_1)b_3 + \nu a_0]\xi^2
$$

$$
+ \sum_{i=3}^{m+2} [\beta_i b_{i+2} + (\chi_i + \delta_i)b_{i+1} + \mu a_{i-3} + \nu a_{i-2}]\xi^i
$$

$$
+ [(\chi_{m+3} + \delta_{m+4})b_{m+4} + \mu a_m]\xi^{m+3}
$$

$$
+ \varepsilon_0 b_0 \xi^n + (\varepsilon + \phi_1)b_1 \xi^{n+1}
$$

$$
+ (\varepsilon + \phi_2 + \varphi_2)b_2 \xi^{n+2} + (\varepsilon + \phi_3 + \varphi_3)b_3 \xi^{n+3}
$$

$$
+ \sum_{i=n+4}^{n+m+4} [(\varepsilon + \phi_{i-n} + \varphi_{i-n})b_{i-n} - \Omega^2 a_{i-n+4}]\xi^i = 0 \tag{7.334}
$$

This leads to a system of equations, which is obtained by setting the coefficients in front of each power of ξ equal to zero. The powers of ξ are $0, 1, 2, \ldots, m+3, n, n+1, \ldots, n+m+4$. We observe that if $n > m+3$, we obtain the maximum possible number of equations, which is $2m+9$; in particular, for $i = 0, \ldots, m+3$ we get $m+4$ equations, while for $i = n, \ldots, n+m+4$, we obtain $m+5$ equations. The case $n \le m+3$ yields $n+m+5$ equations. Hence, we have to distinguish between the two cases: (1) $n > m+3$, (2) $n \le m+3$.

We begin with the case $n > m+3$. Equation (7.334) leads to

from ξ^0: $(n^3 + 9n^2 + 24n + 16)b_2 - (2n^3 + 18n^2 + 52n + 48)b_1 = 0$ $\qquad$ (7.335)

from ξ^1: $3(n^3 + 9n^2 + 24n + 16)b_3 - [2(2n^3 + 18n^2 + 52n + 48)$

$\qquad\qquad + 2(n^3 + 9n^2 + 26n + 24)]b_2 = 0$ $\qquad$ (7.336)

from ξ^2: $6(n^3 + 9n^2 + 24n + 16)b_4 - [3(2n^3 + 18n^2 + 52n + 48)$ $\qquad$ (7.337)

$\qquad\qquad + 6(n^3 + 9n^2 + 26n + 24)]b_3 - \frac{1}{2}\Omega^2 a_0 \left[\frac{1}{2}(n+2)(n+3) - 1\right](n+4) = 0$

from ξ^3: $10(n^3 + 9n^2 + 24n + 16)b_5 - [4(2n^3 + 18n^2 + 52n + 48)$

$\qquad\qquad + 12(n^3 + 9n^2 + 26n + 24)]b_4 + \frac{1}{2}\Omega^2 \left\{\frac{1}{3}(n+2)(n+3)(n+4)a_0\right.$

$\qquad\qquad \left. - \left[\frac{1}{2}(n+2)(n+3) - 1\right](n+4)a_1\right\} = 0$ $\qquad$ (7.338)

$\qquad\qquad\qquad \cdots$

from ξ^i: $\frac{1}{2}(i+1)(i+2)(n^3 + 9n^2 + 24n + 16)b_{i+2}$

$\qquad\qquad - [(i+1)(2n^3 + 18n^2 + 52n + 48) + i(i+1)(n^3 + 9n^2 + 26n + 24)]b_{i+1}$

$\qquad\qquad + \frac{1}{2}\Omega^2\{\frac{1}{3}(n+2)(n+3)(n+4)a_{i-3} - [\frac{1}{2}(n+2)(n+3) - 1]$

$\qquad\qquad\qquad \times (n+4)a_{i-2}\} = 0$ $\qquad$ (7.339)

$\qquad\qquad\qquad \cdots$

from ξ^{m+3}: $[(m+4)(2n^3 + 18n^2 + 52n + 48)$

$\qquad\qquad + (m+3)(m+4)(n^3 + 9n^2 + 26n + 24)]b_{m+4}$

$\qquad\qquad - \frac{1}{6}\Omega^2 a_m (n+2)(n+3)(n+4) = 0$ $\qquad$ (7.340)

from ξ^n: $b_0 = 0$ $\qquad$ (7.341)

from ξ^{n+1}: $b_1 = 0$ $\qquad$ (7.342)

from ξ^{n+2}: $b_2 = 0$ $\qquad$ (7.343)

from ξ^{n+3}: $b_3 = 0$ $\qquad$ (7.344)

from ξ^{n+4}: $[(n^4 + 10n^3 + 35n^2 + 50n + 24) + 4(2n^3 + 18n^2 + 52n + 48)$

$\qquad\qquad + 12(n^2 + 7n + 12)]b_4 - \Omega^2 a_0 = 0$ $\qquad$ (7.345)

$\qquad\qquad\qquad \cdots$

from ξ^{n+i}: $\quad [(n^4 + 10n^3 + 35n^2 + 50n + 24) + i(2n^3 + 18n^2 + 52n + 48)$

$$+ i(i-1)(n^2 + 7n + 12)]b_i - \Omega^2 a_{i-4} = 0 \qquad (7.346)$$

$$\cdots$$

from ξ^{n+m+4}: $\quad [(n^4 + 10n^3 + 35n^2 + 50n + 24) + (m+4)(2n^3 + 18n^2 + 52n + 48)$

$$+ (m+4)(m+3)(n^2 + 7n + 12)]b_{m+4} - \Omega^2 a_m = 0 \qquad (7.347)$$

We obtain immediately $b_0 = b_1 = b_2 = b_3 = 0$. Consequently, Eqs. (7.337) and (7.345) result in a system of two equations with two unknowns, b_4 and r_0:

$$\frac{12(n^3 + 9n^2 + 24n + 16)}{[(n+2)(n+3)/2 - 1](n+4)}b_4 - \Omega^2 a_0 = 0$$

$$[(n^4 + 10n^3 + 35n^2 + 50n + 24) + 4(2n^3 + 18n^2 + 52n + 48) \qquad (7.348)$$

$$+ 12(n^2 + 7n + 12)]b_4 - \Omega^2 a_0 = 0$$

One can check that the above system of two equations with two unknowns imposes the condition that both b_4 and a_0 should vanish, since the determinant differs from zero.

We use this property with two recursive equations (7.339) and (7.346). It can be demonstrated that the coefficients b_i, for $i = 5, \ldots, m+4$, and coefficients a_i, for $i = 1, \ldots, m$, vanish. In the first step, we assume that b_{i+1} and a_{i-3} are zero (for example, we can take $i = 3$ to start the recursive form). In the second step, with the previous assumption, we rewrite Eq. (7.339) with $i = j$ and Eq. (7.346) with $i = j+2$, for $j = 3, \ldots, m+1$:

$$\tfrac{1}{2}(j+1)(j+2)(n^3 + 9n^2 + 24n + 16)b_{j+2} - \tfrac{1}{2}\Omega^2 \left[\tfrac{1}{2}(n+2)(n+3) - 1\right](n+4)a_{j-2} = 0$$

$$\left[(n^4 + 10n^3 + 35n^2 + 50n + 24) + (j+2)(2n^3 + 18n^2 + 52n + 48)\right.$$

$$\left.+(j+1)(j+2)(n^2 + 7n + 12)\right]b_{j+2} - \Omega^2 a_{j-2} = 0$$

$$(7.349)$$

Equation (7.347) results in $b_{j+2} = 0$ and $a_{j-2} = 0$. In the third step we observe that $r_{m-1} = 0$ and $b_{m+3} = 0$. Hence, Eqs. (7.340) and (7.347) constitute a system of two equations with two unknowns: b_{m+4} and a_m. The compatibility of this system demands that both the unknowns are zero. This conclusion contradicts our hypothesis that the inertial coefficient is an mth order polynomial function. Finally, in the case $n > m+3$, we obtain only a trivial solution.

Now, consider the case $n = m+3$. Equation (7.334) yields a system of $2m+8$ equations:

$$\frac{d^2(D(\xi)d^2\psi/d\xi^2)}{d\xi^2} - \Omega^2 R(\xi)\psi(\xi)$$

$$= \beta_0 b_2 + \chi_0 b_1$$

$$+ [\beta_1 b_3 + (\chi_1 + \delta_1)b_2]\xi$$

$$+ [\beta_4 b_2 + (\chi_1 + \delta_1)b_3 + va_0]\xi^2$$

$$+ \sum_{i=3}^{m+2}[\beta_i b_{i+2} + (\chi_i + \delta_i)b_{i+1} + \mu a_{i-3} + va_{i-2}]\xi^i$$

$$+ [(\chi_{m+3} + \delta_{m+4})b_{m+4} + \mu a_m + \varepsilon_0 b_0]\xi^{m+3} + (\varepsilon_1 + \phi_1)b_1\xi^{n+1}$$

$$+ (\varepsilon_2 + \phi_2 + \varphi_2)b_2\xi^{n+2} + (\varepsilon_3 + \phi_3 + \varphi_3)b_3\xi^{n+3}$$

$$+ \sum_{i=n+4}^{n+m+4}[(\varepsilon_{i-n} + \phi_{i-n} + \varphi_{i-n})b_{i-n} - \Omega^2 a_{i-n+4}]\xi^i = 0 \qquad (7.350)$$

We expand Eq. (7.350):

from ξ^0: $(n^3 + 9n^2 + 24n + 16)b_2 - (2n^3 + 18n^2 + 52n + 48)b_1 = 0$ $\qquad$ (7.351)

from ξ^1: $3(n^3 + 9n^2 + 24n + 16)b_3$

$\qquad - [2(2n^3 + 18n^2 + 52n + 48) + 2(n^3 + 9n^2 + 26n + 24)]b_2 = 0$ (7.352)

from ξ^2: $6(n^3 + 9n^2 + 24n + 16)b_4 - [3(2n^3 + 18n^2 + 52n + 48)$

$\qquad + 6(n^3 + 9n^2 + 26n + 24)]b_3$

$\qquad - \frac{1}{2}\Omega^2 a_0[\frac{1}{2}(n+2)(n+3) - 1](n+4) = 0$ $\qquad\qquad$ (7.353)

from ξ^3: $10(n^3 + 9n^2 + 24n + 16)b_5$

$\qquad - [4(2n^3 + 18n^2 + 52n + 48) + 12(n^3 + 9n^2 + 26n + 24)]b_4$

$\qquad + \frac{1}{2}\Omega^2\{\frac{1}{3}(n+2)(n+3)(n+4)a_0 - [\frac{1}{2}(n+2)(n+3) - 1](n+4)a_1\} = 0$
$$\qquad\qquad (7.354)$$

$$\cdots$$

from ξ^i: $\frac{1}{2}(i+1)(i+2)(n^3 + 9n^2 + 24n + 16)b_{i+2}$

$\qquad - [(i+1)(2n^3 + 18n^2 + 52n + 48) + i(i+1)(n^3 + 9n^2 + 26n + 24)]b_{i+1}$

$\qquad + \frac{1}{2}\Omega^2\{\frac{1}{3}(n+2)(n+3)(n+4)a_{i-3}$

$\qquad - [\frac{1}{2}(n+2)(n+3) - 1](n+4)a_{i-2}\} = 0$ $\qquad\qquad$ (7.355)

$$\cdots$$

from ξ^{m+3}: $[(m+4)(2n^3+18n^2+52n+48)$

$$+ (m+3)(m+4)(n^3+9n^2+26n+24)]b_{m+4}$$

$$+ (n^4+10n^3+35n^2+50n-+24)b_0$$

$$- \tfrac{1}{6}\Omega^2 a_m(n+2)(n+3)(n+4) = 0 \tag{7.356}$$

from ξ^{n+1}: $b_1 = 0$ (7.357)

from ξ^{n+2}: $b_2 = 0$ (7.358)

from ξ^{n+3}: $b_3 = 0$ (7.359)

from ξ^{n+4}: $[(n^4+10n^3+35n^2+50n+24)+4(2n^3+18n^2+52n+48)$

$$+ 12(n^2+7n+12)]b_4 - \Omega^2 a_0 = 0 \tag{7.360}$$

$$\cdots$$

from ξ^i: $[(n^4+10n^3+35n^2+50n+24)+i(2n^3+18n^2+52n+48)$

$$+ i(i-1)(n^2+7n+12)]b_i - \Omega^2 a_{i-4} = 0 \tag{7.361}$$

$$\cdots$$

from ξ^{n+m+4}: $[(n^4+10n^3+35n^2+50n+24)+(m+4)(2n^3+18n^2+52n+48)$

$$+ (m+4)(m+3)(n^2+7n+12)]b_{m+4} - \Omega^2 a_m = 0 \tag{7.362}$$

Bearing in mind that $b_3 = 0$, Eq. (7.359) and Eqs. (7.353) and (7.360) result in

$$b_4 = 0 \quad a_0 = 0 \tag{7.363}$$

Like in the previous case, the recursive equations (7.355) and (7.361) may be rewritten as follows:

$$A_1 b_{i+2} + A_2 b_{i+1} + A_3 a_{i-3} + A_4 a_{i-2} = 0 \qquad \text{for } 3 \le i \le m+2$$

$$A_5 b_{i+2} + A_6 a_{i-2} = 0 \qquad \text{for } 2 \le i \le m+2 \tag{7.364}$$

where A_1, A_2, A_3, A_4, A_5 and A_6 are non-zero constants. For $i = 3$ this system leads to $b_5 = 0$ and $a_1 = 0$. Hence, for $i = 4$, we obtain $b_6 = 0$ and $a_2 = 0$. Acting in this way, successively, for $5 \le i \le m+2$, we get for each value of i $b_{i+2} = 0$ and $a_{i-2} = 0$. Hence, all coefficients a_i for $i = 0, \ldots, m$ are zero, concluding that this case results in a trivial solution.

Let us assume that n equals $m+2$. This requirement together with Eq. (7.334) yields a system of $2m + 7$ equations with $m + 6$ unknowns. These equations read

from ξ^0: $(n^3 + 9n^2 + 24n + 16)b_2 - (2n^3 + 18n^2 + 52n + 48)b_1 = 0$ (7.365)

from ξ^1: $3(n^3 + 9n^2 + 24n + 16)b_3 - [2(2n^3 + 18n^2 + 52n + 48)$

$$+ 2(n^3 + 9n^2 + 26n + 24)]b_2 = 0 \qquad (7.366)$$

from ξ^2: $6(n^3 + 9n^2 + 24n + 16)b_4 - [3(2n^3 + 18n^2 + 52n + 48)$

$$+ 6(n^3 + 9n^2 + 26n + 24)]b_3 - \tfrac{1}{2}\Omega^2 a_0 \left[\tfrac{1}{2}(n+2)(n+3) - 1\right](n+4) = 0 \qquad (7.367)$$

from ξ^3: $10(n^3 + 9n^2 + 24n + 16)b_5$

$$- [4(2n^3 + 18n^2 + 52n + 48) + 12(n^3 + 9n^2 + 26n + 24)]b_4$$

$$+ \tfrac{1}{2}\Omega^2 \left\{\tfrac{1}{3}(n+2)(n+3)(n+4)a_0 - \left[\tfrac{1}{2}(n+2)(n+3) - 1\right](n+4)a_1\right\} = 0 \qquad (7.368)$$

$$\ldots$$

from ξ^i: $\tfrac{1}{2}(i+1)(i+2)(n^3 + 9n^2 + 24n + 16)b_{i+2}$

$$- [(i+1)(2n^3 + 18n^2 + 52n + 48) + i(i+1)(n^3 + 9n^2 + 26n + 24)]b_{i+1}$$

$$+ \tfrac{1}{2}\Omega^2 \left\{\tfrac{1}{3}(n+2)(n+3)(n+4)a_{i-3} - \left[\tfrac{1}{2}(n+2)(n+3) - 1\right](n+4)a_{i-2}\right\}$$

$$= 0 \qquad (7.369)$$

$$\ldots$$

from ξ^{m+2}: $(n^4 + 10n^3 + 35n^2 + 50n + 24)b_0 + \tfrac{1}{2}(m+3)(m+4)$

$$\times (n^3 + 9n^2 + 24n + 16)b_{m+4} - [(m+3)(2n^3 + 18n^2 + 52n + 48)$$

$$+ (m+2)(m+3)(n^3 + 9n^2 + 26n + 24)]b_{m+3}$$

$$+ \tfrac{1}{2}\Omega^2 \left\{\frac{1}{3}(n+2)(n+3)(n+4)a_{m-1}\right.$$

$$\left. - \left[\tfrac{1}{2}(n+2)(n+3) - 1\right](n+4)a_m\right\} = 0 \qquad (7.370)$$

from ξ^{m+3}: $-[(n^3 + 9n^2 + 26n + 24)(m+3)(m+4)$

$$+ (2n^3 + 18n^2 + 52n + 48)(m+4)]b_{m+4}$$

$$+ [(2n^3 + 18n^2 + 52n + 48) + (n^4 + 10n^3 + 35n^2 + 50n + 24)]b_1$$

$$- \tfrac{1}{6}\Omega^2(n+2)(n+3)(n+4)a_m = 0 \qquad (7.371)$$

from ξ^{n+2}: $b_2 = 0$ (7.372)

from ξ^{n+3}: $b_3 = 0$ (7.373)

from ξ^{n+4}: $\quad [(n^4 + 10n^3 + 35n^2 + 50n + 24) + 4(2n^3 + 18n^2 + 52n + 48)$

$$+ 12(n^2 + 7n + 12)]b_4 - \Omega^2 a_0 = 0 \tag{7.374}$$

from ξ^{n+i}: $\quad [(n^4 + 10n^3 + 35n^2 + 50n + 24) + i(2n^3 + 18n^2 + 52n + 48)$

$$+ i(i-1)(n^2 + 7n + 12)]b_i - \Omega^2 a_{i-4} = 0 \tag{7.375}$$

from ξ^{n+m+4}: $\quad [(n^4 + 10n^3 + 35n^2 + 50n + 24) + (m+4)(2n^3 + 18n^2 + 52n + 48)$

$$+ (m+4)(m+3)(n^2 + 7n + 12)]b_{m+4} - \Omega^2 a_m = 0 \tag{7.376}$$

From Eq. (7.365), we find that $b_1 = 0$. The system of two equations (7.368) and (7.374) with two unknowns b_4 and r_0 yields

$$b_4 = 0 \quad a_0 = 0 \tag{7.377}$$

Analogously, we have with Eqs. (7.369), for $i = j$, (7.375), for $i = j+2$, for $j = 3, \ldots, m+1$ a system of two equations with two unknowns b_i and a_{i-4}. The solution of this system (with the assumption that the natural frequency differs from zero) is

$$\begin{aligned} b_i &= 0 \quad && \text{for } i = 5, \ldots, m+3 \\ a_i &= 0 \quad && \text{for } i = 1, \ldots, m-1 \end{aligned} \tag{7.378}$$

Equations (7.371) and (7.376) result in

$$b_{m+4} = 0 \quad a_m = 0 \tag{7.379}$$

Equation (7.370) yields

$$b_0 = 0 \tag{7.380}$$

Thus, one cannot obtain a non-trivial solution for $n = m + 2$.

The case $n = m + 1$, likewise, results in a trivial solution. We obtain $b_3 = 0$ as in Eq. (7.373). From Eqs. (7.365) and (7.366) we get $b_1 = b_2 = 0$. Using the recursive equation given by Eq. (7.334), it is easy to demonstrate that the coefficients $a_1 = \cdots = a_m = 0$. Consequently, we have the same trivial solution as in the above cases.

We deal now with the case $n = m$. We obtain the following equations:

from ξ^0: $\quad (n^3 + 9n^2 + 24n + 16)b_2 - (2n^3 + 18n^2 + 52n + 48)b_1 = 0 \tag{7.381}$

from ξ^1: $\quad 3(n^3 + 9n^2 + 24n + 16)b_3 - [2(2n^3 + 18n^2 + 52n + 48)$

$$+ 2(n^3 + 9n^2 + 26n + 24)]b_2 = 0 \tag{7.382}$$

from ξ^2: $\quad 6(n^3 + 9n^2 + 24n + 16)b_4 - [3(2n^3 + 18n^2 + 52n + 48)$

$$+ 6(n^3 + 9n^2 + 26n + 24)]b_3 - \tfrac{1}{2}\Omega^2 a_0 \left[\tfrac{1}{2}(n+2)(n+3) - 1\right](n+4) = 0 \tag{7.383}$$

from ξ^3: $10(n^3 + 9n^2 + 24n + 16)b_5 - [4(2n^3 + 18n^2 + 52n + 48)$

$$+ 12(n^3 + 9n^2 + 26n + 24)]b_4 + \tfrac{1}{2}\Omega^2\left\{\tfrac{1}{3}(n+2)(n+3)(n+4)a_0\right.$$

$$\left. - \left[\tfrac{1}{2}(n+2)(n+3) - 1\right](n+4)a_1\right\} = 0 \qquad\qquad (7.384)$$

$$\cdots$$

from ξ^i: $\tfrac{1}{2}(i+1)(i+2)(n^3 + 9n^2 + 24n + 16)b_{i+2}$

$$- [(i+1)(2n^3 + 18n^2 + 52n + 48) + i(i+1)(n^3 + 9n^2 + 26n + 24)]b_{i+1}$$

$$+ \tfrac{1}{2}\Omega^2\left\{\tfrac{1}{3}(n+2)(n+3)(n+4)a_{i-3}\right.$$

$$\left. - \left[\tfrac{1}{2}(n+2)(n+3) - 1\right](n+4)a_{i-2}\right\} = 0 \qquad\qquad (7.385)$$

$$\cdots$$

from ξ^m: $(n^4 + 10n^3 + 35n^2 + 50n + 24)b_0$

$$+ \tfrac{1}{2}(m+1)(m+2)(n^3 + 9n^2 + 24n + 16)b_{m+2}$$

$$- [(m+1)(2n^3 + 18n^2 + 52n + 48)$$

$$+ m(m+1)(n^3 + 9n^2 + 26n + 24)]b_{m+1}$$

$$+ \tfrac{1}{2}\Omega^2\{\tfrac{1}{3}(n+2)(n+3)(n+4)a_{m-3}$$

$$- \left[\tfrac{1}{2}(n+2)(n+3) - 1\right](n+4)a_{m-2}\} = 0 \qquad\qquad (7.386a)$$

from ξ^{m+1}: $[(n^4 + 10n^3 + 35n^2 + 50n + 24) + (2n^3 + 18n^2 + 52n + 48)]b_1$

$$+ \tfrac{1}{2}(m+2)(m+3)(n^3 + 9n^2 + 24n + 16)b_{m+3}$$

$$- [(m+2)(2n^3 + 18n^2 + 52n + 48)$$

$$+ (m+1)(m+2)(n^3 + 9n^2 + 26n + 24)]b_{m+2}$$

$$+ \tfrac{1}{2}\Omega^2\{\tfrac{1}{3}(n+2)(n+3)(n+4)a_{m-2}$$

$$- \left[\tfrac{1}{2}(n+2)(n+3) - 1\right](n+4)a_{m-1}\} = 0 \qquad\qquad (7.386b)$$

from ξ^{m+2}: $[(n^4 + 10n^3 + 35n^2 + 50n + 24) + 2(2n^3 + 18n^2 + 52n + 48)$

$$+ 2(n^2 + 7n + 12)]b_2 + \tfrac{1}{2}(m+3)(m+4)(n^3 + 9n^2 + 24n + 16)b_{m+4}$$

$$- [(m+3)(2n^3 + 18n^2 + 52n + 48)$$

$$+ (m+2)(m+3)(n^3 + 9n^2 + 26n + 24)]b_{m+3}$$

$$+ \tfrac{1}{2}\Omega^2\{\tfrac{1}{3}(n+2)(n+3)(n+4)a_{m-1}$$

$$- \left[\tfrac{1}{2}(n+2)(n+3) - 1\right](n+4)a_m\} = 0 \qquad\qquad (7.387)$$

from ξ^{m+3}: $[(n^4 + 10n^3 + 35n^2 + 50n + 24) + 3(2n^3 + 18n^2 + 52n + 48)$

$$+ 6(n^2 + 7n + 12)]b_3 - [(m+4)(2n^3 + 18n^2 + 52n + 48)$$

$$+ (m+4)(m+3)(n^3 + 9n^2 + 26n + 24)]b_{m+4}$$

$$+ \tfrac{1}{6}\Omega^2(n+2)(n+3)(n+4)a_m = 0 \qquad\qquad (7.388)$$

from ξ^{m+4}: $[(n^4 + 10n^3 + 35n^2 + 50n + 24) + 4(2n^3 + 18n^2 + 52n + 48)$

$$+ 12(n^2 + 7n + 12)]b_4 - \Omega^2 a_0 = 0 \qquad (7.389)$$

from ξ^{m+i}: $[(n^4 + 10n^3 + 35n^2 + 50n + 24) + i(2n^3 + 18n^2 + 52n + 48)$

$$+ i(i - 1)(n^2 + 7n + 12)]b_i - \Omega^2 a_{i-4} = 0 \qquad (7.390)$$

from ξ^{m+m+4}: $[(n^4 + 10n^3 + 35n^2 + 50n + 24) + (m + 4)(2n^3 + 18n^2 + 52n + 48)$

$$+ (m + 4)(m + 3)(n^2 + 7n + 12)]b_{m+4} - \Omega^2 a_m = 0 \qquad (7.391)$$

This case is different from the previous cases, where $n > m$. Since no b_i term is zero the system of two equations (7.385) and (7.391) does not result in a trivial solution.

We note that our system comprises of $n + m + 5$ equations, or, as $n = m$, $2m + 5$ equations. We have $m + 6$ unknowns, namely $b_0, b_1, \ldots, b_{m+4}$ and Ω^2. To assure the compatibility of all equations and to express the unknowns in terms of an arbitrary constant, our system must have a rank lower than the number of unknowns. According to the very definition of the rank, a matrix is of rank p if it contains minors of order p different from 0, while all minors of order $p + 1$ (if there are such) are zero (Uspenky, 1948). To satisfy this condition, all minors in our system of rank superior to or equal to $m + 6$ must vanish. This leads to $(2m + 5)!/(m + 6)!(m - 1)!$ equations. The coefficients r_i have to satisfy these relations. When the coefficients r_i satisfy these relations, the rank of our system equals $m + 6$. In this case, a solution is derived: from Eqs. (7.390), (7.391) and (7.392), we can express the coefficient Ω^2. Then, by equating each expression for Ω^2, with the relation between coefficients a_i (given by the minors of the system), we are able to express the unknowns b_i in terms of an arbitrary constant. Here, the natural frequency is expressed in terms of b_{m+4}:

$$\omega^2 = [(n^4 + 10n^3 + 35n^2 + 50n + 24) + (m + 4)(2n^3 + 18n^2 + 52n + 48)$$

$$+ (m + 3)(m + 4)(n^2 + 7n + 12)]b_{m+4}/a_m L^4 \qquad (7.392)$$

We can portray the variation of the natural frequency in terms of b_{m+4} and r_m. This is done in Figure 7.17. We note that for $n=0$, we have

$$\omega^2 = [24 + 48(m + 4) + 12(m + 3)(m + 4)]b_{m+4}/a_m L^4$$

$$= 12(m + 5)(m + 6)b_{m+4}/a_m L^4$$

We conclude that a necessary condition to obtain a non-trivial solution is to represent the mode shape by an nth order polynomial function with $n < m+1$. To illustrate this, the Sections 7.4.4–7.4.8 treat different cases for m, namely $1 \leq n \leq m$. The case $n = 0$ has been treated by Elishakoff and Becquet (2001).

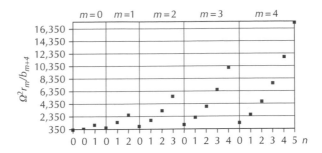

FIGURE 7.17
Variation of $\Omega^2 r_m/b_{m+4}$ for different values of n and m

7.4.4 Constant Inertial Coefficient ($m = 0$)

This case is the classical problem of vibration of a beam with a constant density. However, we are looking for varying flexural rigidity, so that the system possesses the postulated mode shape given in Eq. (7.329). The case for general m, in conjunction with $n = 0$ was studied earlier. Here, we study the case with $m = 0$ and $n > 0$.

Let us first consider the case of a fifth-order polynomial mode shape function ($n = 1$). From Eq. (7.330), with $m = 0$ and $n = 1$, we obtain the following equations:

$$50b_2 - 120b_1 = 0 \tag{7.393}$$

$$-360b_2 + 120b_0 = 0 \tag{7.394}$$

$$300b_4 + 240b_1 - \tfrac{25}{2}\Omega^2 a_0 = 0 \tag{7.395}$$

$$-1200b_4 + 400b_2 + 10\Omega^2 a_0 = 0 \tag{7.396}$$

$$b_3 = 0 \tag{7.397}$$

$$840b_4 - \Omega^2 a_0 = 0 \tag{7.398}$$

We note that, in this case, the number of equations, namely, six, equals the number of unknowns b_0, b_1, b_2, b_3, b_4 and Ω; Eq. (7.398) yields the natural frequency:

$$\Omega^2 = 840b_4/a_0 \tag{7.399}$$

From Eqs. (7.394)–(7.397), we get the following expressions:

$$
\begin{aligned}
b_3 &= 0 \\
b_2 &= -18b_4 \\
b_1 &= 85b_4/2 \\
b_0 &= -54b_4
\end{aligned}
\tag{7.400}
$$

Equation (7.393), in conjunction with Eq. (8.400) leads to

$$b_4 = 0 \tag{7.401}$$

Thus, Eqs. (7.400) and (7.401) lead to a trivial solution. Other cases with $n > 1$ result in trivial solutions, as in the general case (see discussion for the case $n > m$). The present conclusion agrees with the general one made in Section 7.4.3: the case $n > m$ leads only to a trivial solution.

7.4.5 Linearly Varying Inertial Coefficient ($m = 1$)

In this case, the number of unknowns is seven (since $m+6 = 7$). The number of equations is decided by the degree of the polynomial function that represents the mode shape. We start with the case of a fifth-order polynomial mode shape function ($n = 1$). As each coefficient of ξ^i must vanish, we obtain a square system $A\alpha = 0$, with

$$\alpha^T = [b_0 \quad b_1 \quad b_2 \quad b_3 \quad b_4 \quad b_5 \quad \Omega^2]$$

$$A = \begin{bmatrix} 0 & -120 & 50 & 0 & 0 & 0 & 0 \\ 120 & 0 & -360 & 150 & 0 & 0 & 0 \\ 0 & 240 & 0 & -720 & 300 & 0 & -\frac{25}{2}a_0 \\ 0 & 0 & 400 & 0 & -1200 & 500 & 10a_0 - \frac{25}{2}a_1 \\ 0 & 0 & 0 & 600 & 0 & -1800 & 10a_1 \\ 0 & 0 & 0 & 0 & 840 & 0 & -a_0 \\ 0 & 0 & 0 & 0 & 0 & 1120 & -a_1 \end{bmatrix} \tag{7.402}$$

The determinant det(A) of the system must be zero, as a necessary condition to have a non-trivial solution:

$$\det(A) = 425{,}440{,}512 \times 10^8 (a_1 - 3{,}200a_0/2{,}931) = 0$$
$$a_1 = 3{,}200a_0/2{,}931 \tag{7.403}$$

If the coefficients a_i satisfy Eq. (7.403), one can solve the system $A\alpha = 0$. The following results are obtained:

$$b_0 = 65{,}863b_5/1{,}200$$
$$b_1 = 1{,}549b_5/320$$
$$b_2 = 4{,}707b_5/400 \tag{7.404}$$
$$b_3 = -47b_5/3$$
$$b_4 = 977b_5/800$$

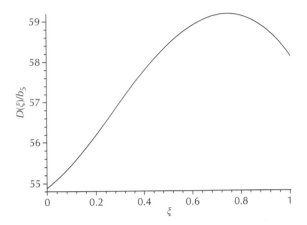

FIGURE 7.18
Variation of $D(\xi)/b_5$, for the beam of linear mass density ($m = 1, n = 1$)

Figure 7.18 depicts the variation of $D(\xi)/b_5$. The function $R(\xi)/a_0$ is linear and positive. Note that, for $n > 1$, only trivial solutions are available, as the solutions derived in Section 7.4.3.

7.4.6 Parabolically Varying Inertial Coefficient ($m = 2$)

This section deals with a material density and a flexural rigidity represented, respectively, by

$$R(\xi) = a_0 + a_1\xi + a_2\xi^2 \quad D(\xi) = b_0 + b_1\xi + b_2\xi^2 + b_3\xi^3 + b_4\xi^4 + b_5\xi^5 + b_6\xi^6$$
$$(7.405)$$

Consider the case of a fifth-order polynomial mode shape function ($n = 1$). This case yields a square system $A\alpha = 0$, with

$$\alpha^T = [b_0 \quad b_1 \quad b_2 \quad b_3 \quad b_4 \quad b_5 \quad b_6 \quad \Omega^2]$$

$$A = \begin{bmatrix} 0 & -120 & 50 & 0 & 0 & 0 & 0 & 0 \\ 120 & 0 & -360 & 150 & 0 & 0 & 0 & 0 \\ 0 & 240 & 0 & -720 & 300 & 0 & 0 & -\frac{25}{2}a_0 \\ 0 & 0 & 400 & 0 & -1200 & 500 & 0 & 10a_0 - \frac{25}{2}a_1 \\ 0 & 0 & 0 & 600 & 0 & -1800 & 750 & 10a_1 - \frac{25}{2}a_2 \\ 0 & 0 & 0 & 0 & 840 & 0 & -2520 & 10a_2 - a_0 \\ 0 & 0 & 0 & 0 & 0 & 1120 & 0 & -a_1 \\ 0 & 0 & 0 & 0 & 0 & 0 & 1440 & -a_2 \end{bmatrix}$$
$$(7.406)$$

We have a non-trivial solution if the determinant $\det(A)$ vanishes:

$$\det(A) = 94{,}894{,}571{,}520 \times 10^9 \left(a_2 - \frac{2{,}931}{4{,}540}a_1 + \frac{160}{227}a_0 \right) = 0$$

$$a_2 = \frac{2{,}931}{4{,}540}a_1 - \frac{160}{227}a_0 \tag{7.407}$$

We write the unknowns in terms of an arbitrary constant, for instance b_6. Thus,

$$b_0 = -(10{,}368a_0 - 17{,}400a_1 + 18{,}281a_2)b_6/112a_2$$

$$b_1 = 3(340a_0 - 282a_1 + 485a_2)b_6/14a_2$$

$$b_2 = -27(32a_0 - 45a_1 + 44a_2)b_6/28a_2 \tag{7.408}$$

$$b_3 = -(564a_1 - 805a_2)b_6/28a_2$$

$$b_4 = 3(4a_0 - 33a_2)b_6/7a_2$$

$$b_5 = 9a_1b_6/7a_2$$

7.4.6.1 Case of a sixth-order polynomial mode shape function ($n = 2$)

The following equations are obtained for the case $m = 2$ and $n = 2$ from Eq. (7.330). This is a system of nine equations with eight unknowns $\alpha^T = [b_0 \ b_1 \ b_2 \ b_3 \ b_4 \ b_5 \ b_6 \ \Omega^2]$:

$$108b_2 - 240b_1 = 0 \tag{7.409}$$

$$324b_3 - 720b_2 = 0 \tag{7.410}$$

$$360b_0 - 1440b_3 + 648b_4 - 27a_0\Omega^2 = 0 \tag{7.411}$$

$$600b_1 - 2400b_4 + 1080b_5 + (20a_0 - 27a_1)\Omega^2 = 0 \tag{7.412}$$

$$900b_2 - 3600b_5 + 1620b_6 + (20a_1 - 27a_2)\Omega^2 = 0 \tag{7.413}$$

$$1260b_3 - 5040b_6 + 20a_2\Omega^2 = 0 \tag{7.414}$$

$$1680b_4 - a_0\Omega^2 = 0 \tag{7.415}$$

$$2160b_5 - a_1\Omega^2 = 0 \tag{7.416}$$

$$2700b_6 - a_2\Omega^2 = 0 \tag{7.417}$$

Equation (7.417) leads to

$$\Omega^2 = 2700b_6/a_2 \tag{7.418}$$

We have a system with the number of equations in excess of the unknowns. To obtain a solution in terms of an arbitrary constant, the rank of this system must be lower than the number of unknowns. According to the definition of the rank, all minors of order 8 must vanish. This requirement results in nine equations:

$$61{,}876{,}343{,}970{,}201{,}600{,}000{,}000a_2 - 35{,}198{,}570{,}373{,}120{,}000{,}000{,}000a_1 = 0$$
$$-7{,}602{,}891{,}200{,}593{,}920{,}000{,}000a_2 + 30{,}718{,}752{,}325{,}632{,}000{,}000{,}000a_1$$
$$-17{,}827{,}847{,}331{,}840{,}000{,}000{,}000a_0 = 0$$
$$-24{,}750{,}537{,}588{,}080{,}640{,}000{,}000a_2 + 14{,}079{,}428{,}149{,}248{,}000{,}000{,}000a_1 = 0$$
$$-1{,}342{,}848{,}315{,}949{,}056{,}000{,}000a_2 - 20{,}351{,}173{,}415{,}731{,}200{,}000{,}000a$$
$$+14{,}262{,}277{,}865{,}472{,}000{,}000{,}000a_0 = 0$$
$$1{,}955{,}029{,}165{,}867{,}008{,}000{,}000a_2 - 7{,}899{,}107{,}740{,}876{,}800{,}000{,}000a_1 \qquad (7.419a)$$
$$+4{,}584{,}303{,}599{,}616{,}000{,}000{,}000a_0 = 0$$
$$35{,}357{,}910{,}840{,}115{,}200{,}000{,}000a_2 - 20{,}113{,}468{,}784{,}640{,}000{,}000{,}000a_1 = 0$$
$$14{,}613{,}349{,}320{,}622{,}080{,}000{,}000a_2$$
$$-23{,}770{,}463{,}109{,}120{,}000{,}000{,}000a_0 26{,}878{,}908{,}284{,}928{,}000{,}000{,}000a_1 = 0$$
$$26{,}955{,}705{,}165{,}742{,}080{,}000{,}000a_1 - 2{,}843{,}678{,}786{,}715{,}648{,}000{,}000a_2$$
$$-17{,}114{,}733{,}438{,}566{,}400{,}000{,}000a_0 = 0$$

These nine conditions are reduced to two relations between the coefficients of the inertial coefficient:

$$a_1 = 24{,}440a_0/36{,}183 \qquad a_2 = 17{,}875a_0/46{,}521 \qquad (7.419b)$$

Equations (7.409)–(7.416) allow the coefficients b_i to be expressed in terms of b_6:

$$b_0 = 182{,}213{,}869b_6/500{,}500$$
$$b_1 = -1{,}377b_6/175$$
$$b_2 = -612b_6/35 \qquad (7.420)$$
$$b_3 = -272b_6/7$$
$$b_4 = 418{,}689b_6/100{,}100$$
$$b_5 = 846b_6/385$$

Figure 7.19 represents the variation of $D(\xi)/b_6$ while Figure 7.20 shows $R(\xi)/r_0$. For $n > 2$, the solutions are trivial, in agreement with the general case.

7.4.7 Cubically Varying Inertial Coefficient ($m = 3$)

In this case, the material density is a third-order polynomial function and, according to Eq. (7.331), the flexural rigidity is represented by a seventh-order polynomial function.

Consider, first, the case of a fifth-order polynomial mode shape function ($n = 1$). Equation (7.332), for $m = 3$ and $n = 1$, results, for each coefficient

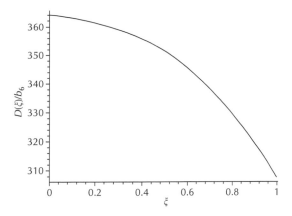

FIGURE 7.19
Variation of $D(\xi)/b_6$ for the beam of parabolically varying mass density ($m = 2, n = 2$)

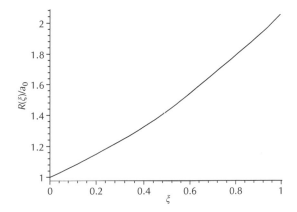

FIGURE 7.20
Variation of $R(\xi)/a_0$ for the beam of parabolically varying mass density ($m = 2, n = 2$)

of ξ, in a square system, $A\alpha = 0$:

$$\alpha^T = [b_0 \quad b_1 \quad b_2 \quad b_3 \quad b_4 \quad b_5 \quad b_6 \quad b_7 \quad \Omega^2]$$

$$A = \begin{bmatrix} 0 & -120 & 50 & 0 & 0 & 0 & 0 & 0 & 0 \\ 120 & 0 & -360 & 150 & 0 & 0 & 0 & 0 & 0 \\ 0 & 240 & 0 & -720 & 300 & 0 & 0 & 0 & -\frac{25}{2}a_0 \\ 0 & 0 & 400 & 0 & -1200 & 500 & 0 & 0 & 10a_0 - \frac{25}{2}a_1 \\ 0 & 0 & 0 & 600 & 0 & -1800 & 750 & 0 & 10a_1 - \frac{25}{2}a_2 \\ 0 & 0 & 0 & 0 & 840 & 0 & -2520 & 1050 & 10a_2 - a_0 - \frac{25}{2}a_3 \\ 0 & 0 & 0 & 0 & 0 & 1120 & 0 & -3360 & 10a_3 - a_1 \\ 0 & 0 & 0 & 0 & 0 & 0 & 1440 & 0 & -a_2 \\ 0 & 0 & 0 & 0 & 0 & 0 & 0 & 1800 & -a_3 \end{bmatrix}$$

$$(7.421)$$

The requirement $\det(A) = 0$ yields

$$a_3 = \frac{8{,}000}{14{,}059}a_0 - \frac{14{,}655}{28{,}118}a_1 + \frac{11{,}350}{14{,}059}a_2 \qquad (7.422)$$

Since the above expression is satisfied, the rank of A is 8 (corresponding to the minor of order 9), according to the definition of the rank of a matrix. Hence, we express the coefficients b_i if, for example, b_7 is taken as a parameter. The solution obtained is as follows:

$$b_0 = -5(10{,}368a_0 - 17{,}400a_1 + 18{,}281a_2 - 29{,}376a_3)b_7/448a_3$$

$$b_1 = (10{,}200a_0 - 8{,}460a_1 + 14{,}550a_2 - 16{,}751a_3)b_7/112a_3$$

$$b_2 = -15(288a_0 - 405a_1 + 396a_2 - 694a_3)b_7/112a_3$$

$$b_3 = -(2{,}820a_1 - 4{,}025a_2 + 4{,}392a_3)b_7/112a_3 \qquad (7.423)$$

$$b_4 = 5(12a_0 - 99a_2 + 143a_3)b_7/a_3$$

$$b_5 = 3(15a_1 - 122a_3)b_7/28a_3$$

$$b_6 = 5a_2b_7/4a_3$$

The natural frequency reads

$$\Omega^2 = 1800b_7/a_3 \qquad (7.424)$$

7.4.7.1 Case of a sixth-order polynomial mode shape function ($n = 2$)

We obtain, in this case, ten equations with nine unknowns $b_0, b_1, b_2, b_3, b_4, b_5, b_6, b_7$ and Ω^2. We can write the system of equations in the form $A\alpha = 0$, with

$$\alpha = [b_0 \quad b_1 \quad b_2 \quad b_3 \quad b_4 \quad b_5 \quad b_6 \quad b_7 \quad \Omega^2]$$

$$A = \begin{bmatrix}
0 & -240 & 108 & 0 & 0 & 0 & 0 & 0 & 0 \\
0 & 0 & -720 & 324 & 0 & 0 & 0 & 0 & 0 \\
360 & 0 & 0 & -1440 & 648 & 0 & 0 & 0 & -27a_0 \\
0 & 600 & 0 & 0 & -2400 & 1080 & 0 & 0 & 20a_0 - 27a_1 \\
0 & 0 & 900 & 0 & 0 & -3600 & 1620 & 0 & 20a_1 - 27a_2 \\
0 & 0 & 0 & 1260 & 0 & 0 & -5040 & 2268 & 20a_2 - 27a_3 \\
0 & 0 & 0 & 0 & 1680 & 0 & 0 & -6720 & 20a_3 - a_0 \\
0 & 0 & 0 & 0 & 0 & 2160 & 0 & 0 & -a_1 \\
0 & 0 & 0 & 0 & 0 & 0 & 2700 & 0 & -a_2 \\
0 & 0 & 0 & 0 & 0 & 0 & 0 & 3300 & -a_3
\end{bmatrix}$$

$$(7.425)$$

The natural frequency is given by the last equation of the system:

$$\Omega^2 = 3300b_7/a_3 \qquad (7.426)$$

To assure the compatibility of the system, and in order to express the unknowns in terms of a parameter, the rank of A has to be lower than 9. To satisfy this, all the minors of A of order 9 must be zero. This leads to ten equations that can be reduced to two relations between the coefficients a_i. We choose to express the other coefficients b_i in terms of b_7. The two relations that assure the compatibility of the system read

$$a_2 = \frac{374{,}497{,}900}{451{,}772{,}199}a_1 - \frac{8{,}818{,}875}{50{,}196{,}911}a_0$$
$$a_3 = \frac{49{,}751{,}625}{50{,}196{,}911}a_1 - \frac{33{,}605{,}000}{50{,}196{,}911}a_0$$

(7.427)

The coefficients of the flexural rigidity read

$$b_0 = (307{,}395a_0 - 239{,}360a_2 + 427{,}356a_3)b_7/1{,}260a_3$$
$$b_1 = -(2{,}860a_0 - 4{,}081a_1 + 3{,}952a_3)b_7/28a_3$$
$$b_2 = -121(25a_1 - 36a_2)b_7/45a_3$$
$$b_3 = -4(3{,}740a_2 - 5{,}427a_3)b_7/315a_3 \qquad (7.428)$$
$$b_4 = (55a_0 - 988a_3)b_7/28a_3$$
$$b_5 = 55a_1b_7/36a_3$$
$$b_6 = 11a_2b_7/9a_3$$

7.4.7.2 Case of a seventh-order polynomial mode shape function ($n = 3$)

If the mode shape is given by a third-order polynomial function, we obtain a system of eleven equations with the nine unknowns $b_0, b_1, b_2, b_3, b_4, b_5, b_6, b_7$ and Ω^2. The system matrix is:

$$A = \begin{bmatrix}
0 & -420 & 196 & 0 & 0 & 0 & 0 & 0 & 0 \\
0 & 0 & -1260 & 588 & 0 & 0 & 0 & 0 & 0 \\
0 & 0 & 0 & -2520 & 1176 & 0 & 0 & 0 & -49a_0 \\
840 & 0 & 0 & 0 & -4200 & 1960 & 0 & 0 & 35a_0 - 49a_1 \\
0 & 1260 & 0 & 0 & 0 & -6300 & 2940 & 0 & 35a_1 - 49a_2 \\
0 & 0 & 1764 & 0 & 0 & 0 & -8820 & 4116 & 35a_2 - 49a_3 \\
0 & 0 & 0 & 1764 & 0 & 0 & 0 & -11760 & 35a_3 \\
0 & 0 & 0 & 0 & 3024 & 0 & 0 & 0 & -a_0 \\
0 & 0 & 0 & 0 & 0 & 3780 & 0 & 0 & -a_1 \\
0 & 0 & 0 & 0 & 0 & 0 & 4620 & 0 & -a_2 \\
0 & 0 & 0 & 0 & 0 & 0 & 0 & 5544 & -a_3
\end{bmatrix}$$

(7.429)

To assure the compatibility of this system, and to express all the unknowns in terms of an arbitrary constant, all the minors of order 9 of the matrix A

must vanish. The resulting 55 equations are reducible to 3 relations. Thus, the coefficients a_i must satisfy the following relations in order to have a non-trivial solution:

$$a_3 = \frac{385}{279}a_0 \quad a_2 = \frac{433{,}895}{174{,}096}a_0 \quad a_1 = \frac{164{,}297}{43{,}524}a_0 \qquad (7.430)$$

With the above expressions for the coefficients a_i, the solution, with b_7 taken as an arbitrary constant, results in

$$
\begin{aligned}
b_0 &= 175{,}566{,}743 b_7 / 245{,}700 \\
b_1 &= -1{,}519 b_7 / 90 \\
b_2 &= -217 b_7 / 6 \\
b_3 &= -155 b_7 / 2 \\
b_4 &= 93 b_7 / 70 \\
b_5 &= 23{,}471 b_7 / 5{,}850 \\
b_6 &= 1{,}127 b_7 / 520
\end{aligned}
\qquad (7.431)
$$

The attendant natural frequency is

$$\Omega^2 = 5544 b_7 / a_3 \qquad (7.432)$$

Figure 7.21 shows $D(\xi)/b_7$, while Figure 7.22 portrays $R(\xi)/a_0$. Other cases, i.e., $n > 3$, yield trivial solutions.

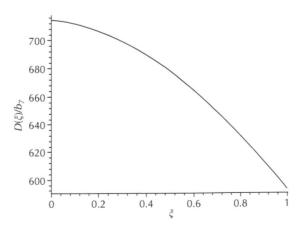

FIGURE 7.21
Variation of $D(\xi)/b_7$ for the beam of cubically varying mass density ($m = 3, n = 3$)

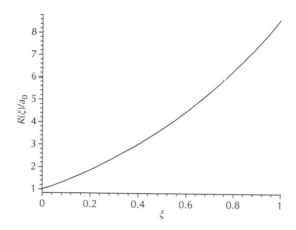

FIGURE 7.22
Variation of $R(\xi)/a_0$ for the beam of cubically mass density ($m = 3, n = 3$)

7.4.8 Inertial Coefficient Represented as a Quadratic ($m = 4$)

The equations for each n are obtained from Eq. (7.332), with $m = 4$. Consider, first, the case of a fifth-order polynomial mode shape function ($n = 1$). This case results in a square system of ten equations with ten unknowns, with the system's matrix A given by:

$$\alpha^{\mathrm{T}} = [b_0 \quad b_1 \quad b_2 \quad b_3 \quad b_4 \quad b_5 \quad b_6 \quad b_7 \quad b_8 \quad \Omega^2]$$

$$A = \begin{bmatrix}
0 & -120 & 50 & 0 & 0 & 0 & 0 & 0 & 0 & 0 \\
120 & 0 & -360 & 150 & 0 & 0 & 0 & 0 & 0 & 0 \\
& 240 & 0 & -720 & 300 & 0 & 0 & 0 & 0 & -\frac{25}{2}a_0 \\
& & 400 & 0 & -1200 & 500 & 0 & 0 & 0 & 10a_0 - \frac{25}{2}a_1 \\
& & & 600 & 0 & -1800 & 750 & 0 & 0 & 10a_1 - \frac{25}{2}a_2 \\
& & & & 840 & 0 & -2520 & 1050 & 0 & 10a_2 - a_0 - \frac{25}{2}a_3 \\
& & & & & 1120 & 0 & -3660 & 1400 & 10a_3 - a_1 - \frac{25}{2}a_4 \\
& & & & & & 1440 & 0 & -4320 & 10a_4 - a_2 \\
& & & & & & & 1800 & 0 & -a_3 \\
& & & & & & & & 2200 & -a_4
\end{bmatrix}$$

$$(7.433)$$

To have a non-trivial solution, the determinant $\det(A)$ must vanish. Consequently,

$$a_4 = -\frac{(70{,}400/193{,}533)a_0}{+(21{,}494/64{,}511)a_1 - (99{,}880/193{,}533)a_2 + (618{,}596/967{,}665)a_3} \tag{7.434}$$

The solution reads

$$b_0 = -(570{,}240a_0 - 957{,}000a_1 + 1{,}005{,}455a_2 - 1{,}615{,}680a_3 + 2{,}118{,}658a_4)b_8/4{,}032a_4$$

$$b_1 = (112{,}200a_0 - 93{,}060a_1 + 160{,}050a_2 - 184{,}261a_3 + 304{,}230a_4)b_8/1{,}008a_4$$

$$b_2 = -(31{,}680a_0 - 44{,}550a_1 + 43{,}560a_2 - 76{,}340a_3 + 93{,}831a_4)b_8/672a_4$$

$$b_3 = -(31{,}020a_1 - 44{,}275a_2 + 48{,}312a_3 - 85{,}940a_4)b_8/1{,}008a_4$$

$$b_4 = (660a_0 - 5{,}445a_2 + 7{,}865a_3 - 9{,}282a_4)b_8/252a_4$$

$$b_5 = (330a_1 - 2{,}684a_3 + 3{,}915a_4)b_8/168a_4$$

$$b_6 = (55a_2 - 442a_4)b_8/36a_4$$

$$b_7 = 11a_3b_8/9a_4$$

$$(7.435)$$

The coefficient Ω^2 is given as

$$\Omega^2 = 2200b_8/a_4 \qquad (7.436)$$

7.4.8.1 Case of a sixth-order polynomial mode shape function ($n = 2$)

The system to solve is $A\alpha = 0$ with the same α as in the case $n = 1$. The matrix A is a 11×10 matrix:

$$A = \begin{bmatrix} 0 & -240 & 108 & 0 & 0 & 0 & 0 & 0 & 0 & 0 \\ 0 & 0 & -720 & 324 & 0 & 0 & 0 & 0 & 0 & 0 \\ 360 & 0 & 0 & -1440 & 648 & 0 & 0 & 0 & 0 & -27a_0 \\ 0 & 600 & 0 & 0 & -2400 & 1080 & 0 & 0 & 0 & 20a_0 - 27a_1 \\ 0 & 0 & 900 & 0 & 0 & -3600 & 1620 & 0 & 0 & 20a_1 - 27a_2 \\ 0 & 0 & 0 & 1260 & 0 & 0 & -5040 & 2268 & 0 & 20a_2 - 27a_3 \\ 0 & 0 & 0 & 0 & 1680 & 0 & 0 & -6720 & 3024 & 20a_3 - a_0 - 27a_4 \\ 0 & 0 & 0 & 0 & 0 & 2160 & 0 & 0 & -8640 & 20a_4 - a_1 \\ 0 & 0 & 0 & 0 & 0 & 0 & 2700 & 0 & 0 & -a_2 \\ 0 & 0 & 0 & 0 & 0 & 0 & 0 & 3300 & 0 & -a_3 \\ 0 & 0 & 0 & 0 & 0 & 0 & 0 & 0 & 3960 & -a_4 \end{bmatrix}$$

$$(7.437)$$

We have one equation in excess of the number of unknowns. To have a non-trivial solution, the rank of the matrix A must be less than 10. Thus, all minors of order must vanish. This leads to a set of eleven equations between the coefficients a_i. This set is reduced to two relations:

$$a_3 = -\frac{122{,}622{,}500}{273{,}428{,}747}a_0 - \frac{14{,}119{,}875}{273{,}428{,}747}a_1 + \frac{343{,}955{,}700}{273{,}428{,}747}a_2$$

$$a_4 = -\frac{121}{196}a_1 + \frac{9{,}306}{8{,}575}a_2 - \frac{48{,}843}{171{,}500}a_3$$

$$(7.438)$$

When the relations between the coefficients a_i are satisfied, the system can be solved. We write the coefficients b_i in terms of an arbitrary constant, chosen here to be b_8:

$$b_0 = (307{,}395a_0 - 239{,}360a_2 + 427{,}356a_3 - 116{,}883a_4)b_8/1{,}050a_4$$

$$b_1 = -3(2{,}860a_0 - 4{,}081a_1 + 3{,}952a_3 - 7{,}144a_4)b_8/70a_4$$

$$b_2 = -2(3{,}025a_1 - 4{,}356a_2 + 4{,}900a_4)b_8/75a_4$$

$$b_3 = -8(3{,}740r_2 - 5{,}427a_3)b_8/525a_4$$

$$b_4 = 3(55a_0 - 988a_3 + 1{,}443a_4)b_8/70a_4 \qquad (7.439)$$

$$b_5 = (11a_1 - 196a_4)b_8/6a_4$$

$$b_6 = 22a_2b_8/15a_4$$

$$b_7 = 6a_3b_8/5a_4$$

The natural frequency in this particular case reads

$$\omega^2 = 3960b_8/L^4a_4 \qquad (7.440)$$

7.4.8.2 Case of a seventh-order polynomial mode shape function ($n = 3$)

This case yields a 12×10 system with the following matrix:

$$A = \begin{bmatrix}
0 & -420 & 196 & 0 & 0 & 0 & 0 & 0 & 0 & 0 \\
0 & 0 & -1{,}260 & 588 & 0 & 0 & 0 & 0 & 0 & 0 \\
0 & 0 & 0 & -2{,}520 & 1{,}176 & 0 & 0 & 0 & 0 & -49a_0 \\
840 & 0 & 0 & 0 & -4{,}200 & 1{,}960 & 0 & 0 & 0 & 35a_0 - 49a_1 \\
0 & 1{,}260 & 0 & 0 & 0 & -6{,}300 & 2{,}940 & 0 & 0 & 35a_1 - 49a_2 \\
0 & 0 & 1{,}764 & 0 & 0 & 0 & -8{,}820 & 4{,}116 & 0 & 35a_2 - 49a_3 \\
0 & 0 & 0 & 2{,}352 & 0 & 0 & 0 & -11{,}760 & 5{,}488 & 35a_3 - 49a_4 \\
0 & 0 & 0 & 0 & 3{,}024 & 0 & 0 & 0 & -15{,}120 & 35a_4 - a_0 \\
0 & 0 & 0 & 0 & 0 & 3{,}780 & 0 & 0 & 0 & -a_1 \\
0 & 0 & 0 & 0 & 0 & 0 & 4{,}620 & 0 & 0 & -a_2 \\
0 & 0 & 0 & 0 & 0 & 0 & 0 & 5{,}544 & 0 & -a_3 \\
0 & 0 & 0 & 0 & 0 & 0 & 0 & 0 & 6{,}552 & -a_4
\end{bmatrix}$$

$$(7.441)$$

To obtain a solution in terms of an arbitrary constant, we demand that all the minors of order superior to 9 vanish. This leads to $C_{12}^2 = 12!/2!/10! = 66$ equations. Some of them turn out to be redundant. Finally, the conditions of its compatibility impose three relations between the coefficients a_i:

$$a_2 = -\frac{161{,}161}{1{,}941{,}552}a_0 + \frac{128{,}777}{188{,}762}a_1$$

$$a_3 = -\frac{12{,}397}{40{,}449}a_0 + \frac{295{,}152}{660{,}667}a_1 \qquad (7.442)$$

$$a_4 = -\frac{12{,}454}{13{,}483}a_0 + \frac{1{,}131{,}624}{4{,}624{,}669}a_1$$

The above system allows us to calculate the natural frequency in terms of b_8:

$$\Omega^2 = 6552 b_8 / a_4 \tag{7.443}$$

We obtain the coefficient b_i in terms of an arbitrary constant b_8:

$$
\begin{aligned}
b_0 &= -(23{,}595 a_0 - 34{,}034 a_1 + 31{,}875 a_4) b_8 / 90 a_4 \\
b_1 &= -104(275 a_1 - 399 a_2) b_8 / 165 a_4 \\
b_2 &= -169(24 a_2 - 35 a_3) b_8 / 33 a_4 \\
b_3 &= -5(1{,}209 a_3 - 1{,}771 a_4) b_8 / 66 a_4 \\
b_4 &= (13 a_0 - 425 a_4) b_8 / 6 a_4 \\
b_5 &= 26 a_1 b_8 / 15 a_4 \\
b_6 &= 78 a_2 b_8 / 55 a_4 \\
b_7 &= 13 a_3 b_8 / 11 a_4
\end{aligned}
\tag{7.444}
$$

7.4.8.3 Case of an eighth-order polynomial mode shape function ($n = 4$)

We obtain a system of thirteen equations with ten unknowns — $b_0, b_1, b_2, b_3, b_4, b_5, b_6, b_7, b_8$ and Ω^2 — with the matrix A being

$$
A = \begin{bmatrix}
0 & -671 & 320 & 0 & 0 & 0 & 0 & 0 & 0 & 0 \\
0 & 0 & -2{,}016 & 960 & 0 & 0 & 0 & 0 & 0 & 0 \\
0 & 0 & 0 & -4{,}032 & 1{,}920 & 0 & 0 & 0 & 0 & -80a_0 \\
0 & 0 & 0 & 0 & -6{,}720 & 3{,}200 & 0 & 0 & 0 & 56a_0 - 80a_1 \\
1{,}680 & 0 & 0 & 0 & 0 & -10{,}080 & 4{,}800 & 0 & 0 & 56a_1 - 80a_2 \\
0 & 2{,}352 & 0 & 0 & 0 & 0 & -14{,}112 & 6{,}720 & 0 & 56a_2 - 80a_3 \\
0 & 0 & 3{,}136 & 0 & 0 & 0 & 0 & -18{,}816 & 8{,}960 & 56a_3 - 80a_4 \\
0 & 0 & 0 & 4{,}032 & 0 & 0 & 0 & 0 & -24{,}192 & 56a_4 \\
0 & 0 & 0 & 0 & 5{,}040 & 0 & 0 & 0 & 0 & -a_0 \\
0 & 0 & 0 & 0 & 0 & 6{,}160 & 0 & 0 & 0 & -a_1 \\
0 & 0 & 0 & 0 & 0 & 0 & 7{,}392 & 0 & 0 & -a_2 \\
0 & 0 & 0 & 0 & 0 & 0 & 0 & 8{,}736 & 0 & -a_3 \\
0 & 0 & 0 & 0 & 0 & 0 & 0 & 0 & 10{,}192 & -a_4
\end{bmatrix}
\tag{7.445}
$$

The compatibility of this system imposes the condition that all minors of order superior to 9 vanish. If so, we obtain a system of rank 9, and we can express all the unknowns in terms of an arbitrary constant. This condition yields $C_{13}^3 = 13!/3!/10! = 286$ relations, reduced to 4 independent relations between coefficients of the material density polynomial function. Yet, we need an arbitrary constant to write the coefficients b_i and b_8 is taken as an arbitrary constant, as in the preceding cases. The relations to be satisfied are

$$a_1 = \frac{3,157}{4,590}a_0 \quad a_2 = \frac{86,276,872}{20,579,265}a_0$$

$$a_3 = \frac{3,303,872}{1,210,545}a_0 \quad a_4 = \frac{2,717}{1,830}a_0$$

$$(7.446)$$

The above expressions assure the compatibility of the system, with the attendant result

$$b_0 = 85,072,969,046b_8/70,509,285$$

$$b_1 = -122,000b_8/3,969$$

$$b_2 = -12,200b_8/189$$

$$b_3 = -1,220b_8/9$$

$$b_4 = 854b_8/627$$

$$(7.447)$$

$$b_5 = 122,549b_8/159,885$$

$$b_6 = 37,528b_8/9,639$$

$$b_7 = 1,216b_8/567$$

The natural frequency is given by

$$\Omega^2 = 10,192b_8/a_4 \quad (7.448)$$

Figure 7.23 depicts $D(\xi)/b_8$ and Figure 7.24 represents $R(\xi)/a_0$. The other cases, corresponding to $n > 4$, result in trivial solutions. This is in perfect agreement with our previous conclusion, which maintains that we obtain only trivial solutions when $n > m$.

7.4.8.4 The particular case, $m = 5$

A final illustration given in this section treats the case of a fifth-order polynomial function for the material density with a ninth-order polynomial for the mode shape. We obtain from Eq. (7.332) the following equations:

$$486b_2 - 1,008b_1 = 0 \qquad (7.449)$$

$$1,458b_3 - 3,024b_2 = 0 \qquad (7.450)$$

$$2,916b_4 - 6,048b_3 - \tfrac{243}{2}\Omega^2 a_0 = 0 \qquad (7.451)$$

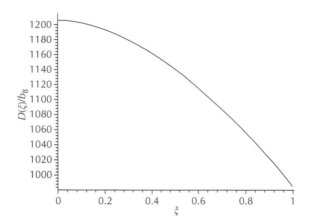

FIGURE 7.23
Variation of $D(\xi)/b_8$ for the beam with quartically varying mass density ($m = 4, n = 4$)

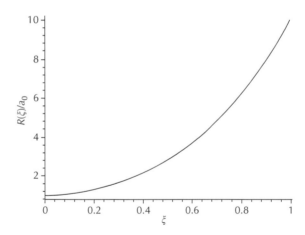

FIGURE 7.24
Variation of $R(\xi)/a_0$ for the beam with quartically varying mass density ($m = 4, n = 4$)

$$4{,}860b_5 - 10{,}080b_4 + 84\Omega^2 a_0 - \tfrac{243}{2}\Omega^2 a_1 = 0 \qquad (7.452)$$

$$7{,}290b_6 - 15{,}120b_5 + 84\Omega^2 a_1 - \tfrac{243}{2}\Omega^2 a_2 = 0 \qquad (7.453)$$

$$10{,}206b_7 - 21{,}168b_6 + 3{,}024b_0 + 84\Omega^2 a_2 - \tfrac{243}{2}\Omega^2 a_3 = 0 \qquad (7.454)$$

$$13{,}608b_8 - 28{,}224b_7 + 4{,}032b_1 + 84\Omega^2 a_3 - \tfrac{243}{2}\Omega^2 a_4 = 0 \qquad (7.455)$$

$$17{,}496b_7 - 36{,}288b_8 + 5{,}184b_2 + 84\Omega^2 a_4 - \tfrac{243}{2}\Omega^2 a_5 = 0 \qquad (7.456)$$

$$-45{,}360b_9 + 6{,}480b_3 + 84\Omega^2 a_5 = 0 \qquad (7.457)$$

$$7{,}920b_4 - \Omega^2 a_0 = 0 \qquad (7.458)$$

$$9{,}504b_5 - \Omega^2 a_1 = 0 \tag{7.459}$$

$$11{,}232b_6 - \Omega^2 a_2 = 0 \tag{7.460}$$

$$13{,}104b_7 - \Omega^2 a_3 = 0 \tag{7.461}$$

$$15{,}120b_8 - \Omega^2 a_4 = 0 \tag{7.462}$$

$$17{,}280b_9 - \Omega^2 a_5 = 0 \tag{7.463}$$

The above equations are a system of fifteen equations with eleven unknowns $b_0, b_1, b_2, b_3, b_4, b_5, b_6, b_7, b_8, b_9$ and Ω. To assure the compatibility of this system, the rank of this system must be equal to 10. This imposes the condition that the minors of order 11 vanish: We obtain $C_{15}^4 = 15!/4!/11! = 1365$ relations. These reduce to five independent equations

$$r_1 = \frac{80}{117}a_0 \quad a_2 = \frac{23{,}680}{50{,}787}a_0$$
$$a_3 = \frac{18{,}171{,}758{,}943}{3{,}947{,}525{,}120}a_0 \quad a_4 = \frac{7{,}709{,}175}{2{,}597{,}056}a_0 \quad a_5 = \frac{3{,}807}{2{,}387}a_0 \tag{7.464}$$

We choose to take b_9 as an arbitrary constant; Eq. (7.463) results in

$$\Omega^2 = 17{,}280b_9/a_5 \tag{7.465}$$

The coefficients b_i are evaluated as follows:

$$b_0 = 2{,}716{,}841{,}089{,}828{,}405b_9/1{,}465{,}478{,}396{,}928$$
$$b_1 = -22{,}599b_9/448 \quad b_2 = -837b_9/8$$
$$b_3 = -217b_9 \quad b_4 = 1{,}736b_9/1{,}269$$
$$b_5 = 347{,}200b_9/445{,}419 \tag{7.466}$$
$$b_6 = 102{,}771{,}200b_9/228{,}499{,}947$$
$$b_7 = 1{,}101{,}519b_9/289{,}408, \ b_8 = 2{,}025b_9/952$$

Equations (7.464) and (7.466) yield

$$D(\xi) = \left(\frac{2{,}716{,}841{,}089{,}828{,}405}{1{,}465{,}478{,}396{,}928} - \frac{22{,}599}{448}\xi - \frac{837}{8}\xi^2 - 217\xi^3 + \frac{1{,}736}{1{,}269}\xi^4 \right.$$
$$\left. + \frac{347{,}200}{445{,}419}\xi^5 + \frac{102{,}771{,}200}{228{,}499{,}947}\xi^6 + \frac{1{,}101{,}519}{289{,}408}\xi^7 + \frac{2{,}025}{952}\xi^8 + \xi^9 \right) b_9$$
$$R(\xi) = \left(1 + \frac{80}{117}\xi + +\frac{23{,}680}{50{,}787}\xi^2 + \frac{18{,}171{,}758{,}943}{3{,}947{,}525{,}120}\xi^3 + \frac{7{,}709{,}175}{2{,}597{,}056}\xi^4 + \frac{3{,}807}{2{,}387}\xi^5 \right) a_0 \tag{7.467}$$

Figure 7.25 depicts $D(\xi)$ and Figure 7.26 depicts $R(\xi)$.

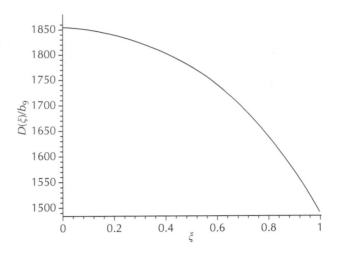

FIGURE 7.25
Variation of $D(\xi)/b_9$ $(m = 5, n = 5)$

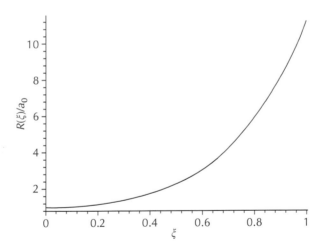

FIGURE 7.26
Variation of $R(\xi)/a_0$ $(m = 5, n = 5)$

7.4.9   Concluding Remarks

In this section we *pre-select* the degree of the polynomial function that represents the mode shape, and construct the inhomogeneous beams that have this specified function as their fundamental mode shape. We show that once the degree m of the polynomial representing the material density is fixed, one cannot choose a degree in the polynomial mode shape arbitrarily, if the polynomial solution is to be valid. In order to obtain a non-trivial solution, it is necessary that the condition $n \leq m$ is satisfied. In this case, the natural frequency of the beam is obtained in a closed-form. It is of interest to investigate if the inequality $n < m$ constitutes a general condition for all sets of boundary conditions.

8

Vibration of Beams in the Presence of an Axial Load

Previous chapters dealt with either buckling or vibration of an inhomogeneous beam. This chapter deals with the vibration of beams in the presence of axial loads. The formula suggesting a linear relationship between the natural frequency squared and the axial load is known to be exact for uniform beams with pinned or guided ends. Its approximate nature for other cases was established by Massonet (1940, 1972) and Galef (1968). In this chapter, we establish, in exact terms, the validity of this linear relationship for a certain class of inhomogeneous beams, under different boundary conditions.

8.1 Closed–Form Solutions for Inhomogeneous Vibrating Beams under Axially Distributed Loading

8.1.1 Introductory Comments

The free vibration of uniform beams subjected to an in-plane load are well studied in the literature. When the in-plane force is a compressive load applied at the ends of the beam, the governing differential equation reads

$$D\frac{d^4W}{dx^4} + P\frac{d^2W}{dx^2} - \rho(x)A\omega^2 W(x) = 0 \qquad (8.1)$$

If the beam is pinned at both its ends, the mode shape can be easily established as

$$W(x) = \sin(\pi x/L) \qquad (8.2)$$

where L is the length. Substitution of Eq. (8.2) into (8.1) leads to the expression for the natural frequency squared:

$$\omega_1^2 = \pi^4 EI/\rho AL^4 - \pi^2 P/\rho AL^2 \qquad (8.3)$$

which can be conveniently re-cast, if the following notations are introduced:

$$\omega_{1,0}^2 = \pi^4 EI/\rho AL^4 \quad P_{cr} = \pi^2 EI/L^2 \quad \alpha = P/P_{cr} \tag{8.4}$$

where $\omega_{1,0}^2$ is the fundamental natural frequency without the axial load P, P_{cr} is the Euler buckling load and, α the ratio between the actual compressive load P and the critical value P_{cr}. Equation (8.3) is rewritten as

$$\omega_1^2 = \omega_{1,0}^2(1 - \alpha) \tag{8.5}$$

indicating that when the actual load approaches its buckling value, the natural frequency vanishes. This is a classical case and reproduced in almost any textbook on vibration; see, e.g., Weaver et al. (1990). For other boundary conditions the linear relationship between the frequency squared, ω_1^2, and the load ratio α does not hold, as was demonstrated by Amba Rao (1967) and Bokaian (1988).

The vibration of uniform beams under in-plane forces was studied in several references. The readers may consult papers by Amba Rao (1967) and Bokaian (1988, 1990), who derived the set of transcendental equations satisfied by the natural frequency and the load ratio, for several combinations of boundary conditions. In an extensive study, Bokaian (1988) studied ten sets of boundary conditions including pinned, guided, free or clamped ends. In particular, Bokaian (1988) showed that Eq. (8.5) holds not only for columns with pinned ends; if each end is pinned or has a guided support Eq. (8.5) is applicable. Guided support is associated with the following boundary conditions

$$\frac{dW}{dx} = 0 \quad \frac{d^3W}{dx^3} = 0 \tag{8.6}$$

For other boundary conditions the relationship $\omega^2(\alpha)$ is non-linear.

We deal here with vibrations of inhomogeneous beams subjected to axially distributed loads. A study by Pilkington and Carr (1970) was devoted to this problem, for the *uniform* beam, and gave an approximate solution. Number of studies treated the vibrations of inhomogeneous columns subjected to axial loading. Theoretical and experimental results were reported by Mok and Murrey (1965). Free vibrations of the beam column with a linear taper were studied by Raju and Rao (1986, 1988), by the Rayleigh method. The Galerkin method with rotational and translational restraints was employed by Yeh and Liu (1989) for the uniform beam column. Williams and Banerjee (1985) studied free vibrations of axially loaded beams with linear or parabolic taper for various sets of boundary conditions. Banerjee and Williams (1985a,b) determined the first five natural frequencies of axially loaded tapered columns for 11 combinations of boundary conditions, within both Bernoulli–Euler and Bresse–Timoshenko beam theories. Other related studies are those by Glück (1973), Kounadis (1975), Sato (1980), Gajewski (1981) and Gottlieb (1987).

The latter investigation is an important contribution for it constructs seven different classes of *inhomogeneous* Bernoulli–Euler beams of continuously varying material density and flexural rigidity; these classes are analytically solvable; moreover, they are isospectral with a *homogeneous* beam in a clamped–clamped configuration. Two beams of the same length are dubbed isospectral in a given boundary condition configuration if they have the same free vibration spectrum. This finding is of prime importance in the context of inverse problems. Gottlieb's paper identified asympathetic sets of data, which would not be suitable for solving the increase problems in a unique manner.

We are unaware of any work prior to that of Guídé and Elishakoff (2001b) — which constitutes a backbone of this section — on the topic of inhomogeneous beams subjected to axially distributed loading, in an *exact, closed-form* setting. Apparently for the first time in the literature, the closed-form solutions are derived for specific beams possessing variable flexural rigidity. A remarkable conclusion is reached. For the uniform beams the linear relationship between the frequency squared, ω_1^2, and the compressive loads ratio α holds for pinned beams with pinned or guided ends only. In the case of a distributed axial load and more complex, *inhomogeneous* beams, one is able to determine the variation of the flexural rigidity that leads to the analogous linear relationship, in the closed form.

8.1.2 Basic Equations

The differential equation of the inhomogeneous vibrating column, which is under a distributed axial load reads

$$\frac{d^2}{dx^2}\left[D(x)\frac{d^2W}{dx^2}\right] - \frac{d}{dx}\left[N(x)\frac{dW}{dx}\right] - R(x)\omega^2 W(x) = 0 \qquad (8.7)$$

where $D(x)$ is the flexural rigidity, $N(x)$ the axial force, $\rho(x)$ the mass density, $R(x) = \rho(x)A(x)$, $A(x)$ is the cross-sectional area, ω the natural frequency, $W(x)$ the buckling mode and x the axial coordinate. With the non-dimensional coordinate $\xi = x/L$, Eq. (8.7) becomes

$$\frac{d^2}{d\xi^2}\left[D(\xi)\frac{d^2\psi}{d\xi^2}\right] - L^2\frac{d}{d\xi}\left[N(\xi)\frac{d\psi}{d\xi}\right] - \rho(\xi)A\omega^2\psi(\xi) = 0 \qquad (8.8)$$

We note that

$$N(\xi) = -q_0 L \sum_{i=0}^{m} g_i(1 - \xi^{i+1})\Big/(i+1) \qquad (8.9)$$

where

$$g_i = q_i/q_0 \quad (i = 1, 2, \ldots, m), \qquad g_0 = 1 \tag{8.10}$$

Therefore, Eq. (8.7) becomes

$$\frac{d^2}{d\xi^2}\left[D(\xi)\frac{d^2\psi}{d\xi^2}\right] + q_0 L^3 \frac{d}{d\xi}\left[\sum_{i=0}^{m}\frac{g_i}{i+1}(1 - \xi^{i+1})\frac{d\psi}{d\xi}\right] - \rho(\xi)A\omega^2\psi(\xi) = 0 \tag{8.11}$$

Considering the degree of the first member of the governing equation, the flexural rigidity $D(\xi)$ is taken to be equal to

$$D(\xi) = \sum_{i=0}^{m+3} b_i \xi^i \tag{8.12}$$

and the inertial coefficient

$$R(\xi) = \rho(\xi)A(\xi) = \sum_{i=0}^{m-1} a_i \xi^i \tag{8.13}$$

Since the degree of the polynomial $R(\xi)$ is $m - 1$, m must be larger than or equal to unity.

8.1.3   Column that is Clamped at One End and Free at the Other

The simplest mode shape that fulfills the boundary conditions reads:

$$\psi(\xi) = 6\xi^2 - 4\xi^3 + \xi^4 \tag{8.14}$$

In this case we first consider two particular cases, namely $m = 1$ and $m = 2$, with the general case following for m larger than 2.

8.1.3.1   *Particular case: constant mass density ($m = 1$)*

Upon substitution of Eqs. (8.12)–(8.14) into Eq. (8.11) we get for the first term

$$\frac{d^2}{d\xi^2}\left[D(\xi)\frac{d^2\psi}{d\xi^2}\right] = (2b_2 + 6b_3\xi + 12b_4\xi^2)(12 - 24\xi + 12\xi^2)$$
$$+ 2(b_1 + 2b_2\xi + 3b_3\xi^2 + 4b_4\xi^3)(-24 + 24\xi)$$
$$+ 24(b_0 + b_1\xi + b_2\xi^2 + b_3\xi^3 + b_4\xi^4) \tag{8.15}$$

The second and third terms become:

$$q_0 L^3 \frac{d}{d\xi} \left[\sum_{i=0}^{m} \frac{g_i}{i+1} (1 - \xi^{i+1}) \frac{d\psi}{d\xi} \right] = q_0 L^3 \{ -(1 + g_1\xi)(12\xi - 12\xi^2 + 4\xi^3)$$

$$+ \left[1 - \xi + \frac{1}{2} g_1(1 - \xi^2) \right] (12 - 24\xi + 12\xi^2) \}$$

(8.16)

$$\omega^2 R(\xi)\psi(\xi) = \omega^2 a_0 (6\xi^2 - 4\xi^3 + \xi^4)$$

(8.17)

We demand that the result of the substitution of Eqs. (8.15)–(8.17) into Eq. (8.11) be valid for any ξ. This leads to the following five linear equations:

$$24b_2 - 48b_1 + 24b_0 + q_0 L^3(12 + 6g_1) = 0$$

$$72b_3 - 144b_2 + 72b_1 - q_0 L^3(48 + 12g_1) = 0$$

$$144b_4 - 288b_3 + 144b_2 + q_0 L^3(48 - 12g_1) - 6\omega^2 a_0 = 0 \qquad (8.18)$$

$$-480b_4 + 240b_3 + q_0 L^3(-16 + 24g_1) + 4\omega^2 a_0 = 0$$

$$360b_4 - 10q_0 L^3 g_1 - \omega^2 a_0 = 0$$

The last equation yields the expression for the natural frequency squared:

$$\omega^2 = (360b_4 - 10q_0 L^3 g_1)/a_0$$

(8.19)

Denoting

$$q_{0,cr} = 36b_4/g_1 L^3$$

(8.20)

we rewrite Eq. (19) as follows:

$$\omega^2 = 360b_4(1 - q_0/q_{0,cr})/a_0$$

(8.21)

This identifies $q_{0,cr}$ as the critical buckling value of the parameter q_0. When $q_0 \equiv 0$ we get Eq. (8.21) to be the natural frequency coefficient of the beam with axially distributed load not taken into account. When q_0 tends to its critical value the natural frequency vanishes. The coefficients b_i in the flexural rigidity are deduced from the remaining equations of the set (8.18):

$$b_0 = 26b_4 + \tfrac{1}{20} q_0 L^3(2 - 13g_1) \quad b_1 = 16b_4 + \tfrac{1}{10} q_0 L^3(2 - 3g_1)$$

$$b_2 = 6b_4 - \tfrac{1}{5} q_0 L^3(1 + g_1) \quad b_3 = -4b_4 + \tfrac{1}{15} q_0 L^3(1 + g_1)$$

(8.22)

8.1.3.2   Particular case: linearly varying mass density ($m = 2$)

Instead of Eq. (8.18) we get the following set:

$$24b_2 - 48b_1 + 24b_0 + q_0L^3(12 + 6g_1 + 4g_2) = 0$$

$$72b_3 - 144b_2 + 72b_1 - q_0L^3(48 + 12g_1 + 8g_2) = 0$$

$$144b_4 - 288b_3 + 144b_2 + q_0L^3(48 - 12g_1 + 4g_2) - 6\omega^2 a_0 = 0$$

$$240b_5 - 480b_4 + 240b_3 + q_0L^3(-16 + 24g_1 - 16g_2) + 2\omega^2(2a_0 - 3a_1) = 0$$

$$-720b_5 + 360b_4 + q_0L^3(-10g_1 + 20g_2) - \omega^2(a_0 - 4a_1) = 0$$

$$504b_5 - 8q_0L^3g_2 - \omega^2 a_1 = 0 \qquad (8.23)$$

with the last equation yielding the natural frequency squared. When we introduce

$$q_{0,cr} = 63b_5/g_2L^3 \qquad (8.24)$$

we express the natural frequency squared as follows:

$$\omega^2 = 504b_5(1 - q_0/q_{0,cr})/a_1 \qquad (8.25)$$

Again, $q_{0,cr}$ is the critical buckling load; when q_0 approaches its buckling value the natural frequency approaches zero. The coefficients b_i in the flexural rigidity are

$$b_0 = \frac{1}{180a_1}\{(6552a_0 + 5112a_1)b_5 + q_0L^3[13a_1g_1 + 18a_1 - (104a_0 + 71a_1)g_2]\}$$

$$b_1 = \frac{1}{90\rho_1}\{(2016a_0 + 1836a_1)b_5 + q_0L^3[13a_1g_1 + 18a_1 - (32a_0 + 19a_1)g_2]\}$$

$$b_2 = \frac{1}{180a_1}\{(1512a_0 + 2232a_1)b_5 - q_0L^3[6a_1g_1 + 36a_1 + (24a_0 + 35a_1)g_2]\}$$

$$b_3 = \frac{1}{45a_1}\{(-252a_0 + 198a_1)b_5 + q_0L^3[-2a_1g_1 + 3a_1 + (4a_0 - 3a_1)g_2]\}$$

$$b_4 = \frac{1}{180a_1}\{(252a_0 - 648a_1)b_5 + q_0L^3[5a_1g_1 + 18a_1 - (4a_0 - 6a_1)g_2]\}$$

$$(8.26)$$

8.1.3.3   General variation of density ($m \geq 3$)

When the parameter m is larger than 2, we get the following set of equations:

from ξ^0: $2(12b_2 - 24b_1 + 12b_0) + 12q_0L^3G = 0$ $\qquad (8.27a)$

from ξ^1: $6(12b_3 - 24b_2 + 12b_1) - 2q_0L^3(12G + 12) = 0$ $\qquad (8.27b)$

from ξ^2: $\quad 12(12b_4 - 24b_3 + 12b_2) + 3q_0 L^3 (4G + 12 - 6g_1) - 6\omega^2 a_0 = 0$

$$(8.27c)$$

from ξ^3: $\quad 20(12b_5 - 24b_4 + 12b_3) - 4q_0 L^3 (4g_2 - 6g_1 + 4)$

$$- \omega^2 (6a_1 - 4a_0) = 0 \qquad (8.27d)$$

from ξ^i: $4 \le i \le m+1$ $\quad (i+1)(i+2)(12b_{i+2} - 24b_{i+1} + 12b_i)$

$$- (i+1)q_0 L^3 \left(\frac{12g_{i-1}}{i} - \frac{12g_{i-2}}{i-1} + \frac{4g_{i-3}}{i-2} \right)$$

$$- \omega^2 (6a_{i-2} - 4a_{i-3} + a_{i-4}) = 0 \qquad (8.27e)$$

from ξ^{m+2}: $\quad (m+3)(m+4)(-24b_{m+3} + 12b_{m+2}) + (m+3)q_0 L^3$

$$\times \left(\frac{12g_m}{m+1} - \frac{4g_{m-1}}{m} \right) - \omega^2 (-4a_{m-1} + a_{m-2}) = 0 \qquad (8.27f)$$

from ξ^{m+3}: $\quad 12(m+5)b_{m+3} - (m+4)q_0 L^3 \dfrac{4g_m}{m+1} - \omega^2 a_{m-1} = 0 \quad (8.27g)$

We note from Eq. (8.27e), that the above system of equations is available when the parameter m is larger than 2, since, in the interval of the index i, the upper limit $m+1$ must be larger than or equal to the lower one, namely 4. The remaining cases, namely $m = 1$ and $m = 2$, are treated separately.

It is instructive to introduce the following quantity:

$$q_{0,\mathrm{cr}} = 3(m+1)(m+5)b_{m+3}/g_m L^3 \qquad (8.28)$$

With this notation the natural frequency squared, obtainable from Eq. (8.27g), is as follows:

$$\omega^2 = 12(m+4)(m+5)b_{m+3}(1 - q_0/q_{0,\mathrm{cr}})/r_{m-1} \qquad (8.29)$$

This equation exhibits the property established for the particular cases $m = 1$ and $m = 2$; namely, when q_0 approaches its critical buckling value, the natural frequency tends to zero. The coefficients describing the flexural rigidity, read

$$b_0 = 5b_4 - 4b_5 - \frac{(m+4)(m+5)}{10\rho_{m-1}}(-12\rho_1 + 23\rho_0)b_{m+3}$$

$$+ \frac{q_0 L^3}{60} \left(16g_2 - \frac{3}{2}g_1 - 5G + 11 + \frac{m+4}{(m+1)\rho_{m-1}}(-24a_1 + 46a_0)g_m \right)$$

$$b_1 = 4b_4 - 3b_5 - \frac{(m+4)(m+5)}{10a_{m-1}}(-9a_1 + 16a_0)b_{m+3}$$

$$+ \frac{q_0 L^3}{30}\left(6g_2 - \frac{3}{2}g_1 + 5G + 1 + \frac{m+4}{(m+1)a_{m-1}}(-9a_1 + 16a_0)g_m\right)$$

$$b_2 = 3b_4 - 2b_5 - \frac{(m+4)(m+5)}{10a_{m-1}}(-6a_1 + 9a_0)b_{m+3}$$

$$+ \frac{q_0 L^3}{60}\left(8g_2 - \frac{9}{2}g_1 - 5G - 7 + \frac{m+4}{(m+1)a_{m-1}}(-12a_1 + 18a_0)g_m\right)$$

$$b_3 = 2b_4 - b_5 - \frac{(m+4)(m+5)}{10a_{m-1}}(3a_1 - 2a_0)b_{m+3}$$

$$+ \frac{q_0 L^3}{30}\left(2g_2 - 3g_1 + 2 - \frac{m+4}{(m+1)a_{m-1}}(3a_1 - 2a_0)g_m\right)$$

for $4 \leq i \leq m+1$,

$$b_i = 2b_{i+1} - b_{i+2} + \frac{(m+4)(m+5)}{(i+1)(i+2)a_{m-1}}(6a_{i-2} - 4a_{i-3} + a_{i-4})b_{m+3}$$

$$+ \frac{q_0 L^3}{3(i+2)}\left[\frac{3g_{i-1}}{i} - \frac{3g_{i-2}}{i-1} + \frac{g_{i-3}}{i-2}\right.$$

$$\left. - \frac{m+4}{(m+1)(i+1)a_{m-1}}(6a_{i-2} - 4a_{i-3} + a_{i-4})g_m\right]$$

$$b_{m+2} = \left[2 + \frac{m+5}{m+3}\left(\frac{a_{m-2}}{a_{m-1}} - 4\right)\right]b_{m+3}$$

$$- \frac{q_0 L^3}{m+1}g_m\left[\frac{1}{m+4} + \frac{1}{3(m+3)}\left(\frac{a_{m-2}}{a_{m-1}} - 4\right)[3pt]\right] + \frac{q_0 L^3}{3m(m+4)}g_{m-1}$$

$$\tag{8.30}$$

8.1.4   Column that is Pinned at its Ends

The simplest polynomial buckling mode that satisfies the boundary conditions is

$$\psi(\xi) = \xi - 2\xi^3 + \xi^4 \tag{8.31}$$

For the pinned column, we present three particular cases, namely $m = 1$, $m = 2$ and $m = 3$, prior to considering the general case $m \geq 4$.

8.1.4.1 *Particular case: constant inertial coefficient ($m = 1$)*

The set of equations governing b_0, b_1, b_2, b_3, b_4 and ω^2, treated as unknowns, reads

$$-24b_1 + 24b_0 - q_0 L^3 = 0$$
$$-72b_2 + 72b_1 - q_0 L^3 (12 + 7g_1) - \omega^2 a_0 = 0$$
$$-144b_3 + 144b_2 + q_0 L^3 (30 + 6g_1) = 0 \qquad (8.32)$$
$$-240b_4 + 240b_3 + q_0 L^3 (-16 + 12g_1) + 2\omega^2 a_0 = 0$$
$$360b_4 - 10q_0 L^3 g_1 - \omega^2 a_0 = 0$$

The last equation of the set (8.32) leads to the following expression for the natural frequency squared:

$$\omega^2 = 360 b_4 \left(1 - q_0/q_{0,\mathrm{cr}}\right)/a_0 \qquad (8.33)$$

which, remarkably, coincides with Eq. (8.21), while the expression for $q_{0,\mathrm{cr}}$ is the same as in Eq. (8.20). The coefficients b_i, however, differ from those given in Eq. (8.22):

$$b_0 = 3b_4 + q_0 L^3 \left(\tfrac{1}{15} - \tfrac{1}{20}g_1\right) \quad b_1 = 3b_4 - \tfrac{1}{40}q_0 L^3 (1 - 2g_1)$$
$$b_2 = -2b_4 - \tfrac{1}{120}q_0 L^3 (17 + g_1) \quad b_3 = -2b_4 + \tfrac{1}{30}q_0 L^3 (2 + g_1) \qquad (8.34)$$

8.1.4.2 *Particular case: linearly varying inertial coefficient ($m = 2$)*

Instead of the set (8.32), we have in this case

$$-24b_1 + 24b_0 - q_0 L^3 = 0$$
$$-72b_2 + 72b_1 - q_0 L^3 (12 + 7g_1 + 4g_2) - \omega^2 a_0 = 0$$
$$-144b_3 + 144b_2 + q_0 L^3 (30 + 6g_1 + 3g_2) - \omega^2 a_1 = 0$$
$$-240b_4 + 240b_3 + q_0 L^3 (-16 + 12g_1) + 2\omega^2 a_0 = 0 \qquad (8.35)$$
$$-360b_5 + 360b_4 + q_0 L^3 (-10g_1 + 10g_2) - \omega^2 (a_0 - 2a_1) = 0$$
$$504b_5 - 8q_0 L^3 g_2 - \omega^2 a_1 = 0$$

yielding the natural frequency squared,

$$\omega^2 = 504 b_5 (1 - q_0/q_{0,\mathrm{cr}})/a_1 \qquad (8.36)$$

Again, Eq. (8.36) coincides with Eq. (8.25), with $q_{0,cr}$ coinciding with Eq. (8.24). The coefficients b_i read

$$b_0 = \frac{1}{240a_1}\{(1008a_0 + 408a_1)b_5 + q_0L^3[8a_1g_1 + 6a_1 - (16a_0 + a_1)g_2]\}$$

$$b_1 = \frac{1}{240a_1}\{(1008a_0 + 408a_1)b_5 + q_0L^3[8a_1g_1 + 16a_1 - (16a_0 + a_1)g_2]\}$$

$$b_2 = \frac{1}{720a_1}\{(-2016a_0 + 1224a_1)b_5 + q_0L^3[-46a_1g_1 - 102a_1 + (32a_0 - 43a_1)g_2]\}$$

$$b_3 = \frac{1}{180a_1}\{(-504a_0 - 324a_1)b_5 + q_0L^3[-4a_1g_1 + 12a_1 + (8a_0 + 3a_1)g_2]\}$$

$$b_4 = \frac{1}{180a_1}\{(252a_0 + 324a_1)b_5 + q_0L^3[5a_1g_1 - (4a_0 - 3a_1)g_2]\}$$

$$(8.37)$$

8.1.4.3 *Particular case: parabolically varying inertial coefficient* ($m = 3$)

When $m = 3$, we obtain the following set of equations that have to be satisfied:

$$-24b_1 + 24b_0 - q_0L^3 = 0$$

$$-72b_2 + 72b_1 - q_0L^3(12 + 7g_1 + 4g_2 + 3g_3) - \omega^2 r_0 = 0$$

$$-144b_3 + 144b_2 + q_0L^3(30 + 6g_1 + 3g_2 + 3g_3) - \omega^2 r_1 = 0$$

$$-240b_4 + 240b_3 + q_0L^3(-16 + 12g_1 - g_3) - \omega^2(-2r_0 + r_2) = 0 \qquad (8.38)$$

$$-360b_5 + 360b_4 + q_0L^3(-10g_1 + 10g_2) - \omega^2(r_0 - 2r_1) = 0$$

$$-504b_6 + 504b_5 + q_0L^3(-8g_2 + 9g_3) - \omega^2(r_1 - 2r_2) = 0$$

$$672b_6 - 7q_0L^3g_2 - \omega^2 r_1 = 0$$

Introducing the critical, buckling value $q_{0,cr}$ of q_0

$$q_{0,cr} = 96b_6/g_3L^3 \qquad (8.39)$$

we write the natural frequency squared as follows:

$$672b_6(1 - q_0/q_{0,cr})/a_2 \qquad (8.40)$$

It is interesting to remark again that when the buckling load q_0 tends to its critical value $q_{0,cr}$ the natural frequency approaches zero. The coefficients b_i

in the flexural rigidity are

$$b_0 = \frac{1}{5{,}040a_2}\{(28{,}224a_0 + 11{,}424a_1 + 5{,}712a_2)b_6 + q_0L^3[336a_2 + 168a_2g_1$$

$$+ 115a_2g_2 + (-294a_0 - 119a_1 + 29a_2)g_3]\}$$

$$b_1 = \frac{1}{5{,}040a_2}\{(28{,}224a_0 + 11{,}424a_1 + 5{,}712a_2)b_6 + q_0L^3[126a_2 + 168a_2g_1$$

$$+ 115a_2g_2 + (-294a_0 - 119a_1 + 29a_2)g_3]\}$$

$$b_2 = \frac{1}{5{,}040a_2}\{(-18{,}816a_0 + 11{,}424a_1 + 5{,}712a_2)b_6 - q_0L^3[714a_2 + 322a_2g_1$$

$$+ 165a_2g_2 + (196a_0 - 119a_1 - 181a_2)g_3]\}$$

$$b_3 = \frac{1}{2{,}520a_2}\{(-9{,}408a_0 - 6{,}048a_1 + 2{,}856a_2)b_6 + q_0L^3[168a_2 - 56a_2g_1$$

$$- 30a_2g_2 + (98a_0 + 63a_1 - 38a_2)g_3]\}$$

$$b_4 = \frac{1}{2{,}520a_2}\{(4{,}704a_0 - 6{,}048a_1 - 4{,}200a_2)b_6 + q_0L^3[70a_2g_1$$

$$- 30a_2g_2 + (-49a_0 + 63a_1 + 25a_2)g_3]\}$$

$$b_5 = \frac{1}{504a_2}\{(672a_1 - 840a_2)b_6 + q_0L^3[8a_2g_2 + (-7a_1 + 5a_2)g_3]\} \qquad (8.41)$$

8.1.4.4 General variation of inertial coefficient

In this case, we get the following set of equations:

from ξ^0: $24(-b_1 + b_0) - q_0L^3 = 0$ $\qquad\qquad\qquad\qquad\qquad\qquad$ (8.42a)

from ξ^1: $72(-b_2 + b_1) - q_0L^3(12G + g_1) - a_0\omega^2 = 0$ $\qquad\qquad\qquad$ (8.42b)

from ξ^2: $144(-b_3 + b_2) + q_0L^3(12G + 18 - g_2) - \omega^2 a_1 = 0$ $\qquad\qquad$ (8.42c)

from ξ^3: $240(-b_4 + b_3) - q_0L^3(g_3 - 12g_1 + 16) - \omega^2(a_2 - 2a_0) = 0$ $\qquad$ (8.42d)

from ξ^i: $4 \le i \le m$ $\quad 12(i+1)(i+2)(-b_{i+1} + b_i)(i+1)q_0L^3$

$$\times \left(\frac{g_i}{i+1} - \frac{6g_{i-2}}{i-1} + \frac{4g_{i-3}}{i-2}\right) - \omega^2(a_{i-1} - 2a_{i-3} + a_{i-4}) = 0$$

$$(8.42e)$$

from ξ^{m+1}: $12(m+2)(m+3)(-b_{m+2} + b_{m+1}) + (m+2)q_0L^3$

$$\times \left(\frac{6g_{m-1}}{m} - \frac{4g_{m-2}}{m-1}\right) - \omega^2(-2a_{m-2} + a_{m-3}) = 0 \qquad (8.42f)$$

from ξ^{m+2}: $\quad 12(m+3)(m+4)(-b_{m+3}+b_{m+2})+(m+3)q_0L^3$

$$\times \left(\frac{6g_m}{m+1}-\frac{4g_{m-1}}{m}\right)-\omega^2(-2a_{m-1}+a_{m-2})=0 \qquad (8.42\text{g})$$

from ξ^{m+3}: $\quad 12(m+5)b_{m+3}-(m+4)q_0L^3\dfrac{4g_m}{m+1}-\omega^2a_{m-1}=0 \qquad (8.42\text{h})$

As in the general case for the clamped–free column, Eq. (8.42e) forces the parameter m to be larger than 3. Hence, the cases $m=1$, $m=2$ and $m=3$ must be studied separately. The natural frequency squared obtainable from Eq. (8.42h), is expressed as follows:

$$\omega^2 = 12(m+4)(m+5)b_{m+3}(1-q_0/q_{0,\text{cr}})/a_{m-1} \qquad (8.43)$$

when $q_{0,\text{cr}}$ coincides with its value in Eq. (8.29). This equation exhibits the property established for the particular cases $m=1, m=2$ and $m=3$; namely, when q_0 approaches its critical buckling value, the natural frequency tends to zero. The coefficients in the flexural rigidity are

$$b_0 = b_4 + \frac{(m+4)(m+5)}{60a_{m-1}}(3a_2+5a_1+4a_0)b_{m+3} + \frac{q_0L^3}{36}\left(\frac{3}{20}g_3+\frac{1}{4}g_2\right.$$

$$-\frac{13}{10}g_1 + 3G - \frac{3}{5} - \frac{m+4}{5(m+1)a_{m-1}}(3a_2+5a_1+4a_0)g_m\Bigg)$$

$$b_1 = b_4 + \frac{(m+4)(m+5)}{60a_{m-1}}(3a_2+5a_1+4a_0)b_{m+3} + \frac{q_0L^3}{36}\left(\frac{3}{20}g_3+\frac{1}{4}g_2\right.$$

$$-\frac{13}{10}g_1 + 3G - \frac{21}{10} - \frac{m+4}{5(m+1)a_{m-1}}(3a_2+5a_1+4a_0)g_m\Bigg)$$

$$b_2 = b_4 + \frac{(m+4)(m+5)}{60a_{m-1}}(3a_2+5a_1-6a_0)b_{m+3} + \frac{q_0L^3}{12}\left(\frac{1}{20}g_3+\frac{1}{12}g_2\right.$$

$$-\frac{3}{5}g_1 - G - \frac{7}{10} - \frac{m+4}{15(m+1)a_{m-1}}(3a_2+5a_1-6a_0)g_m\Bigg)$$

$$b_3 = b_4 + \frac{(m+4)(m+5)}{20a_{m-1}}(a_2-2a_0)b_{m+3}$$

$$+\frac{q_0L^3}{60}\left(\frac{1}{4}g_3-3g_1+4-\frac{m+4}{(m+1)a_{m-1}}(a_2-2a_0)g_m\right)$$

for $4 \le i \le m$,

$$b_i = b_{i+1} + \frac{(m+4)(m+5)}{(i+1)(i+2)a_{m-1}}(a_{i-1} - 2a_{i-3} + a_{i-4})b_{m+3}$$

$$\left| \frac{q_0 L^3}{12(i+2)} \left[\frac{g_i}{i+1} - \frac{6g_{i-2}}{i-1} + \frac{4g_{i-3}}{i-2} \right. \right.$$

$$\left. - \frac{4(m+4)}{(m+1)(i+1)a_{m-1}}(a_{i-1} - 2a_{i-3} + a_{i-4})g_m \right]$$

$$b_{m+1} = \left[1 + \frac{m+5}{m+3}\left(\frac{a_{m-2}}{a_{m-1}} - 2\right) + \frac{(m+4)(m+5)}{(m+2)(m+3)a_{m-1}}(-2a_{m-2} + a_{m-3}) \right] b_{m+3}$$

$$- \frac{q_0 L^3}{m+1} g_m \left[\frac{1}{2(m+4)} + \frac{1}{3(m+3)}\left(\frac{a_{m-2}}{a_{m-1}} - 2\right) \right.$$

$$+ \frac{m+4}{3(m+2)(m+3)} \frac{-2a_{m-2} + a_{m-3}}{a_{m-1}} \right]$$

$$+ \frac{q_0 L^3}{m} g_{m-1} \left[\frac{1}{3(m+4)} - \frac{1}{2(m+3)} \right] + \frac{q_0 L^3}{3(m-1)(m+3)} g_{m-2}$$

$$b_{m+2} = \left[1 + \frac{m+5}{m+3}\left(\frac{a_{m-2}}{a_{m-1}} - 2\right) \right] b_{m+3}$$

$$- \frac{q_0 L^3}{m+1} g_m \left[\frac{1}{2(m+4)} + \frac{1}{3(m+3)}\left(\frac{a_{m-2}}{a_{m-1}} - 2\right) \right] + \frac{q_0 L^3}{3m(m+4)} g_{m-1}$$

$$\tag{8.44}$$

8.1.5 Column that is clamped at its ends

According to the boundary conditions it is convenient to choose the following polynomial function as the buckling mode of the column:

$$\psi(\xi) = \xi^2 - 2\xi^3 + \xi^4 \tag{8.45}$$

We have to consider first the two particular cases $m = 1$ and $m = 2$, prior to presenting the general case for m larger than 2.

8.1.5.1 Particular case: constant mass density ($m = 1$)

To satisfy the governing differential equation the following equations should be valid:

$$4b_2 - 24b_1 + 24b_0 + q_0 L^3(2 + g_1) = 0$$

$$12b_3 - 72b_2 + 72b_1 - q_0 L^3(16 + 6g_1) = 0$$

$$24b_4 - 144b_3 + 144b_2 + q_0L^3(30 + 3g_1) - \omega^2 a_0 = 0$$

$$-240b_4 + 240b_3 + q_0L^3(-16 + 12g_1) + 2\omega^2 a_0 = 0$$

$$360b_4 - 10q_0L^3g_1 - \omega^2 a_0 = 0$$

$$(8.46)$$

The last equation yields the expression for the natural frequency squared, which can be written as follows:

$$\omega^2 = 360b_4(1 - q_0/q_{0,cr})/a_0 \qquad (8.47)$$

where $q_{0,cr}$ is the critical buckling value of the parameter q_0, and when q_0 tends to $q_{0,cr}$ the natural frequency vanishes.

$$b_0 = \frac{11}{18}b_4 + \frac{q_0L^3}{4320}(42 - 49g_1) \quad b_1 = \frac{2}{3}b_4 + \frac{q_0L^3}{144}(10 + 3g_1)$$

$$b_2 = \frac{1}{3}b_4 - \frac{q_0L^3}{720}(162 + 41g_1) \quad b_3 = -2b_4 + \frac{q_0L^3}{30}(2 + g_1)$$

$$(8.48)$$

8.1.5.2   *Particular case: linearly varying inertial coefficient* ($m = 2$)

Instead of Eq. (8.46) we get the following set:

$$4b_2 - 24b_1 + 24b_0 + q_0L^3\left(2 + g_1 + \tfrac{2}{3}g_2\right) = 0$$

$$12b_3 - 72b_2 + 72b_1 - q_0L^3(16 + 6g_1 + 4g_2) = 0$$

$$24b_4 - 144b_3 + 144b_2 + q_0L^3(30 + 3g_1 + 4g_2) - \omega^2 a_0 = 0$$

$$40b_5 - 240b_4 + 240b_3 + q_0L^3\left(-16 + 12g_1 - \tfrac{8}{3}g_2\right) + \omega^2(2a_0 - a_1) = 0$$

$$-360b_5 + 360b_4 + q_0L^3(-10g_1 + 10g_2) - \omega^2(a_0 - 2a_1) = 0$$

$$504b_5 - 8q_0L^3g_2 - \omega^2 a_1 = 0$$

$$(8.49)$$

The last equation yields the natural frequency squared. As in the case of the clamped–free column we obtain

$$\omega^2 = 504b_5(1 - q_0/q_{0,cr})/a_1 \qquad (8.50)$$

Again, $q_{0,cr}$ is the critical buckling load; when q_0 approaches its buckling value, the natural frequency approaches zero. The coefficients in the flexural rigidity are

$$b_0 = \frac{1}{12{,}960a_1}\{(11{,}088a_0 + 4392a_1)b_5$$

$$+ q_0L^3[73a_1g_1 + 126a_1 - (176a_0 + 18a_1)g_2]\}$$

$$b_1 = \frac{1}{2{,}160a_1}\{(2{,}016a_0 + 888a_1)b_5 + q_0L^3[85a_1g_1 + 150a_1 - (32a_0 - 44a_1)g_2]\}$$

$$b_2 = \frac{1}{2{,}160a_1}\{(1{,}008a_0 + 936a_1)b_5$$

$$- q_0L^3[103a_1g_1 - 306a_1 - (16a_0 + 78a_1)g_2]\}$$

$$b_3 = \frac{1}{180a_1}\{(-504a_0 + 24a_1)b_5 + q_0L^3[-4a_1g_1 + 12a_1 + (8a_0 - a_1)g_2]\}$$

$$b_4 = \frac{1}{180a_1}\{(252a_0 - 324a_1)b_5 + q_0L^3[5a_1 - (4a_0 - 3a_1)g_2]\}$$

$$(8.51)$$

8.1.5.3 General case of variation of the inertial coefficient ($m \geq 3$)

Concerning the general case, we get the following set of equations:

from ξ^0: $\quad 2(2b_2 - 12b_1 + 12b_0) + 2q_0L^3G = 0$ $\hspace{2cm}$ (8.52a)

from ξ^1: $\quad 6(2b_3 - 12b_2 + 12b_1) - 2q_0L^3(6G + 2) = 0$ $\hspace{1cm}$ (8.52b)

from ξ^2: $\quad 12(2b_4 - 12b_3 + 12b_2) + 3q_0L^3(4G + 6 - g_1) - \omega^2a_0 = 0$ $\hspace{0.3cm}$ (8.52c)

from ξ^3: $\quad 20(2b_5 - 12b_4 + 12b_3) - 4q_0L^3\left(\tfrac{2}{3}g_2 - 3g_1 + 4\right)$

$$- \omega^2(a_1 - 2a_0) = 0 \hspace{3cm} (8.52\text{d})$$

from ξ: $4 \leq i \leq m+1$ $\quad (i+1)(i+2)(2b_{i+2} - 12b_{i+1} + 12b_i)$

$$- (i+1)q_0L^3\left(\frac{2g_{i-1}}{i} - \frac{6g_{i-2}}{i-1} + \frac{4g_{i-3}}{i-2}\right)$$

$$- \omega^2(a_{i-2} - 2a_{i-3} + a_{i-4}) = 0 \hspace{2cm} (8.52\text{e})$$

from ξ^{m+2}: $\quad (m+3)(m+4)(-12b_{m+3} + 12b_{m+2})$

$$+ (m+3)q_0L^3\left(\frac{6g_m}{m+1} - \frac{4g_{m-1}}{m}\right)$$

$$- A\omega^2(-2\rho_{m-1} + \rho_{m-2}) = 0 \hspace{2cm} (8.52\text{f})$$

from ξ^{m+3}: $\quad 12(m+5)b_{m+3} - (m+4)q_0L^3\frac{4g_m}{m+1} - \omega^2a_{m-1} = 0$ $\hspace{0.5cm}$ (8.52g)

As for the clamped–free column, we introduce

$$q_{0,\mathrm{cr}} = 3(m+1)(m+5)b_{m+3}/g_m L^3 \tag{8.53}$$

Therefore, we can express the natural frequency squared as follows:

$$\omega^2 = 12(m+4)(m+5)b_{m+3}(1 - q_0/q_{0,\mathrm{cr}})/a_{m-1} \tag{8.54}$$

The property that, when q_0 approaches its critical buckling value, the natural frequency tends to zero, stated for the particular cases $m = 1$ and $m = 2$, is still valid in this general case, $m \geq 3$. The coefficients describing the flexural rigidity read

$$b_0 = \frac{19}{36}b_4 - \frac{1}{9}b_5 + \frac{(m+4)(m+5)}{300a_{m-1}}\left(\frac{19}{2}a_1 + a_0\right)b_{m+3}$$

$$+ \frac{q_0 L^3}{360}\left(\frac{11}{6}g_2 - \frac{39}{12}g_1 + 5G - \frac{3}{2} - \frac{m+4}{(m+1)a_{m-1}}\left(4\rho_1 + \frac{13}{3}a_0\right)g_m\right)$$

$$b_1 = \frac{2}{3}b_4 - \frac{5}{36}b_5 - \frac{(m+4)(m+5)}{24a_{m-1}}a_1 b_{m+3}$$

$$+ \frac{q_0 L^3}{360}\left(\frac{5}{2}g_2 - \frac{15}{2}g_1 + 30G - 5 - \frac{m+4}{(m+1)a_{m-1}}(5a_1 + 4a_0)g_m\right)$$

$$b_2 = \frac{5}{6}b_4 - \frac{1}{6}b_5 + \frac{(m+4)(m+5)}{60a_{m-1}}(3a_1 - a_0)b_{m+3}$$

$$+ \frac{q_0 L^3}{360}\left(4g_2 - \frac{21}{2}g_1 - 30G - 21 - \frac{m+4}{(m+1)a_{m-1}}(6a_1 + 2a_0)g_m\right)$$

$$b_3 = b_4 - \frac{1}{6}b_5 + \frac{(m+4)(m+5)}{20a_{m-1}}(a_1 - 2a_0)b_{m+3}$$

$$+ \frac{q_0 L^3}{60}\left(\frac{2}{3}g_2 - 3g_1 + 4 - \frac{m+4}{(m+1)a_{m-1}}(a_1 - 2a_0)g_m\right)$$

for $4 \leq i \leq m+1$,

$$b_i = b_{i+1} - \frac{1}{6}b_{i+2} + \frac{(m+4)(m+5)}{(i+1)(i+2)a_{m-1}}(a_{i-2} - 2a_{i-3} + a_{i-4})b_{m+3}$$

$$+ \frac{q_0 L^3}{12(i+2)}\left[\frac{2g_{i-1}}{i} - \frac{6g_{i-2}}{i-1} + \frac{4g_{i-3}}{i-2}\right.$$

$$\left. - \frac{4(m+4)}{(m+1)(i+1)a_{m-1}}(a_{i-2} - 2a_{i-3} + a_{i-4})g_m\right]$$

$$b_{m+2} = \left[1 + \frac{m+5}{m+3}\left(\frac{a_{m-2}}{a_{m-1}} - 2\right)\right]b_{m+3}$$

$$- \frac{q_0 L^3}{3(m+4)}g_m\left[\frac{3}{2(m+1)} + \frac{m+4}{(m+1)(m+3)}\left(\frac{a_{m-2}}{a_{m-1}} - 2\right)\right]$$

$$+ \frac{q_0 L^3}{3m(m+4)}g_{m-1}$$

$$(8.55)$$

8.1.6 Column that is Pinned at One End and Clamped at the Other

The pre-selected mode shape in accordance with the boundary conditions reads

$$\psi(\xi) = \xi - 3\xi^3 + 2\xi^4 \tag{8.56}$$

We consider three particular cases, namely $m = 1$, $m = 2$ and $m = 3$, prior to the study of the general case when $m \geq 4$.

8.1.6.1 Particular case: constant inertial coefficient ($m = 1$)

The set of equations resulting from the governing Eq. (8.11) is

$$-36b_1 + 48b_0 - q_0 L^3 = 0$$

$$-108b_2 + 144b_1 - q_0 L^3(18 + 10g_1) - \omega^2 a_0 = 0$$

$$-216b_3 + 288b_2 + q_0 L^3(51 + 12g_1) = 0 \tag{8.57}$$

$$-360b_4 + 480b_3 + q_0 L^3(-32 + 18g_1) + 3\omega^2 a_0 = 0$$

$$720b_4 - 20q_0 L^3 g_1 - 2\omega^2 a_0 = 0$$

The last equation of the set (8.57) leads to the following expression for the natural frequency squared:

$$\omega^2 = 360b_4(1 - q_0/q_{0,cr})/a_0 \tag{8.58}$$

This expression coincides with Eq. (8.21), while the expression for $q_{0,cr}$ is the same as in Eq. (8.20). The coefficients b_i in the flexural rigidity are given by

$$b_0 = \frac{159}{128}b_4 + q_0 L^3\left(\frac{331}{7680} - \frac{33}{2560}g_1\right) \quad b_1 = \frac{53}{32}b_4 + \frac{1}{640}q_0 L^3(19 - 11g_1)$$

$$b_2 = -\frac{9}{8}b_4 - \frac{1}{480}q_0 L^3(61 + 11g_1) \quad b_3 = -\frac{3}{2}b_4 + q_0 L^3\left(\frac{1}{15} + \frac{1}{40}g_1\right)$$

$$(8.59)$$

8.1.6.2   Particular case: linearly varying inertial coefficient ($m = 2$)
In the place of set (8.57) we obtain the set

$$-36b_1 + 48b_0 - q_0L^3 = 0$$

$$-108b_2 + 144b_1 - q_0L^3(18 + 10g_1 + 6g_2) - \omega^2 a_0 = 0$$

$$-216b_3 + 288b_2 + q_0L^3(51 + 12g_1 + 7g_2) - \omega^2 a_1 = 0$$

$$-360b_4 + 480b_3 + q_0L^3(-32 + 18g_1) + 3\omega^2 a_0 = 0 \qquad (8.60)$$

$$-540b_5 + 720b_4 + q_0L^3(-20g_1 + 15g_2) - \omega^2(a_0 - 2a_1) = 0$$

$$1008b_5 - 16q_0L^3g_2 - 2\omega^2 a_1 = 0$$

yielding the natural frequency squared:

$$\omega^2 = 504b_5(1 - q_0/q_{0,cr})/a_1 \qquad (8.61)$$

Equation (8.61) always coincides with Eq. (8.25), with $q_{0,cr}$ coinciding with Eq. (8.24). The coefficients b_i read

$$b_0 = \frac{1}{61,440a_1}\{(106,848a_0 + 34,236a_1)b_5$$

$$+ q_0L^3[1,328a_1g_1 + 2,648a_1 - (1,696a_0 - 363a_1)g_2]\}$$

$$b_1 = \frac{1}{46,080a_1}\{(106,848a_0 + 34,236a_1)b_5$$

$$+ q_0L^3[1,328a_1g_1 + 1,638a_1 - (1,696a_0 - 363a_1)g_2]\}$$

$$b_2 = \frac{1}{3,840a_1}\{(-6,048a_0 + 3,804a_1)b_5$$

$$+ q_0L^3[-208a_1g_1 - 488a_1 + (96a_0 - 173a_1)g_2]\}$$

$$b_3 = \frac{1}{960a_1}\{(-2,016a_0 - 972a_1)b_5 + q_0L^3[-16a_1g_1 - 64a_1 + (9a_0 + 32a_1)g_2]\}$$

$$b_4 = \frac{1}{720a_1}\{(1,008a_0 - 972a_1)b_5 + q_0L^3[20a_1g_1 - (16a_0 - 9a_1)g_2]\}$$

$$(8.62)$$

8.1.6.3 Particular case: parabolically varying inertial coefficient ($m = 3$)

Instead of the set (8.57) we have

$$-36b_1 + 48b_0 - q_0 L^3 = 0$$

$$-108b_2 + 144b_1 - q_0 L^3 \left(18 + 10g_1 + 6g_2 + \tfrac{9}{2}g_3\right) - \omega^2 a_0 = 0$$

$$-216b_3 + 288b_2 + q_0 L^3 (51 + 12g_1 + 7g_2 + 6g_3) - \omega^2 a_1 = 0$$

$$-360b_4 + 480b_3 + q_0 L^3 (-32 + 18g_1 - g_3) - \omega^2 (-3a_0 + a_2) = 0 \qquad (8.63)$$

$$-540b_5 + 720b_4 + q_0 L^3 (-20g_1 + 15g_2) - \omega^2 (2a_0 - 3a_1) = 0$$

$$-756b_6 + 1008b_5 + q_0 L^3 \left(-16g_2 + \tfrac{27}{2}g_3\right) - \omega^2 (2a_1 - 3a_2) = 0$$

$$1344b_6 - 14q_0 L^3 g_2 - 2\omega^2 a_1 = 0$$

In the same way as the pinned column, the natural frequency squared is determined as follows:

$$\omega^2 = 672b_6 (1 - q_0/q_{0,\mathrm{cr}})/a_2 \qquad (8.64)$$

The coefficients b_i are

$$b_0 = \frac{1}{3,440,640a_2} \{ (7,977,984a_0 + 2,556,288a_1 + 1,011,528a_2)b_6 + q_0 L^3 [148,288a_2$$
$$+ 74,368a_2 g_1 + 50,760a_2 g_2 + (-83,104a_0 - 26,628a_1 + 28,251a_2)g_3] \}$$

$$b_1 = \frac{1}{2,580,480a_2} \{ (7,977,984a_0 + 2,556,288a_1 + 1,011,528a_2)b_6 + q_0 L^3 [76,608a_2$$
$$+ 74,368a_2 g_1 + 50,760a_2 g_2 + (-83,104a_0 - 26,628a_1 + 28,251a_2)g_3] \}$$

$$b_2 = \frac{1}{645,120a_2} \{ (-1,354,752a_0 + 852,096a_1 + 337,176a_2)b_6 - q_0 L^3 [81,984a_2$$
$$+ 34,944a_2 g_1 + 18,920a_2 g_2 - (14,112a_0 + 8,876a_1 + 17,463a_2)g_3] \}$$

$$b_3 = \frac{1}{5,376a_2} \{ (-150,528a_0 - 72,576a_1 + 37,464a_2)b_6 + q_0 L^3 [3,584a_2 - 896a_2 g_1$$
$$- 360a_2 g_2 + (1,568a_0 + 756a_1 - 447a_2)g_3] \}$$

$$b_4 = \frac{1}{40,320a_2} \{ (75,264a_0 - 72,576a_1 - 37,800a_2)b_6 + q_0 L^3 [1,120a_2 g_1$$
$$- 360a_2 g_2 + (-784a_0 + 756a_1 + 225a_2)g_3] \}$$

$$b_5 = \frac{1}{2,016a_2} \{ (2,688a_1 - 2,520a_2)b_6 + q_0 L^3 [32a_2 g_2 + (-28a_1 + 15a_2)g_3] \}$$

$$(8.65)$$

8.1.6.4   General variation of the inertial coefficient

In this case we get following set of equations:

from ξ^0: $2(-18b_1 + 24b_0) - q_0L^3 = 0$ (8.66a)

from ξ^1: $6(-18b_2 + 24b_1) - q_0L^3(18G + g_1) - a_0\omega^2 = 0$ (8.66b)

from ξ^2: $12(-18b_3 + 24b_2) + q_0L^3(24G + 27 - g_2) - \omega^2 a_1 = 0$ (8.66c)

from ξ^3: $20(-18b_4 + 24b_3) - q_0L^3(g_3 - 18g_1 + 32) - \omega^2(a_2 - 3a_0) = 0$

(8.66d)

from ξ^i: $4 \le i \le m$ $(i+1)(i+2)(-18b_{i+1} + 24b_i)$

$$- (i+1)q_0L^3 \left(\frac{g_i}{i+1} - \frac{9g_{i-2}}{i-1} + \frac{8g_{i-3}}{i-2} \right)$$

$$- \omega^2(a_{i-1} - 3a_{i-3} + 2a_{i-4}) = 0$$ (8.66e)

from ξ^{m+1}: $(m+2)(m+3)(-18b_{m+2} + 24b_{m+1})$

$$+ (m+2)q_0L^3 \left(\frac{9g_{m-1}}{m} - \frac{8g_{m-2}}{m-1} \right) - \omega^2(-3a_{m-2} + 2a_{m-3}) = 0$$

(8.66f)

from ξ^{m+2}: $(m+3)(m+4)(-18b_{m+3} + 24b_{m+2})$

$$+ (m+3)q_0L^3 \left(\frac{9g_m}{m+1} - \frac{8g_{m-1}}{m} \right) - \omega^2(-3a_{m-1} + 2a_{m-2}) = 0$$

(8.66g)

from ξ^{m+3}: $24(m+5)b_{m+3} - (m+4)q_0L^3\frac{8g_m}{m+1} - \omega^2 a_{m-1} = 0$ (8.66h)

The natural frequency squared, obtainable from Eq. (8.66h), is expressed as follows

$$\omega^2 = 12(m+4)(m+5)b_{m+3}(1 - q_0/q_{0,cr})/a_{m-1}$$ (8.67)

when $q_{0,cr}$ coincides with its value in Eq. (8.29). This equation presents the property stated for the particular cases $m = 1$ and $m = 2$; namely, when q_0 approaches its critical buckling value, the natural frequency tends to zero. The coefficients in the flexural rigidity are

$$b_0 = \frac{71}{256}b_4 + \frac{(m+4)(m+5)}{2560a_{m-1}}(27a_2 + 60a_1 + 169a_0)b_{m+3}$$

$$+ \frac{1}{12}q_0L^3 \left(\frac{3}{160}g_3 + \frac{1}{24}g_2 - \frac{17}{80}g_1 + \frac{1}{2}G - \frac{11}{40} \right.$$

$$\left. - \frac{m+4}{(m+1)24a_{m-1}}(24a_2 + 3a_1 - 64a_0)g_m \right)$$

$$b_1 = \frac{27}{64}b_4 + \frac{(m+4)(m+5)}{1920a_{m-1}}(27a_2 + 60a_1 + 169a_0)b_{m+3}$$

$$+ \frac{1}{12}q_0 L^3 \left(\frac{3}{160}g_3 + \frac{1}{24}g_2 - \frac{17}{80}g_1 + \frac{1}{2}G - \frac{21}{40} \right.$$

$$\left. - \frac{m+4}{(m+1)24a_{m-1}}(24a_2 + 3a_1 - 64a_0)g_m \right)$$

$$b_2 = \frac{9}{16}b_4 + \frac{(m+4)(m+5)}{480a_{m-1}}(9a_2 + 20a_1 - 27a_0)b_{m+3}$$

$$+ \frac{1}{12}q_0 L^3 \left(\frac{3}{160}g_3 + \frac{1}{24}g_2 - \frac{27}{80}g_1 - G - \frac{21}{40} \right.$$

$$\left. - \frac{m+4}{8(m+1)a_{m-1}} \frac{8a_2 + a_1 - 24a_0}{} g_m \right)$$

$$b_3 = \frac{3}{4}b_4 + \frac{(m+4)(m+5)}{40a_{m-1}}(a_2 - 3a_0)b_{m+3}$$

$$+ \frac{1}{120}q_0 L^3 \left(\frac{1}{4}g_3 - \frac{9}{2}g_1 + 8 - \frac{m+4}{(m+1)a_{m-1}}(a_2 - 3a_0)g_m \right)$$

for $4 \le i \le m$,

$$b_i = \frac{3}{4}b_{i+1} + \frac{(m+4)(m+5)}{2(i+1)(i+2)\rho_{m-1}}(\rho_{i-1} - 3\rho_{i-3} + 2\rho_{i-4})b_{m+3}$$

$$+ \frac{q_0 L^3}{24(i+2)}\left[\frac{g_i}{i+1} - \frac{9g_{i-2}}{i-1} + \frac{8g_{i-3}}{i-2} \right. \tag{8.68}$$

$$\left. - \frac{4(m+4)}{(m+1)(i+1)a_{m-1}}(a_{i-1} - 3a_{i-3} + 2a_{i-4})g_m \right]$$

$$b_{m+1} = \left[\frac{9}{8} + \frac{3}{8}\frac{m+5}{m+3}\left(2\frac{a_{m-2}}{a_{m-1}} - 3 \right) \right.$$

$$+ \frac{(m+4)(m+5)}{2(m+2)(m+3)a_{m-1}}(-3a_{m-2} + 2a_{m-3}) \bigg] b_{m+3}$$

$$- \frac{q_0 L^3}{24(m+1)}g_m \left[\frac{27}{4(m+4)} + \frac{1}{3(m+3)}\left(2\frac{a_{m-2}}{a_{m-1}} - 3 \right) \right.$$

$$+ \frac{3(m+4)}{(m+2)(m+3)a_{m-1}}(-3a_{m-2} + 2a_{m-3}) \bigg]$$

$$+ \frac{q_0 L^3}{m}g_{m-1}\left[\frac{1}{3(m+4)} - \frac{3}{8(m+3)} \right] + \frac{q_0 L^3}{3(m-1)(m+3)}g_{m-2}$$

$$b_{m+2} = \left[\frac{3}{4} + \frac{1}{2}\frac{m+5}{m+3}\left(2\frac{a_{m-2}}{a_{m-1}} - 3 \right) \right] b_{m+3} + \frac{q_0 L^3}{3m(m+4)}g_{m-1}$$

$$- \frac{q_0 L^3}{24(m+4)}g_m\left[\frac{9}{2(m+1)} + \frac{4(m+4)}{3(m+3)}\left(2\frac{a_{m-2}}{a_{m-1}} - 3 \right) \right]$$

8.1.7   Concluding Remarks

We have established a linear relationship between the natural frequency squared and the load intensity ratio to be valid for uncovered inhomogeneous beams for different sets of boundary conditions. This was made possible by obtaining closed-form solutions for both the buckling intensity and natural frequency.

   Note that such a linear relationship can be obtained albeit approximately, for an inhomogeneous beam. Indeed, within the Boobnov–Galerkin method, instead of the exact mode shape, its approximation $\varphi(\xi)$ is utilized. Therefore, in the general case, the left-hand side of Eq. (8.11) denoted by $\varepsilon(\xi)$, and usually referred to as a residual, does not vanish generally. The residual $\varepsilon(\xi)$ is required to be orthogonal to the coordinate function $\varphi(\xi)$:

$$\langle \varepsilon, \varphi \rangle = 0 \qquad (8.69)$$

where the inner product, identified by angle brackets, is defined as

$$\langle \varepsilon, \varphi \rangle = \int_0^1 \varepsilon(\xi)\varphi(\xi)\, d\xi \qquad (8.70)$$

The approximation for the natural frequency squared ω_{app}^2 is obtained by setting q_0 to be identically zero. For ω_{app}^2, we obtain

$$\omega_{\mathrm{app}}^2 = \frac{1}{\langle \rho A \varphi, \varphi \rangle} \left\langle \frac{d^2}{d\xi^2}\left[D(\xi)\frac{d^2\varphi}{d\xi^2} \right], \varphi \right\rangle \qquad (8.71)$$

Likewise, the approximation of the buckling intensity $q_{0,\mathrm{app}}$ is obtained by dropping the last term in Eq. (8.11) and utilizing the Boobnov–Galerkin procedure. This leads to the following expression:

$$q_{0,\mathrm{app}} = -\left\langle \frac{d}{d\xi}\sum_{i=0}^{m}\frac{g_i}{i+1}(1-\xi^{i+1})\frac{d\varphi}{d\xi}, \varphi \right\rangle^{-1} \left\langle \frac{d^2}{d\xi^2}\left[D(\xi)\frac{d^2\varphi}{d\xi^2} \right], \varphi \right\rangle \qquad (8.72)$$

The approximation of the Boobnov–Galerkin method to the original Eq. (8.11) without neglecting any term leads to the natural frequency squared:

$$\omega_{\mathrm{app}}^2 = \langle \rho A \varphi, \varphi \rangle^{-1}\left[\left\langle \frac{d^2}{d\xi^2}\left[D(\xi)\frac{d^2\varphi}{d\xi^2} \right], \varphi \right\rangle \right.$$

$$\left. + q_0 L^3 \left\langle \frac{d}{d\xi}\sum_{i=0}^{m}\frac{g_i}{i+1}(1-\xi^{i+1})\frac{d\varphi}{d\xi}, \varphi \right\rangle \right] \qquad (8.73)$$

The first term in Eq. (8.73) can be rewritten, in view of Eq. (8.71) as ω_{app}^2/α whereas the second term can be expressed as $\omega_{app,0}^2\alpha$, where $\alpha = q_0/q_{0,app}$, leading to the linear relationship

$$\omega_{app}^2 = \omega_{app,0}^2(1 - \alpha) \tag{8.74}$$

It should be noted that Eq. (8.74) is also derivable by the Rayleigh method (Galef, 1968) (see also Massonet (1940, 1972)). According to Galef, "it is more than a fortunate happenstance that permits the careful analysis and computing effort of Amba–Rao (1967) to be expressed so concisely and accurately." According to Stephen (1989) the "less well-known Galef (1968) formula" is "important for preliminary design calculations and provide useful checks for numerical predictions, particularly the finite element method." Galef's (1968) formula was utilized for the vibrating beam with axially distributed loading by Pilkington and Carr (1970). Bokaian (1988) concluded in his work, that "Galef's formula, previously assumed to be valid for beams with all types of end conditions, is observed to be valid only for a few." Bokaian (1989) suggested a modified Galef's formula as $\omega^2 = \omega_0^2(1 - \gamma\alpha)$, where γ is an empirical corrective coefficient depending on the type of end conditions; depending on the boundary conditions, γ may be greater than equal to or less than unity.

In the present section we established that Galef's formula holds for specific *inhomogeneous* columns under four sets of boundary conditions, with preselected polynomial mode shapes. Thus, its validity is shown to extend to some inhomogeneous columns, under boundary conditions different from the pinned or guided ones, as was previously established by Bokaian (1988) for the *uniform* columns, under axial compression. Also, the linear relationship, which is of an *approximate* nature for the uniform columns with distributed loading (Pilkington and Carr, 1970) turns out to be *exact* for inhomogeneous columns.

The advantage of the present study over the Rayleigh or Boobnov–Galerkin approximations lies in the fact that our derivation is *exact*; moreover, the natural frequency and the buckling load are written in *closed form*. It should be stressed that whereas the approximate relationship in Eq. (7.74) is applicable for any beam, the exact one is valid only for specific beams, constructed in this section.

8.2 A Fifth-Order Polynomial that Serves as both the Buckling and Vibration Modes of an Inhomogeneous Structure

8.2.1 Introductory Comments

Vibration of homogeneous uniform beams in the presence of axial loads is a classical topic included in almost any vibration textbook. For the beam that is pinned at both ends the natural frequency squared turns out to be a linear function of the applied concentrated load. In non-uniform beams, or in beams that are not pinned at both ends, there is no reason to expect the above linear relationship to hold in exact terms. In the present study, surprisingly, such a validity of the linear relationship is established.

It appears to be instructive to list both the vibration and buckling modes of uniform beams under different boundary conditions. The vibration mode shape is denoted by $\psi_V(\xi)$, whereas the buckling mode is designated by $\psi_B(\xi)$, where ξ is a non-dimensional axial coordinate $\xi = x/L$, L being the length of the beam and x the axial coordinate. For the beam that is pinned at both its ends

$$\psi_V(\xi) = \psi_B(\xi) = \sin(m\pi\xi) \tag{8.75}$$

where m is a positive integer, denoting the serial number of the mode shape. For the beam that is clamped at both its ends

$$\psi_V(\xi) = \sinh(\lambda_m\xi) - \sin(\lambda_m\xi) + \alpha_m[\cosh(\lambda_m\xi) - \cos(\lambda_m\xi)]$$
$$\alpha_m = (\sinh\lambda_m - \sin\lambda_m)/(\cos\lambda_m - \cosh\lambda_m) \tag{8.76}$$

where λ_m satisfies the transcendental equation

$$\cos\lambda_m \cosh\lambda_m - 1 = 0 \tag{8.77}$$

with roots found numerically; for example, $\lambda_1 = 4.73$, $\lambda_2 = 7.853$, $\lambda_3 = 10.996$, etc. The mode buckling shape reads

$$\psi_B(\xi) = \cos(2m\pi\xi) - 1 \tag{8.78}$$

For the cantilever beam

$$\psi_V(\xi) = \sin(\lambda_m\xi) - \sinh(\lambda_m\xi) - \beta_m[\cos(\lambda_m\xi) - \cosh(\lambda_m\xi)]$$
$$\beta_m = (\sinh\lambda_m + \sin\lambda_m)/(\cos\lambda_m + \cosh\lambda_m) \tag{8.79}$$

where λ_m satisfies the following transcendental equation

$$\cos\lambda_m \cosh\lambda_m - 1 = 0 \tag{8.80}$$

with $\lambda_1 = 1.875104$, $\lambda_2 = 4.694091$, $\lambda_3 = 7.854757$, etc. The buckling mode reads

$$\psi_B(\xi) = \cos[(2m - 1)\pi\xi] - 1 \tag{8.81}$$

For the beam that is clamped at the left end and pinned at the right end,

$$\psi_V(\xi) = \sinh(\lambda_m \xi) - \sin(\lambda_m \xi) + \gamma_m[\cosh(\lambda_m \xi) - \cos(\lambda_m \xi)]$$
$$\gamma_m = (\sin \lambda_m - \sinh \lambda_m)/(\cos \lambda_m - \cosh \lambda_m) \tag{8.82}$$

where λ_m satisfies the following transcendental equation:

$$\tan \lambda_m = \tanh \lambda_m \tag{8.83}$$

with $\lambda_1 = 3.927$, $\lambda_2 = 7.069$, $\lambda_3 = 10.210$, etc. The buckling mode reads

$$\psi_B(\xi) = \sin(\delta_m \xi) - \delta_m \xi + \delta_m[1 - \cos(\delta_m \xi)] \tag{8.84}$$

where δ_m satisfies the equation

$$\tan \delta_m = \delta_m \tag{8.85}$$

As seen, only in the case of the uniform beam that is pinned at both ends do the vibration and buckling modes coincide, both representing a sinusoidal function.

For the inhomogeneous beams, at first glance, it may appear that there could be no compelling reason for the vibration and buckling mode shapes to coincide. Yet, we will demonstrate that the initial intuition may be wrong under some special circumstances. We postulate *ab initio* that these mode shapes coincide. We construct beams that have this interesting property. Remarkably, it turns out that such a formulation leads to a non-trivial solution, what will be reported in what follows.

8.2.2 Formulation of the Problem

Let us first solve a simple auxiliary problem of bending of homogeneous beams. Consider, first, a uniform beam, pinned at both its ends. It is subjected to the load $q(\xi) = q_0 \xi^n$. The governing differential equation reads

$$\frac{EId^4 W}{dx^4} = q_0 x^n \tag{8.86}$$

We introduce the non-dimensional coordinate

$$\xi = x/L \tag{8.87}$$

where L is the length. Then, Eq. (8.12) becomes

$$\frac{d^4 W}{d\xi^4} = \alpha\xi^n \quad \alpha = q_0 L^{n+4}/EI \tag{8.88}$$

The solution W_n of this equation that satisfies the boundary conditions reads

$$W_n(\xi) = \frac{\alpha L^{n+4}}{(n+1)(n+2)(n+3)(n+4)}\psi(\xi) \tag{8.89}$$

$$\psi_n(\xi) = \xi^{n+4} - \tfrac{1}{6}(n^2 + 7n + 12)\xi^3 + \tfrac{1}{6}(n^2 + 7n + 6)\xi$$

For the beam that is clamped at one end and free at the other the solution of Eq. (8.14) subject to appropriate boundary conditions reads

$$W_n(\xi) = \frac{\alpha}{(n+1)(n+2)(n+3)(n+4)}\psi(\xi) \tag{8.90}$$

$$\psi_n(\xi) = \xi^{n+4} - \tfrac{1}{6}(n+2)(n+3)(n+4)\xi^3 + \tfrac{1}{2}(n+1)(n+3)(n+4)\xi^2$$

For the beam that is clamped at both its ends, we get

$$W_n(\xi) = \frac{\alpha}{(n+1)(n+2)(n+3)(n+4)}\psi(\xi) \tag{8.91}$$

$$\psi_n(\xi) = \xi^{n+4} - (n+2)\xi^3 + (n+1)\xi^2$$

Finally, for the beam that is pinned at one end and clamped at the other, the functions $w(\xi)$ and $\psi(\xi)$ are, respectively,

$$W_n(\xi) = \frac{\alpha}{2(n+1)(n+2)(n+3)(n+4)}\psi(\xi) \tag{8.92}$$

$$\psi_n(\xi) = 2\xi^{n+4} - (n+3)\xi^3 + (n+1)\xi$$

We address the following problem: for an inhomogeneous vibrating beam under axially distributed loading whose exact mode shape (both in vibration and in buckling settings) is $\psi(\xi)$ when $n = 1$, reconstruct the axial rigidity of the beam. Thus, we postulate the fifth-order polynomial to be the exact mode shape of the sought for inhomogeneous beam. We will explore cases of reconstruction of the flexural rigidity $D(\xi)$ for various axial variations of the inertial coefficient, $R(\xi) = \rho(\xi)A$, and the distributed axial load $q(\xi)$.

8.2.3 Basic Equations

The differential equation of the inhomogeneous vibrating column, which is under distributed axial load reads

$$\frac{d^2}{dx^2}\left[D(x)\frac{d^2 W}{dx^2}\right] - \frac{d}{dx}\left[N(x)\frac{dW}{dx}\right] - R(x)\omega^2 W(x) = 0 \qquad (8.93)$$

where $D(x)$ is the flexural rigidity, $N(x)$ the axial force, $R(\xi)$ the inertial coefficient, ω the natural frequency, $W(x)$ the buckling mode and x the axial coordinate. With the non-dimensional coordinate $\xi = x/L$, Eq. (8.93) becomes

$$\frac{d^2}{d\xi^2}\left[D(\xi)\frac{d^2\psi}{d\xi^2}\right] - L^2\frac{d}{d\xi}\left[N(\xi)\frac{d\psi}{d\xi}\right] - R(\xi)\omega^2\psi(\xi) = 0 \qquad (8.94)$$

We consider for the inertial coefficient $R(\xi)$ an mth order polynomial given by

$$R(\xi) = \sum_{i=0}^{m} a_i\xi^i \qquad (8.95)$$

Considering the degree of the remaining member of the governing equation, the flexural rigidity $D(\xi)$ is taken to be equal to

$$D(\xi) = \sum_{i=0}^{m+4} b_i\xi^i \qquad (8.96)$$

and the distributed axial load reads

$$q(\xi) = q_0 \sum_{i=0}^{m+1} g_i\xi^i \qquad (8.97)$$

where

$$g_i = q_i/q_0 \quad (i = 1, 2, \ldots, m), \qquad g_0 = 1 \qquad (8.98)$$

Since the axial load equals

$$N(\xi) = -\int_x^L q(t)\, dt = -L\int_\xi^1 q(t)\, dt \qquad (8.99)$$

we have

$$N(\xi) - q_0 L \sum_{i=0}^{m+2} \frac{g_i}{i+1}(1 - \xi^{i+1}) \qquad (8.100)$$

Therefore, Eq. (8.20) becomes

$$\frac{d^2}{d\xi^2}\left[D(\xi)\frac{d^2\psi}{d\xi^2}\right] + q_0 L^3 \frac{d}{d\xi}\left[\sum_{i=0}^{m+2}\frac{g_i}{i+1}(1-\xi^{i+1})\frac{d\psi}{d\xi}\right] - R(\xi)\omega^2\psi(\xi) = 0$$

$$(8.101)$$

Upon substitution of the polynomial expressions of the modes shapes (8.15)–(8.18) into the differential equation (8.27), we get a polynomial equation of order $m + 5$ in ξ with zero as its right-hand side. We demand, therefore, that every coefficient of ξ, raised to any power, be zero. We get, thus, a set of $m+6$ linear algebraic equations with $m + 7$ unknowns, namely, the coefficients b_i in the flexural rigidity, the natural frequency squared, Ω^2, and the axial load, q_0, which is treated as a control parameter.

8.2.4   Closed-Form Solution for the Pinned Beam

8.2.4.1   Uniform inertial coefficient ($m = 0$)

Substituting the mode shape $\psi(\xi)$ from Eq. (8.15) into the governing differential equation (8.27) we get a system of six linear algebraic equations with seven unknowns, namely the coefficients b_i in the flexural rigidity, the natural frequency squared, Ω^2, and the axial load, q_0, which is treated as a parameter:

$$-40b_1 - \tfrac{7}{3}q_0 L^3 = 0$$
$$-120b_2 + 120b_0 - q_0 L^3\left(20 + \tfrac{37}{3}g_1\right) - \tfrac{7}{3}a_0\Omega^2 = 0$$
$$-240b_3 + 240b_1 + 30q_0 L^3 = 0$$
$$-400b_4 + 400b_2 + q_0 L^3(20 + 30g_1) + \tfrac{10}{3}a_0\Omega^2 = 0 \qquad (8.102)$$
$$600b_3 - 25q_0 L^3 = 0$$
$$840b_4 - 15g_1 q_0 L^3 - a_0\Omega^2 = 0$$

We have to write all unknowns in terms of a single free parameter, say, b_4 and the axial load parameter q_0. In other words, the rank of the set (8.28) must be less than 6. A matrix is of rank p if it contains minors of order p different from 0, while all minors of order $p + 1$ (if there are such) are 0 (Uspensky, 1948). Accordingly, the rank of the matrix of the set (8.28) is less than 6 if all minors of order 6 vanish. Yet, the minor obtained by extraction of the sixth column of the matrix of the set (8.28) equals $-5,806,080,000,000L^3$, which is non-zero. We conclude that the problem has no non-trivial solution.

8.2.4.2   Linearly varying inertial coefficient ($m = 1$)

Substituting the mode shape $\psi(\xi)$ from Eq. (8.4) into the governing differential equation (8.16) we get a homogeneous set of seven linear algebraic equations with eight unknowns, namely the coefficients b_i in the flexural rigidity, the natural frequency squared, Ω^2, and the axial load q_0, which is treated as a parameter:

$$-40b_1 - \tfrac{7}{3}q_0 L^3 = 0$$
$$-120b_2 + 120b_0 - q_0 L^3\left(20 + \tfrac{37}{3}g_1 + \tfrac{20}{3}g_2\right) - \tfrac{7}{3}a_0\Omega^2 = 0$$
$$-240b_3 + 240b_1 + q_0 L^3\left(30 - \tfrac{7}{3}g_2\right) - \tfrac{7}{3}a_1\Omega^2 = 0$$
$$-400b_4 + 400b_2 + q_0 L^3\left(20 + 30g_1 + \tfrac{20}{3}g_2\right) + \tfrac{10}{3}a_0\Omega^2 = 0 \qquad (8.103)$$
$$-600b_5 + 600b_3 - q_0 L^3\left(25 - \tfrac{50}{3}g_2\right) + \tfrac{10}{3}a_1\Omega^2 = 0$$
$$840b_4 - 15g_1 q_0 L^3 - a_0\Omega^2 = 0$$
$$1120b_5 - \tfrac{35}{3}q_0 L^3 g_2 - a_1\Omega^2 = 0$$

We have to write all unknowns in terms of a single free parameter, say b_5, and the axial load parameter q_0. In other words, the rank of the set (8.29) or its matrix must be less than 7. The rank of the matrix of the set (8.29) is less than 7 if all minors of order 7 vanish. We produce the minors, extracting one column of the matrix among eight; therefore we obtain a set of eight determinantal equations for four unknowns, namely the parameters a_0, a_1, g_1 and g_2:

$$2{,}444{,}288a_0 g_2 + 7{,}999{,}488a_0 + 6{,}005{,}760a_1 g_1 + 15{,}353{,}856a_1 + 5{,}117{,}952a_1 g_2 = 0$$
$$a_1 = 0$$
$$1{,}419{,}264a_0 g_2 + 4{,}644{,}864a_0 + 7{,}520{,}256a_1 g_1 + 6{,}580{,}224a_1 + 2{,}193{,}408a_1 g_2 = 0$$
$$3{,}211{,}264a_1 g_2 - 2{,}451{,}456a_1 = 0$$
$$236{,}544a_0 g_2 + 774{,}144a_0 + 2{,}350{,}080a_1 g_1 = 0$$
$$1{,}548{,}288a_1 g_2 + 580{,}608a_1 = 0$$
$$19{,}869{,}696g_2 + 650{,}228{,}096 = 0$$
$$a_1 = 0$$

$$(8.104)$$

The solution of the set (8.104) reads

$$a_1 = 0 \quad g_2 = -\tfrac{36}{11} \qquad (8.105)$$

The solution (8.105a) does not agree with the linearly varying mass density. We conclude that in this case the problem has no non-trivial solution.

8.2.4.3 Parabolically varying inertial coefficient ($m = 2$)

The governing differential equation (8.27) yields a set of eight equations with eight unknowns, namely the coefficients b_i in the flexural rigidity, the natural frequency squared, Ω^2, and the axial load, q_0, which is treated as a

parameter:

$$-40b_1 - \tfrac{7}{3}q_0L^3 = 0$$
$$-120b_2 + 120b_0 - q_0L^3\left(20 + \tfrac{37}{3}g_1 + \tfrac{20}{3}g_2 + 5g_3\right) - \tfrac{7}{3}a_0\Omega^2 = 0$$
$$-240b_3 + 240b_1 + q_0L^3\left(30 - \tfrac{7}{3}g_2\right) - \tfrac{7}{3}a_1\Omega^2 = 0$$
$$-400b_4 + 400b_2 + q_0L^3\left(20 + 30g_1 + \tfrac{20}{3}g_2 + \tfrac{8}{3}g_3\right) + \left(\tfrac{10}{3}a_0 - \tfrac{7}{3}a_2\right)\Omega^2 = 0$$
$$-600b_5 + 600b_3 - q_0L^3\left(25 - \tfrac{50}{3}g_2\right) + \tfrac{10}{3}a_1\Omega^2 = 0$$
$$-840b_6 + 840b_4 - 15q_0L^3(g_1 - g_3) - \left(a_0 - \tfrac{10}{3}a_2\right)\Omega^2 = 0$$
$$1120b_5 - \tfrac{35}{3}q_0L^3g_2 - a_1\Omega^2 = 0$$
$$1440b_6 - 10q_0L^3g_3 - a_2\Omega^2 = 0$$

$$(8.106)$$

In order to have a non-trivial solution, we have to write all unknowns in terms of a single free parameter, say, b_6 and the axial load parameter q_0. Consider the matrix of the set (8.32), where the first seven columns represent the coefficients of the unknowns, b_i, while the remaining ones constitute the coefficients of the natural frequency squared Ω^2 and the axial load q_0, respectively. The matrix of the set (8.32) reads

$$\begin{bmatrix} 0 & -40 & 0 & 0 & 0 & 0 & 0 & 0 & -\tfrac{7}{3}L^3 \\ 120 & 0 & -120 & 0 & 0 & 0 & 0 & -\tfrac{7}{3}a_0 & -L^3(20 + \tfrac{37}{3}g_1 + \tfrac{20}{3}g_2 + 5g_3) \\ 0 & 240 & 0 & -240 & 0 & 0 & 0 & -\tfrac{7}{3}a_1 & L^3(30 - \tfrac{7}{3}g_2) \\ 0 & 0 & 400 & 0 & -400 & 0 & 0 & (\tfrac{10}{3}a_0 - \tfrac{7}{3}a_2) & L^3(20 + 30g_1 + \tfrac{20}{3}g_2 + \tfrac{8}{3}g_3) \\ 0 & 0 & 0 & 600 & 0 & -600 & 0 & \tfrac{10}{3}a_1 & -L^3(25 - \tfrac{50}{3}g_2) \\ 0 & 0 & 0 & 0 & 840 & 0 & 840 & -(a_0 - \tfrac{10}{3}a_2) & -15L^3(g_1 - g_3) \\ 0 & 0 & 0 & 0 & 0 & 1120 & 0 & -a_1 & -\tfrac{35}{3}L^3g_2 \\ 0 & 0 & 0 & 0 & 0 & 0 & 1440 & -a_2 & -10L^3g_3 \end{bmatrix}$$

$$(8.107)$$

which is a matrix with eight rows and nine columns. To obtain the expression for all the unknowns with respect to two parameters b_6 and q_0, we demand that the rank of the matrix (8.33) be less than 8. According to the definition of the rank, the rank of the matrix (33) is less than 8 if all minors of order 8 vanish. We produce the minors, extracting one column of the matrix among 9; therefore, we obtain a set of nine determinantal equations for six unknowns, namely the parameters a_0, a_1, a_2, g_1, g_1 and g_3. The solution of this set reads

$$a_1 = 0 \quad g_2 = -\tfrac{36}{11} \tag{8.108}$$

Upon substitution of the expressions (8.34) into the set (8.32), we get the solution for the natural frequency squared:

$$\Omega^2 = (1440b_6 - 10q_0L^3g_3)/a_2 \tag{8.109}$$

The value of q_0 that corresponds to the vanishing frequency $\Omega^2 = 0$ is identified as the buckling load $q_{0,cr}$:

$$q_{0,cr} = 144h_6/g_3 L^3 \tag{8.110}$$

We, thus, rewrite Eq. (8.35) as follows:

$$\Omega^2 = 1440b_6(1 - q_0/q_{0,cr})/a_2 \tag{8.111}$$

From Eq. (8.25), when $q_0 \equiv 0$, $\Omega^2 = 1440b_6/a_2 \equiv \Omega_0^2$, which is the natural frequency in the *absence* of an axial load q. Therefore, Eq. (8.25) takes the form

$$\Omega_q^2 = \Omega_0^2(1 - q_0/q_{0,cr}) \tag{8.112}$$

where Ω_q^2 is the natural frequency squared in the presence of the axial load. Note that the critical buckling load coincides with that of Elishakoff and Guédé (2001). When $q_0 \equiv 0$, we get Eq. (8.36) to be the natural frequency coefficient of the beam with axially distributed load not taken into account; this solution agrees with that determined by Elishakoff and Guédé (1999) in an other manner. When q_0 tends to its critical value, the natural frequency vanishes when $q_0 < q_{0,cr}$. The coefficients b_i in the flexural rigidity are deduced from the remaining equations of the set (8.32):

$$
\begin{aligned}
b_0 &= \frac{1}{138,600a_2}\{(510,840a_2 + 2,455,200a_0)b_6 \\
&\quad -[209g_3 - 6,325g_1 + 1,470)a_2 + 17,050g_3a_0]q_0L^3\} \\
b_1 &= -\frac{7q_0L^3}{120} \\
b_2 &= \frac{1}{69,300a_2}\{(255,420a_2 - 712,800r_0)b_6 \\
&\quad -[(2,992g_3 + 3,960g_1 - 315)a_2 - 4,950g_3a_0]q_0L^3\} \\
b_3 &= 13q_0L^3/132 \\
b_4 &= \frac{1}{504a_2}\{(-2,376a_2 + 864a_0)b_6 + [(11g_3 + 9g_1)a_2 - 6g_3a_0]q_0L^3\}, \\
b_5 &= -\frac{3q_0L^3}{88}
\end{aligned}
\tag{8.113}
$$

which solve the problem posed for the parabolically varying inertial coefficient.

8.2.4.4 Cubic inertial coefficient (m = 3)

In this case, the governing differential equation yields a set of nine equations with ten unknowns, namely the coefficients b_i in the flexural rigidity, the natural frequency squared, Ω^2, and the axial load q_0, which is again treated as a parameter:

$$-40b_1 - \tfrac{7}{3}q_0L^3 = 0$$
$$-120b_2 + 120b_0 - q_0L^3\left(20 + \tfrac{37}{3}g_1 + \tfrac{20}{3}g_2 + 5g_3 + 4g_4\right) - \tfrac{7}{3}a_0\Omega^2 = 0$$
$$-240b_3 + 240b_1 + q_0L^3\left(30 - \tfrac{7}{3}g_2\right) - \tfrac{7}{3}a_1\Omega^2 = 0$$
$$-400b_4 + 400b_2 + q_0L^3\left(20 + 30g_1 + \tfrac{20}{3}g_2 + \tfrac{8}{3}g_3 + 4g_4\right) + \left(\tfrac{10}{3}a_0 - \tfrac{7}{3}a_2\right)\Omega^2 = 0$$
$$-600b_5 + 600b_3 - q_0L^3\left(25 - \tfrac{50}{3}g_2 + \tfrac{7}{3}g_4\right) + \left(\tfrac{10}{3}a_1 - \tfrac{7}{3}a_3\right)\Omega^2 = 0$$
$$-840b_6 + 840b_4 - 15q_0L^3(g_1 - g_3) - \left(a_0 - \tfrac{10}{3}a_2\right)\Omega^2 = 0$$
$$-1120b_7 + 1120b_5 - q_0L^3\left(\tfrac{35}{3}g_2 - 14g_4\right) - \left(a_1 - \tfrac{10}{3}a_3\right)\Omega^2 = 0$$
$$1440b_6 - 10q_0L^3g_3 - a_2\Omega^2 = 0$$
$$1800b_7 - 9q_0L^3g_4 - a_3\Omega^2 = 0$$

$$(8.114)$$

In order to have a non-trivial solution, we have to express all unknowns in terms of a single parameter, say, b_7 and the axial load parameter q_0. Therefore, we demand the rank of the matrix of (8.114) to be less than 9; thus, all minors of order 9 vanish. This condition produces a set of ten determinantal equations for eight unknowns, namely the parameters a_0, a_1, a_2, a_3, g_1, g_2, g_3 and g_4. The solution of this set reads

$$a_1 = -\tfrac{74}{255}a_3 \quad g_2 = -\tfrac{26}{55}g_4 - \tfrac{36}{11} \tag{8.115}$$

We substitute the expressions (8.41) into the set (8.40) and we get the solution for the natural frequency squared:

$$\Omega^2 = (1800b_7 - 9q_0L^3g_4)/a_3 \tag{8.116}$$

The value of q_0 that corresponds to the vanishing frequency $\Omega^2 = 0$ is identified as the buckling load $q_{0,cr}$:

$$q_{0,cr} = 200b_7/g_4L^3 \tag{8.117}$$

We, thus, rewrite Eq. (8.42) as follows:

$$\Omega^2 = 1800b_7(1 - q_0/q_{0,cr})/a_3 \tag{8.118}$$

From Eq. (8.44), when $q_0 \equiv 0$, $\Omega^2 = 1800b_7/a_3 \equiv \Omega_0^2$. Therefore, Eq. (8.42) takes the form

$$\Omega_q^2 = \Omega_0^2(1 - q_0/q_{0,cr}) \tag{8.119}$$

where Ω_q^2 again denotes the natural frequency squared in the presence of the axial load. Note that when q_0 tends to its critical value the natural frequency vanishes. The coefficients b_i in the flexural rigidity are deduced from the remaining equations of the set (8.40):

$$b_0 = \{(2{,}554{,}200a_2 + 12{,}276{,}000a_0)b_7 + [(2{,}744g_4 + 13{,}354g_3 + 25{,}300g_1$$
$$-5{,}880)a_3 - (61{,}380a_0 + 12{,}771a_2)g_4]q_0L^3\}/554{,}400a_3$$

$$b_1 = -7q_0L^3/120$$

$$b_2 = -\{(7{,}128{,}000a_0 - 2{,}554{,}200a_2)b_7 + [(1{,}176g_4 + 9{,}746g_3 + 31{,}680g_1$$
$$-2{,}520)a_3 + (12{,}771a_2 - 35{,}640a_0)g_4]q_0L^3\}/554{,}400a_3$$

$$b_3 = \frac{259}{51}b_7 + \left(\frac{13}{132} - \frac{35}{1683}g_4\right)q_0L^3$$

$$b_4 = \frac{1}{10{,}080a_3}\{(21{,}600a_0 - 59{,}400a_2)b_7 + [(180g_1 - 110g_3)a_3$$
$$+(297a_2 - 108a_0)g_4]q_0L^3\}$$

$$b_5 = -\frac{82}{17}b_7 + \left(\frac{164}{14{,}025}g_4 - \frac{3}{88}\right)q_0L^3,$$

$$b_6 = [1{,}800a_2b_7 + (10g_3a_3 - 9g_4a_2)q_0L^3]/1{,}440a_3$$

$$(8.120)$$

which complete the solution for the cubically varying inertial coefficient.

8.2.4.5 *Quartic inertial coefficient (m = 4)*

Under these circumstances, the governing differential equation yields a set of 10 equations with 11 unknowns, namely the coefficients b_i in the flexural rigidity, the natural frequency squared, Ω^2, and the axial load q_0, which is treated as a parameter:

$$-40b_1 - \tfrac{7}{3}q_0L^3 = 0$$
$$-120b_2 + 120b_0 - q_0L^3\left(20 + \tfrac{37}{3}g_1 + \tfrac{20}{3}g_2 + 5g_3 + 4g_4 + \tfrac{10}{3}g_5\right) - \tfrac{7}{3}a_0\Omega^2 = 0$$
$$-240b_3 + 240b_1 + q_0L^3\left(30 - \tfrac{7}{3}g_2\right) - \tfrac{7}{3}a_1\Omega^2 = 0$$
$$-400b_4 + 400b_2 + q_0L^3\left(20 + 30g_1 + \tfrac{20}{3}g_2 + \tfrac{8}{3}g_3 + 4g_4 + \tfrac{10}{3}g_5\right) + \left(\tfrac{10}{3}a_0 - \tfrac{7}{3}a_2\right)\Omega^2 = 0$$
$$-600b_5 + 600b_3 - q_0L^3\left(25 - \tfrac{50}{3}g_2 + \tfrac{7}{3}g_4\right) + \left(\tfrac{10}{3}a_1 - \tfrac{7}{3}a_3\right)\Omega^2 = 0$$
$$-840b_6 + 840b_4 - q_0L^3\left(15g_1 - 15g_3 + \tfrac{7}{3}g_5\right) - \left(a_0 - \tfrac{10}{3}a_2 + \tfrac{7}{3}a_4\right)\Omega^2 = 0$$
$$-1120b_7 + 1120b_5 - q_0L^3\left(\tfrac{35}{3}g_2 - 14g_4\right) - \left(a_1 - \tfrac{10}{3}a_3\right)\Omega^2 = 0$$
$$-1440b_8 + 1440b_6 - q_0L^3\left(10g_3 - \tfrac{40}{3}g_5\right) - \left(a_2 - \tfrac{10}{3}a_4\right)\Omega^2 = 0$$
$$1800b_7 - 9q_0L^3g_4 - a_3\Omega^2 = 0$$
$$2200b_8 - \tfrac{25}{3}q_0L^3g_5 - a_4\Omega^2 = 0$$

$$(8.121)$$

In order to have a non-trivial solution we have to write all unknowns in terms of one, say b_7, and the parameter q_0. To obtain expressions for all the unknowns in terms of the two parameters b_7 and q_0, we demand that the rank of the matrix of the set (8.47) be less than 10, leading to the requirement that all minors of order 10 vanish. Therefore, we obtain a set of 11 determinantal equations for 8 unknowns, namely the parameters a_0, a_1, a_2, a_3, g_1, g_2, g_3 and g_4. The solution of this set reads

$$r_1 = -\tfrac{74}{255}r_3 \quad g_2 = -\tfrac{26}{55}g_4 - \tfrac{36}{11} \tag{8.122}$$

Upon substitution of the expressions (8.48) into the set (8.47), we get the solution for the natural frequency squared:

$$\Omega^2 = \left(2200b_8 - \tfrac{25}{3}q_0 L^3 g_5\right)/a_4 \tag{8.123}$$

The value of q_0 that corresponds to the vanishing frequency $\Omega^2 = 0$ is identified as the buckling load $q_{0,\mathrm{cr}}$:

$$q_{0,\mathrm{cr}} = 264b_8/g_5 L^3 \tag{8.124}$$

We, thus, rewrite Eq. (8.37) as follows:

$$\Omega_q^2 = 2200b_8(1 - q_0/q_{0,\mathrm{cr}})/a_4 \tag{8.125}$$

From Eq. (8.51), when $q_0 \equiv 0$, $\Omega^2 = 2200b_8/a_4 \equiv \Omega_0^2$. Therefore, Eq. (8.49) takes the familiar form

$$\Omega_q^2 = \Omega_0^2(1 - q_0/q_{0,\mathrm{cr}}) \tag{8.126}$$

where Ω_q^2 is the natural frequency squared in the presence of the axial load. As expected, when q_0 tends to its critical value, the natural frequency vanishes. The coefficients b_i in the flexural rigidity are

$$\begin{aligned}
b_0 &= \{(10{,}071{,}600a_4 + 28{,}096{,}200a_2 + 135{,}036{,}000a_0)b_8 \\
&\quad + [(45{,}430g_5 + 24{,}696g_4 + 120{,}186g_3 + 227{,}700g_1 - 52{,}920)a_4 \\
&\quad - (511{,}500a_0 + 106{,}425a_2)g_5]q_0 L^3\}/4{,}989{,}600a_4 \\
b_1 &= -7q_0 L^3/120
\end{aligned}$$

$b_2 = \{(10{,}071{,}600a_4 + 28{,}096{,}200a_2 - 78{,}408{,}000a_0)b_8$

$\quad -[(93{,}170g_5 + 10{,}584g_4 + 87{,}714g_3 + 285{,}120g_1)a_4$

$\quad -(297{,}000a_0 - 106{,}425a_2)g_5]q_0L^3\}/4{,}989{,}600a_4$

$b_3 = \{18{,}803{,}400a_3b_8 + [(13{,}923g_4 + 298{,}350)a_4 + 71{,}225g_5a_3]q_0L^3\}/3{,}029{,}400a_4$

$b_4 = \{(183{,}120a_4 - 653{,}400a_2 + 237{,}600a_0)b_8$

$\quad -[(938g_5 - 990g_3 + 1620g_1)a_4 + (2{,}475a_2 - 900a_0)g_5]q_0L^3\}/90{,}720a_4$

$b_5 = \{-5{,}953{,}200a_3b_8 - [(12{,}546g_4 + 34{,}425)a_4 - 22{,}550a_3g_5]q_0L^3\}/1{,}009{,}800a_4$

$b_6 = \{(-10{,}608a_4 + 3{,}960a_2)b_8 + [(18g_3 + 26g_5)a_4 - 15a_2g_5]q_0L^3\}/2{,}592a_4$

$b_7 = [6{,}600a_3b_8 + (27g_4a_4 - 25a_3g_5)q_0L^3]/5400a_4$

$$(8.127)$$

which is the solution for the problem posed for quartic $R(\xi)$.

8.2.4.6 *Quintic inertial coefficient* ($m = 5$)

The governing differential equation yields a set of 11 equations with 12 unknowns, namely the coefficients b_i in the flexural rigidity, the natural frequency squared, Ω^2, and the axial load q_0 which is treated as a parameter:

$-40b_1 - \frac{7}{3}q_0L^3 = 0$

$-120b_2 + 120b_0 - q_0L^3\left(20 + \frac{37}{3}g_1 + \frac{20}{3}g_2 + 5g_3 + 4g_4 + \frac{10}{3}g_5 + \frac{20}{7}g_6\right)$

$\quad -\frac{7}{3}a_0\Omega^2 = 0$

$-240b_3 + 240b_1 + q_0L^3\left(30 - \frac{7}{3}g_2\right) - \frac{7}{3}a_1\Omega^2 = 0$

$-400b_4 + 400b_2 + q_0L^3\left(20 + 30g_1 + \frac{20}{3}g_2 + \frac{8}{3}g_3 + 4g_4 + \frac{10}{3}g_5 + \frac{20}{7}g_6\right)$

$\quad +\left(\frac{10}{3}a_0 - \frac{7}{3}a_2\right)\Omega^2 = 0$

$-600b_5 + 600b_3 - q_0L^3\left(25 - \frac{50}{3}g_2 + \frac{7}{3}g_4\right) + \left(\frac{10}{3}a_1 - \frac{7}{3}a_3\right)\Omega^2 = 0$

$-840b_6 + 840b_4 - q_0L^3\left(15g_1 - 15g_3 - \frac{7}{3}g_5\right) - \left(a_0 - \frac{10}{3}a_2 + \frac{7}{3}a_4\right)\Omega^2 = 0$

$-1120b_7 + 1120b_5 - q_0L^3\left(\frac{35}{3}g_2 - 14g_4 + \frac{7}{3}g_6\right) - \left(a_1 - \frac{10}{3}a_3 + \frac{7}{3}a_5\right)\Omega^2 = 0$

$-1440b_8 + 1440b_6 - q_0L^3\left(10g_3 - \frac{40}{3}g_5\right) - \left(a_2 - \frac{10}{3}a_4\right)\Omega^2 = 0$

$-1800b_9 + 1800b_7 - q_0L^3\left(9g_4 - \frac{90}{7}g_6\right) - \left(a_3 - \frac{10}{3}a_5\right)\Omega^2 = 0$

$2200b_8 - \frac{25}{3}q_0L^3g_5 - a_4\Omega^2 = 0$

$2640b_9 - \frac{55}{7}q_0L^3g_6 - a_5\Omega^2 = 0$

$$(8.128)$$

We have to write all unknowns in terms of b_8 and the axial load parameter q_0. To obtain the expression for all the unknowns with respect to the two parameters b_9 and q_0, the rank of the matrix of the set (8.54) must be less than 10, which means that all minors of order 10 vanish. Hence, we obtain a set of 11 determinantal equations for 8 unknowns, namely the parameters a_0, a_1, a_2, a_3, g_1, g_2, g_3 and g_4. The solution of this set reads

$$a_1 = -\frac{203}{1683}a_5 - \frac{74}{255}a_3 \quad g_2 = -\frac{3}{11}g_6 - \frac{26}{55}g_4 - \frac{36}{11} \tag{8.129}$$

Upon substitution of the expressions (8.55) into the set (8.54), we get the solution for the natural frequency squared:

$$\Omega^2 = (2680b_9 - \tfrac{55}{7}q_0L^3g_6)/a_5 \tag{8.130}$$

The buckling load $q_{0,cr}$ is obtained by setting $\Omega^2 = 0$:

$$q_{0,cr} = 336b_9/g_6L^3 \tag{8.131}$$

We, thus, rewrite Eq. (8.56) as follows:

$$\Omega^2 = 2680b_9(1 - q_0/q_{0,cr})/a_5 \tag{8.132}$$

From Eq. (8.46), when $q_0 \equiv 0$, $\Omega^2 = 2680b_9/a_5 \equiv \Omega_0^2$. Therefore, Eq. (8.132) takes the form

$$\Omega_q^2 = \Omega_0^2(1 - q_0/q_{0,cr}) \tag{8.133}$$

where Ω_q^2 is the natural frequency squared in the presence of the axial load. One can see that when q_0 tends to its critical value the natural frequency vanishes, as it should. The coefficients b_i in the flexural rigidity are

$$b_0 = \{(28{,}200{,}480a_4 + 78{,}669{,}360a_2 + 378{,}100{,}800a_0)b_9$$
$$+[(70{,}560g_6 + 195{,}020g_5 + 57{,}624g_4 + 280{,}434g_3 + 531{,}300g_1 - 123{,}480)a_5$$
$$-(83{,}930a_4 + 234{,}135a_2 + 1{,}125{,}300a_0)g_6]q_0L^3\}/11{,}642{,}400a_5$$
$$b_1 = -7q_0L^3/120$$

$b_2 = \{(28{,}200{,}480a_4 + 78{,}669{,}360a_2 + 219{,}542{,}400a_0)b_9$

$\qquad -[(30{,}240g_6 - 128{,}380g_5 - 24{,}696g_4 - 204{,}666g_3 - 665{,}280g_1 + 52{,}920)a_5$

$\qquad +(83{,}930a_4 + 234{,}135a_2 + 653{,}400a_0)g_6]q_0L^3\}/11{,}642{,}400a_5$

$b_3 = \{(9{,}378{,}600a_5 + 22{,}564{,}080a_3)b_9 - [(19{,}880g_6 + 13{,}923g_4 + 298{,}350)a_5$

$\qquad +67{,}155a_3g_6]q_0L^3\}/3{,}029{,}400a_5$

$b_4 = \{(5{,}640{,}096a_4 - 20{,}124{,}720a_2 + 7{,}318{,}080a_0)b_9$

$\qquad +[(-6{,}272g_5 - 25{,}410g_3 + 41{,}580g_1)a_5$

$\qquad -(16{,}786a_4 + 59{,}895a_2 - 21{,}780a_0)g_6]q_0L^3\}/2{,}328{,}480a_5$

$b_5 = \{(-9{,}378{,}600a_5 + 50{,}006{,}880a_3)b_9 + [(62{,}720g_6 + 87{,}822g_4 + 240{,}975)a_5$

$\qquad -148{,}830a_3g_6]q_0L^3\}/7{,}068{,}600a_5$

$b_6 = \dfrac{1}{332{,}640a_5}\{(609{,}840a_2 - 1{,}633{,}632a_4)b_9 + [(2{,}310g_3 - 1{,}820g_5)a_5$

$\qquad +(4{,}862a_4 - 1{,}815a_2)g_6]q_0L^3\}$

$b_7 = \{(55{,}440a_3 - 147{,}000a_5)b_9 + [(280g_6 + 189g_4)a_5 - 165r_3g_6]q_0L^3\}/37{,}800a_5$

$b_8 = [11{,}088a_4b_9 + (35g_5a_5 - 33a_4g_6)q_0L^3]/9{,}240a_5$

$$(8.134)$$

Equation (8.60) presents the closed-form solution of the problem posed for quintic $R(\xi)$. In an analogous manner, one can produce results for $m \geq 6$.

8.2.5 Closed-Form Solution for the Clamped–Free Beam

8.2.5.1 *Uniform inertial coefficient ($m = 0$)*

In the case of the uniform inertial coefficient, we get a homogeneous set of six linear algebraic equations with seven unknowns, namely the coefficients b_i in the flexural rigidity, the natural frequency squared, Ω^2, and the axial load q_0, which is treated as a parameter:

$$80b_2 + 120b_1 + q_0L^3(40 + 20g_1) = 0$$
$$240b_3 - 360b_2 + 120b_0 - q_0L^3(140 + 30g_1) = 0$$
$$480b_4 - 720b_3 + 240b_1 + q_0L^3(90 - 60g_1) - 20a_0\Omega^2 = 0$$
$$-1200b_4 + 400b_2 + q_0L^3(20 + 70g_1) + 10a_0\Omega^2 = 0$$
$$600b_3 - 25q_0L^3 = 0$$
$$840b_4 - 15g_1q_0L^3 - a_0\Omega^2 = 0$$

$$(8.135)$$

Again, we would like to write all unknowns in terms of a single free parameter, say b_4, and the axial load parameter q_0. In other words, the rank of the set

(8.61) or its matrix must be less than 6, implying that all minors of order 6 vanish. This leads to the following seven equations with two unknowns, a_0 and g_1:

$$-18{,}247{,}768a_0g_1 + 37{,}297{,}152a_0 = 0$$
$$6{,}967{,}296a_0g_1 + 14{,}432{,}256a_0 = 0$$
$$6{,}137{,}856a_0g_1 + 11{,}529{,}216a_0 = 0$$
$$a_0 = 0 \qquad\qquad\qquad\qquad\qquad\qquad\qquad (8.136)$$
$$1{,}078{,}272a_0g_1 + 456{,}192a_0 = 0$$
$$8{,}957{,}952g_1 - 38{,}320{,}128 = 0$$
$$a_0 = 0$$

The solution of this set reads

$$a_0 = 0 \quad g_1 = \tfrac{77}{18} \qquad\qquad\qquad\qquad (8.137)$$

which leads to a trivial solution for the problem since the inertial coefficient vanishes.

8.2.5.2   *Linearly varying inertial coefficient (m = 1)*

In this case, we obtain a set of seven equations with eight unknowns, namely the coefficients b_i in the flexural rigidity, the natural frequency squared, Ω^2, and the axial load q_0, which is treated as a parameter:

$$80b_2 - 120b_1 + q_0L^3\left(40 + 20g_1 + \tfrac{40}{3}g_2\right) = 0$$
$$240b_3 - 360b_2 + 120b_0 - q_0L^3\left(140 + 30g_1 + 20g_2\right) = 0$$
$$480b_4 - 720b_3 + 240b_1 + q_0L^3\left(90 - 60g_1\right) - 20a_0\Omega^2 = 0$$
$$800b_5 - 1200b_4 + 400b_2 + q_0L^3\left(20 + 70g_1 - \tfrac{140}{3}g_2\right) + \Omega^2\left(10a_0 - 20a_1\right) = 0$$
$$-1800b_5 + 600b_3 - q_0L^3\left(25 - 50g_2\right) + 10a_1\Omega^2 = 0$$
$$840b_4 - 15q_0L^3g_1 - a_0\Omega^2 = 0$$
$$1120b_5 - \tfrac{35}{3}q_0L^3g_2 - a_1\Omega^2 = 0$$

$$\qquad\qquad\qquad\qquad\qquad\qquad\qquad\qquad\qquad (8.138)$$

We have to express the unknowns in terms of the parameters b_5 and Ω^2. Therefore, we demand that the rank of the set (8.64) is less than 7, leading to a system of eight determinantal equations with four unknowns, namely a_0,

a_1, g_1 and g_2:

$$
\begin{aligned}
&2{,}487{,}711{,}744 a_0 g_2 + 1{,}233{,}211{,}392 a_1 g_2 + 4{,}177{,}281{,}024 a_0 + 2{,}011{,}226{,}112 a_1 \\
&\quad - 204{,}374{,}016 a_0 g_1 - 1{,}070{,}972{,}928 a_1 g_1 = 0 \\
&-931{,}295{,}232 a_0 g_2 + 181{,}149{,}696 a_1 g_2 - 1{,}616{,}412{,}672 a_0 + 355{,}332{,}096 a_1 \\
&\quad - 780{,}337{,}152 a_0 g_1 + 818{,}159{,}616 a_1 g_1 = 0 \\
&158{,}312{,}448 a_0 g_2 + 692{,}084{,}736 a_1 g_2 - 1{,}291{,}272{,}192 a_0 + 2{,}358{,}429{,}696 a_1 \\
&\quad - 687{,}439{,}872 a_0 g_1 + 218{,}474{,}496 a_1 g_1 = 0 \\
&-140{,}341{,}248 a_1 g_1 + 603{,}832{,}332 a_1 g_2 + 841{,}300{,}992 a_1 + 387{,}072{,}000 a_0 g_2 \\
&\quad - 309{,}657{,}600 a_0 = 0 \\
&30{,}772{,}224 a_0 g_2 + 51{,}093{,}504 a_0 + 120{,}766{,}464 a_0 g_1 - 103{,}265{,}280 a_1 g_1 = 0 \\
&8{,}957{,}952 a_1 g_1 + 37{,}158{,}912 a_1 g_2 - 38{,}320{,}128 a_1 - 77{,}414{,}400 a_0 g_2 = 0 \\
&2{,}584{,}866{,}816 g_2 - 1{,}003{,}290{,}624 g_1 + 4{,}291{,}854{,}336 = 0 \\
&578{,}285{,}568 a_1 - 743{,}178{,}240 a_0 = 0
\end{aligned}
$$

$$(8.139)$$

The solution of the set (8.65) reads

$$
a_0 = \tfrac{249}{320} a_1 \qquad g_1 = \tfrac{77}{18} + \tfrac{371}{144} g_2 \tag{8.140}
$$

Upon substitution of the expressions (8.66) into the set (8.64), we get the solution for the natural frequency squared:

$$
\Omega^2 = \left(1120 b_5 - \tfrac{35}{3} q_0 L^3 g_2\right)/a_1 \tag{8.141}
$$

The value of q_0 that corresponds to the vanishing frequency $\Omega^2 = 0$ leads to the buckling load $q_{0,\mathrm{cr}}$:

$$
q_{0,\mathrm{cr}} = 96 b_5 / g_2 L^3 \tag{8.142}
$$

We rewrite Eq. (8.67) as follows:

$$
\Omega^2 = 1120 b_5 (1 - q_0/q_{0,\mathrm{cr}})/a_1 \tag{8.143}
$$

From Eq. (8.67), when $q_0 \equiv 0$, $\Omega^2 = 1120 b_5/a_1 \equiv \Omega_0^2$. Therefore, Eq. (8.52) takes the form

$$
\Omega_q^2 = \Omega_0^2 (1 - q_0/q_{0,\mathrm{cr}}) \tag{8.144}
$$

where Ω_q^2 is the natural frequency squared in the presence of the axial load. Note that, when q_0 tends to its critical value, the natural frequency vanishes.

The coefficients b_i in the flexural rigidity are

$$b_0 = \frac{16{,}477}{120}b_5 + \left(\frac{4}{9} - \frac{1{,}493}{1{,}280}g_2\right)q_0L^3 \qquad b_1 = \frac{471}{20}b_5 + \left(\frac{2}{3} + \frac{289}{1{,}920}g_2\right)q_0L^3$$

$$b_2 = \frac{1{,}413}{40}b_5 - \left(\frac{41}{72} + \frac{6{,}739}{11{,}520}g_2\right)q_0L^3 \qquad b_3 = -\frac{47}{3}b_5 + \left(\frac{1}{24} + \frac{1}{9}g_2\right)q_0L^3$$

$$b_4 = \frac{83}{80}b_5 + \left(\frac{11}{144} + \frac{811}{23{,}040}g_2\right)q_0L^3$$

$$\tag{8.145}$$

8.2.5.3 *Parabolically varying inertial coefficient* ($m = 2$)

In this case, we obtain a set of eight equations with eight unknowns, namely the coefficients b_i in the flexural rigidity, the natural frequency squared, Ω^2, and the axial load q_0, which is treated as a parameter:

$$80b_2 - 120b_1 + q_0L^3\left(40 + 20g_1 + \tfrac{40}{3}g_2 + 10g_3\right) = 0$$

$$240b_3 - 360b_2 + 120b_0 - q_0L^3(140 + 30g_1 + 20g_2 + 15g_3) = 0$$

$$480b_4 - 720b_3 + 240b_1 + q_0L^3(90 - 60g_1) - 20\Omega^2 a_0 = 0$$

$$800b_5 - 1200b_4 + 400b_2 + q_0L^3\left(20 + 70g_1 - \tfrac{140}{3}g_2 + 5g_3\right)$$
$$\quad + \Omega^2(10a_0 - 20a_1) = 0$$

$$1200b_6 - 1800b_5 + 600b_3 - q_0L^3(25 - 50g_2 - 50g_3) + \Omega^2(10a_1 - 20a_2) = 0$$

$$-2520b_6 + 840b_4 - q_0L^3(15g_1 - 45g_3) - \Omega^2(a_0 - 10a_2) = 0$$

$$1120b_5 - \tfrac{35}{3}g_2 - a_1\Omega^2 = 0$$

$$1440b_6 - 10q_0L^3g_3 - a_2\Omega^2 = 0$$

$$\tag{8.146}$$

We have to express the unknowns in terms of the parameters b_6 and Ω^2. Therefore, we demand that the rank of the set (8.72) be less than 8, leading to a system of nine determinantal equations with six unknowns, namely the parameters a_0, a_1, a_2, g_1, g_2 and g_3. The solution of this set reads

$$a_0 = \tfrac{249}{320}a_1 - \tfrac{227}{160}a_2 \qquad g_1 = \tfrac{77}{18} + \tfrac{371}{144}g_2 - \tfrac{37}{18}g_3 \qquad (8.147)$$

Upon substitution of the expressions (8.73) into the set (8.72), we get the solution for the natural frequency squared:

$$\Omega^2 = (1440b_6 - 10q_0L^3g_3)/a_2 \qquad (8.148)$$

The value of q_0 that corresponds to the vanishing frequency $\Omega^2 = 0$ is the buckling load $q_{0,\text{cr}}$:

$$q_{0,\text{cr}} = 144b_6/g_3L^3 \tag{8.149}$$

We, thus, rewrite Eq. (8.74) as follows:

$$\Omega^2 = 1440b_6(1 - q_0/q_{0,\text{cr}})/a_2 \tag{8.150}$$

From Eq. (8.76), when $q_0 \equiv 0$, $\Omega^2 = 1440b_6/a_2 \equiv \Omega_0^2$. Therefore, Eq. (8.74) takes the form

$$\Omega_q^2 = \Omega_0^2(1 - q_0/q_{0,\text{cr}}) \tag{8.151}$$

where Ω_q^2 is the natural frequency squared in the presence of the axial load. When q_0 tends to its critical value the natural frequency vanishes. The coefficients b_i in the flexural rigidity are

$$
\begin{aligned}
b_0 &= \frac{1}{40{,}320a_2}\{(7{,}118{,}064a_1 - 3{,}546{,}144a_2)b_6 \\
&\quad + [(17{,}920 + 10{,}640g_2 + 20{,}146g_3)a_2 - 49{,}431a_1g_3]q_0L^3\} \\
b_1 &= \frac{1}{6{,}720a_2}\{(203{,}472a_1 + 6{,}048a_2)b_6 \\
&\quad + [(4{,}480 + 2{,}660g_2 - 1{,}162g_3)a_2 - 1{,}413g_3a_1]q_0L^3\} \\
b_2 &= \frac{1}{40{,}320a_2}\{(54{,}432a_2 + 1{,}831{,}248r_1)b_6 \\
&\quad - [(22{,}960 + 8{,}750g_2 - 5{,}222g_3)a_2 + 12{,}717g_3a_1]q_0L^3\} \\
b_3 &= \frac{1}{672a_2}\{(30{,}912a_2 - 13{,}536a_1)b_6 + [(28 - 35g_2 - 168g_3)a_2 + 94g_3a_1]q_0L^3\} \\
b_4 &= \frac{1}{80{,}640a_2}\{(107{,}568a_1 - 1{,}336{,}608a_2)b_6 \\
&\quad + [(6{,}160 + 3{,}710g_2 + 3{,}682g_3)a_2 - 747g_3a_1]q_0L^3\} \\
b_5 &= \frac{1}{672a_2}[864a_1b_6 + (7g_2a_2 - 6g_3a_1)q_0L^3]
\end{aligned}
$$

$$\tag{8.152}$$

8.2.5.4 *Cubic inertial coefficient* (*m* = 3)

The governing differential equation yields a set of nine equations with ten unknowns, namely the coefficients b_i in the flexural rigidity, the natural frequency squared, Ω^2, and the axial load q_0 which is treated as a

parameter:

$$80b_2 - 120b_1 + q_0L^3\left(40 + 20g_1 + \tfrac{40}{3}g_2 + 10g_3 + 8g_4\right) = 0$$
$$240b_3 - 360b_2 + 120b_0 - q_0L^3(140 + 30g_1 + 20g_2 + 15g_3 + 12g_4) = 0$$
$$480b_4 - 720b_3 + 240b_1 + q_0L^3(90 - 60g_1) - 20\Omega^2a_0 = 0$$
$$800b_5 - 1200b_4 + 400b_2 + q_0L^3\left(20 + 70g_1 - \tfrac{140}{3}g_2 + 5g_3 + 4g_4\right)$$
$$\quad + \Omega^2(10a_0 - 20a_1) = 0$$
$$1200b_6 - 1800b_5 + 600b_3 - q_0L^3(25 - 50g_2 + 50g_3) + \Omega^2(10a_1 - 20a_2) = 0$$
$$-2520b_6 + 840b_4 - q_0L^3(15g_1 - 45g_3 + 48g_4) - \Omega^2(a_0 - 10a_2 + 20a_3) = 0$$
$$-3360b_7 + 1120b_5 - q_0L^3\left(\tfrac{35}{3}g_2 - 42g_4\right) - \Omega^2(a_1 - 10a_3) = 0$$
$$1440b_6 - 10g_3 - a_2\Omega^2 = 0$$
$$1800b_7 - 9q_0L^3g_4 - a_3\Omega^2 = 0$$

$$(8.153)$$

We have to express the unknowns in terms of the parameters b_5 and Ω^2. Therefore, we require the rank of the set (8.79) to be less than 9. This leads to a system of ten determinantal equations with eight unknowns, namely the parameters a_0, a_1, a_2, a_3, g_1, g_2, g_3 and g_4. The solution of this set reads

$$a_0 = \frac{249}{320}a_1 - \frac{227}{160}a_2 + \frac{1393}{800}a_3 \quad g_1 = \frac{77}{18} + \frac{371}{144}g_2 - \frac{37}{18}g_3 + \frac{449}{120}g_4 \quad (8.154)$$

Upon substitution of the expressions (8.80) into the set (8.79), we get the solution for the natural frequency squared:

$$\Omega^2 = (1800b_7 - 9q_0L^3g_4)/a_3 \tag{8.155}$$

The value of q_0 that corresponds to the vanishing frequency $\Omega^2 = 0$ is identified as the buckling load $q_{0,\mathrm{cr}}$:

$$q_{0,\mathrm{cr}} = 200b_7/g_4L^3 \tag{8.156}$$

We rewrite Eq. (8.81) as follows:

$$\Omega^2 = 1800b_7(1 - q_0/q_{0,\mathrm{cr}})/a_3 \tag{8.157}$$

From Eq. (8.83), when $q_0 \equiv 0$, $\Omega^2 = 1800b_7/a_3 \equiv \Omega_0^2$. Therefore, Eq. (8.81) takes the form

$$\Omega_q^2 = \Omega_0^2(1 - q_0/q_{0,\mathrm{cr}}) \tag{8.158}$$

where Ω_q^2 is the natural frequency squared in the presence of the axial load. When q_0 tends to its critical value, the natural frequency vanishes. The coefficients b_i in the flexural rigidity are

$$b_0 = \{(444{,}879{,}000a_1 - 221{,}634{,}000a_2 + 651{,}337{,}200a_3)b_7$$
$$+ [(896{,}000 + 532{,}000g_2 - 224{,}000g_3 - 2{,}577{,}966g_4)a_3$$
$$+ (1{,}108{,}170a_2 - 2{,}224{,}395a_1)g_4]q_0L^3\}/2{,}016{,}000a_3$$

$$b_1 = \{(12{,}717{,}000a_1 + 378{,}000a_2 + 18{,}267{,}600a_3)b_7$$
$$+ [(224{,}000 + 133{,}000g_2 - 56{,}000g_3 + 78{,}342g_4)a_3$$
$$- (1{,}890a_2 + 63{,}585a_1)g_4]q_0L^3\}/336{,}000a_3$$

$$b_2 = \{(114{,}453{,}000a_1 + 3{,}402{,}000a_2 + 164{,}408{,}400a_3)b_7$$
$$- [(1{,}148{,}000 + 437{,}500g_2 + 280{,}000g_3 - 1{,}382{,}322g_4)a_3$$
$$+ (572{,}265a_1 + 17{,}010a_2)g_4]q_0L^3\}/2{,}016{,}000a_3$$

$$b_3 = \{(-253{,}800a_1 + 579{,}600a_2 - 395{,}280a_3)b_7$$
$$+ [(420 - 525g_2 + 700g_3 + 1296g_4)a_3$$
$$+ (1{,}269a_1 - 2{,}898a_2)g_4]q_0L^3\}/10{,}080a_3$$

$$b_4 = \{(6{,}723{,}000a_1 - 83{,}538{,}000a_2 + 179{,}780{,}400a_3)b_7 + [(308{,}000 + 185{,}500g_2$$
$$- 280{,}000g_3 - 439{,}422g_4)a_3 + (417{,}690a_2 - 33{,}615a_1)g_4]q_0L^3\}/4{,}032{,}000a_3$$

$$b_5 = \{(5{,}400a_1 - 43{,}920a_3)b_7 + [(144 + 35g_2)a_3 - 27g_4a_1]q_0L^3\}/3{,}360a_3$$

$$b_6 = [1{,}800a_2b_7 + (10g_3a_3 - 9g_4a_2)q_0L^3]/1440a_3$$

$$\text{(8.159)}$$

8.2.5.5 Quartic inertial coefficient ($m = 4$)

The governing differential equation yields a set of 10 equations with 11 unknowns, namely the coefficients b_i in the flexural rigidity, the natural frequency squaredm, Ω^2, and the axial load q_0, which is treated as a parameter:

$$80b_2 - 120b_1 + q_0L^3\left(40 + 20g_1 + \tfrac{40}{3}g_2 + 10g_3 + 8g_4 + \tfrac{20}{3}g_5\right) = 0$$

$$240b_3 - 360b_2 + 120b_0 - q_0L^3(140 + 30g_1 + 20g_2 + 15g_3 + 12g_4 - 10g_5) = 0$$

$$480b_4 - 720b_3 + 240b_1 + q_0L^3(90 - 60g_1) - 20\Omega^2a_0 = 0$$

$$800b_5 - 1200b_4 + 400b_2 + q_0L^3\left(20 + 70g_1 - \tfrac{140}{3}g_2 + 5g_3 + 4g_4 + \tfrac{10}{3}g_5\right)$$
$$+ \Omega^2(10a_0 - 20a_1) = 0$$

$$1200b_6 - 1800b_5 + 600b_3 - q_0L^3(25 - 50g_2 + 50g_3) + \Omega^2(10a_1 - 20a_2) = 0$$

$$1680b_7 - 2520b_6 + 840b_4 - q_0L^3(15g_1 - 45g_3 + 48g_4)$$
$$- \Omega^2(a_0 - 10a_2 + 20a_3) = 0$$

$$2240b_8 - 3360b_7 + 1120b_5 - q_0 L^3 \left(\tfrac{35}{3} g_2 - 42 g_4 + \tfrac{140}{3} g_5 \right)$$
$$-\Omega^2 (a_1 - 10a_3 + 20a_4) = 0$$
$$-4320b_8 + 1440b_6 - q_0 L^3 (10g_3 - 40g_5) - \Omega^2 (a_2 - 10a_4) = 0$$
$$1800b_7 - 9q_0 L^3 g_4 - a_3 \Omega^2 = 0$$
$$2200b_8 - \tfrac{25}{3} q_0 L^3 g_5 - a_4 \Omega^2 = 0 \tag{8.160}$$

We have to express the unknowns in terms of the parameters b_5 and Ω^2. Therefore, we demand the rank of the set (8.86) to be less than 10, leading to a system of 11 determinantal equations with 10 unknowns, namely the parameters a_0, a_1, a_2, a_3, a_4, g_1, g_2, g_3, g_4 and g_5. The solution for this set reads

$$a_0 = \frac{249}{320} a_1 - \frac{227}{160} a_2 + \frac{1393}{800} a_3 - \frac{3183}{1100} a_4$$
$$\tag{8.161}$$
$$g_1 = \frac{77}{18} + \frac{371}{144} g_2 - \frac{37}{18} g_3 + \frac{449}{120} g_4 - \frac{1841}{396} g_5$$

Upon substitution of the expressions (8.87) into the set (8.86), we get the solution for the natural frequency squared:

$$\Omega^2 = \left(2200b_8 - \tfrac{25}{3} q_0 L^3 g_5 \right) / a_4 \tag{8.162}$$

The value of q_0 that corresponds to the vanishing frequency $\Omega^2 = 0$ is identified as the buckling load $q_{0,\mathrm{cr}}$:

$$q_{0,\mathrm{cr}} = 264 b_8 / g_5 L^3 \tag{8.163}$$

We rewrite Eq. (8.88) as follows:

$$\Omega^2 = 2200 b_8 (1 - q_0 / q_{0,\mathrm{cr}}) / a_4 \tag{8.164}$$

From Eq. (8.90), when $q_0 \equiv 0$, $\Omega^2 = 2200 b_8 / a_4 \equiv \Omega_0^2$. Therefore, Eq. (8.88) takes the form

$$\Omega_q^2 = \Omega_0^2 (1 - q_0 / q_{0,\mathrm{cr}}) \tag{8.165}$$

where Ω_q^2 is the natural frequency squared in the presence of the axial load. When q_0 tends to its critical value, the natural frequency vanishes. The coefficients b_i in the flexural rigidity read

$$b_0 = \{(-506{,}576{,}640a_4 + 477{,}647{,}280a_3 - 162{,}531{,}600a_2 + 326{,}244{,}600a_1)b_8$$
$$+ [(1{,}521{,}760g_5 + 407{,}232g_4 - 134{,}400g_3 + 319{,}200g_2 + 537{,}600)a_4$$
$$- (1{,}809{,}270a_3 - 615{,}650a_2 + 1{,}235{,}775a_1)g_5]q_0L^3\}/1{,}209{,}600a_4$$

$$b_1 = \{(-215{,}804{,}160a_4 + 442{,}075{,}920a_3 + 9{,}147{,}600a_2 + 307{,}751{,}400a_1)b_8$$
$$+ [(-2{,}458{,}560g_5 + 3{,}359{,}664g_4 - 1{,}108{,}800g_3 + 2{,}633{,}400g_2 + 4{,}435{,}200)a_4$$
$$- (1{,}674{,}530a_3 + 34{,}650a_2 + 1{,}165{,}725a_1)g_5]q_0L^3\}/6{,}652{,}800a_4$$

$$b_2 = \{(-215{,}804{,}160a_4 + 442{,}075{,}920a_3 + 9{,}147{,}600a_2 + 307{,}751{,}400a_1)b_8$$
$$+ [(2{,}326{,}640g_5 - 1{,}232{,}616g_4 + 616{,}000g_3 - 962{,}500g_2 - 2{,}525{,}600)a_4$$
$$- (1{,}674{,}530a_3 + 34{,}650a_2 + 1{,}165{,}725a_1)g_5]q_0L^3\}/4{,}435{,}200a_4$$

$$b_3 = \{(20{,}625{,}600a_4 - 7{,}246{,}800a_3 + 10{,}626{,}000a_2 - 4{,}653{,}000a_1)b_8$$
$$- [(57{,}700g_5 + 10{,}206g_4 - 10{,}500g_3 + 7{,}875g_2 - 6{,}300)a_4$$
$$- (27{,}450a_3 - 40{,}250a_2 + 17{,}625a_1)g_5]q_0L^3\}/151{,}200a_4$$

$$b_4 = \{(-1{,}181{,}854{,}080a_4 + 1{,}450{,}228{,}560a_3 - 673{,}873{,}200a_2 + 54{,}232{,}200a_1)b_8$$
$$+ [(957{,}120g_5 + 3{,}032{,}568g_4 - 1{,}848{,}000g_3 + 1{,}224{,}300g_2 + 2{,}032{,}800)a_4$$
$$- (5{,}493{,}290a_3 - 2{,}552{,}550a_2 + 205{,}425a_1)g_5]q_0L^3\}/26{,}611{,}200a_4$$

$$b_5 = \{(1{,}879{,}200a_4 - 805{,}200a_3 + 99{,}000a_1)b_8 + [(-5{,}400g_5 - 1{,}134g_4 + 525g_2)a_4$$
$$+ (3{,}050a_3 - 375a_1)g_5]q_0L^3\}/6{,}652{,}800a_4$$

$$b_6 = \{(-10{,}608a_4 + 1{,}320a_2)b_8 + [(26g_5 + 6g_3)a_4 - 5a_2g_5]q_0L^3\}/864a_4$$

$$b_7 = [6{,}600a_3b_8 + (27g_4a_4 - 25g_5a_3)q_0L^3]/5{,}400a_4$$

$$\text{(8.166)}$$

8.2.5.6 Quintic inertial coefficient ($m = 5$)

The governing differential equation yields a set of 11 equations with 12 unknowns, namely the coefficients b_i in the flexural rigidity, the natural frequency squared, Ω^2, and the axial load q_0, which is treated as a parameter:

$$80b_2 - 120b_1 + q_0L^3\left(40 + 20g_1 + \tfrac{40}{3}g_2 + 10g_3 + 8g_4 + \tfrac{20}{3}g_5 + \tfrac{40}{7}g_6\right) = 0$$

$$240b_3 - 360b_2 + 120b_0 - q_0L^3(140 + 30g_1 + 20g_2 + 15g_3 + 12g_4 - 10g_5$$
$$+ \tfrac{60}{7}g_6) = 0$$

$$480b_4 - 720b_3 + 240b_1 + q_0L^3(90 - 60g_1) - 20\Omega^2a_0 = 0$$

$$800b_5 - 1200b_4 + 400b_2 + q_0L^3\left(20 + 70g_1 - \tfrac{140}{3}g_2 + 5g_3 + 4g_4 + \tfrac{10}{3}g_5 + \tfrac{20}{7}g_6\right)$$
$$+ \Omega^2(10a_0 - 20a_1) = 0$$

$$1200b_6 - 1800b_5 + 600b_3 - q_0L^3(25 - 50g_2 + 50g_3) + \Omega^2(10a_1 - 20a_2) = 0$$

$$1680b_7 - 2520b_6 + 840b_4 - q_0L^3(15g_1 - 45g_3 + 48g_4)$$
$$-\Omega^2(a_0 - 10a_2 + 20a_3) = 0$$
$$2240b_8 - 3360b_7 + 1120b_5 - q_0L^3(\tfrac{35}{3}g_2 - 42g_4 + \tfrac{140}{3}g_5)$$
$$-\Omega^2(a_1 - 10a_3 + 20a_4) = 0$$
$$3360b_7 - 4320b_8 + 1440b_6 - q_0L^3(10g_3 - 40g_5 + \tfrac{320}{7}g_6)$$
$$-\Omega^2(a_2 - 10a_4 + 20a_5) = 0$$
$$-5400b_9 + 1800b_7 - q_0L^3(9g_4 - \tfrac{270}{7}g_6) - \Omega^2(a_3 - 10a_5) = 0$$
$$2200b_8 - \tfrac{25}{3}q_0L^3g_5 - a_4\Omega^2 = 0$$
$$2640b_9 - \tfrac{55}{7}q_0L^3g_6 - a_5\Omega^2 = 0$$

$$(8.167)$$

We have to express the unknowns in terms of the parameters b_5 and Ω^2. Therefore, we demand the rank of the set (8.93) to be less than 11, leading to a system of 12 determinantal equations with 12 unknowns, namely the parameters a_0, a_1, a, a_3, a_4, a_5, g_1, g_2, g_3, g_4, g_5 and g_6. The solution of this set reads

$$a_0 = \frac{249}{320}a_1 - \frac{227}{160}a_2 + \frac{1393}{800}a_3 - \frac{3183}{1100}a_4 + \frac{7553}{1760}a_5$$
$$g_1 = \frac{77}{18} + \frac{371}{144}g_2 - \frac{37}{18}g_3 + \frac{449}{120}g_4 - \frac{1841}{396}g_5 + \frac{557}{72}g_6$$

$$(8.168)$$

Upon substitution of the expressions (8.94) into the set (8.93), we get the solution for the natural frequency squared:

$$\Omega^2 = (2680b_9 - \tfrac{55}{7}q_0L^3g_6)/a_5 \qquad (8.169)$$

The value of q_0 that corresponds to the vanishing frequency $\Omega^2 = 0$ is identified as the buckling load $q_{0,\text{cr}}$:

$$q_{0,\text{cr}} = 336b_9/g_6L^3 \qquad (8.170)$$

We rewrite Eq. (8.95) as follows:

$$\Omega^2 = 2680b_9(1 - q_0/q_{0,\text{cr}})/a_5 \qquad (8.171)$$

From Eq. (8.97), when $q_0 \equiv 0$, $\Omega^2 = 2680b_9/a_5 \equiv \Omega_0^2$. Therefore, Eq. (8.95) takes the form

$$\Omega_q^2 = \Omega_0^2(1 - q_0/q_{0,\text{cr}}) \qquad (8.172)$$

where Ω_q^2 is the natural frequency squared in the presence of the axial load. When q_0 tends to its critical value the natural frequency vanishes. The flexural rigidity coefficients b_i read

$$b_0 = \{(30,311,893,920a_5 - 15,602,560,512a_4 + 14,711,536,224a_3$$
$$- 5,005,973,280a_2 + 1,004,833,680a_1)b_9 - [(70,329,490g_6 + 10,192,000g_5$$
$$- 10,452,288g_4 + 3,449,600g_3 - 8,192,800g_2 - 13,798,400)a_5$$
$$- (46,436,192a_4 - 43,784,334a_3 + 14,898,730a_2$$
$$- 29,905,755a_1)g_6]q_0L^3\}/31,046,400a_5$$

$$b_1 = \{(2,165,227,680a_5 - 604,251,648a_4 + 1,237,812,576a_3 + 25,613,280a_2$$
$$+ 861,703,920a_1)b_9 + [(1,798,368g_6 - 7,644,000g_5 + 7,839,216g_4$$
$$- 2,587,200g_3 + 6,144,600g_2 + 10,348,800)a_5 - (3,683,966a_3 + 76,230a_2$$
$$+ 2,564,595a_1)g_6]q_0L^3\}/15,523,200a_5$$

$$b_2 = \{(590,516,640a_5 - 164,795,904a_4 + 337,585,248a_3 + 6,985,440a_2$$
$$+ 235,010,160a_1)b_9 - [(3,350,410g_6 - 960,400g_5 + 784,392g_4 - 392,000g_3$$
$$+ 612,500g_2 + 1,607,200)a_5 + (490,464a_4 - 1,004,718a_3 - 20,790a_2$$
$$- 699,435a_1)g_6]q_0L^3\}/2,822,400a_5$$

$$b_3 = \{(-676,552,800a_5 + 635,268,480a_4 - 223,201,440a_3 + 327,280,800a_2$$
$$- 143,312,400a_1)b_9 + [(1,376,760g_6 + 524,300g_5 - 261,954g_4 + 269,500g_3$$
$$- 202,125g_2 + 161,700)a_5 - (1,890,680a_4 - 664,290a_3 + 974,050a_2$$
$$- 426,525a_1)g_6]q_0L^3\}/3,880,800a_5$$

$$b_4 = \{(8,743,959,840a_5 - 3,309,191,424a_4 + 4,060,639,968a_3 - 1,886,844,960a_2$$
$$+ 151,850,160a_1)b_9 - [(11,088,770g_6 + 8,212,400g_5 - 7,075,992g_4$$
$$+ 4,132,000g_3 - 2,856,700g_2 - 4,743,200)a_5$$
$$- (9,848,784a_4 - 12,085,238a_3 + 5,615,610a_2$$
$$- 451,935a_1)g_6]q_0L^3\}/62,092,800a_5$$

$$b_5 = \{(-45,276,000a_5 + 57,879,360a_4 - 24,800,160a_3 + 3,049,200a_1)b_9$$
$$+ [(86,240g_6 + 44,100g_5 - 29,106g_4 + 13,475g_2)a_5$$
$$- (172,260a_4 - 73,810a_3 + 9,075a_1)g_6]q_0L^3\}/1,293,600a_5$$

$$b_6 = \{(3,843,840a_5 - 1,633,632a_4 + 203,280a_2)b_9$$
$$- [(8,580g_6 + 1,820g_5 - 770g_3)a_5 - (4,862a_4 - 605a_2)g_6]q_0L^3\}/110,880a_5$$

$$b_7 = \{(18,480a_3 - 147,000a_5)b_9 + [(280g_6 + 63g_4)a - 55g_6a_3]q_0L^3\}/12,600a_5$$

$$b_8 = [11,088a_4b_9 + (35g_5a_5 - 33g_6a_4)q_0L^3]/9,240a_5$$

$$(8.173)$$

8.2.6   Closed-Form Solution for the Clamped–Clamped Beam

8.2.6.1   Uniform inertial coefficient ($m = 0$)

In this case, we get a homogeneous set of six linear algebraic equations with seven unknowns, namely the coefficients b_i in the flexural rigidity, the natural frequency squared, Ω^2, and the axial load q_0, which is treated as a parameter:

$$8b_2 - 36b_1 + q_0L^3(4 + 2g_1) = 0$$
$$24b_3 - 108b_2 + 120b_0 - q_0L^3(26 + 9g_1) = 0$$
$$48b_4 - 216b_3 + 240b_1 + q_0L^3(27 - 6g_1) - 2a_0\Omega^2 = 0 \qquad (8.174)$$
$$-360b_4 + 400b_2 + q_0L^3(20 + 28g_1) + 3a_0\Omega^2 = 0$$
$$600b_3 - 25q_0L^3 = 0$$
$$840b_4 - 15g_1q_0L^3 - a_0\Omega^2 = 0$$

We have to write all unknowns in terms of a single free parameter, say b_4, and the axial load parameter q_0. In other words, the rank of the set (8.100) or its matrix must be less than 6. Thus, all minors of order 6 must vanish, leading to seven equations with two unknowns a_0 and g_1:

$$259{,}034{,}112a_0g_1 + 1{,}135{,}088{,}640a_0 = 0$$
$$80{,}123{,}904a_0g_1 + 146{,}810{,}880a_0 = 0$$
$$137{,}106{,}432a_0g_1 + 334{,}679{,}040a_0 = 0$$
$$a_0 = 0 \qquad (8.175)$$
$$41{,}055{,}728a_0g_1 + 4{,}354{,}560a_0 = 0$$
$$46{,}282{,}752g_1 - 365{,}783{,}044 = 0$$
$$a_0 = 0$$

The solution of the set (8.101) reads

$$a_0 = 0 \quad g_1 = -\tfrac{245}{31} \qquad (8.176)$$

which leads to a trivial solution for the problem since the inertial coefficient vanishes.

8.2.6.2   Linearly varying inertial coefficient ($m = 1$)

The governing differential equation yields a set of seven equations with eight unknowns, namely the coefficients b_i in the flexural rigidity, the natural frequency squared, Ω^2, and the axial load q_0, which is treated as a

parameter:

$$8b_2 - 36b_1 + q_0 L^3 \left(4 + 2g_1 + \tfrac{4}{3}g_2\right) = 0$$

$$24b_3 - 108b_2 + 120b_0 - q_0 L^3 (26 + 9g_1 + 6g_2) = 0$$

$$48b_4 - 216b_3 + 240b_1 + q_0 L^3 (27 - 6g_1) - 2a_0 \Omega^2 = 0$$

$$80b_5 - 360b_4 + 400b_2 + q_0 L^3 \left(20 + 28g_1 + \tfrac{4}{3}g_2\right) + \Omega^2 (3a_0 - 2a_1) = 0$$

$$-540b_5 + 600b_3 - q_0 L^3 (25 - 15g_2) + 3a_1 \Omega^2 = 0$$

$$840b_4 - 15q_0 L^3 g_1 - a_0 \Omega^2 = 0$$

$$1120b_5 - \tfrac{35}{3} q_0 L^3 g_2 - a_1 \Omega^2 = 0$$

$$(8.177)$$

We would like to express the unknowns in terms of the parameters b_5 and Ω^2. Therefore, we demand the rank of the set (8.103) to be less than 7, leading to a system of eight determinantal equations with four unknowns, namely a_0, a_1, g_1 and g_2:

$$399{,}940{,}208{,}640 a_0 g_2 + 57{,}267{,}689{,}472 a_1 g_2 + 1{,}271{,}299{,}276{,}800 a_0$$
$$+267{,}753{,}185{,}280 a_1 + 290{,}118{,}205{,}440 a_0 g_1 - 31{,}922{,}408{,}448 a_1 g_1 = 0$$
$$-63{,}204{,}986{,}880 a_0 g_2 + 28{,}154{,}843{,}136 a_1 g_2 - 164{,}428{,}185{,}600 a_0$$
$$+69{,}603{,}287{,}040 a_1 - 89{,}738{,}772{,}480 a_0 g_1 + 43{,}898{,}443{,}776 a_1 g_1 = 0$$
$$-87{,}166{,}679{,}040 a_0 g_2 + 62{,}465{,}292{,}288 a_1 g_2 - 374{,}840{,}524{,}800 a_0$$
$$+254{,}271{,}467{,}520 a_1 - 153{,}559{,}203{,}840 a_0 g_1 + 86{,}200{,}132{,}608 a_1 g_1 = 0$$
$$2{,}175{,}289{,}344 a_1 g_1 + 3{,}128{,}315{,}904 a_1 g_2 + 21{,}920{,}855{,}040 a_1 + 3{,}483{,}648{,}000 a_0 g_2$$
$$-9{,}289{,}728{,}000 a_0 = 0$$
$$139{,}055{,}616 a_0 g_2 + 487{,}710{,}720 a_0 + 459{,}841{,}536 a_0 g_1 - 202{,}673{,}664 a_1 g_1 = 0$$
$$-46{,}282{,}752 a_1 g_1 + 13{,}934{,}592 a_1 g_2 - 365{,}783{,}040 a_1 - 232{,}243{,}200 a_0 g_2 = 0$$
$$11{,}680{,}671{,}744 g_2 + 5{,}183{,}668{,}224 g_1 + 40{,}967{,}700{,}480 = 0$$
$$11{,}349{,}725{,}184 a_1 - 22{,}295{,}347{,}200 a_0 = 0$$

$$(8.178)$$

The solution of the set (8.104) is

$$a_0 = \frac{1629}{3200} a_1 \qquad g_1 = -\frac{245}{31} - \frac{3353}{1488} g_2 \qquad (8.179)$$

Upon substitution of the expressions (8.105) into the set (8.103), we get the solution for the natural frequency squared:

$$\Omega^2 = \left(1120 b_5 - \tfrac{35}{3} q_0 L^3 g_2\right) / a_1 \qquad (8.180)$$

The value of q_0 that corresponds to the vanishing frequency $\Omega^2 = 0$ is identified as the buckling load $q_{0,cr}$:

$$q_{0,cr} = 96b_5/g_2L^3 \tag{8.181}$$

We rewrite Eq. (8.106) as follows:

$$\Omega^2 = 1120b_5(1 - q_0/q_{0,cr})/a_1 \tag{8.182}$$

From Eq. (8.108), when $q_0 \equiv 0, \Omega^2 = 1120b_5/a_1 \equiv \Omega_0^2$. Therefore, Eq. (8.106) takes the form

$$\Omega_q^2 = \Omega_0^2(1 - q_0/q_{0,cr}) \tag{8.183}$$

where Ω_q^2 is the natural frequency squared in the presence of the axial load. We again note the following property: when q_0 tends to its critical value the natural frequency vanishes. The coefficients b_i in the flexural rigidity are

$$b_0 = \frac{100,051}{40,000}b_5 - \left(\frac{3,409}{74,400} + \frac{4,456,681}{119,040,000}g_2\right)q_0L^3$$

$$b_1 = \frac{771}{2,000}b_5 - \left(\frac{303}{1,240} + \frac{131,667}{1,984,000}g_2\right)q_0L^3$$

$$b_2 = \frac{6,939}{4,000}b_5 + \left(\frac{933}{2,480} + \frac{388,997}{3,968,000}g_2\right)q_0L^3$$

$$b_3 = -\frac{47}{10}b_5 + \left(\frac{1}{24} + \frac{1}{30}g_2\right)q_0L^3 \quad b_4 = \frac{543}{800}b_5 - \left(\frac{35}{248} + \frac{112,633}{2,380,800}g_2\right)q_0L^3 \tag{8.184}$$

8.2.6.3 *Parabolically varying inertial coefficient ($m = 2$)*

The governing differential equation yields a set of eight equations with eight unknowns, namely the seven coefficients b_i of the flexural rigidity, the natural frequency squared, Ω^2, and the axial load q_0, which is treated as a parameter:

$$8b_2 - 36b_1 + q_0L^3\left(4 + 2g_1 + \tfrac{4}{3}g_2 + g_3\right) = 0$$

$$24b_3 - 108b_2 + 120b_0 - q_0L^3\left(26 + 9g_1 + 6g_2 + \tfrac{9}{2}g_3\right) = 0$$

$$48b_4 - 216b_3 + 240b_1 + q_0L^3(27 - 6g_1) - 2a_0\Omega^2 = 0$$

$$80b_5 - 360b_4 + 400b_2 + q_0L^3\left(20 + 28g_1 + \tfrac{4}{3}g_2 + 5g_3\right) + \Omega^2(3a_0 - 2a_1) = 0$$

$$120b_6 - 540b_5 + 600b_3 - q_0 L^3 (25 - 15g_2 + 5g_3) + \Omega^2 (3a_1 - 2a_2) = 0$$
$$-756b_6 + 840b_4 - q_0 L^3 (15g_1 - \tfrac{27}{2}g_3) - \Omega^2 (a_0 - 3a_2) = 0$$
$$1120b_5 - \tfrac{35}{3} q_0 L^3 g_2 - a_1 \Omega^2 = 0$$
$$1440b_6 - 10q_0 L^3 g_3 - a_2 \Omega^2 = 0 \tag{8.185}$$

We have to express the unknowns in terms of the parameters b_6 and Ω^2. Therefore, we demand the rank of the set (8.111) to be less than 8, leading to a system of nine determinantal equations with six unknowns, namely the parameters a_0, a_1, a_2, g_1, g_2 and g_3. The solution of this set reads

$$a_0 = \frac{1629}{3200} a_1 - \frac{681}{1600} a_2 \qquad g_1 = -\frac{245}{31} - \frac{3353}{1488} g_2 - \frac{83}{124} g_3 \qquad ' \tag{8.186}$$

Upon substitution of the expressions (8.112) into the set (8.111), we get the solution for the natural frequency squared:

$$\Omega^2 = (1440b_6 - 10q_0 L^3 g_3)/a_2 \tag{8.187}$$

The value of q_0 that corresponds to the vanishing frequency $\Omega^2 = 0$ is identified as the buckling load $q_{0,\mathrm{cr}}$:

$$q_{0,\mathrm{cr}} = 144b_6/g_3 L^3 \tag{8.188}$$

We rewrite Eq. (8.113) as follows:

$$\Omega^2 = 1440b_6 (1 - q_0/q_{0,\mathrm{cr}})/a_2 \tag{8.189}$$

From Eq. (8.115), when $q_0 \equiv 0, \Omega^2 = 1440b_6/a_2 \equiv \Omega_0^2$. Therefore, Eq. (8.113) takes the form

$$\Omega_q^2 = \Omega_0^2 (1 - q_0/q_{0,\mathrm{cr}}) \tag{8.190}$$

where Ω_q^2 is the natural frequency squared in the presence of the axial load. When q_0 tends to its critical value, $q_{0,\mathrm{cr}}$, the natural frequency vanishes. The coefficients b_i in the flexural rigidity are

$$b_0 = \frac{1}{59{,}520{,}000a_2}\{(191{,}411{,}856a_1 - 48{,}249{,}888a_2)b_6$$
$$+ [(286{,}962g_3 - 677{,}550g_2 - 2{,}727{,}200)a_2 - 1{,}329{,}249a_1g_3]q_0L^3\}$$

$$b_1 = \frac{1}{6{,}720a_2}\{(3{,}441{,}744a_1 + 187{,}488a_2)b_6$$
$$- [(1{,}696{,}800 + 432{,}950g_2 + 43{,}862g_3)a_2 + 23{,}901g_3a_1]q_0L^3\}$$

$$b_2 = \frac{1}{13{,}888{,}000a_2}\{(1{,}687{,}392a_2 + 30{,}975{,}696a_1)b_6$$
$$+ [(5{,}224{,}800 + 1{,}612{,}450g_2 + 193{,}242g_3)a_2 - 215{,}109g_3a_1]q_0L^3\}$$

$$b_3 = \frac{1}{6{,}720a_2}\{(30{,}912a_2 - 40{,}608a_1)b_6 + [(280 - 105g_2 - 168g_3)a_2 + 282g_3a_1]q_0L^3\}$$

$$b_4 = \frac{1}{8{,}332{,}800a_2}\{(7{,}271{,}856a_1 - 41{,}434{,}848a_2)b_6$$
$$+ [(1{,}176{,}000 + 335{,}300g_2 - 106{,}302g_3)a_2 - 50{,}499g_3a_1]q_0L^3\}$$

$$b_5 = \frac{1}{672a_2}[864a_1b_6 + (7g_2a_2 - 6g_3a_1)q_0L^3]$$

$$(8.191)$$

8.2.6.4 *Cubic inertial coefficient ($m = 3$)*

The governing differential equation yields a set of nine equations with ten unknowns, namely the coefficients b_i in the flexural rigidity, the natural frequency squared, Ω^2, and the axial load q_0, which is treated as a parameter:

$$8b_2 - 36b_1 + q_0L^3\left(4 + 2g_1 + \tfrac{4}{3}g_2 + g_3 + \tfrac{4}{5}g_4\right) = 0$$

$$24b_3 - 108b_2 + 120b_0 - q_0L^3\left(26 + 9g_1 + 6g_2 + \tfrac{9}{2}g_3 + \tfrac{18}{5}g_4\right) = 0$$

$$48b_4 - 216b_3 + 240b_1 + q_0L^3(27 - 6g_1) - 2a_0\Omega^2 = 0$$

$$80b_5 - 360b_4 + 400b_2 + q_0L^3\left(20 + 28g_1 + \tfrac{4}{3}g_2 + 5g_3 + 4g_4\right) + \Omega^2(3a_0 - 2a_1) = 0$$

$$120b_6 - 540b_5 + 600b_3 - q_0L^3(25 - 15g_2 + 5g_3) + \Omega^2(3a_1 - 2a_2) = 0$$

$$168b_7 - 756b_6 + 840b_4 - q_0L^3\left(15g_1 - \tfrac{27}{2}g_3 + \tfrac{24}{5}g_4\right) - \Omega^2(a_0 - 3a_2 + 2a_3) = 0$$

$$-1008b_7 + 1120b_5 - q_0L^3\left(\tfrac{35}{3}g_2 - \tfrac{63}{5}g_4\right) - \Omega^2(a_1 - 3a_3) = 0$$

$$1440b_6 - 10q_0L^3g_3 - a_2\Omega^2 = 0$$

$$1800b_7 - 9q_0L^3g_4 - a_3\Omega^2 = 0$$

$$(8.192)$$

We have to express the unknowns in terms of the parameters b_5 and Ω^2. Therefore, we demand the rank of the set (8.118) to be less than 9, leading to a system of ten determinantal equations with eight unknowns, namely the parameters $a_0, a_1, a_2, a_3, g_1, g_2, g_3$ and g_4. The solution of this set read

$$u_0 = \frac{1,629}{3,200}u_1 - \frac{681}{1,600}u_2 + \frac{23,263}{80,000}u_3 \quad g_1 - -\frac{245}{31} - \frac{3,353}{1,488}g_2 - \frac{83}{124}g_3 \quad \frac{77,429}{62,000}g_4 \tag{8.193}$$

Upon substitution of the expressions (8.119) into the set (8.118), we get the solution for the natural frequency squared:

$$\Omega^2 = (1800b_7 - 9q_0 L^3 g_4)/a_3 \tag{8.194}$$

The value of q_0 that corresponds to the vanishing frequency $\Omega^2 = 0$ is the buckling load $q_{0,cr}$:

$$q_{0,cr} = 200b_7/g_4 L^3 \tag{8.195}$$

We write Eq. (8.120) as follows:

$$\Omega^2 = 1800b_7(1 - q_0/q_{0,cr})/a_3 \tag{8.196}$$

From Eq. (8.122), when $q_0 \equiv 0$, $\Omega^2 = 1800b_7/a_3 \equiv \Omega_0^2$. Therefore, Eq. (8.120) takes the form

$$\Omega_q^2 = \Omega_0^2(1 - q_0/q_{0,cr}) \tag{8.197}$$

where Ω_q^2 is the natural frequency squared in the presence of the axial load. Note that when q_0 tends to its critical value the natural frequency vanishes. The coefficients b_i in the flexural rigidity are

$b_0 = \{(179,448,615,000a_1 - 45,234,270,000a_2 + 75,566,536,200a_3)b_7$

$\quad - [(2,045,400,000 + 508,162,500g_2 + 36,080,000g_3 + 636,810,741g_4)a_3$

$\quad - (226,171,350a_2 - 897,243,075a_1)g_4]q_0 L^3\}/44,640,000,000a_3$

$b_1 = \{(1,075,545,000a_1 + 58,590,000a_2 + 423,044,600a_3)b_7$

$\quad - [(424,200,000 + 108,237,500g_2 + 10,640,000g_3$

$\quad + 59,701,003g_4)a_3 + (292,950a_2 + 5,377,725r_1)g_4]q_0 L^3\}/1,736,000,000a_3$

$b_2 = \{(9,679,905,000a_1 + 527,310,000a_2 + 3,807,401,400a_3)b_7 + [(1,306,200,000$

$\quad + 403,112,500g_2 + 51,240,000g_3 + 199,496,973g_4)a_3$

$\quad - (48,399,525a_1 + 2,636,550a_2)g_4]q_0 L^3\}/3,472,000,000a_3$

$b_3 = \{(-3,807,000a_1 + 2,898,000a_2 - 1,778,760a_3)b_7 + [(21,000 - 7,875g_2$

$\quad + 3,500g_3 + 5,832g_4)a_3 + (19,035a_1 - 14,490a_2)g_4]q_0 L^3\}/504,000a_3$

$$b_4 = \{(2{,}272{,}455{,}000a_1 - 12{,}948{,}390{,}000a_2 + 9{,}809{,}435{,}400a_3)b_7$$

$$- [(294{,}000{,}000 + 83{,}825{,}000g_2 + 45{,}360{,}000g_3 + 85{,}683{,}777\,g_4)a_3$$

$$- (64{,}741{,}950a_2 - 11{,}362{,}275a_1)g_4]q_0L^3\}/2{,}083{,}200{,}000a_3$$

$$b_5 = \{(27{,}000a_1 - 65{,}880a_3)b_7 + [(216g_4 + 175g_2)a_3 - 135g_4a_1]q_0L^3\}/16{,}800a_3$$

$$b_6 = [1{,}800a_2b_7 + (10g_3a_3 - 9g_4a_2)q_0L^3]/1{,}440a_3$$

$$(8.198)$$

8.2.6.5 Quartic inertial coefficient ($m = 4$)

The governing differential equation yields a set of 10 equations with 11 unknowns, namely the coefficients b_i in the flexural rigidity, the natural frequency squared, Ω^2, and the axial load q_0, which is treated as a parameter:

$$8b_2 - 36b_1 + q_0L^3\left(4 + 2g_1 + \tfrac{4}{3}g_2 + g_3 + \tfrac{4}{5}g_4 + \tfrac{2}{3}g_5\right) = 0$$

$$24b_3 - 108b_2 + 120b_0 - q_0L^3\left(26 + 9g_1 + 6g_2 + \tfrac{9}{2}g_3 + \tfrac{18}{5}g_4 + 3g_5\right) = 0$$

$$48b_4 - 216b_3 + 240b_1 + q_0L^3(27 - 6g_1) - 2a_0\Omega^2 = 0$$

$$80b_5 - 360b_4 + 400b_2 + q_0L^3\left(20 + 28g_1 + \tfrac{4}{3}g_2 + 5g_3 + 4g_4 + \tfrac{10}{3}g_5\right)$$

$$+ \Omega^2(3a_0 - 2a_1) = 0$$

$$120b_6 - 540b_5 + 600b_3 - q_0L^3(25 - 15g_2 + 5g_3) + \Omega^2(3a_1 - 2a_2) = 0$$

$$168b_7 - 756b_6 + 840b_4 - q_0L^3\left(15g_1 - \tfrac{27}{2}g_3 + \tfrac{24}{5}g_4\right) - \Omega^2(a_0 - 3a_2 + 2a_3) = 0$$

$$224b_8 - 1008b_7 + 1120b_5 - q_0L^3\left(\tfrac{35}{3}g_2 - \tfrac{63}{5}g_4 + \tfrac{14}{3}g_5\right) - \Omega^2(a_1 - 3a_3 + 2a_4) = 0$$

$$-1296b_6 + 1440b_6 - q_0L^3(10g_3 - 12g_5) - \Omega^2(a_2 - 3a_4) = 0$$

$$1800b_7 - 9q_0L^3g_4 - a_3\Omega^2 = 0$$

$$2200b_8 - \tfrac{25}{3}q_0L^3g_5 - a_4\Omega^2 = 0$$

$$(8.199)$$

We have to express the unknowns in terms of the parameters b_5 and Ω^2. Therefore, we demand the rank of the set (8.125) to be less than 10, leading to a system of 11 determinantal equations with 10 unknowns, namely the parameters $a_0, a_1, a_2, a_3, a_4, g_1, g_2, g_3, g_4$ and g_5. The solution for this set reads

$$a_0 = \frac{1{,}629}{3{,}200}a_1 - \frac{681}{1{,}600}a_2 + \frac{23{,}263}{80{,}000}a_3 - \frac{1{,}698}{6{,}875}a_4$$

$$(8.200)$$

$$g_1 = -\frac{245}{31} - \frac{3{,}353}{1{,}488}g_2 - \frac{83}{124}g_3 - \frac{77{,}429}{62{,}000}g_4 - \frac{20{,}503}{409{,}240}g_5$$

Upon substitution of the expressions (8.126) into the set (8.127), we get the solution for the natural frequency squared:

$$\Omega^2 = \left(2200b_8 - \tfrac{25}{3}q_0L^3g_5\right)/a_4 \qquad (8.201)$$

The value of q_0 that corresponds to the vanishing frequency $\Omega^2 = 0$ is identified as the buckling load $q_{0,cr}$:

$$q_{0,cr} = 264b_8/g_5L^3 \tag{8.202}$$

We write Eq. (8.127) as follows:

$$\Omega^2 = 2200b_8(1 - q_0/q_{0,cr})/a_4 \tag{8.203}$$

From Eq. (8.129), when $q_0 \equiv 0$, $\Omega^2 = 2200b_8/a_4 \equiv \Omega_0^2$. Therefore, Eq. (8.127) takes the form

$$\Omega_q^2 = \Omega_0^2(1 - q_0/q_{0,cr}) \tag{8.204}$$

where Ω_q^2 is the natural frequency squared in the presence of the axial load. Note that when q_0 tends to its critical value the natural frequency vanishes. The coefficients b_i in the flexural rigidity are

$b_0 = \{(-303{,}770{,}111{,}040a_4 + 609{,}570{,}058{,}680a_3 - 364{,}889{,}778{,}000a_2$

$\quad - 1{,}447{,}552{,}161{,}000a_1)b_8 + [(922{,}614{,}960g_5 - 1{,}709{,}255{,}196g_4$

$\quad - 238{,}128{,}000g_3 - 3{,}353{,}872{,}500g_2 - 13{,}499{,}640{,}000)a_4 - (2{,}308{,}977{,}495a_3$

$\quad - 1{,}382{,}158{,}250a_2 + 54{,}831{,}152{,}125a_1)g_5]q_0L^3\}/294{,}624{,}000{,}000a_4$

$b_1 = \{(-5{,}412{,}242{,}880a_4 + 30{,}713{,}037{,}960a_3 + 4{,}253{,}634{,}000a_2 + 78{,}084{,}567{,}000a_1)b_8$

$\quad - [(619{,}112{,}880g_5 + 3{,}420{,}595{,}332g_4 + 632{,}016{,}000g_3 + 6{,}429{,}307{,}500g_2$

$\quad + 25{,}197{,}480{,}000)a_4 + (116{,}337{,}265a_3 + 16{,}112{,}250a_2 + 295{,}774{,}875a_1)g_5]q_0L^3\}$

$\quad /103{,}118{,}400{,}000a_4$

$b_2 = \{(-5{,}412{,}242{,}880a_4 + 30{,}713{,}037{,}960a_3 + 4{,}253{,}634{,}000a_2 + 78{,}084{,}567{,}000a_1)b_8$

$\quad + [(341{,}707{,}120g_5 + 1{,}442{,}324{,}268g_4 + 338{,}184{,}000g_3 + 2{,}660{,}542{,}500g_2$

$\quad + 8{,}620{,}920{,}000)a_4 - (116{,}337{,}265a_3 + 16{,}112{,}250a_2 + 295{,}774{,}875a_1)g_5]q_0L^3\}$

$\quad /22{,}915{,}200{,}000a_4$

$b_3 = \{(30{,}938{,}400a_4 - 32{,}610{,}600a_3 + 53{,}130{,}000a_2 - 69{,}795{,}000a_1)b_8$

$\quad - [(86{,}550g_5 + 45{,}927g_4 - 52{,}500g_3 + 118{,}125g_2 - 315{,}000)a_4$

$\quad - (123{,}525a_3 - 201{,}250a_2 + 264{,}375a_1)g_5]q_0L^3\}/7{,}560{,}000a_4$

$b_4 = \{(-2{,}971{,}202{,}688a_4 + 4{,}316{,}151{,}576a_3 - 5{,}697{,}291{,}600a_2 + 999{,}880{,}200a_1)b_8$

$\quad + [(1{,}220{,}832g_5 - 13{,}189{,}176g_4 - 16{,}329{,}600g_3 - 30{,}177{,}000g_2 - 105{,}840{,}000)a_4$

$\quad - (16{,}349{,}059a_3 - 21{,}580{,}650a_2 + 3{,}787{,}425a_1)g_5]q_0L^3\}/749{,}952{,}000a_4$

$b_5 = \{(313{,}200a_4 - 402{,}600a_3 + 165{,}000a_1)b_8 + [(-900g_5 - 567g_4 + 875g_2)a_4$

$\quad + (1{,}525a_3 - 625a_1)g_5]q_0L^3\}/84{,}000a_4$

$b_6 = \{(-15{,}912a_4 + 6{,}600a_2)b_8 + [(39g_5 + 30g_3)a_4 - 25a_2g_5]q_0L^3\}/4{,}320a_4$

$b_7 = [6{,}600a_3b_8 + (27g_4a_4 - 25g_5a_3)q_0L^3]/5{,}400a_4$

$$(8.205)$$

8.2.6.6 *Quintic inertial coefficient* ($m = 5$)

The governing differential equation yields a set of 11 equations with 12 unknowns, namely the coefficients b_i in the flexural rigidity, the natural frequency squared, Ω^2, and the axial load q_0, which is treated as a parameter:

$8b_2 - 36b_1 + q_0L^3\left(4 + 2g_1 + \frac{4}{3}g_2 + g_3 + \frac{4}{5}g_4 + \frac{2}{3}g_5 + \frac{4}{7}g_6\right) = 0$

$24b_3 - 108b_2 + 120b_0 - q_0L^3\left(26 + 9g_1 + 6g_2 + \frac{9}{2}g_3 + \frac{18}{5}g_4 + 3g_5 + \frac{18}{7}g_6\right) = 0$

$48b_4 - 216b_3 + 240b_1 + q_0L^3(27 - 6g_1) - 2a_0\Omega^2 = 0$

$80b_5 - 360b_4 + 400b_2 + q_0L^3\left(20 + 28g_1 + \frac{4}{3}g_2 + 5g_3 + 4g_4 + \frac{10}{3}g_5 + \frac{20}{7}g_6\right)$
$\quad + \Omega^2(3a_0 - 2a_1) = 0$

$120b_6 - 540b_5 + 600b_3 - q_0L^3(25 - 15g_2 + 5g_3) + \Omega^2(3a_1 - 2a_2) = 0$

$168b_7 - 756b_6 + 840b_4 - q_0L^3\left(15g_1 - \frac{27}{2}g_3 + \frac{24}{5}g_4\right) - \Omega^2(a_0 - 3a_2 + 2a_3) = 0$

$224b_8 - 1008b_7 + 1120b_5 - q_0L^3\left(\frac{35}{3}g_2 - \frac{63}{5}g_4 + \frac{14}{3}g_5\right) - \Omega^2(a_1 - 3a_3 + 2a_4) = 0$

$288b_9 - 1296b_6 + 1440b_6 - q_0L^3\left(10g_3 - 12g_5 + \frac{32}{7}g_6\right) - \Omega^2(a_2 - 3a_4 + 2a_5) = 0$

$\quad - 1620b_9 + 1800b_7 - q_0L^3\left(9g_4 - \frac{81}{7}g_6\right) - \Omega^2(a_3 - 3a_5) = 0$

$2200b_8 - \frac{25}{3}q_0L^3g_5 - a_4\Omega^2 = 0$

$2640b_9 - \frac{55}{7}q_0L^3g_6 - a_5\Omega^2 = 0$

$$(8.206)$$

We have to express the unknowns in terms of the parameters b_5 and Ω^2. Therefore, we demand the rank of the set (8.132) to be less than 11, leading to a system of 12 determinantal equations with 12 unknowns, namely the parameters $a_0, a_1, a_2, a_3, a_4, a_5, g_1, g_2, g_3, g_4, g_5$ and g_6. The solution of this set reads

$a_0 = \dfrac{1{,}629}{3{,}200}a_1 - \dfrac{681}{1{,}600}a_2 + \dfrac{23{,}263}{80{,}000}a_3 - \dfrac{1{,}698}{6{,}875}a_4 + \dfrac{67{,}809}{352{,}000}a_5$

$g_1 = -\dfrac{245}{31} - \dfrac{3{,}353}{1{,}488}g_2 - \dfrac{83}{124}g_3 - \dfrac{77{,}429}{62{,}000}g_4 - \dfrac{20{,}503}{409{,}240}g_5 - \dfrac{105{,}629}{124{,}000}g_6$

$$(8.207)$$

Upon substitution of the expressions (8.133) into the set (8.132), we get the solution for the natural frequency squared:

$$\Omega^2 = \left(2680b_9 - \tfrac{55}{7}q_0L^3g_6\right)/a_5 \qquad (8.208)$$

The value of q_0 that corresponds to the vanishing frequency $\Omega^2 = 0$ is identified as the buckling load $q_{0,\mathrm{cr}}$:

$$q_{0,\mathrm{cr}} = 336 b_9/g_6 L^3 \tag{8.209}$$

We express Eq. (8.134) as follows:

$$\Omega^2 = 2680 b_9 (1 - q_0/q_{0,\mathrm{cr}})/a_5 \tag{8.210}$$

From Eq. (8.136), when $q_0 \equiv 0$, $\Omega^2 = 2680 b_9/a_5 \equiv \Omega_0^2$. Therefore, Eq. (8.134) takes the form

$$\Omega_q^2 = \Omega_0^2 (1 - q_0/q_{0,\mathrm{cr}}) \tag{8.211}$$

where Ω_q^2 is the natural frequency squared in the presence of the axial load. Note that when q_0 tends to its critical value the natural frequency vanishes. The coefficients b_i in the flexural rigidity are

$$
\begin{aligned}
b_0 = \{&(30{,}311{,}893{,}920 a_5 - 15{,}602{,}560{,}512 a_4 + 14{,}711{,}536{,}224 a_3 - 5{,}005{,}973{,}280 a_2 \\
&+ 1{,}004{,}833{,}680 a_1) b_9 - [(70{,}329{,}490 g_6 + 10{,}192{,}000 g_5 - 10{,}452{,}288 g_4 + 3{,}449{,}600 g_3 \\
&- 8{,}192{,}800 g_2 - 13{,}798{,}400) a_5 - (46{,}436{,}192 a_4 - 43{,}784{,}334 a_3 + 14{,}898{,}730 a_2 \\
&- 29{,}905{,}755 a_1) g_6] q_0 L^3\}/31{,}046{,}400 a_5
\end{aligned}
$$

$$
\begin{aligned}
b_1 = \{&(2{,}165{,}227{,}680 a_5 - 604{,}251{,}648 a_4 + 1{,}237{,}812{,}576 a_3 + 25{,}613{,}280 a_2 + 861{,}703{,}920 a_1) b_9 \\
&+ [(1{,}798{,}368 g_6 - 7{,}644{,}000 g_5 + 7{,}839{,}216 g_4 - 2{,}587{,}200 g_3 + 6{,}144{,}600 g_2 + 10{,}348{,}800) a_5 \\
&- (3{,}683{,}966 a_3 + 76{,}230 a_2 + 2{,}564{,}595 a_1) g_6] q_0 L^3\}/15{,}523{,}200 a_5
\end{aligned}
$$

$$
\begin{aligned}
b_2 = \{&(590{,}516{,}640 a_5 - 164{,}795{,}904 a_4 + 337{,}585{,}248 a_3 + 6{,}985{,}440 a_2 + 235{,}010{,}160 a_1) b_9 \\
&- [(3{,}350{,}410 g_6 - 960{,}400 g_5 + 784{,}392 g_4 - 392{,}000 g_3 + 612{,}500 g_2 + 1{,}607{,}200) a_5 \\
&+ (490{,}464 a_4 - 1{,}004{,}718 a_3 - 20{,}790 a_2 - 699{,}435 a_1) g_6] q_0 L^3\}/2{,}822{,}400 a_5
\end{aligned}
$$

$$
\begin{aligned}
b_3 = \{&(-676{,}552{,}800 a_5 + 635{,}268{,}480 a_4 - 223{,}201{,}440 a_3 + 327{,}280{,}800 a_2 - 143{,}312{,}400 a_1) b_9 \\
&+ [(1{,}376{,}760 g_6 + 524{,}300 g_5 - 261{,}954 g_4 + 269{,}500 g_3 - 202{,}125 g_2 + 161{,}700) a_5. \\
&- (1{,}890{,}680 a_4 - 664{,}290 a_3 + 974{,}050 a_2 - 426{,}525 a_1) g_6] q_0 L^3\}/3{,}880{,}800 a_5
\end{aligned}
$$

$$
\begin{aligned}
b_4 = \{&(8{,}743{,}959{,}840 a_5 - 3{,}309{,}191{,}424 a_4 + 4{,}060{,}639{,}968 a_3 - 1{,}886{,}844{,}960 a_2 + 151{,}850{,}160 a_1) b_9 \\
&- [(11{,}088{,}770 g_6 + 8{,}212{,}400 g_5 - 7{,}075{,}992 g_4 + 4{,}132{,}000 g_3 - 2{,}856{,}700 g_2 - 4{,}743{,}200) a_5 \\
&- (9{,}848{,}784 a_4 - 12{,}085{,}238 a_3 + 5{,}615{,}610 a_2 - 451{,}935 a_1) g_6] q_0 L^3\}/62{,}092{,}800 a_5
\end{aligned}
$$

$$
\begin{aligned}
b_5 = \{&(-45{,}276{,}000 a_5 + 57{,}879{,}360 a_4 - 24{,}800{,}160 a_3 + 3{,}049{,}200 a_1) b_9 \\
&+ [(86{,}240 g_6 + 44{,}100 g_5 - 29{,}106 g_4 + 13{,}475 g_2) a_5 \\
&- (172{,}260 a_4 - 73{,}810 a_3 + 9{,}075 a_1) g_6] q_0 L^3\}/1{,}293{,}600 a_5
\end{aligned}
$$

$$b_6 = \{(3{,}843{,}840a_5 - 1{,}633{,}632a_4 + 203{,}280a_2)b_9$$

$$- [(8{,}580g_6 + 1{,}820g_5 - 770g_3)a_5 - (4{,}862a_4 - 605a_2)g_6]q_0L^3\}/110{,}880a_5$$

$$b_7 = \{(18{,}480a_3 - 147{,}000a_5)b_9 + [(280g_6 + 63g_4)a_5 - 55g_6a_3]q_0L^3\}/12{,}600a_5$$

$$b_8 = [11{,}088a_4b_9 + (35g_5a_5 - 33g_6a_4)q_0L^3]/9{,}240a_5$$

$$(8.212)$$

8.2.7 Closed-Form Solution for the Beam that is Pinned at One End and Clamped at the Other

8.2.7.1 Uniform inertial coefficient (m = 0)

We get a homogeneous set of six linear algebraic equations with seven unknowns, namely the coefficients b_i in the flexural rigidity, the natural frequency squared, Ω^2, and the axial load q_0, which is treated as a parameter:

$$-48b_1 - 2q_0L^3 = 0$$

$$-144b_2 + 240b_0 - q_0L^3(24 + 14g_1) - 2a_0\Omega^2 = 0$$

$$-288b_3 + 480b_1 + 36q_0L^3 = 0$$

$$-480b_4 + 800b_2 + q_0L^3(40 + 44g_1) + 4a_0\Omega^2 = 0 \qquad (8.213)$$

$$1{,}200b_3 - 50q_0L^3 = 0$$

$$1{,}680b_4 - 30g_1q_0L^3 - 2a_0\Omega^2 = 0$$

We have to write all the unknowns in terms of a single free parameter, say b_4, and the axial load parameter q_0. In other words, the rank of the set (8.139) or its matrix must be less than 6. Thus, all minors of order 6 must vanish. Yet the minor obtained by extraction of the sixth column of the matrix of the set (8.139) equals $-74{,}317{,}824{,}000{,}000L^3$, which is non-zero. We conclude that the problem has no non-trivial solution.

8.2.7.2 Linearly varying inertial coefficient (m = 1)

We have a homogeneous set of seven linear algebraic equations with eight unknowns, namely the coefficients b_i in the flexural rigidity, the natural frequency squared, Ω^2, and the axial load q_0, which is treated as a parameter:

$$-48b_1 - 2q_0L^3 = 0$$

$$-144b_2 + 240b_0 - q_0L^3(24 + 14g_1 + 8g_2) - 2a_0\Omega^2 = 0$$

$$-288b_3 + 480b_1 + q_0L^3(36 - 2g_2) - 2a_1\Omega^2 = 0$$

$$-480b_4 + 800b_2 + q_0L^3(40 + 44g_1 + \tfrac{40}{3}g_2) + 4a_0\Omega^2 = 0 \qquad (8.214)$$

$$-720b_5 + 1200b_3 - q_0L^3(50 - 20g_2) + 4a_1\Omega^2 = 0$$

$$1680b_4 - 30g_1q_0L^3 - 2a_0\Omega^2 = 0$$

$$2240b_5 - \tfrac{70}{3}q_0L^3g_2 - 2a_1\Omega^2 = 0$$

We demand that all unknowns be expressed in terms of a single free parameter, say b_5, and the axial load parameter q_0, leading to the condition that the rank of the set (8.140) or its matrix must be less than 7 or all minors of order 7 must vanish. We produce the minors by extracting one column of the matrix among eight; therefore, we obtain a set of eight determinantal equations for four unknowns, namely the parameters a_0, a_1, g_1 and g_2:

$$2{,}397{,}988{,}454{,}400a_0g_2 + 9{,}591{,}953{,}817{,}600a_0 + 15{,}786{,}861{,}133{,}824a_1g_1$$
$$+ 34{,}792{,}632{,}483{,}840a_1 + 11{,}597{,}544{,}161{,}280a_1g_2 = 0$$

$$a_1 = 0$$

$$178{,}362{,}777{,}600a_0g_2 + 713{,}451{,}110{,}400a_0 + 2{,}201{,}166{,}544{,}896a_1g_1$$
$$+ 828{,}396{,}011{,}520a_1 + 2{,}485{,}188{,}034{,}560a_1g_2 = 0$$

$$6{,}341{,}787{,}648a_1g_2 - 16{,}052{,}649{,}984a_1 = 0$$

$$495{,}452{,}160a_0g_2 + 1{,}981{,}808{,}640a_0 + 8{,}875{,}671{,}552a_1g_1 = 0$$

$$5{,}549{,}064{,}192a_1g_2 + 1{,}486{,}356{,}480a_1 = 0$$

$$4{,}161{,}798{,}144g_2 + 16{,}647{,}192{,}576 = 0$$

$$a_1 = 0$$

$$(8.215)$$

The solution of the set (8.141) reads

$$a_1 = 0 \quad g_2 = -4 \tag{8.216}$$

Yet, $a_1 = 0$ implies that the mass density is constant. Thus, for $a_1 \neq 0$ no solution exists.

8.2.7.3 *Parabolically varying inertial coefficient (m = 2)*

The governing differential equation yields a set of eight equations with nine unknowns, namely the coefficients b_i in the flexural rigidity, the natural frequency squared, Ω^2, and the axial load q_0:

$$- 48b_1 - 2q_0L^3 = 0$$

$$- 144b_2 + 240b_0 - q_0L^3(24 + 14g_1 + 8g_2 + 6g_3) - 2a_0\Omega^2 = 0$$

$$- 288b_3 + 480b_1 + q_0L^3(36 - 2g_2) - 2a_1\Omega^2 = 0$$

$$- 480b_4 + 800b_2 + q_0L^3(40 + 44g_1 + \tfrac{40}{3}g_2 + 8g_3) + (4a_0 - 2a_2)\Omega^2 = 0$$

$$- 720b_5 + 1200b_3 - q_0L^3(50 - 20g_2) + 4a_1\Omega^2 = 0$$

$$- 1008b_6 + 1680b_4 - q_0L^3(30g_1 - 18g_3) - (2a_0 - 4a_2)\Omega^2 = 0$$

$$2240b_5 - \tfrac{70}{3}q_0 L^3 g_2 - 2a_1 \Omega^2 = 0$$
$$2880b_6 - 20q_0 L^3 g_3 - 2a_2 \Omega^2 = 0$$

$$(8.217)$$

We have to express the unknowns in terms of the parameters b_6 and Ω^2. Therefore, we demand the rank of the set (8.143) to be less than 8, leading to a system of nine determinantal equations with six unknowns, namely the parameters a_0, a_1, a_2, g_1, g_2 and g_3. The solution of this set eads

$$a_1 = 0 \quad g_2 = -4 \tag{8.218}$$

Upon substitution of the expressions (8.144) into the set (8.143), we get the solution for the natural frequency squared:

$$\Omega^2 = (1440b_6 - 10q_0 L^3 g_3)/a_2 \tag{8.219}$$

The value of q_0 that corresponds to the vanishing frequency $\Omega^2 = 0$ is identified as the buckling load $q_{0,\mathrm{cr}}$:

$$q_{0,\mathrm{cr}} = 144b_6/g_3 L^3 \tag{8.220}$$

We express Eq. (8.145) as follows:

$$\Omega^2 = 1440b_6(1 - q_0/q_{0,\mathrm{cr}})/a_2 \tag{8.221}$$

From Eq. (8.147), when $q_0 \equiv 0$, $\Omega^2 = 1440b_6/a_2 \equiv \Omega_0^2$. Therefore, Eq. (8.145) takes the form

$$\Omega_q^2 = \Omega_0^2(1 - q_0/q_{0,\mathrm{cr}}) \tag{8.222}$$

where Ω_q^2 is the natural frequency squared in the presence of the axial load. Note that when q_0 tends to its critical value the natural frequency vanishes. The coefficients b_i in the flexural rigidity are

$$b_0 = \frac{1}{21{,}000} a_2 \{(23{,}976a_2 + 174{,}240a_0)b_6 + [(183g_3 + 667g_1 - 490)a_2$$

$$- 1{,}210g_3 a_0]q_0 L^3\}$$

$$b_1 = -q_0 L^3/24$$

$$b_2 = \frac{1}{2{,}100a_2}\{(3{,}996a_2 - 12{,}960a_0)b_6 - [(57g_3 + 93g_1 - 32)a_2 - 90g_3 a_0]q_0 L^3\}$$

$$b_3 = q_0 L^3/12$$

$$b_4 = \frac{1}{840a_2}\{-(2{,}376a_2 + 1{,}440a_0)b_6 - [(11g_3 + 15g_1)a_2 - 10g_3a_0]q_0L^3\}$$

$$b_5 = -q_0L^3/24$$

$$(8.223)$$

8.2.7.4 Cubic inertial coefficient (m = 3)

The governing differential equation yields a set of nine equations with ten unknowns, namely the coefficients b_i in the flexural rigidity, the natural frequency squared, Ω^2, and the axial load q_0, which is treated as a parameter:

$$-48b_1 - 2q_0L^3 = 0$$

$$-144b_2 + 240b_0 - q_0L^3(24 + 14g_1 + 8g_2 + 6g_3 + \tfrac{24}{5}g_4) - 2a_0\Omega^2 = 0$$

$$-288b_3 + 480b_1 + q_0L^3(36 - 2g_2) - 2a_1\Omega^2 = 0$$

$$-480b_4 + 800b_2 + q_0L^3(40 + 44g_1 + \tfrac{40}{3}g_2 + 8g_3 + 8g_4) + (4a_0 - 2a_2)\Omega^2 = 0$$

$$-720b_5 + 1{,}200b_3 - q_0L^3(50 - 20g_2 + 2g_4) + \Omega^2(4a_1 - 2a_3) = 0$$

$$-1{,}008b_6 + 1{,}680b_4 - q_0L^3(30g_1 - 18g_3) - (2a_0 - 4a_2)\Omega^2 = 0$$

$$-1{,}344b_7 + 2{,}240b_5 - q_0L^3(\tfrac{70}{3}g_2 + \tfrac{84}{5}g_4) - (2a_1 - 4a_3)\Omega^2 = 0$$

$$2{,}880b_6 - 20q_0L^3g_3 - 2a_2\Omega^2 = 0$$

$$3{,}600b_7 - 18q_0L^3g_4 - 2a_3\Omega^2 = 0$$

$$(8.224)$$

We have to express the unknowns in terms of the parameters b_7 and Ω^2. Therefore, we demand the rank of the set (8.150) to be less than 9, leading to a system of ten determinantal equations with eight unknowns, namely the parameters $a_0, a_1, a_2, a_3, g_1, g_2, g_3$ and g_4. The solution of this set reads

$$r_1 = -\frac{1002}{5225}r_3 \quad g_2 = -4 - \frac{186}{625}g_4 \tag{8.225}$$

Upon substitution of the expressions (8.151) into the set (8.150), we get the solution for the natural frequency squared:

$$\Omega^2 = (1800b_7 - 9q_0L^3g_4)/a_3 \tag{8.226}$$

The value of q_0 that corresponds to the vanishing frequency $\Omega^2 = 0$ is identified as the buckling load $q_{0,cr}$:

$$q_{0,cr} = 200b_7/g_4L^3 \tag{8.227}$$

We write Eq. (8.154) as follows:

$$\Omega^2 = 1800b_7(1 - q_0/q_{0,cr})/a_3 \tag{8.228}$$

From Eq. (8.155), when $q_0 \equiv 0$, $\Omega^2 = 1800b_7/a_3 \equiv \Omega_0^2$. Therefore, Eq. (8.152) takes the form

$$\Omega_q^2 = \Omega_0^2(1 - q_0/q_{0,cr}) \tag{8.229}$$

where Ω_q^2 is the natural frequency squared in the presence of the axial load. Note that when q_0 tends to its critical value the natural frequency vanishes. The coefficients b_i in the flexural rigidity are deduced from the remaining equations of the set (8.150):

$$b_0 = \{(108{,}900{,}000a_0 + 14{,}985{,}000a_2)b_7 + [(74{,}088g_4 + 174{,}750g_3 + 333{,}500g_1$$
$$- 245{,}000)a_3 - (544{,}500a_0 + 74{,}925a_2)g_4]q_0L^3\}/10{,}500{,}000a_3$$

$$b_1 = -q_0L^3/24$$

$$b_2 = \{(4{,}995{,}000a_2 - 16{,}200{,}000a_0)b_7 - [(10{,}584g_4 + 29{,}250g_3 + 93{,}000g_1$$
$$- 35{,}000)a_3 + (24{,}975a_2 - 81{,}000a_0)g_4]q_0L^3\}/2{,}100{,}000a_3$$

$$b_3 = \frac{501}{209}b_7 + \left(\frac{1}{12} - \frac{3{,}887}{391{,}875}g_4\right)q_0L^3$$

$$b_4 = \frac{(36{,}000a_0 - 59{,}400a_2)b_7 + [(300g_1 - 110g_3)a_3 + (297a_2 - 108a_0)g_4]q_0L^3}{16{,}800a_3}$$

$$b_5 = -\frac{3{,}054}{1{,}045}b_7 + \left(\frac{916}{130{,}625}g_4 - \frac{1}{24}\right)q_0L^3,$$

$$b_6 = [1{,}800a_2b_7 + (10g_3a_3 - 9g_4a_2)q_0L^3]/1{,}440a_3 \tag{8.230}$$

8.2.7.5   Quartic inertial coefficient ($m = 4$)

The governing differential equation yields a set of 10 equations with 11 unknowns, namely the coefficients b_i in the flexural rigidity, the natural frequency squared, Ω^2, and the axial load q_0, which is treated as a parameter:

$$-48b_1 - 2q_0L^3 = 0$$

$$-144b_2 + 240b_0 - q_0L^3(24 + 14g_1 + 8g_2 + 6g_3 + \tfrac{24}{5}g_4 + 4g_5) - 2a_0\Omega^2 = 0$$

$$-288b_3 + 480b_1 + q_0L^3(36 - 2g_2) - 2a_1\Omega^2 = 0$$

$$-480b_4 + 800b_2 + q_0L^3(40 + 44g_1 + \tfrac{40}{3}g_2 + 8g_3 + 8g_4 + \tfrac{20}{3}g_5) + (4a_0 - 2a_2)\Omega^2 = 0$$

$$-720b_5 + 1200b_3 - q_0L^3(50 - 20g_2 + 2g_4) + \Omega^2(4a_1 - 2a_2) = 0$$

$$-1008b_6 + 1680b_4 - q_0 L^3 (30g_1 - 18g_3 + 2g_5) - (2a_0 - 4a_2 + 2a_4)\Omega^2 = 0$$
$$-1344b_7 + 2240b_5 - q_0 L^3 (\tfrac{70}{3} g_2 - \tfrac{84}{5} g_4) - (2a_1 - 4a_3)\Omega^2 = 0$$
$$-1728b_8 + 2880b_5 - q_0 L^3 (20g_3 - 16g_5) - (2a_2 - 4a_4)\Omega^2 = 0$$
$$3600b_7 - 18q_0 L^3 g_4 - 2a_3\Omega^2 = 0$$
$$4400b_8 - \tfrac{50}{3} q_0 L^3 g_5 - 2a_4\Omega^2 = 0$$

$$(8.231)$$

We have to express the unknowns in terms of the parameters b_5 and Ω^2. Therefore, we demand the rank of the set (8.157) to be less than 10, leading to a system of 11 determinantal equations with 10 unknowns, namely the parameters $a_0, a_1, a_2, a_3, a_4, g_1, g_2, g_3, g_4$ and g_5. The solution for this set reads

$$a_1 = -\frac{1002}{5225} a_3 \quad g_2 = -4 - \frac{186}{625} g_4 \qquad (8.232)$$

Upon substitution of the expressions (8.158) into the set (8.157), we get the solution for the natural frequency squared:

$$\Omega^2 = (2200b_8 - \tfrac{25}{3} q_0 L^3 g_5)/a_4 \qquad (8.233)$$

The value of q_0 that corresponds to the vanishing frequency $\Omega^2 = 0$ is identified as the buckling load $q_{0,cr}$:

$$q_{0,cr} = 264b_8/g_5 L^3 \qquad (8.234)$$

We rewrite Eq. (8.159) as follows:

$$\Omega^2 = 2200b_8(1 - q_0/q_{0,cr})/a_4 \qquad (8.235)$$

From Eq. (8.161), when $q_0 \equiv 0$, $\Omega^2 = 2200b_8/a_4 \equiv \Omega_0^2$. Therefore, Eq. (8.159) takes the form

$$\Omega_q^2 = \Omega_0^2(1 - q_0/q_{0,cr}) \qquad (8.236)$$

where Ω_q^2 is the natural frequency squared in the presence of the axial load. Note that when q_0 tends to its critical value the natural frequency vanishes. The coefficients b_i in the flexural rigidity are deduced from the remaining equations of the set (8.157):

$$b_0 = \{(12{,}992{,}400a_4 + 54{,}945{,}000a_2 + 399{,}300{,}000a_0)b_8$$
$$+ [(309{,}450g_5 + 222{,}264g_4 + 524{,}250g_3 + 1{,}000{,}500g_1 - 735{,}000a_4)a_4$$
$$- (1{,}512{,}500a_0 + 208{,}125a_2)g_5]q_0L^3\}/31{,}500{,}000a_4$$

$$b_1 = -q_0L^3/24$$

$$b_2 = \{(1{,}443{,}600a_4 + 6{,}105{,}000a_2 - 19{,}800{,}000a_0)b_8$$
$$- [(23{,}950g_5 + 10{,}584g_4 + 29{,}250g_3 + 93{,}000g_1 - 35{,}000)a_4$$
$$- (75{,}000a_0 - 23{,}125a_2)g_5]q_0L^3\}/2{,}100{,}000a_4$$

$$b_3 = \{27{,}555{,}000a_3b_8 + [(19{,}437g_4 + 783{,}750)a_4 - 104{,}375g_5a_3]q_0L^3\}/9{,}405{,}000a_4$$

$$b_4 = \{(57{,}744a_4 - 217{,}800a_2 + 132{,}000a_0)b_8$$
$$- [(258g_5 + 330g_3 - 900g_1)a_4 - (825a_2 - 500a_0)g_5]q_0L^3\}/50{,}400a_4$$

$$b_5 = \{-33{,}594{,}000a_3b_8 - [(71{,}478g_4 + 391{,}875)a_4 - 127{,}250a_3g_5]q_0L^3\}/9{,}405{,}000a_4$$

$$b_6 = \{(-10{,}608a_4 + 6{,}600r_2)b_8 + [(30g_3 + 26g_5)a_4 - 25a_2g_5]q_0L^3\}/4{,}320a_4$$

$$b_7 = [6{,}600a_3b_8 + (27g_4a_4 - 25a_3g_5)q_0L^3]/5{,}400a_4$$

$$(8.237)$$

8.2.7.6　Quintic inertial coefficient (m = 5)

The governing differential equation yields a set of 11 equations with 12 unknowns, namely the coefficients b_i in the flexural rigidity, the natural frequency squared, Ω^2, and the axial load q_0, which is treated as a parameter:

$$-48b_1 - 2q_0L^3 = 0$$

$$-144b_2 + 240b_0 - q_0L^3(24 + 14g_1 + 8g_2 + 6g_3 + \tfrac{24}{5}g_4 + 4g_5 + \tfrac{24}{7}g_6)$$
$$-2a_0\Omega^2 = 0$$

$$-288b_3 + 480b_1 + q_0L^3(36 - 2g_2) - 2a_1\Omega^2 = 0$$

$$-480b_4 + 800b_2 + q_0L^3(40 + 44g_1 + \tfrac{40}{3}g_2 + 8g_3 + 8g_4 + \tfrac{20}{3}g_5 + \tfrac{40}{7}g_6)$$
$$+ (4a_0 - 2a_2)\Omega^2 = 0$$

$$-720b_5 + 1200b_3 - q_0L^3(50 - 20g_2 + 2g_4) + \Omega^2(4a_1 - 2a_2) = 0$$

$$-1008b_6 + 1680b_4 - q_0L^3(30g_1 - 18g_3 + 2g_5) - (2r_0 - 4a_2 + 2a_4)\Omega^2 = 0$$

$$-1344b_7 + 2240b_5 - q_0L^3(\tfrac{70}{3}g_2 + \tfrac{84}{5}g_4 + 2g_6) - (2a_1 - 4a_3 + 2a_5)\Omega^2 = 0$$

$$-1728b_8 + 2880b_5 - q_0L^3(20g_3 - 16g_5) - (2a_2 - 4a_4)\Omega^2 = 0$$

$$-2160b_9 + 3600b_7 - q_0L^3(18g_4 - \tfrac{108}{7}g_6) - (2a_3 - 4a_5)\Omega^2 = 0$$

$$4400b_8 - \tfrac{50}{3}q_0L^3g_5 - 2a_4\Omega^2 = 0$$

$$5280b_9 - \tfrac{110}{7}q_0L^3g_6 - 2a_5\Omega^2 = 0 \qquad (8.238)$$

We have to express the unknowns in terms of the parameters b_5 and Ω^2. Therefore, we demand the rank of the set (8.164) to be less than 11, leading to a system of 12 determinantal equations with 12 unknowns, namely the parameters $a_0, a_1, a_2, a_3, a_4, a_5, g_1, g_2, g_3, g_4, g_5$ and g_6. The solution of this set reads

$$a_1 = -\frac{1{,}002}{5{,}225}a_3 - \frac{603}{11{,}495}a_5 \quad g_2 = -4 - \frac{186}{625}g_4 - \frac{459}{4{,}375}g_6 \qquad (8.239)$$

Upon substitution of the expressions (8.165) into the set (8.164), we get the solution for the natural frequency squared:

$$\Omega^2 = 2680b_9(1 - q_0/q_{0,cr})/a_5 \qquad (8.240)$$

The value of q_0 that corresponds to the vanishing frequency $\Omega^2 = 0$ is identified as the buckling load $q_{0,cr}$:

$$q_{0,cr} = 336b_9/g_6L^3 \qquad (8.241)$$

Equation (8.166) gets the following form:

$$\Omega^2 = 2680b_9(1 - q_0/q_{0,cr})/a_5 \qquad (8.242)$$

From Eq. (8.168), when $q_0 \equiv 0$, $\Omega^2 = 2680b_9/a_5 \equiv \Omega_0^2$. Therefore, Eq. (8.166) takes the form

$$\Omega_q^2 = \Omega_0^2(1 - q_0/q_{0,cr}) \qquad (8.243)$$

where Ω_q^2 is the natural frequency squared in the presence of the axial load. Note that when q_0 tends to its critical value the natural frequency vanishes. The coefficients b_i in the flexural rigidity are deduced from the remaining equations of the set (8.164):

$$b_0 = \{(28{,}200{,}480a_4 + 78{,}669{,}360a_2 + 378{,}100{,}800a_0)b_9$$

$$+ [(70{,}560g_6 + 195{,}020g_5 + 57{,}624g_4 + 280{,}434g_3 + 531{,}300g_1 - 123{,}480)a_5$$

$$- (83{,}930a_4 + 234{,}135a_2 + 1{,}125{,}300a_0)g_6]q_0L^3\}/11{,}642{,}400a_5$$

$$b_1 = -7q_0L^3/120b_1 = -7q_0L^3/120$$

$$b_2 = \{(28{,}200{,}480a_4 + 78{,}669{,}360a_2 + 219{,}542{,}400a_0)b_9$$

$$- [(30{,}240g_6 - 128{,}380g_5 - 24{,}696g_4 - 204{,}666g_3 - 665{,}280g_1 + 52{,}920)a_5$$

$$+ (83{,}930a_4 + 234{,}135a_2 + 653{,}400a_0)g_6]q_0L^3\}/11{,}642{,}400a_5$$

$$b_3 = \frac{1}{3{,}029{,}400a_5}\{(9{,}378{,}600a_5 + 22{,}564{,}080a_3)b_9 - [(19{,}880g_6 + 13{,}923g_4$$

$$+ 298{,}350)a_5 + 67{,}155a_3g_6]q_0L^3\}$$

$$b_4 = \{(5{,}640{,}096a_4 - 20{,}124{,}720a_2 + 7{,}318{,}080a_0)b_9 + [(-6{,}272g_5 - 25{,}410g_3$$

$$+ 41{,}580g_1)a_5 - (16{,}786a_4 + 59{,}895a_2 - 21{,}780a_0)g_6]q_0L^3\}/2{,}328{,}480a_5$$

$$b_5 = \frac{1}{7{,}068{,}600a_5}\{(-9{,}378{,}600a_5 + 50{,}006{,}880a_3)b_9 + [(62{,}720g_6 + 87{,}822g_4$$

$$+ 240{,}975)a_5 - 148{,}830a_3g_6]q_0L^3\}$$

$$b_6 = \frac{1}{332{,}640a_5}\{(609{,}840a_2 - 1{,}633{,}632a_4)b_9 + [(2{,}310g_3 - 1{,}820g_5)a_5$$

$$+ (4{,}862a_4 - 1{,}815a_2)g_6]q_0L^3\}$$

$$b_7 = \{(55{,}440a_3 - 147{,}000a_5)b_9 + [(280g_6 + 189g_4)a_5 - 165a_3g_6]q_0L^3\}/37{,}800a_5$$

$$b_8 = [11{,}088a_4b_9 + (35g_5a_5 - 33a_4g_6)q_0L^3]/9{,}240a_5$$

$$(8.244)$$

8.2.8   Concluding Remarks

We considered a postulate which required the mode shape both in vibration and buckling to coincide and be represented by a fifth-order polynomial. For each of the four sets of boundary conditions considered we studied the cases of uniform and non-uniform material densities. For the non-uniform density case the variation of the inertial coefficient was taken to be linear, parabolic, cubic, quartic or quintic. In an analogous manner one can obtain, in principle, solutions for any value of m, where m is the degree of variation in the mass density.

In perfect analogy, one can postulate the sixth-, seventh-, etc. order polynomials for the mode shapes. Thus, an infinite number of closed-form solutions can be constructed. Throughout this chapter we established the validity of the relationship developed by Massonet (1940, 1972) and Galef (1968) for vibrating beams in the presence of axial loads. What was known, however, as an approximate solution for homogeneous or inhomogeneous beams turns out to be an exact relationship for inhomogeneous beams.

9

Unexpected Results for a Beam on an Elastic Foundation or with Elastic Support

In this chapter, we concentrate on the vibrations of beams on an elastic foundation, with attendant results that are surprising, at first glance. We also discuss closed-form solutions for vibrations of inhomogeneous beams with elastic springs at the ends, namely a rotational spring or a translational one.

9.1 Some Unexpected Results in the Vibration of Inhomogeneous Beams on an Elastic Foundation

9.1.1 Introductory Remarks

Free vibrations of uniform and non-uniform beams on elastic foundations was studied by several authors. Perhaps the most general formulation for variable cross-section beams on an inhomogeneous elastic foundation, containing the characteristics of both the Winkler foundation and the Pasternak foundation, is due to Eisenberger and Clastornik (1987) and Eisenberger (1994). They represented the variations of the cross-sectional area, elastic moduli, and the displacements in terms of infinite power series for the non-dimensional axial coordinate. By equating the coefficients of identical powers, they derived an infinite set of algebraic equations. Once this series was truncated at some specific power, the eigenvalues can be determined with the associated accuracy. This procedure can be performed with arbitrarily large powers of truncation. The method also allowed the formulation of the flexural rigidity and mass matrices with arbitrary precision. Such an approach was dubbed by Eisenberger (1994) as *the exact finite element method*. Indeed, once the regular or exact finite element codes are developed, the vibration analysis of beams on an elastic foundation can be treated as quite a straightforward problem. The immediate question that arises is: is there anything that is fundamentally new that can be contributed to this subject? It appears that the new contribution could lie in obtaining *closed-form* solutions, which have not been reported up to date, to the best of the author's knowledge.

Here, the beams with variable flexural rigidity and inertial coefficient are treated, but with constant cross section. Naturally, it is quite straightforward to apply the Eisenberger–Clastornik approach to beams with variable modulus of elasticity and variable mass density. This is not our task, however. In this section, the objective is defined as obtaining those *closed-form* solutions, which may serve as benchmark problems for other, exact or approximate methods. Also, some seemingly unexpected conclusions will follow.

As far as the variation of the inertial coefficient is concerned, Dinnik (1914) apparently was the first investigator who studied the vibration of such elastic bodies. He considered (Dinnik, 1914) strings with the density $\rho(x) = \rho_1(x/L)^m$, where L is the length and ρ_1 is the density at $x = L$. For the cases of linear ($m = 1$) and conical ($m = 2$) variations of the string he gave an exact solution in terms of Bessel functions. Mikeladze (1951) utilized the method of integral equations for vibrating strings with variable inertial coefficient, namely with binomial variation $\rho(x) = \rho(0)(1 - x/L)^m$ or with exponential variation $\rho(x) = \rho(0)\exp(ax/L)$, where $\rho(0)$ is the density at $x = 0$. Recently, inhomogeneity in the inertial coefficient was treated, for example, by Masad (1996) and Pronsato et al. (1999). They considered the piecewise constant density. Apparently, the only recent work dealing with continuous variation of the inertial coefficient, but with constant modulus of elasticity is that due to Gutierez et al. (1999). In the present section, we study the continuously varying inertial coefficient in conjunction with continuously varying elastic modulus.

9.1.2   Formulation of the Problem

The differential equation that governs the free vibrations of beams on an elastic foundation reads

$$\frac{d^2}{dx^2}\left[D(x)\frac{d^2W}{dx^2}\right] + k_1(x)W - \frac{d}{dx}\left[k_2(x)\frac{dW}{dx}\right] - \rho(x)A(x)\omega^2W = 0 \qquad (9.1)$$

where $D(x)$ is the flexural rigidity, $A(x)$ the area and $\rho(x)$ the material density, which are functions of the axial coordinate x; A is the cross-sectional area, $k_1(x)$ the variable coefficient of the Winkler foundation, $k_2(x)$ the variable coefficient of the Pasternak foundation and $W(x)$ the mode shape associated with the natural frequency ω. The coefficient k_2 was introduced by Pasternak (1954) and, later on, by Vlasov and Leontiev (1966).

We introduce the dimensionless axial coordinate $\xi = x/L$, $0 \leq \xi \leq 1$, where L is the length of the beam. Equation (9.1) is rewritten as

$$\frac{d^2}{d\xi^2}\left[D(\xi)\frac{d^2W}{d\xi^2}\right] + K_1(\xi)W - \frac{d}{d\xi}\left[K_2(\xi)\frac{dW}{d\xi}\right] - R(\xi)\Omega^2W = 0 \qquad (9.2)$$

where

$$K_1(\xi) = k_1(\xi)L^4 \quad K_2(\xi) = k_2(\xi)L^4 \quad \Omega^2 = \omega^2L^4 \quad R(\xi) = \rho(\xi)A(\xi) \qquad (9.3)$$

We are interested in the simplest closed-form solution for the natural frequency coefficient Ω^2. Namely, we ask the following question: What are the conditions, upon whose fulfillment the static deflection of the associated homogeneous beam, denoted here as W_s, will serve as the exact mode shape of the inhomogeneous beam under consideration?

The static deflection W_s of the homogeneous and uniform beam under uniform loading depends upon boundary conditions; it reads as follows:

$$W_s = \xi - 2\xi^3 + \xi^4 \qquad \text{for the } P\text{–}P \text{ beam} \qquad (9.4)$$

$$W_s = 6\xi^2 - 4\xi^3 + \xi^4 \qquad \text{for the } C\text{–}F \text{ beam} \qquad (9.5)$$

$$W_s = 3\xi^2 - 5\xi^3 + 2\xi^4 \qquad \text{for the } C\text{–}P \text{ beam} \qquad (9.6)$$

$$W_s = \xi^2 - 2\xi^3 + \xi^4 \qquad \text{for the } C\text{–}C \text{ beam} \qquad (9.7)$$

$$W_s = 5 - 6\xi^2 + \xi^4 \qquad \text{for the } G\text{–}P \text{ beam} \qquad (9.8)$$

$$W_s = 4\xi^2 - 4\xi^3 + \xi^4 \qquad \text{for the } G\text{–}C \text{ beam} \qquad (9.9)$$

where P stands for the pinned end, C, for the clamped end, F, for the free end and G for the guided end. The first entry is associated with the left end, whereas the second entry indicates the right end.

Our objective is to find closed-form solutions, rather than treat the general problem of determining the natural frequencies of beams with arbitrary variations of $E(x)$ and $\rho(x)$. Yet, the very idea of obtaining closed-form solutions in the polynomial functions, rather than in terms of transcendental functions, appears to be extremely attractive, due to the inherent simplicity of polynomial solutions.

First, we present some general solutions that are valid for any boundary conditions. To do this the static displacement is represented as a fourth-order polynomial whose coefficients are not specified in advance. Once the coefficients are specified, particular boundary conditions, and associated results for natural frequencies are obtained. Such a consideration is most appropriate when the inertial coefficient is represented as a lower order polynomial. The general polynomial variation in the inertial coefficient is studied in detail in subsequent sections.

9.1.3 Beam with Uniform Inertial Coefficient, Inhomogeneous Elastic Modulus and Elastic Foundation

We first consider a particular case $R(\xi) = \text{const} = a_0$. The flexural rigidity varies as follows:

$$D(\xi) = b_0 + b_1\xi + b_2\xi^2 + b_3\xi^3 + b_4\xi^4 \qquad (9.10)$$

The flexural rigidities of the elastic foundation have the following variations:

$$K_1(\xi) = \text{const} = c_0 \tag{9.11}$$

$$K_2(\xi) = d_0 + d_1\xi + d_2\xi^2 \tag{9.12}$$

It can be easily seen that the chosen orders of polynomials in Eqs. (9.10)–(9.12) in conjuction with the constant inertial coefficient are compatible with Eq. (9.2). We consider the coefficients a_0, c_0, d_0, d_1 and d_2 as specified ones and pose the following problem: Find the coefficients b_0, b_1, b_2, b_3 and b_4 so that the governing equation is identically satisfied.

Our aim is to deal with all boundary conditions in a unified manner. Therefore, the postulated mode shape in Eqs. (9.4)–(9.9) is represented as follows:

$$W_s = \alpha_0 + \alpha_1\xi + \alpha_2\xi^2 + \alpha_3\xi^3 + \alpha_4\xi^4 \tag{9.13}$$

where the coefficients α_0, α_1, α_2, α_3 and α_4 are adjusted so as to describe various boundary conditions.

Namely, for the beam that is pinned at both ends (*P–P*):

$$\alpha_0 = 0 \quad \alpha_1 = 1 \quad \alpha_2 = 0 \quad \alpha_3 = -2 \quad \alpha_4 = 1 \tag{9.14}$$

For the *C–F* beam, we have

$$\alpha_0 = 0 \quad \alpha_1 = 0 \quad \alpha_2 = 6 \quad \alpha_3 = -4 \quad \alpha_4 = 1 \tag{9.15}$$

For the *C–P* beam the coefficients read

$$\alpha_0 = 0 \quad \alpha_1 = 0 \quad \alpha_2 = 3 \quad \alpha_3 = -5 \quad \alpha_4 = 2 \tag{9.16}$$

For the *C–C* beam the coefficients read

$$\alpha_0 = 0 \quad \alpha_1 = 0 \quad \alpha_2 = 1 \quad \alpha_3 = -2 \quad \alpha_4 = 1 \tag{9.17}$$

For the guided–pinned (*G–P*) beam the coefficients are

$$\alpha_0 = 5 \quad \alpha_1 = 0 \quad \alpha_2 = -6 \quad \alpha_3 = 0 \quad \alpha_4 = 1 \tag{9.18}$$

Finally, for the guided–clamped (*G–C*) beam,

$$\alpha_0 = 0 \quad \alpha_1 = 0 \quad \alpha_2 = 4 \quad \alpha_3 = -4 \quad \alpha_4 = 1 \tag{9.19}$$

We substitute Eq. (9.13) into the governing differential equation. The result reads

$$[2(2\alpha_2 b_2 + 12\alpha_4 b_0 + 6\alpha_3 b_1) + 6(12\alpha_4 b_1 + 6\alpha_3 b_2 + 2\alpha_2 b_3)\xi$$

$$+ 12(12\alpha_4 b_2 + 6\alpha_3 b_3 + 2\alpha_2 b_4)\xi^2 + 20(12\alpha_4 b_3 + 6\alpha_3 b_4)\xi^3 + 360\alpha_4 b_4 \xi^4]$$

$$+ c_0(\alpha_0 + \alpha_1\xi + \alpha_2\xi^2 + \alpha_3\xi^3 + \alpha_4\xi^4) + [-(2\alpha_2 d_0 + \alpha_1 d_1) - 2(\alpha_1 d_2 + 2\alpha_2 d_1$$

$$+ 3\alpha_3 d_0)\xi - 3(3\alpha_3 d_1 + 2\alpha_2 d_2 + 4\alpha_4 d_0)\xi^2 - 4(3\alpha_3 d_2 + 4\alpha_4 d_1)\xi^3 - 20 d_2 \alpha_4 \xi^4]$$

$$- a_0 \Omega^2 (\alpha_0 + \alpha_1\xi + \alpha_2\xi^2 + \alpha_3\xi^3 + \alpha_4\xi^4) = 0 \tag{9.20}$$

Since this equation must be valid for any ξ, we get the following set of five algebraic equations, which are obtained by equating the coefficients of ξ^i $(i = 0, 1, 2, 3, 4)$ to zero:

$$2(2\alpha_2 b_2 + 6\alpha_3 b_1 + 12\alpha_4 b_0) - (2\alpha_2 d_0 + \alpha_1 d_1) + c_0\alpha_0 - a_0\Omega^2\alpha_0 = 0 \tag{9.21}$$

$$6(12\alpha_4 b_1 + 6\alpha_3 b_2 + 2\alpha_2 b_3) + c_0\alpha_1 - 2(\alpha_1 d_2 + 2\alpha_2 d_1 + 3\alpha_3 d_0) - a_0\Omega^2\alpha_1 = 0 \tag{9.22}$$

$$12(12\alpha_4 b_2 + 6\alpha_3 b_3 + 2\alpha_2 b_4) + c_0\alpha_2 - 3(3\alpha_3 d_1 + 2\alpha_2 d_2 + 4\alpha_4 d_0) - a_0\Omega^2\alpha_2 = 0 \tag{9.23}$$

$$20(12\alpha_4 b_3 + 6\alpha_3 b_4) + c_0\alpha_3 - 4(3\alpha_3 d_2 + 4\alpha_4 d_1) - a_0\Omega^2\alpha_3 = 0 \tag{9.24}$$

$$360\alpha_4 b_4 + c_0\alpha_4 - 20\alpha_4 d_2 - a_0\Omega^2\alpha_4 = 0 \tag{9.25}$$

The set (9.21)–(9.25) contains six unknowns, namely, b_0, b_1, b_2, b_3, b_4 and Ω^2. At this stage we declare b_4 to be a known constant. The question of whether or not this constant can be an arbitrary one will be discussed at a later stage. We thus arrive at five inhomogeneous equations with five unknowns, with an upper triangular matrix with a non-zero determinant equal to $-59,719,680\alpha_4^5 a_0$. This implies that both the coefficients α_4 and a_0 must differ from zero. These two conditions are satisfied by Eqs. (9.14)–(9.19) since neither a_0 nor α_4 vanish.

Thus, the coefficients b_0, b_1, b_2, b_3 and Ω^2 are uniquely determinable from Eqs. (9.21)–(9.25) once the coefficient b_4 is specified, in addition to the given constant inertial coefficient, constant Winkler foundation flexural rigidity, and three coefficients describing the variable Pasternak foundation. We are looking for a physically realizable solution, defined as the one which leads to a positive function $D(\xi)$ in the interval $[0; 1]$ and a positive value for Ω^2.

We get the natural frequency squared from Eq. (9.25). Since $\alpha_4 \neq 0$ and $a_0 \neq 0$, one can divide by α_4 to arrive at

$$\Omega^2 = (360b_4 + c_0 - 20d_2)/a_0 \qquad (9.26)$$

This equation possesses a perplexing property: It is *independent* of the coefficients α_i, for $i = 0, \ldots, 4$, which determine the boundary conditions. Thus, beams with six *different* sets of boundary conditions share the same natural frequency, in the context of the present formulation.

As is seen from Eq. (9.26), the Winkler and Pasternak foundations may counteract in such a manner as to have an equality, $c_0 = 20d_2$, in which case the beam on both foundations has the same natural frequency as the foundationless one. Moreover, if $c_0 < 20d_2$ one may arrive at a paradoxical conclusion that the presence of the elastic foundations may decrease the natural frequency. Note that this conclusion is contrary to Eq. (2.24) in the book by Vlasov and Leontiev (1966), who studied the vibrations of uniform beams on uniform Winkler and Pasternak foundations. Their formula, for the beam pinned at both ends, reads $\omega^2 = [D(\pi/L)^4 + c_0 + d_0(\pi/L)^2]/\rho A$, and suggests that the increase of either Winkler or Pasternak foundation coefficients increases the natural frequency. An additional consideration stemming from Eq. (9.26) is as follows: the Pasternak foundation cannot appear as a sole foundation in our model, for if one formally sets $c_0 = 0$, one obtains a decrease of the natural frequency for any value of the Pasternak foundation coefficient d_2. This is contrary to one's anticipation that the elastic foundation amounts to strengthening the system. In order to avoid the above peculiar results for the inhomogeneous beam, which are often obtained in non-conservative systems, we must assume that the model of the combined Winkler–Pasternak foundation should not be utilized uncritically; it appears that it is applicable when $c_0 > 20d_2$. Formal consideration of Eq. (9.26) may suggest that for large values of d_2 Eq. (9.26) may yield negative natural frequencies squared. This merely implies that the coefficient b_4 cannot be arbitrary. It may take only the values that correspond to positive natural frequency squared, i.e., $b_4 > (20d_2 - c_0)/360$. It is possible that one of the reasons for the unexpected result associated with Pasternak's foundation is the fact that the inertia of the foundation was not included [for its effect, see, e.g., Saito and Murikami (1969)]. The author is unaware if the counteraction phenomenon has been reported in other works, or the possibility of concocting beams whose natural frequencies are unaffected by the elastic foundations. One could naturally argue that in the present formulation, in order to obtain closed-form solutions, the beam's flexural rigidity must depend on the elastic foundation modulus. At first glance, this is an unnatural assumption. Yet, as it often happens, in order to force the system to have a prescribed mode shape, its coefficients must satisfy some relations. Likewise, in order for the beam to have a closed-form solution, it is necessary that the beam's flexural rigidity and the elastic foundation moduli should be interconnected. This situation is not unlike the exact solution derived by Caughey (1964) and Dimentberg (1982) in the context of stochastic mechanics. There, in order for the system with parametric and external excitations to have a closed-form solutions, the systems' inner properties had to depend upon the excitation. In this respect, both our results as well as those of Caughey (1964) and Dimentberg (1982) can be characterized as somewhat artificial ones, by a pessimistic reader. Yet, solutions by Caughey and Ma (1982) and Dimentberg (1982) have been used by Bergman (2000) to check a general computer program for solving the non-linear stochastic dynamic problem. A more detailed discussion of the relevance of the conditions appearing in the closed-form solutions of stochastic problems is given in the Epilogue. To summarize our discussion, the closed-form solutions derived in this section may serve for validation purposes in the general purpose computer codes.

In order to corroborate Eq. (9.26) with attendant seemingly unexpected conclusions, with approximate evaluation, the Boobnov–Galerkin method is utilized in Section 9.1.12.

The coefficients of the flexural rigidity are determined with b_4 again taken as a parameter. The solution of Eqs. (9.21)–(9.25) leads to the following expressions for the coefficients defining the elastic modulus:

$$
\begin{aligned}
b_0 = &-[(258\alpha_2\alpha_3^3\alpha_4 - 140\alpha_2^2\alpha_4^4 + 7200\alpha_0\alpha_4^3 - 1080\alpha_1\alpha_3\alpha_4^2 - 36\alpha_3^4)d_2 \\
&+ (-63\alpha_3^3\alpha_4 + 234\alpha_2\alpha_3\alpha_4^2 - 360\alpha_1\alpha_4^3)d_1 \\
&+ (-600\alpha_2\alpha_4^3 + 180\alpha_3^2\alpha_4^2)d_0 \\
&+ (3{,}360\alpha_2^2\alpha_4^2 - 129{,}600\alpha_0\alpha_4^3 + 1{,}080\alpha_3^4 - 6{,}480\alpha_2\alpha_3^3\alpha_4 \\
&+ 21{,}600\alpha_1\alpha_3\alpha_4^2)b_4]/8{,}640\alpha_4^4
\end{aligned}
\tag{9.27}
$$

$$
\begin{aligned}
b_1 = &[(-360\alpha_1\alpha_4^2 - 12\alpha_3^3 + 78\alpha_2\alpha_3\alpha_4)d_2 + (-21\alpha_3^2\alpha_4 + 64\alpha_2\alpha_4^2)d_1 \\
&+ 60\alpha_3\alpha_4^2d_0 + (-1{,}920\alpha_2\alpha_3\alpha_4 + 7{,}200\alpha_1\alpha_4^2 + 360\alpha_3^3)b_4]/1{,}440\alpha_4^3
\end{aligned}
\tag{9.28}
$$

$$
b_2 = [(12\alpha_3^2 - 70\alpha_2\alpha_4)d_2 + 21\alpha_3\alpha_4d_1 + 60\alpha_4^2d_0 + (1{,}680\alpha_2\alpha_4 - 360\alpha_3^2)b_4]/720\alpha_4^2
\tag{9.29}
$$

$$
b_3 = (-\alpha_3d_2 + 2\alpha_4d_1 + 30\alpha_3b_4)/30\alpha_4
\tag{9.30}
$$

As is seen in this particular case ($m = 0$), for a constant inertial coefficient, an interesting conclusion is obtained: the beam's flexural rigidity, constructed in such a manner as to correspond to the mode shape given by Eq. (9.13), turns to be independent of both the Winkler foundation and the inertial coefficient. Only the coefficients d_0, d_1 and d_2 for the Pasternak foundation are present in the formulation of the coefficients b_i. Figures 9.1–9.6 portray the variation of the flexural rigidity $D(\xi)$, when $\rho(\xi) = 1$ for the six different boundary conditions. We note that to obtain a positive flexural rigidity on $[0; 1]$, we must have $b_4 > 0$.

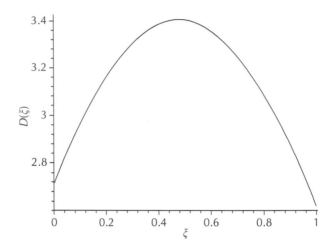

FIGURE 9.1
Variation of $D(\xi)$ for the pinned–pinned beam for $m = 0$, $K_1(\xi) = 1$, $K_2(\xi) = 1 + \xi + \xi^2$ and $b_4 = 1$

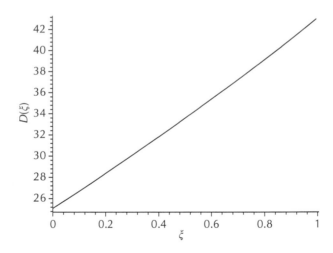

FIGURE 9.2
Variation of $D(\xi)$ for the clamped–free beam for $m = 0$, $K_1(\xi) = 1$, $K_2(\xi) = 1 + \xi + \xi^2$ and $b_4 = 1$

9.1.4 Beams with Linearly Varying Density, Inhomogeneous Modulus and Elastic Foundations

Now consider the case of the linearly varying inertial coefficient,

$$\rho(\xi) = a_0 + a_1\xi \tag{9.31}$$

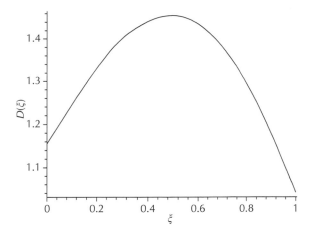

FIGURE 9.3
Variation of $D(\xi)$ for the clamped–pinned beam for $m = 0$, $K_1(\xi) = 1$, $K_2(\xi) = 1 + \xi + \xi^2$ and $b_4 = 1$

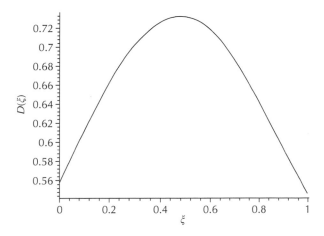

FIGURE 9.4
Variation of $D(\xi)$ for the clamped–clamped beam for $m = 0$, $K_1(\xi) = 1$, $K_2(\xi) = 1 + \xi + \xi^2$ and $b_4 = 1$

with the associated flexural rigidities of the elastic foundations,

$$K_1(\xi) = c_0 + c_1\xi \tag{9.32}$$

$$K_2(\xi) = d_0 + d_1\xi + d_2\xi^2 + d_3\xi^3 \tag{9.33}$$

The coefficients a_i, c_i and d_i in Eqs. (9.31), (9.32) and (9.33) are supposed to be given. We are looking for the beam's elastic modulus as the polynomial

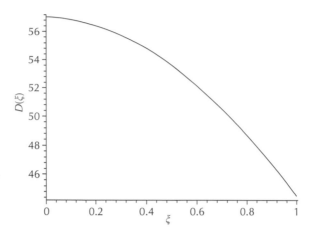

FIGURE 9.5
Variation of $D(\xi)$ for the guided–pinned beam for $m = 0$, $K_1(\xi) = 1$, $K_2(\xi) = 1 + \xi + \xi^2$ and $b_4 = 1$

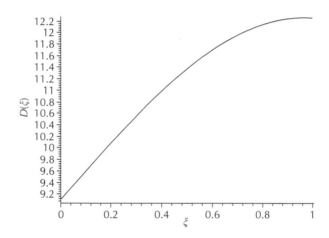

FIGURE 9.6
Variation of $D(\xi)$ for the guided–clamped beam for $m = 0$, $K_1(\xi) = 1$, $K_2(\xi) = 1 + \xi + \xi^2$ and $b_4 = 1$

function of the fifth degree:

$$D(\xi) = b_0 + b_1\xi + b_2\xi^2 + b_3\xi^3 + b_4\xi^4 + b_5\xi^5 \tag{9.34}$$

The linearly varying flexural rigidity in the Winkler foundation, as in Eq. (9.33) was studied in various contexts before. Maliev (1938) and Hetenyi (1955) investigated the bending of a beam resting on a base with linearly varying compressibility. Laterally–loaded pile buried in a soil with flexural rigidity linearly increasing with depth was investigated by Reese and Welch (1975).

The vibrations of a beam on such a foundation were studied by Pavlovic and Wylie (1983).

By substituting the expressions (9.31)–(9.34) into Eq. (9.2), we obtain a system of six equations with seven unknowns, $\alpha^T = [b_0\ b_1\ b_2\ b_3\ b_4\ b_5\ \Omega^2]$. Thus, an infinite number of solutions exists; to pick one of the solutions we have to fix one of the sought for coefficients. To this end, the coefficient b_5 is chosen to be a known constant. Whether or not it can be treated as an arbitrary one will be discussed at a later stage. Hence, we can write our system in the form $A\alpha = B$:

$$A = \begin{bmatrix} 24\alpha_4 & 12\alpha_3 & 4\alpha_2 & 0 & 0 & -a_0\alpha_0 \\ 0 & 72\alpha_4 & 36\alpha_3 & 12\alpha_2 & 0 & -a_0\alpha_1 - a_1\alpha_0 \\ 0 & 0 & 144\alpha_4 & 72\alpha_3 & 24\alpha_2 & -a_0\alpha_2 - a_1\alpha_1 \\ 0 & 0 & 0 & 240\alpha_4 & 120\alpha_3 & -a_0\alpha_3 - a_1\alpha_2 \\ 0 & 0 & 0 & 0 & 360\alpha_4 & -a_0\alpha_4 - a_1\alpha_3 \\ 0 & 0 & 0 & 0 & 0 & -a_1\alpha_4 \end{bmatrix} \quad \alpha = \begin{bmatrix} b_0 \\ b_1 \\ b_2 \\ b_3 \\ b_4 \\ \Omega^2 \end{bmatrix}$$

$$B = \begin{bmatrix} 2\alpha_2 d_0 + \alpha_1 d_1 - \alpha_0 c_0 \\ 2\alpha_1 d_2 + 6\alpha_3 d_0 + 4\alpha_2 d_1 - \alpha_1 c_0 - \alpha_0 c_1 \\ 4\alpha_4 d_0 + 9\alpha_3 d_1 + 6\alpha_2 d_2 + 3\alpha_1 d_3 - \alpha_2 c_0 - \alpha_1 c_1 \\ 16\alpha_4 d_1 + 12\alpha_3 d_2 + 8\alpha_2 d_1 - \alpha_3 c_0 - \alpha_2 c_1 - 40\alpha_2 b_5 \\ 20\alpha_4 d_2 + 15\alpha_3 d_3 - \alpha_4 c_0 - \alpha_3 c_1 - 180\alpha_3 b_5 \\ 24\alpha_4 d_3 - \alpha_4 c_1 - 504\alpha_4 b_5 \end{bmatrix}$$

$$(9.35)$$

In order to obtain a non-trivial solution, the determinant $\det(A)$ must differ from zero:

$$\det(A) = -21{,}499{,}084{,}800\alpha_4^6 a_1 \neq 0 \qquad (9.36)$$

We note that the above condition is satisfied by our hypothesis: a_1 is non-zero since a linearly varying density was postulated; moreover, for each boundary condition the following inequality holds, i.e., $\alpha_4 \neq 0$. Thus, the system $A\alpha = 0$ is solvable. We express the coefficients b_i for each boundary condition as follows:

Pinned–Pinned Beam:

$$b_0 = -[(-5{,}376d_3 + 224c_1 + 112{,}896b_5)a_0$$
$$+ (8{,}160d_0 - 2{,}160d_1 + 1{,}120d_2 + 2{,}520d_3 - 224c_0 - 60{,}480b_5)a_1]/51{,}840a_1$$
$$b_1 = -[(2{,}880d_3 - 120c_1 - 60{,}480b_5)a_0$$
$$+ (768d_1 - 240d_2 - 384d_3 + 120c_0 + 11{,}136b_5)a_1]/8{,}640a_1$$

$$b_2 = [(+1{,}344d_3 - 56c_1 - 28{,}224b_5)a_0$$
$$+ (120d_0 - 280d_2 - 630d_3 + 56c_0 + 15{,}120b_5)a_1]/4{,}320a_1$$
$$b_3 = (48d_1 + 96d_3 - 2{,}784b_5)/720$$
$$b_4 = [(-24d_3 + c_1 + 504b_5)a_0 + (20d_2 + 18d_3 - 648b_5 - c_0)a_1]/360a_1$$

$$(9.37)$$

Clamped–Free Beam:

$$b_0 = -[(89{,}856d_3 - 3{,}744c_1 - 1{,}886{,}976b_5)a_0$$
$$+ (4{,}320d_0 - 9{,}504d_1 - 11{,}808d_2 + 57{,}024d_3$$
$$+ 3{,}744c_0 - 1{,}472{,}256b_5)a_1]/51{,}840a_1$$
$$b_1 = -[(9{,}216d_3 - 384c_1 - 19{,}3536b_5)a_0$$
$$+ (2{,}400d_0 - 288d_1 - 1{,}056d_2 + 6{,}912d_3 + 384c_0$$
$$- 176{,}256b_5)a_1]/8{,}640a_1$$
$$b_2 = [(-1{,}728d_3 + 72c_1 + 36{,}288b_5)a_0$$
$$+ (120d_0 - 504d_1 + 72d_2 - 2{,}160d_3 - 72c_0 + 53{,}568b_5)a_1]/4{,}320a_1$$
$$b_3 = [(192d_3 - 8c_1 - 4{,}032b_5)a_0$$
$$+ (48d_1 - 64d_2 - 144d_3 + 8c_0 + 3{,}168b_5)a_1]/720a_1$$
$$b_4 = [(-24d_3 + c_1 + 504b_5)a_0 + (20d_2 + 36d_3 - c_0 - 1{,}296b_5)a_1]/360a_1$$

$$(9.38)$$

Clamped–Pinned Beam:

$$b_0 = -[(140{,}832d_3 - 5{,}868c_1 - 2{,}957{,}472b_5)a_0$$
$$+ (164{,}160d_0 + 20{,}520d_1 + 16{,}560d_2 + 82{,}215d_3 + 5{,}868c_0$$
$$- 1{,}417{,}500b_5)a_1]/1{,}658{,}880a_1$$
$$b_1 = -[(10{,}080d_3 - 420c_1 - 211{,}680b_5)a_0$$
$$+ (24{,}000d_0 + 3{,}384d_1 + 1{,}680d_2 + 6{,}669d_3 + 420c_0$$
$$- 113{,}076b_5)a_1]/138{,}240a_1$$
$$b_2 = [(-864d_3 + 36c_1 + 18{,}144b_5)a_0$$
$$+ (960d_0 - 2520d_1 - 720d_2 - 1485d_3 - 36c_0 + 23{,}220b_5)a_1]/34{,}560a_1$$
$$b_3 = [(480d_3 - 20c_1 - 10{,}080b_5)a_0$$
$$+ (192d_1 - 160d_2 - 63d_3 + 20c_0 + 252b_5)a_1]/2{,}880a_1$$
$$b_4 = [(-48d_3 + 2c_1 + 1{,}008b_5)a_0 + (40d_2 + 45d_3 - 2c_0 - 1{,}620b_5)a_1]/720a_1$$

$$(9.39)$$

Clamped–Clamped Beam:

$$b_0 = -[(2{,}112d_3 - 88c_1 - 44{,}352b_5)a_0$$
$$+ (3{,}120d_0 + 216d_1 + 136d_2 + 936d_3 + 88c_0 - 17{,}568b_5)a_1]/51{,}840a_1$$
$$b_1 = [(-384d_3 + 16c_1 + 8{,}064b_5)a_0$$
$$+ (-1{,}200d_0 - 120d_1 - 40d_2 - 192d_3 - 16c_0 + 3{,}552b_5)a_1]/8{,}640a_1$$
$$b_2 = -[(96d_3 - 4c_1 - 2{,}016b_5)a_0$$
$$+ (-120d_0 + 252d_1 + 52d_2 + 108d_3 + 4c_0 - 1{,}872b_5)a_1]/4{,}320a_1$$
$$b_3 = -[(-96d_3 + 4c_1 + 2{,}016b_5)a_0$$
$$+ (-48d_1 + 32d_2 + 12d_3 - 4c_0 - 96b_5)a_1]/720a_1$$
$$b_4 = -[(24d_3 - c_1 - 504b_5)a_0 + (-20d_2 - 18d_3 + c_0 + 648b_5)a_1]/360a_1$$

$$(9.40)$$

Guided–Pinned Beam:

$$b_0 = -[(210{,}816d_3 - 8{,}784c_1 - 4{,}427{,}136b_5)a_0$$
$$+ (24{,}480d_0 + 10{,}080d_2 + 8{,}784c_0)a_1]/51{,}840a_1$$
$$b_1 = -(2{,}304d_1 + 10{,}944d_3 - 202{,}176b_5)/8{,}640$$
$$b_2 = [(4{,}032d_3 - 168c_1 - 84{,}672b_5)a_0 + (120d_0 - 840d_2 + 168c_0)a_1]/4{,}320a_1$$
$$b_3 = (48d_1 + 288d_3 - 8{,}352b_5)/720$$
$$b_4 = -[(24d_3 - c_1 - 504b_5)a_0 + (-20d_2 + c_0)a_1]/360a_1$$

$$(9.41)$$

Guided–Clamped Beam:

$$b_0 = -[(33{,}792d_3 - 1{,}408c_1 - 709{,}632b_5)a_0$$
$$+ (12{,}480d_0 + 1{,}728d_1 + 2{,}176d_2 + 29{,}952d_3 + 1{,}408c_0$$
$$- 562{,}176b_5)a_1]/51{,}840a_1$$
$$b_1 = -[(3{,}072d_3 - 128c_1 - 64{,}512b_5)a_0$$
$$+ (2{,}400d_0 + 480d_1 + 320d_2 + 3{,}072d_3 + 128c_0 - 56{,}832b_5)a_1]/8{,}640a_1$$
$$b_2 = -[(384d_3 - 16c_1 - 8{,}064b_5)a_0$$
$$+ (-120d_0 + 504d_1 + 208d_2 + 864d_3 + 16c_0 - 14{,}976b_5)a_1]/4{,}320a_1$$
$$b_3 = -[(-192d_3 + 8c_1 + 4{,}032b_5)a_0$$
$$+ (-48d_1 + 64d_2 + 48d_3 - 8c_0 - 384b_5)a_1]/720a_1$$
$$b_4 = -[(24d_3 - c_1 - 504b_5)a_0 + (-20d_2 - 36d_3 + c_0 + 1{,}296b_5)a_1]/360a_1$$

$$(9.42)$$

It is remarkable that the natural frequency is again independent of the boundary conditions. It is expressed as

$$\Omega^2 = (504b_5 + c_1 - 24d_3)/a_1 \tag{9.43}$$

In order to corroborate this expression, with the seemingly unexpected influence of the Pasternak foundation's coefficient d_3, with that obtained by approximate evaluation, the Boobnov–Galerkin method is applied in Section 9.1.12.

Expression (9.43) has the same form as that for the natural frequency found for $m = 0$. We conclude that the considerations that followed Eq. (9.26) are equally applicable here. Namely, the effect of Winkler and Pasternak foundations may be nullified if $c_1 = 24d_3$, implying that c_1 must exceed $24d_3$, because the inequality $c_1 < 24d_3$ may lead to destabilization of the structure by supporting it using the elastic foundations. Moreover, for a physical realization of the problem, one arrives at the conclusion that b_5 is *not* an arbitrary constant: its range of variation $b_5 \geq (c_1 - 24d_3)/504$ is determined by the non-negativity of the natural frequency.

For the elastic foundations with coefficients

$$K_1(\xi) = 1 + \xi \quad K_2(\xi) = 1 + \xi + \xi^2 + \xi^3 \tag{9.44}$$

and $b_5 = 1$, Figures 9.7–9.12 present the variation of $D(\xi)$ for the different boundary conditions for the linear inertial coefficient $R(\xi) = 1 + \xi$.

In this section, we treated the general solution of a beam with constant and linearly varying functions for the inertial coefficient. It is of interest to treat for each set of boundary conditions *all* possible cases or the polynomially

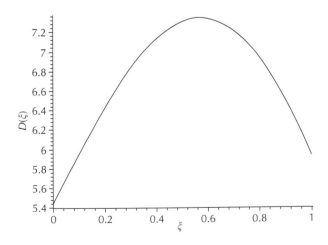

FIGURE 9.7
Variation of $D(\xi)$ for the pinned–pinned beam for $m = 1$, $K_1(\xi) = 1 + \xi$, $K_2(\xi) = 1 + \xi + \xi^2 + \xi^3$ and $b_5 = 1$

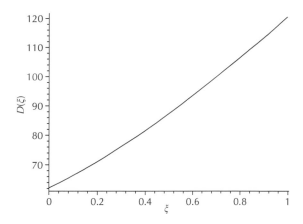

FIGURE 9.8
Variation of $D(\xi)$ for the clamped–free beam for $m = 1$, $K_1(\xi) = 1 + \xi$, $K_2(\xi) = 1 + \xi + \xi^2 + \xi^3$ and $b_5 = 1$

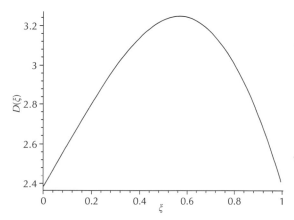

FIGURE 9.9
Variation of $D(\xi)$ for the clamped–pinned beam for $m = 1$, $K_1(\xi) = 1 + \xi$, $K_2(\xi) = 1 + \xi + \xi^2 + \xi^3$ and $b_5 = 1$

varying inertial coefficient. Section 9.1.5 deals with the case of a general order polynomial function that represents the inertial coefficient.

9.1.5 Beams with Varying Inertial Coefficient Represented as an *m*th Order Polynomial

This section deals with an inhomogeneous beam on an elastic foundation, with an inertial coefficient represented by an *m*th order polynomial function:

$$R(\xi) = \sum_{i=0}^{m} a_i \xi^i \tag{9.45}$$

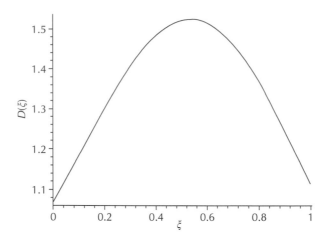

FIGURE 9.10
Variation of $D(\xi)$ for the clamped–clamped beam for $m = 1$, $K_1(\xi) = 1+\xi$, $K_2(\xi) = 1+\xi+\xi^2+\xi^3$ and $b_5 = 1$

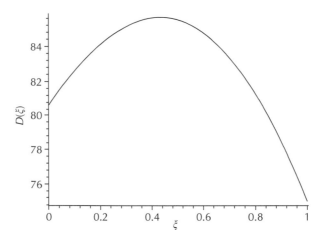

FIGURE 9.11
Variation of $D(\xi)$ for the guided–pinned beam for $m = 1$, $K_1(\xi) = 1+\xi$, $K_2(\xi) = 1+\xi+\xi^2+\xi^3$ and $b_5 = 1$

The Young's modulus, the Winkler foundation coefficients and the Pasternak foundation coefficients are given by the following expressions:

$$D(\xi) = \sum_{i=0}^{p} b_i\xi^i \quad K_1(\xi) = \sum_{i=0}^{q} c_i\xi^i \quad K_2(\xi) = \sum_{i=0}^{r} d_i\xi^i \tag{9.46}$$

We substitute the above expressions into Eq. (9.2), and choose the orders p, q and r to ensure the compatibility of Eq. (9.2), i.e., for each term of it to have the same polynomial degree. Let us consider each term separately: The order of

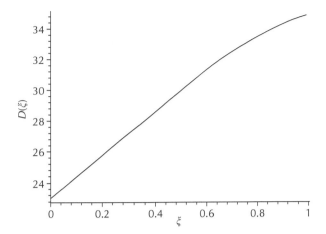

FIGURE 9.12
Variation of $D(\xi)$ for the guided–clamped beam for $m = 1$, $K_1(\xi) = 1 + \xi$, $K_2(\xi) = 1 + \xi + \xi^2 + \xi^3$ and $b_5 = 1$

$\rho(\xi)\Omega^2 W$ is $m + 4$, since W is the fourth order polynomial, while $R(\xi)$ is the mth order polynomial. The expression $d(K_2(\xi)dW/d\xi)/d\xi$ is a polynomial function of degree $r + 2$. The expression $K_1(\xi)W$ is a $(q + 4)$th order polynomial function. The expression $d^2(D(\xi)d^2W/d\xi^2)/d\xi^2$ is a polynomial function of degree p. To achieve the same order in the polynomial for each term of Eq. (9.2) the following conditions must be met:

$$p = m + 4 \quad q = m \quad r = m + 2$$

$$D(\xi) = \sum_{i=0}^{m+4} b_i \xi^i \quad K_1(\xi) = \sum_{i=0}^{m} c_i \xi^i \quad K_2(\xi) = \sum_{i=0}^{m+2} d_i \xi^i \quad (9.47)$$

Now, since the orders of the flexural rigidity, the Winkler foundation and the Pasternak foundation are specified, we evaluate each term of Eq. (9.2) separately:

$$\frac{d^2(D(\xi)d^2W/d^2\xi)}{d\xi^2} = \sum_{i=0}^{m+4} 24(2i+1)\alpha_4 b_i \xi^i + \sum_{i=2}^{m+4} 12(i-1)i\alpha_4 b_i \xi^i$$

$$+ \sum_{i=0}^{m+3} 12(i+1)\alpha_3 b_{i+1}\xi^i + \sum_{i=1}^{m+3} 6i(i+1)\alpha_3 b_{i+1}\xi^i$$

$$+ \sum_{i=0}^{m+2} 2(i+1)(i+2)\alpha_2 b_{i+2}\xi^i \quad (9.48)$$

$$\frac{d(K_2(\xi)dW/d\xi)}{d\xi} = \sum_{i=0}^{m+1}(i+1)\alpha_1 d_{i+1}\xi^i + \sum_{i=0}^{m+2}2(i+1)\alpha_2 d_i\xi^i$$

$$+ \sum_{i=0}^{m+2}3(i+2)\alpha_3 d_i\xi^{i+1} + \sum_{i=0}^{m+2}4(i+3)\alpha_4 d_i\xi^{i+2} \quad (9.49)$$

$$K_1(\xi)W - R(\xi)\Omega^2 W = \sum_{i=0}^{m}(c_i - \Omega^2 a_i)(\alpha_0\xi^i + \alpha_1\xi^{i+1} + \alpha_2\xi^{i+2}$$

$$+ \alpha_3\xi^{i+3} + \alpha_4\xi^{i+4}) \quad (9.50)$$

Equation (9.2), after substituting into it Eqs. (9.48)–(9.50), becomes a polynomial function that must vanish for every value of ξ. This requirement forces each coefficient of any power of ξ to vanish. The order of this polynomial being $m+4$, we obtain $m+5$ equations:

ξ^0: $2(2\alpha_2 b_2 + 6\alpha_3 b_1 + 12\alpha_4 b_0) + c_0\alpha_0 - (d_1\alpha_1 + 2d_0\alpha_2) - \Omega^2 a_0\alpha_0 = 0$
$$(9.51)$$

ξ^1: $6(2\alpha_2 b_3 + 6\alpha_3 b_2 + 12\alpha_4 b_1) + (c_1 - \Omega^2 a_1)\alpha_0 + (c_0 - \Omega^2 a_0)\alpha_1$

$- (2d_2\alpha_1 + 4d_1\alpha_2 + 6d_0\alpha_3) = 0$ $\qquad (9.52)$

ξ^2: $12(2\alpha_2 b_4 + 6\alpha_3 b_3 + 12\alpha_4 b_2) + (c_2 - \Omega^2 a_2)\alpha_0 + (c_1 - \Omega^2 a_1)\alpha_1$

$+ (c_0 - \Omega^2 a_0)\alpha_2 - (3\alpha_1 d_3 + 6\alpha_2 d_2 + 9\alpha_3 d_1 + 12\alpha_4 d_0) = 0$ $\quad (9.53)$

ξ^3: $20(2\alpha_2 b_5 + 6\alpha_3 b_4 + 12\alpha_4 b_3) + (c_3 - \Omega^2 a_3)\alpha_0 + (c_2 - \Omega^2 a_2)\alpha_1$

$+ (c_1 - \Omega^2 a_1)\alpha_2 + (c_0 - \Omega^2 a_0)\alpha_3 - (4\alpha_1 d_4 + 8\alpha_2 d_3$

$+ 12\alpha_3 d_2 + 16\alpha_4 d_1) = 0$ $\qquad (9.54)$

$$\cdots$$

ξ^i: $(i+1)(i+2)(2\alpha_2 b_{i+2} + 6\alpha_3 b_{i+1} + 12\alpha_4 b_i)$

$+ (c_i - \Omega^2 a_i)\alpha_0 + (c_{i-1} - \Omega^2 a_{i-1})\alpha_1 + (c_{i-2} - \Omega^2 a_{i-2})\alpha_2$

$+ (c_{i-3} - \Omega^2 a_{i-3})\alpha_3 + (c_{i-4} - \Omega^2 a_{i-4})\alpha_4$

$- [(i+1)\alpha_1 d_{i+1} + (i+1)(2\alpha_2 d_i + 3\alpha_3 d_{i-1} + 4\alpha_4 d_{i-2})] = 0$ $\quad (9.55)$

$$\cdots$$

ξ^{m+1}: $(m+2)(m+3)(2\alpha_2 b_{m+3} + 6\alpha_3 b_{m+2} + 12\alpha_4 b_{m+1}) + (c_m - \Omega^2 a_m)\alpha_1$

$+ (c_{m-1} - \Omega^2 a_{m-1})\alpha_2 + (c_{m-2} - \Omega^2 a_{m-2})\alpha_3 + (c_{m-3} - \Omega^2 a_{m-3})\alpha_4$

$- [(m+2)\alpha_1 d_{m+2} + (m+2)(2\alpha_2 d_{m+1} + 3\alpha_3 d_m + 4\alpha_4 d_{m-1})] = 0$
$$(9.56)$$

ξ^{m+2} : $(m+3)(m+4)(2\alpha_2 b_{m+4} + 6\alpha_3 b_{m+3} + 12\alpha_4 b_{m+2}) + (c_m - \Omega^2 a_m)\alpha_2$

$\qquad + (c_{m-1} - \Omega^2 a_{m-1})\alpha_3 + (c_{m-2} - \Omega^2 a_{m-2})\alpha_4$

$\qquad - (m+3)(2\alpha_2 d_{m+2} + 3\alpha_3 d_{m+1} + 4\alpha_4 d_m) = 0$ (9.57)

ξ^{m+3} : $(m+4)(m+5)(6\alpha_3 b_{m+4} + 12\alpha_4 b_{m+3})$

$\qquad + (c_m - \Omega^2 a_m)\alpha_3 + (c_{m-1} - \Omega^2 a_{m-1})\alpha_4$

$\qquad - (m+4)(3\alpha_3 d_{m+2} + 4\alpha_4 d_{m+1}) = 0$ (9.58)

ξ^{m+4} : $12(m+5)(m+6)\alpha_4 b_{m+4} + (c_m - \Omega^2 a_m)\alpha_4 - 4(m+5)\alpha_4 d_{m+2} = 0$
(9.59)

The natural frequency is obtained from Eq. (9.59) directly, in terms of the coefficient of the flexural rigidity b_{m+4}, the Winkler foundation coefficient c_m and the Pasternak foundation coefficient d_{m+2}:

$$\omega^2 = [12(m+5)(m+6)b_{m+4} + c_m - 4(m+5)d_{m+2}]/a_m L^4 \qquad (9.60)$$

Note that in the case of no elastic foundation, Eq. (9.60) is reduced to the expression for the natural frequency of an inhomogeneous beam, with $c_m = 0$. For $m = 0$, Eq. (9.60) reduces to Eq. (9.26), whereas for $m = 1$ Eq. (9.43) is obtained. Intriguingly, depending on the coefficients c_m and d_{m+2} in Eq. (9.60), the natural frequency of an inhomogeneous beam on an elastic foundation turns out to be *greater* or *lower* than the associated beam without an elastic foundation. In order to avoid paradoxical conclusion of reducing natural frequency as a result of elastic foundations for the value of $4(m+5)d_{m+2}$ being greater than c_m, we arrive at the range of applicability of this hybrid model: $c_m \geq 4(m+5)d_{m+2}$. Moreover, the coefficient b_{m+4} cannot be chosen arbitrarily. It must satisfy the non-negativity requirement of ω^2 in Eq. (9.60). An additional remarkable phenomenon may take place: one can find a pair of coefficients $(c_m; d_{m+2})$ that results in the *same* natural frequency as the beam without the elastic foundation. This takes place when $c_m = 4(m+5)d_{m+2}$.

We observe that the system (9.51)–(9.59) consists of $m+5$ equations with $m+6$ unknowns, namely b_i for $i = 0, \ldots, m+4$ and the coefficient Ω^2. Hence, to solve the system, we have to choose an arbitrary constant, say b_{m+4}, and express the other unknowns in terms of b_{m+4}.

One observes from Eq. (9.58) that one can express b_{m+3} in term of b_{m+4}. With Eq. (9.57), we obtain b_{m+2}; then b_{m+1} is calculated from Eq. (9.56). The recursive equation (9.55), in conjunction with Eqs. (9.51)–(9.54), allows one to determine all other coefficients b_i. Once all coefficients are expressed in terms of b_{m+4}, we need to know only a single coefficient b_{i*} to calculate the other coefficients b_i, $i* \neq i$.

In the sections 9.1.6, 9.1.7, 9.1.8, 9.1.9, 9.1.10 and 9.1.11 we deal with specific boundary conditions. For each set of boundary condition we distinguish particular cases in which m is fixed at some specific integer value, as well as the general case of m. In all cases, a closed-form solution for the natural frequency is derived.

9.1.6   Case of a Beam Pinned at its Ends

In this case, the coefficients α_i take the following values:

$$\alpha_0 = 0 \quad \alpha_1 = 1 \quad \alpha_2 = 0 \quad \alpha_3 = -2 \quad \alpha_4 = 1 \tag{9.61}$$

and Eqs. (9.51)–(9.59) reduce to

from ξ^0: $24(b_0 - b_1) - d_1 = 0$ (9.62)

from ξ^1: $72(b_1 - b_2) + c_0 - \Omega^2 a_0 - (2d_2 - 12d_0) = 0$ (9.63)

from ξ^2: $144(b_2 - b_3) + c_1 - \Omega^2 a_1 - (3d_3 - 18d_1 + 12d_0) = 0$ (9.64)

from ξ^3: $240(b_3 - b_4) + c_2 - 2c_0 - \Omega^2(a_2 - 2a_0) - (4d_4 - 24d_2 + 16d_1) = 0$
$$\tag{9.65}$$

$$\cdots$$

from ξ^j: $12(i + 1)(i + 2)(b_i - b_{i+1}) - \Omega^2(a_{i-1} - 2a_{i-3} + a_{i-4})$
$$+ c_{i-1} - 2c_{i-3} + c_{i-4} - (i + 1)(d_{i+1} - 6d_{i-1} + 4d_{i-2}) = 0 \tag{9.66}$$

$$\cdots$$

from ξ^{m+1}: $12(m + 2)(m + 3)(b_{m+1} - b_{m+2}) - \Omega^2(a_m - 2a_{m-2} + a_{m-3})$
$$+ c_m - 2c_{m-2} + c_{m-3} - (m + 2)(d_{m+2} - 6d_m + 4d_{m-1}) = 0 \tag{9.67}$$

from ξ^{m+2}: $12(m + 3)(m + 4)(b_{m+2} - b_{m+3}) - \Omega^2(a_{m-2} - 2a_{m-1})$
$$- 2c_{m-1} + c_{m-2} - (m + 3)(-6d_{m+1} + 4d_m) = 0 \tag{9.68}$$

from ξ^{m+3}: $12(m + 4)(m + 5)(b_{m+3} - b_{m+4}) - \Omega^2(a_{m-1} - 2a_m) - 2c_m$
$$+ c_{m-1} - (m + 4)(-6d_{m+2} + 4d_{m+1}) = 0 \tag{9.69}$$

from ξ^{m+4}: $12(m + 5)(m + 6)b_{m+4} + c_m - \Omega^2 a_m - 4(m + 5)d_{m+2} = 0$
$$\tag{9.70}$$

Note that the above system, when there is neither a Winkler foundation nor a Pasternak foundation, reduces to the results obtained in the paper by Candan and Elishakoff (2000). As the natural frequency is independent of the boundary conditions, we find from Eq. (9.70) the expression Eq. (9.60) for the P–P beam.

The system (9.62)–(9.70) has $m + 5$ equations with $m + 6$ unknowns. In order to solve it, we take the coefficient b_{m+4} as an arbitrary parameter. One must differentiate between two separate cases: in Eqs. (9.65) and (9.67), respectively, the coefficients of ξ^3 and ξ^{m+1}, impose the condition $m > 2$. The inertial coefficient is a polynomial function of order greater than 2. Hence, we need to consider particular cases $0 \leq m \leq 2$. It is worth noting that these particular cases, as well as the general case $m \geq 3$ are the same cases as treated by Candan and Elishakoff (2000) for the beam without elastic foundations. In the general case, Eq. (9.66) corresponds to the coefficient of ξ^i, for $4 \leq i \leq m$. This requires that one can use this equation only for $m > 3$.

9.1.6.1   *Particular cases* $0 \leq m \leq 2$

We treat, first, the case of a constant inertial coefficient $R(\xi) = a_0$. Equation (9.2), in this case, results in the following system of five equations with six unknowns:

$$24(b_0 - b_1) - d_1 = 0$$

$$72(b_1 - b_2) + 12d_0 - 2d_2 + c_0 - a_0\Omega^2 = 0$$

$$144(b_2 - b_3) + 18d_1 - 12d_0 = 0 \qquad (9.71)$$

$$240(b_3 - b_4) + 24d_2 - 16d_1 - 2c_0 + 2a_0\Omega^2 = 0$$

$$360b_4 - 20d_2 + c_0 - a_0\Omega^2 = 0$$

Then, the coefficients b_i, $i = \{0, 1, 2, 3\}$ are expressible in terms of b_4, resulting in

$$b_0 = 3b_4 - d_0/12 - d_1/60 - 11d_2/60$$

$$b_1 = 3b_4 - d_0/12 - 7d_1/120 - 11d_2/60$$

$$b_2 = -2b_4 + d_0/12 - 7d_1/120 + d_2/15 \qquad (9.72)$$

$$b_3 = -2b_4 + d_1/15 + d_2/15$$

The natural frequency reads

$$\omega^2 = (360b_4 + c_0 - 20d_2)/L^4 a_0 \qquad (9.73)$$

It is instructive to see the discussion following Eq. (9.26). To recapitulate, we note that b_4 is *not* an arbitrary constant. It should be chosen so as to get a positive value for ω^2 in Eq. (9.73).

Let us now consider the case where the inertial coefficient is a linearly varying function $R(\xi) = a_0 + a_1\xi$. The flexural rigidity $D(\xi)$, the coefficients c_i of the Winkler foundation $K_1(\xi)$ and the coefficients d_i of the Pasternak foundation $K_2(\xi)$, in agreement with Eq. (9.47), read

$$D(\xi) = \sum_{i=0}^{5} b_i \xi^i \quad K_1(\xi) = c_0 + c_1\xi \quad K_2(\xi) = d_0 + d_1\xi + d_2\xi^2 + d_3\xi^3 + d_4\xi^4$$

$$\text{(9.74)}$$

Equation (9.2), in this sub-case, results in a system of six equations with seven unknowns, namely the coefficients b_i of the flexural rigidity and the natural frequency coefficient Ω:

$$24(b_0 - b_1) - d_1 = 0$$

$$72(b_1 - b_2) + 12d_0 - 2d_2 + c_0 - a_0\Omega^2 = 0$$

$$144(b_2 - b_3) + 18d_1 - 12d_0 - 3d_3 + c_1 - a_1\Omega^2 = 0$$

$$240(b_3 - b_4) + 24d_2 - 16d_1 - 2c_0 + 2a_0\Omega^2 = 0 \qquad \text{(9.75)}$$

$$360(b_4 - b_5) + 30d_3 - 20d_2 - 2c_1 + c_0 + (2a_1 - a_0)\Omega^2 = 0$$

$$504b_5 - 24d_3 + c_1 - a_1\Omega^2 = 0$$

The coefficient b_5 is taken as an arbitrary constant. The other unknowns are

$$b_0 = (17a_1 + 42a_0)b_5/10a_1$$
$$\quad - [(48d_3 - 2c_1)a_0 + (23d_3 + 4d_2 + 4d_1 + 20d_0 + 2c_0)a_1]/240a_1$$
$$b_1 = (17a_1 + 42a_0)b_5/10a_1$$
$$\quad - [(48d_3 - 2c_1)a_0 + (23d_3 + 4d_2 + 14d_1 + 20d_0 + 2c_0)a_1]/240a_1$$
$$b_2 = (17a_1 - 28a_0)b_5/10a_1 + [(96d_3 - 4c_1)a_0$$
$$\quad + (-69d_3 - 32d_2 - 42d_1 + 60d_0 + 4c_0)a_1]/720a_1$$
$$b_3 = -(9a_1 + 14a_0)b_5/5a_1 + [(24d_3 - c_1)a_0 + (9d_3 - 8d_2 + 12d_1 + c_0)a_1]/180a_1$$
$$b_4 = -(9a_1 + 7a_0)b_5/5a_1 - [(24d_3 - c_1)a_0 + (-18d_3 - 20d_2 + 12d_1 + c_0)a_1]/360a_1$$
$$\text{(9.76)}$$

The natural frequency is obtained as

$$\omega^2 = (504b_5 + c_1 - 24d_3)/L^4 a_1 \qquad \text{(9.77)}$$

Again, b_5 cannot be assigned any desired value. It must correspond to a positive natural frequency squared. Also, for no destabilization paradox c_1 must exceed $24d_3$.

In the particular case of the parabolically varying inertial coefficient ($m = 2$), the different terms appearing in Eq. (9.2) are given by

$$R(\xi) = \sum_{i=0}^{2} a_i \xi^i \quad D(\xi) = \sum_{i=0}^{6} b_i \xi^i \quad K_1(\xi) = \sum_{i=0}^{2} c_i \xi^i \quad K_2(\xi) = \sum_{i=0}^{4} d_i \xi^i \quad (9.78)$$

The system that is obtained from Eq. (9.2) consists of seven equations with eight unknowns, namely b_i for $i = \{0, 1, \ldots, 6\}$ and the coefficient of the natural frequency Ω:

$$24(b_0 - b_1) - d_1 = 0$$

$$72(b_1 - b_2) + 12d_0 - 2d_2 + c_0 - a_0\Omega^2 = 0$$

$$144(b_2 - b_3) + 18d_1 - 12d_0 - 3d_3 + c_1 - a_1\Omega^2 = 0$$

$$240(b_3 - b_4) + 24d_2 - 16d_1 - 4d_4 + c_2 - 2c_0 + (2a_0 - a_2)\Omega^2 = 0 \qquad (9.79)$$

$$360(b_4 - b_5) + 30d_3 - 20d_2 - 2c_1 + c_0 + (2a_1 - a_0)\Omega^2 = 0$$

$$504(b_5 - b_6) + 36d_4 - 24d_3 - 2c_2 + c_1 + (2a_2 - a_1)\Omega^2 = 0$$

$$672b_6 - 28d_4 + c_2 - a_2\Omega^2 = 0$$

We solve this system by choosing b_6 as a known constant:

$$b_0 = (17a_2 + 34a_1 + 84a_0)b_6/15a_2 - [(1176d_4 - 42c_2)a_0 + (476d_4 - 17c_2)a_1$$
$$+ (304d_4 + 75d_3 + 84d_2 + 84d_1 + 420d_0 + 42c_0 + 17c_1)a_2]/5040a_2$$

$$b_1 = (17a_2 + 34a_1 + 84a_0)b_6/15a_2 - [(1176d_4 - 42c_2)a_0 + (476d_4 - 17c_2)a_1$$
$$+ (304d_4 + 75d_3 + 84d_2 + 294d_1 + 420d_0 + 42c_0 + 17c_1)a_2]/5040a_2$$

$$b_2 = (17a_2 + 34a_1 - 56a_0)b_6/15a_2 + [(784d_4 - 28c_2)a_0 + (-476d_4 + 17c_2)a_1$$
$$+ (-304d_4 - 75d_3 - 224d_2 - 294d_1 + 420d_0 + 28c_0 - 17c_1)a_2]/5040a_2$$

$$b_3 = (17a_2 - 36a_1 - 56a_0)b_6/15a_2 + [(392d_4 - 14c_2)a_0 + (252d_4 - 9c_2)a_1$$
$$+ (-152d_4 - 90d_3 - 112d_2 + 168d_1 + 14c_0 + 9c_1)a_2]/2520a_2$$

$$b_4 = -(25a_2 + 36a_1 - 28a_0)b_6/15a_2 + [(-196d_4 + 7c_2)a_0 + (252d_4 - 9c_2)a_1$$
$$+ (100d_4 - 90d_3 + 140d_2 - 7c_0 + 9c_1)a_2]/2520a_2$$

$$b_5 = -(5a_2 - 4a_1)b_6/3a_2 + [(-28d_4 + c_2)a_1 + (20d_4 + 24d_3 - c_1)a_2]/504a_2$$
$$\qquad (9.80)$$

The natural frequency reads

$$\omega^2 = (672b_6 + c_2 - 28d_4)/L^4 a_2 \qquad (9.81)$$

We again emphasize that b_6 is not an arbitrary constant; rather, $b_6 > (28d_4 - c_2)/672$ in order that the natural frequency squared is a positive quantity. An additional constraint on the coefficients is $c_2 > 28d_4$.

We now proceed to treat the general case for which the degree of the polynomial representing the inertial coefficient function is not fixed.

9.1.6.2 *General case* (*m* > 2)

When the inertial coefficient is an mth order polynomial function, we obtain from Eqs. (9.62)–(9.70) a system of $m + 5$ equations with $m + 6$ unknowns, namely the coefficients of the flexural rigidity b_i and the natural frequency coefficient Ω. As in the particular cases, we take b_{m+4} to be an arbitrary constant. From Eq. (9.70), we obtain the closed-form solution for the natural frequency of a beam on elastic foundations:

$$\omega^2 = (12(m+5)(m+6)b_{m+4} + c_m - 4(m + 50d_{m+2})/L^4 a_m \tag{9.82}$$

From Eqs. (9.63)–(9.69), several expressions of Ω^2 are found. These are listed below:

$$\Omega^2 = 72(b_1 - b_2)/a_0 + c_0/a_0 - (2d_2 + 12d_0)/a_0 \tag{9.83}$$

$$\Omega^2 = 144(b_2 - b_3)/a_1 + c_1/a_1 - (3d_3 - 18d_1 + 12d_0)/a_1 \tag{9.84}$$

$$\Omega^2 = 240(b_3 - b_4)/(a_2 - 2a_0) + (c_2 - 2c_0)/(a_2 - 2a_0) - (4d_4 - 24d_2 + 16d_1)/(a_2 - 2a_0) \tag{9.85}$$

$$\cdots$$

$$\Omega^2 = \frac{12(i+1)(i+2)(b_i - b_{i+1})}{(a_{i-1} - 2a_{i-3} + a_{i-4})} + \frac{(c_{i-1} - 2c_{i-3} + c_{i-4}) - (i+1)(d_{i+1} - 6d_{i-1} + 4d_{i-2})}{(a_{i-1} - 2a_{i-3} + a_{i-4})} \tag{9.86}$$

$$\cdots$$

$$\Omega^2 = \frac{12(m+2)(m+3)(b_{m+1} - b_{m+2})}{(a_m - 2a_{m-2} + a_{m-3})} + \frac{(c_m - 2c_{m-2} + c_{m-3}) - (m+2)(d_{m+2} - 6d_m + 4d_{m-1})}{(a_m - 2a_{m-2} + a_{m-3})} \tag{9.87}$$

$$\Omega^2 = \frac{12(m+3)(m+4)(b_{m+2} - b_{m+3})}{(a_{m-2} - 2a_{m-1})} + \frac{(-2c_{m-1} + c_{m-2}) - (m+3)(-6d_{m+1} + 4d_m)}{(a_{m-2} - 2a_{m-1})} \tag{9.88}$$

$$\Omega^2 = \frac{12(m+4)(m+5)(b_{m+3} - b_{m+4})}{(a_{m-1} - 2a_m)} + \frac{(-2c_m + c_{m-1}) - (m+4)(-6d_{m+2} + 4d_{m+1})}{(a_{m-1} - 2a_m)} \tag{9.89}$$

$$\Omega^2 = [12(m+5)(m+6)b_{m+4} + c_m - 4(m+5)d_{m+2}]/a_m \tag{9.90}$$

Since the expression in Eqs. (9.83)–(9.89) must be compatible, we obtain an expression for the beam's flexural rigidity coefficients, by equating various pairs of equations arising from the above system. From Eqs. (9.89) and (9.90), we obtain b_{m+3}:

$$b_{m+3} = [(m+6)(a_{m-1}/a_m - 2)/(m+4) + 1]b_{m+4}$$
$$+ (a_{m-1}/a_m - 2)[c_m - 4(m+5)d_{m+1}]/12(m+4)(m+5)$$
$$- [(-2c_m + c_{m-1}) - (m+4)(-6d_{m+2} + 4d_{m+1})]/12(m+4)(m+5)$$
$$\tag{9.91}$$

The coefficients b_{m+2} are obtained by equating Eqs. (9.88) and (9.89):

$$b_{m+2} = \left[\frac{m+5}{m+3}\frac{a_{m-2} - 2a_{m-1}}{a_{m-1} - 2a_m} + 1\right]b_{m+3} - \frac{m+5}{m+3}\frac{a_{m-2} - 2a_{m-1}}{a_{m-1} - 2a_m}b_{m+4}$$
$$+ \frac{a_{m-2} - 2a_{m-1}}{a_{m-1} - 2a_m}\frac{(-2c_m + c_{m-1}) - (m+4)(-6d_{m+2} + 4d_{m+1})}{12(m+3)(m+4)}$$
$$- \frac{(-2c_{m-1} + c_{m-2}) - (m+3)(-6d_{m+1} + 4d_m)}{12(m+3)(m+4)} \tag{9.92}$$

Equations (9.87)–(9.88) allow one to express b_{m+1} in terms of b_{m+2} and b_{m+3}:

$$b_{m+1} = \left[\frac{m+4}{m+2}\frac{a_m - 2a_{m-2} + a_{m-3}}{a_{m-2} - 2a_{m-1}} + 1\right]b_{m+2} - \frac{m+4}{m+2}\frac{a_m - 2a_{m-2} + a_{m-3}}{a_{m-2} - 2a_{m-1}}b_{m+3}$$
$$+ \frac{a_m - 2a_{m-2} + a_{m-3}}{a_{m-2} - 2a_{m-1}}\frac{(-2c_{m-1} + c_{m-2}) - (m+3)(-6d_{m+1} + 4d_m)}{12(m+2)(m+3)}$$
$$- \frac{(c_m - 2c_{m-2} + c_{m-3}) - (m+2)(d_{m+2} - 6d_m + 4d_{m-1})}{12(m+2)(m+3)} \tag{9.93}$$

From Eq. (9.86) we obtain the following expression for b_i:

$$b_i = \left[\frac{i+3}{i+1}\frac{a_{i-1} - 2a_{i-3} + a_{i-4}}{a_i - 2a_{i-2} + a_{i-3}} + 1\right]b_{i+1} - \frac{i+3}{i+1}\frac{a_{i-1} - 2a_{i-3} + a_{i-4}}{a_i - 2a_{i-2} + a_{i-3}}b_{i+2}$$
$$+ \frac{a_{i-1} - 2a_{i-3} + a_{i-4}}{a_i - 2a_{i-2} + a_{i-3}}\frac{(c_i - 2c_{i-2} + c_{i-3}) - (i+2)(d_{i+2} - 6d_i + 4d_{i-1})}{12(i+1)(i+2)}$$
$$- \frac{(c_{i-1} - 2c_{i-3} + c_{i-4}) - (i+1)(d_{i+1} - 6d_{i-1} + 4d_{i-2})}{12(i+1)(i+2)} \tag{9.94}$$

where i belongs to the set $\{4, 5, \ldots, m\}$. Equation (9.85) results in

$$b_3 = \left[\frac{3}{2}\frac{a_2 - 2a_0}{a_3 - 2a_1 + a_0} + 1\right]b_4 - \frac{3}{2}\frac{a_2 - 2a_0}{a_3 - 2a_1 + a_0}b_5$$
$$+ \frac{a_2 - 2a_0}{a_3 - 2a_1 + a_0}\frac{(c_3 - 2c_1 + c_0) - 5(d_5 - 6d_3 + 4d_2)}{240}$$
$$- \frac{(c_2 - 2c_0)}{240} + \frac{(d_4 - 6d_2 + 4d_1)}{60} \tag{9.95}$$

Equation (9.84) leads to

$$
\begin{aligned}
b_2 = {} & \left[\frac{5}{3} \frac{a_1}{a_2 - 2a_0} + 1 \right] b_3 - \frac{5}{3} \frac{a_1}{a_2 - 2a_0} b_4 \\
& - \frac{c_1}{144} + \frac{(3d_3 - 18d_1 + 12d_0)}{144} \\
& + \frac{a_1[(c_2 - 2c_0) + (4d_4 - 24d_2 + 16d_1)]}{(a_2 - 2a_0)}
\end{aligned} \tag{9.96}
$$

Finally, from Eq. (9.83), we get

$$
\begin{aligned}
b_1 = {} & [2a_0/a_1 + 1]b_2 - 2a_0 b_3/a_1 - (c_0 - 2d_2 - 12d_0)/72 \\
& + [c_1 - (3d_3 - 18d_1 + 12d_0)]a_0/a_1
\end{aligned} \tag{9.97}
$$

The coefficient b_0 (from Eq. (9.62)) is given by

$$
b_0 = b_1 + d_1/24 \tag{9.98}
$$

In the above expressions for the coefficients of the flexural rigidity, the first term corresponds to the beam without elastic foundations. These terms coincide with those derived by Elishakoff and Candan (2000). Note that we have taken b_{m+4} to be arbitrary constant. The above expressions show that the knowledge of a single coefficient b_i, not necessarily of b_{m+4}, for $i \in \{0, 1, \ldots, m + 4\}$ is sufficient to determine the other coefficients.

This general solution is valid if the coefficients a_i satisfy some relations. These relations are obtained from Eqs. (9.91)–(9.97):

$$
\begin{aligned}
&\text{(a)} \quad a_{m-1} \neq 2a_m \\
&\text{(b)} \quad a_{m-2} \neq 2a_{m-1} \\
&\text{(c)} \quad a_i - 2a_{i-2} + a_{i-3} \neq 0 \qquad \text{for } 3 \leq i \leq m \\
&\text{(d)} \quad a_2 \neq 2a_0 \\
&\text{(e)} \quad a_1 \neq 0 \\
&\text{(f)} \quad a_0 \neq 0
\end{aligned} \tag{9.99}
$$

Section 9.1.12 treats these particular cases.

9.1.7   Beam Clamped at the Left End and Free at the Right End

To have a mode shape that satisfies these boundary conditions, the coefficient α_i in Eq. (9.13) must take the following values:

$$
\alpha_0 = 0 \quad \alpha_1 = 0 \quad \alpha_2 = 6 \quad \alpha_3 = -4 \quad \alpha_4 = 1 \tag{9.100}
$$

Thus, Eqs. (9.51)–(9.59) may be rewritten as

ξ^0: $2(12b_2 - 24b_1 + 12b_0) - 12d_0 = 0$ (9.101)

ξ^1: $6(12b_3 - 24b_2 + 12b_1) - 24(d_1 - d_0) = 0$ (9.102)

ξ^2: $12(12b_4 - 24b_3 + 12b_2) + 6c_0 - 12(3d_2 - 3d_1 + d_0) - 6\Omega^2 a_0 = 0$ (9.103)

ξ^3: $20(12b_5 - 24b_4 + 12b_3) + 6c_1 - 4c_0 - 16(3d_3 - 3d_2 + d_1)$

$\qquad - \Omega^2(6a_1 - 4a_0) = 0$ (9.104)

$$\cdots$$

ξ^i: $(i+1)(i+2)(12b_{i+2} - 24b_{i+1} + 12b_i) + 6c_{i-2} - 4c_{i-3} + c_{i-4}$

$\qquad - 4(i+1)(3d_i - 3d_{i-1} + d_{i-2}) - \Omega^2(6a_{i-2} - 4a_{i-3} + a_{i-4}) = 0$ (9.105)

$$\cdots$$

ξ^{m+3}: $(m+4)(m+5)(-24b_{m+4} + 12b_{m+3}) - 4c_m + c_{m-1}$

$\qquad - 4(m+4)(-3d_{m+2} + d_{m+1}) - \Omega^2(-4a_m + a_{m-1}) = 0$ (9.106)

ξ^{m+4}: $12(m+5)(m+6)b_{m+4} + c_m - \Omega^2 a_m - 4(m+5)d_{m+2} = 0$ (9.107)

The natural frequency is obtained from Eq. (9.107); it is remarkable that the closed-form solution is identical to the natural frequency for a pinned–pinned beam:

$$\omega^2 = \{12(m+5)(m+6)b_{m+4}/a_m + [c_m - 4(m+5)d_{m+2}]/a_m\}/L^4 \quad (9.108)$$

The above system contains $m + 5$ equations with $m + 6$ unknowns, namely b_i where $i \in \{0, 1, \ldots, m+4\}$ and the natural frequency coefficient Ω. In order to solve this system, we again choose the coefficient b_{m+4} to assume an arbitrary value.

Let us consider Eqs. (9.104) and (9.106). To be compatible, these two equations require that $m > 0$. The form of the recursive equation (9.105) gives another condition. Using this, we can express b_i in terms of b_{i+1} and b_{i+2}. But this recursive equation is satisfied for $4 \leq i \leq m + 2$. Hence, to express the coefficient b_{m+2}, when m is not fixed, we need Eq. (9.105) with $i = m + 2$ and Eq. (9.106). Then, we can express the coefficients b_0, b_1, b_2, b_3, b_{m+2} and b_{m+3}. To satisfy this requirement, the condition $m > 1$ must be satisfied. Thus, these conditions lead to two separate cases: (1) $m \leq 1$ and (2) $m > 1$. In the general case, $m > 1$, the recursive equation (9.105) is satisfied for $4 \leq i \leq m + 1$. Consequently, we use it only for $m > 2$.

We begin with the two particular cases $m = 0$ and $m = 1$. The general case is treated thereafter.

9.1.7.1   Particular cases *m* = 0 or *m* = 1

We begin with the simplest case of a constant inertial coefficient. This requirement leads to a flexural rigidity $D(\xi) = b_0 + b_1\xi + b_2\xi^2 + b_3\xi^3 + b_4\xi^4$.
The terms associated with the elastic foundations are: $K_1(\xi) = c_0$ and
$K_2(\xi) = d_0 + d_1\xi + d_2\xi^2$. From the governing equation, we obtain a system of
five equations with six unknowns in terms of b_j and Ω:

$$2(12b_2 - 24b_1 + 12b_0) - 12d_0 = 0$$

$$6(12b_3 - 24b_2 + 12b_1) - 24(d_1 - d_0) = 0$$

$$12(12b_4 - 24b_3 + 12b_2) + 6c_0 - 12(3d_2 - 3d_1 + d_0) - 6a_0\Omega^2 = 0 \qquad (9.109)$$

$$20(-24b_4 + 12b_3) - 4c_0 - 16(-3d_2 + d_1) + 4a_0\Omega^2 = 0$$

$$360b_4 - 20d_2 + c_0 - a_0\Omega^2 = 0$$

The natural frequency reads

$$\omega^2 = (360b_4 + c_0 - 20d_2)/a_0 \qquad (9.110)$$

The coefficients b_0, b_1, b_2 and b_3 are given as

$$b_0 = 26b_4 - 73d_2/60 + 11d_1/60 + d_0/12$$

$$b_1 = 16b_4 - 23d_2/30 + d_1/30 - d_0/6$$

$$b_2 = 6b_4 - 19d_2/60 - 7d_1/60 + d_0/12 \qquad (9.111)$$

$$b_3 = -4b_4 + (2d_2 + d_1)/15$$

Now, consider the second case $m = 1$, when the inertial coefficient is
a linearly varying function while the beam's flexural rigidity, the Winkler
foundation and the Pasternak foundation are, respectively, polynomial functions of order 5, 1 and 3. We take b_5 to be arbitrary constant, and express b_i in
terms of b_{i+1} and b_{i+2}. The system reads

$$2(12b_2 - 24b_1 + 12b_0) - 12d_0 = 0$$

$$6(12b_3 - 24b_2 + 12b_1) - 24(d_1 - d_0) = 0$$

$$12(12b_4 - 24b_3 + 12b_2) + 6c_0 - 12(3d_2 - 3d_1 + d_0) - 6a_0\Omega^2 = 0$$

$$20(12b_5 - 24b_4 + 12b_3) + 6c_1 - 4c_0 - 16(-3d_2 + d_1) - (6a_1 - 4a_0)\Omega^2 = 0$$

$$30(-24b_5 + 12b_4) - 4c_1 + c_0 - 20(-3d_3 + d_2) - (-4a_1 + a_0)\Omega^2 = 0$$

$$504b_5 - 24d_3 + c_1 - a_1\Omega^2 = 0$$

$$(9.112)$$

The unknowns are obtained as

$$b_0 = 2b_1 - b_2 + d_0/2$$
$$b_1 = 2b_2 - b_3 + (d_1 - d_0)/3$$
$$b_2 = 2b_3 - b_4 + 21a_0b_5/a_1 + (3d_2 - 3d_1 + d_0)/12 - a_0d_3/a_1$$
$$\quad + (a_0c_1 - a_1c_0)/24a_1$$
$$b_3 = [120a_1b_4 + (696a_1 - 504a_0)b_5]/60a_1 + 24(a_0 - a_1)d_3/60a_1$$
$$\quad + (4d_1 - 12d_2)/60 - (a_0c_1 - a_1c_0)/60a_1$$
$$b_4 = (504a_0 - 1296a_1)b_5/360a_1 + 12(3a_1 - 2a_0)d_3/360a_1 + d_2/18$$
$$\quad + (a_0c_1 - a_1c_0)/360a_1$$

(9.113)

The natural frequency reads

$$\omega^2 = (504b_5 + c_1 - 24d_3)/a_1 \tag{9.114}$$

We treated the particular cases of an inhomogeneous, clamped–free beam on elastic foundations. The general case, when the degree of the polynomial function representing the inertial coefficient of the beam, m, is greater than unity, is considered in Section 9.1.7.2

9.1.7.2 *General case (m > 1)*

We already derived the closed-form solution for the natural frequency for the general case in Eq. (9.108). The system of equations, given by Eqs. (9.101)–(9.107), contains one unknown in excess of the number of equations. Therefore, we choose b_{m+4} to be arbitrary constant. The compatibility equations, read

$$\Omega^2 = [12(12b_4 - 24b_3 + 12b_2) + 6c_0 - 12(3d_2 - 3d_1 + d_0)]/6a_0 \tag{9.115}$$

$$\Omega^2 = [20(12b_5 - 24b_4 + 12b_3) - (-6c_1 + 4c_0)$$
$$\quad - 16(3d_3 - 3d_2 + d_1)]/(6a_1 - 4a_0) \tag{9.116}$$

$$\cdots$$

$$\Omega^2 = (i+1)(i+2)\frac{12b_{i+2} - 24b_{i+1} + 12b_i}{6a_{i-2} - 4a_{i-3} + a_{i-4}} + \frac{6c_{i-2} - 4c_{i-3} + c_{i-4}}{6a_{i-2} - 4a_{i-3} + a_{i-4}}$$
$$\quad - 4(i+1)\frac{3d_i - 3d_{i-1} + d_{i-2}}{6a_{i-2} - 4a_{i-3} + a_{i-4}} \tag{9.117}$$

$$\cdots$$

$$\Omega^2 = (m+4)(m+5)\frac{-24b_{m+4} + 12b_{m+3}}{-4a_m + a_{m-1}} + \frac{-4c_m + c_{m-1}}{-4a_m + a_{m-1}}$$

$$- 4(m+4)\frac{-3d_{m+2} + d_{m+1}}{-4a_m + a_{m-1}} \tag{9.118}$$

$$\Omega^2 = 12(m+5)(m+6)b_{m+4}/a_m + [c_m - 4(m+5)d_{m+2}]/a_m \tag{9.119}$$

We note that Eqs. (9.115)–(9.119) are valid only if the coefficients of the polynomial function that represents the inertial coefficient satisfy some auxiliary conditions. Indeed, Eq. (9.115) imposes that the condition $a_0 \neq 0$. From Eq. (9.120), we get $a_1 \neq 2a_0/3$. Then, Eq. (9.116) for $i \in \{4, \dots, m+2\}$ leads to: $a_{i-2} \neq 2a_{i-3}/3 + a_{i-4}$. Finally, Eq. (9.118) requires that $a_m \neq a_{m-1}/4$. From Eqs. (9.119) and (9.118), we obtain b_{m+3}:

$$b_{m+3} = \left[\frac{m+6}{m+4}\left(\frac{a_{m-1}}{a_m} - 4\right) + 2\right]b_{m+4} + \frac{c_m - 4(m+5)d_{m+2}}{12(m+4)(m+5)}\left(\frac{a_{m-1}}{a_m} - 4\right)$$

$$+ \frac{4c_m - c_{m-1} + 4(m+4)(-3d_{m+2} + d_{m+1})}{12(m+4)(m+5)} \tag{9.120}$$

The coefficient b_{m+2} is given by Eq. (9.117) with $i = m+2$ and (9.118):

$$b_{m+2} = -\left(2\frac{m+3}{m+5}\frac{6a_m - 4a_{m-1} + a_{m-2}}{-4a_m + a_{m-1}} + 1\right)b_{m+4}$$

$$+ \left(\frac{m+3}{m+5}\frac{6a_m - 4a_{m-1} + a_{m-2}}{-4a_m + a_{m-1}} + 2\right)b_{m+3}$$

$$+ \frac{6a_m - 4a_{m-1} + a_{m-2}}{-4a_m + a_{m-1}}\frac{-4c_m + c_{m-1} - 4(m+4)(-3d_{m+2} + d_{m+1})}{12(m+3)(m+4)}$$

$$- [6c_m - 4c_{m-1} + c_{m-2} - 4(m+3)(3d_{m+2} - 3d_{m+1} + d_m)]/12(m+3)(m+4) \tag{9.121}$$

From Eq. (9.117), for $i \in \{4, \dots, m+1\}$, we obtain

$$b_i = \frac{i+3}{i+1}\frac{6a_{i-2} - 4a_{i-3} + a_{i-4}}{6a_{i-1} - 4a_{i-2} + a_{i-3}}b_{i+3} - \left[2\frac{i+3}{i+1}\frac{6a_{i-2} - 4a_{i-3} + a_{i-4}}{6a_{i-1} - 4a_{i-2} + a_{i-3}} + 1\right]b_{i+2}$$

$$+ \left[\frac{i+3}{i+1}\frac{6a_{i-2} - 4a_{i-3} + a_{i-4}}{6a_{i-1} - 4a_{i-2} + a_{i-3}} + 2\right]b_{i+1}$$

$$+ \frac{6a_{i-2} - 4a_{i-3} + a_{i-4}}{6a_{i-1} - 4a_{i-2} + a_{i-3}}\frac{6c_{i-1} - 4c_{i-2} + c_{i-3} - 4(i+2)(3d_{i+1} - 3d_i + d_{i-1})}{12(i+1)(i+2)}$$

$$- \frac{6c_{i-2} - 4c_{i-3} + c_{i-4}}{12(i+1)(i+2)} + 4(i+1)\frac{3d_1 - 3d_{i-1} + d_{i-2}}{12(i+1)(i+2)} \tag{9.122}$$

From Eqs. (9.116) and (9.117) with $i = 4$, we get

$$b_3 = \frac{3}{2}\frac{6a_1 - 4a_0}{6a_2 - 4a_1 + a_0}b_6 - \left[3\frac{6a_1 - 4a_0}{6a_2 - 4a_1 + a_0} + 1\right]b_5 + \left[\frac{3}{2}\frac{6a_1 - 4a_0}{6a_2 - 4a_1 + a_0} + 2\right]b_4$$

$$+ \frac{6a_1 - 4a_0}{6a_2 - 4a_1 + a_0}\frac{[6c_2 - 4c_1 + c_0 - 4(i+1)(3d_4 - 3d_3 + d_2)]}{240}$$

$$+ \frac{(-6c_1 + 4c_0)}{240} + \frac{16(3d_3 - 3d_2 + d_1)}{240} \tag{9.123}$$

Equation (9.115) results in

$$b_2 = \frac{5}{3}\frac{6a_0}{6a_1 - 4a_0}b_5 - \left[\frac{10}{3}\frac{6a_0}{6a_1 - 4a_0} + 1\right]b_4 + \left[\frac{5}{3}\frac{6a_0}{6a_1 - 4a_0} + 2\right]b_3$$

$$- \frac{6a_0}{6a_1 - 4a_0}\frac{[(-6c_1 + 4c_0) + 16(3d_3 - 3d_2 + d_1)]}{144}$$

$$+ \frac{[-6c_0 + 12(3d_2 - 3d_1 + d_0)]}{144} \tag{9.124}$$

Equations (9.105) and (9.106) lead to

$$b_1 = 2b_2 - b_3 + (d_1 - d_0)/3$$
$$b_0 = 2b_1 - b_2 + d_0/2 \tag{9.125}$$

9.1.8 Case of a Clamped–Pinned Beam

In this case, the mode shape is given by the following polynomial function: $W = 3\xi^2 - 5\xi^3 + 2\xi^4$, corresponding to the following values of the coefficients: $\alpha_0 = 0, \alpha_1 = 0, \alpha_2 = 3, \alpha_3 = -5, \alpha_4 = 2$. Equations (9.51)–(9.59) yield

from ξ^0: $2(6b_2 - 30b_1 + 24b_0) - 6d_0 = 0$ \hfill (9.126)

from ξ^1: $6(6b_3 - 30b_2 + 24b_1) - (12d_1 - 30d_0) = 0$ \hfill (9.127)

from ξ^2: $12(6b_4 - 30b_3 + 24b_2) + 3c_0 - 3\Omega^2 a_0 - (18d_2 - 45d_1 + 24d_0) = 0$ \hfill (9.128)

from ξ^3: $20(6b_5 - 30b_4 + 24b_3) + 3c_1 - 5c_0 - (24d_360d_2 + 32d_1)$

$$- \Omega^2(3a_1 - 5a_0) = 0 \tag{9.129}$$

. . .

from ξ^i: $(i+1)(i+2)(6b_{i+2} - 30b_{i+1} + 24b_i) - \Omega^2(3a_{i-2} - 5a_{i-3} + 2a_{i-4})$

$$+ 3c_{i-2} - 5c_{i-3} + 2c_{i-4} - (i+1)(6d_i - 15d_{i-1} + 8d_{i-2}) = 0$$

$$(9.130)$$

$$\dots$$

from ξ^{m+3}: $(m+4)(m+5)(-30b_{m+4} + 24b_{m+3}) - \Omega^2(-5a_m + 2a_{m-1})$

$$- 5c_m + 2c_{m-1} - (m+4)(-15d_{m+2} + 8d_{m+1}) = 0 \qquad (9.131)$$

from ξ^{m+4}: $12(m+5)(m+6)b_{m+4} + c_m - \Omega^2 a_m - 4(m+5)d_{m+2} = 0$

$$(9.132)$$

Equation (9.130) is valid for $i \in \{4, \dots, m+2\}$. It links the coefficients b_i, b_{i+1} and b_{i+2}. We obtain the same recursive form of equation as for the clamped–free beam. This implies that the general case to be considered is the case $m \geq 2$. Thus, we have to consider two cases: (1) the particular case $m \leq 1$ and (2) the general case $m > 1$. From Eq. (9.132) we obtain the closed-form solution for the natural frequency:

$$\omega^2 = \{12(m+5)(m+6)b_{m+4}/a_m + [c_m - 4(m+5)d_{m+2}]/a_m\}/L^4 \qquad (9.133)$$

This is the same expression as was derived in the case of a pinned–pinned beam as well as in the case of the clamped–free beam.

9.1.8.1   *Particular case: m ≤ 1*

In this case the inertial coefficient and the flexural rigidity read, respectively, as follows:

$$R(\xi) = a_0 \quad D(\xi) = b_0 + b_1\xi + b_2\xi^2 + b_3\xi^3 + b_4\xi^4 \qquad (9.134)$$

The Winkler foundation and the Pasternak foundation flexural rigidities are given, respectively, by

$$K_1(\xi) = c_0 \quad K_2(\xi) = d_0 + d_1\xi + d_2\xi^2 \qquad (9.135)$$

The governing equation results in a system of five equations with six unknowns b_0, b_1, b_2, b_3, b_4 and Ω^2. The system reads

$$12b_2 - 60b_1 + 48b_0 - 6d_0 = 0$$

$$36b_3 - 180b_2 + 144b_1 + 30d_0 - 12d_1 = 0$$

$$72b_4 - 360b_3 + 288b_2 - 24d_0 + 45d_1 - 18d_2 + 3c_0 - 3a_0\Omega^2 = 0 \qquad (9.136)$$

$$-600b_4 + 480b_3 + 60d_2 - 32d_1 - 5c_0 + 5a_0\Omega^2 = 0$$

$$360b_4 + c_0 - 20d_2 - a_0\Omega^2 = 0$$

This system yields the following solution:

$$b_0 = \tfrac{1}{4}(5b_1 - b_2) + \tfrac{1}{8}d_0$$
$$b_1 = \tfrac{1}{4}(5b_2 - b_3) + \tfrac{1}{12}d_1 - \tfrac{5}{24}d_0$$
$$b_2 = \tfrac{1}{4}(14b_4 + 5b_3) - \tfrac{7}{48}d_2 - \tfrac{5}{32}d_1 + \tfrac{1}{12}d_0$$
$$b_3 = -\tfrac{5}{2}b_4 + \tfrac{1}{15}d_1 + \tfrac{1}{12}d_2$$

(9.137)

Specifically, b_4 is taken to be an arbitrary constant; once the expression for b_3 is substituted into the equation for b_2, it is also expressed in term of b_4. Likewise, substituting the expression for b_2 and b_3 into those for b_0 and b_1 we get a final analytical representation for the flexural rigidity coefficients.

In the case of a constant inertial coefficient, the closed-form solution for the natural frequency is obtained from the last of Eqs. (9.140):

$$\omega^2 = (360b_4 + c_0 - 20d_2)/L^4 a_0$$

(9.138)

Note that the discussion following Eq. (9.26) appears to be equally pertinent here.

Now, consider now the case of a linearly varying function for the inertial coefficient. The inertial coefficient, the flexural rigidity, the coefficients of the Winkler foundation and the Pasternak foundation are written, respectively, as follows:

$$R(\xi) = a_0 + a_1\xi \quad D(\xi) = b_0 + b_1\xi + b_2\xi^2 + b_3\xi^3 + b_4\xi^4 + b_5\xi^5$$
$$K_1(\xi) = c_0 + c_1\xi \quad K_2(\xi) = d_0 + d_1\xi + d_2\xi^2 + d_3\xi^3$$

(9.139)

Substitution of Eq. (9.139) into Eq. (9.2) results in a system of six equations with seven unknowns, the coefficient Ω and the coefficients of the flexural rigidity b_j:

$$12b_2 - 60b_1 + 48b_0 - 6d_0 = 0$$

$$36b_3 - 180b_2 + 144b_1 + 30d_0 - 12d_1 = 0$$

$$72b_4 - 360b_3 + 288b_2 - 24d_0 + 45d_1 - 18d_2 + 3c_0 - 3a_0\Omega^2 = 0$$

$$120b_5 - 600b_4 + 480b_3 - 24d_3 + 60d_2 - 32d_1 + 3c_1 - 5c_0 - \Omega^2(3a_1 - 5a_0) = 0$$

$$-900b_5 + 720b_4 + 75d_3 - 40d_2 - 5c_1 + 2c_0 - \Omega^2(-5a_1 + 2a_0) = 0$$

$$504b_4 + c_1 - 24d_3 - a_1\Omega^2 = 0$$

(9.140)

The coefficient b_5 is taken to be a known constant. The coefficients b_i read as follows:

$$b_0 = \tfrac{1}{4}(5b_1 - b_2) + \tfrac{1}{8}d_0$$

$$b_1 = \tfrac{1}{4}(5b_2 - b_3) + \tfrac{1}{12}d_1 - \tfrac{5}{24}d_0$$

$$b_2 = [504a_0b_5 - 24a_1b_4 + 120a_1b_3 + a_0c_1 - a_1c_0 - 24a_0d_3$$
$$\qquad + a_1(6d_2 - 15d_1 + 8d_0)]/96a_1$$

$$b_3 = [(1392a_1 - 2520a_0)b_5 + 600a_1b_4 - 5a_0c_1 + 5a_1c_0$$
$$\qquad + (-48a_1 + 120a_0)d_3 + a_1(-60d_2 + 32d_1)]/480a_1$$

$$b_4 = [(-1620a_1 + 1008a_0)b_5 + 2a_0c_1 - 2a_1c_0 + (45a_1 - 48a_0)d_3$$
$$\qquad + 40a_1d_2]/720a_1$$

(9.141)

The natural frequency is obtained from the last of Eqs. (9.140):

$$\omega^2 = (504b_5 + c_1 - 24d_3)/a_1 \qquad\qquad (9.142)$$

Again, $b_5 > (24d_3 - c_1)/504$; moreover $c_1 > 24d_3$. The two particular cases having been treated, we turn now to the general case $m > 1$.

9.1.8.2 General case

The equations expressing the natural frequency coefficient are obtained from Eqs. (9.128)–(9.132):

$$\Omega^2 = [12(6b_4 - 30b_3 + 24b_2) + 3c_0 - (18d_2 - 45d_1 + 24d_0)]/3a_0 \qquad (9.143)$$

$$\Omega^2 = [20(6b_5 - 30b_4 + 24b_3) + 3c_1 - 5c_0 - (24d_3 + 60d_2 + 32d_1)]/(3a_1 - 5a_0) \qquad (9.144)$$

$$\cdots$$

$$\Omega^2 = (i + 1)(i + 2)(6b_{i+2} - 30b_{i+1} + 24b_i)/(3a_{i-2} - 5a_{i-3} + 2a_{i-4})$$
$$\qquad + [3c_{i-2} - 5c_{i-3} + 2c_{i-4} - (i + 1)(6d_i - 15d_{i-1} + 8d_{i-2})]$$
$$\qquad /(3a_{i-2} - 5a_{i-3} + 2a_{i-4}) \qquad (9.145)$$

$$\cdots$$

$$\Omega^2 = (m + 4)(m + 5)(-30b_{m+4} + 24b_{m+3})/(-5a_m + 2a_{m-1})$$
$$\qquad + [-5c_m + 2c_{m-1} - (m + 4)(-15d_{m+2} + 8d_{m+1})]/(-5a_m + 2a_{m-1}) \qquad (9.146)$$

$$\Omega^2 = [12(m + 5)(m + 6)b_{m+4} + c_m - 4(m + 5)d_{m+2}]/a_m \qquad (9.147)$$

Comparison of Eqs. (9.147) and (9.146) yields

$$
b_{m+3} = \left[\frac{m+6}{m+4} \left(\frac{a_{m-1}}{a_m} - \frac{5}{2} \right) + \frac{5}{4} \right] b_{m+4} + \left(\frac{2a_{m-1}}{a_m} - 5 \right) \frac{c_m - 4(m+5)d_{m+2}}{24(m+4)(m+5)}
$$
$$
+ \frac{[5c_m - 2c_{m-1} + (m+4)(-15d_{m+2} + 8d_{m+1})]}{24(m+4)(m+5)} \tag{9.148}
$$

We obtain b_{m+2} from Eq. (9.145), with $i = m + 2$, and Eq. (9.146):

$$
b_{m+2} = -\left(\frac{5}{4} \frac{m+5}{m+3} \frac{3a_m - 5a_{m-1} + 2a_{m-2}}{-5a_m + 2a_{m-1}} + \frac{1}{4} \right) b_{m+4}
$$
$$
+ \left(\frac{m+5}{m+3} \frac{3a_m - 5a_{m-1} + 2a_{m-2}}{-5a_m + 2a_{m-1}} + \frac{5}{4} \right) b_{m+3} + \frac{3a_m - 5a_{m-1} + 2a_{m-2}}{-5a_m + 2a_{m-1}}
$$
$$
\times \frac{-5c_m + 2c_{m-1} - (m+4)(-15d_{m+2} + 8d_{m+1})}{24(m+3)(m+4)}
$$
$$
- \frac{3c_m - 5c_{m-1} + 2c_{m-2} - (m+3)(6d_{m+2} - 15d_{m+1} + 8d_m)}{24(m+3)(m+4)} \tag{9.149}
$$

From Eq. (9.145), we express b_i in terms of b_{i+1}, b_{i+2} and b_{i+3}. This expression is valid for $i \in \{4, \ldots, m+1\}$:

$$
b_i = \frac{i+3}{4(i+1)} \frac{3a_{i-2} - 5a_{i-3} + 2a_{i-4}}{3a_{i-1} - 5a_{i-2} + 2a_{i-3}} b_{i+3} - \left(\frac{5}{4} \frac{i+3}{i+1} \frac{3a_{i-2} - 5a_{i-3} + 2a_{i-4}}{3a_{i-1} - 5a_{i-2} + 2a_{i-3}} + \frac{1}{4} \right) b_{i+2}
$$
$$
+ \left(\frac{i+3}{i+1} \frac{3a_{i-2} - 5a_{i-3} + 2a_{i-4}}{3a_{i-1} - 5a_{i-2} + 2a_{i-3}} + \frac{5}{4} \right) b_{i+1} + \frac{3a_{i-2} - 5a_{i-3} + 2a_{i-4}}{3a_{i-1} - 5a_{i-2} + 2a_{i-3}}
$$
$$
\times \frac{3c_{i-1} - 5c_{i-2} + 2c_{i-3} - (i+2)(6d_{i+1} - 15d_i + 8d_{i-1})}{24(i+1)(i+2)}
$$
$$
- \frac{[3c_{i-2} - 5c_{i-3} + 2c_{i-4} - (i+1)(6d_i - 15d_{i-1} + 8d_{i-2})]}{24(i+1)(i+2)} \tag{9.150}
$$

Eqs. (9.144) and (9.145) lead to

$$
b_3 = \frac{3}{8} \frac{3a_1 - 5a_0}{3a_2 - 5a_1 + 2a_0} b_6 - \left(\frac{15}{8} \frac{3a_1 - 5a_0}{3a_2 - 5a_1 + 2a_0} + \frac{1}{4} \right) b_5
$$
$$
+ \left(\frac{3}{2} \frac{3a_1 - 5a_0}{3a_2 - 5a_1 + 2a_0} + \frac{5}{4} \right) b_4
$$
$$
+ \frac{3a_1 - 5a_0}{3a_2 - 5a_1 + 2a_0} \frac{[3c_2 - 5c_1 + 2c_0 - 5(6d_4 - 15d_3 + 8d_2)]}{480}
$$
$$
- \frac{[3c_1 - 5c_0 - (24d_3 + 60d_2 + 32d_1)]}{480} \tag{9.151}
$$

Then, b_2 is obtained from Eqs. (9.144) and (9.145):

$$b_2 = \frac{5}{12}\frac{3a_0}{3a_1 - 5a_0}b_5 - \left(\frac{25}{12}\frac{3a_0}{3a_1 - 5a_0} + \frac{1}{4}\right)b_4 + \left(\frac{5}{3}\frac{3a_0}{3a_1 - 5a_0} + \frac{5}{4}\right)b_3$$
$$+ \frac{3a_0}{3a_1 - 5a_0}\frac{[3c_1 - 5c_0 - (24d_3 + 60d_2 + 32d_1)]}{288}$$
$$- \frac{[3c_0 - (18d_2 - 45d_1 + 24d_0)]}{288} \qquad (9.152)$$

The coefficients b_0 and b_1 are derived directly from Eqs. (9.130) and (9.131):

$$b_1 = (5b_2 - b_3)/4 + d_1/12 - 5d_0/24$$
$$b_0 = (5b_1 - b_2)/4 + d_0/8 \qquad (9.153)$$

For $m = 2$, we do not use Eq. (9.150) because it is valid only for $m \geq 3$. The coefficients b_i with $i \in \{0, 1, 2, 3, 4, 5\}$ are given, respectively, by Eqs. (9.157), (9.152), (9.151), (9.149) and (9.152). To be valid, the above expressions imply some relations between the coefficients a_i: (a) Eq. (9.143) is true only if $a_0 \neq 0$; (b) Eq. (9.152) imposes the condition $a_1 \neq 5a_0/3$; (c) Eqs. (9.150) and (9.155) result in $a_{i-1} \neq (5a_{i-2} - 2a_{i-3})/3$ for $i \in \{3, \ldots, m + 1\}$ (d) Eq. (9.149) is valid if $a_{m-1} \neq 5a_m/2$; (e) from Eq. (9.148), we obtain $a_m \neq 0$.

To summarize, if $m > 2$ and the above conditions are satisfied, then the beam has the postulated mode shape $W = 3\xi^2 - 5\xi^3 + 2\xi^4$. The closed-form solution for the natural frequency is given by Eq. (9.138), while the flexural rigidity is given by Eqs. (9.148)–(9.153).

9.1.9   Case of a Clamped–Clamped Beam

This case corresponds to the following mode shape:

$$W = \xi^2 - 2\xi^3 + \xi^4 \qquad (9.154)$$

The values for the coefficients α_i are: $\alpha_0 = 0, \alpha_1 = 0, \alpha_2 = 3, \alpha_3 = -5, \alpha_4 = 2$. The first step is to formulate the system of $m + 5$ equations with $m + 6$ unknowns resulting from the governing equation:

ξ^0: $2(2b_2 - 12b_1 + 12b_0) - 2d_0 = 0$ \qquad (9.155)

ξ^1: $6(2b_3 - 12b_2 + 12b_1) - (4d_1 - 12d_0) = 0$ \qquad (9.156)

ξ^2: $12(2b_4 - 12b_3 + 12b_2) + c_0 - (6d_2 - 18d_1 + 12d_0) - \Omega^2 a_0 = 0$ \quad (9.157)

ξ^3: $20(2b_5 - 12b_4 + 12b_3) + c_1 - 2c_0$

$\qquad - (8d_3 - 24d_2 + 16d_1) - \Omega^2(a_1 - 2a_0) = 0$ (9.158)

$$\cdots$$

ξ^i: $(i+1)(i+2)(2b_{i+2} - 12b_{i+1} + 12b_i) + c_{i-2} - 2c_{i-3} + c_{i-4}$

$\qquad - (i+1)(2d_i - 6d_{i-1} + 4d_{i-2}) - \Omega^2(a_{i-2} - 2a_{i-3} + a_{i-4}) = 0$ (9.159)

$$\cdots$$

ξ^{m+3}: $(m+4)(m+5)(-12b_{m+4} + 12b_{m+3}) - 2c_m + c_{m-1}$

$\qquad - (m+4)(-6d_{m+2} + 4d_{m+1}) - \Omega^2(-2a_m + a_{m-1}) = 0$ (9.160)

ξ^{m+4}: $12(m+5)(m+6)b_{m+4} + (c_m - \Omega^2 a_m) - 4(m+5)d_{m+2} = 0$ (9.161)

Since the number of equations is less than the number of the sought for coefficients, we take one sought for coefficient, for example b_{m+4}, to be a known constant. Equation (9.159) has the same form as Eq. (9.130). It yields a relation between three consecutive coefficients b_i, valid for $i \in \{4, \ldots, m+2\}$. Hence, we have to consider the same particular cases as was done for the 3 clamped–free beam. As a result, we treat two cases: (1) the particular case $m \leq 1$ and (2) the general case $m > 1$.

Before we start with the particular case, we note that Eq. (9.161) allows one to directly get the closed-form solution for the natural frequency. This expression is the same as for the previous sets of boundary conditions (see Eq. (9.133) for example).

9.1.9.1 Particular cases ($m \leq 1$)

The case $m = 0$ corresponds to the classic case of a constant inertial coefficient $R(\xi) = a_0$. The flexural rigidity and the foundation coefficients are written as follows:

$$D(\xi) = \sum_{i=0}^{5} b_i \xi^i \quad K_1(\xi) = c_0 \quad K_2(\xi) = \sum_{i=0}^{2} d_i \xi^i \quad (9.162)$$

From Eq. (9.2), we obtain a system of five equations with six unknowns:

$4b_2 - 24b_1 + 24b_0 - 2d_0 = 0$

$12b_3 - 72b_2 + 72b_1 - 4d_1 + 12d_0 = 0$

$24b_4 - 144b_3 + 144b_2 + c_0 - 6d_2 + 18d_1 - 12d_0 - a_0\Omega^2 = 0$ (9.163)

$-240b_4 + 240b_3 - 2c_0 + 24d_2 - 16d_1 + 2a_0\Omega^2 = 0$

$360b_4 + c_0 - 20d_2 - a_0\Omega^2 = 0$

To assure the compatibility of this system, the coefficients b_i must be expressed as follows:

$$b_0 = -\tfrac{1}{6}b_2 + b_1 + \tfrac{1}{12}d_0$$
$$b_1 = -\tfrac{1}{6}b_3 + b_2 + \tfrac{1}{18}d_1 - \tfrac{1}{6}d_0$$
$$b_2 = \tfrac{7}{3}b_4 + b_3 - \tfrac{7}{72}d_2 - \tfrac{1}{8}d_1 + \tfrac{1}{12}d_0 \tag{9.164}$$
$$b_3 = -2b_4 + \tfrac{1}{15}(d_2 + d_1)$$

The natural frequency reads

$$\omega^2 = (360b_4 + c_0 - 20d_2)/a_0 \tag{9.165}$$

with conclusions following Eq. (9.26) to be equally applicable here. The second case is $m = 1$. Substituting the expressions

$$R(\xi) = a_0 + a_1\xi \quad D(\xi) = \sum_{i=0}^{6} b_i\xi^i \quad K_1(\xi) = c_0 + c_1\xi \quad K_2(\xi) = \sum_{i=0}^{3} d_i\xi^i \tag{9.166}$$

into Eq. (9.2) leads to a system of six equations with seven unknowns:

$$4b_2 - 24b_1 + 24b_0 - 2d_0 = 0$$
$$12b_3 - 72b_2 + 72b_1 - 4d_1 + 12d_0 = 0$$
$$24b_4 - 144b_3 + 144b_2 + c_0 - 6d_2 + 18d_1 - 12d_0 - a_0\Omega^2 = 0$$
$$40b_5 - 240b_4 + 240b_3 + c_1 - 2c_0 - 8d_3 + 24d_2 - 16d_1 - \Omega^2(a_1 - 2a_0) = 0$$
$$-360b_5 + 360b_4 - 2c_1 + c_0 + 30d_3 - 20d_2 - \Omega^2(-2a_1 + a_0) = 0$$
$$504b_5 + c_1 - 24d_3 - \Omega^2 a_1 = 0 \tag{9.167}$$

The coefficient b_5 is taken to be a known parameter. The solution of the above system is

$$b_0 = -\tfrac{1}{6}b_2 + b_1 + \tfrac{1}{12}d_0$$
$$b_1 = -\tfrac{1}{6}b_3 + b_2 + \tfrac{1}{18}d_1 - \tfrac{1}{6}d_0$$

$b_2 = \frac{1}{144}(504a_0b_5 - 24a_1b_4 + 144a_1b_3 + a_0c_1 - a_1c_0 - 24a_0d_3$
$\qquad + 6a_1d_2 - 18a_1d_1 + 12a_1d_0)/a_1$

$b_3 = \frac{1}{120}[(232a_1 - 504a_0)b_5 + 120a_1b_4 - a_0c_1 + a_1c_0 + (-8a_1 + 24a_0)d_3$
$\qquad - 12a_1d_2 + 8a_1d_1]/a_1$

$b_4 = \frac{1}{360}[(-648a_1 + 504a_0)b_5 + a_0c_1 - a_1c_0 + (18a_1 - 24a_0)d_3 + 20a_1d_2]/a_1$
$$\tag{9.168}$$

The closed-form solution for the natural frequency is obtained directly from the last equation in the set (9.167):

$$\omega^2 = (504b_5 + c_1 - 24d_3)/a_1 \tag{9.169}$$

with the conclusions following Eq. (9.43) to be valid in this instance too.

We turn to the general case when the inertial coefficient is described by a polynomial function of degree $m > 1$.

9.1.9.2   General case (m > 1)

In this case, the expressions for the natural frequency squared, arising from Eqs. (9.157)–(9.161), are

$$\Omega^2 = [12(2b_4 - 12b_3 + 12b_2) + c_0 - (6d_2 - 18d_1 + 12d_0)]/a_0 \tag{9.170}$$

$$\Omega^2 = [20(2b_5 - 12b_4 + 12b_3) + c_1 - 2c_0 - (8d_3 - 24d_2 + 16d_1)]/(a_1 - 2a_0) \tag{9.171}$$

$$\Omega^2 = [(i+1)(i+2)(2b_{i+2} - 12b_{i+1} + 12b_i) + c_{i-2} - 2c_{i-3} + c_{i-4}$$
$$\qquad - (i+1)(2d_i - 6d_{i-1} + 4d_{i-2})]/(a_{i-2} - 2a_{i-3} + a_{i-4}) \tag{9.172}$$

$$\Omega^2 = (m+4)(m+5)(-12b_{m+4} + 12b_{m+3}) - 2c_m + c_{m-1}$$
$$\qquad - (m+4)(-6d_{m+2} + 4d_{m+1})/(-2a_m + a_{m-1}) \tag{9.173}$$

$$\Omega^2 = [12(m+5)(m+6)b_{m+4} + c_m - 4(m+5)d_{m+2}]/a_m \tag{9.174}$$

The coefficient b_{m+3} is obtained from Eqs. (9.173) and (9.174):

$$b_{m+3} = \left[\frac{m+6}{m+4}\left(\frac{a_{m-1}}{a_m} - 2\right) + 1\right]b_{m+4} + \left(\frac{a_{m-1}}{a_m} - 2\right)\frac{c_m - 4(m+5)d_{m+2}}{12(m+4)(m+5)}$$
$$\qquad + \frac{2c_m - c_{m-1} + (m+4)(-6d_{m+2} + 4d_{m+1})}{12(m+4)(m+5)} \tag{9.175}$$

Equation (9.172), with $i = m + 2$, and Eq. (9.173) allow us to express b_{m+2} as

$$
\begin{aligned}
b_{m+2} = & -\left(\frac{m+5}{m+3}\frac{a_m - 2a_{m-1} + a_{m-2}}{-2a_m + a_{m-1}} + \frac{1}{6}\right) b_{m+4} \\
& + \left(\frac{m+5}{m+3}\frac{a_m - 2a_{m-1} + a_{m-2}}{-2a_m + a_{m-1}} + 1\right) b_{m+3} + \frac{a_m - 2a_{m-1} + a_{m-2}}{-2a_m + a_{m-1}} \\
& \times \frac{-2c_m + c_{m-1} - (m+4)(-6d_{m+2} + 4d_{m+1})}{12(m+3)(m+4)} \\
& - \frac{c_m - 2c_{m-1} + c_{m-2} - (m+3)(2d_{m+2} - 6d_{m+1} + 4d_m)}{12(m+3)(m+4)}
\end{aligned} \tag{9.176}
$$

Equation (9.172) results in

$$
\begin{aligned}
b_i = & \frac{1}{6}\frac{i+3}{i+1}\frac{a_{i-2} - 2a_{i-3} + a_{i-4}}{a_{i-1} - 2a_{i-2} + a_{i-3}} b_{i+3} - \left(\frac{i+3}{i+1}\frac{a_{i-2} - 2a_{i-3} + a_{i-4}}{a_{i-1} - 2a_{i-2} + a_{i-3}} + \frac{1}{6}\right) b_{i+2} \\
& + \left(\frac{i+3}{i+1}\frac{a_{i-2} - 2a_{i-3} + a_{i-4}}{a_{i-1} - 2a_{i-2} + a_{i-3}} + 1\right) b_{i+1} \\
& + \frac{a_{i-2} - 2a_{i-3} + a_{i-4}}{a_{i-1} - 2a_{i-2} + a_{i-3}} \frac{-2c_{i-2} + c_{i-3} - (i+2)(2d_{i+1} - 6d_i + 4d_{i-1})}{12(i+1)(i+2)} \\
& - \frac{c_{i-2} - 2c_{i-3} + c_{i-4} - (i+1)(2d_i - 6d_{i-1} + 4d_{i-2})}{12(i+1)(i+2)}
\end{aligned} \tag{9.177}
$$

Note that Eq. (9.177) is valid for $i \in \{4, \ldots, m + 1\}$. Equations (9.171) and (9.172), with $i = 4$, lead to the following expression for the coefficient b_3:

$$
\begin{aligned}
b_3 = & \frac{1}{4}\frac{a_1 - 2a_0}{a_2 - 2a_1 + a_0} b_6 - \left(\frac{3}{2}\frac{a_1 - 2a_0}{a_2 - 2a_1 + a_0} + \frac{1}{6}\right) b_5 + \left(\frac{3}{2}\frac{a_1 - 2a_0}{a_2 - 2a_1 + a_0} + 1\right) b_4 \\
& + \frac{1}{240}\frac{a_1 - 2a_0}{a_2 - 2a_1 + a_0}[c_2 - 2c_1 + c_0 - 5(2d_4 - 6d_3 + 4d_2)] \\
& - \frac{1}{240}[c_1 - 2c_0 - (8d_3 - 24d_2 + 16d_1)]
\end{aligned} \tag{9.178}
$$

Equations (9.170) and (9.171) yield

$$
\begin{aligned}
b_2 = & \frac{5}{18}\frac{a_0}{a_1 - 2a_0} b_5 - \left(\frac{5}{3}\frac{a_0}{a_1 - 2a_0} + \frac{1}{6}\right) b_4 + \left(\frac{5}{3}\frac{a_0}{a_1 - 2a_0} + 1\right) b_3 \\
& + \frac{1}{144}\frac{a_0}{a_1 - 2a_0}[c_1 - 2c_0 - (8d_3 - 24d_2 + 16d_1)] \\
& - \frac{1}{144}[c_0 - (6d_2 - 18d_1 + 12d_0)]
\end{aligned} \tag{9.179}
$$

We obtain the coefficients b_0 and b_1 from Eqs. (9.155) and (9.156):

$$b_1 = -\tfrac{1}{6}b_3 + b_2 + \tfrac{1}{18}d_1 - \tfrac{1}{6}d_0$$

$$b_0 = -\tfrac{1}{6}b_2 + b_1 + \tfrac{1}{12}d_0 \tag{9.180}$$

It must be noted that when $m = 2$, we do not need to use Eq. (9.181). This equation is employed only when $m > 2$. Moreover, the expressions given by Eqs. (9.170)–(9.180) are valid only if the coefficients a_i satisfy some auxiliary conditions:

$$\begin{array}{ll}
\text{(a)} & a_{m-1} \neq 2a_m \\
\text{(b)} & a_{i-1} - 2a_{i-2} + a_{i-3} \neq 0 \qquad \text{for } 3 \leq i \leq m+1 \\
\text{(c)} & a_1 \neq 2a_0 \\
\text{(d)} & a_0 \neq 0
\end{array} \tag{9.181}$$

9.1.10 Case of a Guided–Pinned Beam

In this case, the mode shape is given by

$$W = 5 - 6\xi^2 + \xi^4 \tag{9.182}$$

with coefficients α_i taking the values $\alpha_0 = 5, \alpha_1 = 0, \alpha_2 = -6, \alpha_3 = 0, \alpha_4 = 1$. In this case, we obtain a system of $m + 5$ equations with $m + 6$ unknowns, by substituting the previously obtained values of the coefficients α_i into Eqs. (9.51)–(9.59):

from ξ^0: $24(b_0 - b_2) + 5c_0 + 12d_0 - \Omega^2 5a_0 = 0$ \hfill (9.183)

from ξ^1: $72(b_1 - b_3) + 5c_1 + 24d_1 - \Omega^2 5a_1 = 0$ \hfill (9.184)

from ξ^2: $144(b_2 - b_4) + 5c_2 - 6c_0 - (-36d_2 + 12d_0) - \Omega^2(5a_2 - 6a_0) = 0$ \hfill (9.185)

from ξ^3: $240(b_3 - b_5) + 5c_3 - 6c_1 - (-48d_3 + 16d_1) - \Omega^2(5a_3 - 6a_1) = 0$ \hfill (9.186)

from ξ^i: $12(i+1)(i+2)(b_i - b_{i+2}) + 5c_i - 6c_{i-2} + c_{i-4}$

$\qquad - (i+1)(-12d_i + 4d_{i-2}) - \Omega^2(5a_i - 6a_{i-2} + a_{i-4}) = 0$ \hfill (9.187)

from ξ^{m+1}: $12(m+2)(m+3)(b_{m+1} - b_{m+3}) - 6c_{m-1} + c_{m-3}$

$\qquad - (m+2)(-12d_{m+1} + 4d_{m-1}) - \Omega^2(-6a_{m-1} + a_{m-3}) = 0$ \hfill (9.188)

from ξ^{m+2}: $12(m+3)(m+4)(b_{m+2} - b_{m+4}) - 6c_m + c_{m-2}$

$\qquad - (m+3)(-12d_{m+2} + 4d_m) - \Omega^2(-6a_m + a_{m-2}) = 0$ \hfill (9.189) from ξ^{m+3}: $12(m+4)(m+5)b_{m+3} + c_{m-1} - 4(m+4)d_{m+1} - \Omega^2 a_{m-1} = 0$

(9.190)

from ξ^{m+4}: $12(m+5)(m+6)b_{m+4} + c_m - 4(m+5)d_{m+2} - \Omega^2 a_m = 0$

(9.191)

From Eq. (9.191), we get the same closed-form solution for the natural frequency:

$$\omega^2 = \{12(m+5)(m+6)b_{m+4}/a_m + [c_m - 4(m+5)d_{m+2}]/a_m\}/L^4 \quad (9.192)$$

It must be stressed that the discussion following Eq. (9.60) is equally valid here too.

This case is somewhat different from the previous cases because the recursive equation (9.187) allows us to express b_i in terms of b_{i+2}. This equation is valid for $4 \leq i \leq m$. Thus, to express b_m (and hence b_{m-1}), we must use Eq. (9.187), with $i = m$, and Eq. (9.189) and hence Eq. (9.187), with $i = m - 1$, and Eq. (9.188). At this juncture, we have obtained the general expression for b_{m+3} from Eqs. (9.190) and (9.191), for b_{m+2} from Eqs. (9.189) and (9.191), and for b_{m+1} from Eqs. (9.192) and (9.194). The coefficients b_m and b_{m-1} are obtained, respectively, from Eq. (9.187), with $i = m$, and Eq. (9.193), and from Eq. (9.187), with $i = m - 1$, and Eq. (9.188). The coefficient b_3 is calculated from Eqs. (9.186) and (9.187) with $i = 5$. The coefficient b_2 is evaluated with the aid of Eqs. (9.185) and Eq. (9.187) specified for $i = 4$. Then, the remaining coefficients b_1 and b_0 are given, respectively, by Eq. (9.182) and (9.186), and by Eq. (9.183) and (9.185). This implies that prior to facilitating the general case, we have to treat particular cases $m = i$ for $i \in \{0, 1, 2, 3, 4\}$.

9.1.10.1   *Particular cases* $(0 \leq m \leq 4)$

The case of a constant inertial coefficient results in a system of five equations with six unknowns, the coefficients of the flexural rigidity and the natural frequency. To solve this system,

$$24(b_0 - b_2) + 5c_0 + 12d_0 - 5a_0\Omega^2 = 0$$

$$72(b_1 - b_3) + 24d_1 = 0$$

$$144(b_2 - b_4) - 6c_0 + 36d_2 - 12d_0 + 6a_0\Omega^2 = 0 \qquad (9.193)$$

$$240b_3 - 16d_1 = 0$$

$$360b_4 + c_0 - 20d_2 - a_0\Omega^2 = 0$$

One of the coefficients, namely b_4, is taken as a specified parameter:

$$b_0 = b_2 + 75b_4 - \tfrac{25}{6}d_2 - \tfrac{1}{2}d_0$$

$$b_1 = b_3 - \tfrac{1}{3}d_1$$

$$b_2 = -14b_4 + \tfrac{7}{12}d_2 + \tfrac{1}{12}d_0 \tag{9.194}$$

$$b_3 = \tfrac{1}{15}d_1$$

The natural frequency reads

$$\omega^2 = (360b_4 + c_0 - 20d_2)/a_0 \tag{9.195}$$

with the conclusions associated with Eq. (9.26) to be equally pertinent here.

The second case is $m = 1$. In this case, the inertial coefficient is a linear function. This implies that the flexural rigidity, the Winkler foundation and the Pasternak foundation coefficients, respectively, are represented by polynomial functions of degree 5, 1 and 3. The governing equation yields a system of six equations with sevens unknowns:

$$24(b_0 - b_2) + 5c_0 + 12d_0 - 5a_0\Omega^2 = 0$$

$$72(b_1 - b_3) + 5c_1 + 24d_1 - 5a_1\Omega^2 = 0$$

$$144(b_2 - b_4) - 6c_0 + 36d_2 - 12d_0 + 6a_0\Omega^2$$

$$240(b_3 - b_5) - 6c_1 + 48d_3 - 16d_1 + 6a_1\Omega^2 = 0 \tag{9.196}$$

$$360b_4 + c_0 - 20d_2 - a_0\Omega^2 = 0$$

$$504b_5 + c_1 - 24d_3 - a_1\Omega^2 = 0$$

The number of unknowns being higher than the number of equations, we have to take b_5, for example, as a specified constant. Thus, the following system is obtained:

$$b_0 = \tfrac{1}{24}[2520a_0b_5 + 24a_1b_2 + 5a_0c_1 - 5a_1c_0 - 120a_0d_3 - 12a_1d_0]/a_1$$

$$b_1 = 35b_5 + b_3 - \tfrac{5}{3}d_3 - \tfrac{1}{3}d_1$$

$$b_2 = \tfrac{1}{24}[-504a_0b_5 + 24a_1b_4 - a_0c_1 + a_1c_0 + 24a_0d_3 - 6a_1d_2 + 2a_1d_0]/a_1$$

$$b_3 = -\tfrac{58}{5}b_5 + \tfrac{2}{3}d_3 + \tfrac{1}{15}d_1$$

$$b_4 = \tfrac{1}{360}[504a_0b_5 + a_0c_1 - a_1c_0 - 24a_0d_3 + 20d_2a_1]/a_1$$

$$\tag{9.197}$$

The natural frequency is given by

$$\omega^2 = (504b_5 + c_1 - 24d_3)/a_1 \qquad (9.198)$$

Note that this equation coincides with Eq. (9.43); the conclusions following Eq. (9.43) are applicable here too.

The third particular case is $m = 2$. We have the following expressions for the pertinent functions:

$$R(\xi) = a_0 + a_1\xi + a_2\xi^2 \quad D(\xi) = b_0 + b_1\xi + b_2\xi^2 + b_3\xi^3 + b_4\xi^4 + b_5\xi^5 + b_6\xi^6$$

$$K_1(\xi) = c_0 + c_1\xi + c_2\xi^2 \quad K_2(\xi) = d_0 + d_1\xi + d_2\xi^2 + d_3\xi^3 + d_4\xi^4$$

$$(9.199)$$

The governing equation results in a system of seven equations with eight unknowns:

$$24(b_0 - b_2) + 5c_0 + 12d_0 - 5a_0\Omega^2 = 0$$

$$72(b_1 - b_3) + 5c_1 + 24d_1 - 5a_1\Omega^2 = 0$$

$$144(b_2 - b_4) + 5c_2 - 6c_0 + 36d_2 - 12d_0 - \Omega^2(5a_2 - 6a_0) = 0$$

$$240(b_3 - b_5) - 6c_1 + 48d_3 - 16d_1 + 6a_1\Omega^2 = 0 \qquad (9.200)$$

$$360(b_4 - b_6) - 6c_2 + c_0 + 60d_4 - 20d_2 - \Omega^2(-6c_2 + a_0) = 0$$

$$504b_5 + c_1 - 24d_3 - a_1\Omega^2 = 0$$

$$672b_6 + c_2 - 28d_4 - a_2\Omega^2 = 0$$

The unknowns, to ensure the compatibility of the system, must read, once the coefficient b_6 is treated as a specified constant, as follows:

$$b_0 = \tfrac{1}{24}(3360a_0b_6 + 24a_2b_2 + 5a_0c_2 - 5a_2c_0 - 140a_0d_4 - 12a_2d_0)/a_2$$

$$b_1 = \tfrac{1}{72}(3360a_1b_6 + 72a_2b_3 + 5a_1c_2 - 5a_2c_1 - 140a_1d_4 - 24a_2d_1)/a_2$$

$$b_2 = \tfrac{1}{72}[(1680a_2 - 2016a_0)b_6 + 72a_2b_4$$
$$\qquad - 3a_0c_2 + 3a_2c_0 + (-70a_2 + 84a_0)d_4 - 18a_2d_2 + 6a_2d_0]/a_2$$

$$b_3 = \tfrac{1}{120}(-2016a_1b_6 + 120a_2b_5 - 3a_1c_2 + 3a_2c_1 + 84a_1d_4 - 24a_2d_3 + 8a_1d_1)/a_2$$

$$b_4 = \tfrac{1}{360}[(-3672a_2 + 672a_0)b_6 + a_0c_2 - a_2c_0 + (108a_2 - 28a_0)d_4 + 20a_2d_2]/a_2$$

$$b_5 = \tfrac{1}{504}(672a_1b_6 + a_1c_2 - a_2c_1 - 28a_1d_4 + 24a_2d_3)/a_2$$

$$(9.201)$$

The closed-form solution for the natural frequency reads

$$\omega^2 = (672b_6 + c_2 - 28d_4)/a_2 \qquad (9.202)$$

with the following constraints: $b_6 > (28d_4 - c_2)/672$ and $c_2 > 28d_4$.
We now deal with the case $m = 3$. The attendant functions are

$$R(\xi) = \sum_{i=0}^{3} a_i \xi^i \quad D(\xi) = \sum_{i=0}^{7} b_i \xi^i \quad K_1(\xi) = \sum_{i=0}^{3} c_i \xi^i \quad K_2(\xi) = \sum_{i=0}^{5} d_i \xi^i \quad (9.203)$$

Hence, a system of eight equations with nine unknowns is derived from Eq. (9.2). The system reads

$$24(b_0 - b_2) + 5c_0 + 12d_0 - 5a_0\Omega^2 = 0$$

$$72(b_1 - b_3) + 5c_1 + 24d_1 - 5a_1\Omega^2 = 0$$

$$144(b_2 - b_4) + 5c_2 - 6c_0 + 36d_2 - 12d_0 - \Omega^2(5a_2 - 6a_0) = 0$$

$$240(b_3 - b_5) + 5c_3 - 6c_1 + 48d_3 - 16d_1 - \Omega^2(5a_3 - 6a_1) = 0$$

$$360(b_4 - b_6) - 6c_2 + c_0 + 60d_4 - 20d_2 - \Omega^2(-6a_2 + a_0) = 0 \qquad (9.204)$$

$$504(b_5 - b_7) - 6c_3 + c_1 + 72d_5 - 24d_3 - \Omega^2(-6a_3 + a_1) = 0$$

$$672b_6 + c_2 - 28d_4 - a_2\Omega^2 = 0$$

$$864b_7 + c_3 - 32d_5 - a_3\Omega^2 = 0$$

We take the coefficient b_7 to be specified constant and obtain

$$b_0 = \tfrac{1}{24}(4320a_0b_7 + 24a_3b_2 + 5a_0c_3 - 5a_3c_0 - 160a_0d_5 - 12a_3d_0)/a_3$$

$$b_1 = \tfrac{1}{72}(4320a_1b_7 + 72a_3b_3 + 5a_1c_3 - 5a_3c_1 - 160a_1d_5 - 24a_3d_1)/a_3$$

$$b_2 = \tfrac{1}{144}[(4320a_2 - 5184a_0)b_7 + 144a_3b_4 + (5a_2 - 6a_0)c_3 - 5a_3c_2$$
$$+ 6a_3c_0 + (-160a_2 + 192a_0)d_4 - 36a_3d_2 + 12a_3d_0]/a_3$$

$$b_3 = \tfrac{1}{120}[(2160a_3 - 2592a_1)b_7 + 120a_3b_5$$
$$- 3a_1c_3 + 3a_3c_1 + (-80a_3 + 96a_1)d_5 - 24a_3d_3 + 8a_3d_1]/a_3$$

$$b_4 = \tfrac{1}{360}[(-5184a_2 + 864a_0)b_7 + 360a_3b_6 + (-6a_2 + a_0)c_3$$
$$+ 6a_3c_2 - a_3c_0 + (192a_2 - 32a_0)d_5 - 60a_3d_4 + 20a_3d_2]/a_3$$

$$b_5 = \tfrac{1}{504}[(-4680a_3 + 864a_1)b_7 + a_1c_3 - a_3c_1 + (120a_3 - 32a_1)d_4 + 24a_3d_3]/a_3$$

$$b_6 = \tfrac{1}{672}(864a_2b_7 + a_2c_3 - a_3c_2 - 32a_2d_5 + 28a_3d_4)/a_3$$

$$(9.205)$$

The natural frequency is derived from the last equation of the set (9.204):

$$\omega^2 = (864b_7 + c_3 - 32d_5)/a_3 \tag{9.206}$$

Due to the conclusion following Eq. (9.60) we deduce that b_7 must exceed $(32d_5 - c_2)/864$, and $c_3 > 32d_5$.

The last case in the series of particular cases is $m = 4$:

$$R(\xi) = \sum_{i=0}^{4} b_i \xi^i \quad D(\xi) = \sum_{i=0}^{8} b_i \xi^i \quad K_1(\xi) = \sum_{i=0}^{4} c_i \xi^i \quad K_2(\xi) = \sum_{i=0}^{6} d_i \xi^i \tag{9.207}$$

The governing equation yields a system of nine equations with ten unknowns:

$$24(b_0 - b_2) + 5c_0 + 12d_0 - 5a_0\Omega^2 = 0$$
$$72(b_1 - b_3) + 5c_1 + 24d_1 - 5a_1\Omega^2 = 0$$
$$144(b_2 - b_4) + 5c_2 - 6c_0 + 36d_2 - 12d_0 - \Omega^2(5a_2 - 6a_0) = 0$$
$$240(b_3 - b_5) + 5c_3 - 6c_1 + 48d_3 - 16d_1 - \Omega^2(5a_3 - 6a_1) = 0$$
$$360(b_4 - b_6) + 5c_4 - 6c_2 + c_0 + 60d_4 - 20d_2 - \Omega^2(5a_4 - 6a_2 + a_0) = 0$$
$$504(b_5 - b_7) - 6c_3 + c_1 + 72d_5 - 24d_3 - \Omega^2(-6a_3 + a_1) = 0$$
$$672(b_6 - b_8) - 6c_4 + c_2 + 84d_6 - 28d_4 - \Omega^2(-6a_4 + a_2) = 0$$
$$864b_7 + c_3 - 32d_5 - a_3\Omega^2 = 0$$
$$1080b_8 + c_4 - 36d_6 - a_4\Omega^2 = 0$$

$$\tag{9.208}$$

The coefficients b_i are evaluated as follows:

$$b_0 = \tfrac{1}{24}(5400a_0b_8 + 24a_4b_2 + 5a_0c_4 - 5a_4c_0 - 160a_0d_6 - 12a_4d_0)/a_4$$

$$b_1 = \tfrac{1}{72}(5400a_1b_8 + 72a_4b_3 + 5a_1c_4 - 5a_4c_1 - 160a_1d_6 - 12a_4d_1)/a_4$$

$$b_2 = \tfrac{1}{144}[(5400a_2 - 6480a_0)b_8 + 144a_4b_4 + (5a_2 - 6a_0)c_4 - 5a_4c_2 + 6a_4c_0$$
$$+ (-180a_2 + 216a_0)d_5 - 36a_4d_2 + 12a_4d_0]/a_4$$

$$b_3 = \tfrac{1}{240}[(5400a_3 - 6480a_1)b_8 + 240a_4b_5 + (5a_3 - 6a_1)c_4 - 5a_4c_3 + 6a_4c_1$$
$$+ (-180a_3 + 216a_1)d_6 - 48a_4d_3 + 16a_4d_1]/a_4$$

$$b_4 = \tfrac{1}{360}[(5400a_4 - 6480a_2 - 1080a_0)b_8 + 360a_4b_6 + (-6a_2 + a_0)c_4$$
$$+ 6a_4c_2 - a_4c_0 + (-180a_4 + 216a_2 - 36a_0)d_6 - 60a_4d_4 + 20a_4d_2]/a_4$$

$$b_5 = \tfrac{1}{504}[(-6480a_3 + 1080a_1)b_8 + 504a_4b_7 + (-6a_3 + a_1)c_4 + 6a_4c_3 - a_4c_1$$
$$+ (216a_3 - 36a_1)d_6 - 72a_4d_5 + 24a_4d_3]/a_4$$

$$b_6 = \tfrac{1}{672}[(-5808a_4 + 1080a_2)b_8 + a_2c_4 - a_4c_2 + (132a_4 - 36a_2)d_6 + 28a_4d_4]/a_4$$

$$b_7 = \tfrac{1}{864}(1080a_3b_8 + a_3c_4 - a_4c_3 - 36a_3d_6 + 32a_4d_5)/a_4$$

$$(9.209)$$

The natural frequency reads

$$\omega^2 = (1080b_8 + c_4 - 36d_6)/L^4a_4 \qquad (9.210)$$

For its positivity, the coefficient b_8 must be in excess of the quantity $(36d_6 - c_4)/1080$. Additionally, the following inequality must hold: $c_4 > 36d_6$.

9.1.10.2 *General case (m > 4)*

In this case, we have a system of $m + 5$ equations with $m + 6$ unknowns. We rewrite Eqs. (9.180)–(9.190) in order to express the natural frequency coefficient squared, Ω^2, via each contributing equation. This leads to $m + 5$ different expressions of Ω^2, which permit us to express the different coefficients b_i by equating various expressions of Ω^2 with each other. These expressions, named compatibility equations, are

$$\Omega^2 = [24(b_0 - b_2) + 5c_0 + 12d_0]/5a_0 \qquad (9.211)$$

$$\Omega^2 = [72(b_1 - b_3) + 5c_1 + 24d_1]/5a_1 \qquad (9.212)$$

$$\Omega^2 = [144(b_2 - b_4) + 5c_2 - 6c_0 - (-36d_2 + 12d_0)]/(5a_2 - 6a_0) \qquad (9.213)$$

$$\Omega^2 = [240(b_3 - b_5) + 5c_3 - 6c_1 - (-48d_3 + 16d_1)]/(5a_3 - 6a_1) \qquad (9.214)$$

$$\dots$$

$$\Omega^2 = [12(i + 1)(i + 2)(b_i - b_{i+2}) + 5c_i - 6c_{i-2} + c_{i-4}$$
$$- (i + 1)(-12d_i + 4d_{i-2})]/(5a_i - 6a_{i-2} + a_{i-4}) \qquad (9.215)$$

$$\dots$$

$$\Omega^2 = [12(m + 2)(m + 3)(b_{m+1} - b_{m+3}) - 6c_{m-1} + c_{m-3}$$
$$- (m + 2)(-12d_{m+1} + 4d_{m-1})]/(-6a_{m-1} + a_{m-3}) \qquad (9.216)$$

$$\Omega^2 = [12(m+3)(m+4)(b_{m+2} - b_{m+4}) - 6c_m + c_{m-2}$$
$$- (m+3)(-12d_{m+2} + 4d_m)]/(-6a_m + a_{m-2}) \tag{9.217}$$

$$\Omega^2 = [12(m+4)(m+5)b_{m+3} + c_{m-1} - 4(m+4)d_{m+1}]/a_{m-1} \tag{9.218}$$

$$\Omega^2 = [12(m+5)(m+6)b_{m+4} + c_m - 4(m+5)d_{m+2}]/a_m \tag{9.219}$$

We express the coefficient b_{m+3} from Eqs. (9.218) and (9.219):

$$b_{m+3} = \frac{m+6}{m+4}\frac{a_{m-1}}{a_m}b_{m+4} + \frac{a_{m-1}}{a_m}\frac{c_m - 4(m+5)d_{m+2}}{12(m+4)(m+5)} - \frac{c_{m-1} - 4(m+4)d_{m+1}}{12(m+4)(m+5)} \tag{9.220}$$

Equations (9.217) and (9.219) result in

$$b_{m+2} = \left[\frac{(m+5)(m+6)}{(m+3)(m+4)}\frac{-6a_m + a_{m-2}}{a_m} + 1\right]b_{m+4}$$
$$+ \frac{-6a_m + a_{m-2}}{a_m}\frac{c_m - 4(m+5)d_{m+2}}{12(m+3)(m+4)}$$
$$- \frac{-6c_m + c_{m-2} - (m+3)(-12d_{m+2} + 4d_m)}{12(m+3)(m+4)} \tag{9.221}$$

Equations (9.216) and (9.218) lead to

$$b_{m+1} = \left[\frac{(m+4)(m+5)}{(m+2)(m+3)}\frac{-6a_{m-1} + a_{m-3}}{a_{m-1}} + 1\right]b_{m+3}$$
$$+ \frac{-6a_{m-1} + a_{m-3}}{a_{m-1}}\frac{c_{m-1} - 4(m+4)d_{m+1}}{12(m+2)(m+3)}$$
$$- \frac{-6c_{m-1} + c_{m-3} - (m+2)(-12d_{m+1} + 4d_{m-1})}{12(m+2)(m+3)} \tag{9.222}$$

With Eq. (9.215) evaluated for $i = m$ and Eq. (9.217), we get the expression for b_m:

$$b_m = \left[\frac{(m+3)(m+4)}{(m+1)(m+2)}\frac{5a_m - 6a_{m-2} + a_{m-4}}{-6a_m + a_{m-2}} + 1\right]b_{m+2}$$
$$- \frac{(m+3)(m+4)}{(m+1)(m+2)}\frac{5a_m - 6a_{m-2} + a_{m-4}}{-6a_m + a_{m-2}}b_{m+4}$$
$$+ \frac{5a_m - 6a_{m-2} + a_{m-4}}{-6a_m + a_{m-2}}\frac{-6c_m + c_{m-2} - (m+3)(-12d_{m+2} + 4d_m)}{12(m+1)(m+2)}$$
$$- \frac{[5c_m - 6c_{m-2} + c_{m-4} - (m+1)(-12d_m + 4d_{m-2})]}{12(m+1)(m+2)} \tag{9.223}$$

Equations (9.215) taken at $i = m - 1$ and Eq. (9.216) yield

$$
\begin{aligned}
b_{m-1} = {} & \left[\frac{(m+2)(m+3)}{m(m+1)} \frac{5a_{m-1} - 6a_{m-3} + a_{m-5}}{-6a_{m-1} + a_{m-3}} + 1 \right] b_{m+1} - \frac{(m+2)(m+3)}{m(m+1)} \\
& \times \frac{5a_{m-1} - 6a_{m-3} + a_{m-5}}{-6a_{m-1} + a_{m-3}} b_{m+3} + \frac{5a_m - 6a_{m-2} + a_{m-4}}{-6a_m + a_{m-2}} \\
& \times \frac{-6c_{m-1} + c_{m-3} - (m+2)(-12d_{m+1} + 4d_{m-1})}{12m(m+1)} \\
& - \frac{[5c_{m-1} - 6c_{m-3} + c_{m-5} - m(-12d_{m-1} + 4d_{m-3})]}{12m(m+1)}
\end{aligned}
\tag{9.224}
$$

At this stage, the coefficients b_{m+3}, b_{m+2}, b_{m+1}, b_m and b_{m-1} are evaluated. From Eq. (9.215), valid for $i \in \{4, \dots, m\}$, we can express the coefficient b_i for $i \in \{4, \dots, m-2\}$ in terms of b_{i+2} and b_{i+4} by equating Eq. (9.215) evaluated for i and $i + 2$:

$$
\begin{aligned}
b_i = {} & \left[\frac{(i+3)(i+4)}{(i+1)(i+2)} \frac{5a_i - 6a_{i-2} + a_{i-4}}{5a_{i+2} - 6a_i + a_{i-2}} + 1 \right] b_{i+2} \\
& - \frac{(i+3)(i+4)}{(i+1)(i+2)} \frac{5a_i - 6a_{i-2} + a_{i-4}}{5a_{i+2} - 6a_i + a_{i-2}} b_{i+4} \\
& + \frac{5a_i - 6a_{i-2} + a_{i-4}}{5a_{i+2} - 6a_i + a_{i-2}} \frac{5c_{i+2} - 6c_i + c_{i-2} - (i+3)(-12d_{i+2} + 4d_i)}{12(i+1)(i+2)} \\
& - \frac{5c_i - 6c_{i-2} + c_{i-4} - (i+1)(-12d_i + 4d_{i-2})}{12(i+1)(i+2)}
\end{aligned}
\tag{9.225}
$$

The coefficient b_3 is obtained through Eq. (9.214) and Eq. (9.215) specified for $i = 5$:

$$
\begin{aligned}
b_3 = {} & \left(\frac{21}{10} \frac{5a_3 - 6a_1}{5a_5 - 6a_3 + a_1} + 1 \right) b_5 - \frac{21}{10} \frac{5a_3 - 6a_1}{5a_5 - 6a_3 + a_1} b_7 \\
& + \frac{1}{240} \frac{5a_3 - 6a_1}{5a_5 - 6a_3 + a_1} [5c_5 - 6c_3 + c_1 - 6(-12d_5 + 4d_3)] \\
& - \frac{1}{240} [5c_3 - 6c_1 - (-48d_3 + 16d_1)]
\end{aligned}
\tag{9.226}
$$

We obtain the coefficient b_2 from Eqs. (9.213) and (9.215) with $i = 4$:

$$b_2 = \left(\frac{5}{2} \frac{5a_2 - 6a_0}{5a_4 - 6a_2 + a_0} + 1 \right) b_4 - \frac{5}{2} \frac{5a_2 - 6a_0}{5a_4 - 6a_2 + a_0} b_6$$

$$+ \frac{1}{144} \frac{5a_2 - 6a_0}{5a_4 - 6a_2 + a_0} [5c_4 - 6c_2 + c_0 - 5(-12d_4 + 4d_2)]$$

$$- \frac{1}{144} [5c_2 - 6c_0 - (-36d_2 + 12d_0)] \tag{9.227}$$

The coefficients b_0 and b_1 as obtained from Eqs. (9.211)–(9.214):

$$b_1 = \left(\frac{10}{3} \frac{5a_1}{5a_3 - 6a_1} + 1 \right) b_3 - \frac{10}{3} \frac{5a_1}{5a_3 - 6a_1} b_5$$

$$+ \frac{5a_1}{5a_3 - 6a_1} \frac{[5c_3 - 6c_1 - (-48d_3 + 16d_1)]}{72 - [5c_1 + 24d_1]/72}$$

$$b_0 = \left(6 \frac{5a_0}{5a_2 - 6a_0} + 1 \right) b_2 - 6 \frac{5a_0}{5a_2 - 6a_0} b_4 \tag{9.228}$$

$$+ \frac{1}{24} \frac{5a_0}{5a_2 - 6a_0} \frac{[5c_2 - 6c_0 - (-36d_2 + 12d_0)]}{24 - [5c_0 + 12d_0]}$$

We note that Eq. (9.225) is valid if $m > 4$. This explains why the case $m = 4$ belongs to the set of particular cases. Moreover, some relations must be satisfied by the coefficients a_i in order to use the above expressions. From Eqs. (9.219)–(9.226), we get

(a) $a_{m-1} \neq 0$
(b) $a_{m-2} \neq 6a_m$ $a_{m-3} \neq 6a_{m-1}$
(c) $5a_{i+2} - 6a_i + a_{i-2} \neq 0$ for $2 \leq i \leq m - 2$ (9.229)
(d) $a_3 \neq 6a_1/5$ $a_2 \neq 6a_0/5$
(e) $a_1 \neq 0$ $a_0 \neq 0$

9.1.11   Case of a Guided–Clamped Beam

The mode shape and the coefficients α_i in this case are

$$W = 4\xi^2 - 4\xi^3 + \xi^4$$

$$\alpha_0 = 0 \quad \alpha_1 = 0 \quad \alpha_2 = 4 \quad \alpha_3 = -4 \quad \alpha_4 = 1 \tag{9.230}$$

As for other sets of boundary conditions, the governing equation results in a system of $m + 5$ equations with $m + 6$ unknowns. We obtain it from

Eqs. (9.51)–(9.59) by substituting the above coefficients α_i. The system reads

$$\xi^0: \quad 2(8b_2 - 24b_1 + 12b_0) - 8d_0 = 0 \tag{9.231}$$

$$\xi^1: \quad 6(8b_3 - 24b_2 + 12b_1) - (16d_1 - 24d_0) = 0 \tag{9.232}$$

$$\xi^2: \quad 12(8b_4 - 24b_3 + 12b_2) + 4c_0 - (24d_2 - 36d_1 + 12d_0) - \Omega^2 4a_0 = 0 \tag{9.233}$$

$$\xi^3: \quad 20(8b_5 - 24_3b_4 + 12b_3) + 4c_1 - 4c_0 - (32d_3 - 48d_2 + 16d_1)$$
$$- \Omega^2(4a_1 - 4a_0) = 0 \tag{9.234}$$

$$\xi^i: \quad (i+1)(i+2)(8b_{i+2} - 24b_{i+1} + 12b_i) - \Omega^2(4a_{i-2} - 4a_{i-3} + a_{i-4})$$
$$+ 4c_{i-2} - 4c_{i-3} + c_{i-4} - (i+1)(8d_i - 12d_{i-1} + 4d_{i-2}) = 0 \tag{9.235}$$

$$\xi^{m+3}: \quad (m+4)(m+5)(-24b_{m+4} + 12b_{m+3}) - \Omega^2(-4a_m + a_{m-1}) - 4c_m$$
$$+ c_{m-1} - (m+4)(-12d_{m+2} + 4d_{m+1}) = 0 \tag{9.236}$$

$$\xi^{m+4}: \quad 12(m+5)(m+6)b_{m+4} + c_m - \Omega^2 a_m - 4(m+5)d_{m+2} = 0 \tag{9.237}$$

Let us consider Eq. (9.239): it is valid for $i \in \{4, \ldots, m+2\}$. In order to use it in the general case, we must have $m > 1$. Thus, two separate cases must be treated, namely (a) the case $m \leq 1$ and (b) the case $m > 1$. We begin with the particular cases $m \leq 1$.

9.1.11.1   Particular cases $(0 \leq m \leq 1)$

We consider the simplest case of a constant inertial coefficient $R(\xi) = a_0$. The flexural rigidity is represented as $D(\xi) = b_0 + b_1\xi + b_2\xi^2 + b_3\xi^3 + b_4\xi^4$, while the coefficients of the Winkler foundation and the Pasternak foundation, respectively, are given by the following expressions: $K_1(\xi) = c_0, K_2(\xi) = d_0 + d_1\xi + d_2\xi^2$. After substituting these expressions into Eq. (9.2), we obtain a system of five equations with six unknowns:

$$16b_2 - 48b_1 + 24b_0 - 8d_0 = 0$$
$$48b_3 - 144b_2 + 72b_1 - 16d_1 + 24d_0 = 0$$
$$96b_4 - 288b_3 + 144b_2 + 4c_0 - 24d_2 + 36d_1 - 12d_0 - 4a_0\Omega^2 = 0 \tag{9.238}$$
$$-480b_4 + 240b_3 - 4c_0 + 48d_2 - 16d_1 + 4a_0\Omega^2 = 0$$
$$360b_4 + c_0 - 20d_2 - a_0\Omega^2 = 0$$

To be compatible, the unknowns, namely the flexural rigidity coefficients and the coefficient Ω^2 must be expressed in terms of a single parameter. We choose

b_4 to be a specified constant and we obtain

$$
\begin{aligned}
b_0 &= -\tfrac{2}{3}b_2 + 2b_1 + \tfrac{1}{3}d_0 \\
b_1 &= -\tfrac{2}{3}b_3 + 2b_2 + \tfrac{2}{9}d_1 - \tfrac{1}{3}d_0 \\
b_2 &= \tfrac{28}{3}b_4 + 2b_3 - \tfrac{7}{18}d_2 - \tfrac{1}{4}d_1 + \tfrac{1}{12}d_0 \\
b_3 &= -4b_4 + \tfrac{2}{15}d_2 + \tfrac{1}{15}d_1
\end{aligned}
\tag{9.239}
$$

The natural frequency reads

$$
\omega^2 = (360b_4 + c_0 - 20d_2)/L^4 a_0 \tag{9.240}
$$

with attendant discussion following Eq. (9.26).
   The second case is $m = 1$. Thus,

$$
R(\xi) = a_0 + a_1\xi \quad D(\xi) = \sum_{i=0}^{6} b_i\xi^i \quad K_1(\xi) = c_0 + c_1\xi \quad K_2(\xi) = \sum_{i=0}^{3} d_i\xi^i
\tag{9.241}
$$

The governing equation results in a system of six equations with seven unknowns:

$$
16b_2 - 48b_1 + 24b_0 - 8d_0 = 0
$$
$$
48b_3 - 144b_2 + 72b_1 - 16d_1 + 24d_0 = 0
$$
$$
96b_4 - 288b_3 + 144b_2 + 4c_0 - 24d_2 + 36d_1 - 12d_0 - 4a_0\Omega^2 = 0
$$
$$
160b_5 - 480b_4 + 240b_3 + 4c_1 - 4c_0 - 32d_3 + 48d_2 - 16d_1 - (4a_1 - 4a_0)\Omega^2 = 0
$$
$$
-720b_5 + 360b_4 - 4c_1 + c_0 + 60d_3 - 20d_2 - (-4a_1 + a_0)\Omega^2 = 0
$$
$$
504b_5 + c_1 - 24d_3 - \Omega^2 a_1 = 0
$$

$$\tag{9.242}$$

We express the unknowns in terms of b_4:

$$
\begin{aligned}
b_0 &= -\tfrac{2}{3}b_2 + 2b_1 + \tfrac{1}{3}d_0 \\
b_1 &= -\tfrac{2}{3}b_3 + 2b_2 + \tfrac{2}{9}d_1 - \tfrac{1}{3}d_0 \\
b_2 &= (504a_0b_5 + 72a_1b_3 - 24a_1b_4 + a_0c_1 - a_1c_0 - 24a_0d_3 \\
    &\quad + 6a_1d_2 - 9a_1d_1 + 3a_1d_0)/36a_1
\end{aligned}
$$

$$b_3 = [(464a_1 - 504a_0)b_5 + 120a_1b_4 - a_0c_1 + a_1c_0 + (-16a_1 + 24a_0)d_3$$
$$- 12a_1d_2 + 4a_1d_1]/60a_1$$

$$b_4 = [(-1296a_1 + 504a_0)b_5 + a_0c_1 - a_1c_0 + (36a_1 - 24a_0)d_3 + 20a_1d_2]/360a_1$$
$$\tag{9.243}$$

The natural frequency is

$$\omega^2 = (504b_5 + c_1 - 24d_3)/a_1 \tag{9.244}$$

with the discussion following Eq. (9.43) to be applied in this case. Now, we turn our attention to the general case $m > 1$.

9.1.11.2 General case ($m > 1$)

Here, the inertial coefficient is represented by a polynomial function of order $m > 1$. This implies that

$$D(\xi) = \sum_{i=0}^{m+4} b_i \xi^i \quad K_1(\xi) = \sum_{i=0}^{m} c_i \xi^i \quad K_2(\xi) = \sum_{i=0}^{m+2} d_i \xi^i \tag{9.245}$$

From Eqs. (9.231)–(9.235), we derive the compatibility equations:

$$\Omega^2 = [12(8b_4 - 24b_3 + 12b_2) + 4c_0 - (24d_2 - 36d_1 + 12d_0)]/4a_0 \tag{9.246}$$

$$\Omega^2 = [20(8b_5 - 24b_4 + 12b_3) + 4c_1 - 4c_0 - (32d_3 - 48d_2 + 16d_1)](4a_1 - 4a_0) \tag{9.247}$$

$$\cdots$$

$$\Omega^2 = [(i+1)(i+2)(8b_{i+2} - 24b_{i+1} + 12b_i) + 4c_{i-2} - 4c_{i-3} + c_{i-4}$$
$$- (i+1)(8d_i - 12d_{i-1} + 4d_{i-2})]/(4a_{i-2} - 4a_{i-3} + a_{i-4}) \tag{9.248}$$

$$\cdots$$

$$\Omega^2 = \frac{[(m+4)(m+5)(-24b_{m+4} + 12b_{m+3}) - 4c_m + c_{m-1} - (m+4)(-12d_{m+2} + 4d_{m+1})]}{(-4a_m + a_{m-1})} \tag{9.249}$$

$$\Omega^2 = [12(m+5)(m+6)b_{m+4} + c_m - 4(m+5)d_{m+2}]/a_m \tag{9.250}$$

We get b_{m+3} from Eqs. (9.250) and (9.251):

$$b_{m+3} = \left[\frac{m+6}{m+4}\left(\frac{a_{m-1}}{a_m} - 4\right) + 2\right]b_{m+4} + \left(\frac{a_{m-1}}{a_m} - 4\right)\frac{c_m - 4(m+5)d_{m+2}}{12(m+4)(m+5)}$$
$$- \frac{[-4c_m + c_{m-1} - (m+4)(-12d_{m+2} + 4d_{m+1})]}{12(m+4)(m+5)} \tag{9.251}$$

Equation (9.249), with $i = m + 2$, and (9.250) allow us to determine b_{m+2}:

$$b_{m+2} = -\left(2\frac{m+5}{m+3}\frac{4a_m - 4a_{m-1} + a_{m-2}}{-4a_m + a_{m-1}} + \frac{2}{3}\right)b_{m+4}$$

$$+ \left(\frac{m+5}{m+3}\frac{4a_m - 4a_{m-1} + a_{m-2}}{-4a_m + a_{m-1}} + 2\right)b_{m+3}$$

$$+ \frac{4a_m - 4a_{m-1} + a_{m-2}}{-4a_m + a_{m-1}}\frac{-4c_m + c_{m-1} - (m+4)(-12d_{m+2} + 4d_{m+1})}{12(m+3)(m+4)}$$

$$- \frac{[4c_m - 4c_{m-1} + c_{m-2} - (m+3)(8d_{m+2} - 12d_{m+1} + 4d_m)]}{12(m+3)(m+4)} \quad (9.252)$$

The coefficients b_i for $i \in \{4, \ldots, m+1\}$ are obtained from Eq. (9.249):

$$b_i = \frac{2}{3}\frac{i+3}{i+1}\frac{4a_{i-2} - 4a_{i-3} + a_{i-4}}{4a_{i-1} - 4a_{i-2} + a_{i-3}}b_{i+3} - \left(2\frac{i+3}{i+1}\frac{4a_{i-2} - 4a_{i-3} + a_{i-4}}{4a_{i-1} - 4a_{i-2} + a_{i-3}} + \frac{2}{3}\right)b_{i+2}$$

$$+ \left(\frac{i+3}{i+1}\frac{4a_{i-2} - 4a_{i-3} + a_{i-4}}{4a_{i-1} - 4a_{i-2} + a_{i-3}} + 2\right)b_{i+1}$$

$$+ \frac{4a_{i-2} - 4a_{i-3} + a_{i-4}}{4a_{i-1} - 4a_{i-2} + a_{i-3}}\frac{4c_{i-1} - 4c_{i-2} + c_{i-3} - (i+2)(8d_{i+1} - 12d_i + 4d_{i-1})}{12(i+1)(i+2)}$$

$$- \frac{4c_{i-2} - 4c_{i-3} + c_{i-4} - (i+1)(8d_i - 12d_{i-1} + 4d_{i-2})}{12(i+1)(i+2)} \quad (9.253)$$

For Eq. (9.249) for $i = 4$ and Eq. (9.248), we obtain

$$b_3 = \frac{4a_1 - 4a_0}{4a_2 - 4a_1 + a_0}b_6 - \left(3\frac{4a_1 - 4a_0}{4a_2 - 4a_1 + a_0} - \frac{2}{3}\right)b_5 + \left(\frac{3}{2}\frac{4a_1 - 4a_0}{4a_2 - 4a_1 + a_0} + 2\right)b_4$$

$$+ \frac{1}{240}\frac{4a_1 - 4a_0}{4a_2 - 4a_1 + a_0}[4c_2 - 4c_1 + c_0 - 5(8d_4 - 12d_3 + 4d_2)]$$

$$- \frac{1}{240}[4c_1 - 4c_0 - (32d_3 - 48d_2 + 16d_1)] \quad (9.254)$$

Equations (9.247) and (9.248) result in

$$b_2 = \frac{10}{9}\frac{a_0}{a_1 - a_0}b_5 - \left(\frac{10}{3}\frac{a_0}{a_1 - a_0} + \frac{2}{3}\right)b_4 + \left(\frac{5}{3}\frac{a_0}{a_1 - a_0} + 2\right)b_3$$

$$+ \frac{1}{144}\frac{a_0}{a_1 - a_0}[4c_1 - 4c_0 - (32d_3 - 48d_2 + 16d_1)]$$

$$- \frac{1}{144}[4c_0 - (24d_2 - 36d_1 + 12d_0)] \quad (9.255)$$

Then, the coefficients b_0 and b_1 are given by Eqs. (9.231)–(9.232):

$$b_1 = -\tfrac{2}{3}b_3 + 2b_2 + \tfrac{2}{9}d_1 - \tfrac{1}{3}d_0$$
$$b_0 = -\tfrac{2}{3}b_2 + 2b_1 + \tfrac{1}{3}d_0$$

(9.256)

We note that the above expressions are valid if the coefficients a_i satisfy some secondary relations: (a) from Eq. (9.247), we conclude that $a_0 \neq 0$; (b) Eq. (9.256) imposes the condition $a_1 \neq a_0$; (c) Eqs. (9.254) and (9.255) demand that $4a_{i-1} - 4a_{i-2} + a_{i-3} \neq 0$; (d) Eq. (9.253) results in $a_m \neq a_{m-1}/4$; (e) Eq. (9.252) leads to $a_m \neq 0$.

9.1.12 Cases Violated in Eq. (9.99)

We assume in this section that some of the relations in Eq. (9.99) are violated. In this case, we obtain other general expressions for the coefficients b_i as follows.

Let us consider the case in which the inertial coefficients do not satisfy the relation (a) of Eq. (9.99). In this case, using Eq. (9.69), we are able to write an expression for the coefficient b_{m+3}:

$$b_{m+3} = b_{m+4} + \frac{2c_m - c_{m-1} + (m+4)(-6d_{m+2} + 4d_{m+1})}{12(m+4)(m+5)}$$

(9.257)

The coefficient b_{m+2} is also affected. We have to obtain it from Eqs. (9.92) and (9.90), in terms of the coefficients b_{m+3} and b_{m+4}:

$$b_{m+2} = b_{m+3} + \frac{(m+5)(m+6)}{(m+3)(m+4)}\frac{a_{m-2} - 2a_{m-1}}{a_m}b_{m+4}$$
$$+ \frac{a_{m-2} - 2a_{m-1}}{a_m}\frac{c_m - 4(m+5)d_{m+2}}{12(m+3)(m+4)}$$
$$- \frac{-2c_{m-1} + c_{m-2} - (m+3)(-6d_{m+1} + 4d_m)}{12(m+3)(m+4)}$$

(9.258)

The other expressions of the coefficients b_i are unchanged.

If the condition (b) in Eq. (9.99) is not satisfied, i.e., $a_{m-2} = 2a_{m-1}$, then the coefficient b_{m+2} is directly obtained from Eq. (9.68):

$$b_{m+2} = b_{m+3} + \frac{2c_{m-1} - c_{m-2} + (m+3)(-6d_{m+1} + 4d_m)}{12(m+3)(m+4)}$$

(9.259)

The coefficient b_{m+1} is evaluated with the aid of Eqs. (9.87) and (9.90):

$$b_{m+1} = b_{m+2} + \frac{a_m - 2a_{m-2} + a_{m-3}}{a_m} \frac{(m+5)(m+6)}{(m+2)(m+3)} b_{m+4}$$

$$- \frac{(c_m - 2c_{m-2} + c_{m-3}) - (m+2)(d_{m+2} - 6d_m + 4d_{m-1})}{12(m+2)(m+3)}$$

$$+ \frac{a_m - 2a_{m-2} + a_{m-3}}{a_m} \frac{c_m - 4(m+5)d_{m+2}}{12(m+2)(m+3)} \tag{9.260}$$

If the condition (c) in Eq. (9.99) is not valid for $i = j$, then a new expression for the coefficient b_j is derived from Eq. (9.66) evaluated for $i = j$:

$$b_j = b_{j+1} - \frac{c_{j-1} - 2c_{j-3} + c_{j-4} - (j+1)(d_{j+1} - 6d_{j-1} + 4d_{j-2})}{12(j+1)(j+2)} \tag{9.261}$$

The coefficient b_{j-1} is expressed from Eqs. (9.86) with $i = j - 1$ and $i = j + 1$ (we assume that $j > 4$):

$$b_{j-1} = b_j + \frac{a_{j-2} - 2a_{j-4} + a_{j-5}}{a_j - 2a_{j-2} + a_{j-3}} \frac{(j+2)(j+3)(b_{j+1} - b_{j+2})}{j(j+1)}$$

$$+ \frac{a_{j-2} - 2a_{j-4} + a_{j-5}}{a_j - 2a_{j-2} + a_{j-3}} \frac{(c_j - 2c_{j-2} + c_{j-3}) - (j+2)(d_{j+2} - 6d_j + 4d_{j-1})}{12j(j+1)}$$

$$- \frac{(c_{j-2} - 2c_{j-4} + c_{j-5}) - j(d_j - 6d_{j-2} + 4d_{j-3})}{12j(j+1)} \tag{9.262}$$

If $j = 4$, then the expression for the coefficient b_4 is derived from Eq. (9.66) by setting $i = 4$:

$$b_4 = b_5 - [c_3 - 2c_1 + c_0 - 5(d_5 - 6d_3 + 4d_2)]/360 \tag{9.263}$$

The expression for the coefficient b_3 is expressed from Eqs. (9.85) and (9.86) with $i = 5$:

$$b_3 = b_4 + \frac{504}{240} \frac{a_2 - 2a_0}{a_4 - 2a_2 + a_1} (b_5 - b_6)$$

$$+ \frac{a_2 - 2a_0}{a_4 - 2a_2 + a_1} \frac{c_4 - 2c_2 + c_1 - 6(d_6 - 6d_4 + 4d_3)}{240}$$

$$- \frac{[c_2 - 2c_0 - (4d_4 - 24d_2 + 16d_1)]}{240} \tag{9.264}$$

In the particular case $a_2 = 2a_0$, we express directly the coefficient b_3 from Eq. (9.65):

$$b_3 - b_4 + [c_2 + 2c_0 + 4d_4 - 24d_2 + 16d_1]/240 \qquad (9.265)$$

The coefficient b_2 is derived by equating Eq. (9.84) and Eq. (9.86) with $i = 4$:

$$b_2 = b_3 + \frac{360}{144} \frac{a_1}{a_3 - 2a_1 + a_0}(b_4 - b_5) + \frac{a_1}{144(a_3 - 2a_1 + a_0)}$$
$$\times [(c_3 - 2c_1 + c_0) - 5(d_5 - 6d_3 + 4d_2)] + \left(\frac{-c_1 + 3d_3 - 18d_1 + 12d_0}{144}\right)$$
$$(9.266)$$

If $a_1 = 0$, then b_2 is given by Eq. (9.64):

$$b_2 = b_3 + \tfrac{1}{144}(-c_1 + 3d_3 - 18d_1 + 12d_0) \qquad (9.267)$$

The coefficient b_1 is evaluated by equating Eqs. (9.83) and (9.85):

$$b_1 = \frac{240}{72} \frac{a_0}{a_2 - 2a_0}(b_3 - b_4) + \frac{1}{72} \frac{a_0}{a_2 - 2a_0}[c_2 - 2c_0 - (4d_4 - 24d_2 + 16d_1)]$$
$$+ \frac{1}{72}(-c_0 + 2d_2 + 12d_0) \qquad (9.268)$$

When the coefficient $a_0 = 0$, the coefficient b_1 is given by Eq. (9.63):

$$b_1 = b_2 + \tfrac{1}{72}(-c_0 + 2d_2 + 12d_0) \qquad (9.269)$$

We obtain this expression directly from Eq. (9.97) by taking $a_0 = 0$.

In this section, we treated each particular case of violation of Eq. (9.99). We do not deal with the remaining cases when more than one relation in Eq. (9.99) is violated, to save space.

When the conditions in Sections 9.1.7–9.1.11 are violated, one constructs alternative solutions in the manner that was used in the present section.

9.1.13 Does the Boobnov–Galerkin Method Corroborate the Unexpected Exact Results?

In order to corroborate expressions for the natural frequencies obtained in this chapter, it makes sense to contrast the exact solutions with an approximate method. We provide some results obtained by the Boobnov–Galerkin method.

We first refer to Eq. (9.26), which suggests the increase of the natural frequency with the increase of the coefficient c_0 and its decrease with the increase of the coefficient d_2. For the pinned–pinned beam we choose the comparison function as follows:

$$\psi(\xi) = \sin(\pi\xi) \tag{9.270}$$

Substituting it into Eq. (9.2) with the flexural rigidity given in Eqs. (9.27)–(9.30), specified for *P–P* beam, we get

$$\Omega^2 = (360.483b_4 + c_0 - 20.032d_2 - 0.006d_1 - 0.012d_0)/a_0 \tag{9.271}$$

This equation has terms associated with b_4, c_0 and d_2, as the exact equation (9.26) does. Yet Eq. (9.276) also contains terms with d_0 and d_1. The coefficient of b_4 deviates from its exact value 360 by 0.13%, whereas the coefficient of d_2 differs from its exact counterpart by 0.16%. The coefficient of c_0 in Eq. (9.276) is unity as in the exact equation (9.26).

For the clamped–free beam, by taking the following mode shape

$$\psi(\xi) = 4 \sin^2\left(\tfrac{1}{2}\pi\xi\right) + \pi^2\xi^2 \tag{9.272}$$

we obtain the following natural frequency coefficient:

$$\Omega^2 = (376.385b_4 + c_0 - 21.130d_2 - 0.321d_1 - 0.256d_0)/a_0 \tag{9.273}$$

This result differs from the exact result by 7.3% for the coefficient of b_4 and by 5.65% for the coefficient of d_2.

For the clamped–pinned beam, the mode shape is postulated to be

$$\psi(\xi) = \cos\left(\tfrac{1}{2}\pi\xi\right)\sin^2\left(\tfrac{1}{2}\pi\xi\right) \tag{9.274}$$

The natural frequency coefficient is derived:

$$\Omega^2 = (385.759b_4 + c_0 - 21.710d_2 - 0.343d_1 - 0.564d_0)/a_0 \tag{9.275}$$

The coefficients, respectively, in front of b_4 and d_2 differ from the exact values by 7.15% and 8.55%.

For the clamped–clamped beam, the comparison function reads

$$\psi(\xi) = \sin^2(\pi\xi) \tag{9.276}$$

The natural frequency coefficient is evaluated as follows:

$$\Omega^2 = (387.636b_4 + c_0 - 21.780d_2 - 0.361d_1 - 0.723d_0)/a_0 \tag{9.277}$$

The difference between approximate values and the exact ones are, respectively, for b_4 and d_2, 7.67% and 8.9%.

For the guided–pinned beam, we postulate the mode shape as

$$\psi(\xi) = \cos\left(\tfrac{1}{2}\pi\xi\right) \tag{9.278}$$

We obtain the following expression for the natural frequency coefficient:

$$\Omega^2 = (360.483b_4 + c_0 - 20.029d_2 - 0.037d_1 - 0.003d_0)/a_0 \tag{9.279}$$

The coefficient of b_4 deviates from its exact value by only 0.13%, whereas the coefficient of d_2 differs from its counterpart by 0.15%.

For the clamped–guided beam, the mode shape taken and the natural frequency coefficient derived read, respectively, as follows:

$$\psi(\xi) = \sin^2\left(\tfrac{1}{2}\pi\xi\right) \tag{9.280}$$

$$\Omega^2 = (387.636b_4 + c_0 - 21.255d_2 + 0.081d_1 - 0.181d_0)/a_0 \tag{9.281}$$

The value 387.636 differs from its exact counterpart 360 by 0.18% for the coefficient of b_4, while the difference for the coefficient of d_2 is 0.4%.

Now, let us compare the Boobnov–Galerkin results for the linearly varying density with Eq. (9.43). For the P–P beam the natural frequency coefficient reads, by using the comparison function in Eq. (9.270), as follows:

$$\Omega^2 = \frac{1}{a_1}\left(504.676b_5 + \frac{a_1 + 2.003a_0}{a_1 + 2a_0}c_1 - 24.042\frac{a_1 + 1.999a_0}{a_1 + 2a_0}d_3\right)$$
$$+ \frac{1}{a_1 + 2a_0}(-0.027c_0 - 19.31d_0 - 0.012d_1 - 0.102d_2) \tag{9.282}$$

The coefficient 504.676 differs from its exact counterpart 504 by only 0.13%; the coefficient of d_3 in the approximate formula (9.282) is obtained as $-24.042(a_1 + 1.999a_0)/(a_1 + 2a_0)$ vs. the exact value, namely 24. For the values $a_1 = a_0 = 1$ the difference is 0.14%.

For the clamped–clamped beam, we obtain, by employing the comparison function in Eq. (9.276),

$$\Omega^2 = \frac{1}{a_1}\left(542.690b_5 + \frac{a_1 + 2.154a_0}{a_1 + 2a_0}c_1 - 26.215\frac{a_1 + 1.972a_0}{a_1 + 2a_0}d_3\right)$$
$$+ \frac{1}{a_1 + 2a_0}(-0.154c_0 - 98.384d_0 - 0.723d_1 - 0.489d_2) \tag{9.283}$$

For the guided–pinned beam, the natural frequency coefficient reads, via the use of Eq. (9.278) as a comparison function,

$$\Omega^2 = \frac{1}{a_1}\left(542.690\frac{a_1 + 3.296a_0}{a_1 + 3.362a_0}b_5 + \frac{0.999a_1 + 3.367a_0}{a_1 + 3.363a_0}c_1 - 24.591\frac{a_1 + 3.287a_0}{a_1 + 3.363a_0}d_3\right)$$
$$+ \frac{1}{a_1 + 3.363a_0}(0.005c_0 + 1.293d_0 - 0.124d_1 - 0.007d_2) \tag{9.284}$$

For the clamped–guided beam, the Boobnov–Galerkin method leads to the following expression for the natural frequency coefficient, when Eq. (9.281) is utilized as a comparison function:

$$\Omega^2 = \frac{1}{a_1}\left(529.052\frac{a_1 + 1.332a_0}{a_1 + 1.298a_0}b_5 + \frac{a_1 + 1.398a_0}{a_1 + 1.298a_0}c_1 - 24.566\frac{a_1 + 1.366a_0}{a_1 + 1.298a_0}d_3\right)$$
$$+ \frac{1}{a_1 + 1.298a_0}(-0.010c_0 - 7.486d_0 + 0.106d_1 + 0.363d_2) \tag{9.285}$$

We observe that for both constant density as well as linearly varying density the approximate formulas, as the exact ones, exhibit a seemingly unexpected behavior of the decrease of the natural frequency with the increase of the Pasternak foundation coefficients. The phenomenon is explainable by the fact that for the cases for which the closed-form solution is derivable, the beam's flexural rigidity is related to the coefficients of both the Winkler and Pasternak foundations. This leads to the constraints on the combination of the Winkler and Pasternak foundation coefficients in such a manner that the natural frequency increases with the increase of the foundation's effective, combined coefficient.

This implies, for example, that for the beam of constant inertial coefficient, the effective coefficient, of both foundations can be introduced as

$$c_{0,WP} = c_0 - 20d_2 \tag{9.286}$$

Analogously, for the beam with linearly varying inertial coefficient, the effective Winkler–Pasternak foundation coefficient reads

$$c_{1,WP} = c_1 - 20d_3 \tag{9.287}$$

Once these effective coefficients are in place, Eqs. (9.26) and (9.43), are transformed, respectively, into the following expressions:

$$\Omega^2 = (360b_4 + c_{0,WP})/a_0 \tag{9.288}$$
$$\Omega^2 = (504b_5 + c_{1,WP})/a_1 \tag{9.289}$$

For the beam with the inertial coefficient being an mth order polynomial, Eq. (9.60) suggests the introduction of the following effective Winkler–Pasternak foundation coefficient:

$$c_{m,\text{WP}} = c_m - 4(m+5)d_{m+2} \tag{9.290}$$

In these circumstances, Eq. (9.60) is replaced by

$$\Omega^2 = [12(m+5)(m+6)b_{m+4} + c_{m,\text{WP}}]/a_m \tag{9.291}$$

Note that the effective Winkler–Pasternak foundation coefficient depends on the degree of the polynomial representing the beam's inertial coefficient for the derived class of beams for which closed-form solutions were obtained. To arrive at the physically feasible closed-form solutions one cannot retain solely the Pasternak foundation as the model for the elastic foundation. Yet one can have a foundation that is modeled only by its translational nature; the Pasternak foundation property can appear only with the Winkler foundation property, in such a manner that the effective modulus is positive.

Equations (9.288), (9.289) and (9.291) suggest that with the increase of the *effective* Winkler–Pasternak foundation moduli $c_{0,\text{WP}}, c_{1,\text{WP}}$ or $c_{m,\text{WP}}$ the natural frequency increases, in line with one's anticipation.

9.1.14 Concluding Remarks

The inclusion of the elastic foundations of two types to the problem of the beam leads to the modification of the closed-form solution of the fundamental frequency of the beam without elastic foundations. Additional terms, due to the foundations, are introduced in the expression for the natural frequency of foundationless beams. A closed-form solution is obtained when the elastic modulus and foundation have joint parameters. It is remarkable that only the highest coefficients b_{m+4}, c_m, d_{m+2} and a_m, respectively, of the flexural rigidity, the Winkler foundation, the Pasternak foundation and the inertial coefficient appear in the closed-form solution of the natural frequency. It is important to note that the expression for the natural frequency is formally independent of the boundary condition. Yet the flexural rigidities of the beam in these cases are different for each set of boundary conditions. Some unexpected results have been reported on the influence of the elastic foundations on the natural frequencies. Some restrictions have been imposed on the coefficients of the beam's flexural rigidity in terms of the elastic foundation coefficients arising from the conditions of the positivity of the natural frequency and the natural flexural rigidity.

9.2 Closed-Form Solution for the Natural Frequency of an Inhomogeneous Beam with a Rotational Spring

9.2.1 Introductory Remarks

Closed-form solutions for natural frequencies of inhomogeneous beams with pinned supports were derived by Candan and Elishakoff (2000). This section follows their study to deal with beams with an elastic rotational spring. To illustrate the feasibility of the proposed method, we consider free vibrations of a beam with rotational elastic support at the left end, whereas the right end is pinned.

9.2.2 Basic Equations

Consider inhomogeneous beams, governed by the differential equation

$$\frac{d^2}{d\xi^2}\left(D(\xi)\frac{d^2W}{d\xi^2}\right) - R(\xi)\omega^2 W(\xi) = 0 \tag{9.292}$$

where $W(\xi)$ is the displacement, $\xi = x/L$, with x being the axial coordinate, ξ is a non-dimensional axial coordinate, $D(\xi)$ a flexural rigidity and $R(\xi)$ an inertial coefficient function.

We take the inertial coefficient to be an mth order polynomial function:

$$R(\xi) = \sum_{i=0}^{m} a_i \xi^i \tag{9.293}$$

The beam's displacement is postulated to be a fourth-order polynomial function:

$$W(\xi) = \alpha_0 + \alpha_1\xi + \alpha_2\xi^2 + \alpha_3\xi^3 + \xi^4 \tag{9.294}$$

The boundary conditions read

$$W(0) = 0 \quad D(0)\frac{d^2W}{d\xi^2} = (k_1/L)\frac{dW}{d\xi} \tag{9.295}$$

$$W(1) = 0 \quad D(1)W''(1) = 0$$

which corresponds to the rotational spring at $\xi = 0$, of stiffness k_1 times the length squared. In order to have a compatibility of the order in the polynomial expressions, arising from the substitution of Eqs. (9.298) and (9.299) into

Eq. (9.292), we conclude that the flexural rigidity $D(\xi)$ must be represented as a polynomial

$$D(\xi) = \sum_{i=0}^{m+4} b_i \xi^i \tag{9.296}$$

We will consider several particular cases.

Satisfaction of Eq. (9.295) leads to the following final expression for the postulated mode shape:

$$W(\xi) = 3\frac{b_0 L}{3b_0 L + k_1}\xi + \frac{3}{2}\frac{k_1}{3b_0 L + k_1}\xi^2 - \frac{1}{2}\frac{5k_1 + 12b_0 L}{3b_0 L + k_1}\xi^3 + \xi^4 \tag{9.297}$$

We introduce the non-dimensional elastic constant:

$$\beta = k_1/b_0 L \tag{9.298}$$

Note that when β tends to zero, Eq. (9.298) reduces to the mode shape of a pinned–pinned beam,

$$W(\xi) = \xi - 2\xi^3 + \xi^4 \tag{9.299}$$

whereas when β approaches infinity, Eq. (9.298) tends to the mode shape of a clamped–pinned beam,

$$W(\xi) = \tfrac{3}{2}\xi^2 - \tfrac{5}{2}\xi^3 + \xi^4 \tag{9.300}$$

Figure 9.13 portrays the mode shape of the beam for different values of the coefficient β.

9.2.3   Uniform Inertial Coefficient

Consider, first, the case $m = 0$, corresponding to constant inertial coefficient:

$$R(\xi) = a_0 \tag{9.301}$$

The flexural rigidity in Eq. (9.296) is then a fourth-order polynomial. Substitution of Eqs. (9.293), (9.296) and (9.298) into Eq. (9.292) leads to the following polynomial expressions:

$$A_0 + A_1\xi + A_2\xi^2 + A_3\xi^3 + A_4\xi^4 = 0 \tag{9.302}$$

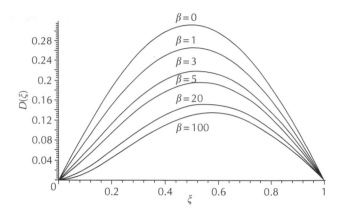

FIGURE 9.13
Mode shapes for different values of the coefficient β

with

$$A_0 = 24b_0 + 6\frac{k_1}{3b_0L + k_1}b_2 - 6\frac{12b_0L + 5k_1}{3b_0L + k_1}b_1 \tag{9.303}$$

$$A_1 = 72b_1 - 18\frac{12b_0L + 5k_1}{3b_0L + k_1}b_2 + 18\frac{k_1}{3b_0L + k_1}b_3 - 3\frac{b_0L}{3b_0L + k_1}a_0\omega^2 \tag{9.304}$$

$$A_2 = 144b_2 - 36\frac{12b_0L + 5k_1}{3b_0L + k_1}b_3 + 36\frac{k_1}{3b_0L + k_1}b_4 - \frac{3}{2}\frac{k_1}{3b_0L + k_1}a_0\omega^2 \tag{9.305}$$

$$A_3 = 240b_3 - 60\frac{12b_0L + 5k_1}{3b_0L + k_1}b_4 + \frac{1}{2}\frac{12b_0L + 5k_1}{3b_0L + k_1}a_0\omega^2 \tag{9.306}$$

$$A_4 = 360b_4 - a_0\omega^2 \tag{9.307}$$

Equation (9.302) is a polynomial function, which is equal to zero for any $\xi \in [0; 1]$. Then, to satisfy this requirement, each coefficient of ξ^i for $i = 0, \ldots, 4$ must vanish. Hence, one obtains a system of five equations with six unknowns, namely b_0, b_1, b_2, b_3, b_4 and ω^2. Thus, we have an infinite number of closed-form solutions. In order to derive the solutions of this system, we have to express the unknowns in terms of an arbitrary constant. It is convenient to choose the coefficient b_0 as such a constant. We use the non-dimensional elastic coefficient β to simplify the analytical expressions:

$$b_1 = 4(2592 + 1872\beta + 444\beta^2 + 35\beta^3)(3 + \beta)b_0/G_1 \tag{9.308}$$

$$b_2 = -48(48 + 12\beta - \beta^2)(3 + \beta)^2 b_0/G_1 \tag{9.309}$$

$$b_3 = -64(3 + \beta)^3(12 + 5\beta)b_0/G_1 \tag{9.310}$$

$$b_4 = 128(3 + \beta)^4 b_0/G_1 \tag{9.311}$$

where

$$G_1 = 31{,}104 + 37{,}152\beta + 15{,}696\beta^2 + 2{,}748\beta^3 + 163\beta^4 \qquad (9.312)$$

Since the coefficients in G_1 are positive, it takes only positive values for $\beta > 0$. The natural frequency is expressed from Eq. (9.307):

$$\omega^2 = 360b_4/a_0$$
$$= 46{,}080(3 + \beta)^4 b_0/a_0 G_1 \qquad (9.313)$$

We note that without elastic support, when $k = 0$, the flexural rigidity and the natural frequency squared read

$$D(\xi) = (1 + \xi - \tfrac{2}{3}\xi^2 - \tfrac{2}{3}\xi^3 + \tfrac{1}{3}\xi^4)b_0$$
$$\omega^2 = 120b_0/a_0 \qquad (9.314)$$

When the elastic coefficient k tends to infinity, the flexural rigidity and the natural frequency squared become

$$D(\xi) = \frac{1}{163}(163 + 140\xi + 48\xi^2 - 320\xi^3 + 128\xi^4)b_0$$
$$\omega^2 = \frac{46{,}080}{163}b_0/a_0 \qquad (9.315)$$

Figure 9.14. shows the dependence of $D(\xi)$ on ξ for various values of the coefficient β.

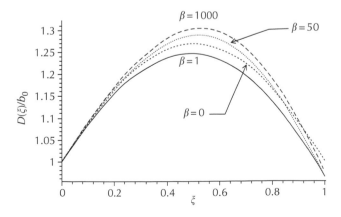

FIGURE 9.14
Variation of the flexural rigidity for linear material density

9.2.4   Linear Inertial Coefficient

In this section, we consider a beam with a linear inertial coefficient:

$$R(\xi) = a_0 + a_1\xi \tag{9.316}$$

The flexural rigidity, given by Eq. (9.296), is a fifth-order polynomial. Thus, the governing equation, with the previous assumptions, can be rewritten as follows:

$$A_0 + A_1\xi + A_2\xi^2 + A_3\xi^3 + A_4\xi^4 + A_5\xi^5 = 0 \tag{9.317}$$

with

$$A_0 = 24b_0 + 6\frac{k_1}{3b_0L + k_1}b_2 - 6\frac{12b_0L + 5k_1}{3b_0L + k_1}b_1 \tag{9.318}$$

$$A_1 = 72b_1 - 18\frac{12b_0L + 5k_1}{3b_0L + k_1}b_2 + 18\frac{k_1}{3b_0L + k_1}b_3 - \frac{3b_0L}{3b_0L + k_1}a_0\omega^2 \tag{9.319}$$

$$A_2 = 144b_2 - 36\frac{12b_0L + 5k_1}{3b_0L + k_1}b_3 + 36\frac{k_1}{3b_0L + k_1}b_4$$
$$- \left(\frac{3k_1}{6b_0L + 2k_1}a_0 + \frac{3b_0L}{3b_0L + k_1}a_1\right)\omega^2 \tag{9.320}$$

$$A_3 = 240b_3 - 60\frac{12b_0L + 5k_1}{3b_0L + k_1}b_4 + 60\frac{k_1}{3b_0L + k_1}b_5$$
$$+ \left(\frac{12b_0L + 5k_1}{6b_0L + 2k_1}a_0 - \frac{3k_1}{6b_0L + 2k_1}a_1\right)\omega^2 \tag{9.321}$$

$$A_4 = 360b_4 - 90\frac{12b_0L + 5k_1}{3b_0L + k_1}b_5 - \left(a_0 - \frac{12b_0L + 5k_1}{6b_0L + 2k_1}a_1\right)\omega^2 \tag{9.322}$$

$$A_5 = 504b_5 - a_1\omega^2 \tag{9.323}$$

Note that in the paper by Elishakoff (2001c) there is a typographical error in the expression for $D(\xi)$ and A_1 (Eqs. 23, 24 and 28) (Storch, 2001a). Equation (9.317) is valid for every ξ within the interval $[0; 1]$. Thus, the coefficients A_i, with $i = 0, \ldots, 5$, must vanish. This requirement leads to a system of six equations with seven unknowns. These are the coefficients b_i with $i = 0, \ldots, 5$ and the natural frequency squared ω^2. We express the unknowns in terms of an arbitrary constant. For convenience, we take the coefficient b_0 to be a parameter. Then, the other coefficients b_i read

$$b_1 = 4(3 + \beta)[(435{,}456 + 459{,}648\beta + 179{,}424\beta^2 + 30{,}744\beta^3 + 1{,}960\beta^4)a_0 +$$
$$+ (176{,}256 + 213{,}408\beta + 91{,}584\beta^2 + 16{,}512\beta^3 + 1{,}047\beta^4)a_1]b_0/G_2 \tag{9.324}$$

$$b_2 = 16(3 + \beta)^2[(-24,192 - 14,112\beta - 1,512\beta^2 + 168\beta^3)a_0 +$$
$$+ (14,688 + 10,368\beta + 2,496\beta^2 + 215\beta^3)a_1]b_0/G_2 \qquad (9.325)$$

$$b_3 = -64(3 + \beta)^3[(2,016 + 1,512\beta + 280\beta^2)a_0$$
$$+ (1,296 + 384\beta - 7\beta^2)a_1]b_0/G_2 \qquad (9.326)$$

$$b_4 = 256(3 + \beta)^4[(84 + 28\beta)a_0 - (108 + 45\beta)a_1]b_0/G_2 \qquad (9.327)$$

$$b_5 = 5120(3 + \beta)^5 a_1 b_0/G_2 \qquad (9.328)$$

where

$$G_2 = (5,22,5472 + 7,983,360\beta + 4,717,440\beta^2 + 1,340,640\beta^3 + 181,272\beta^4$$
$$+ 9,128\beta^5)a_0 + (2,115,072 + 3,265,920\beta + 1,982,880\beta^2 + 584,640\beta^3$$
$$+ 82,560\beta^4 + 4,375\beta^5)a_1 \qquad (9.329)$$

Note that $G_2 = 0$ has only negative roots. It is always positive for physically realizable values of β, namely $\beta > 0$. Figure 9.15 shows the dependence of the flexural rigidity vs. ξ for different values of the elastic coefficient β.

The natural frequency squared reads

$$\omega^2 = 504b_5/a_1 = 2,580,480(3 + \beta)^5 b_0/G_2 \qquad (9.330)$$

Since G_2 is positive so is the natural frequency for $\beta > 0$.

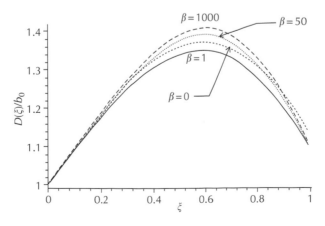

FIGURE 9.15
Variation of the flexural rigidity for linear material density when a_0 and a_1 equal unity

When the elastic coefficient vanishes, the flexural rigidity and the natural frequency read, respectively, as follows:

$$D(\xi) = b_0(1 + \xi) + [(17a_1 - 28a_0)\xi^2 - (28a_0 + 18a_1)\xi^3$$
$$+ (14a_0 - 18a_1)\xi^4 + 10a_1\xi^5]b_0/(42a_0 + 17a_1) \qquad (9.331)$$
$$\omega^2 = 5040b_0/(42a_0 + 17a_1)$$

When the elastic coefficient tends to infinity, the flexural rigidity and the natural frequency tend to the following, respective, expressions:

$$D(\xi) = b_0 + [4(1{,}960a_0 + 1{,}047a_1)\xi + 16(168a_0 + 215a_1)\xi^2 - 64(280a_0 - 7a_1)\xi^3$$
$$+ 256(28a_0 - 45a_1)\xi^4 + 5{,}120a_1\xi^5]b_0/(9{,}128a_0 + 4{,}375a_1)$$
$$\omega^2 = 368{,}640b_0/(1{,}304a_0 + 625a_1)$$

$$(9.332)$$

Note that the expressions in Eq. (9.332) represent simplified versions of Eqs. (41) in the paper by Elishakoff (2001c) (Storch, 2001a).

In an analogous manner one can determine the closed-form solutions for the natural frequencies of beams with parabolic, cubic, quartic, etc. variations of the mass density.

This section presented a simple closed-form solution for the natural frequency and the mode shape of the inhomogeneous beam with a rotational spring. An analogous treatment can be performed for beams with two springs at each end.

9.3   Closed-Form Solution for the Natural Frequency of an Inhomogeneous Beam with a Translational Spring

9.3.1   Introductory Remarks

Since Bernoulli and Euler derived their celebrated equation for vibrating beams over 250 years ago, there were several thousand papers written for determining the spectra of homogeneous or inhomogeneous beams, with or without end springs. It was decided to refrain from citing numerous papers dealing with the topic in the title to save space. Whereas for the homogeneous beams some *exact* and *closed-form* solutions are known [see, e.g., texts by Weaver et al. (1990) and Rao, 1995], it is not so for inhomogeneous beams. Numerous *exact* solutions are available in terms of Bessel functions. Yet, the author is unaware of closed-form solutions for inhomogeneous beams with end springs. The present section generalizes the study by Candan and

Elishakoff (2000) who derived some closed-form solutions for inhomogeneous beams *without* end springs, under different boundary conditions. Here, we *incorporate* the translational spring into the analysis.

9.3.2 Basic Equations

Consider an inhomogeneous beam with a translational spring at one of its ends and clamped at the other. The governing differential equation reads

$$\frac{d^2}{d\xi^2}\left(D(\xi)\frac{d^2W}{d\xi^2}\right) - R(\xi)\omega^2 W = 0 \qquad (9.333)$$

where $W(\xi)$ is the mode shape, $D(\xi) = E(\xi)A(\xi)$ the flexural rigidity, $E(\xi)$ the modulus of elasticity, $R(\xi) = \rho(\xi)A(\xi)$ the inertial coefficient, $\rho(\xi)$ the mass density, $A(\xi)$ the cross-sectional area, ω the natural frequency, x the axial coordinate, $\xi = x/L$ the non-dimensional axial coordinate and L the length.

The associated boundary conditions read

$$\frac{d^2W}{d\xi^2} = 0 \qquad \frac{d}{d\xi}\left(D(\xi)\frac{d^2W}{d\xi^2}\right) = kW \qquad \text{at } \xi = 0$$

$$W = 0 \qquad \frac{dW}{d\xi} = 0 \qquad\qquad\qquad \text{at } \xi = 1 \qquad (9.334)$$

We postulate the mode shape to be expressed as a polynomial function of fourth order:

$$W(\xi) = \alpha_0 + \alpha_1\xi + \alpha_2\xi^2 + \alpha_3\xi^3 + \alpha_4\xi^4 \qquad (9.335)$$

The inertial coefficient $R(\xi)$ is expressed as an mth order polynomial

$$R(\xi) = \sum_{i=0}^{m} a_i\xi^i \qquad (9.336)$$

The second term in Eq. (9.333) is an $(m+4)$th order polynomial. In order that the first term also constitutes a fourth-order polynomial, we express $D(\xi)$ as the $(m+4)$th order polynomial,

$$D(\xi) = \sum_{i=0}^{m+4} b_i\xi^i \qquad (9.337)$$

The objective of this section is defined as follows: construct the beam's flexural rigidity $D(\xi)$ in such a manner that the postulated mode shape in Eq. (9.342) will serve as an exact mode shape. Bearing in mind Eq. (9.342), we get the following explicit expression for the mode shape, by satisfying the boundary conditions specified in Eq. (9.339):

$$W(\xi) = \frac{1}{1 - (k/3b_0)} \left[3 - \left(\frac{k}{6b_0} + 4 \right) \xi + \frac{k}{2b_0} \xi^3 \right] + \xi^4 \tag{9.338}$$

We introduced here the non-dimensional coefficient $\gamma = k/b_0$. Figure 9.16 depicts the variation of the mode shape vs. ξ for $\gamma = 0, 1, 2$. Figure 9.17 shows the mode shape for γ equal to 50, 100 and 10,000. For the latter value, the

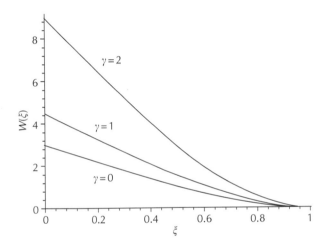

FIGURE 9.16
Variation of the mode shape for different values of $\gamma = k/b_0, \gamma = 0, 1, 2$

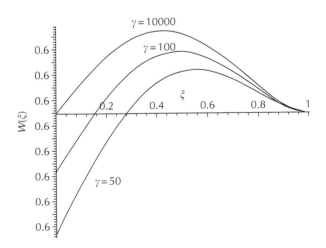

FIGURE 9.17
Variation of the mode shape for different values of $\gamma = k/b_0, \gamma = 50, 100, 10,000$

displacement at the left end is shown to be extremely close to zero, and the mode shape tends to that of a pinned–clamped beam.

Some particular cases follow from Eq. (9.338). First of all, when the elastic spring coefficient k vanishes, the left end of the beam is free. The mode shape reduces to that of the free–clamped beam

$$W(\xi) = 3 - 4\xi + \xi^4 \tag{9.339}$$

When k tends to infinity, the boundary condition at the left end approaches that of the pinned beam; we thus arrive at the pinned–clamped beam with mode shape

$$W(\xi) = \tfrac{1}{2}\xi - \tfrac{3}{2}\xi^3 + \xi^4 \tag{9.340}$$

Our task is to determine the $D(\xi)$ that corresponds to the mode shape in Eq. (9.338). It is natural that the sought for flexural rigidity function depends on the elastic coefficients and the spring coefficient k. These two, in concert, result in a system whose mode shape is as postulated.

9.3.3   Constant Inertial Coefficient

Consider, first, the case when

$$R(\xi) = a_0 \tag{9.341}$$

Substituting Eqs. (9.338) and (9.339) and (9.341) into Eq. (9.333) yields the following polynomial equation:

$$\sum_{i=0}^{4} A_i \xi^i = 0 \tag{9.342}$$

with

$$A_0 = 24b_0 + 6kb_1/(b_0 - k/3) - 3a_0\omega^2/(1 - k/3b_0)$$

$$A_1 = 72b_1 + 18kb_2/(b_0 - k/3) + (4 + k/6b_0)a_0\omega^2/(1 - k/3b_0)$$

$$A_2 = 144b_2 + 36kb_3/(b_0 - k/3) \tag{9.343}$$

$$A_3 = 240b_3 + 60kb_4/(b_0 - k/3) - \tfrac{1}{2}ka_0\omega^2/(b_0 - k/3)$$

$$A_4 = 360b_4 - a_0\omega^2$$

In order for Eq. (9.342) to hold, it is necessary and sufficient that all coefficients A_i vanish:

$$A_i = 0 \quad i = 0, 1, 2, 3, 4 \tag{9.344}$$

Thus, we arrive at five equations with six unknowns. It is convenient to take the coefficient b_0 as a parameter. The coefficients b_i of the flexural rigidity read:

$$b_1 = -\tfrac{4}{3}(3 - \gamma)(53\gamma^3 + 1{,}440\gamma^2 - 10{,}800\gamma + 17{,}280)b_0/G_1$$

$$b_2 = -48(3 - \gamma)^2\gamma^2 b_0/G_1$$

$$b_3 = 64(3 - \gamma)^3\gamma b_0/G_1 \tag{9.345}$$

$$b_4 = \tfrac{128}{3}(3 - \gamma)^4 b_0/G_1$$

where

$$G_1 = 53\gamma^4 - 4{,}320\gamma^3 + 41{,}040\gamma^2 - 138{,}240\gamma + 155{,}520 \tag{9.346}$$

The natural frequency is obtained by the last equations of the sets (9.343) and (9.345) as follows:

$$\omega^2 = 360b_4/a_0 = 15{,}360b_0(3 - \gamma)^4/G_1 \tag{9.347}$$

When the elastic coefficient k vanishes, we get the free–clamped beam, with natural frequency

$$\omega^2 = 360b_4/a_0 = 8b_0/a_0 \tag{9.348}$$

When k tends to infinity, the system becomes a pinned–clamped beam, with

$$\omega^2 = 360b_4/a_0 = \frac{15{,}360}{53}\frac{b_0}{a_0} \tag{9.349}$$

In order for the problem to be physically realizable, it is necessary and sufficient that the flexural rigidity and natural frequency be non-negative.

In order for the natural frequency to be non-negative it is necessary that Eq. (9.351) takes non-negative values. It is of interest, therefore, to find the roots of the denominator:

$$53\gamma^4 - 4{,}320\gamma^3 + 41{,}040\gamma^2 - 13{,}8240\gamma + 155{,}520 = 0 \tag{9.350}$$

The real roots are 2.6023 and 71.1304, implying that these values cannot be taken by the ratio γ, for otherwise, an infinite value for the natural frequency would result. Moreover, in the open interval (2.6023, 71.1304) the natural frequency would be negative. As for the flexural rigidity, it takes negative values in the range (30.2, 71.1304). In order for *both* the natural frequency and the flexural rigidity to take positive values, the region $\gamma \in [2.6003, 71.1304]$ must be excluded from the analytical analysis presented herein.

Figure 9.18 portrays $D(\xi)$ for different values of γ, namely $\gamma = 0, 1, 2, 2.5$. Figure 9.19 depicts $D(\xi)$ for $\gamma = 72, 73, 75$ and 80.

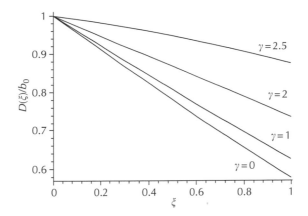

FIGURE 9.18
Variation of the flexural rigidity for $\gamma = 0; 1; 2; 2.5$

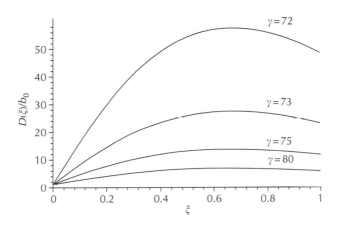

FIGURE 9.19
Variation of the flexural rigidity for $\gamma = 72; 73; 75; 80$

We conclude that the proposed method produces closed-form solutions only outside the region $\gamma \in [2.6023, 71.1304]$ on the positive semi-axis of γ. The flexural rigidity coefficient of the translational spring can still be chosen arbitrarily. Yet, the flexural rigidity of the beam at its origin, b_0, must be chosen outside the interval $[71.1304^{-1}, 2.6023^{-1}] = [0.01406, 0.3843]$. The number of closed-form solutions derived in this section is infinite, since b_0 can belong to the entire positive semi-axis, except the above interval.

9.3.4   Linear Inertial Coefficient

In this case, Eq. (9.338) in conjunction with Eqs. (9.336)–(9.338), leads to a system of six equations with seven unknowns. To solve it, as in the previous case we take the coefficient b_0 to be a constant. Then, the system reads

$$A_0 = 24b_0 + \frac{6kb_1}{(b_0 - k/3)} - \frac{3a_0\omega^2}{(b_0 - k/3)}$$

$$A_1 = 72b_1 + 18\frac{k}{(b_0 - k/3)}b_2 + \left(\frac{4 + k/6b_0}{(1 - k/3b_0)}a_0 - 3\frac{1}{(1 - k/3b_0)}a_1 \right)\omega^2$$

$$A_2 = 144b_2 + \frac{36kb_3}{(b_0 - k/3)} + \frac{(4 + k/6b_0)a_1\omega^2}{(1 - k/3b_0)}$$

$$A_3 = 240b_3 + \frac{60kb_4}{(b_0 - k/3)} - \frac{ka_0\omega^2}{2(b_0 - k/3)}$$

$$A_4 = 360b_4 + 90\frac{k}{(b_0 - k/3)}b_5 - \left(a_0 + \frac{k}{2(b_0 - k/3)}a_1 \right)\omega^2$$

$$A_5 = 504b_5 - a_1\omega^2$$

$$(9.351)$$

The solution follows:

$$b_1 = (4/3)b_0(\gamma - 3)[(951\gamma^4 - 50{,}400\gamma^3 + 498{,}960\gamma^2 - 1{,}814{,}400\gamma + 2{,}177{,}280)a_1$$
$$+ (2{,}968\gamma^4 + 71{,}736\gamma^3 - 846{,}720\gamma^2 + 2{,}782{,}080\gamma - 2{,}903{,}040)a_0]/G_2$$

$$b_2 = -(16/3)(\gamma - 3)^2[(-317\gamma^3 - 10{,}080\gamma^2 + 75{,}600\gamma - 120{,}960)a_1$$
$$+ (504\gamma^3 - 1512\gamma^2)a_0]/G_2$$

$$b_3 = -64b_0(\gamma - 3)^3\gamma(27\gamma a_1 + 56\gamma a_0 - 168a_0)/G_2$$

$$b_4 = (256/3)b_0(\gamma - 3)^4(-27\gamma a_1 + 28\gamma a_0 - 84a_0)/G_2$$

$$b_5 = (5{,}120/3)b_0(\gamma - 3)^5a_1/G_2$$

$$(9.352)$$

where

$$G_2 = [(951\gamma^5 - 50{,}400\gamma^4 + 498{,}960\gamma^3 - 1{,}814{,}400\gamma^2 + 2{,}177{,}280\gamma)a_1$$
$$+ (2{,}968\gamma^5 - 250{,}824\gamma^4 + 3{,}024{,}000\gamma^3 - 14{,}636{,}160\gamma^2$$
$$+ 31{,}933{,}440\gamma - 26{,}127{,}360)a_0]$$

$$(9.353)$$

The natural frequency is obtained from the last of Eqs. (9.351):

$$\omega^2 = 504b_5/a_1 \qquad (9.354)$$

which, upon substitution of b_5 from Eq. (3.357) becomes

$$\omega^2 = 860{,}160 b_0 (\gamma - 3)^5 / G_2 \tag{9.355}$$

This case is similar to the case of the beam with uniform density: we cannot choose the coefficient γ arbitrarily . When γ belongs to the interval [28.2, 63.7637], we derive negative values for the flexural rigidity function. Thus, this region should be excluded from the consideration for physically realizable beams. In order to find regions of variations of γ with attendant non-negative natural frequency, one has to study the real roots of $G_2 = 0$. For the specific case $r_0 = r_1$ the real roots are: $\gamma_1 = 2.3562$, $\gamma_2 = 2.4661$, $\gamma_3 = 63.7637$. Equation (9.358) is positive only when $\gamma > 3$. Since $G_2(0) < 0$, and the expression $(\gamma - 3)^5 < 0$ in the interval $I_1 = [0, \gamma_1]$ we conclude that the natural frequency is positive in this interval. Finally, we obtain a positive natural frequency for $\gamma \in I_1 \cup I_2 \cup I_3$, where $I_2 = [\gamma_2, 3]$, $I_3 = [\gamma_3, \infty]$, $\cup$ denotes union. In order to have *both* the flexural rigidity and the natural frequency non-negative, for the beam with inertial coefficient $R(\xi) = r_0(1 + \xi)$, the coefficient γ must belong to the union of the above intervals, namely $I_1 \cup I_2 \cup I_3$. Figures. 9.20 and 9.21 depict the variation of the flexural rigidity $D(x)$ for different values of γ.

It must be stressed that the coefficient k of the elastic support can be chosen *arbitrarily*; an infinite number of beam flexural rigidities can be constructed so as to have a closed-form solution. The coefficient b_0 can be chosen arbitrarily on the positive semi-axis, except on the union of intervals arising from the above formulation.

In an analogous manner one can consider cases of parabolic, cubic and higher order variations of the inertial coefficient.

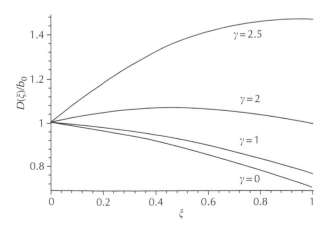

FIGURE 9.20
Variation of the flexural rigidity for different values of γ (0; 1; 2; 2.5)

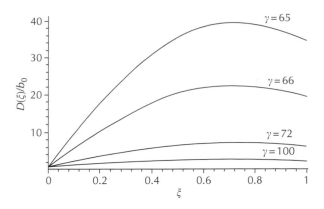

FIGURE 9.21
Variation of the flexural rigidity for different values of γ (65; 66; 72; 100)

To summarize, in this section, a benchmark closed-form solution was constructed for the mode shape and fundamental natural frequency of an inhomogneous beam with a translational end spring. The solution is attractive due to the extreme simplicity of the exact mode shape. Moreover, as in the other sections of this monograph, a rational expression was derived for the natural frequency.

10

Non-Polynomial Expressions for the Beam's Flexural Rigidity for Buckling or Vibration

Previous chapters dealt with the postulated representation of the elastic flexural rigidity in the form of a polynomial. The following question begs an answer: is it possible to determine closed-form solutions for buckling and vibration problems with non-polynomial flexural rigidity? An affirmative reply to this question is provided in this chapter. Some other seemingly provocative questions are posed and elucidated. Most intriguingly, the approximate method of successive approximation is revived here to derive closed-form solutions.

10.1 Both the Static Deflection and Vibration Mode of a Uniform Beam Can Serve as Buckling Modes of a Non-uniform Column

10.1.1 Introductory Remarks

There has been a long quest to find the closed-form solutions for vibration frequencies and buckling loads of non-uniform structures. These were pioneered 240 years ago by Euler (1759) for the cross section whose moment of inertia varies as

$$I(\xi) = I_0(a + b\xi)^m \tag{10.1}$$

where I_0 is the moment of inertia at the origin of coordinates, $\xi = x/L$ is the non-dimensional axial coordinate, a and b are real numbers and m is a real number, chosen so that the moment of inertia is a positive quantity. Euler (1759) studied two particular cases, namely $m = 2$ and $m = 4$, resulting in solutions in elementary functions, whereas Dinnik (1929, 1932) also considered the cases where $m \neq 2$ or $m \neq 4$, the solution being obtained in terms of Bessel functions. Dinnik (1929, 1932) reported some additional exact solutions, including exponentially varying moments of inertia.

In the discussion that ensued, Tukerman (1929) noted that Engesser (1909) presented a method of obtaining an infinite number of closed-form solutions.

Tuckerman (1929) suggested rewriting the buckling equation

$$EI(\xi)\frac{d^2W}{d\xi^2} + PL^2W = 0 \tag{10.2}$$

as follows

$$I(\xi) = -PL^2W/EW'' \tag{10.3}$$

where the prime denotes differentiation with respect to ξ. Substituting an arbitrary function $W(\xi)$ that satisfies the boundary conditions, results in the desired variation of the moment of inertia $I(\xi)$. Engesser (1909) discussed the case of parabolic deflection

$$W = cL(\xi - \xi^2) \tag{10.4}$$

which resulted in the moment of inertia

$$I(\xi) = 4I_0(1 - \xi^2) \tag{10.5}$$

where I_0 is the moment of inertia in the middle cross section, with buckling load being $P_{cr} = 8EI_0/L^2$.

In this chapter, we construct an infinite set of closed-form solutions for two cases of buckling. The first is the Euler's case of buckling of pinned columns whereas the second is for the clamped–free column under its own weight. We pose a seemingly provocative question: can a static deflection curve or a vibration mode of a beam serve as a buckling mode? It is shown that the reply to this question is in the affirmative, once it is made more specific. It turns out that a static displacement of a *uniform* beam may serve as a buckling mode of the *inhomogeneous* column. An analogous conclusion applies to the vibration mode of a uniform beam; it can also be an inhomogeneous column's buckling mode. One ought also mention the studies by Bert (1993, 1995), Stephen (1994), and Xie (1995) who stressed a relationship that exists between the fundamental natural frequency and maximum static deflection for various linear vibratory systems.

10.1.2   Basic Equations

Consider the auxiliary problem of the pinned uniform beam that is subjected to a distributed load of intensity

$$p(\xi) = p_0\xi^n \tag{10.6}$$

The differential equation governing the static deflection $w(\xi)$ reads

$$EI\frac{d^4w}{d\xi^4} = L^4p(\xi) \tag{10.7}$$

Integration and satisfaction of the boundary conditions results in the deflection

$$w(\xi) = \frac{\alpha L^{n+4}}{(n+1)(n+2)(n+3)(n+4)} \psi_1(\xi) \tag{10.8}$$

where

$$\psi_1(\xi) = \xi^{n+4} - \tfrac{1}{6}(n^2 + 7n + 12)\xi^3 + \tfrac{1}{6}(n^2 + 7n + 6)\xi$$
$$(n = 0, 1, 2, \ldots) \tag{10.9}$$

We have thus a countable infinity of functions that can be substituted into Eq. (10.3), to get an infinite sequence of distributions of the elastic modulus, leading to the closed-form solutions.

10.1.3 Buckling of Non-uniform Pinned Columns

Demanding the buckling mode of the non-uniform column to coincide with the static deflection $\psi_1(\xi)$ of the uniform beam, we get from Eq. (10.3)

$$I(\xi) = -PL^2 \psi / E\psi'' \tag{10.10}$$

We introduce *a parent moment of inertia* $I_p(\xi)$ as follows

$$I_p(\xi) = -\psi_1/\psi'' \tag{10.11}$$

Then, Eq. (10.10) is rewritten as

$$I(\xi) = PL^2 I_p(\xi)/E \tag{10.12}$$

For the function $I_p(\xi)$ we get the following expressions, listed here for the first ten values of n:

$$n = 0: \quad I_p(\xi) = \frac{1}{12}(1 + \xi - \xi^2) \tag{10.13}$$

$$n = 1: \quad I_p(\xi) = \frac{1}{20}\left(\frac{7}{3} - \xi^2\right) \tag{10.14}$$

$$n = 2: \quad I_p(\xi) = \frac{1}{30}\left(\frac{4(1+\xi)}{1+\xi+\xi^2} - \xi^2\right) \tag{10.15}$$

$$n = 3: \quad I_p(\xi) = \frac{1}{42}\left(\frac{6}{1+\xi^2} - \xi^2\right) \tag{10.16}$$

$$n = 4: \quad I_p(\xi) = \frac{1}{168}\left(\frac{25(1+\xi)}{1+\xi+\xi^2+\xi^3+\xi^4} - 3\xi^2\right) \tag{10.17}$$

$$n = 5: \quad I_p(\xi) = \frac{1}{72} \left(\frac{11}{1 + \xi^2 + \xi^4} - \xi^2 \right) \tag{10.18}$$

$$n = 6: \quad I_p(\xi) = \frac{1}{90} \left(\frac{14(1 + \xi)}{\sum_{i=0}^{6} \xi^i - \xi^2} \right) \tag{10.19}$$

$$n = 7: \quad I_p(\xi) = \frac{1}{330} \left(\frac{52}{\sum_{i=0}^{3} \xi^{2i} - 3\xi^2} \right) \tag{10.20}$$

$$n = 8: \quad I_p(\xi) = \frac{1}{132} \left(\frac{21(1 + \xi)}{\sum_{i=0}^{8} \xi^i - \xi^2} \right) \tag{10.21}$$

$$n = 9: \quad I_p(\xi) = \frac{1}{156} \left(\frac{25}{\sum_{i=0}^{4} \xi^{2i} - \xi^2} \right) \tag{10.22}$$

$$n = 10: \quad I_p(\xi) = \frac{1}{546} \left(\frac{88(1 + \xi)}{\sum_{i=0}^{10} \xi^i - 3\xi^2} \right) \tag{10.23}$$

Figure 10.1 represents the parent moments of inertia for $n = 0, 1, 2, 3$ and 4, whereas Figure 10.2 portrays the functions for $n = 5, 6, 7, 8, 9$ and 10.

The immediate question that arises is: what are the buckling loads? To answer the question, we rewrite Eq. (10.12) as follows:

$$P = EI(\xi)/L^2 I_p(\xi) \tag{10.24}$$

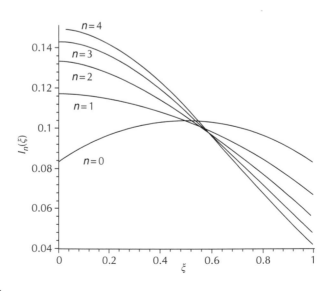

FIGURE 10.1

Parent flexural rigidities $I_n(\xi)$ for the pinned column under a compressive concentrated load, $n = 0, 1, 2, 3, 4$

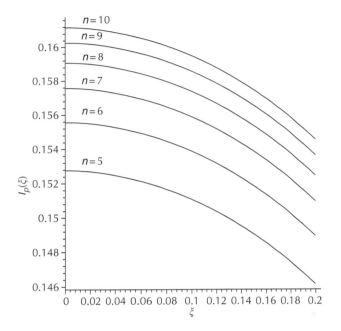

FIGURE 10.2
Parent flexural rigidities for the pinned column under a compressive concentrated load, $n = 5, 6, 7, 8, 9, 10$

We substitute $I(\xi)$ by the actual moment of inertia $I_a(\xi)$ to get

$$P = EI_a(\xi)/L^2 I_p(\xi) \qquad (10.25)$$

Now, consider the actual moment of inertia $I_a(\xi)$ as being *proportional* to the parent moment of inertia $I_p(\xi)$:

$$I_a(\xi) = \gamma I_p(\xi) \qquad (10.26)$$

where γ is the coefficient of proportionality. Then, Eq. (10.24) becomes

$$P = \gamma E/L^2 \qquad (10.27)$$

which is the expression for the buckling load. It depends on the arbitrary coefficient γ. Thus, by choosing γ at will one can achieve any pre-selected value of the buckling load.

As is seen from Eqs. (10.13)–(10.23) only Eqs. (10.13) and (10.14) furnish the polynomial expressions for the flexural rigidity; the other equations are rational expressions.

10.1.4   Buckling of a Column under its Own Weight

Now, consider the buckling of the column under its own weight. The governing differential equation reads (Timoshenko and Gere)

$$EI\frac{d^2W}{dx^2} = \int_x^L q_0[W(u) - W(x)]du \tag{10.28}$$

where q_0 is the load intensity. Equation (10.28) can be rewritten as

$$IW'' = Q\left[\int_\xi^1 W(\gamma)d\gamma - W(\xi)(1 - \xi)\right] \tag{10.29}$$

where

$$Q = q_0L^3/E \tag{10.30}$$

We now resort to the auxiliary problem of the deflection of the uniform beam under the transverse load given in Eq. (10.6). For clamped–free beam we get

$$w(\xi) = \frac{\alpha L^{n+4}}{(n + 1)(n + 2)(n + 3)(n + 4)}\psi_2(\xi) \tag{10.31}$$

where

$$\psi_1(\xi) = \xi^{n+4} - \tfrac{1}{6}(n + 2)(n + 3)(n + 4)\xi^3 + \tfrac{1}{2}(n + 1)(n + 3)(n + 4)\xi^2$$
$$(n = 0, 1, 2, \ldots) \tag{10.32}$$

In these new circumstances, we introduce the following parent moment of inertia:

$$I_{\mathrm{p}}(\xi) = \left[\int_\xi^1 \psi(\gamma)d\gamma - \psi(\xi)(1 - \xi)\right]\Big/\psi'' \tag{10.33}$$

For various n, the parent moments of inertia read

$$n = 0: \quad I_{\mathrm{p}}(\xi) = \frac{(3 + 6\xi - 6\xi^2 + 2\xi^3)}{30} \tag{10.34}$$

$$n = 1: \quad I_{\mathrm{p}}(\xi) = \frac{(26 + 52\xi - 42\xi^2 + 4\xi^3 + 5\xi^4)}{120(2 + \xi)} \tag{10.35}$$

$$n = 2: \quad I_{\mathrm{p}}(\xi) = \frac{71 + 142\xi - 102\xi^2 + 4\xi^3 + 5\xi^4 + 6\xi^5}{210(3 + 2\xi + \xi^2)} \tag{10.36}$$

$$n = 3: \quad I_p(\xi) = \frac{155 + 310\xi - 207\xi^2 + 4\xi^3 + 5\xi^4 + 6\xi^5 + 7\xi^6}{336(4 + 3\xi + 2\xi^2 + \xi^3)} \tag{10.37}$$

$$n = 4: \quad I_p(\xi) = \frac{295(1 + 2\xi) - 375\xi^2 + 4\xi^3 + 5\xi^4 + 6\xi^5 + 7\xi^6 + 8\xi^8}{504(5 + 4\xi + 3\xi^2 + 2\xi^3 + \xi^4)} \tag{10.38}$$

$$n = 5: \quad I_p(\xi) = \frac{511(1 + 2\xi) - 627\xi^2 + 4\xi^3 + 5\xi^4 + 6\xi^5 + 7\xi^6 + 8\xi^8 + 9\xi^8}{720(6 + 5\xi + 4\xi^2 + 3\xi^3 + 2\xi^4 + \xi^5)} \tag{10.39}$$

As is seen, only $n = 0$ yields a polynomial flexural rigidity expression. For $n \geq 1$ the rational, but non-polynomial expressions are derived.

Figure 10.3 depicts the parent moments of inertia for $n = 0, 1, 2, 3, 4$ and 5, while Figure 10.4 shows the dependence of I_p as a function of ξ, for n taking values between 6 and 10.

We are again confronted with the question of the evaluation of the buckling load. In view of the definition (10.30), and substituting $I(\xi) = I_a(\xi)$, Eq. (10.28) can be rewritten for the buckling intensity:

$$Q = q_0 L^3 / E = I_a(\xi)/I_p(\xi) \tag{10.40}$$

We choose the actual moment of inertia $I_a(\xi)$ to be proportional to $I_p(\xi)$:

$$I_a(\xi) = \delta I_p(\xi) \tag{10.41}$$

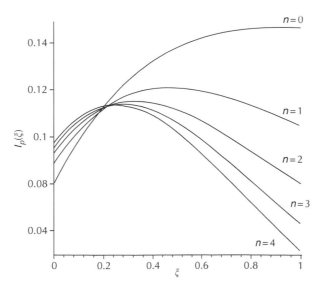

FIGURE 10.3
Parent flexural rigidities for the clamped–free column under its own weight; $n = 0, 1, 2, 3, 4$

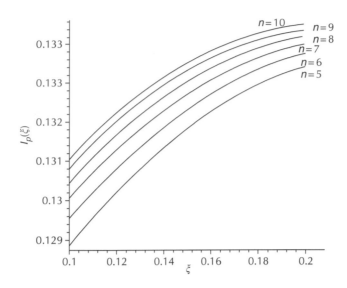

FIGURE 10.4
Parent flexural rigidities of the clamped–free column under its own weight; $n = 5, 6, 7, 8, 9, 10$

where δ is the coefficient of proportionality. Then, Eq. (10.40) takes the form

$$q_0 = \delta E / L^3 \tag{10.42}$$

Thus, the buckling load can be made arbitrary by an proper choice of the parameter δ.

10.1.5 Vibration Mode of a Uniform Beam as a Buckling Mode of a Non-uniform Column

Since the static deflections can serve as the buckling modes, it is natural to ask if the vibration mode of a uniform beam can serve in such a capacity too. For the clamped–free uniform beam the fundamental vibration mode reads (Rao, 1995)

$$\psi(\xi) = \sin \beta_1 \xi - \sinh \beta_1 \xi - r_1(\cos \beta_1 \xi - \cosh \beta_1 \xi) \tag{10.43}$$

where

$$r_1 = (\sin \beta_1 + \sinh \beta_1)/(\cos \beta_1 + \cosh \beta_1),$$
$$\beta_1 = 1.875104068711961166445308241078214162570111733531 \tag{10.44}$$

The parameter β_1 satisfies the characteristic equation $1 + \cos(\beta_1)\cosh(\beta_1) = 0$. The calculation of the parent moment of inertia with this function, via

Eq. (10.33) results in

$$I_p(\xi) = A/\beta_1 M - (1 - \xi)B/\beta_1^2 C \tag{10.45}$$

where

$$
\begin{aligned}
A = & -\{[\cos(\beta_1)]^2 + 2\cos(\beta_1)\cosh(\beta_1) + [\cosh(\beta_1)]^2 + [\sin(\beta_1)]^2 \\
& - [\sinh(\beta_1)]^2 - 2\cos(\beta_1)\cosh(\xi\beta_1) - \cos(\xi\beta_1)\cosh(\beta_1) \\
& - \cosh(\beta_1)\cosh(\xi\beta_1) - \sin(\xi\beta_1)\sinh(\beta_1) + \sinh(\beta_1)\sinh(\xi\beta_1)\} \\
B = & \sin(\xi\beta_1) - \sinh(\xi\beta_1) - \frac{[\sin(\beta_1) + \sinh(\beta_1)][\cos(\xi\beta_1) - \cosh(\xi\beta_1)]}{M} \\
C = & -\left[\sin(\xi\beta_1) + \sinh(\xi\beta_1) - \frac{[\sin(\beta_1) + \sinh(\beta_1)][\cos(\xi\beta_1) - \cosh(\xi\beta_1)]}{M}\right] \\
M = & \cos(\beta_1) + \cosh(\beta_1)
\end{aligned}
$$

$$\tag{10.46}$$

Using Eq. (10.40) again

$$Q = q_0 L^3/E = I_a(\xi)/I_p(\xi) \tag{10.47}$$

and choosing the actual moment of inertia

$$I_a(\xi) = \omega I_p(\xi) \tag{10.48}$$

we arrive at the buckling intensity

$$q = \omega E/L^3 \tag{10.49}$$

Due to the arbitrariness of the positive parameter ω, the buckling parameter can be made as *large* as desired. Figure 10.5 depicts the parent moment of inertia; it should be borne in mind that for an accurate portrayal of it, there is a need of extreme accuracy in the value of the parameter β_1, which is given in Eq. (10.44) upto 50 significant digits. Otherwise, the parent inertial moment figure may present a spurious discontinuity in the vicinity of $\xi = 1$.

10.1.6 Non-uniform Axially Distributed Load

Now, consider the case studied by Dinnik (1955a) in terms of Bessel functions. The column is under an axially distributed load proportional to ξ^t, where t is a positive integer. We use Eq. (10.28) with

$$q(x) = q_0(x/L)^t \tag{10.50}$$

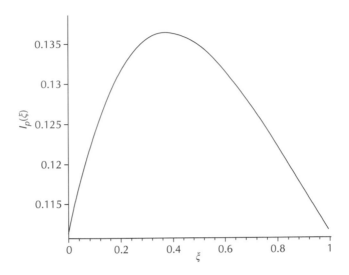

FIGURE 10.5
Parent flexural rigidity of the vibrating non-homogeneous column with the vibration mode of
the uniform beam serving as a buckling mode of the non-uniform column

Instead of Eq. (10.28) we have

$$I\frac{d^2w}{dx^2} = q_0 \int_x^L \left(\frac{u}{L}\right)^t [W(u) - W(x)]du$$

$$= q_0 \int_x^L \left(\frac{u}{L}\right)^t W(u)du - \frac{q_0 W(x)}{t+1}\left(L - \frac{x^{t+1}}{L^t}\right) \qquad (10.51)$$

With $u = \gamma L$ and $\xi = x/L$, Eq. (10.51) can be rewritten as

$$I\frac{d^2w}{dx^2} = Q\left[\int_\xi^1 \gamma^t W(\gamma)d\gamma - \frac{W(\xi)}{t+1}(1-\xi^{t+1})\right] \qquad (10.52)$$

Introducing $W(\xi) = \psi(\xi)$ and defining the parent moment of inertia as

$$I_p(\xi) = \frac{[I_0(\xi) - (\psi(\xi)/t+1)(1-\xi^{t+1})]}{\psi''} \qquad I_0(\xi) = \int_\xi^1 \gamma^t \psi(\gamma)d\gamma \qquad (10.53)$$

we get

$$I(\xi) = QI_p(\xi) \qquad (10.54)$$

Choosing the actual moment of inertia $I(\xi) \equiv I_a(\xi)$ as being proportional to $I_p(\xi)$,

$$I_a(\xi) = \lambda I_p(\xi) \tag{10.55}$$

we obtain

$$q_0 = E\lambda/L^3 \tag{10.56}$$

For ψ, we again use the vibration modes of the uniform clamped–free beams (10.32). The function $I_0(\xi)$ in Eq. (10.53) reads

$$I_0(\xi) = \int_\xi^1 \gamma^t \psi(\gamma) d\gamma = \frac{R_0 + R_1\xi^{3+t} + R_2\xi^{4+t} + R_3\xi^{5+n+t}}{6(3+t)(4+t)(5+n+t)} \tag{10.57a}$$

where

$$
\begin{aligned}
R_0 &= 432 + 822n + 495n^2 + 114n^3 + 9n^4 + 174t + 18t^2 + 2n^4t + 34n^3t + 2n^3t^2 \\
&\quad + 175n^2t + 15n^2t^2 + 317nt + 31nt^2 \\
R_1 &= -(720 + 1284n + 708n^2 + 156n^3 + 12n^4 + 324t + 36t^2 + 3n^4t + 51n^3t \\
&\quad + 3n^3t^2 + 273n^2t + 24n^2t^2 + 549nt + 57nt^2) \\
R_2 &= 360 + 462n + 213n^2 + 42n^3 + 3n^4 + 192t + 24t^2 + n^4t + 17n^3t + n^3t^2 \\
&\quad + 98n^2t + 9n^2t^2 + 232nt + 26nt^2 \\
R_3 &= -(72 + 42t + 6t^2)
\end{aligned}
$$

$$\tag{10.57b}$$

The analytical expressions obtained by the computerized symbolic algebraic code MAPLE, when using as ψ as the vibration mode of the uniform clamped–free beam, are not reproduced here, due to their length. Some parent inertial moments are depicted in Figure 10.6 for various values of $t(t = 0, 12, \ldots, 5)$ and $n = 1$.

10.1.7 Concluding Remarks

We obtained several *infinite series* of closed-form solutions for the buckling loads of non-uniform columns. On the one hand, using an infinite number of static deflections given in Eq. (10.8) we arrived at the infinite number of moments of inertia that lead to the postulated displacement, for the pinned columns.

On other hand, for the clamped–free columns, under their own weight, an *infinite number* of closed-form solutions were obtained by using the postulated mode shapes in Eq. (10.32).

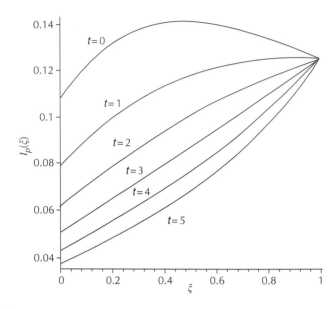

FIGURE 10.6
Parent flexural rigidities for the clamped–free column under a non-uniform axially distributed
load; $t = 0, 1, \ldots, 5$

We also showed that the *vibration* mode of the uniform cantilever in
Eq. (10.43) can serve as the exact *buckling* mode of the column under its own
weight. For the column under an axially distributed load proportional to an
arbitrary positive power of the axial coordinate, we obtained, on the one
hand, an infinite number of closed-form solutions for any integer value of t;
on the other hand, an infinite number of solutions has been found for any
n in Eq. (10.57) at any fixed value of t. A remarkable conclusion is that all
solutions were obtained in an exact, closed-form manner. Moreover, the solutions derived are attractive due to their *simplicity*. The present section provides
new, exact closed-form solutions, in addition to those reported by Tukerman
(1929), Duncan (1937) and Elishakoff and Rollot (1999).

10.2 Resurrection of the Method of Successive Approximations to Yield Closed-Form Solutions for Vibrating Inhomogeneous Beams

10.2.1 Introductory Comments

The method of successive approximations was apparently pioneered by
Picard (1905) and later employed by Hadamard (1932), Hohenemser and
Prager (1933), Kolousek (1954), Ananiev (1948), Nowacki (1963), Birger and
Panovko (1968), Biderman (1972) and others. Presently, it almost never

appears in the texts on vibration, due to the development of the versatile finite element method. The present study attempts to inject new life into it, but not for its original purpose of approximate solutions. Rather, a closed-form solution is obtained by the resurrected method. Moreover, the solutions obtained as intermediate approximations of the mode shape are shown to become closed-form solutions for the *inhomogeneous* beams.

The differential equation that governs the free vibrations of the beams of variable cross section reads

$$\frac{d^2}{dx^2}\left(EI(x)\frac{d^2W}{dx^2}\right) - \omega^2\rho A(x)W(x) = 0 \tag{10.58}$$

Birger and Mavliutov (1986) suggest replacing this equation by the equivalent integral equation. As they say (p. 401) "Eq. (10.58) can be brought to the form of an integral equation, which gives series of advantages for the approximate solution." The amplitude values of the shear force $V_y(x)$ and the bending moment $M_z(x)$ read

$$\frac{d}{dx}\left(EI(x)\frac{d^2W}{dx^2}\right) = -V_y(x) \tag{10.59}$$

$$EI(x)\frac{d^2W}{dx^2} = -M_z(x) \tag{10.60}$$

Birger and Mavliutov (1986) integrate Eq. (10.1) between x and L. For specificity, we consider the cantilever beam. We note that the shearing force at $x = L$ is absent. We get

$$-\frac{d}{dx}\left(EI(x)\frac{d^2W}{dx^2}\right) = \omega^2\int_x^L \rho A(x_1)W(x_1)dx_1 \tag{10.61}$$

We repeat the operation of integration, utilizing the condition $M_z(L) = 0$, to get

$$EI(x)\frac{d^2W}{dx^2} = \omega^2\int_x^L\int_{x_1}^L \rho A(x_2)W(x_2)dx_2dx_1 \tag{10.62}$$

We solve the direct vibration problem. Since it is assumed that the flexural rigidity $EI(x)$ is known, we divide both sides of Eq. (10.5) by it, and integrate the result twice from 0 to x. In view of the conditions

$$W(0) = 0 \qquad \frac{dW(0)}{dx} = 0 \tag{10.63}$$

we obtain for the cantilever beam

$$W(x) = \omega^2 \int_0^x \int_0^{x_1} \frac{1}{EI(x)} \int_{x_2}^L \int_{x_3}^L \rho A(x_4) W(x_4) dx_4 dx_3 dx_2 dx_1 \qquad (10.64)$$

This is a homogeneous integral equation that is equivalent to the differential equation (10.1) with the appropriate boundary conditions. It is an integral equation, since the unknown function appears under the integral sign. In short form, Eq. (10.7) can be written as

$$W = \omega^2 KW \qquad (10.65)$$

where KW is the integral operator, represented by the right-hand side of Eq. (10.7). It is easy to see that $W(x)$ is the solution of Eq. (10.8), as is the function $CW(x)$ where C is an arbitrary constant. Equation (10.8) possesses the trivial solution $W(x) \equiv 0$, but for some values of $\omega^2 = \omega_1^2, \omega_2^2$, etc. it has non-trivial solutions, with $\omega_1, \omega_2, \ldots$ being the natural frequencies. If one uses the following function for $W(x_1)$,

$$W_1(x) = (x/L)^2 \qquad (10.66)$$

one gets the second approximation

$$W_{(2)} = \omega_1^2 KW_{(1)} \qquad (10.67)$$

If $W_{(1)}$ were an exact solution, then the functions $W_{(2)}$ and $W_{(1)}$ would coincide for all values of x. But $W_{(1)}$ is not an exact solution. Hence,

$$W_2(x) \neq W_1(x) \qquad (10.68)$$

Birger and Mavliutov (1986) impose the condition

$$W_{(2)}(L) = W_{(1)}(L) = 1 \qquad (10.69)$$

resulting in

$$\omega_{(1)}^2 = [KW_{(1)}(L)]^{-1} = \left[\int_0^L \int_0^{z_1} \frac{1}{EI(x)} \int_{x_2}^L \int_{x_3}^L \rho A \frac{x_4^2}{L^2} dx_4 dx_3 dx_2 dx_1 \right]^{-1}$$
$$(10.70)$$

According to Birger and Mavliutov (1986), "usually already the first approximation yields an error not exceeding 2–5%. It can be shown that the process of successive approximations always converges to the first natural frequency. Obtaining by this means of consequent frequencies and modes requires performing the orthogonalization process." Collatz (1945) and Ponomarev et al. (1959) utilize another, integral criterion for determining the successive approximations of the natural frequency.

10.2.2 Evaluation of the Example by Birger and Mavliutov

We evaluate a particular example with two objectives in mind: (1) illustration of the method of successive approximations and (2) use of its *intermediate* results in Section 10.2.3 for inhomogeneous beams.

Birger and Mavliutov (1986) consider the example of a blade in the form of a cantilever beam. They write "for construction of the approximate model the cross-section of the blade can be assumed to be constant." For determination of the fundamental vibration frequency, they used the integral method:

$$\omega_{(1)}^2 = [KW_{(1)}(L)]^{-1} = \left[\frac{\rho A}{EI} \int_0^L \int_0^{x_1} \int_{x_2}^L \int_{x_3}^L \left(\frac{x_4}{L}\right)^2 dx_4 dx_3 dx_2 dx_1 \right]^{-1} \quad (10.71)$$

That corresponds to the first approximation in Eq. (10.9). We find

$$KW_{(1)}(x) = \frac{\rho A}{EI} \int_0^x \int_0^{x_1} \int_{x_2}^L \int_{x_3}^L \left(\frac{x_4}{L}\right)^2 dx_4 dx_3 dx_2 dx_1$$

$$= \frac{\rho A}{EI} \frac{1}{L^2} \int_0^x \int_0^{x_1} \left(\frac{1}{4}L^4 - \frac{1}{3}L^3 x_2 + \frac{1}{12}x_2^4 \right) dx_2 dx_1 \quad (10.72)$$

or

$$KW_{(1)}(x) = \rho A L^4 \left[\frac{1}{8}(x/L)^2 - \frac{1}{18}(x/L)^3 + \frac{1}{360}(x/L)^6 \right] / EI \quad (10.73)$$

The value of $KW_{(1)}(x)$ at $x = L$ is

$$KW_{(1)}(L) = \frac{13}{180}\rho A L^4 / EI \quad (10.74)$$

The first approximation of the natural frequency becomes

$$\omega_{(1)}^2 = \frac{180}{13} EI / \rho A L^4 \quad (10.75)$$

For further refinement we calculate

$$W_{(2)}(x) = \omega_{(1)}^2 KW_{(1)}(x) = KW_{(1)}(x)/KW_{(1)}(L)$$

$$= \frac{180}{13} \left[\frac{1}{8}(x/L)^2 - \frac{1}{18}(x/L)^3 + \frac{1}{360}(x/L)^6 \right] \quad (10.76)$$

Now,

$$KW_{(2)}(x) = \frac{\rho A}{EI} \int_0^x \int_0^{x_1} \int_{x_2}^L \int_{x_3}^L \left\{ \frac{180}{13} \left[\frac{1}{8} \left(\frac{x_4}{L}\right)^2 - \frac{1}{18} \left(\frac{x_4}{L}\right)^3 \right. \right.$$

$$\left. \left. + \frac{1}{360} \left(\frac{x_4}{L}\right)^6 \right] \right\} dx_4 dx_3 dx_2 dx_1$$

$$= \frac{\rho A L^4}{EI} \left[\frac{1}{131{,}040} \left(\frac{x}{L}\right)^{10} - \frac{1}{1{,}092} \left(\frac{x}{L}\right)^7 + \frac{1}{208} \left(\frac{x}{L}\right)^6 \right.$$

$$\left. - \frac{71}{1{,}092} \left(\frac{x}{L}\right)^3 + \frac{59}{416} \left(\frac{x}{L}\right)^2 \right] \tag{10.77}$$

leading to the second approximation

$$\omega_{(2)}^2 = [KW_{(2)}(L)]^{-1} = \frac{8190}{661} EI/\rho A L^4 \approx 12.39 EI/\rho A L^4 \tag{10.78}$$

Note that Birger and Mavliutov (1986) do not give an explicit expression for $KW_{(2)}(x)$; they quote a factor of 12.32 in Eq. (10.22). For further refinement, we calculate

$$W_{(3)}(x) = \omega_{(2)}^2 KW_{(2)}(x) = KW_{(2)}(x)/KW_{(2)}(L)$$

$$= \frac{8{,}190}{661} \left[\frac{1}{131{,}040} \left(\frac{x}{L}\right)^{10} - \frac{1}{1{,}092} \left(\frac{x}{L}\right)^7 + \frac{1}{208} \left(\frac{x}{L}\right)^6 \right.$$

$$\left. - \frac{71}{1{,}092} \left(\frac{x}{L}\right)^3 + \frac{59}{416} \left(\frac{x}{L}\right)^2 \right] \tag{10.79}$$

We get

$$KW_{(3)}(x) = \frac{\rho A}{EI} \int_0^x \int_0^{x_1} \int_{x_2}^L \int_{x_3}^L W_{(3)}(x_4) dx_4 dx_3 dx_2 dx_1$$

$$= \frac{\rho A L^4}{EI} \left[\frac{1}{254{,}077{,}824} \left(\frac{x}{L}\right)^{14} - \frac{1}{698{,}016} \left(\frac{x}{L}\right)^{11} + \frac{1}{84{,}608} \left(\frac{x}{L}\right)^{10} \right.$$

$$\left. - \frac{71}{74{,}032} \left(\frac{x}{L}\right)^7 + \frac{413}{84{,}608} \left(\frac{x}{L}\right)^6 - \frac{45{,}541}{698{,}016} \left(\frac{x}{L}\right)^3 + \frac{12{,}031}{84{,}608} \left(\frac{x}{L}\right)^2 \right] \tag{10.80}$$

leading to the third approximation

$$\omega_{(3)}^2 = \frac{15{,}879{,}864}{1{,}284{,}461} \frac{EI}{\rho A L^4} \approx \frac{12.36 EI}{\rho A L^4} \tag{10.81}$$

which nearly coincides with the exact solution (Rao, 1995), with the factor $\approx$ $1.875^4 \approx 12.3596$.

10.2.3 Reinterpretation of the Integral Method for Inhomogeneous Beams

As we saw above, the integral method can be effectively utilized for the approximate solution of eigenvalue problems. Its new twist will be presented here for obtaining *closed-form* solutions by the integral method.

We pose the following problem: find closed-form solutions for beams of variable inertial coefficient, and variable flexural rigidity:

$$\delta(x) = \rho(x)A(x) \quad D(x) = E(x)I(x) \tag{10.82}$$

so that the beam possesses the pre-selected mode shape $\psi(x)$. To this end, we rewrite Eq. (10.5) by identifying the mode shape $W(x)$ with the selected function $\psi(x)$, in conjunction with Eq. (10.26):

$$D(x) = \left[\omega^2 \int_x^L \int_{x_1}^L \delta(x_2)\psi(x_2)dx_2dx_1\right] \Big/ \psi''(x) \tag{10.83}$$

By substituting various functions $\psi_j(x)$ into Eq. (10.27) we obtain appropriate expressions for the flexural rigidity $D(x)$. We adopt functions

$$\psi_j(x) = (x/L)^{j+4} - \tfrac{1}{6}(j+2)(j+3)(j+4)(x/L)^3 + \tfrac{1}{2}(j+1)(j+3)(j+4)(x/L)^2 \tag{10.84}$$

valid for $j \geq 0$. The expressions in Eq. (10.84) are proportional to the displacement of the uniform beam under the load $(x/L)^j$.

Consider a particular example. Let the width $b(x)$ of the cross section of the beam be constant and equal to b, whereas the height varies linearly:

$$h(x) = (h_1 - h_0)x/L + h_0 \tag{10.85}$$

where h_0 is the height at $x = 0$, while h_1 is the height at $x = L$. Then, the cross-sectional area is

$$A(x) = bh(x) = bh_0\left(1 + \frac{h_1 - h_0}{h_0}\frac{x}{L}\right) \tag{10.86}$$

or, with α being the ratio of the heights,

$$\alpha = h_1/h_0 \tag{10.87}$$

we have

$$A(x) = bh(x) = A_0[1 - (1 - \alpha)x/L] \tag{10.88}$$

where A_0 is the cross-sectional area at $x = 0$. The moment of inertia reads

$$I(x) = \tfrac{1}{12}bh(x)^3 = I_0[1 - (1 - \alpha)x/L]^3 \tag{10.89}$$

where

$$I_0 = \tfrac{1}{12}bh_0^3 \tag{10.90}$$

is the moment of inertia at $x = 0$. The modulus of elasticity is written as

$$E(x) = E_0 e(x) \tag{10.91}$$

where E_0 is the modulus of elasticity at $x = 0$. The flexural rigidity becomes

$$D(x) = D_0 e(x)[1 - (1 - \alpha)x/L]^3 \tag{10.92}$$

where D_0 is the flexural rigidity at the origin

$$D_0 = E_0 I_0 \tag{10.93}$$

Let the density be given as a polynomial of mth order

$$\rho(x) = \rho_0 \varphi(x) \tag{10.94}$$

where

$$\varphi(x) = 1 + \beta_1(x/L) + \beta_2(x/L)^2 + \beta_3(x/L)^3 + \cdots + \beta_m(x/L)^m \tag{10.95}$$

Equation (10.27) can be rewritten as

$$(x)[1 - (1 - \alpha)x/L]^3$$
$$= \left[\omega^2 \rho_0 A_0 \int_x^L \int_{x_1}^L \varphi(x)[1 - (1 - \alpha)x/L] \times \psi(x_2) dx_2 dx_1\right] \Big/ D_0 \psi'' \tag{10.96}$$

We define the natural frequency as

$$\omega^2 = \gamma D_0/\rho_0 A_0 L^4 \tag{10.97}$$

where γ is the arbitrary parameter. Since the factor multiplying γ is a known quantity, one can maintain that an arbitrary value of the natural frequency can be obtained by designing a material with given modulus of elasticity. Indeed, in order for the beam to have an arbitrary natural frequency as given in Eq. (10.38), the modulus of elasticity must take the form

$$e(x) = \gamma \tilde{e}(x) \tag{10.98}$$

where $\tilde{e}(x)$ reads

$$\tilde{c}(x) = \frac{E(x)}{E(0)} \qquad E(x) = \frac{1}{L^4} \frac{\left[\int_x^L \int_{x_1}^L \varphi(x_2)[1 - (1-\alpha)x_2/L]\psi(x_2)dx_2dx_1\right]}{\psi''[1 - (1-\alpha)x/L]^3}$$

$$(10.99)$$

The quantity $\tilde{e}(x)$ can be designated as a *parent modulus of elasticity*. The function $e(x) = \gamma\tilde{e}(x)$ then produces a one-parameter *family* of elastic moduli. Let us, now, consider the particular cases.

10.2.4 Uniform Material Density

Let the density of the beam's material be constant:

$$\rho(x) = \rho_0 \qquad \beta_j = 0 \quad j = 1, 2, \ldots, m \qquad (10.100)$$

The parent elastic modulus is

$$\tilde{e}_0(\xi) = \frac{(40 + 142\alpha) + (10 + 102\alpha)\xi - (20 - 62\alpha)\xi^2 - (50 - 22\alpha)\xi^3 + (25 - 18\alpha)\xi^4 - (5 - 5\alpha)\xi^5}{2520[1 - (1-\alpha)\xi]^3}$$

$$(10.101)$$

Note that Eq. (45) in the paper by Elishakoff (2000f) contains a numerical error (Storch, 2001). Re-evaluation of the numerical examples, presented in Sections 10.2.4–10.2.6 was conducted by Ms. Roberta Santoro. As can be seen, we obtain a rational expression for any α, except $\alpha = 1$, corresponding to the beam of uniform cross section, in which case

$$\tilde{e}_0(\xi) = (26 + 16\xi + 6\xi^2 - 4\xi^3 + \xi^4)/360 \qquad (10.102)$$

and the parent modulus of elasticity becomes a polynomial expression. The expressions (10.45) and (10.46) correspond to the function (10.28) with $j = 0$, which is the reason for the subscript 0 in Eqs. (10.45) and (10.46). Figure 10.7 depicts the parent modulus of elasticity for values $\alpha_1 = \frac{1}{3}, \alpha_2 = 0.5$ and $\alpha_3 = 1$.
    Fixing j at unity in Eq. (10.28) yields

$$\tilde{e}_1(\xi) = [115 + 413\alpha + (30 + 298\alpha)\xi - (55 - 183\alpha)\xi^2$$
$$- (140 - 68\alpha)\xi^3 + (55 - 47\alpha)\xi^4 - (2 - 6\alpha)\xi^5$$
$$- (3 - 3\alpha)\xi^6]/\{3360[1 - (1-\alpha)\xi]^3(2 + \xi)\} \qquad (10.103)$$

Figure 10.8 portrays the variation of $\tilde{e}_1(\xi)$ for $\alpha_1 = \frac{1}{3}, \alpha_2 = 0.5$ and $\alpha_3 = 1$.

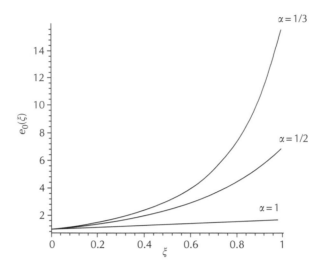

FIGURE 10.7
Variation of the parent modulus of elasticity when $j = 0$

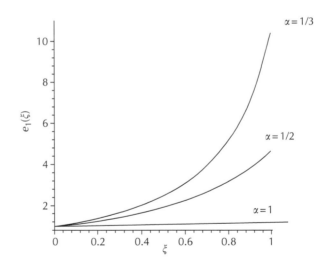

FIGURE 10.8
Variation of the parent modulus of elasticity when $j = 1$

For $j = 2$ in Eq. (10.28) we obtain

$$\tilde{e}_2(\xi) = [805 + 2921\alpha + (215 + 2107\alpha)\xi - (375 - 1302\alpha)\xi^2$$
$$- (965 - 497\alpha)\xi^3 + (335 - 308\alpha)\xi^4 - (3 - 21\alpha)\xi^5$$
$$- (5 - 14\alpha)\xi^6 - (7 - 7\alpha)\xi^7]/\{5120[1 - (1 - \alpha)\xi]^3(3 + 2\xi + \xi^2)\}$$

$$(10.104)$$

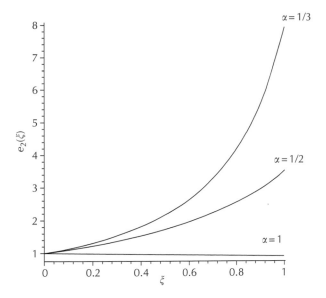

FIGURE 10.9
Variation of the parent modulus of elasticity when $j = 2$

Figure 10.9 shows the function $\tilde{e}_2(\xi)$ for various values of α.
For $j = 3$ one gets

$$\tilde{e}_3(\xi) = [1096 + 3984\alpha - (1895 + 5080\alpha)\xi + 2520\xi^4 - (2142 - 1512\alpha)\xi^5$$
$$+ (420 - 420\alpha)\xi^6 + 5\xi^9 - (4 - 4\alpha)\xi^{10}]$$
$$/\{5120[1 - (1 - \alpha)\xi]^3(\xi - 1)^2(4 + 3\xi + 2\xi^2 + \xi^3)\} \tag{10.105}$$

Figure 10.10 illustrates the dependence of $\tilde{e}_3(\xi)$ on ξ for different values of α.

10.2.5 Linearly Varying Density

Let the material density vary as

$$\rho(x) = \rho_0(1 + \beta x/L) \tag{10.106}$$

whereas the cross-sectional area and the moment of inertia vary as in Eqs. (10.32) and (10.34), respectively. The following results are obtained for the parent modulus of elasticity for $j = 0$:

$$\tilde{e}_0(\xi) = \sum_{j=0}^{6} A_j \xi^j / 10{,}080[1 - (1 - \alpha)\xi]^3$$

$$A_0 = 160 + 568\alpha + 103\beta + 465\alpha\beta \qquad A_1 = -(40 + 408\alpha + 46\beta + 362\alpha\beta)$$

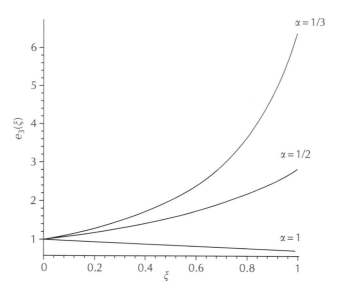

FIGURE 10.10
Variation of the parent modulus of elasticity when $j = 3$

$$A_2 = -(80 - 248\alpha + 11\beta - 259\alpha\beta) \quad A_3 = -(200 - 88\alpha + 68\beta - 156\alpha\beta)$$
$$A_4 = 100 - 72\alpha - 125\beta + 53\alpha\beta \quad A_5 = -(20 - 20\alpha - 70\beta + 50\alpha\beta)$$
$$A_6 = -(15 - 15\alpha)\beta$$

$$(10.107)$$

Again, for a beam with uniform cross section the variation is polynomial, while for values of $\alpha \neq 1$, the rational expression is obtained. For $j = 1, 2$ and 3 we arrive at

$$\tilde{e}_1(\xi) = \sum_{j=0}^{7} B_j \xi^j / 10{,}080[1 - (1 - \alpha)\xi]^3 (2 + \xi)$$

$$B_0 = 345 + 1{,}239\alpha + 223\beta + 1{,}016\alpha\beta$$
$$B_1 = 90 + 894\alpha + 101\beta + 793\alpha\beta$$
$$B_2 = -(165 - 549\alpha + 21\beta - 570\alpha\beta) \quad\quad (10.108)$$
$$B_3 = -(420 - 204\alpha + 143\beta - 347\alpha\beta)$$
$$B_4 = 165 - 141\alpha - 256\beta + 124\alpha\beta$$
$$B_5 = -(6 - 18\alpha - 117\beta + 99\alpha\beta)$$
$$B_6 = -9 + 9\alpha + 5\beta \quad B_7 = -(7 - 7\alpha)\beta$$

$$\tilde{e}_2(\xi) = \sum_{j=0}^{7} C_j \xi^j / 75{,}600[1 - (1-\alpha)\xi]^3 (3 + 2\xi + \xi^2)$$

$$C_0 = 4{,}025 + 14{,}560\alpha + 2{,}608\beta + 1{,}195\alpha\beta$$

$$C_1 = 1{,}075 + 1{,}053\alpha + 1{,}191\beta + 9{,}344\alpha\beta$$

$$C_2 = -(1{,}875 - 6{,}510\alpha + 226\beta - 6{,}736\alpha\beta)$$

$$C_3 = -(4{,}825 - 2{,}485\alpha + 1{,}643\beta - 4{,}128\alpha\beta) \qquad (10.109)$$

$$C_4 = 1{,}675 - 1{,}540\alpha - 3{,}060\beta + 1{,}520\alpha\beta$$

$$C_5 = -(15 - 105\alpha - 1{,}193\beta + 1{,}088\alpha\beta)$$

$$C_6 = -(25 - 70\alpha - 14\beta - 84\alpha\beta)$$

$$C_7 = -(35 - 35\alpha + 21\beta - 56\alpha\beta)$$

$$C_8 = -(28 - 28\alpha)\beta$$

$$\tilde{e}_3(\xi) = \sum_{j=0}^{9} D_j \xi^j / 166{,}320[1 - (1-\alpha)\xi]^3 (4 + 3\xi + 2\xi^2 + 2\xi^3)$$

$$D_0 = 8(1{,}507 + 5{,}478\alpha + 978\beta + 4{,}500\alpha\beta)$$

$$D_1 = 3{,}267 + 31{,}768\alpha + 3{,}592\beta + 28{,}176\alpha\beta$$

$$D_2 = 2(-2{,}716 + 9{,}856\alpha - 320\beta + 10{,}176\alpha\beta)$$

$$D_3 = 14{,}311 + 7{,}656\alpha - 4{,}872\beta + 12{,}528\alpha\beta \qquad (10.110)$$

$$D_4 = 4(1{,}155 - 1{,}100\alpha - 2{,}276\beta + 1{,}176\alpha\beta)$$

$$D_5 = 11 + 176\alpha + 3{,}296\beta - 3{,}120\alpha\beta$$

$$D_6 = 2(-11 + 66\alpha - 6\beta + 72\alpha\beta)$$

$$D_7 = -33 + 88\alpha - 20\beta + 108\alpha\beta$$

$$D_8 = 4(-11 + 11\alpha - 7\beta + 18\alpha\beta) \qquad D_9 = 36(-1+\alpha)\beta$$

Figure 10.11–10.14 depict some of the dependences of the parent modulus of elasticity for various values of α, $\beta = 1$.

10.2.6 Parabolically Varying Density

Let the density vary as

$$\rho(x) = \rho_0[1 + \beta_1 x/L + \beta_2(x/L)^2] \qquad (10.111)$$

The expressions for the parent modulus of elasticity read, for $j = 0$,

$$\tilde{e}_0(\xi) = \sum_{j=0}^{9} E_j \xi^j / 30{,}240[1 - (1-\alpha)\xi]^3 (\xi - 1)^2$$

$$E_0 = [480 + 309\beta_1 + 215\beta_2 + (1{,}704 + 1{,}395\beta_1 + 1{,}180\beta_2)\alpha]$$

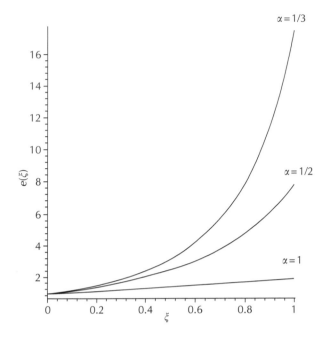

FIGURE 10.11
Variation of the parent modulus of elasticity for the beam with linearly varying material density
($\beta = 1, j = 0$)

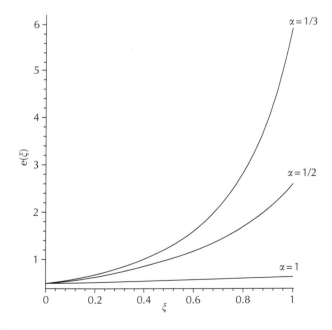

FIGURE 10.12
Variation of the parent modulus of elasticity for the beam with linearly varying material density
($\beta = 1, j = 1$)

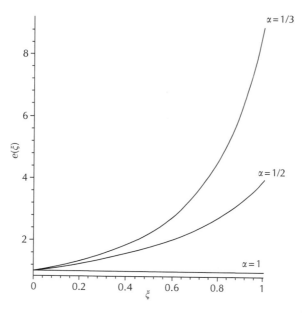

FIGURE 10.13
Variation of the parent modulus of elasticity for the beam with linearly varying material density
($\beta = 1, j = 2$)

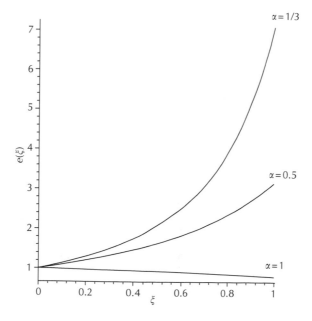

FIGURE 10.14
Variation of the parent modulus of elasticity for the beam with linearly varying material density
($\beta = 1, j = 3$)

$$E_1 = -[840 + 480\beta_1 + 309\beta_2 + (2{,}184 + 1{,}704\beta_1 + 1{,}395\beta_2)\alpha]$$

$$E_2 = 0 \quad E_3 = 0 \quad E_4 = 1{,}260 \quad E_5 = -1{,}260 + 756\beta_1 + 756\alpha$$

$$E_6 = [420 - 840\beta_1 + 504\beta_2 + (504\beta_1 - 336)\alpha]$$

$$E_7 = [-60 + 300\beta_1 - 600\beta_2 + (60 - 240\beta_1 + 360\beta_2)\alpha]$$

$$E_8 = [-45\beta_1 + 225\beta_2 + (45\beta_1 + 180\beta_2)\alpha] \quad E_9 = -(35 - 35\alpha)\beta_2$$

$$(10.112)$$

Again, this is a rational expression for $\alpha \neq 1$ and polynomial expression for $\alpha = 1$. For $j = 1, 2$ and 3 the expression for $\tilde{e}(\xi)$ are

$$\tilde{e}_1(\xi) = \sum_{j=0}^{8} F_j \xi^j / 50{,}400[1 - (1-\alpha)\xi]^3 (2 + \xi)$$

$$F_0 = [1{,}725 - 1{,}115\beta_1 + 778\beta_2 + (6{,}195 + 5{,}080\beta_1 + 4{,}302\beta_2)\alpha]$$

$$F_1 = [450 + 505\beta_1 + 441\beta_2 + (4{,}470 + 3{,}965\beta_1 + 3{,}524\beta_2)\alpha]$$

$$F_2 = -[825 + 105\beta_1 - 104\beta_2 - (2{,}745 + 2{,}850\beta_1 + 2{,}746\beta_2)\alpha]$$

$$F_3 = -[2{,}100 + 715\beta_1 + 233\beta_2 - (1{,}020 + 1{,}735\beta_1 + 1{,}968\beta_2)\alpha]$$

$$F_4 = [825 - 1{,}325\beta_1 - 570\beta_2 - (705 - 620\beta_1 - 1{,}190\beta_2)\alpha]$$

$$F_5 = -[30 - 585\beta_1 + 907\beta_2 - (90 - 495\beta_1 + 412\beta_2)\alpha]$$

$$F_6 = -[45 - 25\beta_1 - 436\beta_2 - (45 + 70\beta_1 - 366\beta_2)\alpha]$$

$$F_7 = -[(35 - 35\alpha)\beta_1 + (21 - 56\alpha)\beta_2] \quad F_8 = -(28 - 28\alpha)\beta_2$$

$$(10.113)$$

$$\tilde{e}_2(\xi) = \sum_{j=0}^{9} G_j \xi^j / 831{,}600[1 - (1-\alpha)\xi]^3 (3 + 2\xi + \xi^2)$$

$$G_0 = [44{,}275 + 28{,}688\beta_1 + 200{,}052\beta_2 + (160{,}160 + 131{,}472\beta_1 + 111{,}420\beta_2)\alpha]$$

$$G_1 = [11{,}825 + 13{,}101\beta_1 + 11{,}416\beta_2 + (115{,}885 + 102{,}784\beta_1 + 91{,}368\beta_2)\alpha]$$

$$G_2 = -[20{,}625 + 2{,}486\beta_1 - 2{,}780\beta_2 - (71{,}610 + 74{,}096\beta_1 + 71{,}316\beta_2)\alpha]$$

$$G_3 = -[53{,}075 + 18{,}073\beta_1 + 5{,}856\beta_2 - (27{,}335 + 45{,}408\beta_1 + 51{,}264\beta_2)\alpha]$$

$$G_4 = [18{,}425 - 33{,}660\beta_1 - 14{,}492\beta_2 - (16{,}940 + 33{,}660\beta_1 - 31{,}212\beta_2)\alpha]$$

$$G_5 = -[165 - 13{,}123\beta_1 + 23{,}128\beta_2 - (1{,}155 - 11{,}968\beta_1 + 11{,}160\beta_2)\alpha]$$

$$G_6 = -[257 + 154\beta_1 - 9{,}816\beta_2 - (770 + 924\beta_1 - 8{,}892\beta_2)\alpha]$$

$$G_7 = -[385 + 1{,}231\beta_1 + 140\beta_2 - (385 + 616\beta_1 + 756\beta_2)\alpha]$$

$$G_8 = -[(308 - 308\alpha)\beta_1 + (196 - 504\alpha)\beta_2] \quad G_9 = -(252 - 252\alpha)\beta_2$$

$$(10.114)$$

$$\tilde{e}_3(\xi) = \sum_{j=0}^{12} H_j \xi^j / 166{,}320[1 - (1-\alpha)\xi]^3(4 - 5\xi + \xi^5)$$

$$H_0 = [12{,}056 + 7{,}824\beta_1 + 5{,}475\beta_2 + (43{,}824 + 36{,}000\beta_1 + 30{,}525\beta_2)\alpha]$$

$$H_1 = -[20{,}845 + 12{,}056\beta_1 + 7{,}824\beta_2 + (55{,}880 + 43{,}824\beta_1 + 36{,}000\beta_2)\alpha]$$

$$H_2 = 0 \quad H_3 = 0 \quad H_4 = 27{,}720 \quad H_5 = -23{,}562 + 16{,}632\beta_1 + 16{,}632\alpha$$

$$H_6 = [4{,}620 - 15{,}708\beta_1 + 11{,}088\beta_2 - (4{,}620 - 11{,}088\beta_1)\alpha]$$

$$H_7 = [(3{,}300 - 3{,}300\alpha)\beta_1 + (11{,}220 + 7{,}920\alpha)\beta_2]$$

$$H_8 = (2{,}475 - 2{,}475\alpha)\beta_2 \quad H_9 = 55 \quad H_{10} = -44 + 44\alpha + 44\beta_1$$

$$H_{11} = [(36 + 36\alpha)\beta_1 + 36\beta_2] \quad H_{12} = -(30 - 30\alpha)\beta_2$$

$$(10.115)$$

10.2.7 Can Successive Approximations Serve as Mode Shapes?

We now pose a somewhat provocative question. The first approximation of the mode shape in Eq. (10.66) does not satisfy all boundary conditions. It satisfies geometric conditions, but not the essential ones. However, two subsequent approximations, utilized in this study are given in Eqs. (10.76) and (10.79), which satisfy all boundary conditions. Can they serve as *exact* mode shapes of some inhomogeneous beams? The reply is in the affirmative. Substituting Eq. (10.76) into (10.96) we get, for the beam with uniform density,

$$\tilde{e}(\xi) = \frac{3(416 + 301\xi + 186\xi^2 + 71\xi^3 - 44\xi^4 + 3\xi^5 + 2\xi^6 + \xi^7)}{416(3 + 2\xi + \xi^2)} \qquad (10.116)$$

Likewise, substitution of Eq. (10.79) into Eq. (10.96) yields

$$\tilde{e}(\xi) = (367{,}392 + 265{,}575\xi + 163{,}758\xi^2 + 61{,}941\xi^3 - 39{,}876\xi^4 + 3{,}270\xi^5$$

$$+ 2{,}112\xi^6 + 954\xi^7 - 204\xi^8 + 3\xi^9 + 2\xi^{10} + \xi^{11})/[122{,}464(3 + 2\xi + \xi^2)]$$

$$(10.117)$$

Likewise, one can obtain closed-form solutions for the beams with linearly or parabolically varying material density. We do not reproduce these formulas here, to save space. Figures 10.15–10.18 depict the variations of the modulus of elasticity; Figures 10.15 and 10.16 are for constant density, whereas Figures 10.17 and 10.18 depict the variations of the modulus of elasticity for the linearly varying density, $\beta = 1$.

10.2.8 Concluding Remarks

As we have demonstrated, a method that was designed and used for decades for *approximate* evaluation of the natural frequencies can be "twisted" to yield

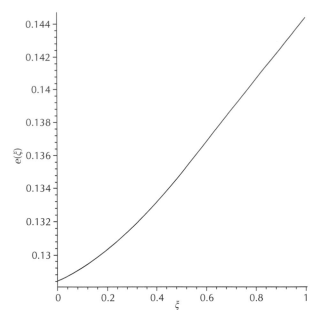

FIGURE 10.15
Variation of the parent modulus of elasticity produced by the second approximation serving as an exact mode shape of a beam with uniform material density

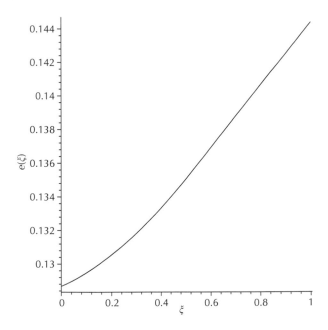

FIGURE 10.16
Variation of the parent modulus of elasticity produced by the third approximation serving as an exact mode shape of a beam with uniform material density

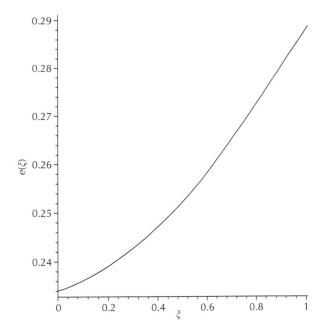

FIGURE 10.17
Variation of the parent modulus of elasticity produced by the second approximation serving as an exact mode shape of a beam with linearly varying material density ($\beta = 1$)

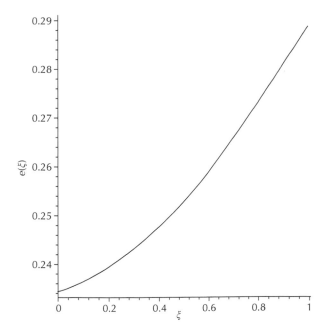

FIGURE 10.18
Variation of the parent modulus of elasticity produced by the third approximation serving as an exact mode shape of a beam with linearly varying material density ($\beta = 1$)

closed-form solutions. This methodology was applied to the longitudinally vibrating bars by Elishakoff et al. (2001b).

10.3 Additional Closed-Form Solutions for Inhomogeneous Vibrating Beams by the Integral Method

10.3.1 Introductory Remarks

In the recent study by Elishakoff (2000f), the method of successive approximations was revived to obtain closed-form solutions for the cantilever beams. Here, the above method is extended to include beams with two other sets of boundary conditions. Namely, we treat beams which are (1) pinned at both ends, (2) pinned at one end and guided at the other and (3) free at both ends. This section closely follows the article by Elishakoff Baruch and Becquet (2002).

The governing equation of the vibrating beam with a variable cross section reads:

$$\frac{d^2}{dx^2}\left(D(x)\frac{d^2W(x)}{dx^2}\right) - \omega^2 \rho(x)A(x)W(x) = 0 \qquad (10.118)$$

in which $D(x)$ denotes the flexural rigidity of the beam, $W(x)$ is the associated mode shape, $\rho(x)$ and $A(x)$ are, respectively, the material density and the cross-sectional area of the beam. In the method of successive approximations one replaces the equation by the associated integral equation. This could be written in the following form:

$$W(x) = \omega^2 KW(x) \qquad (10.119)$$

where K is the integral operator. The operator K depends on the boundary conditions, which will be treated hereinafter.

In this study, following Elishakoff (2000), we treat a particular case where the beam is non-prismatic. It has a constant width b, whereas the height varies linearly:

$$h(x) = h_0[(\alpha - 1)x/L + 1] \quad \alpha = h_l/h_0 \qquad (10.120)$$

where h_0 is the height of the beam at $x = 0$ and h_l is the height of the beam at $x = L$.

The beam's geometry allows one to get expressions for the cross-sectional area and the moment of inertia, respectively:

$$A(x) = bh(x) = A_0[1 + (\alpha - 1)x/L]$$

$$I(x) = bh(x)^3/12 = I_0[1 + (\alpha - 1)x/L]^3 \qquad (10.121)$$

where A_0 and I_0 are, respectively, the cross-sectional area and the moment of inertia at $x = 0$. The modulus of elasticity is given by

$$E(x) = E_0 e(x) \tag{10.122}$$

where E_0 is the modulus of elasticity at $x = 0$. Using Eqs. (10.121) and (10.122), one expresses the flexural rigidity as

$$D(x) = D_0 e(x)[1 + (\alpha - 1)x/L]^3 \tag{10.123}$$

where D_0 is the flexural rigidity at the origin. From Eqs. (10.121) and (10.122), we get

$$D_0 = E_0 I_0 \tag{10.124}$$

The material density is given by

$$\rho(x) = \rho_0 \varphi(x)$$
$$\varphi(x) = 1 + \beta_1(x/L) + \cdots + \beta_i(x/L)^i + \cdots + \beta_m(x/L)^m \tag{10.125}$$

where ρ_0 is the material density at $x = 0$. Note that the terms $e(x), \varphi(x)$ and α are dimensionless.

10.3.2 Pinned–Pinned Beam

10.3.2.1 Determination of the integral operator

By integrating Eq. (10.118) twice with respect to x, one obtains an expression for the bending moment:

$$D(x)\frac{d^2 W(x)}{dx^2} = \omega^2 \int_0^x \int_0^{x_1} \rho(x_2)A(x_2)W(x_2)dx_2dx_1 + A_1 x + A_2 \tag{10.126}$$

The following boundary conditions hold:

$$D(0)\frac{d^2 W(0)}{dx^2} = D(L)\frac{d^2 W(L)}{dx^2} = 0 \tag{10.127}$$

Equation (10.126) in conjunction with Eq. (10.127) allows to determine the constants A_1 and A_2 as follows:

$$A_1 = \omega^2 a_1 \quad A_2 = 0$$

$$a_1 = -\int_0^L \int_0^x \rho(x_1)A(x_1)W(x_1)dx_1dx/L \tag{10.128}$$

Equation (10.126) can be rewritten as follows:

$$\frac{d^2W(x)}{dx^2} = \omega^2 \frac{1}{D(x)} \int_0^x \int_0^{x_1} \rho(x_2)A(x_2)W(x_2)dx_2dx_1 + \omega^2 a_1 \frac{x}{D(x)} \quad (10.129)$$

Then, we integrate Eq. (10.129) twice:

$$
\begin{aligned}
W(x) &= \omega^2 \int_0^x \int_0^{x_1} \frac{1}{D(x_2)} \int_0^{x_2} \int_0^{x_3} \rho(x_4)A(x_4)W(x_4)dx_4dx_3dx_2dx_1 \\
&\quad + \omega^2 a_1 \int_0^x \int_0^{x_1} \frac{x_2}{D(x_2)}dx_2dx_1 + A_3x + A_4
\end{aligned}
\quad (10.130)
$$

We use the two other boundary conditions, i.e.

$$W(0) = W(L) = 0 \quad (10.131)$$

Equation (10.131) permits us to evaluate the two constants A_3 and A_4. These are

$$
\begin{aligned}
A_3 &= -\frac{\omega^2}{L}\Bigg[\int_0^L \int_0^{x_1} \frac{1}{D(x_2)} \int_0^{x_2} \int_0^{x_3} \rho(x_4)A(x_4)W(x_4)dx_4dx_3dx_2dx_1 \\
&\quad + a_1 \int_0^L \int_0^{x_1} \frac{x_2}{D(x_2)}dx_2dx_1 \Bigg]
\end{aligned}
\quad (10.132)
$$

$$A_4 = 0$$

Finally, we rewrite Eq. (10.130) as Eq. (10.119) with $KW(x)$ defined as

$$
\begin{aligned}
KW(x) &= \int_0^x \int_0^{x_1} \frac{1}{D(x_2)} \int_0^{x_2} \int_0^{x_3} \rho(x_4)A(x_4)W(x_4)dx_4dx_3dx_2dx_1 \\
&\quad - \frac{1}{L} \int_0^L \int_0^x \rho(x_1)A(x_1)W(x_1)dx_1dx \int_0^x \int_0^{x_1} \frac{x_2}{D(x_2)}dx_2dx_1 \\
&\quad - \frac{x}{L} \int_0^L \int_0^{x_1} \frac{1}{D(x_2)} \int_0^{x_2} \int_0^{x_3} \rho(x_4)A(x_4)W(x_4)dx_4dx_3dx_2dx_1 \\
&\quad - \frac{x}{L^2} \int_0^L \int_0^x \rho(x_1)A(x_1)W(x_1)dx_1dx \int_0^L \int_0^{x_1} \frac{x_2}{D(x_2)}dx_2dx_1
\end{aligned}
\quad (10.133)
$$

The integral operator will be utilized at a later stage.

10.3.2.2 Obtaining closed-form solutions

Equation (10.129) can be rewritten as

$$
e(x)[1 + (\alpha - 1)x/L]^3
$$
$$
= \left\{ \omega^2 \rho_0 A_0 \left[\int_0^x \int_0^{x_1} \varphi(x_2)[1 + (\alpha - 1)x_2/L]W(x_2)dx_2dx_1 \right. \right.
$$
$$
\left. \left. -(x/L) \int_0^L \int_0^x \varphi(x_1)[1 + (\alpha - 1)x_1/L]W(x_1)dx_1 dx \right] \right\} \bigg/ D_0 W''(x)
$$

$$(10.134)$$

One has to select the mode shape of the beam in order to express the appropriate modulus of elasticity $E(x)$. One can check that the following set of functions

$$
\psi_n(x) = (x/L)^{n+4} - \tfrac{1}{6}(n+3)(n+4)(x/L)^3 + \tfrac{1}{6}(n+1)(n+6)(x/L) \quad (10.135)
$$

satisfies all boundary conditions for the pinned–pinned beam. We postulate that the mode shape coincides with the function $\psi_n(x)$. Figure 10.19 portrays the variation of $\psi_n(x)$ for $0 \le x \le L$ and $0 \le n \le 3$.

At this stage, we also express the natural frequency as

$$
\omega^2 = \gamma D_0 / \rho_0 A_0 L^4 \quad (10.136)
$$

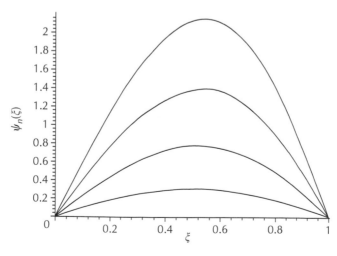

FIGURE 10.19
Variations of $\psi_n(\xi)$ for the beam pinned at both its ends

where γ is an arbitrary positive constant. Then, Eq. (10.134) becomes

$$e(x) = \gamma \left[\int_0^x \int_0^{x_1} \varphi(x_2)[1 + (\alpha - 1)x_2/L]\psi_n(x_2)dx_2dx_1 - (x/L) \right.$$

$$\left. \times \int_0^L \int_0^x \varphi(x_1)[1 + (\alpha - 1)x_1/L]\psi_n(x_1)dx_1dx \right]$$

$$\Big/ L^4[1 + (\alpha - 1)x/L]^3\psi_n''(x) \tag{10.137}$$

Using Eq. (10.137), we rewrite the expression for the modulus of elasticity:

$$e(x) = \gamma\tilde{e}(x)$$

$$\tilde{e}_n(x) = \left[\int_0^x \int_0^{x_1} \varphi(x_2)[1 + (\alpha - 1)x_2/L]\psi_n(x_2)dx_2dx_1 - (x/L) \right.$$

$$\left. \times \int_0^L \int_0^x \varphi(x_1)[1 + (\alpha - 1)x_1/L]\psi_n(x_1)dx_1dx \right]$$

$$\Big/ L^4[1 + (\alpha - 1)x/L]^3\psi_n''(x) \tag{10.138}$$

The quantity $\tilde{e}(x)$ is designated as the *parent modulus of elasticity*. Note that in the following, the subscripts in $\tilde{e}_n(x)$ denote the value n in the function $\psi_n(x)$. For each value of n, we determine the expression for the parent modulus of elasticity $\tilde{e}_n(x)$. We introduce the dimensionless parameter $\xi = x/L$. For $n = 0$, we get from Eq. (10.138):

$$\tilde{e}_0(\xi) = \frac{(25 + 17\alpha)(\xi + 1) + (17 - 45\alpha)\xi^2 - (10 + 18\alpha)\xi^3 + (32 - 18\alpha)\xi^4 + 10(\alpha - 1)\xi^5}{5040[1 + (\alpha - 1)\xi]^3}$$

$$\tag{10.139}$$

For the uniform beam, i.e., $\alpha = 1$, the parent modulus of elasticity reads

$$\tilde{e}_0(\xi) = \tfrac{1}{360}(3 + 3\xi - 2\xi^2 - 2\xi^3 + \xi^4) \tag{10.140}$$

resulting in a polynomial function of order 4 for the parent modulus of elasticity. For $\alpha \neq 1$, we obtain rational expressions for each n. These are not reproduced here due to limitations of space.

When the mode shape is represented by a fifth order polynomial, corresponding to the case $n = 1$ in Eq. (10.135), we obtain from Eq. (10.135) the following expression for the parent modulus of elasticity:

$$\tilde{e}_1(\xi) = [(73 + 51\alpha)(\xi + 1) + (51\alpha - 123)\xi^2 - (25 + 47\alpha)\xi^3$$

$$+ (59 - 47\alpha)\xi^4 + (3 + 9\alpha)\xi^5 + 9(\alpha - 1)\xi^6]/10{,}080[1 + (\alpha - 1)\xi]^3(1 + \xi)$$

$$\tag{10.141}$$

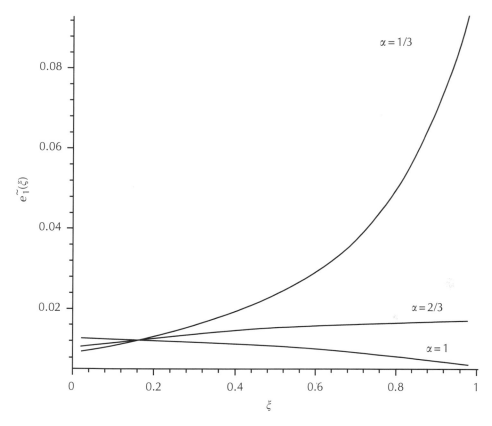

FIGURE 10.20
Dependence of the parent modulus of elasticity on ξ for the pinned–pinned beam (when $n = 1$)

Figure 10.20 portrays the variation of the parent modulus of elasticity $\tilde{e}_1(\xi)$ for various values of the coefficient, namely for $\alpha = \{1, \frac{2}{3}, \frac{1}{3}\}$.

We now treat the case $n = 2$, i.e., the mode shape is a polynomial function of sixth order. The parent modulus of elasticity associated with this case is given by

$$\tilde{e}_2(\xi) = [(128 + 91\alpha)(\xi + 1) + (91\alpha - 208)\xi^2 - (40 + 77\alpha)\xi^3$$
$$+ (86 - 77\alpha)\xi^4 + (2 + 7\alpha)\xi^5 + (2 + 7\alpha)\xi^6 + 7(\alpha - 1)\xi^7]$$
$$/15,120[1 + (\alpha - 1)\xi]^3(1 + \xi + \xi^2) \tag{10.142}$$

Figure 10.21 shows the variation of the function $\tilde{e}_2(\xi)$ within the interval $\xi \in [0; 1]$.

The parent modulus of elasticity for $n = 3$ reads

$$\tilde{e}_3(\xi) = [(139 + 100\alpha)(\xi + 1) + (100\alpha - 221)\xi^2 - (41 + 80\alpha)\xi^3$$
$$+ (85 - 80\alpha)\xi^4 + (1 + 4\alpha)(\xi^5 + \xi^6 + \xi^7) + 4(\alpha - 1)\xi^8]$$
$$/15120[1 + (\alpha - 1)\xi]^3(1 + \xi + \xi^2 + \xi^3) \qquad (10.143)$$

and Figure 10.22 illustrates the dependence of $\tilde{e}_3(\xi)$ on ξ for different values of α.

The following remarkable possibility was uncovered in Section 10.2: that of using the function, generated by the use of the method of successive approximation for *homogeneous beams*, as mode shapes for *inhomogeneous* beams. This is done for the case under study in Section 10.3.2.3.

10.3.2.3 Generating a feasible mode shape by successive approximations

Consider a beam with a constant cross-sectional area, and a constant flexural rigidity. In Eq. (10.119), the following first approximation of the mode shape is utilized:

$$W_{(1)}(x) = \tfrac{16}{5}[(x/L) - 2(x/L)^3 + (x/L)^4] \quad \text{with } W_{(1)}(L/2) = 1 \qquad (10.144)$$

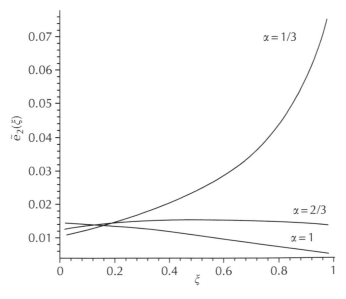

FIGURE 10.21
Dependence of the parent modulus of elasticity on ξ for the pinned–pinned beam (when $n = 2$)

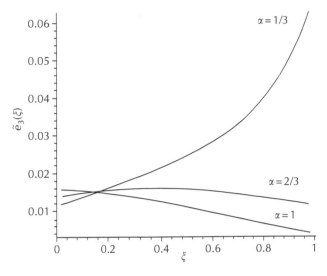

FIGURE 10.22
Dependence of the parent modulus of elasticity on ξ for the pinned–pinned beam (when $n = 3$)

A second approximation of the mode shape is generated by Eq. (10.119):

$$W_{(2)}(x) = \omega_{(1)}^2 K W_{(1)}(x) \tag{10.145}$$

The mode shape $W_{(1)}(x)$ is an approximation of the true mode shape; $W_{(2)}(x)$ does not coincide with $W_{(1)}(x)$ for each value of x. We impose the condition due to Birger and Mavliutov (1986):

$$W_{(2)}(L/2) = W_{(1)}(L/2) = 1 \tag{10.146}$$

By substituting Eq. (10.146) into Eq. (10.145), we obtain a first approximation for the natural frequency:

$$\omega_{(1)}^2 = [K W_{(1)}(L/2)]^{-1}$$
$$= (26{,}880/277) D/\rho A L^4 \approx 97.040 D/\rho A L^4 \tag{10.147}$$

Note that the coefficient is 0.4% lower than the exact coefficient π^4. The second approximation of the mode shape is as follows:

$$W_{(2)}(x) = \omega_{(1)}^2 K W_{(1)}(x) = K W_{(1)}(x)/K W_{(1)}(L/2)$$
$$= (256/1385)[17(x/L) - 28(x/L)^3 + 14(x/L)^5 - 4(x/L)^7 + (x/L)^8] \tag{10.148}$$

resulting with the attendant second approximation for the natural frequency:

$$\omega_{(2)}^2 = [KW_{(2)}(L/2)]^{-1}$$

$$= (52,652,160)/(540,553)D/\rho AL^4 \approx 97.404D/\rho AL^4 \qquad (10.149)$$

Note that the coefficient is 0.005% more than its exact counterpart. From the above equation, we are able to write a third approximation for the mode shape:

$$W_{(3)}(x) = \omega_{(2)}^2 KW_{(2)}(x) = KW_{(2)}(x)/KW_{(2)}(L/2)$$

$$= (4,096/2,702,765)[2,073(x/L) - 3,410(x/L)^3 + 1,683(x/L)^5$$

$$- 396(x/L)^7 + 55(x/L)^9 - 6(x/L)^{11} + (x/L)^{12}] \qquad (10.150)$$

The third approximation for the natural frequency of the corresponding mode shape is

$$\omega_{(3)}^2 = [KW_{(3)}(L/2)]^{-1} = (377,781,680,640/3,878,302,429)D/\rho AL^4$$

$$\approx 97.409D/\rho AL^4$$

$$W_{(4)}(x) = KW_{(3)}(x)/KW_{(3)}(L/2)$$

$$= (65,536/19,391,512,145)[929,569(x/L) - 1,529,080(x/L)^3$$

$$+ 754,572(x/L)^5 - 177,320(x/L)^7 + 24,310(x/L)^9$$

$$- 218,424,310(x/L)^{11} + 140(x/L)^{13} - 8(x/L)^{15} + (x/L)^{16}]$$

$$(10.151)$$

In the third approximation the difference from the exact solution constitutes only $6 \times 10^{-5}\%$ for the natural frequency.

From the successive approximations of the mode shape in the *homogeneous* beam, we can determine the flexural rigidity of the associated *inhomogeneous* beam in order to have a closed-form solution for the natural frequency. For instance, by substituting Eq. (10.148) into Eq. (10.138), we postulate the mode shape to coincide with Eq. (10.148). We obtain the expression for the flexural rigidity. Then, its natural frequency is given in Eq. (10.136). This flexural rigidity reads

$$\tilde{e}(\xi) = [(2,028 + 1,382\alpha)(1 + \xi) + (1,382\alpha - 3,582)\xi^2 - (777 + 1,423\alpha)\xi^3$$

$$+ (1,995 - 1,423\alpha)\xi^4 + (147 + 425\alpha)\xi^5 + (425\alpha - 513)\xi^6$$

$$- (70\alpha + 18)\xi^7 + (92 - 70\alpha)\xi^8 + 18(\alpha - 1)\xi^9]/110,880[1 - (1 - \alpha)\xi]^3$$

$$\times (3 + 3\xi - 2\xi^2 - 2\xi^3 + \xi^4) \qquad (10.152)$$

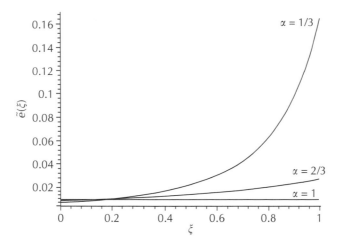

FIGURE 10.23
Dependence of the parent modulus of elasticity on ξ for the pinned–pinned beam

An analogous procedure, by postulating the expressions given in either Eqs. (10.144), (10.150) or (10.151) to serve as exact mode shapes, yields the variations of $\tilde{e}(x)$ presented in Figure 10.23.

10.3.3 Guided–Pinned Beam

10.3.3.1 Determination of the integral operator

The governing equation, upon being integrated once, results in

$$\frac{d}{dx}\left(D(x)\frac{d^2W(x)}{dx^2}\right) = \omega^2 \int_0^x \rho(x_1)A(x_1)W(x_1)dx_1 + A_1 \tag{10.153}$$

On the end that is guided, the shear force vanishes. In the left-hand side of Eq. (10.153), we recognize the expression for the shear force. Then, we obtain directly $A_1 = 0$. Equation (10.153) is integrated one more time:

$$D(x)\frac{d^2W(x)}{dx^2} = \omega^2 \int_0^x \int_0^{x_1} \rho(x_2)A(x_2)W(x_2)dx_2dx_1 + A_2 \tag{10.154}$$

The left-hand side in this equation is the expression for the bending moment. The bending moment is zero on the pinned end $x = L$. Hence,

$$\omega^2 \int_0^L \int_0^x \rho(x_1)A(x_1)W(x_1)dx_1dx + A_2 = 0 \tag{10.155}$$

We express the constant A_2 from Eq. (10.155):

$$A_2 = \omega^2 a_2 \quad a_2 = -\int_0^L \int_0^x \rho(x_1) A(x_1) W(x_1) dx_1 dx \qquad (10.156)$$

In order to obtain the expression for the mode shape $W(x)$, we integrate Eq. (10.154) successively twice, and for each integration, we determine the integration constants with the aid of boundary conditions. The first integration leads to

$$\frac{dW(x)}{dx} = \omega^2 \left[\int_0^x \frac{1}{D(x_1)} \int_0^{x_1} \int_0^{x_2} \rho(x_3) A(x_3) W(x_3) dx_3 dx_2 dx_1 \right.$$
$$\left. + a_2 \int_0^x \frac{1}{D(x_1)} dx_1 \right] + A_3 \qquad (10.157)$$

On the guided end, the remaining boundary condition readse

$$\frac{dW(0)}{dx} = 0 \qquad (10.158)$$

We obtain $A_3 = 0$. The integration of Eq. (10.157) yields the following expression for the mode shape:

$$W(x) = \omega^2 \left[\int_0^x \int_0^{x_1} \frac{1}{D(x_2)} \int_0^{x_2} \int_0^{x_3} \rho(x_4) A(x_4) W(x_4) dx_4 dx_3 dx_2 dx_1 \right.$$
$$\left. + a_2 \int_0^x \int_0^{x_1} \frac{1}{D(x_2)} dx_2 \right] + A_4 \qquad (10.159)$$

The last boundary condition is $W(L) = 0$, resulting in

$$A_4 = \omega^2 a_4$$
$$a_4 = -\int_0^L \int_0^x \frac{1}{D(x_1)} \int_0^{x_1} \int_0^{x_2} \rho(x_3) A(x_3) W(x_3) dx_3 dx_2 dx_1 dx \qquad (10.160)$$
$$- a_2 \int_0^L \int_0^x \frac{1}{D(x_1)} dx_1$$

In this particular case, the integral operator K reads

$$KW(x) = \int_0^x \int_0^{x_1} \frac{1}{D(x_2)} \int_0^{x_2} \int_0^{x_3} \rho(x_4)A(x_4)W(x_4)dx_4dx_3dx_2dx_1$$

$$- \int_0^L \int_0^x \rho(x_1)A(x_1)W(x_1)dx_1dx \int_0^x \int_0^{x_1} \frac{1}{D(x_2)}dx_2$$

$$- \int_0^L \int_0^x \frac{1}{D(x_1)} \int_0^{x_1} \int_0^{x_2} \rho(x_3)A(x_3)W(x_3)dx_3dx_2dx_1dx$$

$$- \int_0^L \int_0^x \rho(x_1)A(x_1)W(x_1)dx_1dx \int_0^L \int_0^x \frac{1}{D(x_1)}dx_1 \qquad (10.161)$$

This integral operator will be utilized later on, to generate candidates for the exact mode shapes.

10.3.3.2   Obtaining closed-form solutions
From Eq. (10.154), we obtain

$$e(x)[1 + (\alpha - 1)x/L]^3 = \left\{ \omega^2 \rho_0 A_0 \left[\int_0^x \int_0^{x_1} \varphi(x_2)[1 + (\alpha - 1)x_2/L]W(x_2)dx_2dx_1 \right.\right.$$

$$\left.\left. - \int_0^L \int_0^x \varphi(x_1)[1 + (\alpha - 1)x_1/L]W(x_1)dx_1dx \right] \right\} \Big/ W''(x)D_0$$

$$(10.162)$$

Analogous to the case of a pinned–pinned beam, we postulate the mode shape to be as follows:

$$\psi_n(x) = (x/L)^{n+4} - \tfrac{1}{2}(n^2 + 7n + 12)(x/L)^2 + \tfrac{1}{2}(n^2 + 7n + 10) \qquad (10.163)$$

Figure 10.24 shows the function $\psi_n(x)$.

One can check that this function satisfies the boundary conditions of a guided–pinned beam for any integer value of n. The natural frequency is taken as in Eq. (10.136) Then, Eq. (10.162) can be rewritten as follows:

$$e(x) = \gamma \left[\int_0^x \int_0^{x_1} \varphi(x_2)[1 + (\alpha - 1)x_2/L]\psi_n(x_2)dx_2dx_1 \right.$$

$$\left. - \int_0^L \int_0^x \varphi(x_1)[1 + (\alpha - 1)x_1/L]\psi_n(x_1)dx_1dx \right] \Big/ L^4[1 + (\alpha - 1)x/L]^3 \psi_n''(x)$$

$$(10.164)$$

We introduce the parent modulus of elasticity $\tilde{e}(x)$:

$$e(x) = \gamma \tilde{e}(x) \qquad (10.165)$$

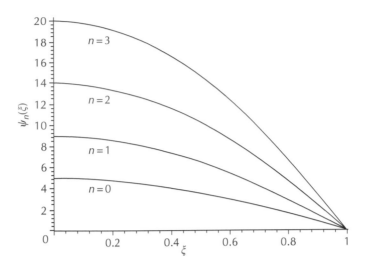

FIGURE 10.24
Variations of $\psi_n(\xi)$ for the guided–pinned beam

$$
\tilde{e}_n(x) = \left[\int_0^x \int_0^{x_1} \varphi(x_2)[1 + (\alpha - 1)x_2/L]\psi_n(x_2)dx_2dx_1 \right.
$$
$$
\left. - \int_0^L \int_0^x \varphi(x_1)[1 + (\alpha - 1)x_1/L]\psi_n(x_1)dx_1dx \right]
$$
$$
\Big/ L^4[1 + (\alpha - 1)x/L]^3\psi_n''(x)
$$

In the following, we express the parent modulus of elasticity in the particular case of a beam with a constant material density ρ_0. This is done for different expressions of the mode shape (corresponding to different values of n), and different types of non-prismatic beams (by taking different values of the coefficient α). The first case is $n = 0$. The parent modulus of elasticity associated with this mode shape reads

$$
\tilde{e}_0(\xi) = (310 + 117\alpha)(1 + \xi) + (117\alpha - 215)\xi^2 - (40 + 58\alpha)\xi^3 + (65 - 58\alpha)\xi^4
$$
$$
+ (2 + 5\alpha)\xi^5 + 5(\alpha - 1)\xi^6]/2520[1 - (1 - \alpha)\xi]^3(1 + \xi) \qquad (10.166)
$$

In the case $\alpha = 1$, we obtain a polynomial function:

$$
\tilde{e}_0(\xi) = \frac{1}{360}(\xi^4 - 14\xi^2 + 61) \qquad (10.167)
$$

Figure 10.25 illustrates the dependence of $\tilde{e}_0(\xi)$ on ξ for $\alpha = \{1, \frac{2}{3}, \frac{1}{3}\}$.

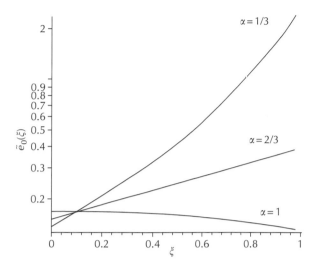

FIGURE 10.25

Dependence of the parent modulus of elasticity on ξ for the guided–pinned beam $(n = 0)$

For $n = 1$, the parent modulus of elasticity $\tilde{e}_1(\xi)$ reads

$$\tilde{e}_1(\xi) = [(449 + 171\alpha)(1 + \xi) + (171\alpha - 307)\xi^2 - (55 + 81\alpha)\xi^3 + (85 - 81\alpha)\xi^4$$
$$+ (1 + 3\alpha)(\xi^5 + \xi^6) + 3(\alpha - 1)\xi^7]/3360[1 - (1 - \alpha)\xi]^3(1 + \xi + \xi^2)$$
$$(10.168)$$

Figure 10.26 portrays the variations of $\tilde{e}_1(\xi)$ for different values of the coefficient α.

For $n = 2$, for the parent modulus of elasticity is given by

$$\tilde{e}_2(\xi) = [(2{,}102 + 805\alpha)(1 + \xi) + (805\alpha - 1{,}426)\xi^2 - (250 + 371\alpha)\xi^3$$
$$+ (380 - 371\alpha)\xi^4 + (2 + 7\alpha)(\xi^5 + \xi^6 + \xi^7)$$
$$+ 7(\alpha - 1)\xi^8]/15{,}120[1 - (1 - \alpha)\xi]^3(1 + \xi + \xi^2 + \xi^3) \qquad (10.169)$$

Figure 10. 27 represents the function $\tilde{e}_2(\xi)$ for different values of α.

We now treat the case $n = 3$. The parent modulus of elasticity $\tilde{e}_3(\xi)$ is

$$\tilde{e}_3(\xi) = [(2{,}149 + 826\alpha)(1 + \xi) + (826\alpha - 1{,}451)\xi^2 - (251 + 374\alpha)\xi^3$$
$$+ (379 - 374\alpha)\xi^4 + (1 + 4\alpha)(\xi^5 + \xi^6 + \xi^7 + \xi^8)$$
$$+ 4(\alpha - 1)\xi^9]/1{,}120[1 - (1 - \alpha)\xi]^3(1 + \xi + \xi^2 + \xi^3 + \xi^4) \qquad (10.170)$$

Figure 10.28 shows the dependence of $\tilde{e}_3(\xi)$ on ξ.

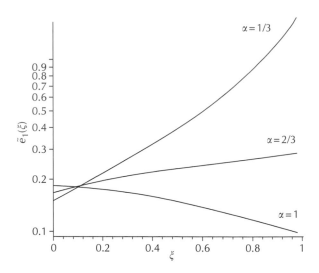

FIGURE 10.26
Dependence of the parent modulus of elasticity on ξ for the guided–pinned beam ($n = 1$)

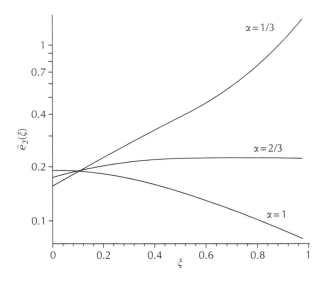

FIGURE 10.27
Dependence of the parent modulus of elasticity on ξ for the guided–pinned beam ($n = 2$)

Analogously one can obtain an infinite sequence of present moduli of elasticity associated with mode shapes coinciding with functions obtained as successive approximations of the displacement functions. In the following, we want to express the parent modulus of elasticity of the inhomogeneous beam via the successive mode shapes generated for a uniform homogeneous beam.

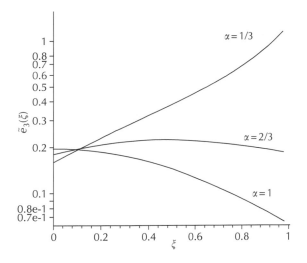

FIGURE 10.28
Dependence of the parent modulus of elasticity on ξ for the guided–pinned beam ($n = 3$)

10.3.3.3 *Generating a feasible mode shape by successive approximations*

We first approximate the mode shape by the following function:

$$W_{(1)}(x) = \tfrac{1}{5}(x/L)^4 - \tfrac{6}{5}(x/L)^2 + 1 \tag{10.171}$$

which satisfies all boundary conditions. Then, Eq. (10.119) allows us to express a second approximation of the mode shape:

$$W_{(2)}(x) = \omega^2 K W_{(1)}(x) \tag{10.172}$$

We impose the following condition:

$$W_{(2)}(0) = W_{(1)}(0) = 1 \tag{10.173}$$

Thus,

$$\begin{aligned}
\omega_{(1)}^2 &= 1/K W_{(1)}(0) \\
&= (1680/277)D/\rho A L^4 \approx 6.065 D/\rho A L^4
\end{aligned} \tag{10.174}$$

and

$$\begin{aligned}
W_{(2)}(x) &= K W_{(1)}(x)/K W_{(1)}(0) \\
&= (1/1385)[1385 - 1708(x/L)^2 + 350(x/L)^4 - 28(x/L)^6 + (x/L)^8]
\end{aligned} \tag{10.175}$$

In like manner, we express the second approximation of the natural frequency and a third approximation of the mode shape, respectively,

$$\omega_{(2)}^2 = 1/KW_{(2)}(0)$$

$$= (3{,}290{,}760/540{,}553)D/\rho AL^4 \approx 6.088D/\rho AL^4 \tag{10.176}$$

$$W_{(3)}(x) = KW_{(2)}(x)/KW_{(2)}(0)$$

$$= (1/2{,}702{,}765)[2{,}702{,}765 - 3{,}334{,}386(x/L)^2 + 685{,}575(x/L)^4$$

$$- 56{,}364(x/L)^6 + 2{,}475(x/L)^8 - 66(x/L)^{10} + (x/L)^{12} \tag{10.177}$$

Note that the exact solution for the natural frequency of a guided–pinned beam is $\pi/2$. The first and second approximations of the natural frequency, given by Eqs. (10.174) and (10.176), respectively, are very close to the exact solution, deviating from the exact value by only 0.4% and 0.005%, respectively.

As in the study by Elishakoff (2000f), we are looking for an inhomogeneous beam that possesses the mode shape given in Eq. (10.176), i.e., $\psi(x) \equiv W_{(2)}(x)$. By substituting Eq. (10.176) in Eq. (10.165), we obtain the expression for the parent modulus of elasticity. This is done for the beam with uniform material density. Figure 10.29 portrays the variation of the parent modulus of elasticity:

$$\tilde{e}(\xi) = 61[(403{,}878 + 151{,}853\alpha)(1 + \xi) + (151{,}853\alpha - 281{,}697)\xi^2$$

$$- (53{,}172 + 76{,}672\alpha)\xi^3 + (87{,}738 - 76{,}672\alpha)\xi^4 + (3{,}192 + 7{,}874\alpha)\xi^5$$

$$+ (7{,}874\alpha - 8{,}358)\xi^6 - (108 + 376\alpha)\xi^7 + (387 - 376\alpha)\xi^8 + (2 + 9\alpha)\xi^9$$

$$+ 9(\alpha - 1)\xi^{1}0]/55{,}440[1 - (1 - \alpha)\xi]^3[61(1 + \xi) - 14(\xi^2 + \xi^3) + \xi^4 + \xi^5] \tag{10.178}$$

for different values of α.

We leave out the exact parent moduli arising from approximations given in Eq. (10.171) or (10.177); these expressions are not reproduced here because they are lengthy.

10.3.4 Free–Free Beam

To discuss this particular case, we note that a beam with free ends has two vanishing natural frequencies. The governing equation becomes, if we set $\omega \equiv 0$ ab initio,

$$\frac{d^2}{dx^2}\left(D(x)\frac{d^2W(x)}{dx^2}\right) = 0 \tag{10.179}$$

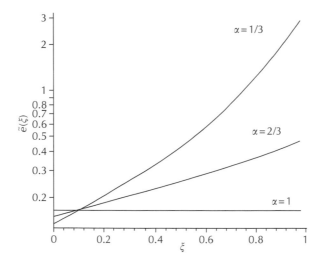

FIGURE 10.29
Dependence of the parent modulus of elasticity on ξ for the guided–pinned beam

Equation (10.179) is integrated twice to yield

$$D(x)\frac{d^2 W(x)}{dx^2} = A_1 x + A_2 \tag{10.180}$$

The constant $A_2 = 0$ because of the boundary condition $W''(0) = 0$ at $x = 0$. Also, $A_1 = 0$, due to the boundary conditions $W''(L) = 0$. One can check that the function

$$W(x) = A_3 x + A_4 \tag{10.181}$$

satisfies the differential equation (10.179).

10.3.4.1 Determination of the integral operator

Let us assume that the natural frequency differs from zero. By integrating the governing equation once we obtain

$$\frac{d}{dx}\left(D(x)\frac{d^2 W(x)}{dx^2}\right) = \omega^2 \int_0^x \rho(x_1)A(x_1)W(x_1)dx_1 + C_1 \tag{10.182}$$

The right term in this equation represents the shear force. It equals zero at $x = 0$, hence $C_1 = 0$. Another integration of Eq. (10.182) leads to

$$D(x)\frac{d^2 W(x)}{dx^2} = \omega^2 \int_0^x \int_0^{x_1} \rho(x_2)A(x_2)W(x_2)dx_2 dx_1 + C_2 \tag{10.183}$$

The term on the right of the Eq. (10.183) represents the bending moment and that must vanish at $x = 0$. Hence $C_2 = 0$. We rewrite Eq. (10.183) as follows:

$$\frac{d^2W(x)}{dx^2} = \omega^2 \frac{1}{D(x)} \int_0^x \int_0^{x_1} \rho(x_2)A(x_2)W(x_2)dx_2dx_1 \tag{10.184}$$

By integrating Eq. (10.184) twice, we obtain

$$W(x) = \omega^2 \int_0^x \int_0^{x_1} \frac{1}{D(x_2)} \int_0^{x_2} \int_0^{x_3} \rho(x_4)A(x_4)W(x_4)dx_4dx_3dx_2dx_1 + C_3x + C_4 \tag{10.185}$$

Equation (10.185) is rewritten as

$$W(x) = \omega^2 \left(\int_0^x \int_0^{x_1} \frac{1}{D(x_2)} \int_0^{x_2} \int_0^{x_3} \rho(x_4)A(x_4)W(x_4)dx_4dx_3dx_2dx_1 \right.$$

$$\left. + E_3x + E_4 \right)$$

$$= \omega^2[f(x) + E_3x + E_4] \tag{10.186}$$

with E_3 and E_4 being two new constants: $E_3 = C_3/\omega^2$, $E_4 = C_4/\omega^2$. Satisfying the boundary conditions at $x = L$ imposes two additional conditions:

$$V_y(L) = 0 \quad \int_0^L \rho(x)A(x)W(x)dx = 0$$

$$M_z(L) = 0 \quad \int_0^L \int_0^x \rho(x_1)A(x_1)W(x_1)dx_1dx = 0 \tag{10.187}$$

Substituting Eq. (10.186) into Eq. (10.187), we obtain the following system of equations for the constants E_3 and E_4:

$$\omega^2 \int_0^L \rho(x)A(x)[f(x) + E_3x + E_4]dx = 0$$

$$\omega^2 \int_0^L \int_0^x \rho(x_1)A(x_1)[f(x_1) + E_3x_1 + E_4]dx_1dx = 0 \tag{10.188}$$

Since the natural frequency is non-zero, Eq. (10.188) leads to the following system:

$$\begin{bmatrix} a_{11} & a_{12} \\ a_{21} & a_{22} \end{bmatrix} \begin{Bmatrix} E_3 \\ E_4 \end{Bmatrix} = \begin{Bmatrix} b_1 \\ b_2 \end{Bmatrix} \tag{10.189}$$

with

$$b_1 = -\int_0^L \rho(x)A(x)f(x)dx \quad b_2 = -\int_0^L \int_0^x \rho(x_1)A(x_1)f(x_1)dx_1dx$$

$$a_{11} = \int_0^L \rho(x)A(x)xdx \quad a_{12} = \int_0^L \rho(x)A(x)dx \quad (10.190)$$

$$a_{21} = \int_0^L \int_0^x \rho(x_1)A(x_1)x_1dx_1dx \quad a_{22} = \int_0^L \int_0^x \rho(x_1)A(x_1)dx_1dx$$

The constants E_3 and E_4 are calculated from Eq. (10.189). Their solutions are denoted by E_3^* and E_4^*, respectively:

$$E_3^* = \frac{b_1a_{22} - b_2a_{12}}{a_{11}a_{22} - a_{12}a_{21}} \quad E_4^* = \frac{b_1a_{11} - b_2a_{21}}{a_{11}a_{22} - a_{12}a_{21}} \quad (10.191)$$

The integral operator K becomes:

$$KW(x) = \int_0^x \int_0^{x_1} \frac{1}{D(x_2)} \int_0^{x_2} \int_0^{x_3} \rho(x_4)A(x_4)W(x_4)dx_4dx_3dx_2dx_1 + B_3^*x + B_4^*$$

$$(10.192)$$

10.3.4.2 Obtaining closed-form solutions

From Eq. (10.183), we can express the flexural rigidity, by substituting Eqs. (10.121), (10.123) and (10.125) into it:

$$e(x)[1 + (\alpha - 1)x/L]^3$$

$$= \left[\omega^2\rho_0A_0\int_0^x \int_0^{x_1} \varphi(x_2) \times [1 + (\alpha - 1)x_2/L]W(x_2)dx_2dx_1\right]\Big/ D_0W''(x)$$

$$(10.193)$$

Let us consider the following functions, parameterized with a positive integer n:

$$\psi_n(x) = \delta_1(x/L)^{n+3} + \delta_2(x/L)^{n+2} + \delta_3(x/L)^{n+1} + \delta_4(x/L)^n \quad (10.194)$$

The above function satisfies the boundary conditions for the uniform beam:

$$\psi_n''(0) = 0 \quad \psi_n'''(0) = 0 \quad \psi_n''(1) = 0 \quad \psi_n'''(1) = 0 \quad (10.195)$$

The first and second conditions are always satisfied by Eq. (10.194). The last two conditions lead to

$$\delta_3 = -[(3n^2 + 15n + 18)\delta_1 + (2n^2 + 6n + 4)\delta_2]/n(n+1)$$
$$\delta_4 = [(2n^2 + 10n + 12)\delta_1 + (n^2 + 3n + 2)\delta_2]/(n-1)n \quad (10.196)$$

We take $\delta_1 = \delta_2 = 1$; thus, we obtain the candidate mode shapes for the free–free beams:

$$\psi_n(x) = \left(\frac{x}{L}\right)^{n+3} + \left(\frac{x}{L}\right)^{n+2} - \frac{5n^2 + 21n + 22}{n(n+1)}\left(\frac{x}{L}\right)^{n+1} + \frac{3n^2 + 13n + 14}{(n-1)n}\left(\frac{x}{L}\right)^n$$

(10.197)

One can check that the above set of functions satisfies the boundary condition (10.195) for any integer $n > 3$. Figure 10.30 portrays the variation of $\psi_n(x)$ for different values of n.

The natural frequency is postulated to be as in Eq. (10.136). Thus, Eq. (10.193) yields

$$e(x) = \gamma \int_0^x \int_0^{x_1} [1 + (\alpha - 1)x_2/L]\psi_n(x_2)dx_2dx_1/L^4\psi_n''(x)[1 + (\alpha - 1)x/L]^3$$

(10.198)

for the constant material density. The parent modulus of elasticity reads

$$\tilde{e}_n(x) = \int_0^x \int_0^{x_1} [1 + (\alpha - 1)x_2/L]\psi_n(x_2)dx_2dx_1/L^4\psi_n''(x)[1 + (\alpha - 1)x/L]^3$$

(10.199)

We first treat the case $n = 4$. For $\alpha = 1$, we obtain

$$\tilde{e}_4(\xi) = \xi^4\frac{56(\alpha-1)\xi^4 + 70\alpha\xi^3 + (927 - 837\alpha)\xi^2 - (2{,}256 - 1{,}140\alpha)\xi + 1{,}596}{30{,}240(7\xi^3 + 5\xi^2 - 31\xi + 19)[1 + (\alpha - 1)\xi]^3}$$

(10.200)

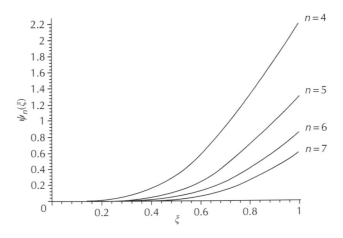

FIGURE 10.30
Variation of $\psi_n(\xi)$ for the free–free beam

The dependence of $\tilde{e}_4(\xi)$ on ξ is shown in Figure 10.31 for different values of α.

The second case is $n = 5$. The flexural rigidity reads

$$\tilde{e}_5(\xi) = \xi^4 \frac{72(\alpha - 1)\xi^4 + 88\alpha\xi^3 + (1{,}034 - 924\alpha)\xi^2 - (2{,}277 - 1{,}089\alpha)\xi + 1{,}452}{110{,}880(4\xi^3 + 3\xi^2 - 18\xi + 11)[1 + (\alpha - 1)\xi]^3}$$

(10.201)

Figure 10.32 portrays the variation of $\tilde{e}_5(\xi)$ for $\alpha = \{1, \frac{2}{3}, \frac{1}{3}\}$.

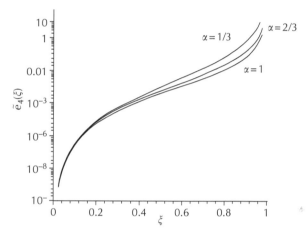

FIGURE 10.31
Dependance of the parent modulus of elasticity on ξ for $n = 4$, for the free–free beam

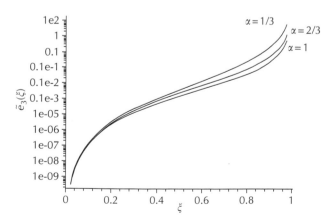

FIGURE 10.32
Dependence of the parent modulus of elasticity on ξ for $n = 5$, for the free–free beam

10.3.4.3   Generating a feasible mode shape by successive approximations

Let us consider, as an auxiliary problem, the free vibrations of the homogeneous beam that has a constant cross section. We first approximate the mode shape by the following function:

$$W_{(1)}(x) = (1/5{,}571)[120(x/L)^{14} - 840(x/L)^{13} + 1{,}820(x/L)^{12} - 48{,}048(x/L)^5$$
$$+ 69{,}160(x/L)^4 - 21{,}458(x/L) + 4{,}817] \qquad (10.202)$$

A second approximation of the mode shape is given by substituting Eq. (10.192) in Eq. (10.119):

$$W_{(2)}(L) = W_{(1)}(L) = 1 \qquad (10.203)$$

From Eqs. (10.86) and (10.87), we obtain a first approximation for the natural frequency:

$$\omega_{(1)}^2 = [KW_{(1)}(L)]^{-1}$$
$$= (13{,}603{,}713{,}480/25{,}321{,}451)D/L^4\rho A \approx (4.814)^4 D/L^4\rho A \qquad (10.204)$$

The second approximation for the mode shape reads

$$W_{(2)}(x) = (1/25{,}321{,}451)[3{,}990(x/L)^{18} - 35{,}910(x/L)^{17} + 101{,}745(x/L)^{16}$$
$$- 38{,}798{,}760(x/L)^9 + 1{,}005{,}244{,}060(x/L)^8 - 436{,}648{,}842(x/L)^5$$
$$+ 490{,}105{,}665(x/L)^4 - 114{,}781{,}014(x/L) + 24{,}850{,}517] \qquad (10.205)$$

From Eq. (10.205), we obtain a second approximation for the mode shape:

$$W_{(2)}^2 = [KW_{(2)}(L)]^{-1}$$
$$= (10{,}762{,}629{,}533{,}040)/(21{,}320{,}380{,}673)D/L^4\rho A \approx (4.740)^4 D/L^4\rho A \qquad (10.206)$$

The third approximation for the mode shape follows:

$$W_{(3)}(x) = (1/21{,}320{,}380{,}673)[9{,}660(x/L)^{22} - 106{,}260(x/L)^{21} + 371{,}910(x/L)^{20}$$
$$- 961{,}015{,}440(x/L)^{13} + 3{,}596{,}527{,}480(x/L)^{10} - 61{,}373{,}420{,}570(x/L)^9$$
$$+ 123{,}996{,}733{,}245(x/L)^8 - 406{,}554{,}351{,}588(x/L)^5$$
$$+ 440{,}102{,}656{,}070(x/L)^4 - 98{,}754{,}906{,}340(x/L) + 21{,}267{,}882{,}506] \qquad (10.207)$$

and the associate approximation of the natural frequency reads

$$W_{(3)}^2 = [KW_{(2)}(L)]^{-1}$$

$$= (698{,}455{,}670{,}847{,}480)/(1{,}393{,}830{,}378{,}763)D/L^4\rho A \approx (4.731)^4 D/L^4\rho A$$

$$(10.208)$$

The fourth approximation for the mode shape and the natural frequency are given in the following:

$$W_{(4)}(x) = (1/555{,}049{,}831{,}310{,}433{,}649)[267{,}344(x/L)^{30} - 4{,}010{,}160(x/L)^{29}$$

$$+ 19{,}382{,}440(x/L)^{29} - 7{,}649{,}866{,}141{,}920(x/L)^{21}$$

$$+ 4{,}624{,}688{,}948{,}880(x/L)^{20} - 408{,}941{,}931{,}861{,}540(x/L)^{17}$$

$$+ 1{,}560{,}622{,}884{,}621{,}570(x/L)^{16} - 51{,}168{,}930{,}690{,}865{,}680(x/L)^{13}$$

$$+ 144{,}017{,}432{,}761{,}722{,}520(x/L)^{12} - 1{,}777{,}388{,}829{,}209{,}193{,}200(x/L)^{9}$$

$$+ 3{,}445{,}010{,}315{,}868{,}040{,}920(x/L)^{8} - 10{,}756{,}739{,}708{,}312{,}742{,}144(x/L)^{5}$$

$$+ 11{,}574{,}459{,}238{,}153{,}405{,}440(x/L)^{4} - 2{,}579{,}321{,}364{,}764{,}022{,}730(x/L)$$

$$+ 555{,}026{,}136{,}832{,}881{,}909] \qquad (10.209)$$

$$\omega_{(4)}^2 = [KW_{(3)}(L)]^{-1}$$

$$= (277{,}877{,}146{,}490{,}240{,}109{,}120/555{,}049{,}831{,}310{,}433{,}649)D/L^4\rho A$$

$$\approx (4.730)^4 D/L^4\rho A \qquad (10.210)$$

We note that Eqs. (10.202), (10.205), (10.207) and (10.209) satisfy the boundary conditions. These mode shapes can serve as the *exact mode shape* for the associated *inhomogeneous* beam's flexural rigidity. For instance, by substituting Eq. (10.205) into Eq. (10.199), we obtain the following parent modulus of elasticity, associated with the mode shape given in Eq. (10.205):

$$\tilde{e}(\xi) = (1/53{,}721{,}360)[1{,}254(\alpha - 1)\xi^{19} + (13{,}860 - 12{,}474\alpha)\xi^{18}$$

$$- (53{,}130 - 39{,}270\alpha)\xi^{17} + 43{,}890\xi^{16} - 38{,}798{,}760(\alpha - 1)\xi^{10}$$

$$- (167{,}187{,}384 - 120{,}628{,}872\alpha)\xi^{9} + 147{,}435{,}288\xi^{8}$$

$$- 1{,}029{,}243{,}699(\alpha - 1)\xi^{6} - (2{,}912{,}657{,}022 - 1{,}540{,}332{,}090\alpha)\xi^{5}$$

$$+ 2{,}156{,}464{,}926\xi^{4} - 1{,}262{,}591{,}154(\alpha - 1)\xi^{2}$$

$$- (3{,}071{,}893{,}682 - 546{,}711{,}374\alpha)\xi + 1{,}640{,}134{,}122]/[(3\xi^{14} - 24\xi^{13}$$

$$+ 60\xi^{12} - 6{,}864\xi^{5} + 13{,}832\xi^{4} - 21{,}458\xi + 14{,}451)(1 + (\alpha - 1)\xi)^{3}]$$

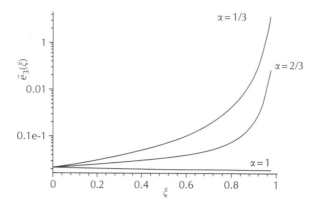

FIGURE 10.33
Dependence of the parent modulus of elasticity on ξ for the second approximation of the mode shape

Figure 10.33 depicts the variation of $\tilde{e}(\xi)$ for $\alpha = \{1, \frac{2}{3}, \frac{1}{3}\}$. One could analogously derive sets of variations of the elastic moduli.

10.3.5   Concluding Remarks

In this section, we derived, for three sets of boundary conditions, new closed-form solutions for the natural frequency of an inhomogeneous beam. These expressions were obtained from turning around the method of successive approximation in such a manner as to obtain *exact, closed-form* solutions. It appears that this method does not lead to analogous solutions when one of the ends is clamped and the other end is not free. The reason is as follows: in order to satisfy the boundary conditions we must express the beam displacement and its slope explicitly. For the inhomogeneous beam these two functions depend on the elastic modulus function. Hence, to satisfy the boundary conditions, the elastic modulus must be specified. This fact prevents the derivation of the function representing the elastic modulus via the mode shape, for the latter depends on the elastic modulus itself.

11

Circular Plates

This chapter is devoted to vibrations of inhomogeneous circular plates in an axisymmetric setting. For three sets of boundary conditions, namely for plates that are either clamped, pinned or free at the circumference, closed-form solutions are obtained for the fundamental natural frequency. Simple polynomial expressions are postulated for the mode shapes in each case, arising from the static displacements of the "corresponding" uniform plates. Inertial coefficients are also represented in a polynomial form. Uniform, linearly, parabolically or cubically varying inertial coefficients are studied with the general case when the inertial coefficient is a polynomial of mth degree.

11.1 Axisymmetric Vibration of Inhomogeneous Clamped Circular Plates: an Unusual Closed-Form Solution

11.1.1 Introductory Remarks

The free vibrations of circular plates attracted investigators nearly two centuries ago (Chladni, 1803; Poisson, 1829). The vibrations of uniform plates have been studied quite extensively already (Gontkevich, 1964; Leissa, 1969). Although the circular plates of variable thickness received much less attention than the uniform plates, still, they have been investigated quite intensively. An exact solution was derived by Conway et al. (1964) by noting an analogy (Conway, 1957) that exists between the free vibration of truncated-cone beams and linearly tapered plates for the special case when Poisson's ratio equals one-third. Conway et al. (1964) derived the natural frequencies for clamped, tapered, circular plates. Series solutions have been provided by Soni (1972) for plates with quadratically varying thickness. The method of Frobenius was utilized by Jain (1972). Frequency parameters of clamped and pinned plates were computed, for the first two modes, for various values of a taper parameter and inplane force, both for linear and parabolic variations of thickness. A perturbation method based on a small parameter was employed by Yang (1993). Zeroth and first-order asymptotic solutions were obtained for the natural frequencies of a clamped plate with linearly varying thickness. Lenox and Conway (1980) obtained an exact expression for the buckling mode for the annular plate; at a later stage they performed the numerical computerized calculations for natural frequencies.

Most of the reported solutions utilized approximate techniques. Laura et al. (1982) used the Rayleigh–Ritz method with polynomial coordinate functions that identically satisfy the external boundary conditions to study the vibrations and elastic stability of polar orthotropic circular plates of linearly varying thickness. Lal and Gupta (1982) utilized Chebychev polynomials to obtain the frequencies of polar orthotropic annular plates of variable thickness. Gorman (1983) and Kanaka Raju (1977) employed the most universal technique, that of the finite element method, for studying the dynamic behavior of polar orthotropic annular plates of variable thickness.

As far as the exact solutions are concerned, one very important paper should be mentioned. Harris (1968) obtained a *closed-form* solution for both the mode shapes and the natural frequencies of the circular plate that is free at its edge. The flexural rigidity was given as $D(r) = D_0(1 - r^2/R^2)^3$, where D_0 is the flexural rigidity at the center, r the polar coordinate and R the radius.

The present section shares with the paper by Harris (1968) the property of offering an exact, closed-form solution the natural frequency. However, whereas Harris (1968) solves a *direct* vibration problem, we pose the *inverse* problem. Namely, we postulate the vibration mode and pursue the following objective: find the variation of the flexural rigidity that leads to the postulated vibration mode. Under some circumstances such a problem has a unique solution with a remarkable by-product: a closed-form expression for the natural frequency. This section complements the paper of Harris (1968), who derived a solution for the circular plates that are free at the boundary, while the present section deals with clamped plates.

The adjective "unusual" utilized in the title of this section (as well as for the entire monograph) was borrowed from the note by Conway (1980) who derived the closed-form expression for the unsymmetrical bending of a particular circular plate resting on a Winkler foundation. He wrote "What is remarkable about this solution is that it is in a closed-form which is far simpler than the constant thickness disk solution which involves Kelvin functions." We consider a vibration case and likewise derive a closed-form solution. But, it appears that the present solution is superior to that by Lenox and Conway (1980) for here we derive not only the closed-form expression for the displacement, but the *natural frequency* too. The unusual aspect characteristic of the study by Conway is preserved in our study: While for the exact solution for homogeneous circular plates the Bessel functions are involved, in the inhomogeneous case the elementary functions turn out to be sufficient.

11.1.2 Basic Equations

The differential equation that governs the free axisymmetric vibrations of the circular plate with variable thickness reads (Kovalenko, 1959)

$$D(r)r^3 \Delta \Delta W + \frac{dD}{dr}\left(2r^3\frac{d^3W}{dr^3} + r^2(2+v)\frac{d^2W}{dr^2} - r\frac{dW}{dr}\right)$$

$$+ \frac{d^2D}{dr^2}\left(r^3\frac{d^2W}{dr^2} + vr^2\frac{dW}{dr}\right) - \rho h\omega^2 r^3 W = 0 \tag{11.1}$$

where Δ is the Laplace operator in polar coordinates,

$$\Delta = \frac{d^2}{dr^2} + \frac{1}{r}\frac{d}{dr} \tag{11.2}$$

D is the flexural rigidity, assumed to vary along the radial coordinate r,

$$D = D(r) = Eh^3/12(1 - v^2) \tag{11.3}$$

h is the thickness, v the Poisson ratio, ρ the material density, r the radial coordinate, φ the circumferential coordinate and W the mode shape. The Poisson ratio v is assumed to be constant. Note that the governing equation reported by Kovalenko (1959) is multiplied here by the term r^3, for further convenience. We are looking for the case when the inertial term

$$\delta(r) = \rho h \tag{11.4}$$

varies along r; likewise, the flexural rigidity is a function of r. We are interested in finding the *closed-form* solution for the natural frequency ω.

We pose the problem as an *inverse* vibration study. Note first that for the uniform circular plate that is under a uniform load, q_0, the displacement can be put in the following form (Timoshenko and Woinowsky-Krieger, 1959):

$$w = \frac{q_0}{64D}(R^2 - r^2)^2 \tag{11.5}$$

where R is the outer radius of the plate. We are interested in determining such a variation of $D(r)$ in Eq. (11.1) that the function

$$W(r) = (R^2 - r^2)^2 \tag{11.6}$$

serves as an exact mode shape. We confine our interest to a circular plate that is clamped along the boundary $r = R$.

11.1.3   Method of Solution

We assume that the inertial term is represented as a polynomial,

$$\delta(r) = \sum_{i=0}^{m} a_i r^i \tag{11.7}$$

Since $W(r)$ is a fourth-order polynomial expression in terms of r, in view of Eq. (11.7) the last term in Eq. (11.1) is a polynomial expression of order $m + 7$. Since the operator $\Delta\Delta$ in Eq. (11.1) involves fourfold differentiation with respect to r, in order that the highest degree of the first term's polynomial expression in $Dr^3 \Delta\Delta W$ be of order $m + 7$, it is necessary and sufficient for the flexural rigidity to be represented as a polynomial of degree $m + 4$. Thus, the sought for flexural rigidity can be put in the form

$$D(r) = \sum_{i=0}^{m+4} b_i r^i \tag{11.8}$$

Further steps involve the substitution of Eqs. (11.6)–(11.8) in the governing differential equation (11.1) and demanding that the polynomial expression so obtained vanishes. This implies that all the coefficients of the powers r^i must be zero, leading, in turn, to the set of algebraic equations in terms of b_i and ω^2. We consider different variations for the inertial term $\delta(r)$ in Eq. (11.4).

11.1.4   Constant Inertial Term ($m = 0$)

In this case, the flexural rigidity is sought for as a fourth-order polynomial:

$$D(r) = b_0 + b_1 r + b_2 r^2 + b_3 r^3 + b_4 r^4 \tag{11.9}$$

We get, instead of the differential Eq. (11.1) the following equation:

$$\sum_{i=0}^{7} c_i r^i = 0 \tag{11.10}$$

where

$$c_0 = 0 \quad c_1 = 0 \quad c_2 = -4(1+\nu)R^2 b_1 \quad c_3 = 64 b_0 - 16(1+\nu)R^2 b_2 - a_0\omega^2 R^4$$

$$c_4 = 12(11+\nu)b_1 - 36(1+\nu)R^2 b_3$$

$$c_5 = 32(7+\nu)b_2 - 64(1+\nu)R^2 b_4 + 2a_0\omega^2 R^2$$

$$c_6 = (340 + 60\nu)b_3 \quad c_7 = 96(5+\nu)b_4 - a_0\omega^2$$

$$\tag{11.11}$$

Since the left-hand side of the differential equation (11.10) must vanish for any r within $[0; R]$, we demand that all the coefficients c_i be zero. This leads to a homogeneous set of six linear algebraic equations for six unknowns. It turns out that the determinant of the matrix of the set derived from Eq. (11.11) is identically zero. Therefore, a non-trivial solution is obtainable. From the requirement $c_7 = 0$, the natural frequency squared is obtained as

$$\omega^2 = 96(5 + v)b_4/a_0 \tag{11.12}$$

Upon substitution of Eq. (11.12) into (11.11), the remaining equations yield the coefficients in the flexural rigidity

$$b_0 = \frac{13 + v}{2}R^4 b_4 \quad b_1 = 0 \quad b_2 = -4R^2 b_4 \quad b_3 = 0 \tag{11.13}$$

Hence, the flexural rigidity reads

$$D(r) = \left(\frac{13 + v}{2}R^4 - 4R^2 r^2 + r^4\right)b_4 \tag{11.14}$$

Figure 11.1 depicts the flexural rigidity for various values of the Poisson ratio v.

11.1.5 Linearly Varying Inertial Term ($m = 1$)

Instead of the set (11.11), here, we get seven linear algebraic equations with seven unknowns:

$$-4(1 + v)R^2 b_1 = 0 \quad 64b_0 - 16(1 + v)R^2 b_2 - a_0\omega^2 R^4 = 0$$

$$12(11 + v)b_1 - 36(1 + v)R^2 b_3 - a_1\omega^2 R^4 = 0$$

$$32(7 + v)b_2 - 64(1 + v)R^2 b_4 + 2a_0\omega^2 R^2 = 0 \tag{11.15}$$

$$(340 + 60v)b_3 - 100(1 + v)R^2 b_5 + 2a_1\omega^2 R^2 = 0$$

$$96(5 + v)b_4 - a_0\omega^2 = 0 \quad (644 + 140v)b_5 - a_1\omega^2 = 0$$

In order to have a non-trivial solution, the determinant of the set (11.15) must vanish,

$$(1 + v)(7 + v)(13{,}765 + 5643v + 728v^2 + 30v^3)a_1 = 0 \tag{11.16}$$

leading to $a_1 = 0$. But this result signifies that the linear inertial coefficient is a constant. Thus, the problem solved in the previous section is re-obtained.

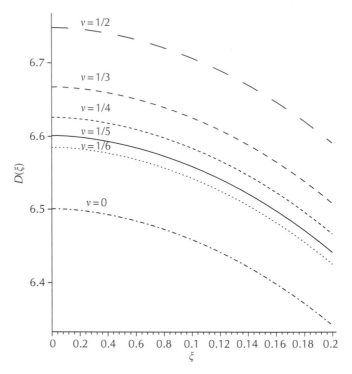

FIGURE 11.1
Variation of the flexural rigidity of the clamped circular plate with constant inertial term, when the Poisson ration takes the values $0; \frac{1}{6}; \frac{1}{5}; \frac{1}{4}; \frac{1}{3}; \frac{1}{2}$

11.1.6 Parabolically Varying Inertial Term ($m = 2$)

For $m = 2$, i.e., the plate whose material density varies parabolically,

$$\delta(r) = a_0 + a_1 r + a_2 r^2 \tag{11.17}$$

the flexural rigidity has to be sought for as a sixth-order polynomial:

$$D(r) = b_0 + b_1 r + b_2 r^2 + b_3 r^3 + b_4 r^4 + b_5 r^5 + b_6 r^6 \tag{11.18}$$

Substitution of Eq. (11.6) in conjunction with Eqs. (11.17) and (11.18) into the governing differential equation (11.1) yields

$$\sum_{i=0}^{9} d_i r^i = 0 \tag{11.19}$$

where

$$d_0 = 0 \quad d_1 = 0 \quad d_2 = -4R^2(1+v)b_1 \quad d_3 = 64b_0 - 16R^2(1+v)b_2 - a_0\omega^2 R^4$$

$$d_4 = 12(11+v)b_1 - 36R^2(1+v)b_3 - a_1\omega^2 R^4$$

$$d_5 = 32(7+v)b_2 - 64R^2(1+v)b_4 + \omega^2(2a_0R^2 - a_2R^4)$$

$$d_6 = 20(17+3v)b_3 - 100R^2(1+v)b_5 + 2a_1\omega^2 R^2 \tag{11.20}$$

$$d_7 = 96(5+v)b_4 - 144R^2(1+v)b_6 - \omega^2(a_0 - 2a_2R^2)$$

$$d_8 = (644+140v)b_5 - a_1\omega^2 \quad d_9 = 64(13+3v)b_6 - a_2\omega^2$$

As in the case of the constant inertial term, we demand that all $d_i = 0$. Thus, we get a set of eight equations with eight unknowns (the seven coefficients b_i and ω^2). The resulting determinantal equation is

$$(1+v)(7+v)a_1(178{,}945 + 114{,}654v + 26{,}393v^2 + 2{,}574v^3 + 90v^4) = 0 \tag{11.21}$$

In order for the homogeneous system to possess a non-trivial solution we must demand that the coefficient a_1 vanishes. We substitute $a_1 = 0$ into the set (11.20). For the natural frequency we arrive at the following expression, obtainable from the requirement $d_9 = 0$:

$$\omega^2 = 64(13+3v)b_6/a_2 \tag{11.22}$$

Then, the coefficients in the flexural rigidity are obtained as

$$b_0 = \frac{R^4 b_6}{12} \frac{(295 + 353v + 61v^2 + 3v^3)R^2 a_2 + (4732 + 2132v + 292v^2 + 12v^3)a_0}{(35 + 12v + v^2)a_2}$$

$$b_1 = 0 \quad b_2 = \frac{R^2 b_6}{3} \frac{(295 + 58v + 3v^2)R^2 a_2 - (728 + 272v + 24v^2)a_0}{(35 + 12v + v^2)a_2}$$

$$b_3 = 0 \quad b_4 = -\frac{b_6}{6} \frac{(95 + 15v)R^2 a_2 - (52 + 12v)a_0}{(5 + v)a_2} \quad b_5 = 0$$

$$\tag{11.23}$$

where b_6 is an arbitrary constant. In order that natural frequency squared is a positive quantity, we demand that the ratio b_6/a_2 be positive. We have two sub-cases: (1) both b_6 and a_2 are positive (2) both are negative. In the former case, namely (1), the necessary condition for $b_0 \geq 0$ positivity of the flexural rigidity $D(r)$ is identically satisfied. In the latter case the above inequality reduces to

$$(295 + 353v + 61v^2 + 3v^3)R^2 a_2 + (4732 + 2132v + 292v^2 + 12v^3)a_0 \geq 0 \tag{11.24}$$

leading to the following inequality

$$\frac{a_0}{|a_2|R^2} \leq \frac{295 + 353v + 61v^2 + 3v^3}{4732 + 2132v + 292v^2 + 12v^3} \qquad (11.25)$$

One can immediately see that

$$\frac{a_0}{|a_2|R^2} < 1 \qquad (11.26)$$

Hence, the associated variation of the inertial coefficient

$$\delta(r) = a_0 + a_2 r^2 \qquad (11.27)$$

takes a negative value at $r = R$. Thus, the possibility that both a_2 and b_6 be negative should be discarded as physically unrealizable. We conclude, therefore, that in Eq. (11.22) both a_2 and b_6 must constitute positive quantities.

11.1.7 Cubic Inertial Term ($m = 3$)

For $m = 3$, the following set of nine linear algebraic equations with nine unknowns is obtained:

$$-4R^2(1+v)b_1 = 0 \quad 64b_0 - 16R^2(1+v)b_2 - a_0\omega^2 R^4 = 0$$
$$12(11+v)b_1 - 36R^2(1+v)b_3 - a_1\omega^2 R^4 = 0$$
$$32(7+v)b_2 - 64R^2(1+v)b_4 + \omega^2(2a_0 R^2 - a_2 R^4) = 0$$
$$20(17+3v)b_3 - 100R^2(1+v)b_5 + \omega^2(2a_1 R^2 - a_3 R^4) = 0 \qquad (11.28)$$
$$96(5+v)b_4 - 144R^2(1+v)b_6 - \omega^2(a_0 - 2a_2 R^2) = 0$$
$$(644 + 140v)b_5 - 196(1+v)R^2 b_7 - \omega^2(a_1 - 2a_3 R^2) = 0$$
$$(832 + 192v)b_6 - a_2\omega^2 = 0 \quad (1044 + 252v)b_7 - a_3\omega^2 = 0$$

The determinantal equation arising from it reads

$$(1+v)(7+v)M = 0 \qquad (11.29)$$

where

$$M = (490{,}685 + 791{,}887v + 367{,}095v^2 + 71{,}999v^3 + 6{,}316v^4$$
$$+ 210v^5)R^2 a_3 + (5{,}189{,}405 + 4{,}577{,}581v + 1{,}567{,}975v^2$$
$$+ 259{,}397v^3 + 20{,}628v^4 + 630v^5)a_1 \qquad (11.30)$$

The solution of Eq. (11.29) is

$$a_1 = -\frac{(7{,}549 + 8{,}931\nu + 1{,}452\nu^2 + 70\nu^3)a_3}{79{,}837 + 36{,}033\nu + 4{,}916\nu^2 + 210\nu^3} \tag{11.31}$$

Upon substitution of Eq. (11.31) into all equations of the set (11.28), the natural frequency squared equals

$$\omega^2 = 36(29 + 7\nu)b_7/a_3 \tag{11.32}$$

and we get the following solution for the coefficients in the flexural rigidity:

$$b_0 = 3[(8{,}555 + 12{,}302\nu + 4{,}240\nu^2 + 514\nu^3 + 21\nu^4)R^2 a_2 + (137{,}228 + 94{,}952\nu$$

$$+ 23{,}392\nu^2 + 2{,}392\nu^3 + 84\nu^4)a_0]R^4 b_7/64a_3(455 + 261\nu + 49\nu^2 + 3\nu^3)$$

$$b_1 = 0$$

$$b_2 = 3[(8{,}555 + 3{,}747\nu + 493\nu^2 + 21\nu^3)R^2 a_2 - (21{,}112 + 12{,}984\nu + 2{,}600\nu^2$$

$$+ 168\nu^3)a_0]R^2 b_7/16a_3(455 + 261\nu + 49\nu^2 + 3\nu^3)$$

$$b_3 = \frac{(7{,}549 + 1{,}382\nu + 70\nu^2)R^4 b_7}{2{,}753 + 578\nu + 30\nu^2}$$

$$b_4 = \frac{3b_7[(1{,}508 + 712\nu + 84\nu^2)a_0 - (2{,}755 + 1{,}100\nu + 105\nu^2)R^2 a_2]}{32a_3(65 + 28\nu + 3\nu^2)}$$

$$b_5 = -\frac{2R^2(4{,}255 + 832\nu + 42\nu^2)b_7}{2{,}753 + 578\nu + 30\nu^2} \qquad b_6 = \frac{9(29 + 7\nu)a_2 b_7}{16a_3(13 + 3\nu)}$$

$$\tag{11.33}$$

where b_7 is an arbitrary constant. For the particular case $\nu = \frac{1}{3}$, the flexural rigidity equals

$$D(r) = \left[\frac{2{,}773}{2{,}464}\frac{R^6 a_2}{a_3} + \frac{235}{16}\frac{R^4 a_0}{a_3} + \left(\frac{8{,}319}{2{,}464}\frac{R^4 a_2}{a_3} - \frac{141}{16}\frac{R^2 a_0}{a_3}\right)r^2 + \frac{72{,}157}{26{,}541}R^4 r^3 \right.$$

$$\left. + \left(-\frac{3{,}525}{896}\frac{R^2 a_2}{a_3} + \frac{141}{64}\frac{a_0}{a_3}\right)r^4 - \frac{9{,}074}{2{,}949}R^2 r^5 + \frac{141}{112}\frac{a_2}{a_3}r^6 + r^7\right]b_7$$

$$\tag{11.34}$$

Note that the paper by Elishakoff (2000c) contains typographical errors in the expressions for b_4 and $D(r)$ (Storch, 2002c).

11.1.8   General Inertial Term ($m \geq 4$)

Now, consider the general expression for the inertial term given in Eq. (11.7), and the flexural rigidity in Eq. (11.8), for $m \geq 4$. Substitution of Eqs. (11.6)–(11.8) into the terms of the differential equation yields

$$r^3 D(r)\Delta\Delta W = 64r^3 \sum_{i=0}^{m+4} b_i r^i \tag{11.35}$$

$$\frac{dD}{dr}\left(2r^3\frac{d^3 W}{dr^3} + r^2(2+v)\frac{d^2 W}{dr^2} - r\frac{dW}{dr}\right) = 4[(17+3v)r^2 - R^2(1+v)]r^2 \sum_{i=1}^{m+4} ib_i r^{i-1} \tag{11.36}$$

$$\frac{d^2 D}{dr^2}\left(r^3\frac{d^2 W}{dr^2} + vr^2\frac{dW}{dr}\right) = 4[(3+v)r^2 - R^2(1+v)]r^3 \sum_{i=2}^{m+4} i(i-1)b_i r^{i-2} \tag{11.37}$$

and

$$-\rho h\omega^2 r^3 W = -\omega^2 r^3(R^2 - r^2)^2 \sum_{i=0}^{m} a_i r^i \tag{11.38}$$

Demanding the sum of (11.35)–(11.38) to be zero, we obtain the equation

$$\sum_{i=0}^{m+7} g_i r^i = 0 \tag{11.39}$$

where the coefficients g_i are

$$g_0 = 0 \quad g_1 = 0 \tag{11.40}$$

$$g_2 = -4R^2(1+v)b_1 \tag{11.41}$$

$$g_3 = 64b_0 - 16R^2(1+v)b_2 - a_0 R^4 \omega^2 \tag{11.42}$$

$$g_4 = 12(11+v)b_1 - 36R^2(1+v)b_3 - a_1 R^4 \omega^2 \tag{11.43}$$

$$g_5 = 32(7+v)b_2 - 64R^2(1+v)b_4 - \omega^2(a_2 R^4 - 2a_0 R^2) \tag{11.44}$$

$$g_6 = 20(17+3v)b_3 - 100R^2(1+v)b_5 - \omega^2(a_3 R^4 - 2a_1 R^2) \tag{11.45}$$

$$\vdots$$

$$g_i = [64 + 4(i-1)(17+3v) + 4(i-2)(i-1)(3+v)]b_{i-1} - 4(i+1)^2$$
$$\times R^2(1+v)b_{i+1} - \omega^2(a_{i-1}R^4 - 2a_{i-3}R^2 + a_{i-5}) \qquad \text{for } 7 \leq i \leq m+3 \tag{11.46}$$

$$\vdots$$

$$g_{m+4} = [64 + 4(m+1)(17 + 3v) + 4m(m+1)(3+v)]b_{m+1}$$
$$- 4R^2(1+v)(m+3)^2 R^2 b_{m+3} - \omega^2(-2a_{m-1}R^2 + a_{m-3}) \qquad (11.47)$$
$$g_{m+5} = [64 + 4(m+2)(17 + 3v) + 4(m+1)(m+2)(3+v)]b_{m+2}$$
$$- 4R^2(1+v)(m+4)^2 R^2 b_{m-4} - \omega^2(-2a_m R^2 + a_{m-2}) \qquad (11.48)$$
$$g_{m+6} = [64 + 4(m+3)(17 + 3v) + 4(m+2)(m+3)(3+v)]b_{m+3} - \omega^2 a_{m-1} \qquad (11.49)$$
$$g_{m+7} = [64 + 4(m+4)(17 + 3v) + 4(m+3)(m+4)(3+v)]b_{m+4} - \omega^2 a_m \qquad (11.50)$$

We demand that all coefficients g_i be zero; thus, we get a set of $m+6$ homogeneous linear algebraic equations for $m+6$ unknowns. In order to find a non-trivial solution the determinant of the set (11.40)–(11.50) must vanish. We expand the determinant along the last column of the matrix of the set, getting a linear algebraic expression with a_i as coefficients. The determinantal equation yields a condition for which the non-trivial solution is obtainable. In this case, the general expression for the natural frequency squared is obtained from the equation $g_{m+7} = 0$, resulting in

$$\omega^2 = [64 + 4(m+4)(17 + 3v) + 4(m+3)(m+4)(3+v)]b_{m+4}/a_m \qquad (11.51)$$

Note that the formulas pertaining to the cases $m = 0$, $m = 2$ and $m = 3$ are formally obtainable from Eq. (11.51) by appropriate substitution.

11.1.9 Alternative Mode Shapes

Let us now pose the following question. In previous sections we postulated the expression given in Eq. (11.6), which is proportional to the deflection of uniform circular plates under distributed loading. Equation (11.6) represents a fourth-order polynomial. A natural question arises: can an inhomogeneous circular plate possess a simpler expression? The simplest polynomial expression, that satisfies the boundary conditions is

$$\psi(r) = (R - r)^2 \qquad (11.52)$$

which represents a second-order polynomial. The third-order polynomial

$$\psi(r) = (R - r)^3 \qquad (11.53)$$

as well as the fourth-order polynomial,

$$\psi(r) = (R - r)^4 \qquad (11.54)$$

also satisfy the boundary conditions. Note that Eq. (11.54) is also a fourth-order polynomial, as is Eq. (11.6), although they are different. Expressions (11.52)–(11.54) are the candidate functions for both Rayleigh–Ritz or Boobnov–Galerkin methods. Thus, in essence, we ask if the coordinate functions utilizable for *approximate* evaluation of natural frequencies of either homogeneous or inhomogeneous plates, can serve as *exact* mode shapes. We consider the candidate mode shape given in Eq. (11.52).

11.1.9.1 Parabolic mode shape

Substitution of Eq. (11.52) into the differential equation (11.1), in conjunction with Eq. (11.9) for a constant inertial term ($m = 0$), yields the equation

$$\sum_{i=0}^{7} e_i r^i = 0 \tag{11.55}$$

where

$$e_0 = -2b_0 R \quad e_1 = 0 \quad e_2 = 2(1+v)b_1 + 2R(1-2v)b_2$$

$$e_3 = 8(1+v)b_2 + 4R(1-3v)b_3 - a_0 R^2 \omega^2$$

$$e_4 = 18(1+v)b_3 + 6R(1-4v)b_4 + 2a_0 R^2 \omega^2 \tag{11.56}$$

$$e_5 = 32(1+v)b_4 - a_0 \omega^2 \quad e_6 = 0 \quad e_7 = 0$$

In order for Eq. (11.55) to be valid for every r, we require $e_i = 0$, for i taking values 0, 2, 3, 4, 5, for the remaining requirements are identically satisfied. We get five equations for six unknowns. Taking b_4 to be an arbitrary positive constant we get the expression for the natural frequency squared:

$$\omega^2 = 32(1+v)b_4/a_0 \tag{11.57}$$

with attendant flexural rigidity coefficients

$$b_0 = 0 \quad b_1 = \frac{R^3(-107 + 155v + 106v^2 + 24v^3)b_4}{18(1 + 3v + 3v^2 + v^3)}$$

$$b_2 = \frac{R^2(107 + 59v + 12v^2)b_4}{18(1 + 2v + v^2)} \quad b_3 = -\frac{5R(7 + 4v)b_4}{9(1 + v)} \tag{11.58}$$

The necessary condition for non-negativity of the flexural rigidity is $b_1 \geq 0$. As Storch (2001c) informs us, "In Eq. (11.58), a necessary and sufficient condition that $b_1 \geq 0$ is that $v \geq 1/2$. For this range of Poisson's ratio we obtain a realizable stiffness distribution (for uniform mass density). For example, for the Poisson's ratio less than 1/2, the plate does not possess the parabolic shape given in Eq. (11.52)." The roots of the equation $b_1 = 0$ are

$$\nu_1 = \tfrac{1}{2} \quad \nu_2 = (-59 + i\sqrt{1655})/24 \quad \nu_3 = (-59 - i\sqrt{1655})/24 \qquad (11.59)$$

The last two roots, as complex numbers, have no physical significance. Only the first root, corresponding to an incompressible material, is acceptable. The associated expression for the flexural rigidity is

$$D(r) = \left[\tfrac{31}{9} R^2 r^2 - \tfrac{10}{3} R r^3 + (r/R)^4 \right] b_4 \qquad (11.60)$$

Note that the last term was communicated to the author by Storch (2002c). Figure 11.2 depicts the variation of the flexural rigidity.

The candidate mode shapes given in Eqs. (11.53) and (11.54) should be investigated separately. It is conjectured that the polynomial function that satisfies the boundary conditions, may not correspond to a physically realizable material density and/or flexural rigidity distribution. For example, for a Poisson's ratio that differs from $\tfrac{1}{2}$, the plate does not possess the parabolic shape given in Eq. (11.52).

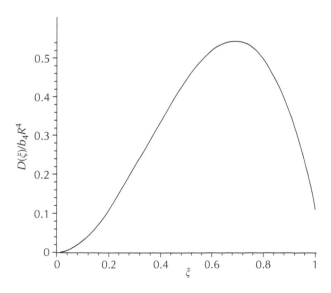

FIGURE 11.2
Variation of the flexural rigidity

11.2   Axisymmetric Vibration of Inhomogeneous Free Circular Plates: An Unusual, Exact, Closed-Form Solution

11.2.1   Introductory Remarks

Harris (1968) obtained closed-form solutions for the circular plate with a free edge. His study appears to be very interesting. He considered a specific case of the variation of the thickness as a function of the radial coordinate r:

$$h(r) = h_0[1 - (r/R)^2] \tag{11.61}$$

leading to the flexural rigidity

$$D(r) = D_0[1 - (r/R)^2]^3 \tag{11.62}$$

where R is the radius of the plate, since he observed that $D(r)$ satisfies the condition

$$D(R) = \frac{dD(R)}{dr} = 0 \tag{11.63}$$

The boundary conditions, demanding that the bending moment,

$$M_r = -D(r)\left(\frac{d^2W}{dr^2} + \frac{v}{r}\frac{dW}{dr}\right) \tag{11.64}$$

and the shearing force,

$$V_r = -D(r)\frac{d}{dr}\nabla^2 W - \frac{dD(r)}{dr}\left(\frac{d^2W}{dr^2} + \frac{v}{r}\frac{dW}{dr}\right) \tag{11.65}$$

are to vanish at the outer boundary, are identically satisfied. This led Harris (1968) to the possibility to choose, *de facto* any expression for the mode shape. He chose the function

$$W(\rho) = \sum_{i=0}^{n} \rho^{2i} \quad \rho = r/R \tag{11.66}$$

and obtained a closed-form expression for the natural frequency. The work by Harris (1968) appears to be the *only one* both for plates with constant thickness, as well as of variable thickness for which both the mode shape and the natural frequency are derived in closed-form. In contrast to the paper by Lennox and Conway (1980) he not only obtained the expression for the mode shape, but also for the eingenvalue. We pursue a somewhat different avenue with the attendant closed-form solution.

11.2.2 Formulation of the Problem

Consider an inhomogeneous circular plate that is free at its edge. We pose the following inverse problem: find a distribution of the flexural rigidity $D(r)$ and the material density $\delta(r)$ (per unit area) so that the free plate will possess the postulated mode shape:

$$W(\rho) = 1 + \alpha\rho^2 + \beta\rho^4 \tag{11.67}$$

where α and β are parameters yet to be determined. If such a plate is found, it is natural to visualize that its flexural rigidity will be dependent upon the parameters α and β; thus,

$$D(\rho) = D(\rho, \alpha, \beta) \tag{11.68}$$

The free parameters α and β will be chosen in such a form that the conditions

$$D(\rho, \alpha, \beta) = 0 \qquad \text{at } \rho = 1 \tag{11.69}$$

and

$$\frac{dD(\rho, \alpha, \beta)}{d\rho} = 0 \qquad \text{at } \rho = 1 \tag{11.70}$$

are satisfied. This leads to two equations for α and β, and thus, to the closed-form solution for the first non-zero natural frequency.

11.2.3 Basic Equations

The governing differential equation, governing the free vibration of the inhomogeneous plate reads

$$D(r)r^3 \Delta\Delta W + \frac{dD}{dr}\left(2r^3\frac{d^3W}{dr^3} + r^2(2+v)\frac{d^2W}{dr^2} - r\frac{dW}{dr}\right)$$

$$+ \frac{d^2D}{dr^2}\left(r^3\frac{d^2W}{dr^2} + vr^2\frac{dW}{dr}\right) - \delta\omega^2 r^3 W = 0 \tag{11.71}$$

We represent the mass density and the flexural rigidity in the following forms, respectively:

$$\delta(r) = \sum_{i=0}^{m} c_i r^i \tag{11.72}$$

$$D(r) = \sum_{i=0}^{m+4} b_i r^i \tag{11.73}$$

Substitution of Eqs. (11.67), (11.72) and (11.73) into the governing differential equation leads to the result

$$\sum_{i=0}^{7} d_i r^i = 0 \tag{11.74}$$

where

$$d_0 = 0 \quad d_1 = 0 \quad d_2 = 2b_1\alpha(1+v) \quad d_3 = 64b_0\beta + 8b_2\alpha(v+1) - a_0\omega^2$$

$$d_4 = 12\beta(11+v) + 18b_3\alpha(1+v) \quad d_5 = 32b_2\beta(7+v) + 32b_4\alpha(1+v) - a_0\alpha\omega^2$$

$$d_6 = 20(17+3v)b_3\beta \quad d_7 = 96(5+v)b_4\beta - a_0\beta\omega^2$$

$$\tag{11.75}$$

Note that in the paper (Elishakoff, 2000c) the expressions for the coefficients d_3 and d_4 contain typographic errors (Storch, 2001b). In order for Eq. (11.74) to be valid, all d_i must vanish. We get six non-trivial equations for five coefficients b_0, b_1, b_2, b_3, b_4 and ω^2, i.e., a total of six unknowns. For non-triviality of the solution, the determinant should vanish. It turns out to be identically zero. Taking b_4 to be an undetermined coefficient, the solution is written as

$$b_0 = -[\alpha^2(1+v) - 6\beta(5+v)]b_4/4\beta^2 \quad b_1 = 0 \quad b_2 = 2\alpha b_4/\beta \quad b_3 = 0$$

$$\tag{11.76}$$

with attendant natural frequency squared,

$$\omega^2 = 96(5+v)b_4/a_0 \tag{11.77}$$

It is interesting to note that Eq. (11.77) coincides with Eq. (11.12), i.e., the expressions for the natural frequencies of the clamped and free circular plates coincide. Equation (11.76) yields the flexural rigidity

$$D(r) = -\frac{[\alpha^2(1+v) - 6\beta(5+v)]b_4}{4\beta^2} + \frac{2\alpha b_4}{\beta}r^2 + b_4 r^4 \tag{11.78}$$

We require that $D(R) = 0$, along with $D'(R) = 0$:

$$-\frac{[\alpha^2(1+v) - 6\beta(5+v)]b_4}{4\beta^2} + \frac{2\alpha b_4}{\beta} + b_4 = 0$$

$$\frac{4\alpha b_4}{\beta} + 4b_4 = 0 \tag{11.79}$$

Solution of Eq. (11.79) yields

$$\alpha = -6/R^2 \quad \beta = 6/R^4 \tag{11.80}$$

Thus, the mode shape of the plate reads

$$W(\rho) = 1 - 6\rho^2 + 6\rho^4 \tag{11.81}$$

whereas the flexural rigidity is

$$D(\rho) = b_4 R^4 (1 - \rho^2)^2 \tag{11.82}$$

11.2.4 Concluding Remarks

This section presents a new closed-form solution for the natural frequency of an inhomogeneous circular plate that is free at its boundary. While Harris (1968) dealt with the flexural rigidity in the form $D(\rho) = D_0 \left(1 - \rho^2\right)^3$, here, we arrive, by using inverse vibration analysis, at a simpler expression for the same, namely $D(\rho) = D_0 \left(1 - \rho^2\right)^2$.

11.3 Axisymmetric Vibration of Inhomogeneous Pinned Circular Plates: An Unusual, Exact, Closed-Form Solution

11.3.1 Basic Equations

Timoshenko and Woinowsky-Krieger (1959) reported the following static displacement of the uniform homogeneous circular pinned plate

$$W = (q/64D)(R^2 - r^2)(\theta R^2 - r^2) \tag{11.83}$$

where the parameter θ depends solely on the Poisson ratio

$$\theta = (5 + v)/(1 + v) \tag{11.84}$$

where q is the uniform pressure and D is the flexural stiffness. We pose the following problem: find the variation of the flexural rigidity of an inhomogeneous circular plate as a function of the polar coordinate r that leads to the following postulated mode shape

$$\psi(r) = (R^2 - r^2)(\theta R^2 - r^2) \tag{11.85}$$

which is proportional to the static displacement in Eq. (11.83).

11.3.2   Constant Inertial Term ($m = 0$)

In this case, the flexural rigidity is sought for as a fourth-order polynomial:

$$D(r) = b_0 + b_1 r + b_2 r^2 + b_3 r^3 + b_4 r^4 \tag{11.86}$$

We then get the differential equation as follows:

$$\sum_{i=0}^{7} c_i r^i = 0 \tag{11.87}$$

where

$$
\begin{aligned}
&c_0 = 0 \quad c_1 = 0 \quad c_2 = -4(3 + v) R^2 b_1 \\
&c_3 = 64 b_0 - 16(3 + v) R^2 b_2 - a_0 \omega^2 R^4 (5 + v)/(1 + v) \\
&c_4 = 12(11 + v) b_1 - 36(3 + v) R^2 b_3 \\
&c_5 = 32(7 + v) b_2 - 64(3 + v) R^2 b_4 + 2 a_0 \omega^2 R^2 (3 + v)/(1 + v) \\
&c_6 = (340 + 60v) b_3 \quad c_7 = 96(5 + v) b_4 - a_0 \omega^2
\end{aligned}
\tag{11.88}
$$

Since the left-hand side of the differential equation must vanish for any r within $[0; R]$, we demand that all the coefficients c_i be zero. This leads to a homogeneous set of six linear algebraic equations for six unknowns. It turns out that the determinant of the matrix of the set derived from Eq. (11.11) is zero. Therefore, a non-trivial solution is obtainable; the natural frequency is obtained from the coefficient c_7:

$$\omega^2 = 96(5 + v) b_4 / a_0 \tag{11.89}$$

Upon substitution of Eq. (11.12) into (11.11), the remaining equations yield the coefficients in the flexural rigidity:

$$b_0 = \frac{57 + 18v + v^2}{2(1 + v)} R^4 b_4 \quad b_1 = 0 \quad b_2 = -4 \frac{3 + v}{1 + v} R^2 b_4 \quad b_3 = 0 \tag{11.90}$$

Hence, the flexural rigidity reads

$$D(r) = \left(\frac{57 + 18v + v^2}{2(1 + v)} R^4 - 4 \frac{3 + v}{1 + v} R^2 r^2 + r^4 \right) b_4 \tag{11.91}$$

Figure 11.3 depicts the flexural rigidity for various values of the Poisson ratio v.

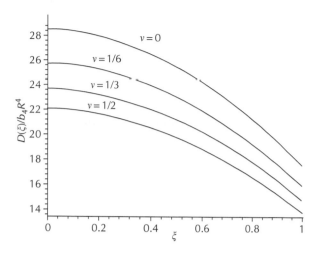

FIGURE 11.3
Variation of the flexural rigidity for different values of the Poisson's coefficient

11.3.3 Linearly Varying Inertial Term ($m = 1$)

Instead of the set (11.11) we get here, seven linear algebraic equations with seven unknowns:

$$-4(3+v)R^2 b_1 = 0$$

$$64b_0 - 16(3+v)R^2 b_2 - a_0\omega^2 R^4 (5+v)/(1+v) = 0$$

$$12(11+v)b_1 - 36(3+v)R^2 b_3 - a_1\omega^2 R^4 (5+v)/(1+v) = 0$$

$$32(7+v)b_2 - 64(3+v)R^2 b_4 + 2a_0\omega^2 R^2 (3+v)/(1+v) = 0 \qquad (11.92)$$

$$(340+60v)b_3 - 100(3+v)R^2 b_5 + 2a_1\omega^2 R^2 (3+v)/(1+v) = 0$$

$$96(5+v)b_6 - a_0\omega^2 = 0$$

$$(644+140v)b_5 - a_1\omega^2 = 0$$

In order to have a non-trivial solution the determinant of the set (11.15) must vanish,

$$-100{,}663{,}296 R^6 \frac{(3+v)(7+v)(5+v)(5{,}546 + 2{,}619v + 364v^2 + 439v^3 + 15v^4)}{1+v} a_1$$

$$= 0 \qquad (11.93)$$

leading to $a_1 = 0$. But this result signifies that the linear inertial coefficient is a constant. Thus, the problem solved in the previous section is obtained.

11.3.4   Parabolically Varying Inertial Term ($m = 2$)

For $m = 2$, i.e., the plate whose material density varies parabolically

$$\delta(r) = a_0 + a_1 r + a_2 r^2 \tag{11.94}$$

the flexural rigidity has to be a sixth-order polynomial

$$D(r) = b_0 + b_1 r + b_2 r^2 + b_3 r^3 + b_4 r^4 + b_5 r^5 + b_6 r^6 \tag{11.95}$$

Substitution of Eq. (11.6) in conjunction with Eqs. (11.12) and (11.13) into the governing differential equation yields

$$\sum_{i=0}^{9} d_i r^i = 0 \tag{11.96}$$

where

$$d_0 = 0 \quad d_1 = 0 \quad d_2 = -4R^2(3+v)b_1$$

$$d_3 = 64b_0 - 16R^2(3+v)b_2 - a_0\omega^2 R^4 \frac{5+v}{1+v}$$

$$d_4 = 12(11+v)b_1 - 36R^2(3+v)b_3 - a_1\omega^2 R^4 \frac{5+v}{1+v}$$

$$d_5 = 32(7+v)b_2 - 64R^2(3+v)b_4 + \omega^2 \left(2a_0 R^2 \frac{3+v}{1+v} - a_2 R^4 \frac{5+v}{1+v} \right) \tag{11.97}$$

$$d_6 = 20(17+3v)b_3 - 100R^2(3+v)b_5 + 2a_1\omega^2 R^2 \frac{3+v}{1+v}$$

$$d_7 = 96(5+v)b_4 - 144R^2(3+v)b_6 - \omega^2 \left(a_0 - 2a_2 R^2 \frac{3+v}{1+v} \right)$$

$$d_8 = (644+140v)b_5 - a_1\omega^2 \quad d_9 = (832+192v)b_6 - a_2\omega^2$$

As in the case of the constant inertial term, we demand that all $d_i = 0$; thus, we get a set of eight equations with eight unknowns (the seven coefficients b_i and ω^2). The resulting determinantal equation is

$$(3+v)(7+v)a_1(360{,}490 + 325{,}523v + 113{,}630v^2$$
$$+ 19{,}024v^3 + 1{,}512v^4 + 45v^5) = 0 \tag{11.98}$$

In order that the homogeneous system has a non-trivial solution, we must demand that the coefficient a_1 vanishes. We substitute $a_1 = 0$ into the set

(11.20). We arrive at the following expression for the natural frequency:

$$\omega^2 = 64(13 + 3\nu)b_6/a_2 \tag{11.99}$$

Then, the coefficients in the flexural rigidity are given by

$$b_0 = R^4 b_6[(3{,}285 + 2{,}670\nu + 744\nu^2 + 82\nu^3 + 3\nu^4)R^2 a_2$$
$$+ (20{,}748 + 14{,}304\nu + 3{,}496\nu^2 + 352\nu^3 + 12\nu^4)a_0]/12(35 + 47\nu + 13\nu^2 + \nu^3)a_2$$

$$b_1 = 0$$

$$b_2 = \frac{R^2 b_6}{3} \frac{(1{,}095 + 525\nu + 73\nu^2 + 3\nu^4)R^2 a_2 - (2{,}184 + 1{,}544\nu + 344\nu^2 + 24\nu^4)a_0}{(35 + 47\nu + 13\nu^2 + \nu^3)a_2}$$

$$b_3 = 0 \quad b_4 = -\frac{b_6}{6} \frac{(285 + 140\nu 15\nu^2)R^2 a_2 - (52 + 64\nu + 12\nu^2)a_0}{(5 + 6\nu + \nu^2)a_2} \quad b_5 = 0$$

$$\tag{11.100}$$

where b_6 is an arbitrary constant. In order for the natural frequency squared to be a positive quantity we demand that the ratio b_6/a_2 is positive. We have two sub-cases: (1) both b_6 and a_2 are positive or (2) both are negative. In the former case, namely (1) the necessary condition for positivity of the flexural rigidity $b_0 \geq 0$ is identically satisfied. In the latter case the above inequality reduces to

$$(3{,}285 + 2{,}670\nu + 744\nu^2 + 82\nu^3 + 3\nu^4)R^2$$
$$+ (20{,}748 + 14{,}304\nu + 3{,}496\nu^2 + 352\nu^3 + 12\nu^4)a_0/a_2 \leq 0 \tag{11.101}$$

leading to the following inequality

$$\frac{a_0}{|a_2|R^2} \geq \frac{3{,}285 + 2{,}670\nu + 744\nu^2 + 82\nu^3 + 3\nu^4}{20{,}748 + 14{,}304\nu + 3{,}496\nu^2 + 352\nu^3 + 12\nu^4} \tag{11.102}$$

For the non-negativity of $\delta(r)$, the following inequality must hold:

$$\frac{a_0}{|a_2|R^2} \geq 1 \tag{11.103}$$

The flexural rigidity reads

$$D(r) = \left[\left(\frac{3595 R^6}{528} + \frac{497 R^4 a_0}{12 a_2} \right) + \left(\frac{719 R^4}{88} - \frac{35 R^2 a_0}{2 a_2} \right) r^2 \right.$$
$$\left. + \left(-\frac{125 R^2}{16} + \frac{7 a_0}{4 a_2} \right) r^4 + r^6 \right] b_6 \tag{11.104}$$

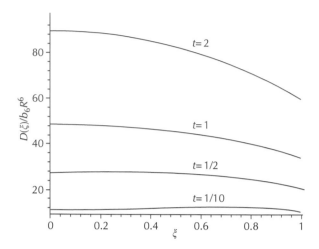

FIGURE 11.4
Variation of the flexural rigidity for different values of $t = a_0/a_2$

Figure 11.4 depicts the variation of $D(r)$ for different values of $t = a_0/a_2$.

11.3.5   Cubic Inertial Term ($m = 3$)

For $m = 3$, the following set of nine linear algebraic equations with nine unknowns is obtained

$$-4R^2(3 + v)b_1 = 0$$

$$64b_0 - 16R^2(3 + v)b_2 - a_0\omega^2 R^4 \frac{5 + v}{1 + v} = 0 \qquad (11.105)$$

$$12(11 + v)b_1 - 36R^2(3 + v)b_3 - a_1\omega^2 R^4 \frac{5 + v}{1 + v} = 0$$

$$32(7 + v)b_2 - 64R^2(3 + v)b_4 + \omega^2 \left(2a_0 R^2 \frac{3 + v}{1 + v} - a_2 R^4 \frac{5 + v}{1 + v}\right) = 0$$

$$20(17 + 3v)b_3 - 100R^2(3 + v)b_5 + \omega^2 \left(2a_1 R^2 \frac{3 + v}{1 + v} - a_3 R^4 \frac{5 + v}{1 + v}\right) = 0 \qquad (11.106)$$

$$96(5 + v)b_4 - 144R^2(1 + v)b_6 - \omega^2 \left(a_0 - 2a_2 R^2 \frac{3 + v}{1 + v}\right) = 0$$

$$(644 + 140v)b_5 - 196(3 + v)R^2 b_7 - \omega^2 \left(a_1 - 2a_3 R^2 \frac{3 + v}{1 + v}\right) = 0 \qquad (11.107)$$

$$(832 + 192v)b_6 - a_2\omega^2 = 0$$

$$(1044 + 252v)b_7 - a_3\omega^2 = 0$$

The determinantal equation arising from it reads

$$(2{,}527{,}200 + 3{,}153{,}105\nu + 1{,}582{,}173\nu^2 + 406{,}618\nu^3 + 56{,}058\nu^4$$
$$+ 3{,}893\nu^5 + 105\nu^6)R^2 a_3 + (10{,}454{,}210 + 11{,}963{,}597\nu + 5{,}573{,}931\nu^2$$
$$+ 134{,}7106\nu^3 + 177{,}016\nu^4 + 11{,}889\nu^5 + 315\nu^6)a_1 = 0 \qquad (11.108)$$

The solution is

$$a_1 = -\frac{(38{,}880 + 31{,}761\nu + 8{,}865\nu^2 + 971\nu^3 + 35\nu^4)R^2 a_3}{160{,}834 + 114{,}773\nu + 28{,}889\nu^2 + 2{,}983\nu^3 + 105\nu^4} \qquad (11.109)$$

We obtain the following expression for the natural frequency squared:

$$\omega^2 = 36(29 + 7\nu)b_7/a_3 \qquad (11.110)$$

and we get the following solution for the coefficients in the flexural rigidity:

$$b_0 = 3[(95{,}265 + 100{,}425\nu + 40{,}266\nu^2 + 7{,}586\nu^3 + 661\nu^4 + 21\nu^5)R^2 a_2$$
$$+ (601{,}692 + 560{,}052\nu + 201{,}512\nu^2 + 34{,}680\nu^3 + 2{,}812\nu^4$$
$$+ 84\nu^5)a_0]R^4 b_7/64a_3 M$$
$$b_1 = 0$$
$$b_2 = 3[(31{,}755 + 22{,}890\nu + 5{,}792\nu^2 + 598\nu^3 + 21\nu^4)R^2 a_2 - (63{,}336 + 60{,}064\nu$$
$$+ 20{,}784\nu^2 + 3{,}104\nu^3 + 168\nu^4)a_0]R^2 b_7/16a_3 M$$
$$b_3 = (64{,}800 + 44{,}295\nu + 105{,}597\nu^2 + 1{,}041\nu^3 + 21\nu^4)R^4 b_7/N$$
$$b_4 = 3b_7[(1{,}508 - 2{,}220\nu + 796\nu^2 + 84\nu^3)a_0$$
$$- (8{,}265 + 6{,}055\nu + 1{,}415\nu^2 + 105\nu^3)R^2 a_2]/a_3(65 + 93\nu + 31\nu^2 + 3\nu^3)$$
$$b_5 = -2R^2(25{,}527 + 20{,}092\nu + 5{,}424\nu^2 + 584\nu^3 + 21\nu^4)b_7/N$$
$$b_6 = 9(29 + 7\nu)a_2 b_7/16a_3(13 + 3\nu)$$
$$M = 455 + 716\nu + 310\nu^2 + 52\nu^3 + 3\nu^4$$
$$N = 5{,}546 + 8{,}165\nu + 2{,}983\nu^2 + 379\nu^3 + 15\nu^4$$
$$(11.111)$$

where b_7 is an arbitrary constant. For the particular case $v = \frac{1}{3}$, the flexural rigidity equals

$$
\begin{aligned}
D(r) = \Bigg[& \frac{1,689,965}{19,712} \frac{R^6 a_2}{a_3} + \frac{3,337}{64} \frac{R^4 a_0}{a_3} + \left(\frac{101,379}{9,856} \frac{R^4 a_2}{a_3} - \frac{705}{32} \frac{R^2 a_0}{a_3} \right) r^2 \\
& + \frac{21,524}{2,295} R^4 r^3 + \left(-\frac{17,625}{1,792} \frac{R^2 a_2}{a_3} + \frac{141}{64} \frac{a_0}{a_3} \right) r^4 \\
& - \frac{389}{51} R^2 r^5 + \frac{141}{112} \frac{a_2}{a_3} r^6 + r^7 \Bigg] b_7
\end{aligned}
\tag{11.112}
$$

11.3.6 General Inertial Term ($m \geq 4$)

Now, consider now the general expression for the inertial term given in Eq. (11.7), and the flexural rigidity in Eq. (11.8), for $m \geq 4$. Substitution of Eqs. (11.6), (11.7) and (11.8) into the terms for the differential equation yields

$$
r^3 D(r) \Delta \Delta W = -64 r^3 \sum_{i=0}^{m+4} b_i r^i
\tag{11.113}
$$

$$
\frac{dD}{dr} \left(2r^3 \frac{d^3 W}{dr^3} + r^2(2 + v) \frac{d^2 W}{dr^2} - r \frac{dW}{dr} \right) = 4[(17 + 3v)r^2 - R^2(3 + v)]r^2 \sum_{i=1}^{m+4} i b_i r^{i-1}
\tag{11.114}
$$

$$
\frac{d^2 D}{dr^2} \left(r^3 \frac{d^2 W}{dr^2} + v r^2 \frac{dW}{dr} \right) = 4[(3 + v)r^2 - R^2(3 + v)]r^3 \sum_{i=2}^{m+4} i(i - 1) b_i r^{i-2}
\tag{11.115}
$$

and

$$
\rho h \omega^2 r^3 W = \omega^2 r^3 (R^2 - r^2)(\theta R^2 - r^2) \sum_{i=0}^{m} a_i r^i
\tag{11.116}
$$

Demanding the sum of (11.31)–(11.44) to be zero, we obtain the following equation

$$
\sum_{i=0}^{m+7} g_i r^i = 0
\tag{11.117}
$$

where the coefficients g_i are

$$
g_0 = 0 \quad g_1 = 0
\tag{11.118}
$$

$$
g_2 = -4(3 + v) b_1
\tag{11.119}
$$

$$g_3 = 64b_0 - 16R^2(3+v)b_2 - \theta a_0 R^4 \omega^2 \tag{11.120}$$

$$g_4 = 12(11+v)b_1 - 36R^2(3+v)b_3 - \theta a_1 R^4 \omega^2 \tag{11.121}$$

$$g_5 = 32(7+v)b_2 - 64R^2(3+v)b_4 - \omega^2 \left[\theta a_2 R^4 - (1+\theta)a_0 R^2\right] \tag{11.122}$$

$$g_6 = 20(17+3v)b_3 - 100R^2(3+v)b_5 - \omega^2 \left[\theta a_3 R^4 - (1+\theta)a_1 R^2\right] \tag{11.123}$$

$$\vdots$$

$$g_i = [64 + 4(i-3)(17+3v) + 4(i-3)(i-4)(3+v)]b_{i-3} - 4(i-1)^2 R^2(3+v)b_{i-1}$$
$$- \omega^2 \left[\theta a_{i-3} R^4 - (1+\theta)a_{i-5} R^2 + a_{i-7}\right] \qquad \text{for } 7 \leq i \leq m+3 \tag{11.124}$$

$$\vdots$$

$$g_{m+4} = [64 + 4(m+1)(17+3v) + 4m(m+1)(3+v)]b_{m+1}$$
$$- 4R^2(m+3)^2(3+v)b_{m+3} - \omega^2 \left[a_{m-3} - (1+\theta)a_{m-1} R^2\right] \tag{11.125}$$

$$g_{m+5} = [64 + 4(m+2)(17+3v) + 4(m+1)(m+2)(3+v)]b_{m+2}$$
$$- 4R^2(m+4)^2(3+v)b_{m+4} - \omega^2 [a_{m-2} - (1+\theta)a_m R^2] \tag{11.126}$$

$$g_{m+6} = [64 + 4(m+3)(17+3v) + 4(m+2)(m+3)(3+v)]b_{m+3} - \omega^2 a_{m-1} \tag{11.127}$$

$$g_{m+7} = [64 + 4(m+4)(17+3v) + 4(m+3)(m+4)(3+v)]b_{m+4} - \omega^2 a_m \tag{11.128}$$

We demand that all coefficients g_i should be zero; thus, we get a set of $m+6$ homogeneous linear algebraic equations for $m+6$ unknowns. In order to find a non-trivial solution, the determinant of the set (11.118)–(11.128) must vanish. We expand the determinant along the last column of the matrix of the set, getting a linear algebraic expression with the a_i as coefficients. The determinantal equation yields a condition for which a non-trivial solution is obtainable. In this case, the general expression of the natural frequency squared is obtained from the equation $g_{m+7} = 0$, resulting in

$$\omega^2 = 4(m+6)[m(3+v) + 4(5+v)]b_{m+4}/a_m \tag{11.129}$$

Note that the formulas pertaining to the cases $m = 0$, $m = 2$ and $m = 3$ are formally obtainable from Eq. (11.129) by appropriate substitution. An alternative mode shape $W = (r - R)[vr - R(v + 2)]$ has been discussed by Elishakoff and Storch (2004).

11.3.7   Concluding Remarks

In this chapter, several axisymmetric vibration problems have been solved in closed-form. The buckling problem has been studied by Elishakoff, Ruta and Stavsky (2004). Non-axisymmetric problems have been considered by Storch and Elishakoff (2004) in the vibration setting, apparently for the first time. No non-axisymmetric solutions are available yet for the buckling problems. Several closed-form solutions exist for the static response of composite elliptic plates. These include papers by Kicher (1969) and Bert (1981a). It appears that closed-form solutions of the polynomial type may be shown to appear in the problems of static behavior, vibrations and buckling of elliptic plates.

Epilogue

This monograph was conceived due to sheer curiosity. The author has been engaged for over fifteen years, in writing, on and off (more off than on), lecture notes on elastic stability, in which, naturally, Duncan's (1937) closed-form solution is reproduced. Duncan's article itself, titled "Galerkin's Method in Mechanics and Differential Equations," seems to be a surprising place to find a closed-form solution, for it is devoted to the Boobnov–Galerkin method, which is an approximate, though an extremely effective, technique. Duncan did not divulge whether he guessed the polynomial mode shape, or used the method of trial and error, or simply used the method that is elucidated in this book without elaborating details. We just asked, as it now appears, the right question, as to whether his solution was the only one. We found that there are an *infinite number* of other solutions of this kind. Then, invariably, different problems yielded exact solutions, in an apparent "domino effect."

Before we complete our study on inhomogeneous structures, it appears instructive to make some pertinent comments on the essence of this work. It is posed as an inverse problem. We postulated throughout this monograph the mode shape, either in buckling or vibration, and designed the structure that possessed exactly this mode shape. Keller (1976) in his insightful article explained the terminology on direct and inverse problems:

> We call two problems **inverses** of one another of the formulation of each involves all or part of the solution of the other. Often, for historic reasons, one of the two problems has been studied extensively for some time, while the other is newer and not so well understood. In such cases, the former is called the **direct problem**, while the latter is called the **inverse problem**" (bold by Keller).

Keller (1976) continues:

> As illustrations, we present the following three inverse problems. The corresponding direct problems, which are their solutions are given in the appendix.

1. What is the question to which the answer is 'Washington, Irving?'
2. What is the question to which the answer is 'Nein, W.?'
3. What is the question to which the answer is 'Chicken Sukiyaki'?

To satisfy the reader's curiosity, Keller's answers from the appendix of his paper are immediately reproduced:

1. What is the capital of the United States, Max?
2. Do you spell your name with a 'V', Herr Wagner?
3. What is the name of the sole surviving Kamikaze pilot?"

Yet, one is advised to provide his/her own possible answers to Keller's insightful questions. According to Keller (1976), "these examples illustrate that inverse problems often have many solutions, and also that some particular solution is preferable to the others."

It can be immediately observed that the solution of some inverse problems may turn out to be *easier* than that of the direct one. One such example is finding a polynomial $p(x)$ of degree n with given roots $x_1, \ldots, x_n$. This constitutes an inverse problem to the direct one of determining the roots $x_1, \ldots, x_n$ of a given polynomial of degree n. In this particular case, the inverse problem is straightforward, since the solution

$$P(x) = c(x - x_1)(x - x_2) \cdots (x - x_n) \tag{EP.1}$$

is immediately derived. Remarkably, this solution is not unique, since the coefficient c can be an arbitrary constant.

An example of the direct problem that is much easier than the inverse one is given by Kirsch (1991): "Find a polynomial Π of degree v that assumes given values $y_1, \ldots, y_n \in \nabla$ at given points $x_1, \ldots, x_n \in \nabla$. This problem is inverse to the direct problem of calculating the given polynomial at given $x_1, \ldots, x_n$. The inverse problem is the *Lagrange interpolation problem*."

Niordson (1967) wrote:

> Ordinarily an eigenvalue problem is formulated so that the potential is given, while its spectrum is unknown. However, a considerable interest is attached to the *inverse* problem: To find the potential and out of some family of functions S, given certain properties of its spectrum. Clearly questions concerning existence and uniqueness of solutions are of great importance in such problems, the fundamental problem however, is to find explicit methods of constructing solutions.

In all problems posed in this monograph the solution of the semi-inverse problem turns out to be much *easier* than that of the direct problem. Yet, we were not interested in the inverse problem *per se*. Our objective was obtaining closed-form solutions, which was achieved by posing an inverse problem. Thus, inverse problem formulation was the *bridge* to our goal rather than the declared destination.

In the problems that were studied, our requirement that the system have a polynomial mode shape resulted in an infinite number of solutions, with arbitrary coefficients appearing in the flexural rigidity of the structure.

Remarkably, Duncan's (1937) unique solution, which is reproduced in the Prologue, may immediately give rise to an infinite set of solutions not unlike the ones reported in the present monograph. Indeed, instead of the flexural rigidity given in Eq. (P.7) we introduce the new flexural rigidity

$$D(x) = [1 - \tfrac{3}{7}(x/L)^2]D_0 b \tag{EP.2}$$

where b is an arbitrary positive parameter. The mode shape is again postulated as given by Duncan (1937) in Eq. (P.8). After substituting Eq. (EP.2) into the governing differential equation (P.5) we get the buckling load

$$D(x) = \tfrac{60}{7}D_0 b/L^2 \tag{EP.3}$$

fully resembling the form of the closed-form solutions derived in this monograph. Equation (EP.3) represents an infinite set of solutions, since the positive parameter b can be chosen arbitrarily. This solution, along with numerous classes of solutions reported in this monograph, depends on the parameter b. This is not unlike the solution of constructing the polynomial with given roots, which depends, as shown in Eq. (EP.1), on the arbitrary coefficient c.

Still, some researchers may be unsettled by the fact that an infinite number of flexural rigidities is generated when the sole requirement is to have a mode shape postulated. They may ask: is it possible to pose a question somewhat differently in order to get a unique solution? The reply is "yes." Indeed, instead of postulating only the mode shape, one can also demand the fundamental natural frequency to constitute a pre-selected value, or that the buckling load be a given quantity.

Indeed, Eq. (2.156) yields an infinite number of flexural rigidities for inhomogeneous clamped–free columns under their own weight:

$$D(\xi) = (\tfrac{3}{2} + 3\xi - 3\xi^2 + \xi^4)b_3 \tag{EP.4}$$

with the corresponding buckling loads in Eq. (2.154):

$$q_{cr} = 15b_3/L^3 \tag{EP.5}$$

However, if in addition to postulating the mode shape

$$W(\xi) = 6\xi^2 - 4\xi^3 + \xi^4 \tag{EP.6}$$

in Eq. (2.151), we demand the buckling load to equal a pre-selected value $\tilde{q}$, then the parameter b_3, which was previously treated as an arbitrary one, is directly found as

$$b_3 = \tilde{q}L^3/15 \tag{EP.7}$$

with the flexural rigidity in Eq. (E.4) becoming

$$D(\xi) = \tfrac{1}{15}\tilde{q}L^3(\tfrac{3}{2} + 3\xi - 3\xi^2 + \xi^4) \tag{EP.8}$$

As is clearly seen, the modified formulation yields a unique solution.

Likewise, Eq. (4.179) gives an infinite number of natural frequencies of the inhomogeneous beams, with the inertial coefficient representing an mth order polynomial. The natural frequency squared

$$\omega^2 = 12(m^2 + 11m + 30)b_{m+4}/a_m L^4 \tag{EP.9}$$

depends on the arbitrary coefficient b_{m+4}. This arbitrariness is removed if we demand the cantilever beam to possess the mode shape (E.6), and to have a pre-selected natural frequency $\tilde{\omega}$. The coefficient b_{m+4} is then given as

$$b_{m+4} = \tilde{\omega}^2 a_m L^4 / 12(m^2 + 11m + 30) \tag{EP.10}$$

Thus, the beam's flexural rigidity, which is proportional to b_{m+4}, becomes a unique function.

Also, naturally, in some problems, the flexural rigidity parameters and other parameters of the system turned out to have common coefficients, as happens for beams on inhomogeneous Winkler or Pasternak foundations. This fact ought not be interpreted as an indication of their interdependence. Indeed, such conditions often arise in closed-form solutions obtained in other fields. For example, in order to obtain closed-form solutions for stochastic problems, some specific conditions often must be met. Let us cite an example. In his book, Nigam (1983) deals with the multi-degree-of-freedom system governed by the following set of equations:

$$\ddot{Y}_j + \beta_j \dot{Y}_j + \partial V / \partial Y_j = Q_j(t) \qquad j = 1, 2, \ldots, n \tag{EP.11}$$

where Y_j are generalized displacements, β_j are damping coefficients, V is the potential function, Q_j are generalized forces and n represents the number of degrees of freedom. Nigam replaces the system (EP.11) by a system of $2n$ first-order equations

$$\dot{Z}_j = Z_{j+n}, \dot{Z}_{j+n} = -\beta_j Z_{j+n} - \partial V / \partial Z_j + Q_j(t) \tag{EP.12}$$

where $Z_j = Y_j$ and $Z_{j+n} = \dot{Y}_j$, $j = 1, 2, \ldots, n$. Then, the Fokker–Planck equation is written for the stationary response, which is not reproduced here. Nigam (1983) notes:

Assume that

$$\beta_j / \Phi_j = \gamma \text{ for every } j, \tag{EP.13}$$

and define

$$H = \frac{1}{2}\sum_{j=1}^{n} z_{j+n}^2 + V(z_1, z_2, \ldots, z_n),$$ (EP.14)

then the solution can be expressed as

$$p(z_1, z_2, \ldots, z_{n+1}, \ldots, z_{2n}) = C \exp[-(\gamma/\pi)H].$$ (EP.15)

where p is the probability density function and Φ_j is the spectral density of $Q_j(t)$. As can be seen, the closed-form solution (EP.15) is obtainable when the ratios between the inner characteristics — damping coefficients of the system, on the one hand, and the spectral densities Φ_j of excitation, on the other — satisfy the condition given in Eq. (EP.13). This does not signify that the inner and external characteristics of the system ought to be proportional, but only indicates that when these characteristics have a common numerical parameter, then, the closed-form solution is derivable. The condition (EP.13) is necessary for the solution to be given by Eq. (EP.15), which was first derived by Ariaratnam (1960) for systems with two degrees of freedom and was extended to multi-degree-of-freedom systems by Caughey (1963). Likewise, it appears instructive to quote Langley (1988):

> Remarkably, equation
>
> $$\ddot{x} + 1/2\gamma[H_y r(H) - H_{yy}/H_y]\dot{x} + H_x/H_y = \sqrt{\gamma}\xi(t)$$ (EP.16)
>
> is identical to the most generally externally excited non-linear oscillator for which an exact solution to the stationary Fokker–Planck equation has appeared in the literature. In the study of the stochastic stability of a class of elastic structures, Caughey and Ma (1982) seemingly happened upon the form of [this] equation..., for which a solution to a stationary Fokker–Planck equation was then found.

For the notations in (E.16) the reader may consult Langley (1988). Note again that the same parameter γ appears both on the left- and right-hand sides of the equation. To recapitulate, the above fact does not imply that the inner properties depend on the external ones. This merely tells us that in order for the closed-form solution to be obtainable the inner and external properties of the system must have common parameters. For other closed-form solutions of this kind the reader may consult works by Dimentberg (1982), Soize (1988) Scheurkogel and Elishakoff (1988) and Lin and Cai (1995).

Therefore, the importance of the derived closed-form solutions is not diminished by the fact that certain conditions must be met. The appearance of conditions is natural too. Indeed, it can be expected that the solution of the inverse problem would depend on either part of or on the entire given data. Thus, if the beam rests on inhomogeneous Winkler or Pasternak foundations with given coefficients, it must come as no surprise that the sought for flexural rigidity of the beam that possesses the pre-selected mode shape, is directly related to the foundation coefficients.

As Eqs. (E.7) and (E.10) show, inhomogeneity allows us to make structures that serve a specific purpose. For homogeneous structures the tapering allows us to achieve such a goal. The tapered structures are known from antiquity. As Gustave Eiffel — the builder of the famous tower — put it (Kunzig, 2003),"The first principle of architectural aesthetics is that the essential lines of a monument should be dictated by a perfect adaptation to its purpose."

A decade later, American architect Louis Sullivan coined what became the dictum: "Form ever follows function." With functionally graded materials (see Section 1.11), one gets additional freedom of tailoring structures via continuous variation of the elastic modulus in one of the directions. It is anticipated that when technology is in place, axial grading will play an ever increasing role in achieving the structural purpose at hand. It is hoped that this monograph provides methods of determining desirable polynomial as well as non-polynomial variations of the elastic modulus in the axial direction.

The following question arises: is there a resemblence in the previous literature to the type of thinking adopted in this book? The connection with the previous work was found via Saint-Venant's semi-inverse method, albeit developed for static problems. As Timoshenko (1953) writes:

> In 1853, Saint-Venant presented his epochmaking memoir on torsion to the French Academy. The committee, composed of Cauchy, Poncelet, Piobert, and Lamé, were very impressed by the work and recommended its publication . . . In the introduction Saint-Venant states that the stresses at any point of an elastic body can be readily calculated if the functions representing the components u, v and w of the displacements are known . . . Saint-Venant then proposes the *semi-inverse method* by which he assumes only some features of the displacements and the forces and determines the remaining features of those quantities so to be satisfied by all the equations of elasticity. He remarks that an engineer, guided by the approximate solution of the elementary strength of materials, can obtain rigorous solutions of practical importance in this way.

Thus, Saint-Venant (1855) postulates the prior knowledge of the two displacement functions $u = \theta z y$ and $v = \theta z x$, and then determines the function $w = \theta \varphi(x, y)$, where $\varphi(x, y)$ is some function of x and y determined later. One should also mention other important semi-inverse solutions, namely by Nemenyi (1951), Truesdell (1953), Truesdell (see Flügge, 1965), Kaloni (1989), Horgan and Villaggio (1996), He (1997, 2000, 2001), Sun and Fu (2001) as well as contributions to the various inverse problems by Hochstadt (1965, 1976 a,b), Costa (1989), Lei (1987), Levine and Costa (1992), Siddiqui (1990), Siddiqui and Kaloni (1986), Villaggio (1979, 1996), Kim and Daniel (2003) and others. We have assumed throughout this monograph, the knowledge of the mode shapes. This provides some connection with Saint-Venant's semi-inverse method, and answers negatively the inquisitive question posed in Ecclesiastes (1:10) some three thousand years ago: "Is there a thing of which it is said, 'See, this is new'? It has been in the ages before us." It appears that the above Ecclesiastes principle has not been violated by this monograph, the eigenvalue problem was found in which the method similar to the present semi-inverse method was utilized (see also the Section 1.10 describing works by Życzkowski and Gajewski). According to Sabatier (1990), "A model is a way of relating parameters that describe a physical system to all possible measurements results."

For other semi-inverse problems readers can consult the works by Horgan and Villaggio (1996), Liu (2000), Sun and Fu (2001) and He (1997, 2000), to name just a few. Naturally, Życzkowski's (1991) approach, elucidated in several of his articles and those of his collaborators, of an assumed exact solution is also a semi-inverse method.

In this monograph we treated models of rods, columns, beams and plates, along with the knowledge of the mode shapes in either buckling or vibration to determine the spatial distribution of flexural rigidities and material densities that are compatible, in an exact manner and in the closed-form with the adopted models.

It is believed that many closed-form solutions are still to be uncovered. For example, we did not touch upon the following models:

1. vibration and buckling of Bresse–Timoshenko beams;
2. vibration and buckling of beams with open cross-section;
3. buckling of circular plates in an non-axisymmetric setting;
4. vibration and buckling of circular and annular plates in non-axisymmetric contexts;
5. vibration and buckling of rectangular plates;
6. vibration and buckling of shells;
7. vibration and buckling of composite structures;
8. closed-form solutions of stochastic boundary-value problems, and many other fascinating topics.

In Dimentberg's terminology (Dimentberg, 2000), the present author does not "possess the keys of *all* secret seifs but found a key that opens *some* seifs." This metaphorical statement reflects the essence of this book quite accurately. Yet, it must be stressed that we do know that the key to *all* "seifs," i.e., finding closed-form solutions to *all* eigenvalue problems, does not exist, at the present time. This is due to the fact that even in uniform beams, columns and plates, often there is an *exact* solution but no *closed-form* one for the eigenvalues. For example, for the clamped–clamped uniform vibrating beam the natural frequencies are found by numerically solving a transcendental equation, namely (1.126); likewise, for a pinned–clamped uniform column, in order to determine the buckling load, first the roots of the transcendental equation (1.195) ought be determined.

It is possible that other types of closed-form solutions are obtainable, for example, in terms of trigonometric and/or exponential functions (see, e.g., Caliò and Elishakoff, 2002), in addition to those expressed via polynomial and rational functions, as developed in this monograph.

Whereas the method of actual construction of inhomogeneous structures with axially graded flexural rigidity and/or inertial coefficients remains beyond the scope of this monograph, earlier progress in science and technology teaches us that the design of inhomogeneous structures with the desired rigidity and inertial functions ought not to be though of as impossible.

As was quoted already, Bulson (1970) posed the following question, in the preface of his book: "In many technical libraries the supports of bookshelves containing works on the stability of structure are in danger of buckling under the weight of literature. Can another work on this subject be justified?" By the very writing of his book Bulson (1970) gave a positive reply to his question. Since then Bažant and Cedolin (1991) composed their *magnum opus* on buckling of structures, and many first-rate texts on vibration of structures appeared. Still, a skeptic may ask: can Bulson's Hamletian question have an affirmative reply *now* too, in the 21st century?

It appears that the answer to this question is an unequivocal and enthusiastic "yes!" Indeed, closest to this book are (a) the exact solutions derived by numerous authors in whose works the solutions are derived in terms of *special* functions, by *clever* substitutions made in the differential equation, as well as (b) those derived by the exact finite element method, where exact buckling loads (or frequencies) "up to the accuracy of the computer" can be derived, in principle. The advantage of the closed-form solutions over the exact ones is obvious, for the former are independent of the accuracy of the computer. On the other hand, the exact finite element method, or Frobenius method, can be applied conceptually to any problem. Here, *specific* classes of "secrets" of the buckling and vibration behavior of rods, columns, beams and plates are uncovered.

It appears to be instructive to tell a story of a famous scholar who was concerned with an extremely complex and previously untreated philosophical problem. After some months he was able to crack the problem and derive an answer to this burning question. How big was his surprise that once, when he was travelling in the countryside, in a village school, a young pupil posed exactly the same question to his teacher. The teacher was able to immediately retort with the right answer. The scholar was flabbergasted that a provincial school could deal so successfully with his most complex problem, which he was so proud of solving. Yet, he was rewarded by a dream where he was told that he should not be depressed by the fact that much younger people could now deal with his problem. Once the new pipeline of thought is opened, all can easily grasp it in the near future. Therefore, many more closed-form solutions ought be anticipated in the future.

It ought to be stressed that the closed-form solutions reported in this book are solutions of approximate governing equations of rods, columns, beams and plates. Fritjof Capra noted (Bennett, 1999):

> You cannot express the full truth in words, you can only approximate it.
> That awareness has come to the fore more and more in science during this century. In this respect, I've been very deeply influenced by my physics mentor,

Geoffrey Chew at the University of California, Berkeley, with whom I worked for over ten years; he places great emphasis on 'approximation' in science. . . . he told. . . that the greatest breakthrough would be in the recognition that whatever we say is limited and approximate. It was Werner Heisenberg who in a sense discovered the concept of approximation in atomic physics. He said that in quantum theory every word, every theory that we might propose clear as it may seen to be, is only an approximation.

This does not imply obviously, that the solution of approximate governing equations must be sought only by approximate means (Chernousko, 1998). In 1905 Lyapunov (see Lyapunov, 1959) wrote:

I think that if in some case it is permissible to use unclear considerations, it is when one ought to establish a new principle, that does not logically follow from what is accepted already, and which is not in logically contradiction with other principles of science, but one cannot set this way when one has to solve a determined problem (of mechanics or of physics), which is posed mathematically in fully exact manner. This problem then becomes a problem of pure analysis and should be solved as such.

Indeed, the present writer recalls a remark made by Academician V. Novozhilov in his paper on the required accuracy: "Who is going to solve the problems of static bending of beams approximately just because the equations of beams are approximate?!" This book is written in that spirit: It is recognized that the equations we use are of approximate nature, in order not to overestimate the obtained results. It is also realized that the infinite number of closed-form solutions reported here are obtained apparently for the first time, so that their importance could be fully estimated in the future.

It is humbly felt that more studies will appear in the literature, hopefully inspired by infinite sets of unusual closed-form solutions presented in this monograph. The best outcome, yet, would be if readers would argue about the method presented in this monograph saying "It isn't unusual," as a popular singer Tom Jones maintained in his 1965 hit. Thus, the monograph's main purpose is to make unusual closed-form solutions part and parcel of seemingly "prosaic" research activities. Moreover, since Bernoulli and Euler proposed, over 250 years ago, differential equations for vibrating rods and beams, no closed-form solutions were reported, to the best of the author's knowledge, for vibrating inhomogeneous beams or columns. The extreme *simplicity* of numerous results obtained in this monograph appears to be the most attractive feature. It is hoped that future textbooks and research monographs on vibration will include some closed-form solutions for inhomogeneous structures reported herein and *inter alia* utilize the solutions derived in this monograph for the validation of numerous approximate techniques.

Appendices

Appendix A: Closed-Form Solutions for Beams ($m < 3$)

In Section 4.1.4 we derived closed-form solutions for the case when the inequality $m \geq 3$ is satisfied. Here, we consider cases in which $m < 3$.

Case 1: $m = 0$

In this sub-case, the expressions for $E(\xi)$ and $\rho(\xi)$ read

$$\rho(\xi) = a_0 \quad E(\xi) = \sum_{i=0}^{4} b_i \xi^i \tag{A.1}$$

By substituting the latter expressions in Eq. (4.5), we obtain

$$-12 \sum_{i=1}^{3} i(i+1)b_{i+1}\xi^i + 12 \sum_{i=2}^{4} i(i-1)b_i\xi^i + 24 \sum_{i=0}^{4} b_i\xi^i - 24 \sum_{i=0}^{3} (i+1)b_{i+1}\xi^i$$

$$+ 48 \sum_{i=1}^{4} ib_i\xi^i - kL^4 a_0(\xi - 2\xi^3 + \xi^4) = 0 \tag{A.2}$$

Equation (A.2) has to be satisfied for any ξ. This requirement yields

$$\begin{aligned}
-24b_1 + 24b_0 &= 0 \\
-72b_2 + 72b_1 - ka_0 &= 0 \\
-144b_3 + 144b_2 &= 0 \\
-240b_4 + 240b_3 + 2kL^4 a_0 &= 0 \\
360b_4 - kL^4 a_0 &= 0
\end{aligned} \tag{A.3}$$

To satisfy the compatibility equations, b_i, where $i = \{0, 1, 2, 3\}$, have to be as follows:

$$b_3 = -2b_4 \quad b_2 = -2b_4 \quad b_1 = 3b_4 \quad b_0 = 3b_4 \tag{A.4}$$

To summarize, if conditions (A.1) are satisfied, where b_i are given by Eq. (A.4), then, the fundamental mode shape is expressed by Eq. (4.45), where the fundamental natural frequency reads

$$\omega^2 = 360 \frac{I}{A} \frac{b_4}{a_0 L^4} \tag{A.5}$$

Case 2: $m = 1$

In this sub-case, the expressions for $E(\xi)$ and $\rho(\xi)$ read

$$\rho(\xi) = a_0 + a_1 \xi \quad E(\xi) = \sum_{i=0}^{5} b_i \xi^i$$

By substituting the latter expressions in Eq. (4.5), we obtain

$$-12 \sum_{i=1}^{4} i(i+1)b_{i+1}\xi^i + 12 \sum_{i=2}^{5} i(i-1)b_i\xi^i + 24 \sum_{i=0}^{5} b_i\xi^i - 24 \sum_{i=0}^{4} (i+1)b_{i+1}\xi^i$$

$$+ 48 \sum_{i=1}^{5} ib_i\xi^i - kL^4 a_0(\xi - 2\xi^3 + \xi^4) - kL^4 a_1\xi(\xi - 2\xi^3 + \xi^4) = 0 \tag{A.6}$$

Equation (A.6) has to be satisfied for any ξ. This requirement yields

$$-24b_1 + 24b_0 = 0$$

$$-72b_2 + 72b_1 - kL^4 a_0 = 0$$

$$-144b_3 + 144b_2 - kL^4 a_1 = 0$$

$$-240b_4 + 240b_3 + 2kL^4 a_0 = 0 \tag{A.7}$$

$$-360b_5 + 360b_4 + 2kL^4 a_1 - kL^4 a_0 = 0$$

$$-504b_5 + kL^4 a_0 = 0$$

To satisfy the compatibility equations, b_i, where $i = \{0, 1, 2, 3, 4\}$, have to be as follows:

$$b_4 = -\frac{9a_1 - 7a_0}{5a_1}b_5 \quad b_3 = -\frac{9a_1 + 14a_0}{5a_1}b_5 \quad b_2 = \frac{17a_1 - 28a_0}{10a_1}b_5$$

$$b_1 = \frac{17a_1 + 42a_0}{10a_1}b_5 \quad b_0 = \frac{17a_1 + 42a_0}{10a_1}b_5 \tag{A.8}$$

To summarize, if conditions (A.13) are satisfied, where b_i are given by Eq. (A.8), then, the fundamental mode shape is expressed by Eq. (4.45), where the fundamental natural frequency reads

$$\omega^2 = 504 \frac{I}{A} \frac{b_5}{a_1 L^4} \tag{A.9}$$

Case 3: $m = 2$

In this sub-case, the expressions for $E(\xi)$ and $\rho(\xi)$ read

$$\rho(\xi) = a_0 + a_1\xi + a_2\xi^2 \quad E(\xi) = \sum_{i=0}^{6} b_i\xi^i \tag{A.10}$$

By substituting the latter expressions in Eq. (4.5), we obtain

$$-12 \sum_{i=1}^{5} i(i+1)b_{i+1}\xi^i + 12 \sum_{i=2}^{6} i(i-1)b_i\xi^i + 24 \sum_{i=0}^{6} b_i\xi^i - 24 \sum_{i=0}^{5} (i+1)b_{i+1}\xi^i$$

$$+ 48 \sum_{i=1}^{6} ib_i\xi^i - kL^4(a_0 + a_1\xi + a_2\xi^2)(\xi - 2\xi^3 + \xi^4) = 0 \tag{A.11}$$

Equation (A.6) has to be satisfied for any ξ. This requirement yields

$$-24b_1 + 24b_0 = 0$$
$$-72b_2 + 72b_1 - kL^4a_0 = 0$$
$$-144b_3 + 144b_2 - kL^4a_1 = 0$$
$$-240b_4 + 240b_3 + 2kL^4a_0 - kL^4a_2 = 0 \tag{A.12}$$
$$-360b_5 + 360b_4 + 2kL^4a_1 - kL^4a_0 = 0$$
$$-504b_6 + 504b_5 + 2kL^4a_2 - kL^4a_1 = 0$$
$$672b_6 - kL^4a_2 = 0$$

To satisfy the compatibility equations, b_i, where $i = \{0, 1, 2, 3, 4, 5\}$ has to be as follows:

$$b_5 = -\frac{5a_2 - 4a_1}{3a_2}b_6 \quad b_4 = -\frac{25a_2 + 36a_1 - 28a_0}{15a_2}b_6 \quad b_3 = \frac{17a_2 - 36a_1 - 56a_0}{15a_2}b_6$$

$$b_2 = \frac{17a_2 + 34a_1 - 56a_0}{15a_2}b_6 \quad b_1 = \frac{17a_2 - 36a_1 + 84a_0}{15a_2}b_6 \quad b_0 = \frac{17a_2 - 36a_1 + 84a_0}{15a_2}b_6 \tag{A.13}$$

To summarize, if conditions (A.27) are satisfied, where b_i are given by Eq. (A.13), then, the fundamental mode shape is expressed by Eq. (4.45), where the fundamental natural frequency reads

$$\omega^2 = 672 \frac{I}{A} \frac{b_6}{a_2 L^4} \tag{A.14}$$

Appendix B: General Case of Mass Density

Substitution of the mode shape in Eq. (4.236) in Eq. (4.234) yields

$$(6\gamma_3\xi + 12\gamma_4\xi^2 + 20\gamma_5\xi^3) \sum_{i=2}^{\mu+4} i(i-1)\beta_i\xi^{i-2} + (12\gamma_3 + 48\gamma_4\xi + 120\gamma_5\xi^2) \sum_{i=1}^{\mu+4} i\beta_i\xi^{i-1}$$

$$+ (24\gamma_4 + 120\gamma_5\xi) \sum_{i=0}^{\mu+4} \beta_i\xi^i - \kappa\Lambda^4(\xi + \gamma_3\xi^3 + \gamma_4\xi^4 + \gamma_5\xi^5) \sum_{i=0}^{\mu} \alpha_i\xi^i = 0 \qquad \text{(B.1)}$$

Adjusting the indices so that the general power in each sum is given by i, we get

$$6\gamma_3 \sum_{i=1}^{m+3} i(i+1)b_{i+1}\xi^i + 12\gamma_4 \sum_{i=2}^{m+4} i(i-1)b_i\xi^i$$

$$+ 20\gamma_5 \sum_{i=3}^{m+5} (i-1)(i-2)b_{i-1}\xi^i + 12\gamma_3 \sum_{i=0}^{m+3} (i+1)b_{i+1}\xi^i$$

$$+ 48\gamma_4 \sum_{i=1}^{m+4} ib_i\xi^i + 120\gamma_5 \sum_{i=2}^{m+5} (i-1)b_{i-1}\xi^i + 24\gamma_4 \sum_{i=0}^{m+4} b_i\xi^i$$

$$+ 120\gamma_5 \sum_{i=1}^{m+5} b_{i-1}\xi^i - kL^4 \sum_{i=1}^{m+4} a_{i=1}\xi^i - kL^4\gamma_3 \sum_{i=3}^{m+3} a_{i-3}\xi^i$$

$$- kL^4\gamma_4 \sum_{i=4}^{m+4} a_{i-4}\xi^i - kL^4\gamma_5 \sum_{i=5}^{m+5} a_{i-5}\xi^i = 0 \qquad \text{(B.2)}$$

For easy identification we explicitly display those recursive relations arising from Eq. (B.2) that do *not* involve a contribution from *each* of the summations. The general recursive relation in which there is a contribution from *each* summation and the range of values of i to which it applies is also given. Thus, from Eq. (B.2) we get

from $i = 0$: $\gamma_3 b_1 + 2\gamma_4 b_0 = 0$ $\qquad\qquad$ (B.3)

from $i = 1$: $120\gamma_5 b_0 + 72\gamma_4 b_1 + 36\gamma_3 b_2 - kL^4 a_0 = 0$ $\qquad\qquad$ (B.4)

from $i = 2$: $240\gamma_5 b_1 + 144\gamma_4 b_2 + 72\gamma_3 b_3 - kL^4 a_1 = 0$ $\qquad\qquad$ (B.5)

from $i = 3$: $400\gamma_5 b_2 + 240\gamma_4 b_3 + 120\gamma_3 b_4 - kL^4(a_2 + \gamma_3 a_0) = 0$ $\qquad$ (B.6)

from $i = 4$: $600\gamma_5 b_3 + 360\gamma_4 b_4 + 180\gamma_3 b_5 - kL^4(a_3 + \gamma_3 a_1 + \gamma_4 a_0) = 0$
$\qquad\qquad\qquad\qquad\qquad\qquad\qquad\qquad\qquad\qquad\qquad\qquad$ (B.7)

from $5 \leq i \leq m+1$: $(i+1)(i+2)[20\gamma_5 b_{i-1} + 12\gamma_4 b_i + 6\gamma_3 b_{i+1}]$

$$- kL^4[a_{i-1} + \gamma_3 a_{i-3} + \gamma_4 a_{i-4} + \gamma_5 a_{i-5}] = 0 \quad \text{(B.8)}$$

from $i = m+2$: $(m+3)(m+4)[20\gamma_5 b_{m+1} + 12\gamma_4 b_{m+2} + 6\gamma_3 b_{m+3}]$

$$- kL^4(\gamma_3 a_{m-1} + \gamma_4 a_{m-2} + \gamma_5 a_{m-3}) = 0 \quad \text{(B.9)}$$

from $i = m+3$: $(m+3)(m+4)[20\gamma_5 b_{m+1} + 12\gamma_4 b_{m+2} + 6\gamma_3 b_{m+3}]$

$$- kL^4(\gamma_3 a_{m-1} + \gamma_4 a_{m-2} + \gamma_5 a_{m-3}) = 0 \quad \text{(B.10)}$$

from $i = m+4$: $(m+5)(m+6)(20\gamma_5 b_{m+3} + 12\gamma_4 b_{m+4})$

$$- kL^4(\gamma_4 a_m + \gamma_5 a_{m-1}) = 0 \quad \text{(B.11)}$$

from $i = m+5$: $[20(m+4)(m+3) + 120(m+4) + 120]\gamma_5 b_{m+4}$

$$- kL^4 \gamma_5 a_m = 0 \quad \text{(B.12)}$$

We get $m+6$ homogeneous equations with $m+6$ unknowns. For a non-trivial solution, the determinant

$$\Delta = \begin{vmatrix} 2\gamma_4 & \gamma_1 & 0 & \cdots & \cdots & \cdots & 0 & 0 \\ 120\gamma_5 & 72\gamma_4 & 36\gamma_3 & 0 & \cdots & \cdots & 0 & -L^4 a_0 \\ 0 & 240\gamma_5 & 144\gamma_4 & 72\gamma_3 & 0 & \cdots & 0 & -L^4 a_1 \\ \ddots & \ddots & \ddots & \ddots & \ddots & \ddots & \ddots & \vdots \\ \ddots & \ddots & \ddots & \ddots & \ddots & \ddots & \ddots & \vdots \\ \ddots & \ddots & \ddots & \ddots & \ddots & \ddots & \ddots & \vdots \\ 0 & \cdots & \cdots & \cdots & 0 & 20m^2 + 260m + 840 & -L^4 \gamma_5 a_m \end{vmatrix} \quad \text{(B.13)}$$

must vanish. Resolving the determinant with respect to the last column we get

$$\Delta = \sum_j (-1)^{m+6+j} m_{j,m+6} D_{j,m+6} \quad \text{(B.14)}$$

These minors have a tridiagonal nature. Each of the minors is easily reducible to the following determinant:

$$D_{m+5} = \begin{vmatrix} r_{11} & 1 & 0 & \cdots & & 0 \\ 1 & r_{22} & 1 & \ddots & & \vdots \\ 0 & \ddots & \ddots & \ddots & & 0 \\ \vdots & \ddots & \ddots & \ddots & & 1 \\ 0 & \cdots & 0 & 1 & r_{m+5,m+5} \end{vmatrix} \quad \text{(B.15)}$$

which, according to Noble and Daniel (1977) can be calculated recursively:

$$D_{m+5} = r_{m+5}D_{m+4} - D_{m+3} \tag{B.16}$$

Thus, the algorithm is given for an analytical calculation of the determinant. For the system parameter for which it equals zero, the last equation (B.12), after solving for k, yields

$$k = (20m^2 + 260m + 840)b_{m+4}/L^4a_m \tag{B.17}$$

It is interesting to compare Eq. (B.17) with the fundamental natural frequency obtained by Candan and Elishakoff (2000):

$$k = (12m^2 + 132m + 360)b_{m+4}/L^4a_m \tag{B.18}$$

For the constant density case, i.e., $m = 0$, the ratio of the second frequency to the first one is not $2:1$ (as in the case of a uniform beam) but $\sqrt{840/360} : 1$.

We now proceed to obtain the coefficients b_i in terms of the a_i and b_{m+4} as follows: substituting for k from Eq. (B.12) into Eq. (B.11) and solving for b_{m+3} we obtain b_{m+3} as a function of b_{m+4}, a_m and a_{m-1}. Substituting for b_{m+3} and k into Eq. (B.10) we obtain b_{m+2} as a function of b_{m+4}, a_m, a_{m-1} and a_{m-2}. Continuing in this way we derive all coefficients b_i in terms of b_{m+4} and the a_i.

Appendix C: Uniformly Distributed Axial Load ($m = 0$)

In this appendix we look for solutions for n larger than 1, since closed-form solutions were derived for $n = 0$, for pinned beams with uniform axial load density. With the general mode shape given in Eq. (5.11), we calculate the first term of the governing equation

$$\frac{d^2}{d\xi^2}\left[D(\xi)\frac{d^2\psi}{d\xi^2}\right] = (n+3)(n+4)[(n+1)(n+2)b_0\xi^n + (n+2)(n+3)b_1\xi^{n+1}$$

$$+ (n+3)(n+4)b_2\xi^{n+2} + (n+4)(n+5)b_3\xi^{n+3}]$$

$$- (n+3)(n+4)(2b_1 + 6b_2\xi + 12b_3\xi^2) \tag{C.1}$$

The second term in Eq. (5.26) reads

$$-q_0L^3\frac{d}{d\xi}\left[N(\xi)\frac{d\psi}{d\xi}\right] = q_0L^3\left[(n+3)(n+4)(\xi^{n+2} - \xi) - (n+4)^2\xi^{n+3}\right.$$

$$\left. + \frac{3}{2}(n+3)(n+4)\xi^2 - \frac{1}{6}(n^2 + 7n + 6)\right] \tag{C.2}$$

We have to collect terms with identical powers of ξ. It is important to know how the power of the first term in Eq. (C.1), namely n, relates to the power of the last term that equals 2. Hence, two special cases arise: $n > 2$ and $n \leq 2$. Let us first concentrate on the case $n = 1$; we obtain a homogeneous set of five linear algebraic equations for five unknowns:

$$-40b_1 - \tfrac{7}{3}q_0 L^3 = 0 \tag{C.3a}$$

$$-120b_2 + 120b_0 - 20q_0 L^3 = 0 \tag{C.3b}$$

$$-240b_3 + 240b_1 + 30q_0 L^3 = 0 \tag{C.3c}$$

$$400b_2 + 20q_0 L^3 = 0 \tag{C.3d}$$

$$600b_3 - 25q_0 L^3 = 0 \tag{C.3e}$$

The determinant of the set (C.3), which equals $6{,}912{,}000{,}000L^3$, is non-zero. Thus, the solution of (C.3) is trivial. Now, consider the case $n = 2$; we deduce from the governing differential equation a set of six equations for five unknowns:

$$-60b_1 - 4q_0 L^3 = 0 \tag{C.4a}$$

$$-180b_2 - 30q_0 L^3 = 0 \tag{C.4b}$$

$$-360b + 360b_0 + 45q_0 L^3 = 0 \tag{C.4c}$$

$$b_1 = 0 \tag{C.4d}$$

$$900b_2 + 30q_0 L^3 = 0 \tag{C.4e}$$

$$1260b_3 - 36q_0 L^3 = 0 \tag{C.4f}$$

We note that a combination of Eqs. (C.4a) and (C.4d) yields $q_0 = 0$, which is a trivial solution for the critical buckling load. We turn now to the case $n > 2$. Equating the difference of the expressions (C.1) and (C.2) to zero for every ξ within the interval $[0; 1]$, we obtain the following set of equations:

from ξ^0: $\quad -2(n+3)(n+4)b_1 - \tfrac{1}{6}(n^2 + 7n + 6)q_0 L^3 = 0 \tag{C.5a}$

from ξ^1: $\quad -6(n+3)(n+4)b_2 - (n+3)(n+4)q_0 L^3 = 0 \tag{C.5b}$

from ξ^2: $\quad 12(n+3)(n+4)b_3 + \tfrac{3}{2}(n+3)(n+4)q_0 L^3 = 0 \tag{C.5c}$

from ξ^n: $\quad b_0 = 0 \tag{C.5d}$

from ξ^{n+1}: $\quad b_1 = 0 \tag{C.5e}$

from ξ^{n+2}: $\quad (n+3)(n+4)b_2 + q_0 L^3 = 0 \tag{C.5f}$

from ξ^{n+3}: $\quad (n+3)(n+5)b_3 - q_0 L^3 = 0 \tag{C.5g}$

We note again that a combination of Eqs. (C.5a) and (C.5e) imposes $q_0 = 0$. We obtain, therefore, a trivial solution.

Appendix D: Linearly Varying Distributed Axial Load ($m = 1$)

In this appendix, we are looking for solutions for n larger than 1, since closed-form solutions were derived for $n = 0$, for pinned beams with linearly varying axial load density. With the general mode shape given in Eq. (5.11), we calculate the first term of the governing equation:

$$\frac{d^2}{d\xi^2}\left[D(\xi)\frac{d^2\psi}{d\xi^2}\right] = (n+3)(n+4)\left[(n+1)(n+2)b_0\xi^n + (n+2)(n+3)b_1\xi^{n+1}\right.$$
$$+ (n+3)(n+4)b_2\xi^{n+2} + (n+4)(n+5)b_3\xi^{n+3}$$
$$\left. + (n+5)(n+6)\xi^{n+4} - (2b_1 + 6b_2\xi + 12b_3\xi^2 + 20b_4\xi^3)\right]$$

$$(D.1)$$

The second term in Eq. (5.26) reads

$$-q_0L^3\frac{d}{d\xi}\left[N(\xi)\frac{d\psi}{d\xi}\right] = q_0L^3\left[(n+3)(n+4)(1+g_1)(\xi^{n+2} - \xi)\right.$$
$$- g_1(n+4)(n+5)\xi^{n+4} + (n+4)^2\xi^{n+3}$$
$$+ 2g_1(n+3)(n+4)\xi^3 + \frac{3}{2}(n+3)(n+4)\xi^2$$
$$\left. - \frac{1}{3}g_1(n^2+7n+6)\xi - \frac{1}{6}(n^2+7n+6)\right] \qquad (D.2)$$

We collect terms with identical powers of ξ. Hence, two special cases arise: $n \leq 3$ and $n > 3$. Let us first concentrate on the case $n = 1$; we obtain a homogeneous set of six linear algebraic equations for six unknowns

$$-40b_1 - \tfrac{7}{3}q_0L^3 = 0 \qquad (D.3a)$$

$$-120b_2 + 120b_0 - q_0L^3(20 + \tfrac{37}{7}g_1) = 0 \qquad (D.3b)$$

$$-240b_3 + 240b_1 + 30q_0L^3 = 0 \qquad (D.3c)$$

$$-400b_4 + 400b_2 + q_0L^3(20 + 30g_1) = 0 \qquad (D.3d)$$

$$600b_3 - 25q_0L^3 = 0 \qquad (D.3e)$$

$$840b_4 - 15q_0L^3g_1 = 0 \qquad (D.3f)$$

The determinant of the set (D.3), which equals $-5{,}806{,}080{,}000{,}000L^3$, is nonzero. Thus, the solution of (D.3) is trivial. Now, consider the case $n = 2$; we deduce from the governing differential equation a set of seven equations for six unknowns

$$-60b_1 - 4q_0L^3 = 0 \tag{D.4a}$$

$$-180b_2 - q_0L^3(30 + 19g_1) = 0 \tag{D.4b}$$

$$-360b_3 + 360b_0 + 45q_0L^3 = 0 \tag{D.4c}$$

$$-600b_4 + 600b_1 + 30q_0L^3g_1 = 0 \tag{D.4d}$$

$$900b_2 + q_0L^3(30 + 15g_1) = 0 \tag{D.4e}$$

$$1260b_3 - 36q_0L^3 = 0 \tag{D.4f}$$

$$1680b_4 - 21q_0L^3g_1 = 0 \tag{D.4g}$$

In order to find a non-trivial solution, the rank of the matrix of the set (D.4) must be less than 6, leading to the requirement that all minors of order 6 be zero. We then get seven inhomogeneous linear equations, among which two are zero, with only one unknown g_1. Some of the determinantal equations coincide with each other; we reduce the equations in terms of the equations involving minors to

$$987{,}614{,}208 + 658{,}409{,}472g_1 = 0$$
$$164{,}602{,}368 - 9{,}258{,}832g_1 = 0 \tag{D.5}$$
$$35{,}271{,}936 + 23{,}514{,}624g_1 = 0$$

The set (D.5) has no solution, since we can easily note that the first equation of (D.5) yields a negative value for g_1 while the second yields a positive one. We conclude that only a trivial solution is associated with the set (D.4):

For $n = 3$, instead of (D.4), we obtain

$$-84b_1 - 6q_0L^3 = 0 \tag{D.6a}$$

$$-252b_2 - q_0L^3(27 + 42g_1) = 0 \tag{D.6b}$$

$$-504b_3 + 63q_0L^3 = 0 \tag{D.6c}$$

$$-840b_4 + 840b_0 + 42q_0L^3g_1 = 0 \tag{D.6d}$$

$$b_1 = 0 \tag{D.6e}$$

$$1764b_2 + q_0L^3(42 + 21g_1) = 0 \tag{D.6f}$$

$$2352b_3 - 49q_0L^3 = 0 \tag{D.6g}$$

$$3024b_4 - 28q_0L^3g_1 = 0 \tag{D.6h}$$

We note that a combination of Eqs. (D.6a) and (D.6e) imposes the condition that $q_0 = 0$. We obtain, therefore, a trivial solution. We now turn to the case $n > 3$. We obtain the following set of equations:

from ξ^0: $-2(n+3)(n+4)b_1 - \frac{1}{6}(n^2 + 7n + 6)q_0 L^3 = 0$ (D.7a)

from ξ^1: $-6(n+3)(n+4)b_2 - [(n+3)(n+4)(1+g_1/2)$

$+ \frac{1}{6}(n^2 + 7n + 6)g_1]q_0 L^3 = 0$ (D.7b)

from ξ^2: $12(n+3)(n+4)b_3 + \frac{3}{2}(n+3)(n+4)q_0 L^3 = 0$ (D.7c)

from ξ^3: $20(n+3)(n+4)b_4 + 2(n+3)(n+4)g_1 q_0 L^3 = 0$ (D.7d)

from ξ^n: $b_0 = 0$ (D.7e)

from ξ^{n+1}: $b_1 = 0$ (D.7f)

from ξ^{n+2}: $(n+3)(n+4)b_2 + q_0 L^3(1+g_1/2) = 0$ (D.7g)

from ξ^{n+3}: $(n+3)(n+5)b_3 - q_0 L^3 = 0$ (D.7h)

from ξ^{n+4}: $(n+3)(n+6)b_4 - g_1/2q_0 L^3 = 0$ (D.7i)

We note again that a combination of Eqs. (D.7a) and (D.7f) leads to $q_0 = 0$. We obtain, therefore, a trivial solution.

Appendix E: Parabolically Varying Distributed Axial Load ($m = 2$)

Here, we look for solutions for n larger than 2, since closed-form solutions were derived for $n = 0$ and $n = 1$, and the case $n = 2$ was proved to be non-realistic, for pinned beams with parabolically varying axial load density. With the general mode shape given in Eq. (5.11), we calculate the first term of the governing equation:

$$\frac{d^2}{d\xi^2}\left[D(\xi)\frac{d^2\psi}{d\xi^2}\right] = (n+3)(n+4)[(n+1)(n+2)b_0\xi^n + (n+2)(n+3)b_1\xi^{n+1}$$

$$[-7pt] \quad + (n+3)(n+4)b_2\xi^{n+2} + (n+4)(n+5)b_3\xi^{n+3}$$

$$+ (n+5)(n+6)b_4\xi^{n+4} + (n+6)(n+7)b_5\xi^{n+5}$$

$$- (2b_1 + 6b_2\xi + 12b_3\xi^2 + 20b_4\xi^3 + 30b_5\xi^4)] \quad (E.1)$$

The second term in Eq. (5.26) reads

$$-q_0 L^3 \frac{d}{d\xi}\left[N(\xi)\frac{d\psi}{d\xi}\right] - q_0 L^3 \left\{\left(1 + \frac{g_1}{2} + \frac{g_2}{3}\right)(n+3)(n+4)(\xi^{n+2} - \xi)\right.$$

$$- (n+4)^2 \xi^{n+3} + \frac{3}{2}(n+3)(n+4)\xi^2 - \frac{1}{6}(n^2 + 7n + 6)$$

$$- \left(\frac{g_1}{2}\right)\left[(n+4)(n+5)\xi^{n+4} - 2(n+3)(n+4)\xi^3\right.$$

$$+ \frac{1}{3}(n^2 + 7n + 6)\xi\right] - \left(\frac{g_2}{3}\right)\left[(n+4)(n+6)\xi^{n+5}\right.$$

$$\left.\left. - \frac{5}{2}(n+4)(n+5)\xi^4 + \frac{1}{2}(n^2 + 7n + 6)\xi^2\right]\right\} \tag{E.2}$$

We collect terms with identical powers of ξ. It is important to establish how the power of the first term n relates to the power, 4, of the last term. Hence, two special cases arise: $n \le 4$ and $n > 4$. Let us first concentrate on the case $n = 3$; we obtain a homogeneous set of nine linear algebraic equations for seven unknowns:

$$-84b_1 - 6q_0 L^3 = 0 \tag{E.3a}$$

$$-252b_2 - q_0 L^3(42 + 27g_1 + 14g_2) = 0 \tag{E.3b}$$

$$-504b_3 + q_0 L^3(63 - 6g_2) = 0 \tag{E.3c}$$

$$-840b_3 + 840b_0 + 42q_0 L^3 g_1 = 0 \tag{E.3d}$$

$$-1260b_5 + 1260b_1 + 35q_0 L^3 g_2 = 0 \tag{E.3e}$$

$$-1764b_2 + q_0 L^3(42 + 21g_1 + 14g_2) = 0 \tag{E.3f}$$

$$2352b_3 - 49q_0 L^3 = 0 \tag{E.3g}$$

$$3024b_4 - 28q_0 L^3 g_1 = 0 \tag{E.3h}$$

$$3780b_5 - 21q_0 L^3 g_2 = 0 \tag{E.3i}$$

In order to find a non-trivial solution, the rank of the matrix of the set (E.3) must be less than 7, or all minors of order 7 must be zero. We then get 55 inhomogeneous linear equations, with only 2 unknowns g_1 and g_2, which has no solution when q_0 is non-zero. Therefore, no solution is also obtainable for the flexural rigidity coefficients and the natural frequency. For $n = 4$, instead of (E.3), we obtain

$$-112b_1 - \tfrac{25}{3}q_0L^3 = 0 \tag{E.4a}$$

$$-336b_2 - q_0L^3(56 + \tfrac{109}{3}g_1 + \tfrac{56}{3}g_2) = 0 \tag{E.4b}$$

$$-672b_3 + q_0L^3(84 - \tfrac{25}{3}g_2) = 0 \tag{E.4c}$$

$$-1120b_4 + 56q_0L^3g_1 = 0 \tag{E.4d}$$

$$-1680b_5 + 1680b_0 + \tfrac{140}{3}q_0L^3g_2 = 0 \tag{E.4e}$$

$$b_1 = 0 \tag{E.4f}$$

$$3136b_2 + q_0L^3(56 + 28g_1 + \tfrac{56}{3}g_2) = 0 \tag{E.4g}$$

$$4032b_3 - 64q_0L^3 = 0 \tag{E.4h}$$

$$5040b_4 - 36q_0L^3g_1 = 0 \tag{E.4i}$$

$$6160b_5 - \tfrac{80}{3}q_0L^3g_2 = 0 \tag{E.4j}$$

We note that a combination of Eqs. (E.4a) and (E.4f) requires $q_0 = 0$. We obtain, therefore, a trivial solution. We now turn to the case $n > 4$. We obtain the following set of equations:

from ξ^0: $\quad -2(n+3)(n+4)b_1 - \tfrac{1}{6}(n^2 + 7n + 6)q_0L^3 = 0 \tag{E.5a}$

from ξ^1: $\quad -6(n+3)(n+4)b_2 - [(n+3)(n+4)(1 + g_1/2 + g_2/3)$

$$+ \tfrac{1}{6}(n^2 + 7n + 6)g_1]q_0L^3 = 0 \tag{E.5b}$$

from ξ^2: $\quad 12(n+3)(n+4)b_3 + \left[\tfrac{3}{2}(n+3)(n+4) - (g_2/6)(n^2 + 7n + 6)\right]q_0L^3$

$$= 0 \tag{E.5c}$$

from ξ^3: $\quad 20(n+3)(n+4)b_4 + 2(n+3)(n+4)g_1q_0L^3 = 0 \tag{E.5d}$

from ξ^4: $\quad 30(n+3)(n+4)b_5 + \tfrac{5}{6}(n+3)(n+4)g_2q_0L^3 = 0 \tag{E.5e}$

from ξ^n: $\quad b_0 = 0 \tag{E.5f}$

from ξ^{n+1}: $\quad b_1 = 0 \tag{E.5g}$

from ξ^{n+2}: $\quad (n+3)(n+4)b_2 + q_0L^3(1 + g_1/2 + g_2/3) = 0 \tag{E.5h}$

from ξ^{n+3}: $\quad (n+3)(n+5)b_3 - q_0L^3 = 0 \tag{E.5i}$

from ξ^{n+4}: $\quad (n+3)(n+6)b_4 - g_1q_0L^3/2 = 0 \tag{E.5j}$

from ξ^{n+5}: $\quad (n+3)(n+7)b_5 - g_2q_0L^3/3 = 0 \tag{E.5k}$

We note again that a combination of Eqs. (E.5a) and (E.5g) imposes the conditions $q_0 = 0$. We obtain, therefore, a trivial solution.

Appendix F: Cubically Varying Distributed Axial Load ($m = 3$)

Here, we look for solutions for n larger than 2, since closed-form solutions were derived for n equal to zero, one or two, for pinned beams with cubic axial load density. With the general mode shape given in Eq. (5.11), we calculate the first and second terms of the governing equation in Eq. (5.26):

$$
\frac{d^2}{d\xi^2}\left[D(\xi)\frac{d^2\psi}{d\xi^2}\right] = (n+3)(n+4)[(n+1)(n+2)b_0\xi^n + (n+2)(n+3)b_1\xi^{n+1}
$$

$$
+ (n+3)(n+4)b_2\xi^{n+2} + (n+4)(n+5)b_3\xi^{n+3}
$$

$$
+ (n+5)(n+6)b_4\xi^{n+4} + (n+6)(n+7)b_5\xi^{n+5}
$$

$$
+ (n+7)(n+8)\xi^{n+6} - (2b_1 + 6b_2\xi + 12b_3\xi^2 + 20b_4\xi^3
$$

$$
+ 30b_5\xi^4 + 42b_6\xi^5)] \tag{F.1}
$$

$$
-q_0L^3\frac{d}{d\xi}\left[N(\xi)\frac{d\psi}{d\xi}\right] = q_0L^3\Bigg\{\left(1 + \frac{g_1}{2} + \frac{g_2}{2}\right)(n+3)(n+4)(\xi^n + 2 - \xi)
$$

$$
- (n+4)^2\xi^{n+3} + \frac{3}{2}(n+3)(n+4)\xi^2 - \frac{1}{6}(n^2+7n+6)
$$

$$
- \left(\frac{g_1}{2}\right)\Big[(n+4)(n+5)\xi^{n+4} - 2(n+3)(n+4)\xi^3
$$

$$
+ \frac{1}{3}(n^2+7n+6)\xi\Big] - \left(\frac{g_2}{3}\right)\Big[(n+4)(n+6)\xi^{n+5}
$$

$$
- \frac{5}{2}(n+4)(n+5)\xi^4 + \frac{1}{2}(n^2+7n+6)\xi^2\Big]
$$

$$
- \left(\frac{g_3}{4}\right)\Big[(n+4)(n+7)\xi^{n+6} - 3(n+3)(n+4)\xi^5
$$

$$
+ \frac{2}{3}(n^2+7n+6)\xi^3\Big]\Bigg\} \tag{F.2}
$$

Here, two special cases arise: $n > 5$ and $n \le 5$. Let us first concentrate on the case $n = 3$; we obtain a homogeneous set of ten linear algebraic equations for eight unknowns:

$$
-84b_1 - 6q_0L^3 = 0 \tag{F.3a}
$$

$$
-252b_2 - q_0L^3\left(42 + 27g_1 + 14g_2 + \tfrac{21}{2}g_3\right) = 0 \tag{F.3b}
$$

$$
-504b_3 + q_0L^3(63 - 6g_2) = 0 \tag{F.3c}
$$

$$
-840b_4 + 840b_0 + q_0L^3(42g_1 - 6g_3) = 0 \tag{F.3d}
$$

$$-1260b_5 + 1260b_1 + 35q_0L^3g_2 = 0 \tag{F.3e}$$

$$-1764b_6 + 1764b_2 + q_0L^3(42 + 21g_1 + 14g_2 + 42g_3) = 0 \tag{F.3f}$$

$$2352b_3 - 49q_0L^3 = 0 \tag{F.3g}$$

$$3024b_4 - 28q_0L^3g_1 = 0 \tag{F.3h}$$

$$3780b_5 - 21q_0L^3g_2 = 0 \tag{F.3i}$$

$$4620b_6 - \tfrac{35}{2}q_0L^3g_3 = 0 \tag{F.3j}$$

In order to find a non-trivial solution, the rank of the matrix of the set (F.3) must be less than 8, or all minors of order 8 must be zero. We then get a set of 45 inhomogeneous linear equations, with only 3 unknowns g_1, g_2 and g_3, which has no solution when q_0 is non-zero. Therefore, no solution is obtainable for the flexural rigidity coefficients and the natural frequency also. For $n = 4$, instead of the set (F.3), we obtain

$$-112b_1 - \tfrac{25}{3}q_0L^3 = 0 \tag{F.4a}$$

$$-336b_2 - q_0L^3\left(56 + \tfrac{100}{3}g_1 + \tfrac{56}{3}g_2 + 14g_3\right) = 0 \tag{F.4b}$$

$$-672b_3 + q_0L^3\left(84 - \tfrac{25}{3}g_2\right) = 0 \tag{F.4c}$$

$$-1120b_4 + q_0L^3\left(56g_1 - \tfrac{25}{3}g_3\right) = 0 \tag{F.4d}$$

$$-1680b_5 + 1680b_0 + \tfrac{140}{3}q_0L^3g_2 = 0 \tag{F.4e}$$

$$-2352b_3 + 2352b_1 + 42q_0L^3g_3 = 0 \tag{F.4f}$$

$$3136b_2 + q_0L^3\left(56 + 28g_1 + \tfrac{56}{3}g_2 + 14g_3\right) = 0 \tag{F.4g}$$

$$4032b_3 - 64q_0L^3 = 0 \tag{F.4h}$$

$$5040b_4 - 36q_0L^3g_1 = 0 \tag{F.4i}$$

$$6160b_5 - \tfrac{80}{3}q_0L^3g_2 = 0 \tag{F.4j}$$

$$7392b_6 - 22q_0L^3g_3 = 0 \tag{F.4k}$$

whose solution is trivial, with all unknowns equal to zero. When $n = 5$, we have 12 equations with, again, 8 unknowns

$$-144b_1 - 11q_0L^3 = 0 \tag{F.5a}$$

$$-432b_2 - q_0L^3(72 + 47g_1 + 24g_2 + 18g_3) = 0 \tag{F.5b}$$

$$-864b_3 + q_0L^3(108 - 11g_2) = 0 \tag{F.5c}$$

$$-1,440b_4 + q_0L^3(72g_1 - 11g_3) = 0 \tag{F.5d}$$

$$-2{,}160b_5 + 60q_0L^3g_2 = 0 \tag{F.5e}$$

$$-3{,}024b_6 + 3{,}024b_0 + 54q_0L^3g_3 = 0 \tag{F.5f}$$

$$b_1 = 0 \tag{F.5g}$$

$$5{,}184b_2 + q_0L^3(72 + 36g_1 + 24g_2 + 18g_3) = 0 \tag{F.5h}$$

$$6{,}480b_3 - 81q_0L^3 = 0 \tag{F.5i}$$

$$7{,}920b_4 - 45q_0L^3g_1 = 0 \tag{F.5j}$$

$$9{,}504b_5 - 33q_0L^3g_2 = 0 \tag{F.5k}$$

$$11{,}232b_6 - 27q_0L^3g_3 = 0 \tag{F.5l}$$

We note that combining Eq. (F.5a) and (F.5g) we get $q_0 = 0$, which leads to all unknowns being zero, which represents a trivial solution. We turn now to the case $n > 5$. We obtain the following set of equations:

from ξ^0:
$$-2(n+3)(n+4)b_1 - \tfrac{1}{6}(n^2 + 7n + 6)q_0L^3 = 0 \tag{F.6a}$$

from ξ^1:
$$-6(n+3)(n+4)b_2 - [(n+3)(n+4)(1 + g_1/2 + g_2/3 + g_3/4)$$
$$+ \tfrac{1}{6}(n^2 + 7n + 6)g_1]q_0L^3 = 0 \tag{F.6b}$$

from ξ^2:
$$12(n+3)(n+4)b_3 + [3/2(n+3)(n+4)$$
$$- (g_2/6)(n^2 + 7n + 6)]q_0L^3 = 0 \tag{F.6c}$$

from ξ^3:
$$20(n+3)(n+4)b_4 + [2(n+3)(n+4)g_1$$
$$- (g_3/6)(n^2 + 7n + 6)]q_0L^3 = 0 \tag{F.6d}$$

from ξ^4:
$$30(n+3)(n+4)b_5 + \tfrac{5}{6}(n+3)(n+4)g_2q_0L^3 = 0 \tag{F.6e}$$

from ξ^5:
$$42(n+3)(n+4)b_6 + \tfrac{3}{4}(n+3)(n+4)g_3q_0L^3 = 0 \tag{F.6f}$$

from ξ^n:
$$b_0 = 0 \tag{F.6g}$$

from ξ^{n+1}:
$$b_1 = 0 \tag{F.6h}$$

from ξ^{n+2}:
$$(n+3)(n+4)b_2 + q_0L^3(1 + g_1/2)$$
$$+ (g_2/3) + (g_3/4) = 0 \tag{F.6i}$$

from ξ^{n+3}:
$$(n+3)(n+5)b_3 - q_0L^3 = 0 \tag{F.6j}$$

from ξ^{n+4}:
$$(n+3)(n+6)b_4 - g_1q_0L^3/2 = 0 \tag{F.6k}$$

from ξ^{n+5}:
$$(n+3)(n+7)b_5 - g_2q_0L^3/3 = 0 \tag{F.6l}$$

from ξ^{n+6}:
$$(n+3)(n+8)b_6 - g_3q_0L^3/4 = 0 \tag{F.6m}$$

We note again that a combination of Eqs. (F.6a) and (F.6h) imposes the condition $q_0 = 0$. We obtain, therefore, a trivial solution.

Appendix G: Constant Inertial Coefficient ($m = 0$)

In this appendix, we look for solutions for n larger than 2, since closed-form solutions were derived for $n = 0$ and $n = 1$, for pinned beams with constant mass density. With the general mode shape given in Eq. (5.88), we calculate the first term of the governing equation, namely Eq. (5.89):

$$\frac{d^2}{d\xi^2}\left[D(\xi)\frac{d^2\psi}{d\xi^2}\right] = (n+3)(n+4)[(n+1)(n+2)b_0\xi^n + (n+2)(n+3)b_1\xi^{n+1}$$
$$+ (n+3)(n+4)b_2\xi^{n+2} + (n+4)(n+5)b_3\xi^{n+3}$$
$$+ (n+5)(n+6)b_4\xi^{n+4}] - (n^2 + 7n + 12)$$
$$\times (2b_1 + 6b_2\xi + 12b_3\xi^2 + 20b_4\xi^3) \tag{G.1}$$

The second term in Eq. (5.89) reads

$$\Omega^2 R(\xi) = \Omega^2 r_0[\xi^{n+4} - \tfrac{1}{6}(n^2 + 7n + 12)\xi^3 + \tfrac{1}{6}(n^2 + 7n + 6)\xi] \tag{G.2}$$

We have to collect terms with identical powers of ξ. It is important to know how the power of the first term in Eq. (G.1), namely n, relates to the power of the last term, which equals 3. Hence, two special cases arise: $n > 3$ and $n \le 3$. Let us first concentrate on the case $n = 2$; we obtain the following set of seven equations for six unknowns:

$$b_1 = 0 \tag{G.3a}$$
$$-180b_2 - 4r_0\Omega^2 = 0 \tag{G.3b}$$
$$-b_3 + b_0 = 0 \tag{G.3c}$$
$$-600b_4 + 600b_1 + 5r_0\Omega^2 = 0 \tag{G.3d}$$
$$b_2 = 0 \tag{G.3e}$$
$$b_3 = 0 \tag{G.3f}$$
$$1680b_4 - r_0\Omega^2 = 0 \tag{G.3g}$$

From Eqs. (G.3b) and (G.3e), $r_0\Omega^2$ is zero. Hence, all unknowns vanish, yielding a trivial solution.

Now, consider the case $n = 3$; we deduce from the governing differential equation a set of eight homogeneous equations for six unknowns:

$$b_1 = 0 \tag{G.4a}$$
$$-252b_2 - 6r_0\Omega^2 = 0 \tag{G.4b}$$
$$b_3 = 0 \tag{G.4c}$$
$$-840b_4 + 840b_0 + 7r_0\Omega^2 = 0 \tag{G.4d}$$

$$b_1 = 0 \tag{G.4e}$$

$$b_2 = 0 \tag{G.4f}$$

$$b_3 = 0 \tag{G.4g}$$

$$3024b_4 - r_0\Omega^2 = 0 \tag{G.4h}$$

As for the case $n = 2$, from Eqs. (G.4b) and (G.4f), $r_0\Omega^2$ is zero, leading to a trivial solution.

We now turn to the case $n > 3$. Equating the difference of the expressions (G.1) and (G.2) to zero for every ξ within the interval $[0;1]$, we obtain the following set of equations:

from ξ^0: $\quad b_1 = 0$ $\tag{G.5a}$

from ξ^1: $\quad 6(n^2 + 7n + 12)b_2 + \frac{1}{6}(n^2 + 7n + 6)r_0\Omega^2 = 0$ $\tag{G.5b}$

from ξ^2: $\quad b_3 = 0$ $\tag{G.5c}$

from ξ^3: $\quad 20(n^2 + 7n + 12)b_2 - \frac{1}{6}(n^2 + 7n + 12)r_0\Omega^2 = 0$ $\tag{G.5d}$

from ξ^n: $\quad b_0 = 0$ $\tag{G.5e}$

from ξ^{n+1}: $\quad b_1 = 0$ $\tag{G.5f}$

from ξ^{n+2}: $\quad b_2 = 0$ $\tag{G.5g}$

from ξ^{n+3}: $\quad b_3 = 0$ $\tag{G.5h}$

from ξ^{n+4}: $\quad (n + 3)(n + 4)(n + 5)(n + 6)b_4 - r_0\Omega^2 0$ $\tag{G.5i}$

Substituting $b_2 = 0$ from Eq. (G.5g) into (G.5b), we get $r_0\Omega^2 = 0$, which in view of the rest of the equations results in all unknowns being zero; i.e., in the case $n > 3$, only a trivial solution is derived.

Appendix H: Linearly Varying Inertial Coefficient ($m = 1$)

We look for solutions for n larger than 0, since closed-form solutions were already derived for $n = 0$. With the general mode shape given in Eq. (5.88), we calculate the first term of the governing equation, Eq. (5.89):

$$\frac{d^2}{d\xi^2}\left[D(\xi)\frac{d^2\psi}{d\xi^2}\right] = (n + 3)(n + 4)[(n + 1)(n + 2)b_0\xi^n + (n + 2)(n + 3)b_1\xi^{n+1}$$
$$+ (n + 3)(n + 4)b_2\xi^{n+2} + (n + 4)(n + 5)b_3\xi^{n+3}$$
$$+ (n + 5)(n + 6)b_4\xi^{n+4} + (n + 6)(n + 7)b_5\xi^{n+5}]$$
$$- (n^2 + 7n + 12)(2b_1 + 6b_2\xi + 12b_3\xi^2 + 20b_4\xi^3 + 30b_5\xi^4)$$
$$\tag{H.1}$$

while the second term in Eq. (5.89) is

$$\Omega^2 R(\xi) = \Omega^2 \big[r_0 \xi^{n+4} - \tfrac{1}{6} r_0 (n^2 + 7n + 12)\xi^3 + \tfrac{1}{6} r_0 (n^2 + 7n + 6)\xi$$
$$+ r_1 \xi^{n+5} - \tfrac{1}{6} r_1 (n^2 + 7n + 12)\xi^4 + \tfrac{1}{6} r_1 (n^2 + 7n + 6)\xi^2 \big] \qquad \text{(H.2)}$$

It is important to establish how the power of the first term in Eq. (H.1), namely n, relates to the power of the last term, which equals 4. Hence, two special cases arise: $n > 4$ and $n \le 4$. Concerning the particular case $n = 1$, we obtain the following set of seven equations for seven unknowns:

$$b_1 = 0 \qquad \text{(H.3a)}$$

$$-120b_2 + 120b_0 - \tfrac{7}{3} r_0 \Omega^2 = 0 \qquad \text{(H.3b)}$$

$$-240b_3 + 240b_1 - \tfrac{7}{3} r_1 \Omega^2 = 0 \qquad \text{(H.3c)}$$

$$-400b_4 + 400b_2 + \tfrac{10}{3} r_0 \Omega^2 = 0 \qquad \text{(H.3d)}$$

$$-600b_5 + 600b_3 + \tfrac{10}{3} r_1 \Omega^2 = 0 \qquad \text{(H.3e)}$$

$$840b_4 - r_0 \Omega^2 = 0 \qquad \text{(H.3f)}$$

$$1120b_5 - r_1 \Omega^2 = 0 \qquad \text{(H.3g)}$$

The determinant of the system equals $-1,316,044,800,000,000r_1$. Since the system is homogeneous, to find a non-trivial solution, the determinant must vanish. This condition yields $r_1 = 0$; i.e., the mass density is constant; this contradicts the description of the problem.

Now, consider the case $n = 2$; we deduce from the governing differential equation a set of eight homogeneous equations for seven unknowns:

$$b_1 = 0 \qquad \text{(H.4a)}$$

$$-180b_2 - 4r_0 \Omega^2 = 0 \qquad \text{(H.4b)}$$

$$-360b_3 + 360b_0 - 4r_1 \Omega^2 = 0 \qquad \text{(H.4c)}$$

$$-600b_4 + 600b_1 + 5r_0 \Omega^2 = 0 \qquad \text{(H.4d)}$$

$$-900b_5 + 900b_2 + 5r_1 \Omega^2 = 0 \qquad \text{(H.4e)}$$

$$b_3 = 0 \qquad \text{(H.4f)}$$

$$1680b_4 - r_0 \Omega^2 = 0 \qquad \text{(H.4g)}$$

$$2160b_5 - r_1 \Omega^2 = 0 \qquad \text{(H.4h)}$$

In order to find a non-trivial solution, the rank of the matrix of the set (H.4) must be less than 7. In view of the definition of the rank, all eight minors of order 7 of the matrix of (H.4) should vanish. This leads to a set of eight linear algebraic equations for two unknowns r_0 and r_1. The solution of this set turns out to be $r_1 = r_0 = 0$, yielding a trivial solution for the set (H.4).

Now, consider the case $n = 3$; we obtain from the governing differential equation a set of nine equations for seven unknowns:

$$b_1 = 0 \tag{H.5a}$$

$$-252b_2 - 6r_0\Omega^2 = 0 \tag{H.5b}$$

$$-504b_3 - 6r_1\Omega^2 = 0 \tag{H.5c}$$

$$-840b_4 + 840b_0 + 7r_0\Omega^2 = 0 \tag{H.5d}$$

$$-1260b_5 + 1260b_1 + 7r_1\Omega^2 = 0 \tag{H.5e}$$

$$b_2 = 0 \tag{H.5f}$$

$$b_3 = 0 \tag{H.5g}$$

$$3024b_4 - r_0\Omega^2 = 0 \tag{H.5h}$$

$$3780b_5 - r_1\Omega^2 = 0 \tag{H.5i}$$

From Eqs. (H.5b) and (H.5f) we determine that $r_0\Omega^2$ is zero, and from Eqs. (H.5c) and (H.5g) we have $r_1\Omega^2$ is zero. Hence, all unknowns are zero.

For the case $n = 4$, we have a set of ten equations for seven unknowns

$$b_1 = 0 \tag{H.6a}$$

$$-336b_2 - \tfrac{25}{3}r_0\Omega^2 = 0 \tag{H.6b}$$

$$-672b_3 - \tfrac{25}{3}r_1\Omega^2 = 0 \tag{H.6c}$$

$$-1120b_4 + \tfrac{28}{3}r_0\Omega^2 = 0 \tag{H.6d}$$

$$-1680b_5 + 1680b_0 + \tfrac{28}{3}r_1\Omega^2 = 0 \tag{H.6e}$$

$$b_1 = 0 \tag{H.6f}$$

$$b_2 = 0 \tag{H.6g}$$

$$b_3 = 0 \tag{H.6h}$$

$$5040b_4 - r_0\Omega^2 = 0 \tag{H.6i}$$

$$6160b_5 - r_1\Omega^2 = 0 \tag{H.6j}$$

As for the case $n = 3$, Eqs. (H.6b) and (H.6g) yield $r_0\Omega^2 = 0$, and Eqs. (H.6c) and (H.6h) yield $r_1\Omega^2 = 0$, which leads to a trivial solution.

We turn to the case $n > 4$. Stipulating the difference between the expressions (H.1) and (H.2) to be zero for every ξ within the interval $[0; 1]$, we obtain the following set of equations:

$$\text{from } \xi^0: \quad b_1 = 0 \tag{H.7a}$$

$$\text{from } \xi^1: \quad 6(n^2 + 7n + 12)b_2 + \tfrac{1}{6}(n^2 + 7n + 6)r_0\Omega^2 = 0 \tag{H.7b}$$

$$\text{from } \xi^2: \quad 12(n^2 + 7n + 12)b_3 + \tfrac{1}{6}(n^2 + 7n + 6)r_1\Omega^2 = 0 \tag{H.7c}$$

$$\text{from } \xi^3: \quad 20(n^2 + 7n + 12)b_4 - \tfrac{1}{6}(n^2 + 7n + 6)r_0\Omega^2 = 0 \tag{H.7d}$$

$$\text{from } \xi^4: \quad 30(n^2 + 7n + 12)b_5 - \tfrac{1}{6}(n^2 + 7n + 6)r_1\Omega^2 = 0 \tag{H.7e}$$

$$\text{from } \xi^n: \quad b_0 = 0 \tag{H.7f}$$

$$\text{from } \xi^{n+1}: \quad b_1 = 0 \tag{H.7g}$$

$$\text{from } \xi^{n+2}: \quad b_2 = 0 \tag{H.7h}$$

$$\text{from } \xi^{n+3}: \quad b_3 = 0 \tag{H.7i}$$

$$\text{from } \xi^{n+4}: \quad (n+3)(n+4)(n+5)(n+6)b_4 - r_0\Omega^2 = 0 \tag{H.7j}$$

$$\text{from } \xi^{n+5}: \quad (n+3)(n+4)(n+6)(n+7)b_5 - r_1\Omega^2 = 0 \tag{H.7k}$$

Substituting $b_2 = 0$ from Eq. (H.7h) into (H.7b), we get $r_0\Omega^2 = 0$; likewise, substituting $b_3 = 0$ from Eq. (H.7i) into (H.7c), we get $r_1\Omega^2 = 0$, which in view of the rest of the equations results in all unknowns being zero; i.e., in the case $n > 4$, only a trivial solution is derived.

Appendix I: Parabolically Varying Inertial Coefficient ($m = 2$)

We are looking for solutions for n larger than 3, since closed-form solutions were derived for $n = 0, 1, 2, 3$ in Section 5.2.6. With the general mode shape given in Eq. (5.88), we calculate the left-hand side of equation (5.89):

$$\frac{d^2}{d\xi^2}\left[D(\xi)\frac{d^2\psi}{d\xi^2} \right] - \Omega^2 R(\xi) = (n+3)(n+4)\sum_{i=0}^{6}(i+b+1)(i+n+2)b_i\xi^{i+n}$$
$$- (n^2 + 7n + 12)\sum_{i=1}^{6} i(i+1)b_i\xi^{i-1}$$
$$- \Omega^2[r_0\xi^{n+4} - \tfrac{1}{6}r_0(n^2 + 7n + 12)\xi^3$$
$$+ \tfrac{1}{6}r_0(n^2 + 7n + 6)\xi + r_1\xi^{n+5}$$
$$- \tfrac{1}{6}r_1(n^2 + 7n + 12)\xi^4 + \tfrac{1}{6}r_1(n^2 + 7n + 6)\xi^2 + r_2\xi^{n+6}$$
$$- \tfrac{1}{6}r_2(n^2 + 7n + 12)\xi^5 + \tfrac{1}{6}r_2(n^2 + 7n + 6)\xi^3] \tag{I.1}$$

We have to establish how the parameter n, relates to the power of the last term, which equals 5. Hence, two special cases arise: $n > 5$ and $n \leq 5$. Consider now the case $n = 4$; we obtain from the governing differential equation a set of 11 homogeneous equations for 8 unknowns:

$$b_1 = 0 \tag{I.2a}$$

$$-336b_2 - \tfrac{25}{3}r_0\Omega^2 = 0 \tag{I.2b}$$

$$-672b_3 - \tfrac{25}{3}r_1\Omega^2 = 0 \tag{I.2c}$$

$$-1120b_4 + \tfrac{28}{3}r_0\Omega^2 - \tfrac{25}{3}r_2\Omega^2 = 0 \tag{I.2d}$$

$$-1680b_5 + 1680b_0 + \tfrac{28}{3}r_1\Omega^2 = 0 \tag{I.2e}$$

$$-2352b_6 + 2352b_1 + \tfrac{28}{3}r_2\Omega^2 = 0 \tag{I.2f}$$

$$b_2 = 0 \tag{I.2g}$$

$$b_3 = 0 \tag{I.2h}$$

$$5040b_4 - r_0\Omega^2 = 0 \tag{I.2i}$$

$$6160b_5 - r_1\Omega^2 = 0 \tag{I.2j}$$

$$7392r_6 - r_2\Omega^2 = 0 \tag{I.2k}$$

It is easily seen that this system has no solution different from the trivial solution $b_0 = b_1 = \cdots = b_6 = 0$.

When $n = 5$, we get the following set

$$b_1 = 0 \tag{I.3a}$$

$$-432b_2 - 11r_0\Omega^2 = 0 \tag{I.3b}$$

$$-864b_3 - 11r_1\Omega^2 = 0 \tag{I.3c}$$

$$-1{,}440b_4 + 12r_0\Omega^2 - 11r_2\Omega^2 = 0 \tag{I.3d}$$

$$-2{,}160b_5 + 12r_1\Omega^2 = 0 \tag{I.3e}$$

$$-3{,}024b_6 + 3{,}024b_0 + 12r_2\Omega^2 = 0 \tag{I.3f}$$

$$b_1 = 0 \tag{I.3g}$$

$$b_2 = 0 \tag{I.3h}$$

$$b_3 = 0 \tag{I.3i}$$

$$7{,}920b_4 - r_0\Omega^2 = 0 \tag{I.3j}$$

$$9{,}504b_5 - r_1\Omega^2 = 0 \tag{I.3k}$$

$$11{,}232r_6 - r_2\Omega^2 = 0 \tag{I.3l}$$

As in the previous case, one can easily see that only the trivial solution is obtainable.

For the general case $n > 5$, we obtain the following set of equations:

from ξ^0: $b_1 = 0$ \hfill (I.4a)

from ξ^1: $6(n^2 + 7n + 12)b_2 + \frac{1}{6}(n^2 + 7n + 6)r_0\Omega^2 = 0$ \hfill (I.4b)

from ξ^2: $12(n^2 + 7n + 12)b_3 + \frac{1}{6}(n^2 + 7n + 6)r_1\Omega^2 = 0$ \hfill (I.4c)

from ξ^3: $20(n^2 + 7n + 12)b_4 - \frac{1}{6}(n^2 + 7n + 6)r_0\Omega^2 + \frac{1}{6}(n^2 + 7n + 6)r_2\Omega^2 = 0$ \hfill (I.4d)

from ξ^4: $30(n^2 + 7n + 12)b_5 - \frac{1}{6}(n^2 + 7n + 6)r_1\Omega^2 = 0$ \hfill (I.4e)

from ξ^5: $42(n^2 + 7n + 12)b_6 - \frac{1}{6}(n^2 + 7n + 6)r_2\Omega^2 = 0$ \hfill (I.4f)

from ξ^n: $b_0 = 0$ \hfill (I.4g)

from ξ^{n+1}: $b_1 = 0$ \hfill (I.4h)

from ξ^{n+2}: $b_2 = 0$ \hfill (I.4i)

from ξ^{n+3}: $b_3 = 0$ \hfill (I.4j)

from ξ^{n+4}: $(n + 3)(n + 4)(n + 5)(n + 6)b_4 - r_0\Omega^2 = 0$ \hfill (I.4k)

from ξ^{n+5}: $(n + 3)(n + 4)(n + 6)(n + 7)b_5 - r_1\Omega^2 = 0$ \hfill (I.4l)

from ξ^{n+6}: $(n + 3)(n + 4)(n + 7)(n + 8)b_6 - r_2\Omega^2 = 0$ \hfill (I.4m)

It can be proved that the only solution of the set (I.4) is the trivial one.

Appendix J: Cubic Inertial Coefficient ($m = 3$)

We are looking for solutions for n larger than 2, since closed-form solutions were derived for $n = 0, 1, 2$ in Section 5.2.7. With the general mode shape given in Eq. (5.88), we calculate the left-hand side of Eqn. (5.89):

$$\frac{d^2}{d\xi^2}\left[D(\xi)\frac{d^2\psi}{d\xi^2}\right] - \Omega^2 R(\xi) = (n + 3)(n + 4)\sum_{i=0}^{7}(i + b + 1)(i + n + 2)b_i\xi^{i+n}$$

$$- (n^2 + 7n + 12)\sum_{i=1}^{7}i(i + 1)b_i\xi^{i-1}$$

$$- \Omega^2\sum_{i=0}^{3}r_i\xi^i\left[\xi^{n+4} - \frac{1}{6}(n^2 + 7n + 12)\xi^3\right.$$

$$\left. + \frac{1}{6}(n^2 + 7n + 6)\xi\right] \hfill (J.1)$$

We have to establish how the parameter n relates to the power of the last term, which equals 6. Hence, two special cases arise: $n > 6$ and $n \leq 6$. Now consider the case $n = 3$; we deduce from the governing differential equation a set of 11 homogeneous equations for 8 unknowns:

$$b_1 = 0 \tag{J.2a}$$

$$-252b_2 - 6r_0\Omega^2 = 0 \tag{J.2b}$$

$$-504b_3 - 6r_1\Omega^2 = 0 \tag{J.2c}$$

$$-840b_4 + 840b_0 + 7r_0\Omega^2 - 6r_2\Omega^2 = 0 \tag{J.2d}$$

$$-1260b_5 + 1260b_1 + 7r_1\Omega^2 - 6r_3\Omega^2 = 0 \tag{J.2e}$$

$$-1764b_6 + 1764b_0 + 7r_2\Omega^2 = 0 \tag{J.2f}$$

$$-2352b_7 + 1764b_3 + 7r_3\Omega^2 = 0 \tag{J.2g}$$

$$3024b_4 - r_0\Omega^2 = 0 \tag{J.2h}$$

$$3780b_5 - r_1\Omega^2 = 0 \tag{J.2i}$$

$$4620b_6 - r_2\Omega^2 = 0 \tag{J.2j}$$

$$5544b_7 - r_3\Omega^2 = 0 \tag{J.2k}$$

It is easily seen that this system has no solution different from the trivial solution $b_0 = b_1 = \cdots = b_7 = 0$.

When $n = 4$, we get the following set:

$$b_1 = 0 \tag{J.3a}$$

$$-336b_2 - \tfrac{25}{3}r_0\Omega^2 = 0 \tag{J.3b}$$

$$-672b_3 - \tfrac{25}{3}r_1\Omega^2 = 0 \tag{J.3c}$$

$$-1120b_4 + \tfrac{28}{3}r_0\Omega^2 - \tfrac{25}{3}r_2\Omega^2 = 0 \tag{J.3d}$$

$$-1680b_5 + 1680b_0 + \tfrac{28}{3}r_1\Omega^2 - \tfrac{25}{3}r_3\Omega^2 = 0 \tag{J.3e}$$

$$-2352b_6 + 2352b_1 + \tfrac{28}{3}r_2\Omega^2 = 0 \tag{J.3f}$$

$$-3136b_7 + 3136b_2 + \tfrac{28}{3}r_3\Omega^2 = 0 \tag{J.3g}$$

$$b_3 = 0 \tag{J.3h}$$

$$5040b_4 - r_0\Omega^2 = 0 \tag{J.3i}$$

$$6160b_5 - r_1\Omega^2 = 0 \tag{J.3j}$$

$$7392b_6 - r_2\Omega^2 = 0 \tag{J.3k}$$

$$8736b_7 - r_3\Omega^2 = 0 \tag{J.3l}$$

As in the previous case, one can easily see that only the trivial solution is obtainable.

Concerning $n = 5$, the following set of equations is derived from Eq. (J.1):

$$b_1 = 0 \tag{J.4a}$$

$$-432b_2 - 11r_0\Omega^2 = 0 \tag{J.4b}$$

$$-864b_3 - 11r_1\Omega^2 = 0 \tag{J.4c}$$

$$-1{,}440b_4 + 12r_0\Omega^2 - 11r_2\Omega^2 = 0 \tag{J.4d}$$

$$-2{,}160b_5 + 12r_1\Omega^2 - 11r_3\Omega^2 = 0 \tag{J.4e}$$

$$-3{,}024b_6 + 3{,}024b_0 + 12r_2\Omega^2 = 0 \tag{J.4f}$$

$$-4{,}032b_7 + 4{,}032b_1 + 12r_3\Omega^2 = 0 \tag{J.4g}$$

$$b_2 = 0 \tag{J.4h}$$

$$b_3 = 0 \tag{J.4i}$$

$$7{,}920b_4 - r_0\Omega^2 = 0 \tag{J.4j}$$

$$9{,}504b_5 - r_1\Omega^2 = 0 \tag{J.4k}$$

$$11{,}232r_6 - r_2\Omega^2 = 0 \tag{J.4l}$$

$$13{,}104b_7 - r_3\Omega^2 = 0 \tag{J.4m}$$

As in the previous cases, it could be shown that the set (J.4) has only a trivial solution.

When $n = 6$, we get

$$b_1 = 0 \tag{J.5a}$$

$$-540b_2 - 14r_0\Omega^2 = 0 \tag{J.5b}$$

$$-1{,}080b_3 - 14r_1\Omega^2 = 0 \tag{J.5c}$$

$$-1{,}800b_4 + 15r_0\Omega^2 - 14r_2\Omega^2 = 0 \tag{J.5d}$$

$$-2{,}700b_5 + 15r_1\Omega^2 - 14r_3\Omega^2 = 0 \tag{J.5e}$$

$$-3{,}780b_6 + 15r_2\Omega^2 = 0 \tag{J.5f}$$

$$-5{,}040b_7 + 5{,}040b_0 + 15r_3\Omega^2 = 0 \tag{J.5g}$$

$$b_1 = 0 \tag{J.5h}$$

$$b_2 = 0 \tag{J.5i}$$

$$b_3 = 0 \tag{J.5j}$$

$$11{,}880b_4 - r_0\Omega^2 = 0 \tag{J.5k}$$

$$14{,}040b_5 - r_1\Omega^2 = 0 \tag{J.5l}$$

$$16{,}380r_6 - r_2\Omega^2 = 0 \tag{J.5m}$$

$$18{,}900b_7 - r_3\Omega^2 = 0 \tag{J.5n}$$

which has a trivial solution.

For the general case $n > 6$, we deduce the following set of equations:

from ξ^0: $b_1 = 0$ $\hspace{5em}$ (J.6a)

from ξ^1: $6(n^2 + 7n + 12)b_2 + \frac{1}{6}(n^2 + 7n + 6)r_0\Omega^2 = 0$ $\hspace{3em}$ (J.6b)

from ξ^2: $12(n^2 + 7n + 12)b_3 + \frac{1}{6}(n^2 + 7n + 6)r_1\Omega^2 = 0$ $\hspace{3em}$ (J.6c)

from ξ^3: $20(n^2 + 7n + 12)b_4 - \frac{1}{6}(n^2 + 7n + 6)r_0\Omega^2 + \frac{1}{6}(n^2 + 7n + 6)r_2\Omega^2 = 0$
$\hspace{31em}$ (J.6d)

from ξ^4: $30(n^2 + 7n + 12)b_5 - \frac{1}{6}(n^2 + 7n + 12)r_1\Omega^2 + \frac{1}{6}(n^2 + 7n + 6)r_3\Omega^2 = 0$
$\hspace{31em}$ (J.6e)

from ξ^5: $42(n^2 + 7n + 12)b_6 - \frac{1}{6}(n^2 + 7n + 6)r_2\Omega^2 = 0$ $\hspace{3em}$ (J.6f)

from ξ^6: $56(n^2 + 7n + 12)b_7 - \frac{1}{6}(n^2 + 7n + 6)r_3\Omega^2 = 0$ $\hspace{3em}$ (J.6g)

from ξ^n: $b_0 = 0$ $\hspace{5em}$ (J.6h)

from ξ^{n+1}: $b_1 = 0$ $\hspace{5em}$ (J.6i)

from ξ^{n+2}: $b_2 = 0$ $\hspace{5em}$ (J.6j)

from ξ^{n+3}: $b_3 = 0$ $\hspace{5em}$ (J.6k)

from ξ^{n+4}: $(n + 3)(n + 4)(n + 5)(n + 6)b_4 - r_0\Omega^2 = 0$ $\hspace{3em}$ (J.6l)

from ξ^{n+5}: $(n + 3)(n + 4)(n + 6)(n + 7)b_5 - r_1\Omega^2 = 0$ $\hspace{3em}$ (J.6m)

from ξ^{n+6}: $(n + 3)(n + 4)(n + 7)(n + 8)b_6 - r_2\Omega^2 = 0$ $\hspace{3em}$ (J.6n)

from ξ^{n+7}: $(n + 3)(n + 4)(n + 8)(n + 9)b_7 - r_3\Omega^2 = 0$ $\hspace{3em}$ (J.6o)

It could be demonstrated that the only solution of the set (J.6) is the trivial one.

References

Abbassi M.M. (1958) Buckling of Struts of Variable Bending Rigidity, *Journal of Applied Mechanics*, **25**, 537–540.

Abbassi M.M. (1960) The Second Approximation for Buckling Loads of Tapered Struts, *Journal of Applied Mechanics*, **27**, 211–212.

Abel N.H. (1826) Résolution d'un problème de méchanique, *Z. reine angew. Math.*, **1**, 97–101.

Abid Mian M. and Spencer A.J.M. (1998) Exact Solutions for Functionally Graded Laminated Elastic Materials, *Journal of Mechanics of Physics and Solids*, **46**, 2283–2295.

Aboudi J., Pindera M.J. and Arnold S.M. (1999) Higher-Order Theory for Functionally Graded Materials, *Composites: Part B*, **30**, 777–832.

Abramovich H., Eisenberger M. and Shulepov O. (1996) Vibrations and Buckling of Non-Symmetric Laminated Composite Beams via the Exact Finite Element Method, *AIAA Journal*, **34**, 1064–1069.

Abrate S. (1995) Vibration of Non-Uniform Rods and Beams, *Journal of Sound and Vibration*, **185**, 703–716.

Adali S. (1979) Optimal Shape and Non-Homogeneity of a Non-Uniformely Compressed Columns, *International Journal of Solids and Structures*, **15(12)**, 935–949.

Adali S. (1981) Optimization and Stability of Columns under Opposing Distributed and Concentrated Loads, *Journal of Structural Mechanics*, **9(1)**, 1–28.

Adali S. (1982) Design of Beams on Winkler-Pasternak Foundations for Minimum Dynamic Response and Maximum Eigenfrequency, *Journal of Mech. Theor. Appl.*, **1(6)**, 975–993.

Adams S.G. (1997) *Maple Talk*, Prentice Hall, Upper Saddle River.

Adams R.A. and Doyle J.F. (2000) Force Identification in Complex Structures, in *Recent Advances in Structural Dynamics*, pp. 456–477, University of Southampton, U.K.

Adelman N.T. and Stavsky Y. (1975) Vibrations of Radially Polarized Composite Piezoceramic Cylinders and Disks, *Journal of Sound and Vibration*, **43(1)**, 37–44.

Afsar A.M. and Skeine H. (2002) Inverse Problems of Material Distributions for Prescribed Apparent Fracture Toughness in FGM Elastic Media, *Composites Science and Technology*, **62(7–8)**, 1063–1077.

Ainola L.Ia. (1971) On the Inverse Problem of Natural Vibrations of Elastic Shells, *Journal of Applied Mechanics and Mathematics — PMM*, **35**, 317–322.

Åkesson B.A. (1976) PFVIBAT-A Computer Program for Plane Frame Vibration Analysis by an Exact Method, *International Journal for Numerical Methods in Engineering*, **10(6)**, 1221–1231.

Åkesson B. and Tägnfors H. (1978) PFVIBAT-II Computer Program for Plane Frame Vibration Analysis, *Publication No 25*, Division of Solid Mechanics, Chalmers University of Technology, Gotthenburg, Sweden.

Åkesson B., Tägnfors H. and Johannesson O. (1972) Böjvängande Balkar Och Ramar, Almqvist and Wisksell, Stockholm (in Swedish).

Akulenko L.D. and Nesterov S.V. (1998) Effective Solution of the Buckling Problem for a Nonuniform Beam, *Mechanics of Solids*, **33(2)**, 163–169.

Akulenko L.D. and Nesterov S.V. (1999) The Natural Oscillations of Distributed Inhomogenous Systems Described by Generlized Boundary-Value Problem, *Journal of Applied Mechanics and Mathematics — PMM*, **63(4)**, 617–625.

Akulenko L.D., Kostin G.V. and Nesterov S.V. (1995) Numerical — Analytic Method for Studying Natural Vibrations of Nonuniform Rods, *Mechanics of Solids*, **30(5)**, 173–182.

Albul A.V., Banichuck N.V. and Barsuk A.A. (1980) Optimisation of Stability of Elastic Rods under Thermal Loads, *Mechanics of Solids*, **15(3)**, 120–125.

Alfutov N.A. (1998) On the Stability of an Elastic Self-Gravitating Bar (The Problem of V.I. Feodosiev), *Mechanics of Solids*, **33(3)**, 82–86.

Alfutov N.A. and Popov B.G. (1998) Stability of a Self-Gravitating Bar, *Izvestiya AN SSSR, Mekhanika Tverdogo Tela*, **23(5)**, 177–180.

Ali M.A. et al. (1994) Influence of Linear Depth Taper on the Frequencies of Free Torsional Vibration of a Cantilever Beam, *Computers and Structures*, **51(5)**, 541–545.

Ali R. (1983) Finite Difference Method in Vibration Analysis, *The Shock and Vibration Digest*, **15(3)**, 3–7.

Alifanov O.M. (1983) Mechanics of Solving Ill-Posed Inverse Problems, *Journal of Engineering Physics*, **45(5)**, 1237–1245.

Alifanov O.M. (1994) *Inverse Heat Transfer Problems*, Springer Verlag, Berlin.

Alison H. (1979) Inverse Unstable Problems and Some of Their Applications, *Mathematical Scientist*, **4**, 9–30.

Alvarez S.I., Faccadenti de Iglesisas G.M. and Laura P.A.A. (1998) Vibrations of Elastically Restrained, Non-Uniform Beam with Translational and Rotational Springs, and a Tip Mass, *Journal of Sound and Vibrations*, **120**, 465–471.

Amazigo J.C. (1976) Buckling of Stochastically Imperfect Structures, in *Buckling of Structures* (Budiansky B., ed.), pp. 172–182, Springer, Berlin.

Amba Rao C.L. (1967) Effect of End Conditions on the Lateral Frequencies of Uniform Straight Columns, *Journal of Acoustical Society of America*, **42**, 900–901.

Ambarzumian V. (1929) Über eine Frage der Eigenwerttheorie, *Zeitschrift für Physik*, **53**, 690–695 (in German).

Ananiev I.V. (1948) *Handbook on Analysis of Free Vibrations of Elastic Systems*, pp. 92–94, GITTL Publishers, Moscow (in Russian).

Anderson L.E. (1990) Algorithm for Solving Inverse Problems for Sturm-Liouville Equations, in *Inverse Methods in Action* (Sabatier P.C., ed.), Springer, Berlin.

Anderson R.G., Irons B.M. and Zienkiewicz O.C. (1968) Vibrations and Stability of Plates Using Finite Elements, *International Journal of Solids and Structures*, **4(10)**, 1031–1035.

Anger G. (1985) On the Relationship between Mathematics and Its Applications: A Critical Analysis by Means of Inverse Problems, *Inverse Problems*, **1**, L7–L11.

Anger G. (1990) *Inverse Problems in Differential Equations*, Akademie Verlag, Berlin.

Anger G. (1998a) Basic Principles of Inverse Problems in Natural Sciences and Medicine, *Zeitschrift für angewandte Mathematik und Mechanik*, **78**, 5189–5196.

Anger G. (1998b) Inverse Problems: Regularization of Ill-Posed Problems, *Zeitschrift für angewandte Mathematik und Mechanik*, **78**, 5235–5237.

Anger G. (2000) Personal Communication.

Anger G., Gorenflo R., Jochmann H., Moritz H. and Webers W. (1993) *Inverse Problems: Principles and Applications in Geophysics, Technology and Medicine*, Akademie Verlag, Berlin.

Antyufeev V.S. (2000) *Monte Carlo Methods for Solving Inverse Problems of Radiation Transfer*, VSP, Zeist, The Netherlands.

Appl F.C. and Zorowski C.F. (1959) Upper and Lower Bounds of Special Eigenvalues, *Journal of Applied Mechanics*, **26**, 246–250.

Arbabi F. (1991) Buckling of Variable Cross-Section Columns: Integral Equations Approach, *Journal of Structural Engineering*, **117**, 2426–2441.

Arbabi F. and Li F. (1990) A Macroelement for Variable-Section Beams, *Computers and Structures*, **37(4)**, 553–559.

Arbabi F. and Li F. (1991) Buckling of Variable Cross-Section Columns: Integral-Equation Approach, *Journal of Structural Engineering*, **117(8)**, 2426–2441.

Archer J.S. (1965) Consistent Matrix Formulations for Structural Analysis Using Finite-Element Techniques, *AIAA Journal*, **3(10)**, 1010–1918.

Ariaratnam S.T. (1960) Random Vibration of Nonlinear Suspensions, *Journal of Mechanical Engineering Sciences*, **2(3)**, 195–201.

Aristizabel-Ochoa, J.D. (1993) Statics, Stability and Vibration of Non-Prismatic Beams and Columns, *Journal of Sound and Vibration*, **162**, 441–455.

Arridge S.R. and Schweiger M. (1997) Image Reconstruction on Optical Tomography, *Philosophical Transactions: Biological Sciences*, **352(1354)**, 716–726.

Ashley H. (1982) On Making Things the Best — Aeronautical Uses of Optimization, *Journal of Aircraft*, **19(1)**, 5–28.

Atanackovic T.M. (1997) Sect. 2.11, Inverse Problem: The Shape of a Piston Ring, in *Stability Theory of Elastic Rods*, pp. 89–94, World Scientific, Singapore.

Auciello N.M. (1995) A Comment on: A Note on Vibrating Tapered Beams, *Journal of Sound and Vibration*, **187(4)**, 724–726.

Auciello N.M. (2001) On the Transverse Vibrations of Non-Uniform Beams with Axial Loads and Elastically Restrained Ends, *International Journal of Mechanical Sciences*, **43**, 193–208.

Auciello N.M. and Ercolano A. (1997) Exact Solution for the Transverse Vibration of a Beam a Part of which is a Taper Beam and Other Part is a Uniform Beam, *International Journal of Solids and Structures*, **34(17)**, 2115–2129.

Avdonin S., Lenhart S. and Protopopescu V. (2002) Solving the Dynamical Inverse Problem for the Schrödinger Equation by the Boundary Control Method, *Inverse Problems*, **18**, 34–61.

Babich V.M. (1961a) Fundamental Solutions of the Dynamic Equations of Elasticity for Nonhomogeneous Media, *Journal of Applied Mechanics and Mathematics — PMM*, **25(1)**, 49–60.

Babich V.M. (1961b) Fundamental Solutions of Dynamic Equations of the Theory of Elasticity of Inhomogeneous Medium, *Journal of Applied Mechanics and Mathematics — PMM*, **25(1)** (in Russian).

Bahai H., Farahani K. and Djoudi M.S. (2002) Eigenvalue Inverse Formulation for Optimizing Vibrating Behaviour of Truss and Continuous Structures, *Computers and Structures*, **80**, 2397–2403.

Baistow L. and Stedman E.W. (1914a) Critical Loads for Long Struts of Varying Sections, *Engineering* (London), **98**, 403.

Baistow L. and Stedman E.W. (1914b) The Design of a Strut of Uniform Strength, *British Adv. Comm. for Aero. R. & M.*, No. 158.

Baker T.S. (1966) A Modified Ritz Method, *Journal of Applied Mechanics*, **33(1)**, 224–226.

Balas M. and Johnson C.D. (1998) Identification of Unknown Force/Moment Distributions in Distributed Parameter Systems, *Proceedings Southeastern Conference on Theoretical and Applied Mechanics (SECTAM)*, Florida Atlantic University.

Banerjee J.R. and Williams F.W. (1985a) Exact Bernoulli-Euler Dynamic Stiffness Matrix for a Range of Tapered Beams, *International Journal for Numerical Methods in Engineering*, **21**, 2289–2302.

Banerjee J.R. and Williams F.W. (1985b) Further Flexural Vibration Curves for Axially Loaded Beams with Linear and Parabolic Taper, *Journal of Sound and Vibration*, **102(3)**, 315–327.

Banerjee J.R. and Williams F.W. (1986) Exact Bernoulli–Euler Static Stiffness Matrix for a Range of Tapered Beam-Columns, *International Journal for Numerical Methods in Engineering*, **23(9)**, 1615–1628.

Banichuk N.V. (1974) Optimization of Stability of a Rod with Elastic Attachment, *Mechanics of Solids*, No. **4**, 150–154.

Banichuk N.V. (1982) Current Problems in the Optimization of Structures, *Mechanics of Solids*, **17(2)**, 95–107.

Bapat C.N. (1995) Vibration of Rods with Uniformly Tapered Sections, *Journal of Sound and Vibration*, **185**, 185–189.

Barakat R. and Baumann E. (1968) Axisymmetric Vibrations of a Thin Circular Plate Having Parabolic Thickness Variation, *Journal of Acoustical Society of America*, **44(2)**, 641–643.

Barcilon V. (1974) On the Uniqueness of Inverse Eigenvalue Problems, *Geophys. J. R Astron. Soc.*, **38**, 287–298.

Barcilon V. (1976) Inverse Problem of a Vibrating Beam, *Journal of Applied Mathematics and Physics*, **27**, 347–358.

Barcilon V. (1979) On the Multiplicity of Solutions of the Inverse Problem for Vibrating Beam, *SIAM Journal of Applied Mechanics*, **37**, 605–613.

Barcilon V. (1982) Inverse Problem for the Vibrating Beam in the Free-Clamped Configuration, *Philosophical Transactions of the Royal Society of London*, Ser. A., **304**, 211–252.

Barcilon V. (1983) Explicit Solution of the Inverse Problem for a Vibrating String, *Journal of Applied Mathematical Analysis and Applications*, **93**, 222–234.

Barling W.H. and Webb H.A. (1918) Design of Aeroplane Struts, *Journal of Royal Aeronautical Society*, **22**, 313.

Barnes D. (1988) Shape of the Strongest Column is Arbitrarily Close to the Shape of the Weakest Column, *Quarterly of Applied Mathematics*, **44(3)**, 583–588.

Barnes E.R. (1977) Shape of the Strongest Column and Some Related Extremal Eigenvalue Problems, *Quarterly of Applied Mathematics*, **34(4)**, 393–439.

Bartero M. and Boccacci P. (1998) *Introduction to Inverse Problems in Imaging*, Institute of Physics Publishing, Bristol.

Baruch M. (1970) Undestructive Determination of the Buckling Load of an Elastic Bar, *AIAA Journal*, 2274–2275.

Baruch M. (1971) Determination of the Stiffness of an Elastic Bar, *AIAA Journal*, 1638–1639.

Baruch M. (1973) Integral Equations for Nondestructive Determination of Buckling Loads for Elastic Plates and Bars, *Israel Journal of Technology*, **11(1–2)**, 1–8.

Baruch M. (1998) Model Data Are Insufficient for Identification of Both Mass and Stiffness Matrices, *AIAA Journal*, **35(11)**, 1797–1798.

Baruch M. (2000) Private Communication, 6 February.

Baruch M., Elishakoff I. and Catellani G. (2003) Solution of Semi-Inverse Buckling Problems Yields a Flexural Rigidity as a Rational Function, *International Journal of Structural Stability and Dynamics*, **3(3)**, 307–334.

Batdorf S.B. (1969) On the Application of Inverse Differential Operators to the Solution of Cylinder Buckling and Other Problems, *AIAA/ASME 10th Structures, Structural Dynamics, and Materials Conference*, New Orleans, LA, pp. 386–391, ASME Press, New York.

Bathe K.J. and Wilson E.L. (1973) Solution Methods for Eigenvalue Problems in Structural Mechanics, *International Journal for Numerical Methods in Engineering*, **6**, 213–216.

Baumeister J. (1973) *Stable Solution of Inverse Problems*, Vieweg, Braunschweig.

Baumeister J. (1991) *Stable Solution of Inverse Problems*, Friedrich Vieweg & Sohn, Braunschweig.

Bažant Z.P. and Cedolin L. (1991) *Stability of Structures*, Oxford University Press, New York.

Beck J.L. (1979) Determining Models of Structures from Earthquake Records, *Ph. D. Dissertation*, School of Engineering, Caltech, Pasadena.

Beck M. (1952) Die Knicklast des einseitig eingespannten tangential gedrückten Stabes, *ZAMP*, **3**, 225–228 (in German) (Erratum: **3**, 476–477, 1952).

Becquet R. and Elishakoff I. (2001a) Class of Analytical Closed-Form Polynomial Solutions for Guided–Pinned Inhomogeneous Beams, *Chaos, Solitons and Fractals*, **12**, 1509–1534.

Becquet R. and Elishakoff I. (2001b) Class of Analytical Closed-Form Polynomial Solutions for Clamped–Guided Inhomogeneous Beams, *Chaos, Solitons and Fractals*, **12**, 1657–1678.

Beliakow Th. (1930) *Das Prinzip der "virtuellen" Biegelinie als Grundlage des Knickproblems*, Kharkov (in German).

Beliakow Th. (1931) Über einige schwierige Knickaufgaben, *Bauingenieur*, **11**, 188 (in German).

Bellman R. and Casti J. (1971) Differential Quadrature and Long-Term Integration, *Journal of Mathematical Analysis and Applications*, **34**, 235–238.

Bellman R., Kashef B.G. and Casti J. (1972) Differential Quadrature: A Technique for the Rapid Solution of Nonlinear Partial Differential Equations, *Journal of Computational Physics*, **10**, 40–52.

Bennett H.Z. (ed.) (1999) *Fritjof Capra in Conversation with Michael Toms*, Aslan Publishing, p.18, Lower Lake, CA.

Bérand P. (1952) Transplantation et isospectralité, *Math. Ann.*, **292**, 547–559 (in French).

Bergman L.A. (2000) Private Communication, June 18.

Bernshtein S.A. (1941) *Basics of Dynamics of Beams*, "Stroyizdat" Publishers, Moscow (in Russian).

Bert C.W. (1961) Nonhomogenous Polar Orthotropic Circular Disks of Varying Thickness, *Ph.D. Dissertation*, Ohio State University.

Bert C.W. (1962) Complete Stress Function for Nonhomogeneous Anisotropic, Plane Problems in Continuum Mechanics, *Journal of Aerospace Sciences*, **29**, 756–757.

Bert C.W. (1981a) Closed-Form Solution of an Arbitrarily Laminated, Anisotropic, Elliptic Plate under Uniform Pressure, *Journal of Elasticity*, **11**, 337–340.

Bert C.W. (1984a) Use of Symmetry On Applying the Rayleigh–Schmidt Method to Static and Free-Vibration Analysis, *Industrial Mathematics*, **34(1)**.

Bert C.W. (1984b) Improved Technique for Estimating Buckling Loads, *Journal of Engineering Mechanics*, **110(12)**, 1655–1665.

Bert C.W. (1987a) Techniques for Estimating Buckling Loads, in *Civil Engineering Practice*, **1** *Structures* (P.N. Cheremisinoff, N.P. Cheremisinoff and S.L. Cheng, eds.), pp. 489–499, Technomic, Lancaster.

Bert C.W. (1987b) Application of a Version of the Rayleigh Technique to Problems of Bars, Beams, Columns, Membranes and Plates, *Journal of Sound and Vibration*, **119(2)**, 317–326.

Bert C.W. (1993) Relationship between Fundamental Natural Frequency and Maximum Static Deflection for Various Linear Vibratory Systems, *Journal of Sound and Vibration*, **162(3)**, 547–557.

Bert C.W. (1995) Effect of Finite Thickness on the Relationship between Fundamental Natural Frequency and Maximum Static Deflection of Beams, *Journal of Sound and Vibration*, **186(4)**, 691–693.

Bert C.W. (2000a) Private Communication, March 31.

Bert C.W. (2000b) Private Communication, April 19.

Bert C.W. and Malik M. (1996) Differential Quadrature Method in Computational Mechanics: A Review, *Applied Mechanics Reviews*, **49(1)**, 1–27.

Bert C.W. and Niedenfuhr F.W. (1963) Stretching of a Polar-Orthotropic Disk of Varying Thickness Under Arbitrary Body Forces, *AIAA Journal*, **1(6)**, 1385–1390.

Bert C.W., Wang X. and Striz A.G. (1994) Static and Free Vibrational Analysis of Beams and Plates by Differential Quadrature Method, *Acta Mechanica*, **102**, 11–24.

Bert C.W., Jang S.K. and Striz A.G. (1988) Two New Approximate Methods for Analyzing Free Vibration of Structural Components, *AIAA Journal*, **26**, 612–618.

Bhat R.B. (1984) Obtaining Natural Frequencies of Elastic Systems by using an Improved Strain Energy Formulation in the Rayleigh–Ritz Method, *Journal of Sound and Vibration*, **93**, 314–320.

Bhat R.B. (1985) Natural Frequencies of Rectangular Plates using Characteristic Orthogonal Polynomials in the Rayleigh–Ritz Method, *Journal of Sound and Vibration*, **102(3)**, 493–499.

Bhat R.B. (1996) Effect of Normal Mode Contents in Assumed Deflection Shapes in Rayleigh–Ritz Method, *Journal of Sound and Vibration*, **189(3)**, 407–419.

Bhat R.B. (2000) Private Communication, 7 March.

Biderman V.L. (1972) *Applied Theory of Mechanical Vibrations*, pp. 207–217, Vysshaiya Shkola Publishers, Moscow (in Russian).

Bielak J. (1969) Base Moment for a Class of Linear Systems, *Journal of the Engineering Mechanics Division*, **95(5)**, 1053–1061.

Bieniek M., Spillers W.R. and Freudenthal A.M. (1962) Nonhomogeneous Thick-Walled Cylinder Under Internal Pressure, *AIAA Journal*, 1249–1255.

Biezeno C. (1924) Graphical and Numerical Methods for Solving Stress Problems, *Proceedings of the First International Congress for Applied Mechanics*, Delft, The Netherlands.

Biezeno C.B. and Grammel R. (1953) *Technische Dynamik*, pp. 114–117, Springer, Berlin (in German).

Birger I.A. (1961) Nonuniformly Heated Columns with Variable Parameters of Elasticity, *Strength Analysis*, **7**, Mashinostroenie Publishing House, Moscow (in Russian).

Birger I.A. and Mavliutov R.R. (1986) *Strength of Materials*, pp. 401–403, 407–408, Nauka Publishers, Moscow (in Russian).

Birger I.A. and Panovko Ya. G. (eds.) (1968) *Strength. Stability. Vibrations*, **3**, pp. 311–312, "Mashinostroenie" Publishers, Moscow (in Russian).

Birman V. (1995) Buckling of Functionally Graded Hybrid Composite Plates, *Proceedings of the 10th Conference on Engineering Mechanics*, **2**, 1199–1202.

Birman V. (1997) Stability of Functionally Graded Shape Memory Alloy Sandwich Panels, *Smart Materials and Structures*, **6**, 282–286.

Bishop R.E.D. (1960) *The Mechanics of Vibration*, Cambridge University Press, Cambridge.

Blasius H. (1914) Träger kleinster Durchbiegung und Stäbe grösster Knickfestingkeit bei gegebenem Materialverbrauch, *ZAMP*, **62**, 182–197 (in German).

Bleich F. (1952) *Buckling Strength of Metal Structures*, McGraw-Hill, New York.

Blevins R.D. (1984) *Formulas for Natural Frequency and Mode Shape*, R.E. Krieger, Malabar.

Bokaian A. (1988) Natural Frequencies of Beams under Compressive Axial Loads, *Journal of Sound and Vibration*, **126(1)**, 49–65.

Bokaian A. (1989) Author's reply, *Journal of Sound and Vibration*, **131**, 351.

Bokaian A. (1990) Natural Frequencies of Beams under Tensile Axial Loads, *Journal of Sound and Vibration*, **142(1)**, 481–498.

Boley D. and Golub G.H. (1987) A Survey of Matrix Inverse Eigenvalue Problems, *Inverse Problems*, **3**, 595–622.

Bolotin V.V. (1953) Integral Equations for Restrained Torsion and Stability of Thin Walled Bars, *Journal of Applied Mechanics and Mathematics — PMM*, **17**, 245–248 (in Russian).

Bolotin V.V. (1964) *The Dynamic Stability of Elastic Systems*, Holden-Day, San Francisco.

Bolt B.A. (1980) What Can Inverse Theory Do for Applied Mathematics and the Sciences? *Gazette of the Australian Mathematical Society*, **7**, 69–78.

Bolt B.A. and Uhrhammer R. (1975) Resolution Techniques for Density and Heterogeneity in the Earth, *Geophysics Journal of the Royal Astro. Soc.*, **42**, 419–435.

Boobnov I.G. (1913) Recommendation on the Work of Pr. S.P. Timoshenko "About Stability of Elastic Systems," *Sb. S.-Peterburgskogo Instituta Putei Soobschenia*, No. 81, 33–36 (in Russian).

Boobnov I.G. (1956) *Selected Works*, pp. 136–139, "Sudpromgiz" Publishers, Leningrad (in Russian).

Borg G. (1946) Eine Umkehrung der Sturm-Liouvilleschen Eigenwertaufgabe, *Acta Mathematica*, **78**, 1–96 (in German).

Borwein P. and Erdélyi T. (1995) *Polynomials and Polynomial Inequality*, Springer, New York.

Boyce E.W. and Goodwin B.E. (1964) Random Transverse Vibration of Elastic Beams, *SIAM Review*, **12(3)**, 619–629.

Boyd J.E. (1920) Theory of Certain Tapered Struts, Ch. 7, *Bureau of Standards Technologic Papers*, No. **152**, 24–35.

Brach R.M. (1968) On the Extremal Fundamental Frequencies of Vibrating Beams, *International Journal of Solids and Structures*, **4**, 667–674.

Braun S.G. and Ram Y.M. (1987) Structural Parameter Identification in the Frequency Domain: The Use of Overdetermined Systems, *Journal of Dynamic Systems, Measurement and Control*, **109**, 120–123.

Broeck van den, J.A. (1941) Columns Subject to Uniformly Distributed Transverse Loads, Illustrating a New Method of Column Analysis, *Engineering Journal*, **24**, 115.

Brush D.O. and Almroth B.O. (1975) *Buckling of Bars, Plates and Shells*, McGraw-Hill, New York.

Bryan G.H. (1888) Application of the Energy Test to the Collapse of a Thin Long Pipe under External Pressure, *Proceedings of the Cambridge Philosophical Society*, **6**, 287–292.

Buckgeim A.L. (2000) *Introduction to the Theory of Inverse Problems*, VSP, Zeist, The Netherlands.

Bufler H. (1963a) Die Inomogene elastische Schicht, *Zeitschrift für angewandte Mathematik und Mechanik*, **43**, Zonder Heft (Special Issue) (in German).

Bufler H. (1963b) Elastische Schicht und elastische Halbraum bei mit Tiefe stetig abnehmenden Elastizitätmodul, *Ingenieur-Archiv*, **32(6)** (in German).

Bulson P.S. (1970) *The Stability of Plates*, Chatto & Windus, London.

Bültmann W. (1944) Die Knickfestigkeit des geraden Stabes mit veränderlicher Druckkraft, *Stahlbau*, **17(10–11)**, 49–50 (in German).

Bültmann W. (1951) Der Knickfestigkeit des geraden Stabes mit veränderlicher Druckkraft bei elastischer Einspannung, *Stahlbau*, **28**, 50–52, (in German).

Bürgermeister G. and Steup H. (1957) *Stabilitätstheorie*, pp. 192–198, Akademie-Verlag, Berlin (in German).

Busby H.R. and Trujillo D.M. (1987) Solution of an Inverse Problem using an Eigenvalue Reduction Technique, *Computers and Structures*, **25(1)**, 109–117.

Busby H.R. and Trujillo D.M. (1993) An Inverse Problem for a Plate under Impulse Loading, in *Inverse Problems in Engineering: Theory and Practise* (N. Zabaka, K. Woodbury and M. Raynaud, eds.), pp. 155–161, ASME Press, New York.

Buser P., Conway J., Doyle P. and Semmler K. (1994) Some Planar Isospectral Domains, *Internat. Math. Res. Notices*, 391–400.

Calió I. (2000) Personal Communication, June 8.

Calió I. and Elishakoff I. (2002) Can Harmonic Functions Constitute Closed-Form Buckling Modes of Inhomogeneous Columns? *AIAA Journal*, **40(2)**, 2532–2537.

Calió I. and Elishakoff I. (2004a) Closed-form Trigonometric Solutions for Inhomogeneous Beam-Columns on Elastic Foundation, *International Journal of Structural Stability and Dynamics* (to appear).

Calió I. and Elishakoff I. (2004b) Can a Trigonometric Function Serve Both as the Vibration and the Buckling Mode of an Inhomogeneous Structure? *Mechanics Based Design of Structures and Machines* (to appear).

Calió I. and Elishakoff I. (2004c) Closed-Form Solutions for Axially Graded Beam-Columns, *Journal of Sound and Vibration* (to appear).

Candan S. and Elishakoff I. (2000) Infinite Number of Closed-Form Solutions for Reliabilities of Stochastically Nonhomogeneous Beams, in *Proceedings, International Conference on Applications of Statistics & Probability* (R.E. Melchers and M.G.Stewart, eds.), pp. 1059–1066, Balkema Publishers, Rotterdam.

Candan S. and Elishakoff I. (2001a) Apparently First Closed-Form Solution for Frequencies of Deterministically and/or Stochastically Inhomogeneous Simply Supported Beams, *Journal of Applied Mechanics*, **68**, 176–185.

Candan S. and Elishakoff I. (2001b) Constructing the Axial Stiffness of Longitudinally Vibrating Rod from Fundamental Mode Shape, *International Journal of Solids and Structures*, **38**, 3443–3452.

Carnegie W. and Thomas J. (1967) Natural Frequencies of Long Tapered Cantilevers, *The Aeronautical Quarterly*, **18**, 309–320.

Carr J.B. (1969) The Torsional Vibrations of Uniform Thin Walled Beams of Open Section, *The Aeronautical Journal of the Royal Aeronautical Society*, **73**, 672–674.

Carter W.J. (1973) Optimal Column, *ASME Paper 73-Mat-DD*.

Case J. (1918) An Approximate Graphical Treatment of Some Strut Problems, *Engineering*, **116**, 699–700.

Catellani G. and Elishakoff I. (2004) Apparently First Closed-Form Solutions of Semi-Inverse Buckling Problems Involving Distributed and Concentrated Loads, *Thin-Walled Structures* (to appear).

Caughey T.K. (1963) Derivation and Application of the Fokker–Planck Equation to Discrete Nonlinear Dynamic Systems Subjected to White Random Excitation, *Journal of Acoustical Society of America*, **35(11)**, 1683–1692.

Caughey T.K. (1964) On the Response of a Class of Nonlinear Oscillators to Stochastic Excitation, *Proc. Colloq. Intern. Du Centre National de la Recherche Scientifique*, Marseille, **148**, 393–402.

Caughey T.K. (1971) Nonlinear Theory of Random Vibrations, *Advances in Applied Mechanics*, **11**, 209–253.

Caughey T.K. and Ma F. (1982) The Exact Steady-State Solution of a Class of Non-Linear Stochastic Systems, *International Journal of Non-Linear Mechanics*, **17**, 137–142.

Cha P.D. (2002) Specifying Nodes at Multiple Locations for any Normal Mode of a Linear Elastic Structure, *Journal of Sound and Vibration*, **250(5)**, 923–934.

Cha P.D. and Dym C.L. (1997) Identifying Nodes and Antinodes of an Axially Vibrating Bar with Lumped Masses, *Journal of Sound and Vibration*, **203**, 533–535.

Cha P.D. and Pierre C. (1999) Imposing Nodes to the Normal Mode of a Linear Elastic Structure, *Journal of Sound and Vibration*, **219(4)**, 669–687.

Cha P.D., Dym C.L. and Wong W.C. (1998) Identifying Nodes and Anti-Nodes of Complex Structures with Virtual Elements, *Journal of Sound and Vibration*, **211**, 249–264.

Chajes A. (1974) *Principles of Structural Stability Theory*, Prentice Hall, Englewood Cliffs.

Chan A.M. and Horgan C.O. (1998) End Effects in Anti-Plane Shear for an Inhomogeneous Isotropic Linearly Elastic Semi-Infinite Strip, *Journal of Elasticity*, **51**, 227–242.

Chang C. and Sun C.T. (1989) Determining Transverse Impact Force on a Composite Laminate by Signal Deconvolution, *Experimental Mechanics*, **29(4)**, 414–419.

Chang Sh. (1982) The Fundamental Frequency of an Elastic System and an Improved Displacement Function, *Journal of Sound and Vibration*, **81**, 229–302.

Chapman S.J. (1995) Drums That Sound the Same, *American Mathematical Monthly*, **102**, 124–138.

Charkraverty, S., Bhat R.B. and Stiharu I. (1999) Recent Research on Vibration of Structures using Boundary Characteristics Polynomials on the Rayleigh–Ritz Method, *The Shock and Vibration Digest*, **31(3)**, 187–194.

Chen D.Y. (1998) Finite Element Analysis of the Lateral Vibration of Thin Annular and Circular Plates with Variable Thickness, *Journal of Vibration and Acoustics*, **120(3)**, 747–753.

Chen D.Y. and Ren B.S. (1998) Finite Element Analysis of the Lateral Vibration of Thin Annular and Circular Plates with Variable Thickness, *Journal of Vibration and Acoustics*, **120(3)**, 747–752.

Chen M.F. and Maniatty D.M. (1995) An Inverse Technique for the Optimization of Some Forming Processes, *Simulation of Materials Processing* (Chen M.F. et al. eds.), pp. 545–550, Balkema, Rotterdam.

Chen R.S. (1997) Evaluation of Natural Vibration Frequency of a Compression Bar with Varying Cross-Section by using the Shooting Method, *Journal of Sound and Vibration*, **201**, 520–525.

Chen Y. (1963) On the Displacements of Nonhomogeneous Thick Cylinder Under Uniform Pressure, *Journal of the Franklin Institute*, **286(2)**.

Chen Y.Z., Cheung Y.K. and Xie J.R. (1989) Buckling Loads of Columns with Varying Cross-Sections, *Journal of Engineering Mechanics*, **115(3)**, 662–667.

Cheney M. (1997) Inverse Boundary-Value Problems, *American Scientist*.

Cheng Z.Q. and Batra R.C. (2000) Three-Dimensional Thermo-Elastic Deformations of a Functionally Graded Elliptic Plate, *Composites: Part B*, **31**, 97–106.

Cheng Z.Q. and Kitpornchai S. (1999) Membrane Analogy of Buckling and Vibration of Inhomogenous Plates, *Journal of Engineering Mechanics*, **125(11)**, 1293–1297.

Chenot J.L., Massoni E. and Fourment L. (1996) Inverse Problems in Finite Element Simulation of Metal Forming Processes, *Engineering Computations*, **13(2–4)**, 190–225.

Chermousko F.L. (1988) Private Communication.

Cheung C-K., Keongh G.E. and May M. (1998) *Getting Started with Maple®*, Wiley, New York.

Chladni E.F.F. (1803) *Die Akustik*, Leipzig (in German) (French translation 1809, *Traité d'Acoustique*, Courcier, Paris).

Cho S-M. (2000) A Sub-Domain Inverse Method for Dynamic Crack Propagation Problems, *MS Thesis*, Purdue University.

Choudhury P. (1957) Stresses in an Elastic Layer with Varying Modulus of Elasticity, *Calcutta Mathematical Society*, **49(2)**, 99–104.

Christiano P. and Salmela L. (1971) Frequencies of Beams with Elastic Warping Restraint, *Journal of Structural Design*, **97**, 1835–1840.

Chu T.H. (1949) Determination of Buckling Loads by Frequency Measurements, *Thesis*, California Institute of Technology.

Chuvikin G.M. (1951) *Stability of Frames and Columns*, "Gosstroyizdat" Publishers, Moscow (in Russian).

Chwalla E. (1934) Zur Berechnung gedrungener Stäbe mit beliebig veränderlichem Querschnitt, *Stahlbau*, 121 (in German).

Cipra B. (1992) You Can't Hear the Shape of a Drum, *Science*, **255**, 1642–1643.

Clausen T. (1851) Über die Form architecktonischer Saeulen, *Bull. Physico-Math. de l' Acad. (Petersburg)*, **9**, 368–379 (in German).

Clough R.W. (1971) Analysis of Structural Vibrations and Dynamic Response, in *Recent Advances in Mathematical Methods in Structural Analysis and Design* (Gallagher R.H., Tamada T. and Oden J.T., eds.), pp. 25–45, University of Alabama Press, Huntsville.

Collatz L. (1945) *Eigenwertaufgaben mit Technischen Anwendungen*, Akademische Verlagsgesellschaft, Leipzig (in German) (see also an English translation, Chelsea Publishing Co., New York, 1948).

Collins J.D. and Thomson W.T. (1969) The Eigenvalue Problem for Structural Systems with Uncertain Parameters, *AIAA Journal*, **7(4)**, 642–648.

Colton D., Ewing R. and Rundell W. (1990) *Inverse Problems in Partial Differential Equations*, SIAM, Philadelphia.

Conn J.F.C. (1944) Vibration of a Truncated Wedge, *Aircraft Engineering*, **16**, 103–105.

Constantinescu A. (1993) A Numerical Investigation of the Elastic Moduli in an Inhomogeneous Body, *Inverse Problems in Engineering: Theory and Practice* (Zabaras N. et al., eds.), pp. 77–84, ASME Press, New York.

Constantinescu A. (1995) On the Identification of Elastic Moduli from Displacement–Force Boundary Measurement, *International Journal of Inverse Problems in Engineering Mechanics*, **1**, 293–313.

Constantinescu A. (1998) On the Identification of Elastic Moduli in Plates, in *Inverse Problems in Engineering Mechanics* (Tanaka M. and Dulikravich G.S., eds.), pp. 205–214, Elsevier, Amsterdam.

Conway H.D. (1957) An Analogy Between the Flexural Vibrations of a Cone and a Disc of Linearly Varying Thickness, *Zeitschrift für angewandte Mathematik und Mechanik*, **37(9, 10)**, 406–407.

Conway H.D. (1958) Some Special Solutions for Flexural Vibrations of Discs of Varying Thickness, *Ingenieur-Archiv*, **26**, 408–410.

Conway H.D. (1959) Analysis of Plane Stress in Polar Coordinates and with Varying Thickness, *Journal of Applied Mechanics*, **26**, 437–439.

Conway H.D. (1965) A General Solution for Plane Stress in Polar Coordinates with Varying Modulus of Elasticity, *Revue Roumaine des Sciences Techniques. Sâerie de Mecanique*, **10(1)**, 109–112.

Conway H.D. (1980) An Unusual Closed-Form Solution for a Variable-Thickness Plate on an Elastic Foundation, *Journal of Applied Mechanics*, **47**, 204.

Conway H.D. (1981) An Unusual Closed-Form Solution for a Variable-Thickness Plate on an Elastic Foundation, *Journal of Applied Mechanics*, **47**, 204.

Conway H.D. and Dubil J.F. (1965) Vibration Frequencies of Truncated Cone and Wedge Beams, *Journal of Applied Mechanics*, **32**, 923–925.

Conway H.D. and Farnham K.A. (1965) The Free Flexural Vibrations of Triangular, Rhombic and Parallelogram Plates and Some Analogies, *International Journal of Mechanical Sciences*, **7**, 811–816.

Conway H.D., Becker E.C.H. and Dubil J.F. (1964) Vibration Frequencies of Tapered Bars and Circular Plates, *Journal of Applied Mechanics*, **31**, 329–331.

Cortinez V.H. and Laura P.A.A. (1994) An Extension of Timoshenko's Method and Its Application to Buckling and Vibration Problems, *Journal of Sound and Vibration*, **169**, 141–144.

Costa G.B. (1989) Polynomial Solutions of Certain Classes of Ordinary Differential Equations, *International Journal of Mathematical Education in Science and Technology*, **20(1)**, 1–11.

Costa G.B. and Levine L.E. (1993) Families of Separable Partial Differential Equations, *International Journal of Mathematical Education in Science and Technology*, **24(5)**, 631–635.

Courant R. and Hilbert D. (1953) *Methods of Mathematical Physics*, Wiley, New York.

Cox S.J. (1992) The Shape of the Ideal Column, *The Mathematical Intelligencer*, **14**, 16–24.

Cox S.J. and McCarthy C.M. (1998) The Shape of the Tallest Column, *SIAM Journal of Mathematical Analysis*, **29(3)**, 547–554.

Cox S.J. and Overton M.L. (1992) On the Optimal Design of Column Against Buckling, *SIAM Journal of Mathematical Analysis*, **23(2)**, 287–235.

Cranch E.T. and Adler A.A. (1956) Bending Vibrations of Variable Section Beams, *Journal of Applied Mechanics*, **23**, 103–108.

Crandall S. (1956) *Engineering Analysis: A Survey of Numerical Procedures*, McGraw Hill, New York (second edition: R. Krieger Publishing, Malabar, FL, 1986).

Crandall S.H. (1995) Rayleigh's Influence on Engineering Vibration Theory, *Journal of the Acoustical Society of America*, **98(3)**, 1269–1272.

Craver W.L. and Egle D.M. (1972) A Method for Selection of Significant Terms in the Assumed Solution in a Rayleigh – Ritz Analysis, *Journal of Sound and Vibration*, **22**, 133–142.

Craver W.L. Jr. and Jampala P. (1993) Tansverse Vibrations of a Linearly Tapered Cantilever Beam with Constraining Springs, *Journal of Sound and Vibration*, **166**, 521–529.

Darboux B.E.J. (1882) Sur la représentations sphérique des surfaces, *C. R. Acad. Sci., Paris*, **93**, 1343–1345 (in French).

Darboux G. (1915) *Leçons sur la Théorie Générale des surfaces et les Applications Géométriques du Calcul Infinitésimal*, **2**, "Gauthier Villars" Publishers, Paris (in French).

Darnley T.R. (1918) Design of a Strut of Uniform Strength, *British Adv. Comm. for Aero.*, T. 1186.

Datta P.K. and Nagraj C.S. (1989) Dynamic Instability Behaviour of Tapered Bars with Flaws Supported on an Elastic Foundation, *Journal of Sound and Vibration*, **131(2)**, 227–229.

Davies R., Henshell R.D. and Warburton G.B. (1972) Consistent Curvature Beam Finite Elements for In-Plane Vibration, *Journal of Sound and Vibration*, **25(4)**, 561–576.

De Rosa M.A. (1989) Stability and Dynamics of Beams on Winkler Elastic Foundations, *Earthquake Engineering and Structural Dynamics*, **18**, 377–388.

De Rosa M.A. (1990) Stability Analysis of Timoshenko Beams on Variable Winkler Soil, *Mechanics Research Communications*, **17(4)**, 255–261.

De Rosa M.A. (1995) Free Vibrations of Timoshenko Beams on Two-Parameter Elastic Foundation, *Computers and Structures*, **57**, 151–156.

De Rosa M.A. (1996) Free Vibration of Tapered Beams with Flexible Ends, *Computers and Structures*, **60**, 197–202.

De Rosa M.A. and Auciello N.M. (1996) Free Vibrations of Tapered Beams with Flexible Ends, *Computers and Structures*, **60**, 197–205.

De Rosa M.A. and Franciosi C. (1996) Optimized Rayleigh Method and Mathematica in Vibration and Buckling Problems, *Journal of Sound and Vibration*, **191**, 795–808.

Denisov A.M. (1999) *Elements of the Theory of Inverse Problems*, VSP International Science Publishers, Zeist, The Netherlands.

Descloux J. and Tolley M. (1983) An Accurate Algorithm for Computing the Eigenvalues of a Polygonal Membrane, *Computer Methods in Applied Mechanics and Engineering*, **39**, 37–53.

Deutsch E. (1953) Einfache Berechnung der Knicklast gerader Stäbe mit beliebig veränderlichem Trägheitsmoment, *Der Stahlbau*, 224–226 (in German).

Dhaliwal R.S. and Singh B.H. (1978) On the Theory of Elasticity of a Nonhomogeneous Medium, *Journal of Elasticity*, **8**, 211–219.

Dimarogonas A.D. (1996) *Vibrations for Engineers*, Prentice Hall, Paramus.

Dimentberg M.F. (1982) An Exact Solution to a Certain Non-Linear Random Vibration Problem, *International Journal of Non-Linear Mechanics*, **17**, 231–236.

Dimentberg M.F. (2000) Personal Communication, 13 June.

Dimitrov N. (1953) Ermittlung konstanter Ersatz-Trägheitsmomente für Druck-stäbe mit veränderlichem Querschnitten, *Der Bauingenieur*, **28**, 208–211 (in German).

Ding Z. and Cheung Y.K. (1998) Eigenfrequencies of Tapered Beams with Intermediate Point Supports, *International Journal of Space Structures*, **13**, 87–95.

Dinnik A.N. (1912) Buckling under Own Weight, *Proceedings of Don Polytechnic Institute*, **1**, Part 2, p. 19 (in Russian).

Dinnik A.N. (1913) On Bending of Columns of Variable Cross-Section, *Izvestiya Donskogo Politekhnicheskogo Instituta*, **1**, 390–404 (in Russian).

Dinnik A.N. (1914) On Bending of Columns, Whose Stiffness Varies by a Binomial Law, *Izvestiya Ekaterinoslavskogo Gornogo Instituta* (in Russian).

Dinnik A.N. (1928) On Bending of Columns, Whose Stiffness Varies by a Cubic Law, *Izvestiya Ekaterinoslavskogo Gornogo Instituta*, **15** (in Russian).

Dinnik A.N. (1929) Design of Columns of Varying Cross-Sections, *Transactions of ASME, Applied Mechanics*, Paper APMS1-11, **51(1)**, 105–114.

Dinnik A.N. (1932) Design of Columns of Varying Cross-Sections, *Transactions of the ASME, Applied Mechanics*, Paper APM-54-16, 165–171.

Dinnik A.N. (1955a) *Prodol'nyi Izgib. Kruchenie (Bending in Presence of Axial Forces. Torsion)*, Academy of USSR Publishing, pp. 96–130 (in Russian).

Dinnik A.N. (1955b) *Selected Works*, Academy of Ukraine Publishing, Kiev, p. 74, Eq. (5.2), p. 134 Eq. (13) (in Russian).

Dondorff J. (1907) Die Knickfestigkeit des geraden Stabes mit veränderlichem Quer-schnitt und veränderlickem Druck, ohne und mit Querstuetzen, *Dissertation*, Aachen (in German).

Donnell L.H. (1933) Stability of the Thin-Walled Tubes under Torsion, *NACA Rep. No. 479*.

Downs B. (1977) Transverse Vibrations of Cantilever Beams Having Unequal Breadth and Depth Taper, *Journal of Applied Mechanics*, **44**, 737–742.

Downs B. (1978) Reference Frequencies for the Validation of Numerical Solutions of Transverse Vibration of Non-uniform Beams, *Journal of Sound and Vibration*, **61**, 71–78.

Downs B. (1980) Vibration Analysis of Continuous Systems by Dynamic Discretization, *Journal of Mechanical Design*, **102**, 391–398.

Doyle J.F. (1993) Force Identification from Dynamic Responses of a Bi-Material Beam, *Experimental Mechanics*, **33**, 64–69.

Doyle J.F. (2002) Reconstructing Dynamic Events from Time-Limited Spatially Dis-tributed Data, *International Journal for Numerical Methods in Engineering*, **53**, 2721–2734.

Driscoll T.A. (1997) Eigenmodes of Isospectral Drums, *SIAM Review*, **39(1)**, 1–17.

Du Z-H. (1961) Plane Problem of Elasticity for Inhomogeneous Medium, in *Problems of Mechanics of Continuous Media*, USSR Academy of Sciences Press, Leningrad (in Russian).

Duffing G. (1918) *Erzwungene Schwingungen bei veränderlicher Eigenfrequenz*, Braunsch-weig (in German).

Duncan W.J. (1937) Galerkin's Method in Mechanics and Differential Equations, *Aeronautical Research Committee Reports and Memoranda, No. 1798*.

Duncan W.J. (1938) The Principles of the Galerkin's Method, *Aeronautical Research Committee, Reports and Memoranda, Technical Report No. 1848*.

Duncan W.J. (1952) Multiply-Loaded and Continuously Loaded Struts, *Engineering*, **174**, 180–182, 202–203.

Durban D. and Baruch M. (1972) Buckling of Elastic Bars with Varying Stiffness and Non-Ideal Boundary Conditions, *Israel Journal of Technology*, **10(4)**, 341–349.

Duren W.L. (1994) The Most Urgent Problem for Mathematics Professors, *Notices of American Math. Soc.*, **41(6)**, 582–586.

Dyka C.T. and Carney III, J.F. (1979) Vibration of Annular Plates of Variable Thickness, *Journal of Engineering Mechanics Division*, **105**, 361–370.

Eisenberger M. (1990) Exact Static and Dynamic Stiffness Matrices for General Variable Cross Section Members, *AIAA Journal*, **28**, 1105–1109.

Eisenberger M. (1991a) Exact Solution of General Variable Cross-Section Members, *Computers and Structures*, **41(4)** 765–772.

Eisenberger M. (1991b) Exact Longitudinal Vibration Frequencies of a Variable Cross-Section Rod, *Applied Acoustics*, **34**, 123–130.

Eisenberger M. (1991c) Buckling Loads for Variable Cross-Section Members with Variable Axial Forces, *International Journal of Solids and Structures*, **27**, 135–143.

Eisenberger M. (1991d) Exact Solutions for General Variable Cross-Section Members, *Computers and Structures*, **41(4)**, 765–772.

Eisenberger M. (1992) Buckling Loads for Variable Cross Section Bars in a Nonuniform Thermal Field, *Mechanics Research Communications*, **19(1)**, 259–266.

Eisenberger M. (1994) Vibration Frequencies for Beams on Variable One- and Two-Parameter Elastic Foundations, *Journal of Sound and Vibration*, **176**, 577–584.

Eisenberger M. (1995a) Non-Uniform Torsional Analysis of Variable and Open Cross-Section Bars, *Thin-Walled Structures*, **21**, 93–105.

Eisenberger M. (1995b) Dynamic Stiffness Matrix for Variable Cross-Section Timoshenko Beams, *Communications in Applied Numerical Methods*, **11**, 507–513.

Eisenberger M. (1997a) Dynamic Stiffness Vibration Analysis of Non-Uniform Members, *International Symposium on Vibrations on Continuous Systems* (Leissa A.W., organizer), Estes Park, Colorado, August 11–15, 1997, pp. 13–15.

Eisenberger M. (1997b) Torsional Vibrations of Open and Variable Cross-Section Bars, *Thin-Walled Structures*, **28 (3/4)**, 269–278.

Eisenberger M. (2000a) Private Communication, 20 February.

Eisenberger M. (2000b) Private Communication, 24 April.

Eisenberger M. and Clastornik J. (1987) Vibration and Buckling of a Beam on a Variable Winkler Elastic Foundation, *Journal of Sound and Vibration*, **115**, 233–241.

Eisenberger M. and Cohen M. (1995) Flexural-Torsional Buckling of Variable and Open Cross Section Members, *Journal of Engineering Mechanics*, **121**, 244–254.

Eisenberger M. and Reich Y. (1989) Static, Vibration, and Stability Analysis of Non-Uniform Beams, *Computers and Structures*, **31**, 567–573.

Eisenberger M., Abramovich H. and Shulepov O. (1995) Dynamic Stiffness Matrix for Laminated Beams using a First Order Shear Deformation Theory, *Composite Structures*, **31**, 265–271.

Eisner E. (1966) Inverse Design for Flexural Vibrations, *Journal of the Acoustical Society of America*, **40(4)**.

El-Nashie M.S. and Al Athel S. (1977) A Simple Finite Element Mechanical Model for the Numerical Estimation of Buckling Loads, *International Journal of Mechanical Engineering Education*, **5**, 295–305.

El-Nashie M.S. and Hussein A. (2000) On the Eigenvalue of Nuclear Reaction and Self-Weight Buckling, *Chaos, Solitons and Fractals*, **11(5)**, 815–818.

Elhay S. and Ram Y.M. (2002) An Affine Inverse Eigenvalue Problem, *Inverse Problems*, **18**, 455–466.

Elishakoff I. (1979) Buckling of Stochastically Imperfect Finite Column on a Nonlinear Elastic Foundation — a Reliability Study, *Journal of Applied Mechanics*, **46**, 411–416.

Elishakoff I. (1980) Hoff's Problem in a Probabilistic Setting, *Journal of Applied Mechanics*, **47**, 403–408.

Elishakoff I. (1987a) A Remark on the Adjustable Parameter Versions of the Rayleigh's Method, *Journal of Sound and Vibration*, **118**, 163–165.

Elishakoff I. (1987b) A Variant of the Rayleigh or Galerkin Methods with Variable Parameter as a Multiplier, *Journal of Sound and Vibration*, **114**, 159–163.

Elishakoff I. (1987c) Application of Bessel and Lommel Functions, and the Undetermined Multiplier Galerkin Method Version, for Instability of Non-Uniform Column, *Journal of Sound and Vibration*, **115(1)**, 182–186.

Elishakoff I. (1999) *Probabilistic Theory of Structures*, pp. 333–339, Dover Publications (second edition), New York (first edition, Wiley, 1983).

Elishakoff I. (2000a) Closed-Form Solution for the Generalized Euler's Problem, *Proceedings of the Royal Society of London*, **496**, No. 2002, 2409–2417.

Elishakoff I. (2000b) Axisymmetric Vibration of Inhomogeneous Clamped Circular Plates: An Unusual Closed-Form Solution, *Journal of Sound and Vibration*, **233(4)**, 727–738.

Elishakoff I. (2000c) Axisymmetric Vibration of Inhomogeneous Free Circular Plates: An Unusual Closed-Form Solution, *Journal of Sound and Vibration*, **234(1)**, 167–170.

Elishakoff I. (2001d) Inverse Buckling Problem for Inhomogeneous Columns, *International Journal of Solids and Structures*, **38(3)**, 457–464.

Elishakoff I. (2000e) Both Static Deflection and Vibration Mode of Uniform Beam Can Serve As a Buckling Mode of a Non-Uniform Column, *Sound and Vibration*, **232(2)**, 477–489.

Elishakoff I. (2000f) Resurrection of the Method of Successive Approximations to Yield Closed-Form Solutions for Vibrating Inhomogeneous Beams, *Journal of Sound and Vibration*, **234(2)**, 349–362.

Elishakoff I. (2000g) Axisymmetric Vibration of Inhomogeneous Free Circular Plates: An Unusual Exact, Closed-Form Solution, *Journal of Sound and Vibration*, **234(1)**, 167–170.

Elishakoff I. (2000h) Euler's Problem Reconsidered — 222 Years Later, *Meccanica*, **35**, 375–380.

Elishakoff I. (2000i) A Selective Review of Direct, Semi-Inverse and Inverse Eigenvalue Problems for Structures Described by Differential Equations with Variable Coefficients, *Archive of Computational Methods in Engineering*, **7(4)**, 451–526.

Elishakoff I. (2001a) Some Unexpected Results in Vibration of Non-Homogeneous Beams on Elastic Foundation, *Chaos, Solitons and Fractals*, **12**, 2177–2218.

Elishakoff I. (2001b) Inverse Buckling Problem for Inhomogeneous Columns, *International Journal of Solids and Structures*, **38(3)**, 457–464.

Elishakoff I. (2001c) Apparently First Closed-Form Solution for Frequency of Beam with Rotational Spring, *AIAA Journal*, **39**, 183–186.

Elishakoff I. (2002) A Closed-Form Solution for the Generalized Euler Problem, *Proceedings of the Royal Society of London*, **456**, 2409–2417.

Elishakoff I. and Becquet R. (2000a) Closed-Form Solutions for Natural Frequency for Inhomogeneous Beams with One Sliding Support and the Other Clamped, *Journal of Sound and Vibration*, **238(3)**, 540–546.

Elishakoff I. and Becquet R. (2000b) Closed-Form Solutions for Natural Frequency for Inhomogeneous Beams with One Sliding Support and the Other Pinned, *Journal of Sound and Vibration*, **238(3)**, 529–539.

Elishakoff I. and Bert C.W. (1988) Comparison of Rayleigh's Noninteger-Power Method with Rayleigh–Ritz Method, *Computer Methods in Applied Mechanics and Engineering*, **67**, 297–309.

Elishakoff I. and Candan S. (2000) Infinite Number of Closed-Form Solutions Exist for Frequencies and Reliabilities of Stochastically Nonhomogeneous Beams, *Applications of Probability and Statistics* (Melchers R. and Stewart S., ed.), Balkema Publishers, Rotterdam.

Elishakoff I. and Candan S. (2001) Apparently First Closed-Form Solution for Vibrating Inhomogeneous Beams, *International Journal of Solids and Structures*, **38**, 3411–3441.

Elishakoff I. and Endres J. (2004) Extension of the Euler's Problem to Axially Graded Columns: 260 Years Later, *Journal of Intelligent Material Systems and Structures* (to appear).

Elishakoff I. and Guédé Z. (2001) A Remarkable Nature of the Effect of Boundary Conditions on Closed-Form Solutions for Vibrating Inhomogeneous Bernoulli–Euler Beams, *Chaos, Solitons and Fractals*, **12**, 659–704.

Elishakoff I. and Guédé Z. (2001) Novel Closed-Form Solutions in Buckling of Inhomogeneous Columns under Distributed Variable Loading, *Chaos, Solitons and Fractals*, **12**, 1075–1089.

Elishakoff I. and Guédé Z. (2004) Analytical Polynomial Solutions for Vibrating Axially Graded Beams, *Mechanics of Advanced Materials and Structures — An International Journal*, **11(6)**, 517–533.

Elishakoff I. and Lee L. (1986) On Equivalence of the Galerkin and Fourier Series Methods for One Class of Problems, *Journal of Sound and Vibration*, **109**, 174–177.

Elishakoff I. and Meyer D. (2004) Inverse Vibration Problem for Inhomogeneous Circular Plate with Translational Spring, *Journal of Sound and Vibration* (to appear).

Elishakoff I. and Pellegrini F. (1987a) Application of Bessel and Lommel Functions, and the Undetermined Multiplier Galerkin Method Version, for Stability of Non-Uniform Columns, *Journal of Sound and Vibration*, **115**, 182–186.

Elishakoff I. and Pellegrini F. (1987b) Exact and Effective Approximate Solutions of Some Divergent-Type Non-Conservative Problems, *Journal of Sound and Vibration*, **114(1)**, 143–147.

Elishakoff I. and Rollot O. (1999) New Closed-Form Solutions for Buckling of a Variable Stiffness Column by Mathematica®, *Journal of Sound and Vibration*, **224(1)**, 172–182.

Elishakoff I. and Storch J. (2004) An Unusual Exact, Closed-Form Solution for Axisymmetric Vibration of Inhomogeneous Simply Supported Circular Plates *Journal of Sound and Vibration* (to appear).

Elishakoff., Baruch M. and Becquet R. (2001b) Turning Around a Method of Successive Iterations to Yield Closed-Form Solutions for Vibrating Inhomogeneous Bars, *Meccanica*, **36**, 573–586.

Elishakoff., Baruch M. and Becquet R. (2002) A New Twist of the Method of Successive Approximations to Yield Closed-Form Solutions for Inhomogeneous Vibrating Beams by Integral Method, *Meccanica*, **37**, 142–166.

Elishakoff I., Gentilini C. and Santoro R. (2004b) Conventional and Non-Conventional Problems for the Strength of Materials Course in a Single Package (to appear).

Elishakoff I., Li Y.W. and Starnes J.H. Jr. (2001a) *Non-Classical Problems in the Theory of Elastic Stability*, Cambridge University Press, New York.

Elishakoff I., Ren Y.J. and Shinozuka M. (1995) Some Exact Solutions for Bending of Beams with Spatially Stochastic Stiffness, *International Journal of Solids and Structures*, **32**, 2315–2327 (Corrigendum: **33**, p. 3491, 1996).

Elishakoff I., Ruta G. and Stavsky Y. (2004a) Semi-Inverse Method for Buckling of Radially Graded Circular Plates (submitted for publication).

Elishakoff I. and Tang J. (1988) Buckling of Polar Orthotropic Circular Plates on Elastic Foundation by Computerized Symbolic Algebra, *Computer Methods in Applied Mechanics and Engineering*, **68**, 229–247.

Engelhardt H. (1954) Die einheitliche Behandlung der Stabknickung mit Berücksichtung des Stabeigengewichte in den Eulerfällen 1 bis 4 als Eigenwertproblem, *Der Stahlbau*, **23**(4), 80–84 (in German).

Engesser F. (1893) Über die Berechnung auf Knickfestigkeit beanspruchten Stabe aus Schweiss — und Guseisen, *Zeitschrift Oesterreichisches Ing. und Archit. Ver.*, **45**, 506–508 (in German).

Engesser F. (1889) Über Knickfestigkeit gerader Stäbe, *Zeitschrift Arch. — und Ingen. in Hannover*, **35** (in German).

Engesser F. (1909) Über die Knickfestigkeit von Stäben veränderlichen Trägheitsmomentes, *Zeitschrift der Österreichischer Ingenieur und Architekten. Verein.*, Nr. **34**, 506–508, (in German).

Erdelyi A. (ed.) (1953) *Higher Transcendental Functions*, **20**, McGraw-Hill, New York.

Erdelyi A., Magnus W., Oberhethinger F. and Tricomi F.G. (1954) *Higher Transcendental Functions*, McGraw-Hill, New York.

Erdogan F. (1995) Fracture Mechanics of Functionally Graded Materials, *Composites Engineering*, **5**, 753–770.

Ericksen J.L. (1977) Semi-Inverse Methods in Finite Elasticity Theory, *Finite Elasticity*, AMD, **27**, pp. 11–21, ASME Press, New York.

Ermopoulos J. Ch. (1986) Buckling of Tapered Bars under Stepped Axial Loads, *Journal of Structural Engineering*, **112**(6), 1346–1354.

Erugin N.P. (1952) Construction of the Entire Set of System of Differential Equations, that Have a Given Integral Curve, *Journal of Applied Mechanics and Mathematics — PMM*, **16**(6), 659–670.

Euler L. (1744) Methodus Inveniendi Lineas Curvas Maximi Minimive Proprietate Gaudentes, Marcum Michaelem Bousquet, Lausanne and Geneva (in Latin).

Euler L. (1759a) On the Strength of Columns, *Acad. Roy. Sci. Belles Lettres Berlin Méns.*, **13**, 252 (English translation by van den Broek, J.A. (1997) *J. Phys.*, **15**, 315–318).

Euler L. (1759b) Sur la force des colonnes, *Memoires de L'Academie des Sciences et Belles-Lettres* (Berlin), **13**, 252–282 (in French).

Euler L. (1778a) Determinatio onerum, quae columnae gestare valent, *Acta Academiae Scientiarum Petropolitanae*, **1**, 121–145 (see also *Leonhardi Euleri Opera Omnia*, **17**, Part 2, pp. 232–251, Bern 1982) (in Latin).

Euler L. (1778b) Examen insignio puradoxi in theoria columnarum occurentio, *Acta Academiae Scientiarum Petropolitanae*, **1**, 146–162 (see also *Leonhardi Euleri Opera Omnia*, **17**, Part 2, pp. 252–265, Bern 1982) (in Latin).

Euler L. (1778c) De altitudine columnarum sub propio pondere corruentium, *Acta Academiae Scientiarum Petropolitanae*, **1**, 163–193 (see also *Leonhardi Euleri Opera Omnia*, **17**, Part 2, pp. 266–293, Bern 1982) (in Latin).

Even C. and Pieranski P. (1999) On "Hearing the Shape of Drums": An Experimental Study using Vibrating Smetic Films, *Europhysics Letters*, **47**(5), 531–537.

Ewins D.J. and Inman D.J. (2000) *Structural Dynamics @ 2000, Current Status and Future Directions*, Research Studies Press Ltd., Baldock.

Fadle J. (1962a) Eine Anwendung des Verfahrens der schrittweisen Naherung auf den Knickstab mit stufenweise veraenderlichem Querschmitt, *Ingenieur-Archiv*, **32**, 26–32 (in German).

Fadle J. (1962b) Eine Naherungsloesung fuer die Knicklast eines Staebes, dessen Druckkraft und dessen Tragheitsmoment sich nach Potenzgezetzen andern, *Zeitschrift für angewandte Mathematik und Mechanik*, **42(6)**, 257–259 (in German).

Fadle J. (1963) Eine Anwendung der Methode der Problemumkehrung auf die Knickung mit geradem Staebes mit veraenderlichem Querschmitt, *Ingenieur-Archiv*, **32**, 373–387 (in German).

Fang T., Leng X.L. and Song C.Q. (2003) Chebyshev Polynomial Approximation for Dynamical Response Problem of Random System, *Journal of Sound and Vibration*, **266**, 198–206.

Fauconneau G. and Laird W.M. (1967) Bounds for the Natural Frequencies of a Simply Supported Beam Carrying a Uniformly Distributed Axial Load, *Journal of Mechanical Engineering Science*, **9**, 149–155.

Feldman F. and Aboudi J. (1997) Buckling Analysis of Functionally Graded Plates Subjected to Uniaxial Loading, *Composite Structures*, **38**, 29–36.

Felipich C. and Rosales M.B. (2002), A Further Study about the Behaviour of Foundation Piles and Beams on a Winkler–Pasternak Soil, *International Journal of Mechanical Sciences*, **44**, 21–36.

Feodosiev V.I. (1996) *Selected Problems and Questions in Strength of Materials*, pp. 39, 190–191 Nauka Publishers, Moscow (in Russian).

Fettis H.E. (1952) Torsional Vibration Modes of Tapered Bars, *Journal of Applied Mechanics*, **19(2)**, 220–222.

Filipich C.P. (1989) Transverse Vibrations of a Tapered Beam Embedded in a Non-Homogeneous Winkler Foundation, *Applied Acoustics*, **26(1)**, 67–72.

Filipich C.P., Laura P.A.A., Sonenblum M. and Gil E. (1988) Transverse Vibrations of a Stepped Beam Subjected to an Axial Force and Embedded in a Non-Homogeneous Winkler Foundation, *Journal of Sound and Vibration*, **126**, 1–8.

Finlayson B.A. (1972) *The Method of Weighted Residuals and Variational Principles with Application in Fluid Mechanics, Heat and Mass Transfer*, Academic Press, New York.

Flavin J.N. (1976) A Bound for the Deflection of a Beam-Column,*Journal of Applied Mechanics*, **43(1)**, 181–182.

Flügge S. (1965) Section 59, Semi-inverse Methods: Reduction to Certain Static Deformation that Depend Upon Material Properties, in *Encyclopedia of Physics*, **III/3**, *The Non-Linear Field Theories of Mechanics*, pp. 199–203.

Forys A.S. (1996) Optimization of an Axially Loaded Beam on a Foundation, *Journal of Sound and Vibration*, **178**, 607–613.

Forys A.S. (1997) Two-Parameters Optimization of an Axially Loaded Beam on a Foundation, *Journal of Sound and Vibration*, **199**, 522–530.

Fourment L. and Chenot J.L. (2000) Inverse Methods Applied to Metal Forming Processes, *Computational Mechanics for the Twenty-First Century* (Topping B.H.V., ed.), pp. 127–144, Saxe-Colurg Publications, Edinburgh, UK.

Franciosi C. and Masi A. (1993) Free Vibrations of Foundation Beams on Two-Parameter Elastic Soil, *Computers and Structures*, **47**, 419–426.

Fraser W.B. and Budiansky B. (1969) The Buckling of Column with Random Initial Deflection, *Journal of Applied Mechanics*, **36**, 232–240.

Freudenthal A.M. (1966) *Introduction to the Mechanics of Solids*, Wiley, New York.

Fried I. (1971) Accuracy of Finite Element Eigenproblems, *Journal of Sound and Vibration*, **18**, 289–295.

Fried I. (1979) *Numerical Solution of Differential Equations*, p. 3, Academic Press, New York.

Friedland S. (1975) On Inverse Multiplicative Eigenvalue Problems for Matrices, *Linear Algebr. Appl.*, **12**, 127–137.

Friedland S. (1977) Inverse Eigenvalue Problems, *Linear Algebra App.*, **17**, 15–51.

Friedland S., Nocedal J. and Overton M.L. (1987) The Formulation and Analysis of Numerical Methods for Inverse Eigenvalue Problems, *SIAM J. Numerical Analysis*, **24**, 634–667.

Frisch-Fay R. (1966) On the Stability of a Strut Under Uniformly Distributed Axial Forces, *International Journal of Solids and Structures*, **2**, 361–369.

Frisch-Fay R. (1980) Buckling of Masonry Pier under Its Own Weight, *International Journal of Solids and Structures*, **16**, 445–450.

Friswell M.J. and Mottershead J.E. (1995) *Finite Element Model Updating Structural Dynamics*, Kluwer Academic Publishers, Dordrecht, The Netherlands.

Fukui Y. and Yamanaka N. (1992) Elastic Analysis for Thick-Walled Tubes of Functionally Graded Material, *JSME International Journal, Ser. I: Solid Mechanics, Strength of Materials*, **35(4)**, 379–385.

Fulcher L.P. (1985) Study of Eigenvalues of Nonuniform Strings, *American Journal of Physics*, **53**, 730–735.

Gaines J.H. and Volterra E. (1966) Transverse Vibrations of Cantilever Bars of Variable Cross-section, *Journal of Accoustical Society of America*, **39**, 674–679.

Gajewski A. (1981) A Note on Unimodal and Bimodal Optimal Design of Vibrating Compressed Columns, *International of Journal Mechanical Sciences*, **23(1)**, 11–16.

Gajewski A. (1985) Bimodal Optimisation of a Column in an Elastic Medium, with Respect to Buckling and Vibration, *International Journal of Mechanical Sciences*, **27(1–2)**, 45–53.

Gajewski A. and Palej R. (1974) Stability and Shape Optimization of an Elastically Clamped Bar Under Tension, *Rozprawy Inżynierskie*, **22(2)**, 265–279 (in Polish).

Gajewski A. and Życzkowski M. (1965a) Calculation of Elastic Stability of Circular Plates with Variable Thickness by an Inverse Method, *Bulletin de L'Academic Polonaise des Sciences*, **14(5)**, 303–312.

Gajewski A. and Życzkowski M. (1965b) Computation of the Elastic and Elastic–Plastic Stability of Circular Plates with Variable Rigidity by Means of the Inverse Method, *Rozprawy Inżynierskie*, **3(13)**, 582–622 (in Polish).

Gajewski A. and Życzkowski M. (1966) Elastic–Plastic Buckling of Some Circular and Annular Plates of Variable Thickness, *Bulletin de L'Academic Polonaise des Sciences*, **14(5)**, 313–320.

Gajewski A. and Życzkowski M. (1969) Optimum Shaping of a Bar Compressed by a Pole-Oriented Force, *Rozprawy Inżynierskie*, **17(2)**, 299–329 (in Polish).

Gajewski A. and Życzkowski M. (1970a) Optimum Design of Elastic Columns Subjected to a General Conservative Behavior of Loading, *Zeitschrift für angewandte Mathematik und Physik*, **21(5)**, 806–818.

Gajewski A. and Życzkowski M. (1970b) Influence of Longitudinal Non-Homogeneity on the Optimal Shape of a Bar Compressed by Force Pointing Towards the Pole, *Bulletin de L'Academic Polonaise des Sciences*, **18(1)**, 19–27.

Gajewski A. and Życzkowski M. (1988) *Optimal Structural Design under Stability Constraints*, Kluwer Academic Publishers, Dordrecht.

Galea Y. (1981) Flambement des poteaux a inertie variable, *Construct. Metall.*, **1**, 21–46 (in French).

Galef A.E. (1968) Bending Frequencies of Compressed Beams, *Journal of Acoustical Society of America*, **44(8)**, 643.

Galerkin B.G. (1915a) Beams and Plates, *Vestnick Inzhenerov*, **1(19)**, 897–908.

Galerkin B.G. (1915b) Rods and Plates, Series Occuring in Some Problems of Elastic Equilibrium of Rods and Plates, *Vestnik Inzhenerov i Tekhnikov*, **19**, 897–908 (English translation: 63-18294 Clearinghouse Fed. Sci. Tech. Info.)

Galerkin B.G. (1952) *Collection of Works*, **1**, pp. 168–195, Academy of USSR, Moscow (in Russian).

Galiullin A.S. (1981) *The Inverse Problem in Dynamics*, Nauka Publishers, Moscow (in Russian).

Gallagher R. and Lee C. (1970) Matrix Dynamic and Instability Analysis with Non-Uniform Elements, *International Journal for Numerical Methods in Engineering*, **2**, 265–275.

Gallagher R.H. and Padlog J. (1963) Discrete Element Approach to Structural Instability Analysis, *AIAA Journal*, **1**, 1437–1439.

Galletly G.D. (1959) Circular Plates on a Generalized Elastic Foundation, *Journal of Applied Mechanics*, **81**, 297–301.

Gantmakher F.P. and Krein M.G. (1950) *Oscillation Matrices and Kernels and Small Vibrations of Mechanical Systems*, GITTL Publishing, Moscow (in Russian) (English Translation by the U.S. Atomic Energy Commission, Washington D. C., 1961).

Gao Y. and Randall R.B. (1999) Reconstruction of Diesel Engine Cylinder Pressure Using a Time-Domain Smoothing Technique, *Mechanical Systems and Signal Processing*, **13(5)**, 709–722.

Gatewood B.E. (1954) Buckling Loads for Columns with Variable Cross-Sections, *Journal of Aerospace Sciences*, 287–288.

Gel'fand I.M. and Levitan B.M. (1951) On the Determination of a Differential Equation from Its Spectral Function, *Izvestiya AN SSSR, Ser. Mathematika*, **15**, 309–360 (in Russian) (English Translation, *Am. Math. Soc. Translations Ser 2.*, **1**, 253–304).

Gellert M. and Glück J. (1972) The Influence of Axial Load on Eigenfrequencies of a Vibrating Lateral Restraint Cantilever, *International Journal of Mechanical Sciences*, **14**, 723–728.

Gere J.M. (1954) Torsional Vibrations of Beams of Thin Walled Open Cross Section, *Journal of Applied Mechanics*, **21**, 381–387.

Gere J.M. and Carter W.O. (1962) Critical Buckling Loads for Tapered Columns, *Journal of Structural Division*, **88**, 1–11.

Gerver M.L. (1970) The Inverse Problem for the Vibrating String Equation, *Izvestiya AN SSSR, Phys. Solid Earth*, No. **8**, 463–471.

Ghani Razaqpur A. and Shah K.R. (1991) Exact Analysis of Beams on Two-Parameter Elastic Foundations, *International Journal of Solids and Structures*, **27**, 435–454.

Girijavallabhan C.V (1969) Buckling Loads of Nonuniform Columns, *Journal of Structural Division*, **95**, 2419–2431.

Gladwell G.M.L. (1984a) The Inverse Problem for the Euler–Bernoulli Beam, *Inverse Problems of Acoustic and Elastic Waves* (Santosa F., Pao Y.H., Symes W.W. and Holland C., eds.), pp. 346–347, SIAM, Philadelphia.

Gladwell G.M.L. (1984b) The Inverse Problem for the Vibrating Beam, *Proceedings, Royal Society of London, Series A*, **393**, 277–295.

Gladwell G.M.L. (1986a) *Inverse Problems in Vibration*, Martinus Nijhoff Publishers, Dordrecht.

Gladwell G.M.L. (1986b) Inverse Problems in Vibration, *Applied Mechanics Reviews*, **39(7)**, 1013–1018.

Gladwell G.M.L. (1986c) Inverse Mode Problem for Lumped-Mass Systems, *Quarterly Journal of Mechanics and Applied Mathematics*, **39**, 297–307.

Gladwell G.M.L (1991) Application of Schur's Algorithm to an Inverse Eigenvalue Problem, *Inverse Problem*, **7**, 557–565.

Gladwell G.M.L. (1995a) On Isospectral Spring-Mass System, *Inverse Problems*, **11**, 591–602.

Gladwell G.M.L. (1995b) Inverse Problems for Finite Element Models, *Inverse Problems*, **13**, 311–322.

Gladwell G.M.L. (1996) Inverse Problems in Vibration II, *Applied Mechanics Reviews*, **49(10)**, Part 2, S25–S33.

Gladwell G.M.L. (1998) Total Positivity and the QR Algorithm, *Linear Algebra and Applications*, **271**, 257–271.

Gladwell G.M.L. (1999a) Inverse Finite Element Vibration Problems, *Journal of Sound and Vibration*, **221(2)**, 309–324.

Gladwell G.M.L. (1999b) Inverse Problems in Vibration: A Personal Story, *Euromech 401: Inverse Methods in Structural Dynamics*, Abstracts (Mottershead J.E. and Friswell M.I., chairmen), The University of Liverpool, 6–8 September.

Gladwell G.M.L. (2000a) Private Communication, February 1.

Gladwell G.M.L. (2000b) Private Communication, May 1.

Gladwell G.M.L. (2002a) Isospectral Systems, *Prikladnaya Mekhanika*, **38(5)**, 3–11 (in Russian).

Gladwell G.M.L. (2002b) Isospectral Vibrating Beams, *Proceedings of the Royal Society of London*, **458(2002)**, 2691–2703.

Gladwell G.M.L. and Coen S. (1981) Inverse Problem in Elastostatics, *IMA Journal of Applied Mathematic*, **27(4)**, 407–421.

Gladwell G.M.L. and Dods S.R.A. (1987) Examples of Reconstruction of an Euler–Bernoulli Beam from Spectral Data, *Journal of Sound and Vibration*, **119**, 81–94.

Gladwell G.M.L. and Gbadeyan J.A. (1985) On the Inverse Problem of the Vibrating String or Rod, *Quarterly Journal of Mechanics and Applied Mathematics*, **38**, 169–175.

Gladwell G.M.L. and Morassi A. (1995) On Isospectral Rods, Beams and Strings, *Inverse Problems*, **11**, 533–554.

Gladwell G.M.L. and Morassi A. (1999) Estimating Damage in a Rod from Change in Node Positions, *Inverse Problems in Engineering*, **7(3)**, 215–233.

Gladwell G.M.L. and Movashedy M. (1995) Reconstruction of a Mass-Spring System from Spectral Data — I: Theory, *Inverse Problems in Engineering*, **1**, 179–189.

Gladwell G.M.L., England A.H. and Wang D. (1987) Examples of Reconstruction of an Euler–Bernoulli Beam from Spectral Data, *Journal of Sound and Vibration*, **119**, 81–94.

Gladwell G.M.L., Willems N.B., He B. and Wang D. (1989) How Can We Recognize an Acceptable Mode Shape for a Vibrating Beam? *Quarterly Journal of Mechanics and Applied Mathematics*, **42**, 303–316.

Glück J. (1973) Vibration Frequencies of an Elastically Supported Cantilever Column with Variable Cross-Section and Distributed Axial Load, *Earthquake Engineering and Structural Dynamics*, **1(4)**, 371–376.

Glushkov G.S. (1939) Stress Determination in Rotating Disks Having Different Elastic Properties in Two Directions, *Trudy Moskovskogo Stanko-Instrumentalnogo Instituta*, **3** (in Russian).

Goel R.P. (1974) Transverse Vibrations of Tapered Beams, *Journal of Sound and Vibration*, **47**, 1–7.

Goel R.P. (1976) Transverse Vibrations of Tapered Beams, *Journal of Sound and Vibration*, **47**, 1–7.

Golecki J. (1959) On the Foundations of the Theory of Elasticity of Plane Incompressible Non-Homogeneous Bodies, *Non-Homogeneity in Elasticity and Plasticity* (Olszak W., ed.), Pergamon Press.

Gonga Rao H.V.S. and Spyrakos C.C. (1986) Closed-Form Series Solutions of Boundary Value Problems with Variable Properties, *Computers and Structures*, **3**, 211–215.

Gontkevich V.S. (1964)*Free Vibrations of Plates and Shells*, "Naukova Dumka" Publishing, Kiev (in Russian).

Gorbunov B.N. (1936) Analysis of Stability of Columns and Arches by the Method of Successive Approximations, *Issledovaniya no Teorii Sooruzhenii (Investigations in the Theory of Structures)*, "Gosstroyizdat" Publishers, Moscow (in Russian).

Gordon C. (1989) When You Can't Hear the Shape of a Minifold, *Mathematical Intelligencer*, **11**, 39–47.

Gordon C., Webb D. and Wolpert S. (1992a) Isospectral Plane Domains and Surfaces via Riemannian Orbifolds, *Inventiones. Mathematicae*, **110**, 1–22.

Gordon C., Webb D. and Wolpert S. (1992b) One Can't Hear the Shape of a Drum, *Bulletin of the American Mathematical Society*, **27**, 134–138.

Gorman D.J. (1974) *Free Vibration Analysis of Beams and Shafts*, Wiley, New York.

Gorman D.G. (1983) Natural Frequencies of Transverse Vibration of Polar Orthotropic Variable Thickness Annular Plate, *Journal of Sound and Vibration*, **86**, 47–60.

Gorr G.V. and Zyra A.V. (1998) Polynomial Solutions in a Problem of the Motion of a Gyrostat with a Fixed Point, *Mechanics of Solids*, **33(6)**, 7–14.

Gottlieb H.P.W. (1985) Exact Vibration Solutions for some Irregularly Shaped Membranes and Simply Supported Plates, *Journal of Sound and Vibration*, **103**, 333–339.

Gottlieb H.P.W. (1986) Historical Comment on the Modification of the Galerkin Method, *Journal of the Acoustical Society of America*, **80**, 348–349.

Gottlieb H.P.W. (1987) Isospectral Euler–Bernoulli Beams with Continuous Density and Rigidity Functions, *Proceedings of the Royal Society of London A*, **413**, 235–250.

Gottlieb H.P.W. (1988) Eigenvalues of the Laplacian for Rectilinear Regions, *Journal of the Australian Mathematical Society Series B*, **29**, 270–281.

Gottlieb H.P.W. (1989) On Standard Eigenvalues of Variable-Coefficient Heat and Rod Equations, *Journal of Applied Mechanics*, **56**, 146–148.

Gottlieb H.P.W. (1991) Inhomogeneous Clamped Circular Plates with Standard Vibration Spectra, *Journal of Applied Mechanics*, **58(3)**, 729–730.

Gottlieb H.P.W. (1992) Axisymmetric Isospectral Annular Plates and Membranes. *IMA Journal of Applied Mechanics*, **49**, 182–192.

Gottlieb H.P.W. (1993) Inhomogeneous Annular Plates with Exactly Beam-Like Radial Spectra, *IMA Journal of Applied Mathematics*, **50**, 107–112.

Gottlieb H.P.W. (2000a) Comments on "On 'Hearing the Shape of the Drums': An Experimental Study using Vibrating Smectic Films," *Europhysics Letters*, **50(2)**, 280–281.

Gottlieb H.P.W. (2000b) Personal Communication.

Gottlieb H.P.W. (2002) Isospectral Strings, *Inverse Problems*, **18**, 971–978.

Gottlieb H.P.W. and McManus J.P. (1998) Exact Shared Modal Functions and Frequencies for Fixed and Free Isospectral Membrane Shapes Formed from Triangles, *Journal of Sound and Vibration*, **212(2)**, 253–284.

Graf K.F. (1975) *Wave Motion in Elastic Solids*, Ohio University Press, Columbus, Ohio.

Gran Olson R. (1942a) Elastische Knickung gerader Stäbe von exponentiell veränderlichem Querschnitt unter dem Einfluss ihres Eigengewichtes, *Ingenieur-Archiv*, **13**, 162–174 (in German).

Gran Olson R. (1942b) Elastische Knickung gerader Stäbe, die als Saeulen von konstanter Druckspannung ausgebildet sind, *Ingenieur-Archiv*, 162 (in German).

Granger S. and Perotin L. (1999a) An Inverse Method for the Indentification of a Distributed Random Excitation Action on a Vibrating Structure. Part 1 Theory, *Mechanical Systems and Signal Processing*, **13(1)**, 53–65.

Granger S. and Perotin L. (1999b) An Inverse Method for the Determination of a Distributed Random Excitation Acting as a Vibrating Structure, *Mechanical Systems and Signal Processing*, **13(1)**, Part I: 53–65; Part II: 67–81.

Grediac M. (1996) On the Direct Determination of Invariant Parameters Governing Anisotropic Plate Bending Problems, *International Journal of Solids Structures*, **33(27)**, 3969–3982.

Grediac M. and Paris P-A. (1996) Direct Identification of Elastic Constants of Anisotropic Plates by Model Analysis: Theoretical and Numerical Aspects, *Journal of Sound and Vibration*, **193(3)**, 401–415.

Grediac M. and Vautrin A. (1990) A New Method for Determination of Bending Rigidities of Anisotropic Plates, *Journal of Applied Mechanics*, **57**, 964–968.

Grediac M. and Vautrin A. (1993) Mechanical Characterization of Anisotropic Plates: Experiments and Results, *European Journal of Mechanics, A/Solids*, **12(6)**, 819–838.

Green C.D. (1969) *Integral Equation Methods*, Barnes and Noble, New York.

Greenhill A.G. (1881) Determination of the Greatest Height Consistent with Stability that a Vertical Pole must or Can Be Made, and of the Greatest Height to which a Tree of Given Proportions can Grow, *Proceedings, Cambridge Philosophical Society*, **4**, 765–774.

Grigoliuk E.I. (1971) About the Boobnov's Method, *Bull. AN SSSR, Mechanics of Solids*, No. **3**, 205 (in Russian).

Grigoliuk E.I. (1975) About the Boobnov's Method. Towards Sixtieth Anniversary of Its Creation, *Investigations in the Theory of Plates and Shells* (Galimov K.Z., ed.), pp. 3–41, Kazan' University Press (in Russian).

Grigoliuk E.I. (1996) *Boobnov's Method*, pp. 1–58, Institute of Mechanics, Moscow State University (in Russian).

Grigoliuk E.I. and Kabanov V.V. (1969) *Stability of Circular Cylindrical Shells*, p.40 Itogi Nauki, VINITI Publishing, Moscow (in Russian).

Grigoliuk E.I. and Selezov I.T. (1973) *Non-Classical Theories of Vibration of Beams, Plates and Shells*, VINITI Press (in Russian).

Groetsch C.W. (1993) *Inverse Problems in the Mathematical Sciences*, View e.g., Braunschweig.

Groetsch C.W. (1999) *Inverse Problems. Activities for Undergraduates*, The Mathematical Association of America, Washington, DC.

Grossi R.O. and Bhat R.B. (1991) A Note on Vibrating Tapered Beams, *Journal of Sound and Vibration*, **147**, 174–178.

Grossi R.O. and Bhat R.B. (1995) Author's Reply, *Journal of Sound and Vibration*, **187(4)**, 727.

Gu Y., Nakamura T., Prchlik L., Sampath S. and Wallace J. (2003) Micro-Indentation and Inverse Analysis to Characterize Elastic–Plastic Graded Materials, *Materials Science and Engineering*, **345(1–2)**, 223–233.

Guédé Z. and Elishakoff I. (2001a) A Fifth Order Polynomial That Serves as Both Buckling and Vibration Mode of an Inhomogenous Structure, *Chaos, Solitons and Fractals*, **12(7)**, 1267–1298.

Guédé Z. and Elishakoff I. (2001b) Apparently First Closed-Form Solution for Inhomogeneous Vibrating Beams Under Axial Loading, *Proceedings of the Royal Society of London*, **457**, 623–649.

Gupta A.K. (1985) Vibration of Tapered Beams, *Journal of Structural Engineering*, **111**, 19–36.

Gupta K.K. and Lawson C.L. (1999) Structural Vibration Analysis by a Progressive Simultaneous Iteration Method, *Proceedings of the Royal Society of London, Series A*, **455(1989)**, 3415–3423.

Gupta R.S. and Rao S.S. (1978) Finite Element Eigenvalue Analysis of Tapered and Timoshenko Beams, *Journal of Sound and Vibration*, **56**, 187–200.

Gupta U.S. and Ansari A.H. (1998) Free Vibration of Polar Orthotropic Circular Plates of Variable Thickness with Elastically Restrained Edge, *Journal of Sound and Vibration*, **213(3)**, 429–445.

Gupta U.S., Lal R. and Verma C.P. (1986) Buckling and Vibrations of Polar Orthotropic Annular Plates of Variable Thickness, *Journal of Sound and Vibration*, **104**, 357–369.

Gürgöze M. (1999) Identifying Nodes and Anti-Nodes of a Longitudinally Vibrating Rod Restrained by a Linear Spring In-Span, *Journal of Sound and Vibration*, **219(3)**, 550–557.

Gürgöze M. and Ÿnceoðlu S. (2000) On the Vibrations of an Axially Vibrating Elastic Rod with Distributed Mass Added in Span, *Journal of Sound and Vibrations*, **230(1)**, 187–194.

Gurushankar G.V. (1975) Thermal Stresses in a Rotating, Nonhomogeneous Anisotropic Disk of Varying Thickness and Density, *Journal of Strain Analysis*, **10(3)**, 137–142.

Guttierez R.H., Laura P.A.A., Bambil J.V.A. and Hodges D. (1998) Axisymmetric Vibrations of Solid Circular and Annular Membranes with Continuously Varying Density, *Journal of Sound and Vibration*, **212**, 611–622.

Hadamard J. (1932) Le problem de Cauchy et les équations aux dérivées partielles linéaire hyperboliques, Paris (in French).

Haftka R.T., Gürdal Z. and Kamat M.P. (1990) *Elements of Structural Optimisation*. Kluwer Academic Publishers (second edition), Dordrecht.

Hald O.H. (1978) Inverse Sturm–Liouville Problem with Symmetric Potentials, *Acta Mathematica*, **141**, 263–291.

Hald O.H. (1980) Inverse Eigenvalue Problem for the Mantle, *Geophys. J.R. Astr. Soc.*, **62**, 41–48.

Hald O.H. (1983) Inverse Eigenvalue Problems for Mantle, *Numerical Treatment in Inverse Problems on Differential and Integral Equations*(Deuflhard P. and Hairer E., eds.), pp. 146–149, Birkhäuser, Boston.

Hald O.H. and McLaughlin J.R. (1989) Solutions of Inverse Nodal Problems, *Inverse Problems*, **5**, 304–347.

Hald O.H. and McLaughlin J.R. (1996) Inverse Nodal Problem: Finding the Potential from Nodal Lines, *AMS Memoir*, **119(572)**.

Hald O.H. and McLaughlin J.R. (1998a) Inverse Problems Using Nodal Position Results, Algorithms and Bounds, in *Proceedings of the Centre for Mathematical Analysis*, **17** (Anderssen R.S. and Newsam G.N., eds.), pp. 32–58, Australian National University, Special Program in Inverse Problems.

Hald O.H. and McLaughlin J.R. (1998b) Inverse Problems: Recovery of BV Coefficients from Nodes, *Inverse Problems*, **14**, 245–273.

Harris C.O. (1942) A Suggestion for Columns of Varying Sections, *Journal of Aeronautical Sciences*, **9**, 97–99.

Harris C.Z. (1968) The Normal Modes of a Circular Plate of Variable Thickness, *Quarterly Journal of Mechanics and Applied Mathematics*, **21**, 32–36,

Hart G.C. and Collins J.D. (1970) The Treatment of Randomness in Finite Element Modeling, *SAE Paper* No. 700842.

Hartmann F. (1936) Der allgemeine Fall der Knickung des geraden Baustahlstabes mit veränderlichem Querschnitt, Abh. IVBH, **4**, p. 319, Leemann, Zürich (in German).

Hauger W. (1966) Die Knicklasten elastischer Stäbe unter gleichmässig verteilten und linear veränderlichen tangentialen Druckkäften, *Ingenieur-Archiv*, **35**, 221–229 (in German).

He J.H. (1997) Semi-Inverse Method of Establishing Generalized Variational Principles for Fluid Mechanics with Emphasis on Turbomachinery Aerodynamics, *International Journal of Turbo Jet-Engines*, **14(1)**, 23–28.

He J.H. (2000) Semi-Inverse Method and General Variational Principles with Multi-Variables in Elasticity, *Applied Mathematics and Mechanics*, **21(7)**, 797–808.

He J.H. (2001) Comments on Derivation and Transformation of Variational Principles with Emphasis on Inverse and Hybrid Problems in Fluid Mechanics: A Systematic Approach, *Acta Mechanica*, **149(1–4)**, 247–249.

Heidevrecht A.C. (1967) Vibration of Non-Uniform Simply-Supported Beams, *Journal of Engineering Mechanics Division*, **93**, No. EM2, 1–15.

Heim v., J.P.G. (1838) *Über Gleichgewicht elastischer fester Körper*, Stuttgart (in German).

Henky H. (1920) Über die angenäherte Lösung von Stabilitätsproblemen im Raum mittels der elastischen Gelenkkette, *Der Eisenbau*, **7**, 437–452 (in German).

Henky H. (1927) Eine wichtige Vereinfachung der Methode von Ritz zur angenäehrten Behandlung von variations Problemen, *Zeitschrift für angewandte Mathematik und Mechanik*, **4(1)** (in German).

Hensel E. (1991) *Inverse Theory and Applications for Engineers*, Prentice Hall, Englewood Cliffs.

Hetenyi M. (1955) *Beams on Elastic Foundations*, The Michigan Press, Ann Arbor, MI.

Hochstadt H. (1965) On the Determination of a Hill's Equation from Its Spectrum, *Arch. Rational Mech. Anal.*, **19**, 353–362.

Hochstadt H. (1973) The Inverse Sturm–Liouville Problem, *Communications on Pure and Applied Mathematics*, **26**, 715–729.

Hochstadt H. (1975) Well-Posed Inverse Spectral Problem, *Proceedings of the National Academy of Sciences of the U.S.A.*, **72**, 2496–2497.

Hochstadt H. (1976a) Determination of the Density of a Vibrating String from Spectral Data, *Journal of Mathematical Analysis and Applications*, **55**, 673–685.

Hochstadt H. (1976b) An Inverse Problem for a Hill's Equation, *Journal of Differential Equations*, **20**, 55–60.

Hochstadt H. and Lieberman B. (1978) An Inverse Sturm–Liouville Problem with Mixed Given Data, *SIAM Journal of Applied Mathematics*, **34**, 676–680.

Hodges D.H. (1997) Improved Approximations via Rayleigh's Quotient, *Journal of Sound and Vibration*, **199(1)**, 155–164.

Hoff N.Y. and Soong T.C. (1965) Buckling of Circular Cylindrical Shells in Axial Compression, *International Journal of Mechanical Sciences*, **7**, 489–520.

Hohenemser K. and Prager W. (1933) *Dynamik der Stabwerke (Dynamics of Structures)*, pp. 107–110, Springer Verlag, Berlin (in German).

Hollandsworth P.E. and Busby H.R. (1989) Impact Force Identification using the General Inverse Technique, *International Journal of Impact Engineering*, **8(4)**, 315–322.

Horgan C.O. (2000) Private Communication, March 21.

Horgan C.O. and Chan A.M. (1999a) Vibration of Inhomogeneous Strings, Rods and Membranes, *Journal of Sound and Vibration*, **225**, 503–513.

Horgan C.O. and Chan A.M. (1999b) The Pressurized Hollow Cylinder or Disk Problem for Functionally Graded Isotropic Linearly Elastic Material, *Journal of Elasticity*, **55**, 43–59.

Horgan C.O. and Villaggio P. (1996) A Semi-Inverse Shape Optimization Problem in Linear Anti-Plane Shear, *Journal of Elasticity*, **45**, 53–60.

Houbolt J.C. and Andreson R.A. (1948) Calculations of Uncoupled Modes and Frequencies in Bending on Torsion of Nonuniform Beams, *NACA TN — 1552*.

Houmat A. (1994) Flexural and Longitudinal Vibrations of Beams by Boundary Collocation, *European Journal of Mechanics, B/Fluids*, **13(2)**, 269–276.

Housner G.W. and Keightley W.O. (1963) Vibrations of Linearly Tapered Beams, Part 1, *Transactions of ASCE*, **128**, 1020–1048.

Hua D. (1995) About an Inverse Eigenvalue Problem Arising in Vibration Analysis, *Mathematical Modeling and Numerical Analysis*, **29(4)**, 421–434.

Huang C-H. and Shih W-Y. (1999) An Inverse Problem in Estimating Interfacial Cracks in Bimaterials by Boundary Element Technique, *International Journal for Numerical Methods in Engineering*, **45**, 1547–1567.

Huang N.C. and Sheu C.Y. (1968) Optimal Design of an Elastic Column of Thin-Walled Cross Section, *Journal of Applied Mechanics*, **35**, 285.

Hung N.V. (1968) Refinement of Energy Method of Stability Investigation of Columns Compressed by Distributed Forces, *Izvestiya VUZOV, Mashinostroenie*, No. **2**, 35–39 (in Russian).

Hutchinson J.W. and Niordson F.I. (1972) Designing Vibrating Membranes, in *Continuum Mechanics and Related Problems of Analysis*, pp. 581–590, Nauka Publishers, Moscow.

Ibrahim R.A. (1987) Structural Dynamics with Parameter Uncertainties, *Applied Mechanics Reviews*, **40(3)**, 309–328.

Imanuvilov D.Yu. and Yamamoto M. (2003) Determination of a Coefficient on an Acoustic Equation with a Single Measurement, *Inverse Problems*, **19**, 157–171.

Ince E.L. (1956) *Ordinary Differential Equations*, Dover, New York, Chapter 7.

Inman D.J. (1995) *Engineering Vibration*, Prentice Hall, Upper Saddle River.

Inoue N. (1957) A Technique for Computing Natural Frequencies of Trapezoidal Vibrations, *Journal of the Aeronautical Sciences*, **24(3)**, 239–240.

Iremonger M.J. (1980) Finite Difference Buckling Analysis of Non-Uniform Columns, *Computers and Structures*, **12(5)**, 741–748.

Irmay S. (1997) A Family of Analytical Solutions of the Navier–Stokes Equations in Space, *International Journal of Engineering Science*, **35(12/13)**, 1261–1263.

Isaacson E.L. and Trubowitz E. (1983) The Inverse Sturm–Liouville Problem, *Communication in Pure and Applied Mathematics*, Part 1, **36**, 767–787; Part 2 (1984), **37**, 1–11.

Ishida R. (1993) Proposal of Constructive Algorithms for Optimal Design and Application to the Design of the Strongest Columns, *Transactions of the JSME, Part A*, **60(579)**, 2672–2677.

Ishida R. (1994) Strongest Column Design by Constructive Algorithm, *Transactions of the JSME, Part A*, **80(579)**, 2672–2677.

Ishida R. and Sugiyama Y. (1993) Proposal of Constructive Algorithms for Optimal Design and Application to the Design of the Strongest Columns, *Transactions of the Japan Society of Mechanical Engineers*, Series A, **59(566)**, 248–253.

Issa M.S. (1990) Free Vibrations of Curved Timoshenko Beams on Pasternak Foundations, *International Journal of Solids and Structures*, **26(1)**, 1243–1252.

Iwamoto T. (1971) The Effect of End Fixity on the Stability of Structures, *Ph. D. Thesis*, Georgia Institute of Technology, Atlanta, GA.

Jacquelin E., Bennani A. and Mamelin P. (2003) Force Reconstruction: Analysis and Regularization of a Deconvolution Problem, *Journal of Sound and Vibration*, **265**, 81–107.

Jain R.K. (1972) Vibrations of Circular Plates of Variable Thickness under an Inplane Force, *Journal of Sound and Vibration*, **23(4)**, 406–414.

Jang S.K., Bert C.W. and Striz A.G. (1989) Application of DQ to Static Analysis of Structural Components, *International Journal for Numerical Methods in Pioneering*, **28(3)**, 561–577.

Janus M. and Waszczyszyn Z. (1991) Exact Finite Elements in the Static and Stability Analysis of Bar Structures, *X Polish Conference, Computer Methods in Mechanics*, 293–300.

Jaroszewicz J. and Zorij L. (1985) Free Transverse Vibrations of Cantilever Beam with Variable Cross-Section, *Eng. Trans.*, **33(4)**, 537–547.

Jaroszewicz J. and Zorij L. (1987) Effects of Variable Cross-Section of Certain Cantilever Beams and Shafts, *Eng. Trans.*, **35(3)**, 523–534.

Jaroszewicz J. and Zorij L. (1994) Analysis of Bending Curve and Critical Load of a Variable Cross-Section Beam by Means of the Influence Method, *J. Theor. Appl. Mech.*, **2(32)**, 429–437.

Jaroszewicz J. and Zorij L. (1996) Critical Euler Load for Cantilever Tapered Beam, *J. Theor. Appl. Mech.*, **4(34)**, 843–851.

Jaroszewicz J. and Zorij L. (1999) Effect of Nonhomogeneous Material Properties on Transverse Vibrations of Elastic Cantilevers, *International Applied Mechanics*, **35(6)**, 633–640.

Jasinski F.S. (1894) Recherches sur la flexion des pièces comprimées, *Annales des Ponts et Chaus.*, **7**, 233–364 (in French).

Javeheri R. and Eslami M.R. (2002) Buckling of Functionally Graded Plates Under In-Plane Compressive Loading, *Zeitschrift für angewandte Mathematik und Mechanik*, **82(4)**, 277–283.

Jin F. and Ye T.Q. (1999) Instability Analysis of Prismatic Members by Wavelet-Galerkin Method, *Advances in Engineering Software*, **30**, 361–367.

Jin Z-H. and Batra R.C. (1996) Some Basic Fracture Mechanics Concepts in Functionally Graded Materials, *Journal of Mechanics of Physics and Solids*, **44**, 1221–1235.

Joga Rao C.V. and Vijayakumar K. (1983) On Admissible Functions for Flexural Vibration and Buckling of Annular Plates, *Journal of the Aeronautical Society of India*, **15**, 1–5.

Johnson C.D. (1994) A General Solution of the Inverse-System Realization Problem for Linear Dynamical Systems with Waveform-Structured Inputs, in *Proceedings, Southeastern Symposium on System Theory*, p. 512, IEEE Computer Society Press.

Johnson C.D. (1996) Identification of Unknown, Time-Varying Forces/Moments in Dynamics and Vibration Problems Using a New Approach to Deconvolution, in *Proceedings, 67th Shock and Vibration Symposium*, Monterey, CA.

Johnson C.D. (1998) Identification of Unknown Static Load-Distributions of Beams from Static Deflection Measurements: The Inverse Problem in Beam Theory, *Developments in Theoretical and Applied Mechanics*, **19**, 159–171, Boca Raton.

Joseph K.T. (1992) Inverse Eigenvalue Problem in Structural Design, *AIAA Journal*, **30(12)**, 2890–2896.

Jubb J.E.M., Phillips I.G. and Becker H. (1975) Interrelation of Structural Stability, Stiffness Residual Stress and Natural Frequency, *Journal of Sound and Vibration*, **39(1)**, 121–134.

Justine T. and Krishnan A. (1980) Effect of Support Flexibility on Fundamental Frequency of Beams, *Journal of Sound and Vibration*, **68**, 310–312.

Kac M. (1966) Can One Hear the Shape of a Drum, *American Mathematical Monthly*, **73**, 1–23.

Kac M. (1987) *Enigmas of Chance. An Autobiography*, University of California Press, Berkeley.

Kachalov A., Kurylev Y. and Lassas M. (2001) *Inverse Boundary Spectral Problems*, Chapman & Hall/CRC, Boca Raton.

Kaipio J.P., Kolehmainen V., Voulkonen M. and Somersalo E. (1999) Inverse Problems with Structural Prior Information, *Inverse Problems*, **15**, 713–729.

Kaloni P.N. (1989) Some Remarks on "Useful Theorem for the Second Order Fluid", *Journal of Non-Newtonian Fluid Mechanics*, **31**, 115–121.

Kaloni P.N. and Huschilt K. (1984) Semi-Inverse Solutions of a Non-Newtonian Fluid, *International Journal of Non-Linear Mechanics*, **19**, 373–381.

Kameswara Rao C. and Appala Saytam A. (1975) Torsional Vibrations and Stability of Thin-Walled Beams on Continuous Elastic Foundation, *AIAA Journal*, **13**, 232–234.

Kammer D.C. and Steltzner A.D. (2000) Structural Identification using Inverse System Dynamics, *Journal of Guidance, Control, and Dynamics*, **23(5)**, 820–825.

Kammer D.C. and Steltzner A.D. (2001) Structural Identification of Mir Using Inverse System Dynamics and Mir/Shuttle Docking Data, *Journal of Vibration and Acoustics*, **123(2)**, 230–237.

Kanaka Raju K. (1977) Large Amplitude Vibrations of Circular Plates with Varying Thickness, *Journal of Sound and Vibration*, **50**, 399–403.

Kaneswara Rao C. and Mirza S. (1989) Torsional Vibrations and Stability of Thin-Walled Beams on Continuous Elastic Foundations, *Thin Walled Structures*, **7**, 73–82.

Kantorovich L.V. and Krylov V.I. (1958) *Approximate Methods of Higher Analysis*, Interscience, New York.

Kanwal R.P. (1971) *Linear Integral Equations: Theory and Technique*, Academic Press, New York.

Karabalis D.L. and Beskos D.E. (1983) Static, Dynamic and Stability Analyses of Structures Composed of Tapered Beams, *Computers and Structures*, **16(6)**, 731–748.

Karas K. (1928) Über die Knickung gerader Stäbe durch Eigengewicht and Einzellasten, *Zeitschrift für Bauwesen*, No. 9/10, 246–256 (in German).

Kardomateas G.A. (1990) Bending of a Cylindrically Orthotropic Curved Beam with Linearly Distributed Elastic Constants, *Quarterly Journal of Mechanics and Applied Mathematics*, **43(1)**, 43–55.

Karihaloo B.L. and Niordson F.I. (1973) Optimum Design of Vibrating Cantilevers, *Journal of Optimization Theory and Application*, **11**, 638–654.

Katz M.M. (1957) Free Vibrations of a Beam Carrying Equal Masses, *Issledovaniya no Teorii Sooruzhenii (Investigation in the Theory of Structures)*, **7**, Moscow (in Russian).

Kazar (Kezerashvili) M., Rand O. and Rovenskii V. (2002) Polynomial Solutions for Problems in Anisotropic Elasticity with Symbolic Computational Tools, *Annual Aeronautical Conference*, p. 74, Technion–Israel Institute of Technology.

Keller J.B. (1960) The Shape of the Strongest Column, *Arch. Rational Mech. Anal.*, **5(4)** 275–285.

Keller J.B. (1976) Inverse Problems, *American Mathematical Monthly*, **83**, 107–118.

Keller J.B. and Niordson F.I. (1966) The Tallest Column, *J. Math. and Mech.*, **16(5)**, 433–446.

Kerr A.D. (1988) Stability of Water Tower, *Ingenieur-Archiv*, **58**, 428–436.

Khvingiya M.V. (1958) Transverse Vibrations and Axial Bending of Conical Springs, Compressed by Axial Loads, *Izvestiya VUZOV, Mashinostroenie*, **3(4)**, 43–51 (in Russian).

Kicher T.P. (1969) The Analysis of Unbalanced Cross-Plied Elliptic Plates under Uniform Pressure, *Journal of Composite Materials*, **3**, 424–432.

Kiessling F. (1930) Eine Methode zur approximativen Berechnung einseitig eingespannter Druckstäbe mit veränderlichen Querchnitt, *Zeitschrift für angewandte Mathematik und Mechanik*, 594–602 (in German).

Kim C.S. and Dickinson S.M. (1988) On the Analysis of Vibrating Slender Beams Subject to Various Complicating Effects, *Journal of Sound and Vibration*, **122**, 441–455.

Kim S.K. and Daniel I.M. (2003) Solution to Inverse Heat Conduction Problem in Nanoscale using Sequential Method, *Numerical Heat Transfer, Part B*, **44**, 439–456.

Kim Y.Y. and Kapania R.K. (2003) Neural Networks for Inverse Problems in Damage Identification and Optical Imaging, *AIAA Journal*, **41(4)**, 732–740.

Kirchhoff G. (1882) *Gesammelte Abhandlungen*, Leipzig (in German).

Kirchhoff G.R. (1879) Über die Transversalschwingungen eines Stabes von veranderlichem Querschnitt, *Monatsberichte der Koniglichen Preussischen Akademie der Wissenschafter zu Berlin*, 815–828 (in German).

Kirsch A. (1991) *An Introduction to the Mathematical Theory of Inverse Problems*, Springer, Berlin.

Klein B. (1955) Rapid Estimation of the Elastic–Plastic Euler Bukling Loads for Simply Supported Tapered Columns under Varying Axial Loading, *Journal of Aeronautical Sciences*, **22(12)**, 873–874.

Klein L. (1974) Transverse Vibrations of Non-Uniform Beam, *Journal of Sound and Vibration*, **37**, 491–505.

Kliushnikov V.D. and Khvostunkov K.A. (1996) On the Stability of a Self-Gravitating Bar, *Mechanics of Solids*, **31(2)**, 179–181.

Kliushnikov V.D. and Khvostunkov K.A. (1996) Stability of Self-Gravitating Rod, *Mechanics of Solids*, **31(2)**, 157–159.

Klosowicz B. (1968) The Inhomogeneous Spherical Pressure Vessel of Maximum Rigidity, *Bull. Acad. Polonaise Sci., Ser. Sci. Techn.*, **16(7)**.

Knobel R. and Lowe B.D. (1993) Inverse Sturm–Liouville for an Impedance, *ZAMP*, **44**, 433–450.

Kochanov Yu.P. (1964) Plane Problem of Elasticity Theory for the Plate of Variable Thickness, *Inzhenernyi Zhurnal*, **4(2)** (in Russian).

Kodnar R. (1964) Die Äquivalenz der Methode Galerkins mit die Methode von Ritz für Nichtlineare Operatoren, *Zeitschrift für angewandte Mathematik und Mechanik*, **44(2)**, 579–583 (in German).

Koenig H.A. (1973) Transient Response of Non-Uniform, Non-Homogeneous Beams, *International Journal of Mechanical Sciences*, **15(5)**, 339–413.

Koiter W.T. (1996) Unrealistic Follower Forces, *Journal of Sound and Vibration*, **194**, 636–638.

Koizumi M. (1993) The Concept of FGM, in *Ceramic Transactions, Functionally Graded Materials* (Holt J.B., Koizumi M., Mirai F. and Munir Z.A., eds.), **37**, 3–10.

Koizumi M. (1997) FGM Activities in Japan, *Composites: Part B*, **28B**, 1–4.

Kolchin G.B. (1967) About Representing the Stress Function of Inhomogeneous Strip in the Form of Polynomials, *Proceedings of the Kishinev Polytechnic Institute*, **7** (in Russian).

Kolchin G.B. (1971) *Analysis of Structural Elements Made of Elastic Inhomogeneous Materials*, "Kartia Moldobeniaske" Publishers, Kishinev (in Russian).

Kolchin G.B. and Faverman E.A. (1972) *Theory of Elasticity of Inhomogeneous Bodies*, "Shtiinza" Publishers, Kishinev (in Russian).

Kollbrunner C.F. and Meister M. (1955) *Knicken*, Springer, Berlin (in German).

Kolousek V. (1954) *Dynamika Stavebnich Konstrukci (Dynamics of Building Constructions)*, Praha (in Czech) (see also Russian translation, Moscow (1965) pp. 18–19).

Korbut B.A. and Ogurtsov B.I. (1965) About the Boobnov–Galerkin Method for Systems of Equations, *Prikladnay, Mekhanika*, **1(10)**, 138–140 (in Russian).

Korenev B.G. and Rabinovich I.M. (1972) *Handbook on Structural Dynamics*, pp. 149–212, "Stroyizdat" Publishers, Moscow, (in Russian).

Kounadis A.N. (1975) Beam-Column of Varying Cross-Section under Lateral Harmonic Loads, *Ingenieur-Archiv*, **44(1)**, 43–51.

Kovalenko A.D. (1959) *Circular Plates of Variable Thickness*, "Naukova Dumka" Publishing, Kiev (in Russian).

Kovalenko A.D., Kostiuk E.D. and Vovkodav I.F. (1970) Bending of Circular Plate of Variable Thickness Along the Circumference, *Prikladnaya Mekhanika*, **6(9)**, 52–58 (in Russian).

Köylüoğlu H., Cakmak A.S., and Nielsen S.A.K. (1994) Response of Stochastically Loaded Bernoulli–Euler Beams with Randomly Varying Bending Stiffness, in *Structural Safety and Reliability* (Schuëller, G.I. Shinozuka, M. and Yao J.T.P., eds.), pp. 267–274, Bakema, Rotterdam.

Krätzig W.B. and Niemann H.J. (eds.) (1996) *Dynamics of Civil Engineering Structures*, Balkema, Rotterdam.

Krein M.G. (1933) On the Spectrum of a Jacobian Matrix, in Connection with the Torsional Oscillations of Shafts, *Matematicheskii Sbornik*, **40**, 455–466 (in Russian).

Krein M.G. (1934) On Nodes of Harmonic Oscillations of Mechanical Systems of a Certain Special Type, *Matematicheskii Sbornik*, **41(2)**, 339–348 (in Russian).

Krein M.G. (1952) About the Inverse Problems for a Non-Homogeneous String, *Doklady Akademii Nauk SSSR (Proceedings of the USSR Academy of Sciences)*, **LXXXII(5)**, 669–672 (in Russian).

Krein M.G. (1953) On Some Cases of the Effective Determination of the Density of an Inhomogeneous String by Its Spectral Function, *Doklady Akademii Nauk SSSR*, **93(4)**.

Krenk S. (1981) Theories of Elastic Plates via Orthogonal Polynomials, *Journal of Applied Mechanics*, **48**, 900–904.

Krishna Murthy A.B. and Prabhakaran K.R. (1969) Vibrations of Tapered Cantilever Beams and Shafts, *Aeronautical Quarterly*, **20**, 171–177.

Kronberg V.A. and Nudelman Ya.L. (1961) Method of Superposition in Problems of Lateral Bending With Own Weight Taken Into Account, *Izvestiya VUZOV, Stroitelstvo i Arkhitektura*, No. **4**, 12–22 (in Russian).

Krynicki E. and Mazurkiewicz Z. (1962) Free Vibration of a Simply Supported Bar with a Linearly Variable Height of Cross-Section, *Journal of Applied Mechanics*, **29**, 497–501.

Krynicki E. and Mazurkiewicz Z. (1966) *Frames Made of Variable-Rigidity Bars*, "PWN" Publishers, Warsaw (in Polish).

Kubo Sh. (ed.) (1992) *Inverse Problems*, Atlanta Technology Publications, Atlanta.

Kubo Sh., Ohji K. and Konishi K. (1992) Finite Element Based Inversion Schemes for Estimating Distributions of Elastic Constants, in *Inverse Problems* (Sh. Kubo, ed.), pp. 212–225, Atlanta Technology Publication; Atlanta.

Kumar B.M. and Sujith R.I. (1997) Exact Solutions for the Longitudinal Vibration of Non-Uniform Rods, *Journal of Sound and Vibration*, **207(5)**, 721–729.

Kunzig R. (2003) Engineering Mastery Spawned a New Kind of Beauty in Eiffel's Creation, U.S. News and World Report, June 30/July 7, pp. 54–55.

Kurdimov A.A. (1951) Application of the Method of Successive Approximations for Finding the Forms and Frequencies of Transverse Vibrations of the Ship with Shear Taken into Account, *Proceedings of Leningrad Shipbuilding Institute*, Vol. 8 (in Russian).

Labropulu F. (2000) A Few More Exact Solutions of a Second Grade Fluid via Inverse Method, *Mechanics Research Communications*, **27(6)**, 713–720.

Lagrange (1770–1773) Sur la force des colonnes, *Mélanges de Philosophie et de Matématique de la Société Royale de Turin (Miscellanea Taurenensia)*, **5**.

Lagrange (1868) *Oeuvres de Lagrange*, **2**, 125–170 (in French).

Lai E.K. and Ananthasuresh G.K. (1998) Design for Desired Mode Shapes — Preliminary Results, *ASME Design Engineering Technical Conferences*, Atlanta, Georgia, September 13–16, DET C98/DAC-5632.

Lai E. and Ananthasuresh G.K. (2002) On the Design of Bars and Beams for Desired Mode Shapes, *Journal of Sound and Vibration*, **254(2)**, 393–406.

Lal R. and Gupta U.S. (1982) Axisymmetric Vibrations of Polar Orthotropic Annular Plates of Variable Thickness, *Journal of Sound and Vibration*, **83**, 229–240.

Lamm P.K. (1993) Inverse Problems and Ill-Posedness, *Inverse Problems in Engineering: Theory and Practice* (Zabaras N. et al., eds.), pp. 1–10, ASME Press, New York.

Lancaster P. and Maroulas J. (1987) Inverse Eigenvalue Problems for Damped Vibrating Systems, *Journal of Mathematical Analysis and Application*, **123(1)**, 238–261.

Langley R.S. (1988) Application of the Principle of Detailed Balance to the Random Vibration of Non-Linear Oscillators, *Journal of Sound and Vibration*, **125(1)**, 85–92.

Lashchenkov B. Ya. (1961) Determination of the Beam's Free Vibrations Frequencies with Arbitrary Variation of Mass and Stiffness, *Proceedings of MIIT*, **134**, Moscow (in Russian).

Laura P.A.A (1985) The Computer and the Usefulness of Old Ideas, *Stock and Vibration Digest*, **18(3)**, 3–5.

Laura P.A.A. (2000) Private Communication, January 31.

Laura P.A.A. and Cortinez V.H. (1986a) Optimization of Eigenvalues in the Case of Vibrating Beams with Point Masses, *Journal of Sound and Vibration*, **108**, 346–347.

Laura P.A.A. and Cortinez V.H. (1986b) Optimization of Eigenvalues when using the Galerkin Method, *American Institute of Chemical Engineers Journal*, **32**, 1025–1026.

Laura P.A.A. and Cortinez V.H. (1987) Vibrating Beam Partially Embedded in Winkler-Type Foundation, *Journal of Engineering Mechanics*, **113**, 143–147.

Laura P.A.A. and Cortinez V.H. (1988) Rayleigh and Galerkin's Methods: Use of a Variable Parameter as a Multiplier vs Minimization with Respect to an Exponential Parameter, *Journal of Sound and Vibration*, **124**, 388–389.

Laura P.A.A., Cortinez V.H., Ercoli L. and Palluzzi V.H. (1985) Analytical and Experiment Investigation on a Vibrating Beam with Free Ends and Indeterminate Supports, *Journal of Sound and Vibration*, **102**, 595–598.

Laura P.A.A. and Guttierez R. (1985) Vibration of Nonuniform Plates on Elastic Foundation, *Journal of Engineering Mechanics*, **111(9)**, 1185–1196.

Laura P.A.A. and Guttierrez R.H. (1986) Vibrations of an Elastically Restrained Cantilever Beam of Varying Cross-Section with Tip Mass of Finite Length, *Journal of Sound and Vibration*, **108**, 123–131.

Laura P.A. and Saffell B.F., Jr. (1967) Study of Small-Amplitude Vibrations of Clamped Rectangular Plates Using Polynomial Approximations, *Journal of the Acoustical Society of America*, **41(4)**, 836–839.

Laura P.A.A., Paloto J.C. and Santos R.D. (1975) A Note on the Vibration and Stability of a Circular Plate Elastically Restrained Against Rotation, *Journal of Sound and Vibration*, **41(2)**, 177–180.

Laura P.A.A., Avalos D.R. and Galles C.D. (1982) Vibrations and Elastic Stability of Polar Orthotropic Circular Plates of Linearly Varying Thickness, *Journal of Sound and Vibration*, **82**, 151–156.

Laura P.A.A. and Cortinez V.H. (1985) Transverse Vibrations of a Cantilever Beam Carrying a Concentrated Mass at Its Free End and Subjected to a Variable Axial Force, *Journal of Sound and Vibration*, **103**, 596–599.

Laura P.A.A. et al. (1996) Free Vibrations of Beams of Bilinearly Varying Thickness, *Ocean Engineering*, **23(1)**, 1–6.

Laura P.A.A., Guttiérrez R.H. and Pistonesi C. (2001) Comments on "Natural Frequencies of Transverse Vibrations of Non-Uniform Circular and Annular Plates," *Journal of Sound and Vibration*, **230(2)**, 447–448.

Law C.K. and Tsay J. (2001) On the Well-Posedness of the Inverse Nodal Problem, *Inverse Problems*, **17**, 1493–1512.

Lee B.K. and Oh S.J. (1994) Free Vibrations and Buckling Loads of Beams-Columns on Elastic Foundations, in *Proceeding of the International Conference on Vibration Engineering* (Zheng Z.C., ed.), pp. 73–77, International Academic Publishers, Beijing.

Lee C.J. and McLaughlin J.R. (1997) Finding the Density of a Membrane with Nodal Lines, *Inverse Problems in Wave Propagation* (Chavent G. et al., eds.), Springer-Verlag, Berlin.

Lee H.C. (1963) A Generalized Minimum Principle and Its Application to the Vibration of a Wedge with Rotary Inertia and Shear, *Journal of Applied Mechanics*, **30**, 176–180.

Lee H.C. and Bishopp K.E. (1964) Application of Integral Equations to the Flexural Vibration of a Wedge with Rotary Inertia and Shear, *Journal of the Franklin Institute*, **277**(4), 327–336.

Lee S.Y. and Kuo Y.H. (1991) Deflection and Stability of Elastically Restrained Nonuniform Beam, *Journal of Engineering Mechanics*, **117**(3), 674–692.

Lee S.Y. and Kuo Y.H. (1992) Exact Solution for the Analysis of General Elastically Restrained Non-Uniform Beams, *Journal of Applied Mechanics*, **59**, 205–212.

Lee S.Y., Ke H.Y. and Kuo Y.H. (1990) Analysis of Non-Uniform Beam Vibration, *Journal of Sound and Vibration*, **142**, 15–29.

Lee T.W. (1976) Transverse Vibrations of a Tapered Beam Carrying a Concentrated Mass, *Journal of Applied Mechanics*, **43**, 366–367.

Lei Z. (1987) A Kind of Inverse Problem of Matrices and Its Numerical Solution, *Mathematica Numerica Sinica*, **9**, 431–437.

Lei Z. (1989) The Solvability Conditions for the Inverse Problem of Symmetric Nonnegative Definite Matrices, *Mathematica Numerica Sinica*, **11**, 337–343.

Leipholz H.H.E. (1967a) Über die Befreiung der Anzatzfunktionen des Ritzschen und Galerkinschen Verfahrens von den Randbedingungen, *Ingenieur-Archiv*, **36**, 251–261 (in German).

Leipholz H.H.E. (1967b) Über die Wahl der Anzatzfunktionen bei der Durchführung des Verfahrens von Galerkin, *Acta Mechanica*, **36**, 295–317.

Leipholz H.H.E. (1976) Use of Galerkin's Method for Vibration Problems, *Shock and Vibration Problems*, **8**, 3–18.

Leipholz H.H.E. and Bhalla K. (1977) On the Solution of the Stability Problem of Elastic Rods Subjected to Triangularly Distributed, Tangential Follower Forces, *Ingenieur-Archiv*, **46**, 115–124.

Leissa A.W. (1969) *Vibration of Plates*, NASA SP 160.

Leissa A.W. (1973) The Free Vibration of Rectangular Plates, *Journal of Sound and Vibration*, **31**, 257–293.

Leissa A.W. (1981) Plate Vibration Research: 1976–1980, *The Shock and Vibration Digest*, **10**, 19–36.

Leissa A.W. (1987) Recent Studies in Plate Vibrations: 1981–85, Part II, Complicating Effects, *The Shock and Vibration Digest*, **19**, 10–24.

Leissa A.W. (2000a) Private Communication, February 14.

Leissa A.W. (2000b) Private Communication, April 18.

Leissa A.W. and Narita Y. (1980) Natural Frequencies of Simply Supported Circular Plates, *Journal of Sound and Vibration*, **70**(2), 221–229.

Leissa A.W. and Shihada S.M. (1995) Convergence of the Ritz Method, *Applied Mechanics Reviews*, **11**, S90–S95.

Leissa A.W. and Vagins M. (1976) Stress Optimization in Nonhomogenous Materials, *Developments in Theoretical and Applied Mechanics*, **8**, pp. 13–22.

Leissa A.W. and Vagins M. (1978) The Design of Orthotropic Materials for Stress Optimization, *International Journal of Solids and Structures*, **14**, 517–526.

Lekhnitskii S.G. (1964) Uniaxial Stress State of Orthotropic Beam with Variable Modulus of Elasticity, *Mekhanika i Mashinostroenie*, No. 1 (in Russian).

Lekhnitskii S.G. (1965) Some Cases of Torsion of Beams with Variable Modulus of Elasticity, *Investigations in Elasticity and Plasticity*, **4**, pp. 72–85, Leningrad University Press, Leningrad.

Lenox T.A. and Conway H.D. (1980) An Exact, Closed-Form Solution for the Flexural Vibration of a Thin Annular Plate Having a Parabolic Thickness Variation, *Journal of Sound and Vibration*, **68(2)**, 231–239.

Leung A.Y.T., Zhou W.E., Lim C.W., Yuen R.K.K. and Lee U. (2001) Dynamic Stiffness for Piecewise Non-Uniform Timoshenko Column by Power Series — Part i: Conservative Axial Force, *International Journal for Numerical Methods in Engineering*, **51(5)**, 505–530.

Levine L.E. and Costa G.B. (1992) A Generalization of the Steady-State Heat Equation for a Solid Sphere, *International Journal of Mathematical Education in Science and Technology*, **25**, 155–156.

Levinson N. (1949) The Inverse Sturm–Liouville Problem, *Matematisk Tidsskrift*, 25–30.

Levitan B.M. (1964) On the Determination of a Sturm–Liouville Equation by Spectra, *Izvestiya AN SSSR, Seriya Mathematicheskaya*, *Amer. Math. Soc. Transl. Ser. 2*, **68(2)**, 1–20.

Li Q.S. (1999a) Flexural Free Vibration of Cantilevered Structures of Variable Stiffness and Mass, *Structural Engineering and Mechanics*, **8(3)**, 243–256.

Li Q.S. (1999b) Flexural Free Vibration of Cantilevered Structures of Variable Stiffness and Mass, *Structural Engineering and Mechanics*, **8(3)**, 243–256.

Li Q.S. (2000) Exact Solutions for Longitudinal Vibration of Multi-Step Bars with Varying Cross-Section, *Journal of Vibration and Acoustics*, **122(2)**, 183–187.

Li Q.S. (2001) Classes of Exact Solutions for Several Static and Dynamic Problems for Non-Uniform Beams, *Structural Engineering and Mechanics*, **12(1)**, 85–100.

Li Q.S., Cao H. and Li G. (1995) Calculation of Free Vibration of High Rise Structures, *Asian Journal of Structural Engineering*, **1(1)**, 17–25.

Li Q.S., Fang J.Q. and Jeary A.P. (1998) Free Longitudinal Vibration Analysis of Tall Buildings and High-Rise, *The Structural Design of Tall Buildings*, **1**, 167–176.

Li Q.S., Cao H. and Li G. (1999) Static and Dynamic Analysis of Straight Bars with Variable Cross-Section, *Computers and Structures*, **59(6)**, 1185–1191.

Li Q.S., Fang J.Q. and Jeary A.P. (2000a) Free Vibration Analysis of Cantilevered Tall Structures under Various Axial Loads, *Engineering Structures*, **22(5)**, 525–534.

Li Q.S., Li G.Q. and Liu D.K. (2000b) Exact Solutions for Longitudinal Vibration of Rods Coupled by Translational Springs, *International Journal of Mechanical Sciences*, **42(6)**, 1135–1152.

Li W.L. (2000) Free Vibrations of Beams with General Boundary Conditions, *Journal of Sound and Vibration*, **237**, 709–725.

Liew K.M. and Yang B. (1999) Three-Dimensional Solutions for Free Vibrations of Circular Plates by a Polynomials-Ritz Analysis, *Computer Methods in Applied Mechanics and Engineering*, **175(1–2)**, 189–201.

Lightfoot E. (1980) Exact Straight-Line Elements, *Journal of Strain Analysis*, **15**, 89–96.

Likar O. (1969) Knicklast am schmalen Fuss eingespsanter, em Korf beweglicher, konischer Stabe, *Beton Stahlbetonbau*, **64(9)**, 220–221 (in German).

Lim T.W. and Pilkey W.D. (1992) A Solution to the Inverse Dynamics Problems for Lightly Damped Flexible Structures using a Model Approach, *Computers and Structures*, **43(1)**, 53–59.

Lin Y.K. (1976) *Probabilistic Theory of Structural Mechanics*, p. 271, Krieger, Malabar.

Lin Y.K. and Cai G.Q. (1995) *Probabilistic Structural Dynamics*, McGraw Hill, New York.

Lindberg G.M. (1963) Vibration of Non-Uniform Beams, *Aeronautical Quarterly*, **14**, 387–395.

Lindberg G.M. and Olson M.D. (1970) Convergence Studies of Eigenvalue Solutions using Two Finite Plate Bending Elements, *International Journal for Numerical Methods in Engineering*, **2**, 99–116.

Liu G.L. (2000) Derivation and Transformation of Variational Principles with Emphasis on Inverse and Hybrid Problems in Fluid Mechanics: A Systematic Approach, *Acta Mechanica*, **140**, 73–89.

Liu G.R. and Han X. (2003) *Computational Inverse Techniques in Nondestructive Evaluation*, CRC Press, Boca Raton.

Liu G.R., Han X. and Lam K.Y. (2001) Material Characterization of FGM Plates using Elastic Waves and Inverse Procedure, *Journal of Composite Materials*, **35(11)**, 954–971.

Liu G.R., Han X. and Ohyoshi T. (2002) Computation Inverse Techniques for Material Characterization Using Dynamic Response, *International Journal of the Society of Materials Engineering for Resources*, **10(1)**, 26–33.

Lizarev A.D. (1959) Free Vibrations of Beams with Elastically Clamped Ends, *Izvestiya VUZOV, Stroitelstvo i Arkhitektura*, No. **10** (in Russian).

Lizarev A.D. (1960) Free Vibrations and Stability of Compressed Beams with Elastically Clamped Ends, *Stroitelnaya Mekhanika i Raschet Sooruzhenii (Strucutral Mechanics and Analysis of Constructions)*, No. **4** (in Russian).

Lizarev A.D. and Kuz'mentsov V.P. (1980) Free Vibrations of Annular Plates of Variable Thickness, *Problemy Prochnosti* (Strength Problems), No. **4**, 96–99 (in Russian).

Lockshin A. (1930) Über die Knickung eines doppelwandigen Druckstabes mit parabolish veränderlicher Querschnittshöhe, *Zeitschrift für angewandte Mathematik und Mechanik*, **10**, 160–166 (in German).

Louis A.K. (1989) *Inverse und schlecht gestellte Probleme*, Teubner, Stuttgart (in German).

Lowe B. (1993) Construction of an Euler–Bernoulli Beam from Spectral Data, *Journal of Sound and Vibration*, **163**, 165–171.

Loy C.T., Lam K.Y. and Reddy J.N. (1999) Vibration of Functionally Graded Cylindrical Shells, *International Journal of Mechanical Sciences*, **41(3)**, 309–324.

Lu H.X., Feng M.H., Lark R.J. and Williams F.W. (1999) The Calculation of Critical Buckling Loads for Externally Constrained Structures, *Communications in Numerical Methods in Engineering*, **15(3)**, 193–201.

Lu L.W. et al. (1983) *Stability Theory of Metal Struts*, Building Industry Press of China, Beijing (in Chinese).

Lurie A.I. (1936) Towards a Resolution of the Problem of Equilibrium of a Variable Thickness Plate, *Proceedings of the Leningrad Industrial Institute*, No. **6** (in Russian).

Lurie H. (1951) A Note on the Buckling of Struts, *Journal of the Royal Aeronautical Sciences*, **55**, 181–184.

Lurie H. (1952) Lateral Vibrations as Related to Structural Stability, *Journal of Applied Mechanics*, **19**, 195–204.

Lübbig H. (ed.) (1995) *The Inverse Problem*, Akademie Verlag, Berlin.

Lyapunov A.M. (1959) About One Problem of Chebyshev, *Collected Papers*, **3**, p. 209 (in Russian).

Ma C.K. and Lin D.C. (2000) Input Force Estimation of a Cantilever Beam, *Inverse Problems in Engineering*, **8**, 511–528.

Ma C.K., Tuan P.C., Lin D.C. and Liu C.S. (1998) A Study of an Inverse Method for the Estimation of Impulsive Loads, *International Journal of Systems Sciences*, **29**, 663–672.

Ma C.-K., Chang J.-M. and Lin D.-C. (2003) Input Force Estimation of Beam Structures by an Inverse Method, *Journal of Sound and Vibration*, **259(2)**, 387–407.

Ma L.S. and Wang T-J. (2003) Nonlinear Bending and Post-Buckling of a Functionally Graded Circular Plate Under Mechanical and Thermal Leadings, *International Journal of Solids and Structures*, **40**, 3311–3330.

Mabie H.H. and Rogers C.B. (1964) Transverse Vibrations of Double-Tapered Cantilever Beams with End Loads, *Journal of the Acoustical Society of America*, **36(3)**, 463–469.

Mabie H.H. and Rogers C.B. (1968) Transverse Vibrations of Tapered Cantilever Beams with End Support, *Journal of the Acoustical Society of America*, **44(6)**, 1739–1741.

Mabie H.H. and Rogers C.B. (1972) Transverse Vibrations of Double-Tapered Cantilever Beams, *Journal of the Acoustical Society of America*, Part 2, **51(5)**, 1771–1774.

Mabie H.H. and Rogers C.B. (1974) Transverse Vibrations of Double-Tapered Canteliver Beams with End Support and with End Mass, *Journal of the Acoustical Society of America*, **55(5)**, 986–991.

Maeve McCarty C. (2000) Recovery of a Density from the Eigenvalues of a Nonhomogeneous Membrane, *Inverse Problems in Engineering: Theory and Practice* (Woodbury K.A., ed.), pp. 155–161, The ASME Press, New York.

Magrab E.B. (1979) *Vibration of Elastic Structural Members*, Sijthoff and Noordhoff, Alphen a/d Rijn.

Mahmoud M.A. and Abu Kiefa M.A. (1999) Neural Network Solution of the Inverse Vibration Problem, *NDT&E International*, **32(2)**, 91–99.

Majima O. and Hayaski K. (1987) Elastic Buckling and Flexural Vibration of Variable-Thickness Annular Plates under Non-Uniform in Plane Forces, *JSME International Journal*, **30(270)**, 1890–270.

Maksimov V.I. (2002) *Dynamical Inverse Problems of Distributed Systems*, VSP, Zeist, The Netherlands.

Makushin V.M. (1962) Effective Application of Energy Method of Elastic Stability of Columns and Plates, *Raschety na Prochnost (Strength Analysis)*, **8**, "Mashinostroenie" Publishing, Moscow (in Russian).

Makushin V.M. (1963) Critical Value of Uniformly Distributed Axial Loads for Some Cases of Boundary Conditions of Compressed Columns, in *Raschety na Prochnost (Strength Analysis)*, **9**, pp. 253–269, Mashinostroenie Publishing, Moscow (in Russian).

Makushin V.M. (1964) Approximate Investigation of the Stability of Columns, Compressed by the Uniformly Distributed Axial Forces (Method of Elastic Deflections), in *Raschety na Prochnost (Strength Analysis)*, **10**, pp. 174–210, Mashinostroenie Publishing, Moscow (in Russian).

Makushin V.M. and Petrov B.P. (1967) Approximate Investigation of Stability of Single-Span Columns, Subjected to Concentrated and Distributed Axial Forces, *Izvestiya VUZOV, Mashinostroenie*, No. **4**, 25–30 (in Russian).

Makushin V.M. and Petrov B.P. (1967) On the Question of Single-Span Weighting Columns, *Izvestiya VUZOV, Mashinostroenie*, No. **5**, 27–33 (in Russian).

Maliev A.S. (1938), Beams on Elastic Foundation with Variable Coefficient of Compression along the Length, *Proceeding of the Leningrad Institute of Industrial Building Engineering*, **6**, 9–34 (in Russian).

Malik M. and Dang H.H (1998) Vibration Analysis of Continuous Systems by Differential Transformation, *Applied Mathematics and Computation*, **96**, 17–26.

Man T.B. (1982) On the Construction of the Programmed Motion, *Hardonic Journal*, **5**, 1802–1811.

Manicka Selvam V.K. (1997) Critical Load Evaluation of Nonprismatic Columns, *Journal of Structural Engineering*, **24**, 175–179.

Manicka Selvam V.K. (1998) Critical Loads of Nonprismatic Cantilever Columns with Intermediate Concentrated and Distributed Axial Loads, *Journal of Structural Engineering*, **24(4)**, 221–224.

Manickarajah D., Xie Y.M. and Steven G.P. (2000) Optimisation of Columns and Frames Against Buckling, *Computers and Structures*, **75**, 45–54.

Manoach E., Karagiozoua D. and Hadjikov L. (1991) An Inverse Problem for an Initially Heated Circular Plate under a Pulse Loading, *Zeitschrift für angewandte Mathematik und Mechanik*, **10**, 413–416.

Manohar C.S. and Ibrahim R.A. (1997) Progress in Structural Dynamics with Stochastic Parameter Variations: 1987–1998, *Applied Mechanics Reviews*, **52(3)**, 177–197.

Manohar C.S., Keane A.J. (1993) Axial Vibrations of Stochastic Rods, *Journal of Sound and Vibration*, **165**, 341–359.

Martin A.I. (1956) Some Integrals Relating to the Vibration of a Cantilever Beam and Approximation for the Effect of Taper on Overtone Frequencies, *Aeronautical Quarterly*, **7(2)**, 109–124.

Martin M.T. and Doyle J.F. (1996) Impact Force Identification from Wave Propagation Responses, *International Journal of Impact Engineering*, **18**, 65–77.

Masad J.A. (1996) Free Vibration of a Non-Homogeneous Rectangular Membrane, *Journal of Sound and Vibration*, **195**, 674–678.

Massey F.J., Jr. (1951) The Kolmgorov–Smirnov Test for Goodness of Fit, *Journal of American Statistical Association*, **253**, 67–68.

Massonet C. (1940) Les Relations entre les Modes Normaux de Vibration et la Stabilité des Systemes Elastiques, *Bull Cours et Lab. Const. Gneie Civil*, Liège, **I**, 353 (in French).

Massonet C. (1972) An Accurate Formula for Assessing the Vibration Frequency of Structures Axially Loaded, *International Journal of Mechanical Sciences*, **14**, 729–734.

Matevosian R.R. (1961) *Stability of Complex Column Systems (Qualitative Theory)*, "Gosstroyizdat" Publishers, Moscow (in Russian).

Mathews P.M. (1958) Vibrations of a Beams on Elastic Foundation, *Zeitschrift für angewandte Mathematik und Mechanik*, **38**, 105–115; **39**, 13–19 (1959).

Matsunaga H. (1996a) Free Vibration and Stability of Thin Elastic Beams Subjected to Axial Forces, *Journal of Sound and Vibration*, **191**, 917–993.

Matsunaga H. (1996b) Buckling Instabilities of Thick Elastic Beams Subjected to Axial Stresses, *Computers and Structures*, **59**, 859–869.

Matsunaga H. (1999) Vibration and Buckling of Deep Beam-Columns on Two-Parameter Elastic Foundations, *Journal of Sound and Vibrations*, **228(2)**, 359–376.

Matsuno Y. (2002) Exactly Solvable Eigenvalue Problems for a Nonlocal Nonlinear Schrödinger Equation, *Inverse Problems*, **18**, 1101–1125.

Maurizi M.J. (1988) Further Note on the Free Vibrations of Beams Resting on an Elastic Foundation, *Journal of Sound and Vibration*, **124(1)**, 191–193.

Mazurkiewicz Z. (1964) Buckling of Straight Bars with Arbitrarily Varying Flexural Rigidities and under Various Boundary Conditions, *Bull. Acad. Pol., Sci. Ser. Sci. Tech.*, Warsaw, **12(9)**, 445–454.

Mazurkiewicz Z. (1965) Buckling of Straight Bars of Variable Flexural Rigidity, *Rozpr. Inzyn.*, **13(3)**, 636 (in Polish).

McCarthy C.M. (1999) The Tallest Column-Optimality Revisited, *Journal of Computational and Applied Mathematics*, **101**, 27–37.

McLaughlin J.R. (1976) An Inverse Problem of Order Four, *SIAM Journal of Mathematical Analysis*, **7**, 646–461.

McLaughlin J.R. (1984) On Construction Solutions to an Inverse Euler–Bernoulli Beam Problems, *Inverse Problems of Acoustic and Elastic Waves* (Santosa F., Pao Y.H., Symes W.W. and Holland C., eds.), pp. 341–345, SIAM, Philadelphia.

McLaughlin J.R. (1988) Inverse Spectral Theory using Nodal Points as Data — a Uniqueness Result, *Journal of Differential Equations*, **73**, 354–362.

McLaughlin J.R. (1995) Formulas for Finding Coefficients from Nodes/Nodal Lines, in *Proceedings of the ICM*, Birkhäuser, Basel, Switzerland.

McLaughlin J.R. (1998) Good Vibrations, *American Scientist*, **86**, 342–349.

McLaughlin J.R. and Hald O.H. (1995) Formula for Finding a Potential from Nodal Lines, *Bull. Amer. Math. Soc.*, **32**, 241–253.

Meirovitch L. (1980) *Computational Methods in Structural Dynamics*, Kluwer Academic Publishers, Dordrecht.

Meirovitch L. (1986) *Elements of Vibration Analysis*, p. 272, McGraw-Hill, New York.

Meirovitch L. (1997) *Principles and Techniques of Vibrations*, Prentice Hall, Upper Saddle River, NJ.

Melerski E.S. (2000) *Design Analysis of Beams, Circular Plates and Cylindrical Tanks on Elastic Foundations*, Balkema, Rotterdam.

Meyers M.K. (1986) Note on the Strongest Fixed–Fixed Column, *Quarterly Journal of Applied Mathematics*, **44(3)**, 583–588.

Michaels J.E. and Pao Y-H (1985) The Inverse Source Problem for an Oblique Force on an Elastic Plate, *Journal of the Acoustical Society of America*, **77(6)**, 2005–2011.

Miesse C.C. (1949) Determination of the Buckling Load for Columns of Variable Stiffness, *Journal of Applied Mechanics*, 406–410.

Mikeladze Sh.E. (1951) *New Methods of Integrating Differential Equations*, GITTL Publishing, Moscow (in Russian).

Mikhlin S.G. (1934) Some Cases of the Plane Problem of the Theory of Elasticity for an Inhomogeneous Medium, *Journal of Applied Mechanics and Mathematics — PMM*, **1(2)** (in Russian).

Mikhlin S.G. (1950) Some Sufficient Conditions for Convergence of the Galerkin Method, *Uchenye Zapiski LGU, Seriya Matematicheskaya*, **135(21)** (in Russian).

Mikhlin S.G. (1957) *Integral Equations*, Pergamon Press, Oxford.

Mishiku M. and Teodosiu I. (1966) Solution of an Elastic Static Plane Problem for Nonhomogeneous Isotropic Bodies by Means of the Theory of Complex Variables, *Journal of Applied Mechanics and Mathematics — PMM*, **30(2)**, 379–387.

Mitelman M. (1970) Critical Load of Fixed-Ended Column with Symmetrically Variable Cross-Section (Extension of Engesser-Vianello Method), *Israel Journal of Technology*, **8(4)**, 353–356.

Mohr G.A. (2000) Polynomial Solutions for Thin Plates, *International Journal of Mechanical Sciences*, **42**, 1197–1204.

Mok C-H. and Murrey E.A. Jr. (1965) Free Vibrations of a Slender Bar With Nonuniform Characteristics, *Journal of the Acoustical Society of America*, **40**, 385–389.

Morassi A. (1993) Crack-Induced Changes in Eigenparameters of Beam Structures, *Journal of Engineering Mechanics*, **119(9)**, 1798–1803.

Morassi A. (2000), Personal Communication, July 18.

Morley A. (1914) Critical Loads for Ideal Long Columns, *Engineering* (London), **97**, 566–568.

Morley A. (1917) Critical Loads for Long Tapering Struts, *Engineering* (London), **104**, 295–298.

Morrow J. (1905) On the Lateral Vibration of Bars, *Philosophical Magazine*, **10**, 113–125.

Mosegaard K. and Sambridge M. (2002) Monte Carlo Analysis of Inverse Problems, *Inverse Problems*, **18**, R29–R54.

Moskalenko V.N. (1973) Method of Variable Parameters in Problems on Free Vibrations of Plates, *Theoria Obolochek i Plastin (Theory of Shells and Plates)*, 524–527, "Nauka" Publishers, Moscow (in Russian).

Mota Soares C.M., Moreira de Freitas M., Araujo A.L. and Pedersen P. (1993) Identification of Material Properties of Composite Plate Structures, *Composite Structures*, **25**, 277–285.

Mottershead J.E. and Lallement G. (1999) Vibration Nodes, and the Cancellation of Poles and Zeros by Unit-Rank Modifications to Structures, *Journal of Sound and Vibration*, **222(5)**, 831–851.

Mottershead J.E., Mares C. and Friswell M.I. (2000) Assignment of Vibration Nodes, in *Proceedings of IMAC-XVIII: A Conference on Structural Dynamics*, pp. 227–233.

Mou Y., Han R.P.S. and Shah A.H. (1997) Exact Dynamic Stiffness Matrix for Beams of Arbitrarily Varying Cross Sections, *International Journal of Numerical Methods in Engineering*, **40(2)**, 233–250.

Mullagulov M. Kh. (1981) Dynamic Equations of the Method of Initial Parameters in the Problems of Column Stability, *Izvestiya VUZOV, Stroitelstvo i Arkhitektura*, No. 3, 43–48.

Muravskii G.B. (1967) On the Elastic Foundation Models, *Stroitelnaya Mekhanika i Raschet Sooruzhenii*, **6**, 14–17 (in Russian).

Naguleswaran S. (1990) The Vibration of Euler–Bernoulli Beam of Constant Depth and with Convex Parabolic Vibration in Breadth, in *Proceedings of the Australian Noise and Vibration Conference*, pp. 204–209.

Naguleswaran S. (1992) Vibration of an Euler–Bernoulli Beam of Constant Depth and with Linearly Varying Breadth, *Journal of Sound and Vibration*, **153**, 509–522.

Naguleswaran S. (1994a) A Direct Solution for the Transverse Vibration of Euler–Bernoulli Wedge and Cone Beams, *Journal of Sound and Vibration*, **172(3)**, 289–304.

Naguleswaran S. (1994b) Vibration in Two Principal Planes of a Non-uniform Beam of Rectangular Cross-Section, One Side of Which Varies as a Square Root of the Axial Coordinate, *Journal of Sound and Vibration*, **172(3)**, 305–319.

Naidu N.R. (1995) Vibrations of Initially Stressed Uniform Beams on a Two-Parameter Elastic Foundation, *Computers and Structures*, **57(5)**, 941–943.

Najafizadeh M.M. and Eslami M.R. (2002a) First-Order Based Thermoelastic Stability of Functionally Graded Material Circular Plates, *AIAA Journal*, **40(7)**, 1444–1450.

Najafizadeh M.M. and Eslami M.R. (2002b) Buckling Analysis of Circular Plates Under Uniform Radial Compression, *International Journal of Mechanical Sciences*, **44**, 2479–2493.

Nakagari S., Takabatake H., Tani S. (1987) Uncertain Eigenvalue Analysis of Composite Laminated Plates by the Stochastic Finite Element Method, *ASME Journal for Industry*, **109**, 9–12.

Nakamura G. and Uhlmann G. (1995) Inverse Problems at the Boundary for an Elastic Medium, *SIAM Journal of Mathematical Analysis*, **26**, 263–279.

Nakamura T. and Lin P-S. (1988) Design of Beams for Specified Fundamental Natural Period and Mode, *Journal of Structural Engineering*, A1JB, 95–104 (in Japanese).

Nakamura T. and Sampath S. (2001) Determiniation of FGM Properites by Inverse Analysis, *Ceramic Transactions*, **114**, 521–528.

Napetvaridze S.G., Khvoles A.R. and Vekua T.P. (1973) The Inverse Problem of Engineering Seismology, *Fifth World Conference on Earthquake Engineering Session 1D: Dynamics of Soils and Soil Structures*, Rome, Italy, pp. 1–15.

Nashed Z. (ed.) (1976) *Generalized Inverses and Applications*, Academic Press, New York.

Natke H.G. (1989) *Baudynamik*, p. 156 B.G. Teubner, Stuttgart, (in German).

Natke H.G. and Cempel C. (1991) Fault Detection and Localisation in Structures: A Discussion, *Journal of Mechanical Systems and Signal Processing*, **5(5)**, 345–356.

Nemenyi P.F. (1951) Recent Developments in Inverse and Semi-Inverse Methods in the Mechanics of Continua, *Advances in Applied Mechanics*, **2**, New York, 1951.

Nemirovskii Y.V. (2001) Inverse Problems of the Mechanics of Thin-Walled Composite Structures, *Mechanics of Composite Materials*, **37(5–6)**, 421–430.

Nesterenko V.F. (2001) *Dynamics of Heterogeneous Materials*, Springer Verlag, New York.

Neuringer J. and Elishakoff I. (2000) Natural Frequency of an Inhomogeneous Rod May Be Independent of Nodal Parameters, *Proceedings of the Royal Society of London*, **456**, 2731–2740.

Neuringer J. and Elishakoff I. (2001) Inhomogeneous Beams That May Possess a Prescribed Polynomial Second Mode, *Chaos, Solitons and Fractals*, **12**, 881–896.

Newberry A.L., Bert C.W. and Striz A.G. (1987) Noninteger-Polynomial Finite Element Analysis of Column Buckling, *Journal of Engineering Mechanics*, **113**, 873–878.

Newmark N.M. (1943) Numerical Procedure for Computing Deflections, Moments and Buckling Loads, *Transactions of ASCE*, **108**, 1161–1188.

Ng T.Y., Lam K.M., Liew K.M. and Reddy J.N. (2001) Dynamic Stability of Analysis of Functionally Graded Cylindrical Shells Under Periodic Axial Loading, *International Journal of Solids and Structures*, **38**, 1295–1309.

Niblett L.T. (1983) Frequencies of Transverse Vibrations of Truncated Conical Cantilevers, *Journal of Sound and Vibrations*, **89(1)**, 141–143.

Nicholson J.W. (1917) The Lateral Vibration of Bars of Variable Section, *Proceedings of the Royal Society of London* (Series A), **93(A654)**, 506–519.

Nicholson J.W. (1920) The Lateral Vibration of Sharply Pointed Bars, *Proceedings of the Royal Society of London*, (Series A), **97**, 171–181.

Niedenfuhr F.W. (1952) Note on the Determination of Bending Frequencies by Rayleigh's Method, *Journal of the Aeronautical Sciences*, **19**, 849–850.

Nielsen J.C.O. (1991) Eigenfrequencies and Eigenmodes of Beam Structures on an Elastic Foundation, *Journal of Sound and Vibration*, **145(3)**, 479–487.

Nigam N.C. (1983) *Introduction to Random Vibrations*, MIT Press, Cambridge, pp. 258–259.

Nikolai E.L. (1955) Lagrange's Problem on Most Favorable Form of the Column, in *Selected Works in Mechanics*, pp. 9–44, GITTL Publishing, Moscow.

Nikolai E. (1955) On Euler's Works on Buckling, in *Papers on Mechanics*, pp. 437–454, "Gostekhizdat" Publishers, Moscow.

Niordson F.I. (1965) On the Optimal Design of a Vibrating Beam, *Quarterly Applied Mechanics*, **23(1)**, 47–53.

Niordson F.I. (1967) A Method for Solving Inverse Eigenvalue Problems, *Recent Progress in Applied Mechanics*, pp. 373–382, Wiley, New York.

Niordson F.J. (1968) Inverse Problem of Natural Frequencies of Elastic Plates, *III All-Union Congress on Theoretical and Applied Mechanics*, Abstract Report, pp. 230–231, Moscow.

Niordson F.I. (1970) About Inversing the Problem of Eigenvalues for the Plate Vibration Problem, *Problems of Mechanics of Solid Deformable Body* (Sedov L.I., ed.), pp. 287–294, "Sudostroenie" Pulishers, Leningrad (in Russian).

Niordson F. (2000) Personal Communication, September 12.

Niordson F.I. (2001) Early Numerical Computations in Engineering, *Applied Mechanics Reviews*, **54(6)**, R17–R19.

Noble B. and Daniel J. (1977) *Applied Linear Algebra*, NJ p. 204, exercise 6.30, Prentice-Hall, Englewood-Cliffs.

Novotvortsev B.I. (1933) Method of Successive Approximations for Application to Study Free Vibrations of Engineering Structures, *Proceedings of the Seismological Institute AN SSSR*, **83** (in Russian).

Nowinski I. and Turski S. (1953) On the Theory of Elasticity of Isotropic Non-Homogeneous Bodies, *Arch. Mech. Stos.*, **5(1)**.

Nowacki W. (1963) *Dynamics of Elastic Systems*, pp. 340–343, Wiley, New York.

Nudelman I.L. (1936) Integral Equations in the Theory of Longitudinal Bending, *Journal of Applied Mechanics and Mathematics — PMM*, **3** (in Russian).

Oberkampf W.L. (2001) What Are Validation Experiments? *Experimental Techniques*, May/June, 35–40.

Ochi M.K. (1989) *Applied Probability and Stochastic Processes*, p. 150, Wiley-Interscience, New York.

Odeh F. (1975) Shape of the Strongest Column with a Follower Load, *Journal of Optimization Theory and Applications*, **15(1)**, 103–118.

Ogarkov B.I. and Katz Yu.S. (1967) Determination of Stresses in Anisotropic Ring with Modulus of Elasticity That Varies Along Radius, *Mashinostroenie*, No. **9**, 40–46 (in Russian).

Ohkami Y. and Tanaka H. (1998) Estimation of the Force and Location of an Impact Exerted on a Spacecraft, *JSME International Journal, Series C*, **41(4)**, 829–835.

Okubo H. and Seika (1963) Stresses in Heterogeneous Plates, *Ingenieur-Archiv*, **32**, 246–254.

Olhoff N. (1970) Optimal Design of Vibrating Circular Plates, *International Journal of Solids and Structures*, **6**, 139–156.

Olhoff N. (1976a) Optimization of Vibrating Beams with Respect to Higher Order Natural Frequencies, *Journal of Structural Mechanics*, **4**, 87–122.

Olhoff N. (1976b) A Survey of the Optimal Design of Vibrating Structural Elements, Part 1. Theory, *Shock and Vibration Digest*, No. **8**, 3–10.

Olhoff N. (1977) Maximizing Higher Order Eigenfrequencies of Beams with Constraints on the Design Geometry, *Journal of Structural Mechanics*, **5**, 107–134

Olhoff N. and Rasmussen S.H. (1977) On Single and Bimodal Optimum Buckling Loads of Clamped Columns, *International Journal of Solids and Structures*, **13(7)**, 605–614.

Olszak W. (ed.) (1959) *Non-Homogeneity in Elasticity and Plasticity*, Pergamon Press, London.

Olszak W. and Rychlewski J. (1961) Nichthomogenitätsprobleme im elastischen und vorplastischen Bereich, *Österreichische Ingenieur-Archiv*, **15**, 130–152 (in German).

Ono A. (1919) On the Stability of Long Struts of Variable Section, *Memoirs, College of Engineering*, Kyushu Imperial University, Fukuoka, Japan, **1(5)**, 395–406.

Ono A. (1925) Lateral Vibrations of Tapered Bars, *Journal of the Society of Mechanical Engineering in Japan*, **28(10)**, 429–441.

Öry H. and Lindert H.W. (1993a) Calculation of Rotor Blade Air Loads from Measured Structural Response Data, *Zeitschrift für Flugwissenschaften und Weltraumsforschung*, **17**, 225–234.

Pan V.Y. (1998) Solving Polynomials with Computers, *American Scientist*.

Panichkin V.I. (1968) Stability of Cantilever Column in the Form of Rotated Body Under Its Own Weight, *Izvestiya VUZOV, Mashinostroenie*, No. **4**, 50–56 (in Russian).

Panovko Ya. G. and Gubanova I. (1965) *Stability and Oscillations of Elastic Systems: Paradoxes, Fallacies and New Concepts*, Consultant's Bureau, New York.

Papkovich P.F. (1933) Essay on Development and Modern State of the Problem of Ship Vibrations, *Journal of Applied Mechanics and Mathematics — PMM*, **1(1)** (in Russian).

Passer W. (1936) Beitrag zur Berechnung von Knickstäben mit veränderlichem Querschnitt, *Bautechnik*, 418–419, (in German).

Pasternak P.L. (1954) *On a New Method of Analysis of an Elastic Foundation by Beams of Two Foundation Constants*, Gosudarstvennoe Izdatel'stvo Literatury no Stroitelstuu i Arkhitekture (in Russian).

Pavlovic M.V. and Wylie G.B. (1983) Vibration of Beams on Non-Homogeneous Elastic Foundations, *Earthquake Engineering and Structural Dynamics*, **11**, 797–808.

Pedersen N. (2000) Design of Cantilever Probes for Atomic Force Microscopy (AFM), *Engineering Optimization*, **32**, 373–392.

Penney J.E. (1997) Vibration Modes of Tapered Beams, *American Mathematical Monthly*, **54**, 391–394.

Penney J.E. and Reed J.R. (1971) An Integral Equation Approach to the Fundamental Frequency of Vibrating Beams, *Journal of Sound and Vibration*, **19**, 393–400.

Perkins K.A.R. (1966) The Effect of Support Flexibility on the Natural Frequencies of a Uniform Cantilever, *Journal of Sound and Vibration*, **4**, 1–8.

Peterson I. (1994) Beating of Fractal Drum: How a Drum's Shape Affects Its Sound, *Sci. News*, **146**, 184–185.

Petyt M. (1990) *Introduction to Finite Element Vibration Analysis*, Cambridge University Press (second edition, 1999).

Pflüger A. (1964) *Stabilitäts probleme der Elastostatik*, Springer Verlag, Berlin, p. 351 (in German).

Pflüger A. (1975) Personal Communication.

Picard E. (1905) *Traité d'analyse*, **2**, 340, **3**, 88 (in French).

Piché R. (1993) Bounds for the Maximum Deflection and Curvature of Nonuniform Beam Columns, *International Journal of Solids and Structures*, **30(12)**, 1565–1577.

Pieczara J. (1987) Statics and Buckling of Elastic Shells of Revolution by Means of Exact Finite Elements, *Ph. D. Thesis*, Cracow University of Technology, Poland (in Polish).

Pietrzak J. (1994) Column Optimization Revisited, *Structural Engineering Review*, **6(2)**, 129–134.

Pilkington D.F. and Carr J.B. (1970) Vibration of Beams Subjected to End and Axially Distributed Loading, *Journal of Mechanical Engineering Science*, **12(1)**, 70–72.

Plaut R.H. (1986) Bimodal Optimization of Compressed Columns on Elastic Foundations, *Journal of Applied Mechanics*, **53(1)**, 130–134.

Plotnikov M.M. (1959) About Stresses in Thick Inhomogeneous Anisotropic Tube, *Mashinostroenie*, No. **8** (in Russian).

Plotnikov M.M. (1960) Towards Analysis of Thick-Walled Inhomogeneous Tube, *Mashinostroenie*, No. **12** (in Russian).

Plotnikov M.M. (1963) Elastic Properties and Stress State in Inhomogeneous Cylinders, *Mashinostroenie*, No. **6**, 19–28 (in Russian).

Plotnikov M.M. (1966) Towards Analysis of Inhomogeneous Cylinder Whose Elasticity Modulus Is a Piece-Wise Smooth Function of the Radius, *Mashinostroenie*, No. **12**, 33–36 (in Russian).

Plotnikov M.M. (1967a) General Case of Hyperbolicity of Function E, and Stresses in Inhomogeneous Anisotropic Cylinder, *Mashinostroenie*, No. **11**, 32–35 (in Russian).

Plotnikov M.M. (1967b) About Stresses in One Problem of Inhomogeneous Anisotropic Cylinder, *Mashinostroenie*, No. **8**, 28–31 (in Russian).

Plotnikov M.M. (1968) About the Influence of the Poisson Ration on the Stress Field of the Inhomogeneous Anisotropic Cylinder, *Mashinostroenie*, No. **3** (in Russian).

Pnueli D. (1972a) An Iterative Procedure to Obtain Exact Buckling Criteria for Columns under Combined Action of Continuous Load and Concentrated Forces, *Israel Journal of Technology*, **10(4)**, 253–256.

Pnueli D. (1972b) Buckling of Columns with Variable Moment of Inertia, *Journal of Applied Mechanics*, **39(4)**, 1139–1141.

Pnueli D. (1972c) Two Shift Theorems Leading to Both Upper and Lower Bounds of Eigenvalues, *Journal of Engineering Mathematics*, **6(1)**, 47–51.

Pnueli D. (1999) Lower Bounds do Eigenvalues for Vibrations and Buckling of Structures by Analytical Applications of *a Priori* Known Qualitative Information, *Chaos, Solutions and Fractals*, **10(11)**, 1783–1806.

Pohl K. (1933) Näherungslvesungen für besondere Fälle von Knickbelastung, *Stahlbau*, 137 (in German).

Poisson S.D. (1829) Memoires de l'Academie Royales des Sciences de l'Institut de la France, L'Équilibre et le mouvement des corps élastiques, Ser. **2, 8**, 357.

Polskii N.I. (1949) Convergence of the B.G. Galerkin Method, *Doklady AN USSR, Otdelenie Fiziko-Matematicheskykh i Khimicheskykh Nauk*, No. **6** (in Russian).

Polyanin A.D. and Manzhirov A.V. (1998) *Handbook of Integral Equations*, CRC Press, Boca Raton.

Pomraning G.C. (1966) A Numerical Study of the Method of Weighted Residuals, *Nuclear Science Engineering*, **24**, 293–301.

Ponomarev S.D., Biderman V.L., Likhachev K.K., Makuskin V.M., Malinin N. N. and Feodosiev V.I. (1959) *Strength Analysis in Mechanical Engineering*, **3**, pp. 346–355, "GITTL" Publishers, Moscow (in Russian).

Popovich V.E. (1962) Application of the Successive Approximations to Problems of Stability and Vibrations of Elastic Systems, *Izvestiya VUZOV, Aviazionnaya Tekhnika*, No. **4**, 103–110.

Pörschel J. and Trubowitz E. (1987) *Inverse Spectral Theory*, Academic Press, London.

Pörschl Th. (1938) Über eine Methode zur angenäherten Lösung nichtlinearer Differentialgleichungen mit Anwendungen auf die Berechnung der Durchbiegung bei der Knickung gerader Stäbe, *Ingenieur-Archiv*, **2**, 34–41 (in German).

Pradhan S.C., Loy C.T., Lam K.Y. and Reddy J.N. (2000) Vibration Characteristics of Functionally Graded Cylindrical Shells Under Various Boundary Conditions, *Applied Acoustics*, **61**, 111–129.

Pranger W.A. (1989) A Formula for the Mass Density of a Vibrating String in Terms of the Trace, *Journal of Mathematical Analysis and Application*, **141**, 399–404.

Praveen G.N. and Reddy J.N. (1998) Nonlinear Transient Thermoelastic Analysis of Functionally Graded Ceramic–Metal Plates, *International Journal of Solids and Structures*, **35(33)**, 4457–4476.

Prezemieniecki J.S. (1968) Discrete-Element Methods for Stability Analysis of Complex Structures, *The Aeronautical Journal*, **72**, 1077–1086.

Pronsato M.E., Laura P.A.A. and Juan A. (1999) Transverse Vibrations of a Rectangular Membrane and Discontinuously Varying Density, *Journal of Sound and Vibration*, **222**, 341–344.

Putty M.W. and Najabi K. (1994) A Micromachined Vibrating Ring Gyroscope, *Technical Digest, Sensors and Actuators Workshop*, Hilton Head Islands, SC, 13–16 June, pp. 213–220.

Rabinovich I.M. (1949) *Achievements of Structural Mechanics of Beam Systems in the USSR*, Moscow (in Russian).

Radeş M. (1970) Steady-State Response of a Finite Beam on a Pasternak-Type Foundation, *International Journal of Solids and Structures*, **6**, 739–756.

Radeş M. (1972) Dynamic Analysis of an Inertial Foundation Model, *International Journal of Solids and Structures*, **8**, 1363–1372.

Radomski B. (1938) Compression Struts with Nonprogressively Variable Moment of Inertia, *Technical Memorendum No. 861*, National Advisory Committee for Aeronautics (NACA).

Rainville E.D. (1960) *Special Functions*, Chapter 6, Macmillan, New York.

Rajagopal K.R. and Gupta A.S. (1981) One Class of Exact Solutions to the Equations of Motion of a Second Grade Fluid, *International Journal of Engineering Science*, **19**, 1009–1014.

Rajasekhariah H.L. (1965) Beam Column under Axially Distributed Load, *UNICIV Rep. R7*, The University of New South Wales, Sydney.

Rajendran S. and Prathab G. (1991) Convergence of Eigenvalues of a Cantilever Beam with 8- and 20-Node Hexahedral Elements, *Journal of Sound and Vibration*, **227(3)**, 667–668.

Raju K.K. (1990) Effect of Elastic Foundation on the Mode Shape in Stability and Vibration Problems of Tapered Columns/Beams, *Journal of Sound and Vibration*, **136(1)**, 171–175.

Raju K.K. and Rao G.V. (1986) Free Vibration Behavior of Prestressed Beams, *Journal of Structural Engineering*, **112(2)**, 433–437.

Raju K.K. and Venkateswara Rao G. (1988) Free Vibration Behavior of Tapered Beam Columns, *Journal of Engineering Mechanics*, **114(5)**, 889–892.

Raju P.N. (1962) Vibrations of Annular Plates, *Journal of Aeronautical Society of India*, **14(2)**, 37–52.

Ram G.S. (1963) Euler Load of a Stepped Column — An Exact Formula, *AIAA Journal*, **1(1)**, 211–212.

Ram G.S. and Rao G.V.E. (1951) Buckling of an n-Section Column, *Journal of Aerospace Sciences*, **19(1)**, 66–67.

Ram Y.M. (1993) Inverse Eigenvalue Problem for Modified Vibrating System, *SIAM Journal of Applied Mathematics*, **53**, 1762–1775.

Ram Y.M. (1994a) Enlarging Spectral Gap by Structural Modification, *Journal of Sound and Vibration*, **176**, 225–234.

Ram Y.M. (1994b) An Inverse Mode Problem for the Continuous Model of an Axially Vibrating Rod, *Journal of Applied Mechanics*, **61**, 624–628.

Ram Y.M. (1994c) Inverse Mode Problems for the Discrete Model of Vibrating Beam, *Journal of Sound and Vibration*, **169(2)**, 239–252.

Ram Y.M. (1998) Pole Assignment for the Vibrating Rod, *Quarterly Journal of Mechanics and Applied Mathematics*, **51(3)**, 461–476.

Ram Y.M. (2000) Private Communication, January 31.

Ram Y. (2002) Nodal Control of a Vibrating Rod, *Mechanical Systems and Signal Processing*, **16(1)**, 69–81.

Ram Y.M. and Elhay S. (1998) Constructing the Shape of a Rod from Eigenvalues, *Communications in Numerical Methods in Engineering*, **14**, 597–608.

Ram Y.M. and Elishakoff I. (2004) Can One Reconstruct the Cross-Section of an Axially Vibrating Rod from One of its Mode Shapes? *Proceedings of the Royal Society of London*, **460**, 1583–1596.

Ram Y.M. and Gladwell G.M.L. (1994) Constructing a Finite Element Model of a Vibrating Rod from Eigendata, *Journal of Sound and Vibration*, **169**, 229–237.

Rao J.S. (1965) The Fundamental Flexural Vibration of a Cantilever Beam of Rectangular Cross-Section with Uniform Taper, *Aeronautical Quarterly*, **14**, 139–144.

Rao S.S. (1995) *Mechanical Vibrations*, pp. 502–519, 527, Addison-Wesley, Reading, MA.

Ratzensdorfer J. (1936) *Die Knickfestigkeit von Stäben und Stabwerken*, Springer, Wien (in German).

Ratzersdorfer J. (1943) Determination of the Buckling Load of Struts by Successive Approximations, *The Journal of the Royal Aeronautical* Society, **47**, 103–105.

Rayleigh J.W.S. (1873) Some General Theorems Relating to Vibrations, *Proceedings of The London Mathematical Society*, **4**, 357–368.

Rayleigh J.W.S. (1894) *The Theory of Sound*, pp. 112–113, Macmillan and Co., London (Dover Publications, 1945).

Rayleigh J.W.S. (1899a) *Scientific Papers*, **1**, pp. 170–181, Cambridge University Press, Cambridge.

Rayleigh J.W.S. (1899b) On the Calculation of the Frequency of System in Its Gravest Mode, with an Example from Hydrodynamics, *Philosophical Magazine*, Ser. 5, **47**, 556–572.

Rayleigh J.W.S. (1911) On the Calculation of Chladni's Figures for a Square Plate, *Philosophical Magazine*, Ser. 6, **22**, 225–229.

Reddy J.N. (2000) Analysis of Functionally Graded Plates, *International Journal for Numerical Methods in Engineering*, **47**, 663–684.

Reddy J.N., Wang C.M. and Kitiporrnchai S. (1999) Axisymmetric Bending of Functionally Graded Circular and Annular Plates, *European Journal of Mechanics: A/Solids*, **18**, 185–199.

Reddy T.Y. and Srinath H. (1974) Elastic Stresses in a Rotating Anisotropic Annular Disk of Variable Thickness and Variable Density, *International Journal of Mechanical Sciences*, **16**, 85–89.

Reese L.C. and Welch R.C. (1975) Lateral Loading of Deep Foundations in Stiff Clay, *Journal of Geotechnical Division*, **101**, 633–649.

Reich Y. and Eisenberger M. (1989) Optimal Shape Design of Columns for Buckling, in *Proceedings of the Sessions Related to Steel Structures at Structures Congress'89*, pp. 677–685, San Francisco.

Reinitzhuber F. (1953) Approximate Formulas for the Buckling of Struts with Linearly Variable Axial Loading, *Publ. Int. Bridge Struct. Engrg.*, **13**, 309–319 (in German).

Reinitzhuber F. (1955) Das Knicken gerader Stäbe mit linear veränderlicher Längskraft im elastischem und unelastischen Bereich, *Bautechnik-Archiv*, No. **11**, Berlin (in German).

Rezeka S.F. (1989) Torsional Vibrations of Nonprismatic Hollow Shaft, *Journal of Vibration, Stress and Reliability in Design*, **111**, 486–489.

Rietsch E. (1977) The Maximum Entropy Approach to Inverse Problems, *J. Geophys.*, **42**, 489–506.

Rissoné R.F. and Williams J.J. (1965) Vibrations of Non-Uniform Cantilever Beams, *The Engineer*, Sept. 24, 497–506.

Ritz W. (1908) Über eine neue Methode zur Lösung gewisser Randwertaufgaben, *Goettingen Nachrichtten Math.-Phys.*, **K1**, 236–248 (in German).

Ritz W. (1909) Über eine neue Methode zur Lösung gewisser Variationsprobleme der mathematischen Physik, *Zeitschrift für die reine eind angewandte Mathematik*, **135(1)**, 1–61 (in German).

Ritz W. (1911) *Gesammelte Werke*, pp. 251–264, Gauthier-Willars, Paris (in German).

Rizos P.F., Aspragathos N. and Dimarogonas A.D. (1990) Identification of Crack Location and Magnitude in a Cantilever Beam from the Vibration Modes, *Journal of Sound and Vibration*, **138**, 381–388.

Robertson R.E. (1951) Vibration of a Clamped Circular Plate Carrying Concentrated Mass, *Journal of Applied Mechanics*, **18(4)**, 349–352.

Rok Shibar (1994) Solutions of Inverse Problems in Elastic Wave Propagation with Artificial Neural Networks, *Dissertation*, Cornell University.

Romanov V.G. (1984) *Inverse Problems of Mathematical Physics*, Nauka Publishers, Moscow (in Russian).

Rooney F.T. and Ferrari M. (1995) Torsion and Flexure of Inhomogeneous Elements, *Composites Engineering*, **5**, 901–011.

Ross R.G. (1971) Synthesis of Stiffness and Mass Matrices from Experimental Vibration Modes, *SAE Paper 710787*, pp. 2627–2635.

Rostovzev N.A. (1964) Towards the Theory of Elasticity of Inhomogeneous Media, *Journal of Applied Mechanics and Mathematics — PMM*, **28(4)**, 601–611 (in Russian).

Rundel W. and Sacks P.E. (1992) Reconstruction of Sturm-Liouville Operators, *Inverse Problems*, **8**, 457–482.

Ruta P. (1999) Application of Chebyshev Series to Solution of Non-Prismatic Beam Vibration Problems, *Journal of Sound and Vibration*, **227(2)**, 449–467.

Rzhanitsyn A.R. (1955) *Stability of Equilibrium of Elastic Systems*, "GITTL" Publishers, Moscow (in Russian), p. 291.

Sabatier P.C. (1977) Positivity Constraints in Linear Inverse Problems, 1) General Theory, *Geophys. J. Royal Astro. Soc.*, **48**, 415–441.

Sabatier P.C. (ed.) (1978) *Applied Inverse Problems*, Springer, Berlin.

Sabatier P.C. (1985) Inverse Problems — An Introduction, *Inverse Problems*, **1**.

Sabatier P.C. (1990) Modelling or Solving Inverse Problems? *Inverse Problems in Action* (Sabatier P.C., ed.), pp. 1–14, Springer, Berlin.

Sabatier P.C. (ed.) (1990) *Inverse Problems in Action*, Springer, Berlin.

Sabatier P.C. (1996) Nonuniqueness in Inverse Problems, *Journal of Inverse and Ill-Posed Problems*, **4**, 707–717.

Sabatier P.C. (2000) Past and Future of Inverse Problems, *Journal of Mathematical Physics*, **41**, 4082–4124.

Sabatier P.C. (2001) Should We Study Sophisticated Inverse Problems? *Inverse Problems*, **17**, 1219–1223.

Sacks P.E. (1988) The Iterative Method for the Inverse Dirichlet Problem, *Inverse Problems*, **4**, 1055–1067.

Saint-Venant J.C. (1855) *Mém. acad. savants é trangers*, **XIV**, 233–560.

Saito H. and Murikami T. (1969) Vibrations of Infinite Beam on an Elastic Foundation with Consideration of Mass of Foundation, *Bull. JSME*, **12(50)**, 200–205.

Sakhnovich L. (2001) Half-Inverse Problems on the Finite Interval, *Inverse Problems*, **17**, 527–532.

Sakiyama T. (1985) A Method of Analyzing the Bending Vibration of Any Type of Tapered Beams, *Journal of Sound and Vibration*, **101(9)**, 267–270.

Sakiyama T. (1986) A Method of Analysis of the Elastic Buckling of Tapered Columns, *Computers and Structures*, **23(1)**, 119–120.

Salunkhe A.L. and Majumdar P.M. (1998) Identification Approach to Estimate Buckling Load of Damaged Composite Plates, *AIAA Journal*, **36(8)**, 1479–1485.

Salvadori M.G. (1951) Numerical Computation of Buckling Loads by Finite Differences, *Transactions ASCE*, **116**, 590–624.

Salzar R.S. (1995) Functionally Graded Metal Matrix Composite Tubes, *Computer Engineering*, **5(7)**, 851–900.

Samanta B.K. (1964) Note on Non-Homogeneous Rotating Circular Disk, *Indian Journal of Theoretical Physics*, **12(2)**, 41–48.

Sanger D.J. (1968) Transverse Vibrations of a Class of Non-Uniform Beams, *Journal of Mechanical Engineering Science*, **10**, 111–120.

Sato K. (1980) Transverse Vibrations of Linearly Tapered Beams with Ends Restrained Elastically Against Rotation Subjected to Axial Force, *International Journal of Mechanical Science*, **22**, 109–115.

Scheurkogel A. and Elishakoff I. (1988) Nonlinear Random Vibration of a Two Degree-of-Freedom Systems, *Non-Linear Stochastic Engineering Systems* (Ziegler F. and Schuëller G.I., eds.), pp. 285–299, Springer, Berlin.

Schile R.D. (1963) Bending of Nonhomogeneous Bars, *International Journal of Mechanical Sciences*, No. **5**.

Schile R.D. (1967) Some Problems in Nonhomogeneous Elasticity, *Ph.D. Dissertation*, Rensselaer Polytechnic Institute.

Schile R.D. and Sierakowski R.L. (1964) On the Axially Symmetric Deformation of Nonhomogeneous Elastic Material, *Journal of Franklin Institute*, **278(5)**.

Schmidt R. (1980) Some Inconsistencies Arising in Applications of Galerkin's Method to Bifurcation Problems of Elastic Stability, *Industrial Mathematics*, **30(2)**, 121–133.

Schmidt R. (1981) A Variant of the Rayleigh Ritz Method, *Industrial Mathematics*, **31**, 37–46.

Schmidt R. (1982) Estimation of Buckling Loads and Other Eigenvalues via a Modification of the Rayleigh–Ritz Method, *Journal of Applied Mechanics*, **49**, 639–640.

Schmidt R. (1983) Modification of Timoshenko's Technique for Estimating Buckling loads, *Industrial Mathematics*, **33(2)**, 169–173.

Schmidt R. (1985) Towards Resurrecting the Original Rayleigh Method, *Industrial Mathematics*, **35(1)**, 69–73.

Schimdt R. (1989) Accurate Fundamental Frequencies of Vibrating Beams via a Ritz–Rayleigh–Sturn–Liouville Approach, *Industrial Mathematics*, Part 1, **39**, 37–46.

Schmidt R. (1990) On the Uses of the Boobnov–Galerkin Method in Eigenvalue Problems, *Industrial Mathematics*, Part 2, **40**, 139–147.

Schmidt R. (1996) Rayleigh's Own Method Versus the Two-Term Rayleigh–Ritz Method, *Industrial Mathematics*, **46(1)**, 1–8.

Schmidt R. (1997) Upper and Lower Bounds to Bifurcational Buckling Loads, Fundamental Frequencies, and Other Eigenvalues via a Two-Term Ritz Method, *Industrial Mathematics*, **47(2)**, 49–57.

Schuëller G.I. (ed.) (1991) *Structural Dynamics — Recent Advances*, Springer, New York.

Schwerin E. (1927) Über Transversalschwingungen von Stäben veränderlichen Querschnitten, *Zeit. Techn. Phys.*, **8**, 270 (in German).

Segal A.I. (1959) Simplification of the Boobnov–Galerkin Method, *Investigations in the Theory of Structures* (A.G. Gvozdev et al., eds.), pp. 235–239, "Gosstroiizdat" Publishers, Moscow (in Russian).

Segal A.I. and Baruch M. (1980) A Nondestructive Dynamic Method for the Determination of the Critical Load of Elastic Columns, *Experimental Mechanics*, **20**, 285–288.

Segall A. and Springer G.S. (1986) A Dynamic Method for Measuring the Critical Loads of Elastic Flat Plates, *Experimental Mechanics*, **26(4)**, 354–359.

Seide P. (1975) Accuracy of Some Numerical Methods for Column Buckling, *Journal of Engineering Mechanics*, **101**, 545–560.

Sekhniashvili E.A. (1950a) Determination of Free Vibration Frequencies of Beams of Variable Stiffness, *Communications of the Georgian Academy of Sciences*, **11(3)** (in Russian).

Sekhniashvili E.A. (1950b) Determination of Free Vibration Frequencies of Variable Stiffness and Arbitrarily Varying Mass via the Method of Adjoint Beam, *Communications of the Georgian Academy of Sciences*, **11(8)** (in Russian).

Sekhniashvili E.A. (1966) *Vibrations of Elastic Systems*, "Sabtchota Sakartvelo" Publishers, Tbilisi (in Russian).

Selvadori A.P.S. (1979) *Elastic Analysis of Soil–Foundation Interaction*, Elsevier, Amsterdam.

Sen B. (1935) Note on Stresses in Some Rotating Disks of Varying Thickness, *Philosophical Magazine*, **19(130)**, 1121–1125.

Sen Gupta A.M. (1949) Stresses in Some Aelotropic and Isotropic Circular Disks of Varying Thickness Rotating About the Central Axis, *Bull. Calcutta Math. Society*, **41**, 129–139.

Settler K. (1938) *Das "Durchbiegungsverfahren" zur Lösung von Stabilitätsproblemen im elastischen und unelastischen Bereich*, Helsinki (in German).

Seyranian A.P. (1983) On a Certain Solution of a Problem of Lagrange, *Doklady AN SSSR*, **271(3)**, 337–340.

Seyranian A.P. (1984) On a Problem of Lagrange, *Mechanics of Solids*, **19(2)**, 100–111.

Seyranian A.P. (1995) New Solutions to Lagrange's Problem, *Phys. Dokl.*, **40**, 251–253.

Seyranian A.P. and Privalova O.G. (2003) The Lagrange Problem on an Optimal Column: Old and New Results, *Structural Multidisciplinary Optimization*, **25**, 393–410.

Sezer M. and Kaynak M. (1996) Chebychev Polynomial Solutions of Linear Differential Equations, *International Journal of Mathematical Education in Science and Technology*, **27(4)**, 607–618.

Sezer M. and Keşan C. (2000) Polynomial Solutions of Certain Differential Equations, *International Journal of Computer Mathematics*, **76**, 93–104.

Shabana A.A. (1994) *Computational Dynamics*, Wiley, New York.

Shaffer B.W. (1967) Orthotropic Annular Disks in Plane Stress, *Journal of Applied Mechanics*, 1027–1029.

Shaker F.J. (1975) Effect of Axial Load on Mode Shape and Frequencies of Beams, NASA Lewis Research Center, *Report NASA-TN*-8109.

Shakhnazarov D.I. (1940) An Equation that Determines Periods of Vibration of Variable Stiffness Beams in a Particular Case, *Journal of Applied Mechanics and Mathematics — PMM*, **4(3)** (in Russian).

Shen C.L. (1993) On the Nodal Sets of the Eigenfactors of Certain Homogeneous and Nonhomogeneous Membranes, *SIAM Journal of Mathematical Analysis*, **24**, 1277–1282.

Shen C.L. and Tsai T.M. (1995) On the Uniform Approximation of the Density Function of a String Equation using Eigenvalues and Nodal Points and Some Related Inverse Nodal Problems, *Inverse Problems*, **11**, 1113–1123.

Sherbourne A.N. and Pandey M D. (1991) Differential Quadrature Method in the Buckling Analysis of Beams and Composite Plates, *Computers and Structures*, **40**, 903–913.

Sherman D.I. (1959) On the Problem of Plane Strain in Nonhomogeneous Media, *Nonhomogeneity in Elasticity and Plasticity* (Olszak W., ed.), Pergamon Press, London.

Sheu C.Y. (1968) Elastic Minimum Weight Design for Specified Fundamental Frequency, *International Journal of Solids and Structures*, **4**, 953–958.

Shevchenko Y.N. (1958) General Solution of the Theory of Elasticity Problem with a Variable Modulus, *Reports of the Academy of Sciences of the Ukrainian SSR*, No. **10**, 1054–1057 (in Ukrainian).

Shevchenko Y.N. (1958) General Solution of the Theory of Elasticity with Variable Modulus of Elasticity, *Doklady Akademii Nauk Ukrainskoi SSR*, No. **10** (in Russian).

Shinozuka M. and Astill C.J. (1972) Random Eigenvalue Problems in Structural Analysis, *AIAA Journal*, **10**, 456–462.

Shinozuka M. and Yamazaki F. (1988) Stochastic Finite Element Analysis: An Introduction, in *Stochastic Structural Dynamics* (Ariaratnam S.T., Schuëller G.I. and Elishakoff I., eds.), pp. 241–292, Elsevier Applied Science, London.

Siddiqui A.M. (1990) Some More Inverse Solutions of a Non-Newtonian Fluid, *Mechanics Research Communication*, **17(3)**, 157–163.

Siddiqui A.M. and Kaloni P.N. (1986) Certain Inverse Solutions of Non-Newtonian Fluids, *International Journal of Nonlinear Mechanics*, **21**, 459–470.

Silver G. (1951) Critical Loads on Variable-Section Columns in the Elastic Range, *Journal of Applied Mechanics*, 414–420.

Simitses G.J., Kamat M.P. and Smith C.V. (1973) Strongest Column by the Finite Element Displacement Method, *AIAA Journal*, **11(9)**, 1231–1232.

Simonian S.S. (1981a) Inverse Problems in Structural Dynamics, Part 1: Theory, *International Journal for Numerical Methods in Engineering*, **17**, 357–365.

Simonian S.S. (1981b) Inverse Problems in Structural Dynamics, Part 2: *International Journal for Numerical Methods in Engineering*, **17**, 367–386.

Simonov N.I. (1957) *Euler's Applied Methods of Analysis*, GITTL, Moscow (in Russian).

Sing R. and Sascena V. (1995) Axisymmetric vibration of circular plate with double linear variable thickness, *Journal of Sound and Vibration*, **179**(1), 879–897.

Singer J. (1962) On the Equivalence of the Galerkin and Rayleigh–Ritz Methods, *Journal of Royal Aeronautical Society*, **66**, 592–597.

Singh B. and Hassan S.M. (1998) Transverse Vibration of a Circular Plate with Arbitrary Thickness Variation, *International Journal of Mechanical Sciences*, **40**(11), 1089–1104.

Singh S.R. and Chakraverty S. (1991) Transverse Vibration of Circular and Elliptic Plates with Variable Thickness, *Indian Journal of Pure and Applied Mathematics*, **22**(9), 787–803.

Singh S.R. and Chakraverty S. (1992) Transverse Vibration of Circular and Ellipical Plates with Quadratically Varying Thickness, *Applied Mathematics Modeling*, **16**, 269–274.

Sinha D.K. (1968) Deformations of an Inhomogeneous Piezoelectric Thick Disk, *Bull. Acad. Polonaise Sci., Ser. Sci. Techn.*, **16**(6).

Sinha J.K. and Friswell M.I. (2002) Model Updating: A Tool for Reliable Modeling, Design Modification and Diagnosis, *The Shock and Vibration Digest*, **34**(1), 27–35.

Smolansky U. (1993) On Hearing the Shape of a Drum, and Related Problems, *Contemporary Physics*, **34**, 297–302.

Soize C. (1988) Steady-State Solution of Fokker Planck Equation in High Dimension, *Probabilistic Engineering Mechanics*, **3**, 196–206.

Soize C. (1999) Personal Communication, November 10 .

Soize C. (2000a) A Nonparametric Model of Random Uncertainties for Reduced Matrix Models in Structural Mechanics, *Probabilistic Engineering Mechanics*, **15**, 277–294.

Soize C. (2000b) Private Communication, June 18.

Sol H. (1986) Identification of Anisotropic Plate Rigidities using Free Vibration Data, *Ph. D. Thesis*, Free University of Brussels.

Soni S.R. (1972) Vibrations of Elastic Plates and Shells of Variable Thickness, *Ph.D. Thesis*, University of Roorkee.

Soni S.R. and Amba Rao C.L. (1975) Axisymmetric Vibrations of Annular Plates of Variable Thickness, *Journal of Sound and Vibration*, **38**(4), 465–473.

Spencer A.J.M. (1998) A Stress Function Formulation for a Class of Exact Solutions for Functionally Graded Plates, in *Proceedings of the IUTAM Symposium on Transformation Problems in Composite and Active Materials* (Bahei-El-Din Y.A. and Dvorak G.I., eds.), Kluwer Academic Publishers, Dordrecht.

Spillers W.R. and Meyers M.K. (1997) The Tadjbakhsh-Keller Problem, *Iranian Journal of Science and Technology*, Transaction B, **21**(3), 209–214.

Spyrakos C.C. and Chen C.C. (1990) Power Series Expansions of Dynamic Stiffness Matrices for Tapered Bars and Shafts, *International Journal for Numerical Methods in Engineers*, **30**, 259–270.

Sridhar S. and Kudrolli A. (1994) Experiments on Not "Hearing the Shape" of Drums, *Physical Review Letters*, **72**, 2175–2178.

Srinivasan A.V. (1964) Buckling Load of Bars with Variable Stiffness: A Simple Numerical Method, *AIAA Journal*, **2**(1), 139–140.

Srivasta H.M. and Buschman R.G. (1992) *Theory and Application of Convolution Integral Equations*, Kluwer, Dordrecht.

Starek L. (2001) Inverse Eigenvalue Problems in Vibration of Mechanical Systems, *Strojnicky Casopis*, **52**(1), 1–20 (in Slovak).

Starek L. and Inman D.J. (1991) On the Inverse Vibration Problem with Rigid Body Modes, *Journal of Applied Mechanics*, **58**, 1101–1104.

Starek L. and Inman D.J. (1997) A Symmetric Inverse Vibration Problem for Non-Proportional Underdamped Systems, *Journal of Applied Mechanics*, **64**, 601–605.

Starek L., Inman D.J. and Kress A. (1992) A Symmetric Inverse Vibration Problem, *Vibration and Acoustics*, **114**, 564–568.

Starek L., Inman D.J. and Pilkey D.J. (1995) A Symmetric Positive Definite Inverse Vibration Problem with Underdamped Modes, in *Proceedings of the Design Engineering Technical Conference*, DE-Vol. 84–3, Vol. 3 — Part C, pp. 1089–1094, ASME Press, New York.

Stavsky Y. (1964) Thermoelastic Field Equations of Inhomogeneous Media, *Israel Journal of Technology*, **2(1)**.

Stavsky Y. (1965) On the Theory of Symmetrically Heterogeneous Plates Having the Same Thickness Variation of the Elastic Moduli, in *Topics in Applied Mechaics* (Abir D., Ollendorff F. and Reiner M., eds.), pp. 105–116, Elsevier.

Steinberg L.G. (1995) Inverse Spectral Problems for Inhomogeneous Elastic Cylinders, *Journal of Elasticity*, **38**, 133–151.

Stephen N.G. (1989) Beam Vibration Under Compressive Axial Load — Upper and Lower Bound Approximation, *Journal of Sound and Vibration*, **131**, 345–350.

Stephen N.G. (1994) On the "Relationship between Fundamental Natural Frequency and Maximum Static Deflection for Various Linear Vibratory Systems," *Journal of Sound and Vibration*, **170(2)**, 285–287.

Stevens K.K. (1987) Force Identification Problems: An Overview, in *Proceedings of SEM Spring Meeting*, pp. 838–844, Houston.

Stewart I. (1992) Beating out the Shape of a Drum, *New Scientist*, **1825**, 26–30.

Stewartson K. and Waechter R.T. (1971) On Hearing the Shape of a Drum: Further Results, *Proceedings of the Cambridge Philosophical Society*, **69**, 353–363.

Stodola A. (1927) *Steam and Gas Turbines*, p. 1145, McGraw-Hill, New York.

Storch J. (2001a) Private Communication, February 14.

Storch J. (2001b) Private Communication, December 28.

Storch J. (2002a) Private Communication, January 1.

Storch J. (2002b) Private Communication, March 26.

Storch J. (2002c) Private Communication, March 27.

Storch J. (2002d) Private Communication, June 23.

Storch J. (2002e) Private Communication, June 23.

Storch J. and Elishakoff I. (2004) Apparently First Closed-Form Solutions of Inhomogeneous Circular Plates in 200 Years After Chladni, *Journal of Sound and Vibration*, **276**, 1108–1114.

Strang G. and Fix G. (1969) Fourier Series Analysis of the Finite Element Method in Ritz–Galerkin Theory, *Studies in Applied Mathematics*, **XLVIII**, 265–275.

Struble D.E. (1971) Boundary Effects in Structural Stability: A New Approach, *Ph. D Thesis*, Georgia Institute of Technology, Atlanta, GA.

Strutt J.W. (Baron Rayleigh), (1894) *The Theory of Sound*, Macmillan, New York (reprinted by Dover Publications, 1945).

Subramanian G. and Raman A. (1996) Isospectral Systems for Tapered Beams, *Journal of Sound and Vibration*, **198**, 257–266.

Sugiyama Y., Ashida K. and Kawagoe H. (1978) Buckling of Long Columns under Their Own Weight, *Bulletin of the JSME*, **21(158)**, 1228–1235.

Sugiyama Y., Langthjem M.A. and Ryu B.J. (1999) Realistic Follower Forces, *Journal of Sound and Vibration*, **225**, 779–782.

Sun H. and Fu M. (2001) Semi-Inverse Method for Generalized Variational Inequalities in Elasticity with Finite Displacement and Friction, *Chinese Journal of Mechanical Engineering*, **37(8)**, 12–17 (in Chinese).

Suppinger E.W. and N.J. Taleb (1956) Free Lateral Vibrations of a Beam of Variable Cross-Section, *ZAMP*, **3(6)**, 7.

Suresh S. and Mortensen A. (1998) *Fundamentals of Functionally Graded Materials*, IOM Publications, London.

Suzuki T. (1985) Gelfand–Levitan's Theory, Deformation Formulas and Inverse Problems, *J. Fac. Sci. Univ. Tokyo*, **IA(32)**, 227–271.

Svirskii I.B. (1968) *Methods of Boobnov–Galerkin-Type and Successive Approximations*, Nauka Publishers, Moscow (in Russian).

Swenson G. (1952) Analysis of Non-Uniform Columns and Beams by a Simple D. C. Network Analyzer, *Journal of Aeronautical Sciences*, **19**, 273–276.

Sysoev V.I. (1961) Towards the Problem of Transverse Vibrations of Beams of Variable Cross Section, *Issledovaniya no Teorii Sooruzhenii (Investigations in the Theory of Structures)* (in Russian).

Szidarowski S. (1964) Two New Practical Methods to Determine the Critical Loads for a Compressed Bar of Variable Flexaral Rigidity, *Acta Tech. Hung.*, **46**, 261–285.

Szyszkowski W. and Watson L.G. (1988) Optimisation of the Buckling Load of Columns and Frames, *Engineering Structures*, **10**, 249–256.

Tada Y. and Wang L. (1995) Reinvestigation on Optimisation of Clamped–Clamped Columns and Symmetry of Corresponding Eigenfunctions, *JSME Int. Journal*, **38(1)**, 38–43.

Tadjbakhsh I. and Keller J.B. (1962) Strongest Columns and Isoperimetric Inequalities for Eigenvalues, *Journal of Applied Mechanics*, **29(1)**, 159–164.

Tai H., Katayama T. and Sekiya T. (1982) Buckling Analysis by Influence Function, *Mathematics Research Communications*, **9(3)**, 139–144.

Takewaki I. (1999) Efficient Inverse Frequency Design of Tapered Beams Including Shear Deformations, *Computer Methods in Applied Mechanics and Engineering*, **179(1–2)**, 67–79.

Takewaki I. (2000) *Dynamic Structural Design: Inverse Problem Approach*, WIT Press, Southampton.

Takewaki I. and Nakamura T. (1995) Hybrid Inverse Mode Problems for FEM Shear Models, *Journal of Engineering Mechanics*, **110**, 873–880.

Taleb N. (1955) Free Lateral Vibrations of Beams with Non-Uniform Cross Sections, *Ph. D. Dissertation*, Princeton University.

Taleb N.J. and Suppiger E.J. (1961) Vibration of Stepped Beams, *Journal of Aerospace Sciences*, **28(4)**, 295–298.

Tanaka M. and Bui H.D. (1993) *Inverse Problems in Engineering Mechanics*, Springer, Berlin.

Tarantola A. and Valette B. (1982) Inverse Problems= Quest for Information, *J. Geophys.*, **50**, 159–170.

Tarantola A. (1987) *Inverse Problem Theory*, Elsevier, Amsterdam.

Taylor J.E. (1968) Optimum Design of a Vibrating Bar with Specified Minimum Cross-Section, *AIAA Journal*, 1379–1381.

Tchentsov N.G. (1936) Columns of Minimum Weight, *Trudy TSAGI*, **265** (in Russian).

Tebede N. and Tall L. (1973) Linear Stability Analysis of Beam-Columns, *Journal of Structural Division*, 2439–2457.

Temple G. and Bickley W.G. (1933) *Rayleigh's Principle and Its Applications to Engineering*, Oxford University Press, New York (also, Dover Publications, 1956).

Teodorescu P.P. (1964) Über das kinetische Problem nichthomogener elastischer Körper, *Bull. Acad. Polonaise Sci, Ser. Sci. Techn.*, **12(2)** (in German).

Ter-Mkrtchian L.N. (1961) Some Problems of the Theory of Elasticity of Inhomogeneous Elastic Bodies, *Journal of Applied Mechanics and Mathematics — PMM*, **26(6)** (in Russian).

Tesfaye E. and Broome T.H. (1977) Effect of Weight on Stability of Masonry Walls, *Journal of Structural Division*, **103**, 961–967.

Thambiratnam D. (1996) Free Vibration Analysis of Beams on Elastic Foundation, *Computers and Structures*, **60(6)**, 971–9680.

Theodorescu P.P. and Predeleanu M. (1959) Über das eben Problem nichthomogener élastischer Körper, *Acta Techn. Acad. Scient. Hung.*, No. **3–4** (in German).

Thomas J. and Dokumaci E. (1973) Improved Finite Elements for Vibration Analysis of Tapered Beams, *Aeronautical Quarterly*, **24**, 39–46.

Thomson W.T. (1949) Matrix Solution of the *n*-Section Column, *Journal of Aerospace Sciences*, 623–624.

Thomson W.T. (1950) Critical Load of Columns of Varying Cross-Section, *Journal of Applied Mechanics*, **18**, 132–134.

Tikhonov A.N. and Arsenin V.Y. (1977) *Solutions of Ill-Posed Problems*, Wiley, New York.

Timoshenko S.P. (1910) On the Stability of Elastic Systems, Application of New Methodology to Stability Investigation of Some Bridge Structures, *Izvestiya Kievskogo Politekhnicheskogo Instituta*, **10**, Book No. 4 (in Russian).

Timoshenko S.P. (1953) *History of Strength of Materials*, McGraw-Hill, New York.

Timoshenko S.P. and Gere J.M. (1961) *Theory of Elastic Stability*, McGraw-Hill, New York.

Timoshenko S.P. and Woinowsky-Krieger S. (1959) *Theory of Plates and Shells*, McGraw-Hill Book Company, New York.

To C.W.S. (1979) Higher Order Tapered Finite Elements for Vibration Analysis, *Journal of Sound and Vibration*, **63**, 33–50.

To C.W.S. (1981) A Linearly Tapered Beam Finite Element Incorporating Shear Deformation and Rotary Inertia for Vibration Analysis, *Journal of Sound and Vibration*, **78(4)**, 475–484.

Todhunter I. and Pearson K. (1893) *A History of the Theory of Elasticity and of the Strength of Materials*, **2**, Part 2, pp. 92–98, Cambridge University Press, Cambridge.

Töllke F. (1930) Über die Bemessung von Druckstäben mit veränderlichem Querschnitt, *Bauingenieur*, 500 (in German).

Tong P. and Pian T.H.H. (1967) The Convergence of the Finite Element Method in Solving Linear Elastic Problems, *International Journal of Solids and Structures*, **3**, 865–879.

Tong P., Pian T.H.H. and Bucciarelli L.L. (1971) Mode Shapes and Frequencies by the Finite Element Method using Consistent and Lumped Matrices, *Computers and Structures*, **1**, 623–638.

Truesdell C. (1953) A New Chapter in the Theory of Elastica, *Proceedings of the First Midwestern Conference on Solid Mechanics*, Sponsored by the College of Engineering and the Panel on Fluid and Solid Mechanics of the University of Illinois, pp. 52–54.

Trujillo D.M. and Busby H.R. (1997) *Practical Inverse Analysis in Engineering*, CRC Press, New York.

Tsai C.Z., Wu E. and Luo B.H. (1998) Forward and Inverse Analysis for Impact on Sandwich Panels, *AIAA Journal*, **36(11)**, 2130–2136.

Tu Y.O. and Handelman G. (1961) Lateral Vibrations of a Bar under Initial Linear Axial Stress, *Journal of SIAM*, **9**(3), 455.

Tukerman L.B. (1929) Discussion of A. Dinnik's (1929) Paper, *Transactions ASME, Applied Mechanics*, **51**(1), p. 112.

Turton F.J. (1942) Pinned–Pinned Slender Solid Struts with Parabolic Taper, *Journal of Royal Aeronautical Society*, **46**, 146.

Tutuncu N. and Ozturk M. (2001) Exact Solutions for Stresses in Functionally Graded Pressure Vessels, *Composites: Part B*, **32**, 683–686.

Ueda S. (1988) The Inverse Problem for the String or Rod Clamped at Both Sides, *JSME International Series*, Series I, **31**(4), 686–689.

Uspensky J.V. (1948) *Theory of Equations*, McGraw-Hill Book Company, New York.

Vainberg D.V. (1952) Analog Between Problems of Plain Stress and of the Bending of Circular Plate of Variable Thickness Subjected to Unsymmetric Loading, *Journal of Applied Mechanics and Mathematics — PMM*, **16**(6) (in Russian).

Vanmarcke E.H. and Grigoriu M. (1983) Stochastic Finite Element Analysis of Simple Beams, *Journal of the Engineering Mechanics Division*, **109**(5), 1203–1214.

Varley E. and Seymour B.A. (1988) A Method of Obtaining Exact Solutions to PDEs with Variable Coefficients, *Studies in Applied Mathematics*, **78**, 183–225.

Vasilenko A.T. (1999) Solution of the Problem of Bending of an Inhomogeneous Curved Beam, *Strength of Materials*, **31**(5), 505–509.

Velte W. and Villaggio P. (1982) Are the Optimum Problems in Structural Design Well Posed? *Arch. Rat. Mech. Anal.*, **78**, 199–211.

Venkataraman S. and Sankar B.V. (2003) Elasticity Analysis and Optimization of a Functionally Graded Plate with Hole, Paper AIAA-2003-1466, *44th AIAA/ASME/ASCEAHS/ASC Structures, Structural Dynamics and Materials Conference*, April 7–11, 2003, Norfolk, VA.

Vianello L. (1898) Graphische Untersuchung der Knickfestigkeit gerader Stabe, *Zeitschrift Verein Deutscher Ing.*, **42**, 1436–1443 (in German).

Villaggio P. (1979) Inverse Boundary Value Problems in Structural Optimization, in *Free Boundary Problems*, **2**, pp. 587–599, Inst. Naz. Alta Matem.

Villaggio P. (1996) The Pillar of Best Efficiency, *Journal of Elasticity*, **42**, 79–89.

Villaggio P. (1997) *Mathematical Models for Elastic Structures*, see Section 32, Inverse Problems in Strings, Cambridge University Press, Cambridge.

Villaggio P. and Velte W. (1982) Are the Optimum Problems in Structural Design Well Posed? *Arch. Rational Mech. Anal.*, **78**(3), 199–211.

Vlasov V.Z. and Leontiev N.N. (1966) *Beams, Plates and Shells on Elastic Foundations*, Israel Program for Scientific Translations, Jerusalem, Israel.

Vogel C.R. (2002) *Computational Methods for Inverse Problems*, SIAM, Philadelphia.

Volmir A.S. (1967) *Stability of Deformable Systems*, Nauka Publishers, Moscow (in Russian).

Volmir A.S. (ed.) (1984) *Collection of Problems in Strength of Materials*, Problem 7.34, p. 30, Nauka Publishers, Moscow (in Russian).

Volvich S.I. (1953) Stability of Columns by Arbitrary Axial Loading, *Proceedings of the Saratov Automotive-Transport Institute*, **12** (in Russian).

Vorovich I.I. (1975) Boobnov–Galerkin Method, Its Development and Role in Applied Mathematics, in *Advances of Mechanics of Deformable Continua*. (Dedicated to the 100th Anniversary of the Birth of B.G. Galerkin) (Ishlinskii A. Ju., ed.), pp. 121–133, Nauka, Moscow (in Russian).

Wah T. and Calcote L.R. (1970) *Structural Analysis by Finite Difference Calculus*, Van Nostrand Reinhold, New York.

Wang D.J., He B.C. and Wang Q.S. (1991) On the Construction of the Euler–Bernoulli Beam via Two Sets of Modes and the Corresponding Frequencies, *Acta Mechanica Sinica*, **22(4)**, 479–483 (in Chinese).

Wang H.C. (1967) Generalized Hypergeometric Function Solution on Transverse Vibration of a Class of Non-uniform Beams, *Journal of Applied Mechanics*, **34**, 702–708.

Wang Q.S. and Wang D.J. (1994) An Inverse Mode Problem for Continuous Second-Order Systems, *Proceedings of the International Conference on Vibration Engineering, ICVE'99*, pp. 167–170.

Ward P.F. (1913) The Transversal Vibrations of a Rod of Varying Cross-Section, *Philosophical Magazine*, **25**, 85–106.

Warner W.H. (2000) Optimal Design of Elastic Rods under Axial Gravitational Load using the Maximum Principle, *International Journal of Solids and Structures*, **37(19)**, 2709–2726.

Waszczyszyn Z. (1978) Exact Finite Elements for Thin Elastic Shells of Revolution, *Theor. And Appl. Mech. Bulgarian Acad. Sci*, **9**, 101–104.

Waszczyszyn Z. and Pieczara J. (1990) Exact Finite Elements for the Linear Buckling Analysis of Undimensioned Structural Problems, *Discretization Methods in Structural Mechanics* (Kuhn G. and Mang H., eds.), pp. 307–316, Springer, Berlin.

Weaver W. Jr. and, Johnston P.R. (1987) *Structural Dynamics by Finite Elements*, Prentice Hall, Englewood Cliffs, NJ.

Weaver W. Jr., Timoshenko S.P. and Young D.H. (1990) *Vibration Problems in Engineering*, Wiley (fourth edition), New York, pp. 395–401.

Webb H.A. and Long E.D. (1919) Struts of Conical Taper, *Journal of Royal Aeronautical Society*, **23**, 179.

Weber M. (1999) On the Reconstruction of a String from Spectral Data, *Mathematische Nachrichten*, **197**, 135–156.

Weidenmuller H. (1994) Why Different Drums Sound the Same, *Physics World*, **7**, 22–23.

Weiner J.H. (2002) *Statistical Mechanics of Elasticity*, Dover Publication, Mineola, New York.

White W.T. (1948) On Integral Equations Approach to Problems of Vibrating Beams, Part, I and II, *Journal of Franklin Institute*, **245**, 25–36, 117–133.

Wiesensel G.N. (1989) Natural Frequency Information for Circular and Annular Plates, *Journal of Sound and Vibration*, **133(1)**, 129–137.

Wilken J.A. (1927) The Bending of Columns of Varying Cross Section, *Philosophical Magazine*, Series 7, **3**, 418, **13**, 845.

Willers F.A. (1941) Das Knicken Schwerer Gestaenge, *Zeitschrift für angewandte Mathematik und Mechanik*, **21**, 43–51 (in German).

Williams F.W. and Banerjee J.R. (1985) Flexural Vibration of Axially Loaded Beams with Linear or Parabolic Taper, *Journal of Sound and Vibration*, **99(1)**, 121–138.

Williams F.W. and Wittrick W.H. (1983) Exact Buckling and Frequency Calculations Surveyed, *Journal of Structural Engineering*, **109**, 167–187.

Wilson J.F. (1971) Stability Experiments on the Strongest Columns and Circular Arches, *Experimental Mechanics*, **11(7)**, 303–308.

Wittmeyer H. (1951) A Simple Method for the Approximate Computation of All the Natural Frequencies of a Member with Variable Cross-Section, *Technical Note KTH-AERO TN13*, Royal Institute of Technology, Division of Aeronautics, Stockholm, Sweden.

Wittrick W.H. (1985) Some Observations on the Dynamic Equations of Prismatic Members in Compression, *International Journal of Mechanical Sciences*, **27**(6), 375–382.

Wolfram S. (1996) *The Mathematica® Book*, Cambridge University Press, Cambridge.

Woodbury K.A. (2002) *Inverse Engineering Handbook*, CRC Press, Boca Raton.

Wrinch D.M. (1922) On the Lateral Vibration of Bars of Conical Type, *Proceedings of the Royal Society of London* (series A), **101**, 493–508.

Wrinch D.M. (1923) On the Lateral Vibration of Rods of Variable Cross-Section, *Philosophical Magazine*, **46**, Series 6, 273–291.

Wu L., Wang Q-S. and Elishakoff I. (2004) Semi-Inverse Method for Axially Graded Beams with an Anti-Symmetric Vibration Mode, *Journal of Sound and Vibration* (to appear).

Xie J. (1992) Numerical Model for Analysis of Flexible Beams-Columns on Elastic Foundations, *Comput. Geotech.*, **13**, 51–62.

Xie Y.M. (1995) Further Comments on "Relationship between Fundamental Natural Frequency and Maximum Static Deflection for Various Linear Vibratory Systems," *Journal of Sound and Vibration*, **186**(4), 689–693.

Xie Y.M. and Steven G.P. (1993) A Simple Evolutionary Procedure for Structural Optimisation, *Computers and Structures*, **49**(5), 885–896.

Yamamoto M. (1988a) Inverse Spectral Problem for Systems of Ordinary Differential Equations of First Order, *J. Fac. Sci. Univ. Tokyo*, **35**, 519–546.

Yamamoto M. (1988b) Continuous Dependence Problem in an Inverse Spectral Problem for Systems of Ordinary Differential Equations of First Order, *Scientific Papers of the College of Arts and Sciences, University of Tokyo*, **38**, 69–130.

Yamamoto M. (1990) Inverse Eigenvalue Problem for a Vibration of a String with Viscous Drag, *Y. Mathematical Analysis and Applications*, **152**, 20–34.

Yamaoka H., Yuki M., Tahara K., Irisawa T., Watanabe R. and Kawasaki A. (1993) Fabrication of Functionally Gradient Materials by Slurry Stacking and Sintering Process, *Ceramic Transactions*, **34**, 165–172.

Yang J.S. (1993) The Vibration of a Circular Plate with Varying Thickness, *Journal of Sound and Vibration*, **165**(1), 178–184.

Yang J. and Shen H-S. (2003) Free Vibration and Parametric Resonance of Shear Deformable Functionally Graded Cylindrical Panels, *Journal of Sound and Vibration*, **261**, 871–893.

Yang J.S. and Xie Z. (1984) Peturbation Method in the Problem of Large Deflections of Circular Plates with Non-uniform Thickness, *Applied Mathematics and Mechanics*, **5**, 1237–1242.

Yang K.Y. (1990) The Natural Frequencies of a Non-uniform Beam with a Tip Mass and with Translational and Rotational Springs, *Journal of Sound and Vibration*, **137**, 339–341.

Yang X.F. (1997) A Solution of the Inverse Nodal Problem, *Inverse Problems*, **13**, 203–213.

Yao Z. and Qu S. (1998) Identification of the Material Parameters of Laminated Plates, *Inverse Problems in Engineering Mechanics* (Tanaka M. and Dulikravich G.S., eds.), pp. 179–185, Elsevier, Amsterdam.

Yatram A.L. and Awadalla E.S. (1967) A Direct Matrix Method for the Elastic Analysis of Structures, *International Journal of Mechanical Sciences*, **9**, 315–321.

Yeh F.H. and Liu W.H. (1989) Free Vibrations of Nonuniform Beams with Rotational and Translational Restraints and Subjected to an Axial Force, *Journal of the Acoustical Society of America*, **85**, 1368–1371.

Yeh K Y. (1994) Analysis of High-Speed Rotating Discs with Variable Thickness and Inhomogeneity, *Journal of Applied Mechanics*, **61(1)**, 186–192.

Yen C.-S. and Wu E. (1995) On the Inverse Problem of Rectangular Plates Subjected to Elastic Impact, *Journal of Applied Mechanics*, Part I, 692–698.

Ylinen A., (1938) *Die Knickfestigkeit eines zentrisch gedrückten Stabes im elastischen und unelaastischen Bereich*, Helsinki(in German).

Yoakimidis N.I. (1994) Symbolic Computations for the Solution of Inverse/Design Problems, *Computers and Structures*, **53**, 63–68.

Yokoyama T. (1991) Vibrations of Timoshenko Beam-Columns on Two-Parameter Elastic Foundations, *Earthquake Engineering and Structural Dynamics*, **20**, 355–370.

Yu Y.Y. and Lai J.L. (1967) Application of Galerkin's Method to the Dynamic Analysis of Structures, *AIAA Journal*, **5(4)**, 795–800.

Yurko V. (2002) *Method of Spectral Mappings in the Inverse Problem Theory*, VSP, Zeist, The Netherlands.

Zaslavsky A. (1965) Column Buckling Loads by the Conjugate Beam Method, *Struct. Engineer*, **43**, 183–188.

Zavriev K.S. (1942) Generalized Method of Successive Approximations for Investigation of Free Vibrations of Elastic Systems, *Proceedings of the Energy Institute of Georgia*, **4** (in Russian).

Zavriev K.S. (1945) Free Vibrations of Beams on Elastic Foundation, *Communications of the Georgian Academy of Sciences* (in Russian).

Zhou D. (1993) A General Solution Vibrations of Beams on Variable Winkler Elastic Foundation, *Computers and Structures*, **47(1)**, 83–90.

Zhou D. (2002) Discussion on "Free Vibrations of Beams with General Boundary Conditions," *Journal of Sound and Vibration*, **257(3)**, 589–592.

Zhu W.Q. and Wu W.Q. (1991) A Stochastic Finite Element Method for Real Eigenvalue Problem, in *Stochastic Structural Dynamics 1* (Elishakoff I. and Lin Y.K., eds.), **2**, pp. 337–351, Springer Verlag, Berlin.

Zinkiewicz O.C. (1970) The Finite Element Method: From Intuition to Generality, *Applied Mechanics Reviews*, **23**, 249–256.

Zienkiewiez O.C. and Schimming B. (1962) Torsion of Non-Homogeneous Bars with Axial Symmetry, *International Journal of Mechanical Sciences*, **4**.

Zubov L.M. (2001) Semi-Inverse Solutions in Non-linear Theory of Elastic Shells, *Arch. Mech.*, **53(4–5)**, 599–610.

Zweiling K. (1953) *Gleichgewicht und Stabilität*, Berlin (in German).

Życzkowski M. (1954) Elastic–Plastic Buckling of Some Non-Prismatic Bars, *Rozprawy Inżynierskie*, **2(2)**, 233–289 (in Polish).

Życzkowski M. (1955) Elastisch-plastische Knickung einiger nichtprizmatisher Stäbe, *Bulletin de L'Academic Polonaise des Sciences*, **3(3)**, 129–137 (in German).

Życzkowski M. (1956a) Calculation of Critical Forces for Elastic Non-Prismatic Bars by Partial Interpolation Method, *Rozprawy Inżynierskie*, **4(3)**, 367–412 (in Polish).

Życzkowski M. (1956b) On the Choice of Optimum Shape of Axially Compressed Bars, *Rozprawy Inżynierskie*, **4(4)**, 441–456 (in Polish).

Życzkowski M. (1956c) Anwendung der Methode der "Voraussetzung einer exacten Gleichung" bei einigen Problemen der elastisch-plastischen Stabilität, *Bulletin de L'Academic Polonaise des Sciences*, **4(1)**, 37–43 (in German).

Życzkowski M. (1956d) The Problem of the Most Suitable Form for Axially Compressed Bars, *Rozprawy Inżynierskie*, **53**, 443–456 (in Polish).

Życzkowski M. (1956e) Calculation of Critical Forces for Elastic Non-Prismatic Bars by Partial Interpolation Method, *Rozprawy Inżynierskie*, **4(3)**, 367–412 (in Polish).

Życzkowski M. (1956f) On the Choice of Optimum Shape of Axially Compressed Bars, *Rozprawy Inżynierskie*, **4(4)**, 441–456 (in Polish).

Życzkowski M. (1956g) The Problem of the Most Suitable Form for Axially Compressed Bars, *Rozprawy Inżynierskie*, **53**, 443–456 (in Polish).

Życzkowski M. (1956h) Anwendung der Methode der "Voraussetzung einer exacten Gleichung" bei einigen Problemen der elastisch-plastischen Stabilität, *Bulletin de L'Academic Polonaise des Sciences*, **4(1)**, 37–43 (in German).

Życzkowski M. (1968) Optimale Formen des dünnwandigen geschlossenen Querschnittes eines Balkens bei Berücksichtigung von Stabilitätsbedingungen, *Zeitschrift für angewandte Mathematik und Mechanik*, **48(7)**, 455–462 (in German).

Życzkowski M. (ed.) (1991) *Strength of Structural Elements*, p. 298, Elsevier, Amsterdam.

Życzkowski M. (2002) Plasticity, Stability, and Structural Optimization, *Applied Mechanics Reviews*, **55(3)**, R17–R26.

Życzkowski M. (2003) Private Communication, March 17.

Author Index

711

Subject Index

Printed and bound by CPI Group (UK) Ltd, Croydon, CR0 4YY

23/10/2024

01778238-0019